Climate Change Adaptation in the Built Environment

Chamindi Malalgoda · Dilanthi Amaratunga ·
Richard Haigh · Shavindree Nissanka ·
Nishara Fernando · Ruben Paul Borg · Mo Hamza
Editors

Climate Change Adaptation in the Built Environment

Transdisciplinary and Innovative Learning

Springer

Editors
Chamindi Malalgoda
Global Disaster Resilience Centre
University of Huddersfield
Huddersfield, UK

Dilanthi Amaratunga
Global Disaster Resilience Centre
University of Huddersfield
Huddersfield, UK

Richard Haigh
Global Disaster Resilience Centre
University of Huddersfield
Huddersfield, UK

Shavindree Nissanka
Global Disaster Resilience Centre
University of Huddersfield
Huddersfield, UK

Nishara Fernando
Department of Sociology
University of Colombo
Colombo, Sri Lanka

Department of Quantity Surveying, Faculty of Engineering
Sri Lanka Institute of Information Technology
Malabe, Sri Lanka

Mo Hamza
Division of Risk Management and Societal Safety
Lund University
Lund, Sweden

Ruben Paul Borg
Faculty for the Built Environment
University of Malta
Msida, Malta

ISBN 978-3-031-75825-6 ISBN 978-3-031-75826-3 (eBook)
https://doi.org/10.1007/978-3-031-75826-3

This Springer imprint is published by the registered company Springer Nature Switzerland AG
The registered company address is: Gewerbestrasse 11, 6330 Cham, Switzerland

International Scientific/Peer-Review Committee

100 keynote speeches in around 40 countries. Among many leadership roles, she is the joint chief editor of the *International Journal of Disaster Resilience in the Built Environment*. She is a member of the European Commission and UNDRR's European Science and Technology Advisory Group (ESTAG) representing the UK, a Steering Committee member of the UK Frontiers of Development programme, a Steering Committee member of the UK Alliance for Disaster Research, motivation of which is to bring together the UK's rich and diverse disaster research community to facilitate collaboration and partnership. She is a fellow of the Royal Institution of Chartered Surveyors (RICS), a fellow of the Royal Geographical Society, and a fellow and a chartered manager of the Chartered Management Institute, UK.

Prof. Richard Haigh is Professor of Disaster Resilience at the University of Huddersfield, aligned to their School of Applied Sciences. He is also Co-Director of the Global Disaster Resilience Centre. His research interests include multi-hazard early warning, disaster risk governance, and resilience in the built environment. He is Editor-In-Chief of the *International Journal of Disaster Resilience in the Built Environment* and Co-Founder of the International Conference on Building Resilience series. He is also an expert member of Working Group 1 of Intergovernmental Coordination Group on the Indian Ocean Tsunami Warning and Mitigation System (ICG/IOTWMS) and was a UK Advocate for the UN campaign on Making Cities Resilient. Since 2014, Richard and his team have secured over 30 external research grants with 288 international partners in 57 countries, with a grant value in excess of £30m. Richard and his team won the 2019 Newton Prize for Indonesia, celebrating the best research innovation partnership for the project Mainstreaming Integrated Disaster Risk Reduction and Climate Change Adaption Strategies into Coastal Urban Agglomeration Policy. Richard has given over 100 invited speeches and keynote presentations for audiences in twenty-five countries across North America, South America, Europe, Asia, Africa, and Australasia. He has also published over 60 peer-reviewed journal articles, 2 edited books, 15 book chapters, and over 20 reports for a variety of stakeholders.

Ms. Shavindree Nissanka Shavindree Nissanka is currently a Senior Lecturer at the Department of Quantity Surveying, Faculty of Engineering, Sri Lanka Institute of Information Technology. While doing her PhD as a doctoral researcher affiliated with the Global Disaster Resilience Centre (GDRC), University of Huddersfield, UK, she worked as the lead researcher for the research collaboration Built Environment Learning for Climate Adaptation (BEACON), co-funded by the EU Erasmus programme involving six partner institutions from the UK, Malta, Sweden, Spain, and Sri Lanka). She has also contributed to other research initiatives led by GDRC in the fields of built environment learning for climate adaptation and disaster resilience in smart cities and infrastructure resilience such as developing disaster resilient building codes for Sri Lanka, Development of the Standard Operating Procedures for the Tsunami Early warning system in Sri Lanka and developing a guide for built environment professionals done by GDRC and Royal Institute of Chartered Surveyors (RICS). She has presented her work at many international conferences and has also published in the research areas of disaster resilient properties: built environment discourse, disaster resilient building codes in Sri Lanka, employment challenges in mega construction projects in Sri Lanka, and integration of the concept of disaster resilience for sustainable construction. She recently guest edited a special issue on "Climate change adaptation in the built environment" in the International Journal of Disaster Resilience in the Built Environment. In the year 2023, she has been named as one of the 50 rising stars in Environmental, Social and Governance categories at the "Women of the Future" program.

Prof. Nishara Fernando is Professor in Sociology attached to the Department of Sociology at the University of Colombo, Sri Lanka. He has been an academic for 24 years since 1999. Professor Fernando is a former Head of the Department of Sociology (2016–2019) and served as the director of the Social Policy Analysis and Research Center (SPARC) at the Faculty of Arts, University of Colombo, from 2015 to 2021. He earned his Bachelor's degree in Sociology and a Master of Philosophy Degree in Sociology from the University of Colombo, and a Postgraduate Diploma in Sociology from the University of La Trobe, Australia. He obtained his Ph.D. from the University of Bonn, Germany. His expertise encompasses a profound engagement with urban studies, particularly focusing on vulnerable urban groups and the intricacies of disaster and development-induced forced relocation. Professor Fernando was honored with the President's Award for Scientific Research in 2018. Throughout his 24 years of service, Prof. Fernando has made significant contributions as a principal investigator and country lead for numerous international sociological and multidisciplinary research projects. These projects were funded by organizations such as ERASMUS+, the Swiss National Science Foundation, UKGCRF, GIZ, IOM, ILO, the World Bank, Save the Children, the Bill Gates Foundation, and SOS Children's Village, among others. He currently holds the position of Senior Research Fellow at the National Centre for Advanced Studies (NCAS) in Sri Lanka.

Prof. Ruben Paul Borg is Structural and Materials Engineer and Academic at the University of Malta. He is Fulbright Scholar and Visiting Professor at the University of Missouri S&T USA. He is visiting academic in various International Institutions. He holds a Bachelor of Architecture and Civil Engineering (Hons.), University of Malta, Specialisation Degree in Reinforced Concrete Structures, Politecnico di Milano Italy, and Ph.D. Materials Engineering University of Sheffield, UK. Prof. Borg was elected Board of Directors member of iiSBE International Initiative for a Sustainable Built Environment, Chairman of ECCE European Council of Civil Engineers for Knowledge and Technology and Representative of the Institution of Civil Engineers UK. Prof. Borg is active in fib International Federation for Structural Concrete, TG7.8 convener, RILEM scientific committees, and appointed expert on European Committee for Standardization CEN TC350 Sustainability of Construction Works. He served as Board Member of the Executive Committee of the Building Industry Consultative Council, Government of Malta, and Chairman for Research and Innovation. Prof. Borg served on the Management Committee of various research projects in materials engineering, sustainability of construction and resilience, and is

Challenges of Built Environment's Stakeholders in Climate Change Adaptation

Analysis and Prediction of Flood Events in Colombo District, Sri Lanka Using Standard Precipitation Index (SPI) and Rainfall Anomaly Index (RAI)

Analysis of the UK Homeowners' Practices Prompting Housing Retrofits: A Practice Theory Approach

The Cultural Landscape of Portovenere, Cinque Terre, and the Islands (Palmaria, Tino and Tinetto) in Northern Italy: How to Tackle Climate Change in the Case of a Coastal UNESCO Site

Lifelong Learning of Coastal Communities: An Investigation on Nature-Based Solutions to Mitigate Climate Change Effects on Coastal Built Environments

Skill and Knowledge Gaps in Climate Change Adaptation: A Case of Sri Lanka

University Educational Contribution on the Improvement of Perceptions and Behavioral Aspects of Students for Climate Change Mitigation; A Case Study in Sri Lanka

Developing a Joint Master's Degree Programme in Agroecosystem Resilience for Sri Lanka

STEM Teaching Tools on Climate Education: A Situational Audit on Climate Pedagogy

Spatial–Temporal Trend Analysis of Seasonal and Annual Rainfall (1997–2022) Using Nonparametric Analysis with Climatic Zone Changes in Sri Lanka

Enhancing University-Enterprise Collaborations: A Framework for Disaster Resilience in Asia

Climate Change Education in Universities: A Sri Lankan Perspective

An Innovative Pedagogy in Climate Change Education: A Case of Problem-Based Teaching and Learning

Digital Competencies, Disaster Risk Reduction, and Education: An Introduction to the Digital Competency Framework to Develop Digital Pedagogical Competencies of Educators in Disaster Risk Reduction

Importance of Sustainable Utilization Practices and Emerging Technologies Towards Food Security

Unveiling the Green Trail: Harnessing RFID Technology for Sustainable Vegetable Traceability

The Resilience of Agro Ecosystems Through Higher Education; Identification of Key Thematic Areas

Green Consumer Behavior Index (GCBI): Evaluation of Consumer's Green Behaviour Towards Sustainable Consumption and Eco-conscious Society

Risk and Vulnerability of Food System to Climate Change: A Case Study of Upcountry Vegetable Supply Chains in Sri Lanka

Building Resilience for Adaptation to Climate Change in High Value Marine Fisheries: A Case on Lobster Fisheries, Sri Lanka

Economics, Marketing and Value Chain Enhancement of the Cottage Industry Based in Highly Vulnerable Agro-Ecosystem: A Case of Kithul (*Caryota urens*)

List of Tables

Analysis and Prediction of Flood Events in Colombo District, Sri Lanka Using Standard Precipitation Index (SPI) and Rainfall Anomaly Index (RAI)

Climate Adaptation in Built Environment: Urban Agriculture and Food Security with Innovative Approaches

Analysis of the UK Homeowners' Practices Prompting Housing Retrofits: A Practice Theory Approach

The Cultural Landscape of Portovenere, Cinque Terre, and the Islands (Palmaria, Tino and Tinetto) in Northern Italy: How to Tackle Climate Change in the Case of a Coastal UNESCO Site

Lifelong Learning of Coastal Communities: An Investigation on Nature-Based Solutions to Mitigate Climate Change Effects on Coastal Built Environments

Skill and Knowledge Gaps in Climate Change Adaptation: A Case of Sri Lanka

University Educational Contribution on the Improvement of Perceptions and Behavioral Aspects of Students for Climate Change Mitigation; A Case Study in Sri Lanka

Spatial–Temporal Trend Analysis of Seasonal and Annual Rainfall (1997–2022) Using Nonparametric Analysis with Climatic Zone Changes in Sri Lanka

Climate Change Education in Universities: A Sri Lankan Perspective

An Innovative Pedagogy in Climate Change Education: A Case of Problem-Based Teaching and Learning

Economics, Marketing and Value Chain Enhancement of the Cottage Industry Based in Highly Vulnerable Agro-Ecosystem: A Case of Kithul (*Caryota urens*)

Governance for Climate Change Adaptation

People-Centered Policy Guidelines for Sustainable Relocation of Communities Displaced by Climate Change: A Case Study of Sri Lanka

Nishara Fernando and Malith De Silva

1 Introduction

Climate change is causing unprecedented environmental challenges worldwide, leading to the displacement of vulnerable communities (Dasgupta et al., 2022; McAdam, 2020; Schwarz, 2023; Tabe, 2019). Rising sea levels, extreme weather events, prolonged droughts, and other climate-related phenomena have triggered migrations and forced relocation of populations from their homes (Ahmad et al., 2022; Clarke et al., 2022; Rahman, 2023; Yin, 2023). In Sri Lanka, a country highly susceptible to climate change impacts, climate-induced displacement poses a significant threat to the livelihoods and well-being of its inhabitants (Manawadu & Wijeratne, 2022; Sarker, 2023; Fernando et al., 2023). Sustainable relocation policies are essential in mitigating the adverse effects of climate-induced displacement and ensuring the well-being of affected communities (Chandra et al., 2023; Lankenau Ahumada, 2021). Poorly planned relocations can exacerbate existing vulnerabilities, disrupt social structures, and lead to increased poverty and marginalization. Hence, formulating people-centered policy guidelines is crucial to foster sustainable development and social equity in the face of climate-induced displacement. By examining existing policies and past relocation efforts, the study seeks to identify policy gaps and propose comprehensive guidelines that prioritize the needs and aspirations of affected communities.

N. Fernando (✉)
Department of Sociology, University of Colombo, Colombo, Sri Lanka
e-mail: nishara.fernando@soc.cmb.ac.lk

M. De Silva
Federation of Sri Lankan Local Government Authorities, Colombo, Sri Lanka

C. Malalgoda et al. (eds.), *Climate Change Adaptation in the Built Environment*,
https://doi.org/10.1007/978-3-031-75826-3_1

2 Climate-Induced Displacement and Relocation

Climate-induced displacement and relocation lack universally agreed definitions (Warner, 2010). It is often perceived as a straightforward causal process, where people are forced to leave their homes due to the direct impacts of climate change (UNESCO, 2020). However, in reality, climate-induced displacement is a complex and multi-causal phenomenon, and understanding the relationship between environmental changes and human mobility can be challenging. Factors such as population growth, underdevelopment, weak governance, and inadequate urban planning can diminish the resilience of local populations, rendering households more susceptible to climate stressors (Nansen Initiative, 2015).

The consequences of climate impacts can adversely affect livelihoods, particularly for those whose subsistence relies on agriculture. Such impacts can also disrupt opportunities and social networks, ultimately triggering migration. Additionally, gender, economic, political, and cultural elements can intertwine with environmental factors, influencing migratory decisions (IOM, 2012). In 2010, the Conference of the Parties (COP) to the United Nations Framework Convention on Climate Change (UNFCCC) emphasized the importance of enhancing understanding, coordination, and cooperation regarding climate change-induced displacement, migration, and planned relocation (UNFCCC, 2011).

Robert McLeman identifies two distinct categories of drivers for climate induced forced relocation: climate processes and climate events (Brown, 2008). Climate processes pertain to gradual changes, including sea-level rise, agricultural land salinization, desertification, and the growing scarcity of water and food, leading to long-term environmental challenges. In contrast, climate events encompass sudden and dramatic hazards like monsoon floods, glacial lake outburst floods, storms, hurricanes, and typhoons, causing acute and immediate impacts on communities (Brown, 2008).

Climate-induced forced relocation has profound and wide-ranging impacts on affected communities, touching various dimensions of their lives. Primarily, the process leads to social disruptions as communities are uprooted from their familiar environments, resulting in the loss of social networks, cultural heritage, and traditional ways of life (Iuchi & Mutter, 2020). The breakdown of social cohesion and community ties can lead to increased psychological distress and mental health challenges among the relocated population (Holmes & Marcelli, 2020). Economically, climate-forced relocation often leads to the loss of livelihoods, particularly for those whose subsistence depends on agriculture or localized economic activities (Adger et al., 2005). The lack of adequate compensation and alternative income opportunities can perpetuate poverty and exacerbate existing inequalities (Cornish, 2020; Kumar et al., 2021).

Moreover, health-related repercussions are observed due to changes in living conditions and exposure to new environmental risks in relocation sites (Cunsolo Willox et al., 2012). The increased vulnerability to infectious diseases, mental health issues, and limited access to healthcare facilities can negatively impact the well-being of displaced communities (Bürgin et al., 2022; Hirani & Richter, 2019; McMichael, 2020).

3 Need for a People-Centered Relocation

The concept of people-centered policies emerged as a response to the shortcomings of earlier relocation practices that neglected the needs and rights of the affected populations (Barnett & Webber, 2010). It signifies a shift from top-down approaches to more inclusive, participatory, and rights-based strategies that aim to empower the displaced communities and ensure their active involvement in decision-making processes (Black et al., 2011).

People-Centered relocation policies gained significant attention following the Indian Ocean Tsunami of 26th December 2004. In response to this catastrophic event, the United Nations International Strategy for Disaster Reduction (UNISDR) strongly recommended the establishment of national level people-centered DRR policies (UNISDR, 2015). The Sendai Framework for Disaster Risk Reduction also recognizes the importance of involving citizens and non-professionals in dealing with multiple hazards (UNISDR, 2015).

The authors belief that policies can only be effective when the beneficiaries involved perceive them as viable and relevant to their circumstances. Therefore, having reliable information and insights from affected communities becomes a foundational step in ensuring people-centered policies (Ajulor, 2018; Gold, 2014). A qualitative method such as Participatory Research is highly recommended by researchers as the appropriate approach for gathering people-centered evidence (Freudenberg & Tsui, 2014; Salimi et al., 2012). According to Salimi et al. (2012), participatory research addresses the challenge of engaging people in policymaking and can lead to social transformation in the design, implementation, and evaluation of policies, particularly in understanding health disparities and determinants.

Furthermore, it is crucial to distinguish between people-centered policy formulation and people-centered policy implementation. Although these terms are often treated separately, policy implementation is an integral component of the policy development process, aimed at actualizing the formulated policy (Ajulor, 2018; Gold, 2014). A people-centered policy approach recognizes the significant impact of beneficiaries in implementing policies. Encouraging the active participation of beneficiaries in the formulation process, such as being members of committees, allows them to define problems and propose solutions beyond expert knowledge. By involving beneficiaries in the formulation process, the policy becomes more attuned to their needs and realities.

4 Existing Policies and Guidelines Related to Involuntary/voluntary Relocation

The study delved into prominent international guidelines such as the World Bank Operational Policy on Involuntary Resettlement (2002), the Involuntary Resettlement Safeguards (2012) of the Asian Development Bank, and The Nansen Initiative (2015). On a national level, the research explored the National Involuntary Resettlement Policy (2001) and the Land Acquisition Act (1950). By exploring these significant documents, the study sought to gain a comprehensive understanding of the principles and measures guiding involuntary relocation and resettlement at both international and national levels.

5 International Guidelines

The **World Bank Operational Policy on Involuntary Resettlement** (OP 4.12) was adopted in 2002 as part of the World Bank's safeguards policies and was updated in 2013 (The World Bank, 2015). Its formulation was prompted by the recognition of the significant social and economic impacts experienced by communities displaced involuntarily due to World Bank-funded projects. The policy aimed to ensure that involuntary resettlement is carried out in a manner that respects the rights and interests of affected communities while promoting sustainable development. The key pillars of the World Bank Operational Policy on Involuntary Resettlement include, (1) Entitlements and Compensation: The policy emphasizes that affected individuals and communities should be provided with adequate compensation for lost assets, land, and livelihoods. Compensation should reflect the full replacement cost of affected assets, and displaced households should be assisted in restoring their standard of living; (2) Consultation and Participation: It highlights the importance of meaningful consultation with affected communities throughout the resettlement process. Communities should have the opportunity to express their preferences, needs, and concerns, and their views should be taken into account in project planning and implementation; (3) Livelihood Restoration and Improvement: The policy promotes efforts to restore or improve the livelihoods of displaced households. This may include support for income-generating activities, vocational training, and access to essential services; (4) Resettlement Planning and Implementation: The policy requires the preparation of comprehensive resettlement plans that address the needs of affected communities. The plans should be implemented prior to displacement and should include measures to minimize adverse impacts and ensure the sustainability of resettlement outcomes and (5) Grievance Redress Mechanism: It mandates the establishment of a grievance redress mechanism to address and resolve concerns and disputes arising from the resettlement process. The mechanism should be accessible, transparent, and capable of providing timely solutions to grievances (The World Bank, 2015).

The **World Bank's Environmental and Social Framework (ESF),** approved on August 4, 2016, represents a significant milestone in promoting sustainable development, environmental protection, and social inclusivity. Comprising a Vision for Sustainable Development along with 10 Environmental and Social Standards (ESSs), an Environmental and Social Policy for Investment Project Financing (IPF), and associated Directives, the ESF sets forth a comprehensive framework that guides both the Bank and Borrowers in their efforts to balance development with environmental and social considerations (The World Bank, 2016).

One of the pivotal aspects of the ESF is its attention to land acquisition and involuntary relocation, critical issues often faced in development projects. The framework is designed to enhance protections for individuals and communities affected by land acquisition, displacement, and resettlement. The ESF promotes a risk-based approach, tailored to the complexities of each project, and ensures that safeguards are in place to address the potential adverse impacts of land acquisition and relocation.

Under the ESF, the emphasis on stakeholder engagement and consultation is noteworthy. This proactive approach empowers individuals and communities to voice their concerns, contribute to decision-making, and seek resolution for grievances. Responsive grievance mechanisms are a key component, providing a platform for affected parties to raise issues and find solutions collaboratively.

Furthermore, the ESF underscores the importance of strengthening national environmental and social management systems and institutions (The World Bank, 2016). This capacity building approach promotes sustainable practices and empowers local stakeholders to take a more active role in the development process, ultimately enhancing project outcomes and community well-being.

Within the framework of the **United Nations Framework Convention on Climate Change** (UNFCCC), the strategic work stream (d) of the five-year work plan of the Executive Committee is dedicated to enhancing cooperation and facilitation concerning human mobility, migration, displacement, and planned relocation (UNFCCC, n.d). In response to the urgent need for integrated approaches to address the adverse impacts of climate change on human mobility, the Executive Committee has taken significant steps to address this complex challenge.

The UNFCCC's strategic work stream (d) focuses on addressing loss and damage associated with climate change through three key pillars (UNFCC, n.d). Firstly, it aims to enhance understanding by addressing knowledge gaps, collecting and sharing data, and highlighting best practices. Secondly, it emphasizes dialogue, coordination, and synergy-building among stakeholders within and beyond the Convention, fostering a comprehensive response. Lastly, the work stream emphasizes actionable support by providing technical guidance, recommendations, and expertise mobilization to enable countries to effectively manage and mitigate loss and damage (UNFCC, n.d).

The Involuntary Resettlement Safeguards of the Asian Development Bank (ADB) were introduced in 2012 as part of its Safeguard Policy Statement. The initiation of these safeguards was driven by the recognition of the potential adverse impacts that ADB-funded projects might have on affected communities, particularly when it involves involuntary resettlement (Asian Development Bank, 2012).

The key pillars of the ADB's Involuntary Resettlement Safeguards revolve around the principles of fairness, consultation, and restoration (Asian Development Bank, 2012). Firstly, the safeguards emphasize the fair and prompt compensation for the loss of assets and livelihoods resulting from resettlement (Asian Development Bank, 2012). Secondly, meaningful consultation and participation of affected communities are central to the safeguards. This participatory approach aims to foster community ownership and minimize negative impacts on their social and cultural well-being (Sapkota, 2017). Lastly, the Involuntary Resettlement Safeguards underscore the importance of livelihood restoration and improvement. This assistance might include vocational training, access to credit facilities, or job placement programs, fostering sustainable development and economic resilience (Asian Development Bank, 2012).

The Nansen Initiative, launched in 2015, is a landmark initiative focused on addressing the challenges faced by individuals and communities displaced across borders due to natural disasters, including the adverse effects of climate change (The Nansen Initiative, 2015). The Nansen Initiative was initiated as a state-led process with the support of the Governments of Norway and Switzerland, in collaboration with the Norwegian Refugee Council and the Office of the United Nations High Commissioner for Refugees (UNHCR) (The Nansen Initiative, 2015).

The key pillars of The Nansen Initiative are centered around its "Protection Agenda," which outlines principles and guidelines to address the protection needs of people displaced across borders in the context of natural disasters and climate change (McAdam, 2016). The initiative aims to enhance understanding, coordination, and cooperation among governments, international organizations, and civil society in providing protection and assistance to those affected by cross-border displacement.

The Nansen Initiative encourages international cooperation and collaboration to enhance the protection of displaced populations and emphasizes the significance of including affected communities in decision-making processes (McAdam, 2016).

6 Land Acquisition Act of 1950 (LAA)

The Land Acquisition Act (LAA) serves the purpose of establishing comprehensive procedures and guidelines governing the acquisition of private land and servitudes. Furthermore, the Act encompasses all aspects related to the acquisition process. Whenever the state intends to take over private land, it is imperative that they adhere to the procedural framework outlined within the LAA (LAA, 1950).

Senanayakea (2019) notes that the Land Acquisition Act (LAA) provides a comprehensive framework that governs the different stages of the land acquisition process. This encompassing framework involves a detailed procedure that includes multiple steps such as notifying affected parties, evaluating suitable compensation, and resorting to legal actions if required (Senanayakea, 2019).

7 National Involuntary Resettlement Policy (NIRP)

The National Involuntary Resettlement Policy (NIRP) was developed in the year 2000 and subsequently approved by the Cabinet in 2001. The primary goal of this policy was to address issues concerning involuntary resettlement resulting from development projects in both the public and private sectors. The NIRP aimed to ensure that affected communities are not adversely impacted and are provided with the necessary support to restore their standard of living and successfully integrate into their new environment (Godamunne, 2012). Additionally, the policy enforces that compensation for land should be equivalent to its replacement value. According to Godamunne (2012), a fundamental aspect of the policy is the identification and support of vulnerable groups, aiming to enhance their quality of life. Quoting Muggah (2008) Godamunne notes that NIRP is a "soft law", a normative framework that exists to address deficiencies in existing legislation and lacks direct legal standing in a court of law.

Above discussed policies display a number of shortcomings and pitfalls. These shortcomings include lack of participation of communities affected by or vulnerable to relocation in the formation of policies and guidelines. As a result, the policies and guidelines represent expert opinions on needs of relocated communities, instead of representing the true needs of affected communities. Further the failure to incorporate community opinions and needs in formulating policies may also interrupt proper implementation due to lack of acceptance of communities. Moreover, it is evident that policies have also failed to integrate ground level officials, specially from the provincial administrative and management services to the formation and implementation of policies that can further contribute to failure of policies and guidelines.

8 Methodology

8.1 Data Collection

The study utilized secondary and primary data. Secondary data was collected from both international and national guidelines and policies that regulates involuntary relocation and resettlement. International guidelines explored under the study included World Bank Operational Policy on Involuntary Resettlement (2002), Involuntary Resettlement Safeguards (2012) of Asian Development Bank and The Nansen Initiative (2015). National policies included the National Involuntary Resettlement Policy (2001) and Land Acquisition Act (1950).

Primary data was collected from four relocation cases using semi-structured interviews, namely, the post-tsunami relocation programme, Aranayake and Meeriyabadda landslide relocation programmes and Lunawa Environment Improvement and Community Development Project.

Due to the lack of theories and concepts gauging people centeredness of policies, (Leyns & De Maeseneer, 2013) the authors took a constructivist approach towards analyzing the data.

Constructivism posits that knowledge and meaning are products of social construction (Hjelm, 2019). Constructivist thinkers derive insights about social phenomena from social facts legitimized by collective agreement among social agents. Constructivists interpret social facts inductively through the dialectical interplay of knowledge and reality (Barbehön, 2020).

Glaser and Strauss (Morse et al., 2021) coined the term "grounded theory" to describe this approach to knowledge-building. It provides a systematic and rigorous iterative process for developing a theory rooted in empirical data (Salvini, 2019).

This study embraces the constructivist grounded theory approach, acknowledging the role of subjectivity in researchers' recognition and interpretation of collected data sets. Accordingly, data gathered from secondary and primary data sources were analyzed separately using thematic analysis method with a specific focus on identifying weaknesses related to people centeredness of relocation policies and programmes. The data was later triangulated to further improve the accuracy and clarity of the findings.

9 Selected Relocation Programmes

9.1 Aranayake Relocation Programme

The landslide that occurred in May 2016 at Samasarakanda mountain in the Aranayaka Divisional Secretariat Division (DSD) of Kegalle District in the Sabaragamuwa Province of Sri Lanka was a devastating natural disaster. The devastating landslide resulted in the displacement of 513 families and the complete destruction of 72 houses (Fernando & Jayasinghe, 2023). In the aftermath of the landslide, efforts were undertaken to relocate 512 families directly impacted by the disaster, as well as those residing in areas at risk of potential future landslides. This study has utilized data sourced from four resettlement villages: Duldeniya, Panapura Watta, Wasanthagama, and China Friendship Village.

9.2 Post-Tsunami Relocation Programme in the Akmeemana DS Division in Galle

The devastating Indian Ocean Tsunami that struck on December 26, 2004, claimed the lives of over 35,000 people in the country and rendered thousands without homes. Responding to the crisis, the Government of Sri Lanka introduced buffer zone regulations in 2005, resulting in the relocation of more than 70,000 families (Fernando,

2010). This study focuses on the outcomes of three specific relocated communities: Eco village, Salzburg village, and Tea Garden village.

9.3 Lunawa Environment Improvement and Community Development Project

The Lunawa Environmental Improvement and Community Development Project (LEICDP) emerged in 2000 as a response to the pressing challenges faced by a significant portion of the urban poor residing in underserved settlements (Jayakody et al., 2022).

The programme relocated 870 households and the three relocation options extended to these households included: settling in donor-built housing in relocation sites—Hike Terrace, Green View Garden, Riverside Garden and Lake View Garden, self-relocation to lands purchased by the families, or on-site relocation after regularizing their plots (Fernando & Punchihewa, 2013).

9.4 Meeriyabadda Landslide Relocation Programme

On October 29th, 2014, an extensive landslide occurred at the Meeriyabedda tea estate situated within the Haldummulla Divisional Secretariat Division of Sri Lanka's Badulla district (Dissanayake et al., 2018). Tragically, this event claimed 37 lives, resulted in the complete destruction of 70 houses, and rendered more than 275 individuals homeless. In response to this devastating landslide, a relocation initiative was undertaken to provide new housing for the victims (Maheshika & Sangasumana, 2017). The relocation effort aimed to move the victims to government-constructed houses in the Makaldeniya Division of the Maskeliya plantation, located 11 km away from the original settlement (Fernando & Jayasinghe, 2023).

10 Results and Discussions

By meticulously assessing these frameworks, policies and case studies, the research team has developed a series of recommendations which are discussed below in detail.

(a) **Ensure the participation of affected and vulnerable communities in the development of involuntary relocation policies for climate change induced relocation and resettlement**

The examination of international and national guideline and policy studies underscores a consistent trend in their formulation, specifically concerning climate-induced relocation and resettlement. These policies have often been crafted by expert panels well-versed in various fields such as relocation, resettlement, disaster risk reduction, law, financial management, and public administration. However, a noticeable void has been identified in the integration of voices from the communities directly impacted by climate-induced displacement. This lack of direct representation from the displaced and relocated communities in the policy-making process results in policies that inadequately address the genuine needs of these affected populations.

Integrating the insights of displaced and relocated communities serves to address several crucial aspects that could otherwise go unnoticed. Firstly, their lived experiences offer invaluable insights that cannot be replicated solely by external experts. Secondly, involving these communities ensures that policies respect their cultural identities and ways of life, thereby mitigating further marginalization and facilitating a smoother transition to new environments.

To effectively integrate affected and vulnerable communities into policy development for climate-induced relocation and resettlement, a participatory approach is essential. This approach entails establishing platforms for meaningful engagement at every stage of policy formulation, from initial planning to implementation and evaluation. Organizing public hearings, focus group discussions, consultations, and participatory workshops becomes instrumental in soliciting their input, addressing their concerns, and incorporating their suggestions.

However, in the process of integrating these communities, certain dynamics must be carefully navigated. One significant challenge pertains to potential power dynamics during consultations. Special attention should be given to ensuring equitable representation and amplification of marginalized groups, including women, minorities, children, and the elderly.

(b) **Placing greater emphasis on livelihood restoration and economic empowerment in the climate induced relocation and resettlement contexts**

Within the realm of climate-induced relocation and resettlement, it becomes imperative to prioritize the restoration of livelihoods and the promotion of economic empowerment. One critical aggravating factor is the considerable increase in the commuting distance, resulting from the relocation's geographical shift. Situated at a range of 10 to 20 km from the relocated individuals' workplaces, the new settlement's location has significantly heightened fuel and transportation costs, thereby imposing an additional financial burden on those affected. This predicament is further compounded by a notable proportion (85%) of respondents expressing dissatisfaction with the insufficient space available in their new settlement to sustain income-generating activities previously undertaken. An illustrative example is the inadequacy of garden space for animal husbandry, a pivotal income source for a significant segment of the reloated communities in Akmeemana, Galle. Moreover, a majority (85%) of respondents reported a decrease in income subsequent to their relocation.

A key challenge faced by the relocated individuals across all surveyed settlements pertains to their mobility. In the case of Tsunami resettlements, the necessity of commuting to Galle city early in the morning poses a formidable challenge as public bus services are sparse before 6:00 a.m. Similarly, returning from Galle after 9:00 p.m. presents a similar predicament. Moreover, the employment opportunities available in the newly established areas are comparatively fewer than those present in their previous locations. Nearly 94% of the resettlers interviewed, expressed their struggle in securing employment within the relocation sites. Furthermore, the monthly household expenditures of the involuntarily relocated individuals have witnessed an increase. This rise can be predominantly attributed to two key factors: escalating transportation costs and rising expenses related to essential utilities such as electricity, water, and gas.

These findings indisputably emphasize the importance of placing a more pronounced emphasis on the restoration of livelihoods within the context of climate-induced relocation and resettlement. Therefore, prioritizing comprehensive livelihood restoration and economic empowerment efforts is not only crucial in mitigating these challenges but also in cultivating self-reliance, fostering social stability, and facilitating the integration of affected communities.

To effectively realize these goals in the climate-induced relocation and resettlement context, policies and strategies should extend beyond the realm of mere restoration. The integration of the following key components into policy frameworks is essential to achieve this overarching objective:

(i) **Designing and implementing skill development programmes that are specifically tailored to the needs and aspirations of the relocated population**

These programmes should align with local market demands, enabling individuals to access viable job opportunities or establish their own enterprises. The programmes should consider prevailing socio-economic conditions, cultural nuances, and available resources. These skill development initiatives can equip individuals with competencies that directly translate into meaningful employment opportunities or entrepreneurial ventures.

(ii) **Ensuring access to financial resources (micro credit, grants, or loans)**

Facilitating financial inclusion is a vital component of effective people-centered relocation policies. For instance, access to micro credit initiatives can play a pivotal role in providing small-scale entrepreneurs with the capital needed to establish or scale up their businesses. By offering accessible and manageable loans, often without requiring traditional collateral, microcredit enables individuals to venture into income-generating activities that might have been otherwise unattainable.

(iii) **Facilitating market access for relocated individuals' products and services**

This strategic focus empowers the relocated community to effectively engage in economic activities, capitalize on their skills and resources, and contribute to local economic growth. Developing direct connections between relocated individuals and potential buyers, distributors, and suppliers is crucial. Creating platforms for networking, trade fairs, and business exhibitions allow entrepreneurs to showcase their offerings and forge valuable relationships within the broader business eco-system.

(c) **Ensuring effective implementation of policies**

In 2015, World Bank Group President, Jim Yong Kim, expressed deep concern over the World Bank findings, highlighting major problems in overseeing projects involving relocation. He had emphasized that the World Bank had failed to implement plans effectively and establish strong tracking systems to ensure policy compliance (The World Bank, 2015). The acknowledgment by the President of the World Bank Group underscores the significance of ensuring the efficient enforcement of climate change induced relocation policies and directives. This can be achieved by integrating monitoring and assessment mechanisms into involuntary resettlement policies.

On a national level, it is essential to recognize that the National Involuntary Resettlement Policy (NIRP) operates as a "soft law," a normative structure designed to address gaps in prevailing legislation, yet it lacks enforceability in a court of law. Moreover, it has come to light that implementing agencies in Sri Lanka have chosen to selectively apply the NIRP in specific cases. The relocation of displaced individuals from Sampur area serves as an instance of a development initiative that neglected to adhere to the principles of the NIRP. In 2006, portions of Sampur were designated as a Special Economic Zone (SEZ), resulting in the displacement of families owning land in that region. Subsequently, the area was reclassified as a high security zone (HSZ), and in 2012, a Gazette notification was issued, announcing the transformation of this land into a Special Zone for Heavy Industries (Fonseka & Jegatheeswaran, 2013). This declaration displaced majority of the Individuals that lived within the designated zones. These individuals received a standardized compensation and relocation package, which falls short of addressing the requirements of affected individuals and did not meet the benchmarks outlined by the NIRP (Fonseka & Jegatheeswaran, 2013).

Achieving proper implementation of involuntary relocation policies in relocation programmes demands a comprehensive and strategic approach encompassing several critical components. One fundamental pillar is establishing a robust monitoring and evaluation system (Lian et al., 2022). This entails defining clear and measurable indicators to track progress and assess outcomes systematically.

Effective interagency coordination stands as another cornerstone of successful implementation. Given the intricate nature of relocation policies, cohesive collaboration among diverse government entities, departments, and sectors is essential. Instituting mechanisms for regular communication, information exchange, and collective decision-making prevents duplication of efforts and gaps in execution.

Thirdly ensuring compliance with legal and policy frameworks is of paramount importance. Striving to align with both national and international laws and standards, governments should safeguard human rights, particularly those of vulnerable or marginalized groups. In the Sri Lankan context, mandating strict adherence to the National Involuntary Resettlement Policy (NIRP) is not only advisable but also imperative for the effective and equitable implementation of involuntary relocation programmes.

In essence, ensuring proper implementation of involuntary relocation policies demands a holistic approach that encompasses diligent monitoring, effective coordination, legal compliance, and the capacity to adapt.

(d) **Treating climate-induced relocation and resettlement as a long-term process**

The exploration of climate-induced relocation and resettlement cases presented in this paper brings to light a spectrum of enduring issues and challenges faced by communities subjected to involuntary relocation. Particularly within the context of relocation sites following climatic events such as tsunamis, respondents highlight an array of persistent negative impacts on livelihoods, daily routines, and overall quality of life. For instance, the relocation process disrupts various sources of income, including fishing, local transportation services, food vending, and tourism-related activities. Consequently, a substantial number of resettled individuals not only slide into poverty but also experience economic and social marginalization. Under these unfavorable conditions, a significant portion of the relocated population is compelled to return to areas designated as buffer zones.

Adopting a perspective that treats climate-induced relocation and resettlement as a long-term process necessitates a departure from the short-sighted approaches, underlining the importance of comprehensive strategies that extend beyond immediate challenges. While short-term strategies might address initial concerns effectively, they frequently fall short in cultivating sustainable solutions that ensure the well-being and sustainable development of relocated communities over an extended period.

A long-term perspective facilitates the establishment of sustainable livelihoods, social cohesion, and community resilience, thereby ensuring the well-being of relocated individuals well beyond the initial phases of displacement. Furthermore, adopting a long-term outlook on climate-induced relocation and resettlement enables effective adaptation to evolving circumstances and unforeseen challenges. Communities grapple with changing dynamics, including shifts in employment opportunities, market demands, and environmental conditions. A long-term approach allows for the flexibility to recalibrate strategies and responses as needed, ensuring that the provided support remains pertinent and impactful over the years.

(e) **Integrating ground level officials into climate-induced relocation and resettlement programmes**

A significant revelation arising from the examined case studies is the notable absence of local-level officials within the climate-induced relocation and resettlement process. This includes officials from District and Divisional Secretariats, as well as Grama Niladari (GN) officers. Interviews conducted with GN officers from Tsunami relocation sites revealed that they were excluded from the relocation activities. Consequently, their understanding of the relocation process, related issues, and challenges faced by relocated communities remains limited. This lack of involvement has seemingly contributed to a lack of consideration for the well-being of the relocated populations.

One of the primary advantages of incorporating local-level officials is their ability to serve as bridges of information between the implementing agencies and the affected communities. Their familiarity with the community's social structure, livelihood patterns, and resource distribution can facilitate more precise and contextually appropriate planning. Local-level officials can play a pivotal role in ensuring that the relocation process aligns with the aspirations and priorities of the displaced communities. This inclusivity can help mitigate potential conflicts, encourage community ownership, and ultimately contribute to the long-term success and resilience of the relocation efforts.

Early engagement of officials stands as a pivotal step, entailing the involvement of local officials from the outset of relocation planning. Their insights at this initial stage shape decision-making, anchoring relocation strategies in local realities and enhancing their efficacy. Capacity building emerges as another crucial facet, involving the provision of training and capacity-building initiatives for local officials. The establishment of joint task forces or committees constitutes another impactful strategy.

11 Conclusion

In conclusion, this research highlights the pressing need to address shortcomings of existing relocation policies and recommends a series of factors that should be taken in to consideration while formulating and implementing community centered policies and guidelines. It proposes prioritizing the integration of affected and vulnerable communities in policy development, ensuring that their voices, experiences, and perspectives are central to the formulation process. Moreover, the study underscores the importance of placing a heightened emphasis on livelihood restoration and economic empowerment through skill development, access to financial resources, and facilitating market linkages.

Furthermore, it portrays the significance of effective implementation of relocation policies to ensure that the rights and well-being of displaced communities are upheld. Lastly, recognizing and integrating local-level officials emerge as a pivotal element in successful relocation policies. Their intimate understanding of the local context, expertise, and their role as conduits of information can greatly enhance the effectiveness and relevance of policy decisions. Nevertheless, it must be emphasized here that these recommendations cannot be generalized to each and every relocation scenario in Sri Lanka. Cultural, geographical, social and political elements associated with relocation and resettlement may vary from one relocation case to another. While acknowledging this limitation authors suggests that by contextualizing the recommendations this particular limitation can be addressed. Furthermore, further research should be carried out to identify further elements which can further improve the 'people-centeredness' of relocation policies, especially in relation to introducing inter and intra disciplinary approaches, assuring continued community participation and integration of lay people as experts of sustainable relocation.

References

Adger, W. N., Arnell, N. W., & Tompkins, E. L. (2005). Successful adaptation to climate change across scales. *Global Environmental Change, 15*(2), 77–86.

UNISDR (United Nations International Strategy for Disaster Reduction). (2015). *Sendai framework for disaster risk reduction 2015–2030*. UNISDR.

Ahmad, M. M., Yaseen, M., & Saqib, S. E. (2022). Climate change impacts of drought on the livelihood of dryland smallholders: Implications of adaptation challenges. *International Journal of Disaster Risk Reduction, 80*, 103210.

Ajulor, O. V. (2018). The challenges of policy implementation in Africa and sustainable development goals. *PEOPLE: International Journal of Social Sciences, 3*(3), 1497–1518.

Asian Development Bank. (2012). Safeguard Policy Statement. Retrieved from https://www.adb.org/documents/safeguard-policy-statement#:~:text=The%20SPS%20aims%20to%20promote,adverse%20project%20impacts%20on%20the

Barbehön, M. (2020). Reclaiming constructivism: Towards an interpretive reading of the 'social construction framework.' *Policy Sciences, 53*(1), 139–160.

Barnett, J. R., & Webber, M. (2010). *Accommodating migration to promote adaptation to climate change* (5270). World Bank Policy Research Working Paper.

Black, R., Adger, W. N., Arnell, N. W., Dercon, S., Geddes, A., & Thomas, D. (2011). The effect of environmental change on human migration. *Global Environmental Change, 21*, S3–S11.

Brown, O. (2008). *Migration and climate change. IOM migration research series no. 31* (vol. 7). Geneva: International Organization for Migration.

Bürgin, D., Anagnostopoulos, D., Vitiello, B., Sukale, T., Schmid, M., & Fegert, J. M. (2022). Impact of war and forced displacement on children's mental health—multilevel, needs-oriented, and trauma-informed approaches. *European Child and Adolescent Psychiatry, 31*(6), 845–853.

Chandra, A., McNamara, K. E., Clissold, R., Tabe, T., & Westoby, R. (2023). Climate-induced non-economic loss and damage: Understanding policy responses, challenges, and future directions in Pacific small island developing states. *Climate, 11*(3), 74.

Clarke, B., Otto, F., Stuart-Smith, R., & Harrington, L. (2022). Extreme weather impacts of climate change: An attribution perspective. *Environmental Research: Climate, 1*(1), 012001.

Cornish, G. E. (2020). *Settling into a New Place: Livelihood recovery and belongingness of households forced to relocate in Yangon, Myanmar*.

Cunsolo Willox, A., Harper, S. L., Ford, J. D., Landman, K., Houle, K., & Edge, V. L. (2012). "From this place and of this place:" Climate change, sense of place, and health in Nunatsiavut, Canada. *Social Science and Medicine, 75*(3), 538–547.

Dasgupta, S., Wheeler, D., Bandyopadhyay, S., Ghosh, S., & Roy, U. (2022). Coastal dilemma: Climate change, public assistance and population displacement. *World Development, 150*, 105707.

Dissanayake, P., Hettiarachchi, S., & Siriwardana, C. (2018). Increase in disaster risk due to inefficient environmental management, land use policies and relocation policies. Case studies from Sri Lanka. *Procedia Engineering, 212*, 1326–1333.

Fernando, N. (2010). *Forced Relocation after the Indian ocean tsunami 2004 case study of vulnerable population in three relocation settlements in Galle, Sri Lanka*.

Fernando, N., & Jayasinghe, N. (2023). Disaster-induced relocation of vulnerable households: Evidence from planned relocation in Sri Lanka. *Rebuilding communities after displacement: Sustainable and resilience approaches* (pp. 245–268). Springer International Publishing.

Fernando, N., & Punchihewa, A. G. (2013). *Relocating the displaced strategies for sustainable relocation*. Colombo, Sri Lanka: Friedrich Ebert Stiftung.

Fernando, R. L. S., Siriwardhana, M. K., Kularathna, E. A. I., & Geethamali, H. K. (2023). Disaster, displacement and resilient livelihoods in Sri Lanka: Policy imperatives. In *Disaster, displacement and resilient livelihoods: Perspectives from South Asia* (pp. 163–191). Emerald Publishing Limited.

Fonseka, B., & Jegatheeswaran, D. (2013). *Politics, policies, and practices with land acquisitions and related issues in the North and East of Sri Lanka*. Centre for Policy Alternatives.
Freudenberg, N., & Tsui, E. (2014). Evidence, power, and policy change in community-based participatory research. *American Journal of Public Health, 104*(1), 11–14.
Godamunne, N. (2012). *Development and displacement: The national involuntary resettlement policy (NIRP) in practice*.
Gold, J. (2014). *International delivery: Centres of government and the drive for better policy implementation*. Mowat Centre for Policy Innovation.
Hirani, S. A. A., & Richter, S. (2019). Maternal and child health during forced displacement. *Journal of Nursing Scholarship, 51*(3), 252–261.
Hjelm, T. (2019). Rethinking the theoretical base of Peter L. Berger's sociology of religion: Social construction, power, and discourse. *Critical Research on Religion, 7*(3), 223–236.
Holmes, L. M., & Marcelli, E. A. (2020). Neighborhood social cohesion and serious psychological distress among Brazilian immigrants in Boston. *Community Mental Health Journal, 56*, 149–156.
IOM. (2012). *International dialogue on migration 18: Climate change, environmental degradation and migration*. Geneva, Switzerland.
Iuchi, K., & Mutter, J. (2020). Governing community relocation after major disasters: An analysis of three different approaches and its outcomes in Asia. *Progress in Disaster Science, 6*, 100071.
Jayakody, C., Malalgoda, C. I., Amaratunga, D., Haigh, R., Liyanage, C., Hamza, M., Witt, E., & Fernando, N. (2022). Addressing housing needs of the displaced people promoting resilient and sustainable communities. *International Journal of Disaster Resilience in the Built Environment, 13*(3), 368–385.
Kumar, A., Singh, N., Cooper, S., Mdee, A., & Singhal, S. (2021). Infrastructural violence: Five axes of inequities in water supply in Delhi, India. *Frontiers in Water, 3*, 727368.
Land Acquisition Act of Sri Lanka. (1950). https://faolex.fao.org/docs/pdf/srl13617.pdf
Lankenau Ahumada, R. (2021). *Weathering the storm: Harnessing climate migration from Central America to advance the sustainable development agenda*.
Leyns, C. C., & De Maeseneer, J. (2013). Conceptualizing person-and people centeredness in primary health care: A literature review. *International Journal of Person Centered Medicine, 3*(1), 13–22.
Lian, H., Shi, G., & Xu, J. (2022). A Study on updating the model for monitoring and evaluation of involuntary resettlement based on the experience of China. *Processes, 10*(2), 225.
Maheshika, H., & Sangasumana, P. (2017). Issues and challenges of post landslide management in Sri Lanka (a case study of Meeriyabedda landslide in Badulla District). *International Journal of Scientific and Research Publications, 7*(12).
Manawadu, L., & Wijeratne, V. P. I. S. (2022). Human mobility response to natural disasters and environmental change. In C. Change (Ed.), *Disaster and Adaptations: Contextualising Human Responses to Ecological Change* (pp. 229–242). Springer International Publishing.
McAdam, J. (2016). From the Nansen initiative to the platform on disaster displacement: Shaping international approaches to climate change, disasters and displacement. *The University of New South Wales Law Journal, 39*(4), 1518–1546.
McAdam, J. (2020). Protecting people displaced by the impacts of climate change: The UN human rights committee and the principle of non-refoulement. *American Journal of International Law, 114*(4), 708–725.
McMichael, C. (2020). Human mobility, climate change, and health: Unpacking the connections. *The Lancet Planetary Health, 4*(6), e217–e218.
Morse, J. M., Bowers, B. J., Clarke, A. E., Charmaz, K., Corbin, J., & Porr, C. J. (2021). The maturation of grounded theory. In *Developing grounded theory* (pp. 3–22). Routledge.
Muggah, R. (2008). *Relocation failures in Sri Lanka: A short history of internal displacement and resettlement*. Macmillan.
Rahman, M. M. (2023). *Impact of climate change on the Kuakata sea shore of Bangladesh*. Available at SSRN 4427286.

Salimi, Y., Shahandeh, K., Malekafzali, H., Loori, N., Kheiltash, A., Jamshidi, E., Frouzan, A. S., & Majdzadeh, R. (2012). Is community-based participatory research (CBPR) useful? A systematic review on papers in a decade. *International Journal of Preventive Medicine, 3*(6), 386.

Salvini, A. (2019). The methodological convergences between symbolic interactionism and constructivist grounded theory. *Przegląd Socjologii Jakościowej, 15*(3), 10–29.

Sapkota, N. (2017). *Involuntary resettlement and sustainable development: Conceptual framework, reservoir resettlement policies, and experience of the Yudongxia Reservoir.*

Sarker, M. N. I. (2023). Livelihood resilience of climate-induced displaced people in South Asia: Implications for Bangladesh. In *Disaster, displacement and resilient livelihoods: Perspectives from South Asia* (pp. 81–98). Emerald Publishing Limited.

Schwarz, J. (2023). Climate change, human displacement, and STEM education: Toward a more transdisciplinary and inclusive culture of science. In *Migration, displacement, and higher education: Now what?* (pp. 219–231). Cham: Springer International Publishing.

Senanayakea, A. C. (2019). *A review of existing laws and policies applicable to disaster induced resettlement in Sri Lanka.* In *Capacity building for research and innovation in disaster resilience* (p. 3).

Tabe, T. (2019). Climate change migration and displacement: Learning from past relocations in the Pacific. *Social Sciences, 8*(7), 218.

The Nansen Initiative. (2015). Agenda for the protection of cross-border displaced persons in the context of disasters and climate change ('Protection Agenda Vol 1').

The World Bank. (2015). *World bank acknowledges shortcomings in resettlement projects, announces action plan to fix problems.* Retrieved from https://www.worldbank.org/en/news/press-release/2015/03/04/world-bank-shortcomings-resettlement-projects-plan-fix-problems

The World Bank. (2016). *Environmental and social framework.* Retrieved from The World Bank https://thedocs.worldbank.org/en/doc/837721522762050108-0290022018/original/ESFFramework.pdf

UNESCO. (2020). *The impact of climate displacement on the right to education; examining the educational vulnerabilities linked to climate displacement and the implications for national law and policy.* https://reliefweb.int/attachments/842f4a05-072d-35e5-9f81-b685b3c7d557/374966eng.pdf

UNFCCC, D. (2011). 1/CP. 17 in Report of the conference of the parties on its seventeenth session. In *United Nations framework convention on climate change.*

United Nations Framework Convention on Climate Change. (n.d.). *Migration, displacement and human mobility.* Retrieved from https://unfccc.int/wim-excom/areas-of-work/migration-displacement-and-human-mobility?gclid=Cj0KCQjwoeemBhCfARIsADR2QCvk3nfWnedr8sgbgnEEUWbIsuNPWXAl1AHtR5eZvxB4KlWPcLdJgcoaAvAFEALw_wcB

Warner, K. (2010). Global environmental change and migration: Governance challenges. *Global Environmental Change, 20*(3), 402–413.

Yin, J. (2023). Rapid decadal acceleration of sea level rise along the US East and Gulf coasts during 2010–22 and its impact on hurricane-induced storm surge. *Journal of Climate, 36*(13), 4511–4529.

Disaster and Climate Risk-Sensitive Business Continuity Plans for SMEs in the Post-COVID19 Context—Key Lessons from Sri Lanka

A. M. Aslam Saja, K. Suresh, S. Weerasekara, A. L. F. Nadhiya, and Minuri Perera

1 Introduction

Much like elsewhere in the Global South, the Small and Medium Enterprise (SME) sector plays a vital role in Sri Lanka's economic growth and socio-economic development (Central Bank of Sri Lanka, 2021). The SME sector accounts for 75% of total enterprises and 45% of total establishment employment, and contributes up to 52% of the GDP in Sri Lanka (Bandara et al. 2020). The World Bank (2022) predicts that by 2030, the world would need to create around 600 million new jobs to accommodate the expanding labour force. Accordingly, given typically high levels of population growth in developing countries the role of SMEs as an employment provider becomes extremely vital. Importantly 7 out of 10 jobs are generated by SMEs in most developing markets (World Bank, 2022). In addition, despite the comparatively smaller size of the organization, the level of contribution to the economy by the SMEs is substantial due to several reasons such as, SMEs are contributed to increasing the income levels of low-income families, providing employment opportunities for the majority of employment, acting as a breeding ground for large organizations, increasing competition among firms.

A. M. Aslam Saja (✉) · A. L. F. Nadhiya
South Eastern University of Sri Lanka, Oluvil 32360, Sri Lanka
e-mail: saja.aslam@seu.ac.lk

K. Suresh
Eastern University of Sri Lanka, Batticaloa, Sri Lanka

S. Weerasekara
Central Bank of Sri Lanka, Colombo, Sri Lanka

M. Perera
University of East Anglia, Norwich, United Kingdom

C. Malalgoda et al. (eds.), *Climate Change Adaptation in the Built Environment*,
https://doi.org/10.1007/978-3-031-75826-3_2

Despite the importance of SMEs to economies, SMEs across the world constantly endure difficulties due to climate induced disasters, pandemics and political instability, conflicts, etc. In the context of the impact of climate change, growing climate threats are especially challenging for small and medium businesses as they face more financial constraints than larger firms. Investing on effective adaptation strategies and mitigation technologies and proper risk management are some of the efforts that the governments should take to reduce the cost of a disaster. However, given the absence of an enabling business environment to promote SME growth in most of the developing nations and the resultant low adaptive capacities, SMEs of such economies are prone to immense risk. In Sri Lanka, most SMEs have been facing simultaneous disasters alongside the COVID-19 pandemic and the economic crisis that followed suit, alongside climate change related risks that are increasing in both frequency and severity. The nexus between disaster risk reduction, climate change adaptation, and risk-sensitive development as an integrated approach to sustainable development interventions in all levels of risk and development governance had been lacking since the adoption of the SFDRR in 2015 (Saja et al., 2020).

New integrated and mainstreamed models for sustainable and risk-informed development approaches as opposed to the existing stand-alone approaches can increase aid effectiveness and become more efficient for resilience investment decision making. For example, a DRR (Disaster Risk Reduction) and CCA (Climate Change Adaptation) integration model centred on an ecosystem based DRR is found to be helpful in four key realms: Development Planning; Disaster Risk Reduction; Climate Change Adaptation; and Ecosystem Management (Zarouk et al., 2021). However, most often these key 'realms' are not integrated into key functions of local governance and projects implemented by development initiatives, although the local authorities and development partners have an important role to play within said realms.

In general, the business plans for the SMEs are developed using the economic layer of a business framework. Yet, the social and environmental layers of the business plans are also important parameters to be included in the sustainable business plans. When business plans consider three key sustainability factors (social, economic, and environmental) and convert them into key risk-informed strategies, it will help SMEs build resilience to disaster and climate risks as well as sustainability components. For example, a detailed business canvas model was developed for handloom SMEs (Wanniarachchi et al., 2020) covering social, economic, and environmental factors. Such a holistic business plan will help in identifying the critical factors—such as the seven components of sustainable SMEs (product, market, profitability, environment, manpower, machinery, and materials)—that may be affected in the event of disasters.

Three key disaster management phases are aligned with five project management phases. In a pre-disaster situation, a number of mitigation and preparedness strategies are undertaken. This includes three project management steps: Understanding risks; Planning and prioritisation; and Mitigation and preparedness. When a disaster strikes, a response phase is initiated. This phase needs a response and a recovery strategy. Once this immediate aftermath of a disaster phase is over, a medium to long-term recovery period starts, which needs long-term resilience actions. The three disaster management phases need to be integrated with DRR and CCA measures for SMEs. This paper intends to present the key lessons from the development of a disaster and climate risk-sensitive business continuity plans for SME sector in the post-COVID19 context using case studies undertaken in the handloom and food/vegetable processing sectors in Sri Lanka, which will assist to make the SME sector in Sri Lanka more resilient and to increase their contribution to the economic growth.

In the post-COVID-19 context, the disaster and climate risk-sensitive business continuity plans have become much more essential for SMEs to ensure resilience and sustainable operations. Yet, the focus of research on the risk-sensitive business continuity plans for SMEs has been mostly on the large scale industries and businesses. Further, there are number of workable and realistic adaptation strategies proposed that SMEs can incorporate into their business plans to mitigate the risks of climate change and disasters in the pre-COVID19 period (Kato & Charoenrat, 2017; Ingirige et al., 2008; Verbano & Venturini, 2013). In this context, the key objective of the study is to explore the level of integration of disaster and climate risks in the existing business continuity plans for the selected SMEs in the post-COVID-19 context using handloom and food/vegetable processing SMEs in Sri Lanka as the case study.

2 Review of Relevant Literature

There has been a considerable body of literature that provides statistical evidence to prove a positive relationship between the existence of SMEs and the overall economic growth of a country. The lack of a standard and a uniform definition of SMEs has been a bottleneck for the sector's growth. For this study, the SME definition developed by the National Policy Framework for Small and Medium Enterprise (SME) Development—which uses the number of employees and annual turnover as parameters as shown in Table 1 is adopted.

Table 1 Small and medium enterprise (SME) categorisation based on criteria in different sectors

Size sector	Criteria	Medium [Rs. Mn—Million Sri Lankan Rupees]	Small
Manufacturing sector	Annual turnover	Rs. Mn. 251–750	Rs. Mn. 16–250
	No. of employees	51–300	11–50
Service sector	Annual turnover	Rs. Mn. 251–750	Rs. Mn. 16–250
	No. of employees	51–200	11–50

Source National policy framework for small and medium enterprise (SME) development

2.1 SMEs' Vulnerability to Disaster and Climate Risks

Although climate change is a systemic global issue, its effects are expected to be felt disproportionately by developing and least developed countries (Micale et al., 2018). With the possibility of limiting the rise of global temperatures to 1.50 °C above pre-industrial levels narrowing each passing year, albeit possible still, cities across the globe, which house most small- and medium-scale enterprises, are expected to be at the core of the climate crisis (Axa & UNEP Finance Initiative, 2015). In particular, given SMEs' inadequate resources, liquidity, skilled labour; low technical capacity; and poor management, they are understood as less equipped to deal with climate crises as opposed to their larger scale counterparts (Alam et al., 2022).

In Asia, a majority of the economic losses that have occurred as a result of disasters have reportedly been borne by the private sector (Surianto et al., 2019). As per Loyaza et al. (2012) less developed countries tend to be more exposed to disasters as they are less diversified and tend to rely more heavily on the agricultural sector. In addition, lack of pre and post-disaster mitigation and management policies in the developing countries contributes to greater damage to these economies. Given that Asia has been identified as a climate hotspot—albeit with regional variations—and has rapidly been privatising as opposed to the past, the climate implication borne by the private sector is naturally high (ibid). In relation to Sri Lanka, a study conducted by Dale et al. (2022), with respect to SMEs exposed to flood risks in Sri Lanka, recognises that deficiencies in Sri Lanka's enabling business environment have resulted in information asymmetries that have barred them from adopting adaptation strategies suitable for flood risk management. Furthermore, their capacity to cope with flood exposure has been obstructed by the absence of financial support available before and after extreme weather events have taken place (ibid).

The Sendai Framework for Disaster Risk Reduction 2015–2030 (SFDRR) was established to set disaster risk reduction priorities for the 2015–2030 period globally. It is the first major agreement since its 1994 and 2005–2105 precursors, i.e. Yokohama Strategy (1994) and the Hyogo Framework for Action (Goniewicz & Burkle, 2019). In its core principle is the need to "enhance disaster preparedness for effective response and to "Build Back Better" in recovery, rehabilitation, and reconstruction" (Surianto et al., 2019, p. 2213) particularly as post-disaster structural recovery may reinforce existing vulnerabilities among SMEs if they are not extended with adequate external support (Chatterjee et al., 2018).

Therefore, SFDRR identifies business and industry as one of the four major groups to be impacted by disasters and aims to promote linkages between large-scale enterprises and small- and medium-scale enterprises as a means of safeguarding the latter (Chatterjee et al., 2015). However, Haraguchi et al. (2016) critique that the success of the framework depends on its ability to convert its statements into actionable agendas.

2.2 Disaster and Climate Risk-Sensitive Business Continuity Plans of SME

The effects of climate risk exposure of SMEs can vary from short- to medium- and long-term. Also, it can be varied according to the different disaster types. Weerasekara et al. (2021) highlighted that in their study which investigated the impact of climate induced disasters on different economic sectors in Sri Lanka over 1997 and 2018 that droughts, floods, landslides, and strong winds have created different impact on different economic sectors, i.e. while drought show a negative sign for agriculture, industry and services sectors, strong winds show a positive impact on the agricultural and industrial sectors. However, landslides do not show any significant impact on any sector and a negative sign is present only in agricultural sector. This has been previously confirmed by (Klomp & Valckx, 2014; Loayza et al., 2012). Similarly, different sub-economic sectors that the SME operates can have different impact from different disasters. For example, the textile and garment sub-sectors are subject to a generally negative impact from different disasters while paper and machinery sub-sectors show a positive impact on industrial sector growth (Weerasekara et al., 2021).

Also, the impact of disasters can vary over time and can produce negative or positive impacts depending on the time (Loayza et al., 2012; Mohan et al., 2018). For instance, while flood might damage the SMEs operate in agricultural sector in the short term however it can be positively impact in the long term given soil nutrients increase after flood so as production. Drawing on the climate vulnerability of SMEs in Philippines, Ballesteros and Domingo (2015) demonstrate that exposure to climate catastrophes can compromise business continuity and recovery as physical damages to key inputs such as capital and labour occur. Despite Philippines being highly exposed and vulnerable to climate risks, its government has failed to effectively translate the DRR framework into business plans (Ballesteros and Domingo 2015).

Among the bottlenecks identified to successful implementation of DRR is primarily the underestimation of the likelihood of disaster and climate crisis occurrence (Goniewicz & Burkle, 2019). Furthermore, drawing on evidence from Poland, the authors denote that the dissonance between central administration, local governments, and the private sector have prevented a coherent disaster risk management strategy from being adopted. Nevertheless, the vitality of governments in mobilising SMEs for disaster preparedness and disaster management through the development of policies and legislations; the provision of necessary infrastructure; and direct intervention in pre- and post-disaster have been witnessed in Southeast Asia (Mavrodieva et al., 2019).

Drawing on the private sector, Shaw (2018) iterates that the private sector's role in disaster risk management is two-pronged. Corporate social responsibility on the one hand, and the creation of a corporate community interface to ensure innovative technical solutions for disaster risk reduction and management on the other, are at the core of the private sector's engagement in the disaster risk reduction framework (ibid). Therefore, partnerships between the private and public sectors should focus on minimising risks to infrastructure; securing livelihoods, supply chains, and local markets; strengthening risk insurance are crucial precursors to disaster risk management (Mavrodieva et al., 2019).

Lee et al. (2019) recognizes that public private partnerships (PPP) in disaster risk management should entail several key interventions including cross-cutting collaborations on science and technology, information sharing and dissemination, building awareness regarding risks, and communicating them across borders. The authors contend that these interventions will not only enhance regional resilience to disasters but will also promote inclusive development (ibid). Such integrated efforts can ensure national security, economic growth, environmental sustainability, and foster livelihood opportunities at the macro level, and build adaptive capacity of SMEs and disaster preparedness, at the micro level (Lee et al., 2016).

For companies of all sizes, Business Continuity Plans (BCPs) are essential because they aid in anticipating and responding to unforeseen hazards and disasters (Fani & Subiadi, 2019). Creating a business continuity strategy is a necessary step in keeping organisations stable during disruptions (Dey, 2016). An organisation can plan for its long-term survival by implementing a business continuity plan. Businesses without a BCP are taking a high stake with their clients and supply chain actors (Fani & Subiadi, 2019).

SMEs are especially susceptible to the effects of disasters, and they frequently experience compromised business continuity and recovery (Ballesteros, 2015). SMEs should also incorporate backup plans into their business continuity strategies and think about the possible effects of extreme weather occurrences (Jones & Ingirige, 2008). In order to mitigate potential climate and disaster risks, small and medium-sized enterprises (SMEs) must establish a disaster risk management plan. This can be achieved through business education and the sharing of best practices (Sarmiento et al., 2016). A business continuity and disaster recovery plan should be included in the strategic plans of SMEs. These can be created with the aid of tools like business modelling, risk analysis, business impact analysis, and mitigation strategies (Ghormley, 2009).

2.3 *Risk Mitigation Measures*

Plans for SMEs' business continuity that are responsive to climate change and disaster risks have been improved by a number of initiatives. In order to guarantee efficient business continuity in a particular area, Baba et al. (2014) recommends the use of Area Business Continuity Management, a coordinated strategy including several stakeholders. Turgay and Aydin (2023) offers SMEs a comprehensive, step-by-step guide on how to create and execute a framework for risk reduction that includes risk ranking, treatment choices, and integration into the organization's culture. In light of climate change, Botzen and van den Bergh (2009) highlights the significance of a robust risk management plan that incorporates damage mitigation, effective risk sharing, and catastrophe risk prevention. With an emphasis on knowledge sharing and best practices for SMEs, Sarmiento et al. (2016) emphasizes the significance of business education in fostering catastrophe resilience.

3 Methodology

3.1 Study Framework

Figure 1 shows the study framework used for assessing the risk-sensitive SMEs in the post-COVID19 context.

In the first phase of this study, a Post COVID19—Climate and Disaster Risk Assessment (CDRA) was conducted through a participatory assessment with a sample of 10 Small and Medium Enterprises (SMEs) in the North and Eastern Provinces of Sri Lanka. The SME samples were selected to take into account the geographical range and the scales of operation to capture diverse perspectives. A number of one-on-one interviews with the SME owners were conducted using a pre-defined semi-structured questionnaire.

A participatory workshop was held followed by a field-level study of 10 selected SMEs, to raise awareness on the climate resilience building of SMEs. Around 25 SMEs participated in this workshop, which helped to validate the CDRAs developed. The level of risk and ranking of risk were calculated using a risk score = f (Likelihood × Impact Matrix) through the CDRA exercise.

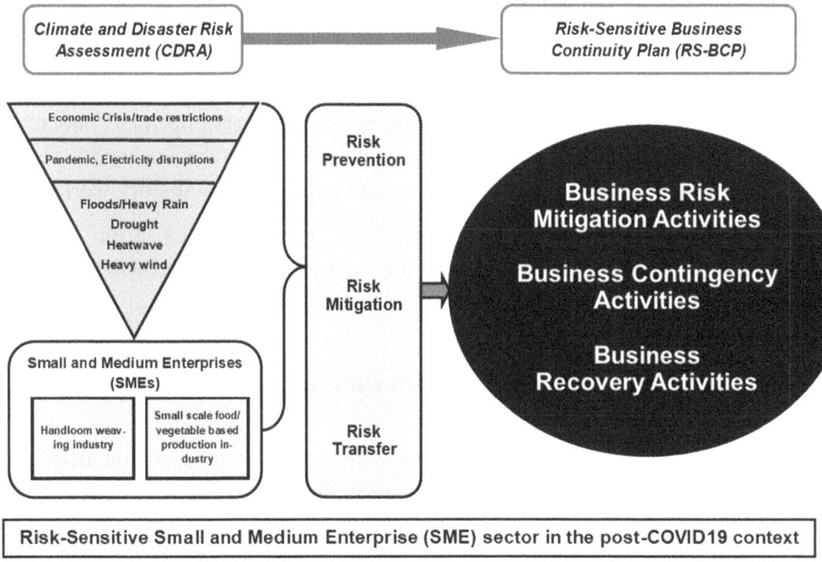

Fig. 1 Study framework for risk-sensitive SME sectors in the post-COVID19 context

In the next phase, key information required to understand the lessons and changes in practices for risk management strategies and activities that were practiced by SMEs to manage disaster risks were extracted through four participatory mapping exercises: community resource map, seasonal calendar, trend analysis, and SME product mobility map. From these exercises, the key components of the post COVID19—Climate and Disaster Risk-Sensitive Business Continuity Plan (RS-BCP) were identified in consultation with the SMEs.

In the final phase, the key findings were then validated with the key stakeholders working with SMEs. These stakeholders included the state departments, district disaster management center, National Enterprise Development Authority (NEDA), and private insurance companies.

3.2 Data Collection

The study settings and participants were chosen using four essential components of a qualitative exploratory case study based research (Creswell & Creswell, 2017):

(1) The research setting (the case study location): Small and Medium Enterprises in the North and Eastern Provinces of Sri Lanka (Handloom and Food/Vegetable Processing Industries were selected for this study since they were identified as the most effected SMEs due to COVID19 pandemic).
(2) Events—the interviews with the participants; one-on-one interviews were conducted using a semi-structured questionnaire.
(3) Actors—in the first phase: 10 selected SME owners and in the second phase: 25 supply chain actors of the selected SMEs (Handloom and Food/Vegetable Processing Industries).
(4) Procedures— the subjects of the interviews: The owners of the Small and Medium Enterprises were interviewed in the initial phase of the data collection. In the second phase of the study, 25 supply chain actors of the selected sectors of the SMEs provided inputs and validation through a participatory workshop.

The data on the current implementation of the BCPs that exist in the selected SMEs are captured during the one-on-one interviews with the SME owners.

3.3 Data Analysis and RS-BCP Production Steps

Data collected from one-on-one interviews with the SME owners and from the participatory workshops were analysed through manual coding of transcripts.

A detailed field-level assessment was carried out using purposefully selected SMEs for evaluating the level of disaster and climate risk integration in SMEs, the availability of disaster mitigation/contingency/continuity plans, and the impact on

SMEs in the absence of such risk-sensitive approaches through a multi-stakeholder engagement.

The following seven steps methodology was used to conduct the participatory Climate and Disaster Risk Assessment of 10 selected SMEs:

- Identification of all types of disaster risks (including climate-induced and non-climate related) that are of key concern or important for SMEs. The SMEs are exposed to different disaster and climate risks and to potential social and economic shocks.
- Assessment of the likelihood that the disaster risks will happen in the scale of 1—with low probability and 5—high probability in the geography where the SME is located.
- Based on the vulnerability and exposure of the SME to the particular risk prioritised based on its concern to SME and the likelihood of occurrence, the assessment of the potential impact of the risk against four key criteria of business impact: financial impact, impact on the brand/reputation for the product, competitiveness, and market survival of the product and the impact on the internal operation of the SMEs due to identified risks.
- Computation of the level of risk.
- Ranking of risk calculated using a risk score = f (Likelihood × Impact Matrix) for each of the prioritised risks that are of key concern to the SMEs. The prioritisation of risks was done using a disaster risk matrix (Fig. 2).
- The inputs from the SMEs during the workshop were analysed and key themes were generated for the identification of key technical gaps in the integration of DRR and CCA measures in the SME sector to develop RS—BCP.
- A participatory validation of Disaster Risk Reduction and Climate Change Adaptation (DRR and CCA) measures identified through CDRA and RS-BCP were carried out with multi-key stakeholder consultations at the sub-national level (district).

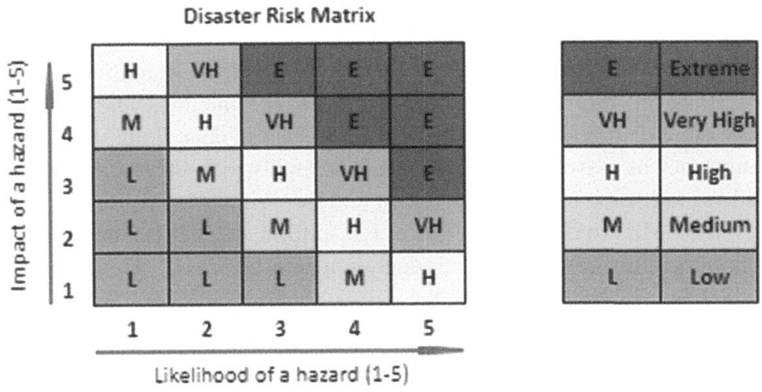

Fig. 2 Risk prioritisation chart for disaster and climate risks

The initial five steps were carried out through one-on-one interviews with the SME owners using a semi-structured questionnaire in March 2022. The disaster risk ranking and the status of current implementation of business continuity planning in the existing SMEs were obtained based on the interview data analysis and it was presented in a workshop with wider participation of SMEs who were not part of the interview data collection phase of the study. The data analysis was finalised with the data obtained from the workshop participants and presented in stakeholder consultation for further validation of the findings. The external validation with key SME supply chain actors can help to generalise the key findings from the case study interviews (Yin, 2003). The key results and analysis from these data collection processes are presented below.

4 Key Results and Analysis

4.1 CDRA Results [Risk Matrix for SMEs]

The major climate induced disaster risks to the SMEs in the study area are floods, heavy rain, heavy wind, heat waves, and drought as shown in Table 2. Further to the past and future disaster/climate change risks, post-COVID19 economic crisis, impact of recent pandemic, and electricity disruptions were also top-ranked socio-economic shocks in the post COVID19 context in Sri Lanka. These risks largely impact the production/supply of raw materials and continuous operation of the factories due to lack of access and availability of employees, raw materials, and supply restrictions.

4.2 Disaster-Disruptive/-Tolerance/-Recovery Period of SMEs

The average tolerance period of non-business operational activity is around 2–4 weeks due to a short-term disruption and average stock availability is around 4 weeks. However, the recovery period has been 2–3 months, when disasters or social/economic disruptions occurred in the past. These data along with other key information such as seasonal variations, product/supply mobility dynamics, and key stakeholders' influence in the supply chain and the community resources available will assist the SMEs to develop a disaster risk-informed SME business plan in the next stage integrating climate change adaptation measures.

Table 2 Risk ranking calculated using a risk score = f (likelihood × impact matrix) for SMEs

Identified hazards [number of SMEs identified]	Likelihood (1–5)	Potential impact (1–5)	Risk score	Level of risk	Ranking
Flood [8]	4	2.75	11	Very high	2
Heavy wind [6]	4	2	8	High	4
Heat wave [5]	3	1.75	5.25	Medium	6
Heavy rain [5]	5	2	10	Very high	3
Water scarcity [4]	3	2.5	7.5	High	5
Drought [2]	4	2	8	High	4
Cyclone [2]	1	4.25	4.25	Medium	7
Tsunami [2]	1	4.25	4.25	Medium	7
Pandemic (COVID19) [10]	2	4	8	High	4
Electricity disruption [1]	2	5	10	Very high	3
Economic crisis [10]	5	3.75	18.75	Extreme	1

4.3 Key Components of RS-BCP Identified for Handloom SME

4.3.1 Business Risk Mitigation Plan [Before a Disaster Occurs]

Key risk mitigation measures identified by the SME stakeholders for potential climate/disaster risks (heavy wind, flood, heavy rain, crop damage and heat waves) and other socio-economic risks (electricity disruption and raw material shortage) in their geographical areas are summarised in Table 3.

4.3.2 Business Contingency Plan [During/Immediately After a Disaster Event]

Key contingency plan measures identified by the SME stakeholders Table 4.

4.3.3 Business Recovery Plan [Recovery After the Disaster Impact] Measures Proposed by SMEs

The Business Recovery Plan includes activities through which SMEs establish the capabilities necessary to protect their assets and continue key business processes after a disaster. All the activities in the Business Risk Management Plan and Business

Table 3 Key risk mitigation measures identified by the SME stakeholders

Potential risk	Measures identified by the handloom SME stakeholders	Measures identified by the food/veg. processing SME stakeholders
Heavy wind	Assess the strength of the weaving center roof structure and reinforce it with strong materials and structure [RM]	Advance the drying process before the heavy wind period [RM/RP]
Flood	Raise the foundation of the weaving center above the past flood level [RM]	Identify the past flood level/ raise the building foundation above the flood level [RM] Enroll in agriculture insurance schemes [RT] Plan for stock alternative raw materials and alternative production [RP] Identify alternative routes to reach customers for supply of products [RM]
Heavy rain	Install windows/similar structures for closing during rainy seasons [RM]	Advance the drying process before the heavy rain period [RM/RP]
Heat waves	Install windows/similar structure for opening during heat waves/install cooling appliances such as fans [RM]	–
Raw material shortage/input price increase	Based on the seasonal calendar, stock enough yarn and other raw materials at least for a month [RP] Establish a network with new potential suppliers who can help during crisis situations [RM]	Sourcing adequate raw materials in advance to manage the price fluctuations [RM]
Crop diseases	–	Obtain crop insurance [RT] Use of prescribed pesticide with the expert advice [RM]
Electricity disruption	Plan the working hours according to the scheduled power cuts [RP] Maximize the electricity disruption period for activities that do not require electricity [sales promotion, acquisition of raw materials] [RP]	Plan the working hours according to the scheduled power cuts [RP] Maximize the electricity disruption period for activities that do not require electricity [sales promotion, acquisition of raw materials] [RP]

Table 4 Key contingency plan measures identified by the handloom/SME stakeholders

Potential risk	Measures identified by the handloom SME stakeholders	Measures identified by the food/veg. processing SME stakeholders
Heavy wind	Relocate all easily movable machineries and equipment to a safer place with strong roofing [brick buildings]	Install and use a heat room/ drying machine
Flood	Relocate all easily movable machineries and equipment to a building in a higher ground	– Identify a safer location for an emergency storage facility in the higher ground with no-flood history – Use of alternative routes for mobility plan for remote work possibilities for employees if accessibility continues to be difficult
Heavy rain	Relocate all easily movable machineries and equipment to a safer place [brick buildings with window closures]	Install and use a heat room/ drying machine
Heat waves	Use cooling appliances and request all employees to follow heat wave protection methods [drinking more water] Shift working schedule to night or morning and avoid peak heat wave time periods such as noon	Reduce the outdoor working hours
Raw material shortage/input price increase	Restructure the selling price based on the increased production cost without losing customer base Explore alternative products to minimize the production cost	Collective advocacy with the relevant departments for uninterrupted supply of raw materials/
Electricity disruption	Changes in the working time for employees (allocation of works shifts based on the power cut schedule)	– Install a generator and store enough fuel in a safer place for a minimum of three days of operation – Explore the feasibility of using solar panels to power the equipment and the factory

RP Risk prevention, *RM* risk mitigation, *RT* risk transfer

Contingency Plan will help the SMEs to continue the key business processes after a disaster without or with minimal interruptions. The to-do list for SMEs derived from this study consultations are mostly common to all risks. This includes the following themes:

> *Theme 1—Location of the SME:—If the current location of the SME is affected, assess the damage to the SME. Decide on whether it is possible to rehabilitate the damages within a shorter period of time or better to move the SME temporarily to a new location if the rehabilitation will take longer time. Therefore, all SMEs need to identify an alternative location for temporary operation in the vicinity.*
>
> *Theme 2—Alternative Market Opportunities for SMEs:—Reach the alternative potential market opportunities identified if the existing markets are affected and not reachable for a longer period.*
>
> *Theme 3—Market Demand for SME products:—Analyse the market demand and revise the business strategy to avoid further losses [sale/price restructuring].*
>
> *Theme 4—Insurance Facility for SMEs: —Record the damages accurately and activate the insurance cover if there is a damage to the SMEs.*
>
> *Theme 5—Target Customer: —Prioritise the customer base which largely depends on particular SME products and start delivering according to the priority plan to avoid the loss of important/large priority customer base.*
>
> *Theme 6—Local human resources:—Train additional people from family and neighbours who have the potential to become employees to be employed during disaster situations.*
>
> *Theme 7—Production Facilities or Machineries:—Encourage SMEs to install multiple machineries (at least two, if there is only one) to avoid any interruption in the production chain due to breaks/damages to the machines as a result of any disaster or the failure of the machine itself. This has to be incorporated into their annual financial plan and expansion plan.*
>
> *Theme 8—SME Expansion Plan:—An expansion plan as well as diversification of products based on the market needs to be included in the annual plan. When the SMEs are impacted by the disasters, the diversification plan can be operationalized and in the normal time the expansion plan can be pursued including new business plan expansions to popularize the product in different parts of Sri Lanka and overseas to expand the customer base [For example, introducing a new design in the handloom products/introducing an alternative food/vegetable product with value additions].*

5 Discussion

5.1 *Integration of Science-Based Knowledge and Know-How in Promoting Risk-Sensitive SME Operation*

The importance of recognizing, ranking, and reducing risks has been highlighted as a key process when dealing with constrained resources in the context of climate change and disasters (Verbano & Venturini, 2013). There has been a lot of scope to integrate the science-based knowledge on disaster risk reduction in the SME Sector. The project risk management requires a dynamic, adaptable strategy that is specific to the needs of small and medium-sized enterprises (Testorelli et al., 2020). Further, a dynamic IT based risk-management framework to take into account newly emerging technical developments is necessary for SMEs (Al-Dosari & Fetais, 2023). However, the focus on risk-integrated SME business plans has been realized after the long period of lockdowns due to COVID19 pandemic and the current economic crisis.

The SMEs highlighted that the key lessons to face future disaster and climate change risks were learned through COVID19 impacts on their businesses, as the SMEs have been the most affected sector due to consequent risks such as COVID19 pandemic, annual floods, and droughts. The scientific knowledge on effective risk communication, risk transfer, and risk reduction measures need to be enhanced for the SMEs, who realised their importance more than ever before, since the COVID19 pandemic and economic crisis.

5.1.1 Key Technical Gaps in Integrating DRR and CCA in the SME Plans

Due to an absence of reliable data on climate change and potential disasters, timely risk communication, and risk-based investment and decision-making were found to be the key gaps in DRR and CCA measures. It was found that a few numbers of SMEs have converted their business plans into risk-sensitive business plans with the support of non-government organisations. In order to improve the results of sustainable planning of SMEs, a problem analysis model integrating enablers like policy, programme, and practice is necessary (Busayo & Kalumba, 2020).

All the SMEs assessed in this study had explored the opportunity to introduce advanced and innovative technologies to link the new supply and market opportunities during disasters and social/economic disruptions. The disaster resilience of SMEs needs to be promoted through innovative initiatives and engagement of local disaster management stakeholders such as district disaster management centers. Social safety programs for protecting the SMEs that are highly vulnerable to disaster and climate risks are not available and alternative livelihood opportunities during disaster events are not promoted.

The supply of key ingredients for food and vegetable based SME products is mostly disrupted during flood session where the crops are damaged and the crops die due to lack of water scarcity during the drought session. Therefore, SMEs need to be provided with timely disaster early warning/weather forecasting information by the District Disaster Management Centre (DDMC). Further, SME committees need to be formed at the DS and district levels and SME stakeholder help desk is required to support SMEs to reduce disaster and socioeconomic risks. A supplier database and a service mobile app can be initiated at the district/provincial levels to integrate DRR into SME BCPs.

The capacity building of SMEs is very limited to regularly assess and monitor their safety protocols/practices and the implementation of standard operation procedures to the identified disaster and climate risks such as heavy rain/heavy wind/heat waves.

There is lack of guidance by the Department of Industries in partnership with the District DMC for SMEs to implement Standard Operating Procedures (SOPs) as part of their BCPs. The SMEs that practice/follow SOPs and implement BCPs are not promoted as models for other SMEs.

5.1.2 Cascading Risks and Impacts

The long-term effects of catastrophic weather events, especially for the most vulnerable communities need to be assessed (Butsch et al., 2023) to understand the probability of cascading disaster events. The seasonal variations in demand and production impact the business negatively. For example, during the annual rainy season, milk production had dropped sharply. However, the cascading risks in addition to the regular/annual disaster risks driven by climate change worsened during 2020–2022 in Sri Lanka due to socio-economic shocks and the COVID19 global pandemic.

This study found that the COVID19 pandemic and subsequent economic crisis had created serial impacts on SMEs to recover to the pre-COVID19 level. Hence, the post-COVID19 scenario highlighted the need for a new approach to DRR taking into account the multi-faceted cascading and systemic risks.

6 Key Recommendations

The key recommendation for integrating climate change measures in the development of Disaster Risk-Sensitive SMEs is to frame them in three risk management strategies:

Risk Prevention (measures to avoid existing or potential new risks)
Risk Mitigation (mitigation and preparedness for existing or potential new risks)
Risk Transfer (transferring the financial consequences of future risks)

The following key risk management strategies for potential disaster/climate risks in the SME sector are recommended from this study along with the details of required support and resources for each identified risk as detailed in Table 5.

Further, a Business Contingency Plan [activities to prepare for disasters in the event of its occurrence to reduce the impact on SMEs for each of the prioritised risks] and a Business Recovery Plan [activities to recover better quickly from the impact of disasters that had impacted the SMEs] need to be incorporated in the overall Disaster-Risk Sensitive SME Development Planning process.

Table 5 Resources required for handloom SME to implement identified risk management strategies

#	Potential risk	Risk management strategy	Business element interrupted	Resources required	External support needed
Handloom SMEs					
1	Shortage of yarn	Stock enough yarn for a number of months based on the seasonal requirements Identify new suppliers	Production	Adequate investment from savings or loan	Support to find out new suppliers Exploring new raw materials for alternative products (technical advice from experts)
2	Heavy wind	Reinforce the roof and building's structure to withstand heavy wind	Weaving/ production	Additional investment/ loan	Timely information from disaster management center about heavy wind
3	Pandemic (COVID19)	Recruit workers not impacted by COVID19 Implement health and safety procedures in the worksite	Weaving/ production/ workforce	Additional trained workers	Training new workers by the rural industries/ institutions
4	Flood/heavy rain	Raise the foundation of the store and weaving center Advocate authorities for constructing proper drainage Identification of alternative access routes for continuous supply/sale	Transport production	Additional investment/ loan	Support from the government officers and local authorities to complete the drainage system
Food and vegetable processing SMEs					
1	Ingredient/input price increase	Increased stock piling	Production reduced	Inputs/ Ingredients support from alternative sources	Support from the consumer affairs authority

(continued)

Table 5 (continued)

#	Potential risk	Risk management strategy	Business element interrupted	Resources required	External support needed
2	Flood	Higher ground for the factory Relocation/ evacuation to a safer place Renovate and resume	Ingredients/ production/sale	Construct the building with a higher foundation	District disaster Management Centre
3	Electricity interruption	Alternative energy sources [use of solar energy] Have a back-up power bank such as diesel generator	Production reduced	Solar system installation or buy a backup power bank	Financial support from government agencies or Industry Development Board
4	Heavy rain	Open drying will be impacted, need to use a dryer machine to dry the vegetables and other ingredients/ products	Production reduced	Heat room [Dryer]	Need to get loan or from the NGOs
5	Heavy wind	Open drying will be interrupted, so a drying machine is needed	Production reduced	Heat room [Dryer]	Need to get loan or from the NGOs
6	Crop diseases	Use of pesticides	Loss of ingredients and production	Pesticides	Agriculture department

7 Concluding Remarks

This study aimed to identify the key lessons in the implementation of disaster and climate risk-sensitive business plans of the handloom and food/vegetable processing sector SMEs in the post-COVID19 context in Sri Lanka. The Climate and Disaster Risk Assessment (CDRA) among SMEs revealed that flood, heavy rain and drought are the high priority disaster and climate risks for the handloom and food/vegetable processing SMEs. While the average tolerance period of non-business operational activity is around 2–4 weeks, the average stock availability is around 4 weeks and the recovery period has been 2–3 months.

Further, innovative disaster risk-sensitive strategies including risk mitigation and business continuity/recovery measures need to be identified as most of the studied SMEs have lacked such a plan based on a systematic and scientifically grounded assessments. The COVID19 pandemic followed by the post-economic crisis in Sri Lanka has provided a window for SMEs to develop such risk-sensitive SMEs operational plans. This study further highlighted the importance of creating a holistic plan to promote Disaster Risk-Sensitive SMEs to evade not only the emerging climate induced disaster risks but also help the SMEs to withstand socio-economic shocks in the future.

Acknowledgements Authors would like to thank SMEs who took part in this study. The contents in this publication are the sole responsibility of the authors and in no way to be taken to reflect the views of the organizations supporting the SMEs took part in this study. All authors have contributed to the writing of the manuscript. The first author conceived the idea and wrote the first draft of the manuscript, other authors have particularly contributed for collecting, analysing the data and reviewing the manuscript.

Appendix: One-on-One Semi-structured Interview Questions

A. Determining Business continuity objectives, scope, and team

A.1. Objective

- Why is your organization introducing business continuity initiative?

  For People; For Business Activities; For Local Community.

A.2. Scope

- Which part of your organization will be covered by business continuity plans?

  At the office or plant; at particular divisions/departments; beyond its own premise e.g. suppliers

A.3. Leader and Members

- Who will serve as leader of your business continuity activities and its members

  Further questions based on the responses.

B. Risk Assessment—Know Your Disaster Scenarios

1. Identify all kinds of hazards your enterprise is exposed to (internal and external threats)—make it specific to your geographic location/industry!
2. Evaluate the likelihood that the hazard will happen
3. Based on the vulnerability and exposure of the enterprise, evaluate the potential impact of the hazard

Further questions based on the responses.

C. Business Impact Analysis.

C.1. Identify prioritized activities/processes or critical factors in the business [Critical elements or single stop failure]

- What is the critical product or service?
- Which is the most critical element in your business that you can't continue your business without it
- Which business activity makes a priority
- Do you have any alternative pathways [Do things differently] / Stocks/ Resource banks/ Alternative workforce etc.

C.2. If risk happens: What is the maximum tolerable period? What is your recovery time period?

References

Alam, A., Du, A. M., Rahman, M., Yazdifar, H., & Abbasi, K. (2022). SMEs respond to climate change: Evidence from developing countries. *Technological Forecasting and Social Change, 185*.

Al-Dosari, K., & Fetais, N. (2023). Risk-management framework and information-security systems for small and medium enterprises (SMEs): A meta-analysis approach. Electronics.

AXA and UNEP Finance Initiative. (2015). Business unusual: Why the climate is changing the rules for our cities and SMEs. AXA Group, Paris, France. Available at: https://www.unepfi.org/psi/wp-content/uploads/2015/10/business_unusual.pdf

Baba, H., Watanabe, T., Nagaishi, M., & Matsumoto, H. (2014). Area business continuity management, a new opportunity for building economic resilience. *Procedia. Economics and Finance, 18*, 296–303.

Ballesteros, M. M., & Domingo, S. N. (2015). *Building Philippine SMEs resilience to natural disasters*. PIDS Discussion Paper Series.

Bandara, E. W. N. D., Perera, K. H., & Madushika, H. P. K. (2020). Study on the impact of national policy framework on the performance of small and medium enterprises (SMEs) in Gampaha District. *SSRN Electronic Journal*.

Botzen, W. J., & van den Bergh, J. C. (2009). Managing natural disaster risks in a changing climate. *Environmental Hazards, 8*, 209–225.

Busayo, E.T., & Kalumba, A.M. (2020). Coastal climate change adaptation and disaster risk reduction: A review of policy, programme and practice for sustainable planning outcomes. *Sustainability*.

Butsch, C., Beckers, L., Nilson, E., Frassl, M. A., Brennholt, N., Kwiatkowski, R., & Söder, M. (2023). Health impacts of extreme weather events—cascading risks in a changing climate. *Journal of Health Monitoring, 8*, 33–56.

Central Bank of Sri Lanka. (2021). Annual Report of the Monetary Board 2021 (Annual Report of the Monetary Board of the Central Bank, pp. 83–86). Central Bank of Sri Lanka. https://www.cbsl.gov.lk/sites/default/files/cbslweb_documents/publications/annual_report/2021/en/13_Box_04.pdf

Chatterjee, R., Okazaki, K., & Shaw, R. (2018). Understanding recovery process of small- and medium-scale enterprises after 2015 Nepal earthquake and impact on resilience building. In *Science and technology in disaster risk reduction in Asia* (pp. 253–272). Elsevier.

Chatterjee, R., Shiwaku, K., Das Gupta, R., Nakano, G., & Shaw, R. (2015). Bangkok to Sendai and beyond: Implications for disaster risk reduction in Asia. *International Journal of Disaster Risk Science, 6*(2), 177–188.

Creswell, J. W., & Creswell, J. D. (2017). *Research design: Qualitative, quantitative, and mixed methods approaches*. Sage Publications.

Dale, T. W., Charlery, L., Gao, J., & Schaer, C. (2022). Enabling private sector adaptation to climate change: Factors supporting and limiting adaptation amongst Sri Lankan SMEs. *Mitigation and Adaptation Strategies for Global Change, 27*(6), 1–30.

Dey, C. K. (2016). Strategies to reduce supply chain disruptions in Ghana [Doctoral Dissertation, Walden University]. Walden Dissertations and Doctoral Studies Collection. Walden University. https://scholarworks.waldenu.edu/dissertations

Fani, S. V., & Subiadi, A. P. (February 2019). *Trend of business continuity plan: A systematic literature review* (pp. 13–15). Sidoarjo, Indonesia.

Ghormley, Y. (2009). *Business continuity and disaster recovery plans*.

Goniewicz, K., & Burkle, F. M., Jr., (2019). Challenges in implementing Sendai framework for disaster risk reduction in Poland. *International Journal of Environmental Research and Public Health, 16*(14).

Haraguchi, M., Lall, U., & Watanabe, K. (2016). Building private sector resilience: Directions after the 2015 Sendai framework. *Journal of Disaster Research, 11*(3), 535–543.

Ingirige, B., Jones, K., & Proverbs, D. (January, 2008). *Investigating SME resilience and their adaptive capacities to extreme weather events: A literature review and synthesis*. Kandalama, Sri Lanka.

Jones, K. A., & Ingirige, B. (2008). *Extreme weather events and business continuity planning* (s.l., s.n.).

Kato, M., & Charoenrat, T. (2017). Business continuity management of small and medium sized enterprises: Evidence from Thailand. *International Journal of Disaster Risk Reduction*.

Klomp, J., & Valckx, K. (2014). Natural disasters and economic growth: A meta-analysis. *Global Environmental Change, 26*, 183–195.

Lee, Y., Watanabe, K., & Li, W.- S. (2016). *Enhancing regional digital preparedness on natural hazards to safeguard business resilience in the Asia-Pacific* (pp. 170–182).

Lee, Y., Watanabe, K., & Li, W.-S. (2019). Public private partnership operational model—A conceptual study on implementing scientific-evidence-based integrated risk management at regional level. *Journal of Disaster Research, 14*(4), 667–677.

Loayza, N. V., Olaberria, E., Rigolini, J., & Christiaensen, L. (2012). Natural disasters and growth: Going beyond the averages. *World Development, 40*(7), 1317–1336.

Mavrodieva, A. V., Budiarti, D. S., Yu, Z., Pasha, F. A., & Shaw, R. (2019). Governmental incentivization for SMEs' engagement in disaster resilience in Southeast Asia. *International Journal of Disaster Risk Management, 1*(1), 32–50.

Micale, V., Tonkonogy, B., & Mazza, F. (2018). *Understanding and increasing finance for climate adaptation in developing countries*.

Mohan, P. S., Ouattara, B., & Strobl, E. (2018). Decomposing the macroeconomic effects of natural disasters: A national income accounting perspective. *Ecological Economics, 146*(1), 1–9.

Saja, A. A., Sahid, M. L., & Sutharshanan, M. (2020). Implementing Sendai framework priorities through risk-sensitive development planning—A case study from Sri Lanka. *Progress in Disaster Science*.

Sarmiento, J. P., Hoberman, G., Jerath, M., & Jordao, G. F. (2016). Disaster risk management and business education: The case of small and medium enterprises. *AD-minister*, 73–90.

Shaw, R. (2018). Role of private sectors in disaster risk reduction: Potential and challenges. *Journal of Disaster Research, 13*(7), 1207–1212.

Strategy, Y. (1994, May). Plan of action for a safer world-guidelines for natural disaster prevention, preparedness and mitigation. In: World conference on natural disaster reduction (pp. 23–27). Yokohama, Japan.

Surianto, S., Alim, S., Nindrea, R. D., & Trisnantoro, L. (2019). Regional policy for disaster risk management in developing countries within the sendai framework: A systematic review. *Open Access Macedonian Journal of Medical Sciences, 7*(13).

Testorelli, R., Ferreira de Araújo Lima, P., & Verbano, C. (2020). Fostering project risk management in SMEs: An emergent framework from a literature review. *Production Planning and Control*, 1304–1318.

Turgay, S., & Aydın, A. (2023). Risk mitigation for SMEs: A step-by-step guide to implementing an effective framework. *Financial Engineering and Risk Management.*

Verbano, C., & Venturini, K. (2013). Managing risks in SMEs: A literature review and research agenda. *Journal of Technology Management and Innovation, 8*, 186–197.

Wanniarachchi, T., Dissanayake, K., & Downs, C. (2020). Improving sustainability and encouraging innovation in traditional craft sectors: The case of the Sri Lankan handloom industry. *Research Journal of Textile and Apparel, 24*(2), 111–130.

Weerasekara, S., Wilson, C., Lee, B., Hoang, V. N., Managi, S., & Rajapaksa, D. (2021). The impacts of climate induced disasters on the economy: Winners and losers in Sri Lanka. *Ecological Economics,* 185.

Word Bank. (2022). Small and medium enterprises (SMEs) finance, improving SMEs' access to finance and finding innovative solution to unlock sources of capital. https://www.worldbank.org/en/topic/smefinance

Yin, R. K. (2003). *Designing case studies. Qualitative research methods.*

Zarouk, Z., Mallawatantri, A., Saja, A. M. A., Attanayake, K., Piyadasa, R. U. K. (2021). A conceptual model for community-driven ecosystem-based disaster risk reduction. In: Amaratunga, D., Haigh, R., Dias, N. (Eds.), *Multi-hazard early warning and disaster risks*. Cham: Springer. https://doi.org/10.1007/978-3-030-73003-1_16

Challenges in Integrating the Paris Agreement, Sustainable Development Goals, and Sendai Framework into Coastal Built Environments: A Review of Sri Lankan Context

Deshan Pasindu, Bawantha Rathnayaka, Dilum Rajapaksha, Chandana Siriwardana, and Lalith Rajapakse

1 Introduction

Sri Lanka, with its breathtaking coastline, stands at the intersection of sustainable development, climate change adaptation, and disaster risk reduction. The 2030 Agenda for Sustainable Development, Paris Agreement, and Sendai Framework for Disaster Risk Reduction (SFDRR) all serve as global roadmaps towards resilient and sustainable futures; aligning national-level policy processes with these international frameworks has the promise of increasing coherence, efficiency, effectiveness and ultimately moving Sri Lanka closer toward climate-resilient development (Daze et al., 2018).

This paper delves into the complex challenges and opportunities presented when incorporating the Paris Agreement, Sustainable Development Goals (SDGs), and Sendai Framework into coastal built environments in Sri Lanka. Coastal cities present special dynamics when faced with climate change, and their dense populations serve as vital economic hubs for trade, fishing, tourism, etc. They face two primary threats: being extremely susceptible to natural and climate change-related risks (Kumar & Taylor, 2015). While their resilience holds keys to national economic prosperity, their resilience is vitally essential in national terms.

Climate Change-induced Sea level rise poses serious threats, from direct inundation, flood and storm damage, loss of wetlands, erosion, saltwater intrusion, rising water tables, and saltwater intrusion to coastal systems. Therefore, an integrated

D. Pasindu (✉) · B. Rathnayaka · D. Rajapaksha · L. Rajapakse
Department of Civil Engineering, University of Moratuwa, Moratuwa, Sri Lanka
e-mail: dpasindu96@gmail.com

C. Siriwardana
School of Built Environment, Massey University, Palmerston North, New Zealand

C. Malalgoda et al. (eds.), *Climate Change Adaptation in the Built Environment*,
https://doi.org/10.1007/978-3-031-75826-3_3

approach to managing climate change and disaster risk in coastal built environments is undeniable. This paper examined the progress of Sri Lanka with each global agenda individually while also exploring their combined application within its context in Sri Lanka.

Examining experiences and strategies in Sri Lanka to unify these agendas, this research seeks to unearth insights that may inform and influence similar efforts elsewhere worldwide. As climate change and disaster risk challenges increase exponentially worldwide, Sri Lanka exemplifies aligning national aspirations with international commitments and offering valuable lessons that other coastal regions could draw from its journey toward synergy among global frameworks.

2 Literature Review

The Paris Agreement, Sustainable Development Goals (SDGs), and Sendai Framework are global initiatives to combat climate change, foster sustainable development, and mitigate disaster risk. Implementing these frameworks within coastal built environments presents many unique challenges that must be resolved. This literature review examined both challenges faced in doing this, with Sri Lanka being used as an example context.

2.1 Paris Agreement

The Paris Agreement, built upon the United Nations Framework Convention on Climate Change (UNFCCC), is a historic international accord designed to combat climate change and its impact. This recognizes the necessity of collaboration among research groups and integrating multiple sciences to meet climate change challenges, specifically regarding coastal built environments where various factors such as sea-level rise, extreme weather events, and coastal erosion interact. Such integrative efforts must consider multiple issues. For instance, rising sea-level levels affect cities, and extreme events along coasts that cause coastal erosion. (Doukas et al., 2018) Highlight the need to move from stand-alone integrated assessment modelling frameworks towards truly integrative processes involving multiple actors and methods. The main goal of the agreement is to limit global warming to below 2 °C or preferably to 1.5 °C compared to pre-industrial levels.

2.1.1 Country Statement of Commitment to the Agreement

As a signatory to the Paris Agreement, Sri Lanka has pledged to combat climate change and decrease greenhouse gas emissions. The country recognizes the urgent need to address the impacts of climate change, particularly in coastal built environments. This agreement reflects equity and adheres to the principle of common yet differentiated responsibilities and respective capabilities when understanding different national circumstances (The Paris Agreement I UNFCCC, 2015). Climate change impacts are expected to have tremendous negative consequences in developing nations because their populations live under lower socio-economic development than elsewhere. It is commonly held that developed nations contribute the most to climate change; however, developing nations are expected to experience its most damaging consequences due to high populations living under such circumstances (Demel et al., 2019). Deliberations over climate issues should consider issues related to social justice and equity (IPCC, 2022). The Convention aims to strengthen the global response to the threat of climate change in the context of sustainable development and efforts to eradicate poverty (The Paris Agreement I UNFCCC, 2015). Sri Lanka signed the United Nations Framework Convention on Climate Change (UNFCCC) in November 1993 and the Paris Agreement in April 2016 before submitting its initial Nationally Determined Contributions (NDCs), detailing how they intended to implement the Paris Agreement in September 2016 (Sri Lanka Updated Nationally Determined Contributions, 2021).

2.1.2 Institutional Structure/Mechanisms Responsible for Implementing the Paris Agreement

Sri Lanka has taken several steps to address climate change by introducing national policies, strategies, and actions. Figure 1 highlights key national policies implemented by Sri Lanka (Ministry of Environment, 2021).

National policies regarding climate change seek to mainstream climate change sector-wise, and some sectors have already integrated climate change risks and commitments into their agendas. These key sectors include power, Urban planning, Waste, transport, industry, coastal and marine, forestry, water, health, tourism and recreation, biodiversity, agriculture, livestock, and fisheries (Climate Change Secretariat, 2016; Ministry of Environment, 2021). Climate change risks and commitments have already been included in several of the above key sectors to achieve the targets through the Paris Agreement strategic plan/action plan or policies. Some essential strategies are as follows (Ministry of Environment, 2021).

- National Energy Policy and strategies of Sri Lanka (2019)
- Long-Term Electricity Generation Expansion Plan (2018–2037)
- The National Policy on Waste Management (2019)
- The National Policy on Sustainable Consumption and Production for Sri Lanka (2019)
- Coastal Zone and Coastal Resource Management Plan (2018)
- National REDD+ Investment Framework and Action Plan (NRIFAP) (2017)
- Strategic Action Plan for Adaptation of Irrigation and Water Resources Sector for Climate Change (2018)

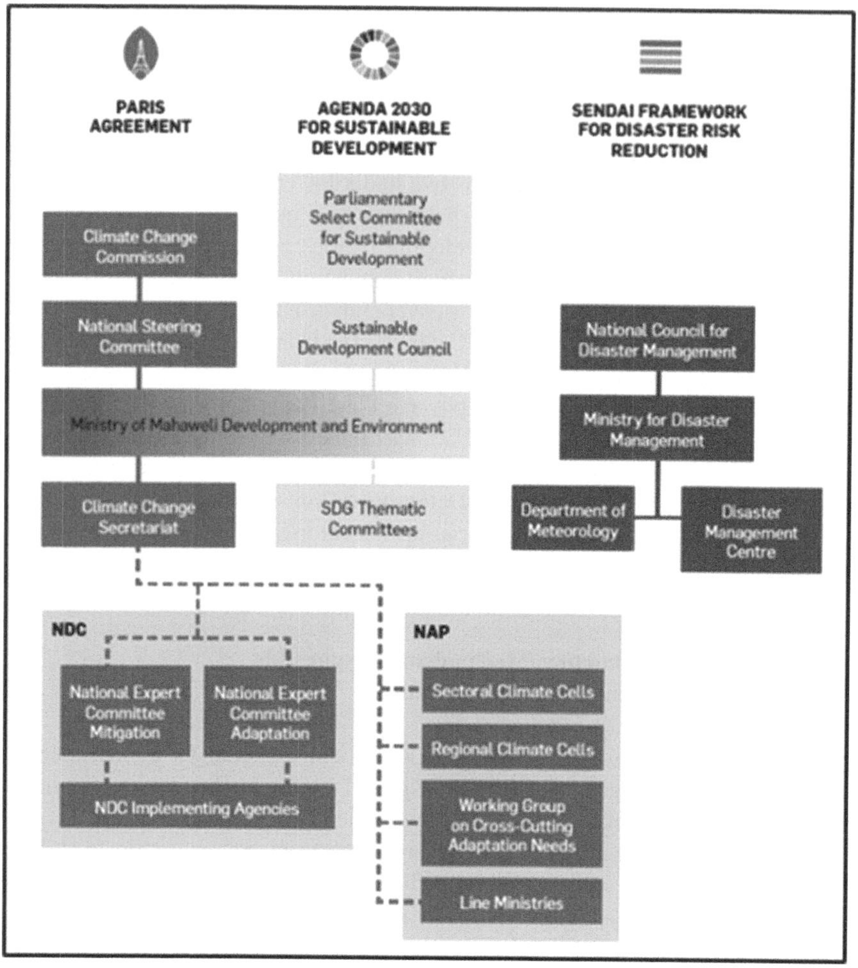

Fig. 1 Key national policies implemented by Sri Lanka

- National Policy on Natural Gas (2019)
- National Policy on Disaster Management (2013)
- Sri Lanka Disaster Management Plan (2018–2030)

2.1.3 National Statement on Progress in Implementing the Paris Agreement

The updated National Developmental Climate Agreements guide development, particularly post-COVID economic recovery, and livelihood needs, along a low-emission trajectory to attain high income and human development within a decade (Gunathilaka & Samarakoon, 2022). Their conceptualization is comprehensive yet compatible with national needs (Sri Lanka Updated Nationally Determined Contributions, 2021a).

The NDCs are grounded on recent analyses of climate risks and vulnerability, such as the third draft of National Communication, draft Climate Change Risk in Sri Lanka—Sector Risk Profile, Marginal Abatement Cost Curve analysis for energy, draft Low Carbon Development Strategy document, draft National Disaster Management Plan document and National REDD+ Investment Framework and Action Plan from 2021 (Sri Lanka updated Nationally Determined Contributions, 2021a).

However, few studies examine the implementation and results of implemented policies and their outcomes. Renewable energy policy changes especially require strict monitoring to reap expected benefits without incurring losses; adaptation financing needs cannot solely be covered through public investment alone (Kumarasiri & Dissanayake, 2021).

Sri Lanka has taken steps to secure climate finance from various climate funds and multilateral agencies, but this has fallen short of meeting demand. Financing needs for adaptation sector resilience building should become apparent through the National Adaptation Plan (NAP) and NDC costing exercises conducted in the future (Secretariat, 2015). Such activities must focus on building environmental requirements such as wind turbines, solar panels, and power grids for storage needs (Sri Lanka updated Nationally Determined Contributions, 2021a).

The NDCs demonstrate how the power sector has already invested in renewable energy by offering supportive policy instruments such as feed-in tariffs, net metering, and net accounting (Ministry of Environment, 2021). Energy efficiency has been incentivized and encouraged through financial incentives, with key municipalities seeing significant investments and increases in waste-to-energy assets and waste composting measures. Regulations have been implemented to force large-scale waste producers to invest in on-site waste treatment and management. Some more extensive industrial production facilities are striving for "carbon neutrality," while the industrial sector has pledged to incorporate "green or eco" concepts when investing in new industrial parks (Sri Lanka updated Nationally Determined Contributions, 2021a).

2.2 *The Sustainable Development Goals*

2.2.1 Country Statement of Commitment to the SDGs

In 2015, Sri Lanka and all United Nations member states unanimously embraced the 17 Sustainable Development Goals (SDGs) and 2030 Agenda (Sustainable Sri Lanka 2030 Vision and Strategic Path, 2023). Under this global framework, each nation had to outline its vision for 2030 (Munasinghe, 2018). Sri Lanka had an advantage due to its solid performance under Millennium Development Goals in areas like health, education, and poverty reduction—these positions made Sri Lanka well suited to pursue these SDGs successfully. The resolution of the three-decades-long conflict also provided an ideal platform (Government of the Democratic Socialist Republic of Sri Lanka, 2018).

Under the direction of President Maithripala Sirisena in 2017, Sri Lanka adopted its "Sustainable Sri Lanka 2030 Vision," an ambitious strategic document (Government of the Democratic Socialist Republic of Sri Lanka, 2018). This vision sought to establish Sri Lanka as an upper middle-income hub of sustainable economic development within the Indian Ocean region (Munasinghe, 2019a, 2019b). Furthermore, this document highlighted the need for "Balanced Inclusive Green Growth," coherent policy frameworks, and efficient implementation mechanisms (Munasinghe, 2019a).

Since committing to the SDGs, Sri Lanka has taken various steps to advance their implementation. A Parliamentary Select Committee for Sustainable Development was created at the highest legislative level to coordinate activities among relevant ministries and provide expert guidance. Established in 2015, Sri Lanka's Ministry of Sustainable Development and Wildlife is the central entity responsible for monitoring SDG progress across Sri Lanka. Since 2017, Sri Lanka passed Sustainable Development Act No. 19 of 2017, and an institutional apparatus for driving SDG implementation has been created utilizing existing public institutions' stems (Government of the Democratic Socialist Republic of Sri Lanka, 2018).

Sri Lanka has developed several national policy frameworks demonstrating its commitment to sustainable development principles, including Vision 2025 and the Blue Green Budget 2018. These frameworks prioritize social protection schemes for all sectors of society and prioritize environmental management as a key priority, striking an equilibrium between economic growth and environmental protection. Vision 2025 foresees Sri Lanka as being transformed into a knowledge-based, competitive social market economy that prioritizes sustainable development as an essential aspect of agricultural and sustainable development policies. The Blue-Green Budget of 2018 underscores this goal. It confirms the government's dedication to realizing the ocean's full economic potential while upholding environmentally friendly development practices (Government of the Democratic Socialist Republic of Sri Lanka, 2018).

In January 2016, Sri Lanka initiated "Sri Lanka NEXT: A Blue Green Era" to express their commitment to the Paris Agreement on Climate Change and Sustainable Development Goals (Sri Lanka NEXT: A Blue-Green Era—Bio Energy Association of Sri Lanka, 2019). This program seeks to foster sustainable development in Sri Lanka through Blue Green Economic initiatives and raise awareness among stakeholders and the public on achieving economic and environmental sustainability (Munasinghe, 2018). The current Central Government in Sri Lanka was elected in August 2020 and immediately introduced a National Policy Framework called "Vistas of Prosperity and Splendour" concerning SDGs under Chap. 2. This framework details plans to reorganize cabinet ministries and develop an effective approach to meeting SDGs (Sustainable Development Goal | Sustainable Development Council of Sri Lanka, 2021). In 2020, The Sustainable Development Council Sri Lanka (SDCSL) also released their draft "National Policy and Strategy on Sustainable Development".

2.2.2 Reflections on Goal 9—Nationally and in Coastal Zones if Applicable, with a Focus on Opportunities and Constraints of Meeting the Goal's Targets

Goal 9 of the Sustainable Development Goals focuses on strengthening infrastructure, supporting inclusive and sustainable industrialization, and driving innovation. Two cabinet ministries, the Ministry of Technology (MoT) and the Ministry of Industries (MoI), are directly aligned to SDG 9. In its Sustainable Sri Lanka 2030 Vision report, the Presidential Expert Committee envisaged an environment free from pollution, a high quality of life for its people, and inclusive green growth (Sustainable Sri Lanka 2030 Vision and Strategic Path, 2023). It was advocated for sustained economic development that is socially equitable and ecologically sound while simultaneously contributing to peace and stability (Government of the Democratic Socialist Republic of Sri Lanka, 2018). Blue-Green Development Strategy is one pathway towards meeting this vision (Sri Lanka NEXT: A Blue-Green Era—Bio Energy Association of Sri Lanka, 2019). Targets established included shifting toward renewable energy sources, planning for pollution-free transport with long-range transport systems, and implementing zonal arrangements following UDA plans while optimizing land usage to develop housing schemes, condominiums, and commercial areas (Sustainable Sri Lanka 2030 Vision and Strategic Path, 2023). Implementing green building guidelines for public sector buildings, green public procurement plans, and the development of eco-villages have all been proposed (Munasinghe, 2019a). In its report, the Presidential Expert Committee indicated that replacing non-renewable raw materials and energy sources with greener alternatives presents opportunities to open new lines of production and service provision (Government of the Democratic Socialist Republic of Sri Lanka, 2018). In addition, sustainable resource extraction causes less environmental damage while contributing to ecosystem resilience.

NDCs of Sri Lanka for 2021 aim to strengthen human settlements and infrastructure resilience by mainstreaming climate adaptation strategies into national, subnational, and local physical planning processes (Government of the Democratic Socialist Republic of Sri Lanka, 2018). The NDC proposal to establish a climate-resilient built environment includes details that include reviewing and updating climate-resilient design strategies, reviewing, and revising existing rules and regulations to protect vulnerable habitats against climate change, and including sustainability concepts into Architectural Engineering curricula (Government of the Democratic Socialist Republic of Sri Lanka, 2018). Nationally Determined Contributions (NDCs) suggest designs that take account of future shoreline loss and erosion when making design recommendations to mitigate SLR/ER, demarcate protected areas to allow shifting urban densification, and implement risk management plans ('Final_NDC_IP_-_10072023', 2021). Sustainable Sri Lanka 2030 Vision report highlighted how lack of access to social security and rising health, education, and transport costs create a sense of social exclusion among informal sector workers (Global Assessment Report on Disaster Risk Reduction, 2019). They advise implementing evidence-based social policies with adequate funding to combat social insecurity and exclusion (Sustainable Development Goal | Sustainable Development Council of Sri Lanka,

2021). This report emphasizes the necessity of constitutional reform to devolve governance responsibilities to provincial and local authorities to decrease centralized control (Government of the Democratic Socialist Republic of Sri Lanka, 2018). Addressing existing disparities may prove essential for sustainable industrialization in Sri Lanka.

2.2.3 Reflections on Goal 11—Nationally and in Coastal Zones if Applicable, with a Focus on Opportunities and Constraints of Meeting the Goal's Targets

Goal 11 of the Sustainable Development Goals (SDGs) seeks to foster inclusive, safe, resilient, and sustainable cities and human settlements. The urban environments in Sri Lanka face various climate-related threats ranging from heatwaves energy demand spikes for cooling purposes to more frequent natural disasters like flooding, drought, and landslides (Sri Lanka Updated Nationally Determined Contributions, 2021b).

The coastal communities in Sri Lanka face similar issues related to sea-level rise, particularly regarding their water sources (Climate Change & Its Impact on Coastal Wetlands of Sri Lanka, 2018). Studies indicate that coastal communities are confronting issues like soil salinization, threats to agricultural lands, and disruption of freshwater fishing locations due to sea-level rise (Perera et al., 2018).

COVID-19 revealed shortcomings in urban planning, specifically regarding congestion and inadequate waste disposal systems in Sri Lanka. Urbanization, as an integral aspect of economic development, has resulted in increased demands on land, water, and resources (Munasinghe, 2019a). Sri Lanka does have an official National Level Plan for physical development. However, many citizens remain unfamiliar with it (Munasinghe, 2019a). In addition to that, ongoing work to update the National Physical Plan through the National Physical Planning Department to meet government objectives for development continues (Munasinghe, 2019a, 2019b). The revised plan comprises four overarching policies: resource conservation, resource, and infrastructure optimisation, prioritising liveable settlement areas for development, and exploring and exploiting potential areas (Sustainable Development Goal | Sustainable Development Council of Sri Lanka, 2021).

National Determined Contributions (NDCs) support the adoption and use of CCS's Guideline for Climate Resilient Human Settlement and Infrastructure and advise incorporating disaster risk reduction (DRR) planning practices into urban and human settlement planning (Government of the Democratic Socialist Republic of Sri Lanka, 2018). NDCs also recommend infrastructure design considerations, including runoff systems, slope stability measures, soil conservation techniques, and disaster risk management plans (Sri Lanka Updated Nationally Determined Contributions, 2021c). Integration of adaptation measures such as urban zoning, disaster risk analysis, forest park establishment and management, groundwater recharge systems, air passage/wind corridor planning and wetland planning, and wise wetland management may increase resilience in urban environments (Sustainable Sri Lanka 2030 Vision and Strategic Path, 2023).

2.3 *The Sendai Framework (2015–2030)*

2.3.1 Country Statement of Commitment to the Sendai Framework

Asia, where 80% of natural disasters take place annually, has adopted the Sendai Framework's approach to disaster risk management with increasing success, seeking to address root-cause risks such as poor urban planning practices, noncompliance with building codes, and ecosystem damage (Institutional Disaster Management Plan, 2019).

The Sendai Framework for Disaster Risk Reduction 2015–2030, adopted at the United Nations Third World Conference on Disaster Risk Reduction held in Sendai, Japan, in 2015, commits nations to protect development gains from natural hazards such as natural disasters. Aligning with other international agreements such as the Paris Climate Agreement, Addis Ababa Action Agenda on Financing Development New Urban Agenda, Sustainable Development Goals, etc., countries can implement clear DRR strategies (Disaster Risk Reduction in Sri Lanka, 2019a).

The Sendai Seven Campaign, initiated in 2016 by the UN Secretary-General, promotes seven goals of the Sendai Framework over seven years. 2021s target will focus on increasing international cooperation for implementation (Disaster Risk Reduction in Sri Lanka, 2019a), while International Day for Disaster Risk Reduction 2021 focused on "ONLY TOGETHER… WE CAN SAVE THE PLANET!".

Sri Lanka recognizes the need to align interventions with post-2015 global frameworks as climate change continues to bring devastating devastation, such as The Sendai Framework's emphasis on collaborative, coordinated, inclusive, risk-informed decision-making—something aligned well with Sri Lanka's focus on disaster risk reduction (Sendai_Framework_for_Disaster_Risk_Reduction_2015-2030, 2015).

Sri Lanka has taken steps, as detailed by the UNSDF 2018–2022, to create an action plan aligned with the Sendai Framework and revise its National Disaster Management Plan. As per this year's UNSDF allocations, DRR integration into various sectorial development plans (Executive Board of the United Nations Development Programme, the United Nations Population Fund, and the United Nations Office for Project Services Draft country program document for Sri Lanka (2018–2022), 2017).

2.3.2 Brief Statement on the Country's Meeting of the Framework's Four Priorities

Priority 1. Understanding Disaster Risks

Understanding disaster risk as the priority in the Sendai Framework for disaster risk reduction is paramount. Policies and practices related to disaster risk must be based on an in-depth knowledge of this subject matter, including aspects such as vulnerability, capacity, and exposure of individuals and assets (What is the Sendai Framework for Disaster Risk Reduction? | UNDRR, 2015).

To address this priority, a risk profile has been developed with support from technical institutions and academia in Sri Lanka with specific emphasis on five hazards: floods, droughts, landslides, cyclones, and coastal hazards such as tsunamis, sea surges, sea-level rise, and coastal erosion (What is the Sendai Framework for Disaster Risk Reduction? | UNDRR, 2015). Hazard profiles were then created, such as landslide maps by the National Building Research Organization (NBRO), coastal hazard profiles by the Coast Conservation and Coastal Resources Management Department, flood hazard maps by the Irrigation Department as well as drought/cyclone profiles developed by Agriculture Department/University of Peradeniya/DMC in 2015 (Hazard Profiles of Sri Lanka, 2015).

The Sri Lanka Disaster Resource Network (SLDRN) was upgraded to increase risk assessment capabilities and analyse future risk trends. Furthermore, the Risk Info platform, also known as Sri Lanka Disaster Risk Information Platform under Open DRI, was also launched with support from GFDRR and the World Bank, and this platform compiles geospatial hazard data from government agencies and stakeholders in an easily understandable format (Disaster Risk Reduction in Sri Lanka, 2019b).

These risk data sources serve as decision-making tools and introduce the concept of National Spatial Data Infrastructure (NSDI) (Disaster Management Centre, 2017). Substantial research and field studies on landslide risk were carried out under the direction of NBRO as the national focal point for landslide risk management, including vulnerability databases to assess disaster impacts on people as well as strategies for creating local hazard maps at the community level (Baseline Study for National Spatial Data Infrastructure Final Report Baseline Study for National Spatial Data Infrastructure Contents, 2017).

Priority 2. Strengthening Disaster Risk Governance to Manage Disaster Risks

Sri Lanka has strengthened its efforts to construct a comprehensive multi-hazard disaster management system since the 2004 Indian Ocean tsunami. The tsunami prompted extensive reforms, culminating in the passage of the Disaster Management Act of 2005. The National Council for Disaster Management (NCDM) and its operational headquarters, the Disaster Management Centre, were also mandated (DMC). Since then, the country's policies have been updated regularly. Table 1 summarises the adopted national policy framework for Sri Lanka (Siriwardana et al., 2018).

Sri Lanka recently approved and established the Sustainable Development Strategy and a National Council for Sustainable Development to encourage fair and environmentally sustainable economic growth via poverty eradication, social development, and increased competitiveness (Munasinghe, 2019b). In 2014, the Sri Lanka Comprehensive Disaster Management Program (SLCDMP) for 2014–2018 was approved by the NCDM to safeguard Sri Lanka by mitigating disaster and climate risks while simultaneously mitigating impacts to people, property, and

Table 1 Summary of the policy and legislation introduced in Sri Lanka

Implementation	Legislations/policy	Scope	Purpose
NDMC, disaster management centre, and the ministry of disaster management	Disaster Management Act (2005)	National, provinces, districts	The formation of the NDMC and the DMC was mandated. The act aims to improve disaster response by coordinating stakeholders and localizing operations
Disaster management centre	The Roadmap for Disaster Risk Management (2006–2016)	National, provinces, districts	Shifting the emphasis from reactive to proactive actions regarding preparedness, response, mitigation, and risk reduction
Ministry of disaster management, disaster management centre	National Policy on Disaster Management (2010)	National	To achieve "how things should be," it is intended to supplement the essential parts of other acts and plans for various ministries, departments, and local governments (such as land-use planning)
Disaster management centre, stakeholders, private sector	Sri Lanka National Disaster Management Plan (2013–2017)	National, provinces, districts, NGOs, CSOs, community leaders	Continuation of the 2006–2016 strategy to establish a multi-sectoral, inter-ministerial, and inter-agency DRM framework based on all stages of disasters and taking international accords into account as well as frameworks
Disaster management centre, stakeholders	National Emergency Operation Plan (NEOP) (2013)	National, provinces, districts	NEOPs, created as part of the 2013–2017 plan, are meant to specify and identify the roles of key stakeholders in various response scenarios to improve coordination

(continued)

Table 1 (continued)

Implementation	Legislations/policy	Scope	Purpose
Partnership strategy	Comprehensive Disaster Management Programme (2014–2018)	National, provinces, NGOs, private sector	Concentrate on risk reduction and the creation of conducive settings. The interventions are based on multi-hazard risk data and follow international guidelines
The government, private sector, community-based organizations	National Adaptation Plan for Climate Change Impacts in Sri Lanka (2016–2025)	National, Provinces, Districts and communities	Based on risk data, identifies sectoral climate adaptation needs and the overarching national plan, roles, cross-cutting concerns, and important players. A National Adaptation Fund (NAF) is also being considered
Partnership strategy	National Disaster Risk Management Plan (2018–2030)-Draft version	National, provinces, districts, communities, non-governmental organizations, private sectors, community based organizations	It is based on many of the same ideas as the 2014–2018 program but with a greater emphasis on actual, concrete implementation at all levels and monitoring and assessment. The importance of participation is emphasized

the economy and this being accomplished through creating an enabling environment conducive to participatory disaster risk reduction strategies based on multi-hazard assessments (Sri Lanka Comprehensive Disaster Management Programme (SLCDMP), 2014). Mahinda Chintana Development Policy Framework and Public Investment Strategy of 2014–2016 have led this initiative, while Disaster Management Policy, National Climate Change Adaptation Strategy for Sri Lanka 2011–2016, and National Action Plan for Disability provided necessary assistance (National Adaptation Plan for Climate Change Impacts in Sri Lanka Climate Change Secretariat Ministry of Mahaweli Development and Environment 2016, 2016).

Priority 3. Investing in Disaster Risk Reduction for Resilience

Sri Lanka, being an island nation vulnerable to flooding, droughts, and landslides, must invest in strengthening resilience against hydrological-meteorological hazards. Initial investments focused on creating robust monitoring systems for hydrological and meteorological hazards and strengthening hazard modelling capabilities for timely early warnings. Various ongoing programs supported by foreign aid agencies demonstrate the significance of investing in flood and drought prevention within spatial planning strategies and regional development plans.

Integration of flood risk considerations into urban development is demonstrated by initiatives like the Metro Colombo Urban Development Program (MCUDP), which seeks to revitalize both economically and physically the Metro Colombo area. Such projects also play an integral role in strengthening capacities at key agencies like the Ministry of Défense and Urban Development and Urban Development Authority (UDA) and development professionals involved with risk-aware infrastructure planning and facility development.

Similar to its counterpart program by the Global Facility for Disaster Reduction and Recovery (GFDRR) for mainstreaming disaster risk management at an urban level, it is an investment in human capital. It aims to enhance government technical proficiency in conducting flood and landslide risk analyses while encouraging the integration of risk reduction into urban planning (GFDRR, 2014). District and divisional officers have participated in capacity-building and training to develop their technical knowledge of disaster risk assessment and incorporate DRM principles seamlessly into development planning (The World Bank & GFDRR, 2012). In 2015, 300 local government officials received training on incorporating disaster risk reduction (DRR) into local government (DMC, 2015). Such initiatives showcase a commitment to investing in human resources, essential when integrating DRR into regional development strategies.

The National Disaster Management Plan 2013–2017 offers guidance to relevant agencies on feasible DRR mainstreaming actions and emphasizes the significance of integrating DRR across different sectors. This includes disaster-sensitive physical design and regional structural planning, development oversight by UDA and local governments, Disaster Impact Assessment (DIA) for new projects, integration during disaster reconstruction efforts, and DRR through Natural Resource Control—a practical manual for development professionals at all levels.

Priority 4. Enhancing Disaster Preparedness for Effective Response to "Build Back Better" in Recovery, Rehabilitation, and Reconstruction

Institutional procedures have been established for comprehensive early warning systems with nationwide coverage. This system comprises technical agencies such as the Meteorological Department Geological Survey and Mines Bureau, responsible for monitoring hazards and issuing early warnings through the Disaster Management Centre's (DMC) Emergency Operations Centre. Provincial Councils, District

Divisions, Local Authorities, GN Divisions, local law enforcement personnel, military personnel, volunteers, and Community-Based Organisations (CBOs) all play an integral part in this network. An integrated citizen-centered approach, including bottom-up and top-down mechanisms, has been implemented for localized hazards through collaborative networks of stakeholders, including local government, private sector entities, the Red Cross, non-governmental organizations (NGOs), and civil society (DMC, 2014).

The National Emergency Operations Plan (NEOP), created with technical assistance from UNDP Sri Lanka, outlines management arrangements, incident command systems, operational protocols, and coordination mechanisms that ensure an effective response to disasters or emergencies. In disaster scenarios, this document also sets forth the roles and responsibilities of various stakeholders (DMC, 2017). Disaster relief services were overseen by the National Disaster Relief Services Centre (NDRSC), while infrastructure restoration and reconstruction fall under the purview of the Ministry of Economic Development. The National Policy for Disaster Risk Management (DRM) (2013) places particular importance on restoring essential services. It takes an "improve-back-better" approach when considering medium and long-term rebuilding and rehabilitation efforts (DMC, 2014). Sri Lanka Red Cross operates in all 25 districts. It works closely with army and navy troops to support search and rescue operations, which are essential in emergency response and relief efforts. In 2016 and 2017, the Ministry of National Policies and Economic Affairs, in collaboration with the Ministry of Disaster Management, conducted Post-Disaster Needs Assessments (PDNAs) following significant disasters such as flooding or landslides with support from international partners like UNDP, EU and World Bank.

2.3.3 Brief Statement on the Country's Meeting of the Framework's Seven Targets

The Sendai Seven Campaign ("7 Targets, 7 Years") was initiated by the UN Secretary-General in 2016 with the primary purpose of raising awareness for each year of this advocacy initiative to implement and encourage implementation of the Sendai Framework to save lives, reduce disaster losses, and enhance management of disaster risk. Each of its seven targets for disaster reduction is designated for each year of this campaign.

- 2016—Target (a): Substantially reduce global disaster mortality by 2030, aiming to lower the average per 100,000 global mortality rates in the decade 2020–2030 compared to the period 2005–2015
- 2017—Target (b): Substantially reduce the number of people affected globally by 2030, aiming to lower the average global figure per 100,000 in the decade 2020–2030 compared to the period 2005–2015
- 2018—Target (c): Reduce direct disaster economic loss concerning the global gross domestic product (GDP) by 2030

- 2019—Target (d): Substantially reduce disaster damage to critical infrastructure and disruption of essential services, among them health and educational facilities, including through developing their resilience by 2030
- 2020—Target (e): Substantially increase the number of countries with national and local DRR strategies by 2020
- 2021—Target (f): Substantially enhance international cooperation with developing countries through adequate and sustainable support to complement their national actions for implementation of the present framework by 2030
- 2022—Target (g): Substantially increase availability and access to multi-hazard early warning systems and disaster risk information and assessments to people by 2030.

The global targets may be clustered into two groups: targets from (a) to (d) concern specific outcomes and targets from (e) to (g) concern inputs or enablers. Together, they assess progress toward achieving the Sendai Framework's expected outcome and develop the necessary tools and means to achieve it.

2.3.4 Implications of Implementing the Sendai Framework on Risk Reduction in Coastal Zones

The 2004 Indian Ocean Tsunami provided a stark reminder of Sri Lanka's coastal area's vulnerability (Tsunami disaster—Sri Lanka Summary situation report 26–31 Dec 2004—Sri Lanka I ReliefWeb, 2004). The coastal regions of Sri Lanka that are vulnerable to coastal intrusion are densely populated, with many urban cities, including Colombo, the capital city, located there (Nianthi et al., 2015). Moreover, many industries and industrial hubs that could be more vulnerable to climate change-related hazards are situated along the coastal belt zone.

Sri Lanka's Disaster Risk Reduction (DRR) and climate change adaptation efforts depend on reducing vulnerability across various sectors while strengthening the capacity of its inhabitants, particularly those living in less favourable economic circumstances (Secretariat, 2015).

Multiple studies have confirmed that coastal cities and communities are particularly susceptible to adverse climate change impacts (Hippola et al., 2020; Citation Jayasekara et al., 2021; Pitigala Liyana Arachchi et al., 2022).

Furthermore, these vulnerabilities are compounded by various contributing factors—including:

- Lack of trust in authorities who disseminate Early Warning (EW)
- Limited knowledge of evacuation routes and shelters in residential areas
- Lack of efficient and sustainable resilience mechanisms focused on the coastal communities
- Lack of efficiencies and effectiveness of national policies and frameworks related to coastal hazards and lack of alignment with the post-2015 global standards.
- Lack of capacity and preparedness of the coastal communities
- Lack of exposure and awareness of modern technology

- The administration process also disregards indigenous knowledge regarding the EW mechanism
- Lack of interest of the coastal communities in evacuation planning and safety drill

3 Discussion

3.1 Opportunities and Constraints of Implementing the Paris Agreement

The objectives of the Paris Agreement are to limit the global mean temperature increase to well below 2 °C above pre-industrial levels and aim for 1.5 °C, opening many possibilities to both developed and developing nations (Key aspects of the Paris Agreement | UNFCCC, 2019). Sri Lanka stands to benefit greatly from these opportunities as a developing nation. However, current greenhouse gas (GHG) emissions in Sri Lanka may seem modest. Still, their development trajectory indicates significant increases that would necessitate active strategies today to mitigate these emissions (National Adaptation Plan for Climate Change Impacts in Sri Lanka Climate Change Secretariat Ministry of Mahaweli Development and Environment 2016, 2016; Highlights of the Year, 2019).

Sri Lanka ranks second out of 181 countries affected by climate change according to the Global Climate Risk Index 2019 (Eckstein et al., 2018) according to its climate change impact vulnerability index scorecard in 2017 as measured by political, geographical, and social aspects, placing them 100th of 181 on World Bank Group's ND-GAIN Index 2020 evaluation system that evaluates countries based on both their vulnerability to global threats such as climate change as well as readiness to build resilience against these hazards (Implementation of the Sendai Framework for Disaster Risk Reduction 2015–2030 Report of the Secretary-General, 2016).

Sri Lanka understands the significance of climate adaptation due to the effects of global climate change, making this issue of paramount importance regarding their social, economic, environmental, and technical benefits. Various opportunities are available for implementing agreements across multiple dimensions (Ranasinghe & Jongejan, 2018).

However, despite these opportunities, there remains an alarming gap between policy formulation and effective implementation across many sectors in Sri Lanka. Research conducted by the Asia–Pacific Network for Global Change Research highlights crucial challenges related to policy implementation, such as institutional coordination, capacity-building finance adoption, technology adoption, and socio-economic factors in Bangladesh, Nepal, and Sri Lanka (innovative-financing-for-development-in-ap, 2017). Their findings highlighted improved coordination with sector ministries/entities and the need for centralized coordination bodies (Wijenayake, 2020).

Additionally, this research draws attention to issues surrounding unauthorized construction projects, land use, and marine crimes caused by a lack of policies and weak enforcement of existing laws. Even though laws exist to protect forest cover, their area has significantly declined as land demands for industry and settlement have increased considerably (Wijenayake, 2020). This study also stresses the need for capacity-building initiatives, mainstreaming climate action into ministry policy-making processes, and increasing collaboration between authorities and local stakeholders (Wijenayake, 2020). Addressing issues associated with duplicate institutions, duplicated laws, inadequate research, financial and technological gaps, and data deficiencies present additional obstacles to progress. The Asia–Pacific Network for Global Change Research's 2019 report offers solutions such as private-sector funding, public–private partnerships, and regional cooperation knowledge-sharing among nations (APNGCR, 2019).

3.2 *Concluding Statement on the SDGs*

Sri Lanka ranked 87 out of 165 nations based on its overall performance with an index country score of 68.1/100 and a statistical performance index rating of 72.6/100 according to the Sustainable Development Report, 2021 (The Sustainable Development Goals Report, 2021).

Once again, here we go with more jargon-laden rhetoric about why, more than ever before, we need to focus on education for economic well-being. The Intergovernmental Panel on Climate Change (IPCC) has highlighted sustainable development as having the power to mitigate systemic vulnerabilities, improve adaptive capacity within communities, and strengthen livelihood security among marginalized groups (IPCC, 2018). Oliver-Smith (1996) observed that when risks and disasters are considered integral parts of environmental and human systems, they serve as litmus tests of adaptation and sustainability. Scholars have repeatedly stressed the need to integrate Disaster Risk Reduction (DRR) and Climate Change Adaptation (CCA) since both can play critical roles in mitigating vulnerability and creating more resilient environments (Dias et al., 2018, 2019). Integrating sustainable development goals (SDGs) and disaster risk reduction and community change adaptation measures (DRR/CCA) into society-level projects seems pragmatic and necessary since addressing existing sustainability and inclusivity shortcomings will also increase resilience within societies/communities. Research by APN in 2019 advocates for tightening up on the linkages between SDGs and Nationally Determined Contributions (NDCs) to generate synergies and establish interlinked pathways linking climate action with sustainable development.

3.3 Opportunities and Constraints of Implementing the Sendai Framework (2015–2030)

Sri Lanka has historically been exposed to many hazards, from weather events, hurricanes, cyclones, and monsoonal rains leading to floods and landslides (Disaster | UNDRR, 2023) to natural catastrophes (Disaster | UNDRR, 2023). Climate Change Secretariat, 2016, and Disaster Management Centre (2014) have reported an increased frequency of drought events due to climate change. Additionally, Japan faces other risks, including lightning strikes, coastal erosion, epidemics, and pandemics (Iizuka, 2020). Notably, COVID-19's pandemic impact in Sri Lanka was profound (Fernando et al., 2021). Due to these challenges, disaster management and Disaster Risk Reduction (DRR) have emerged as top national priorities (Randil et al., 2020). This provides an excellent opportunity to implement the Sendai Framework for Disaster Risk Reduction (SFDRR), thus meeting existing DRR and disaster management challenges (Disaster | UNDRR, 2023).

Due to an increasing frequency of hydro-meteorological hazards, it has become necessary to implement disaster preparedness and response plans at all administrative levels, from local governments down. An inclusive yet well-coordinated disaster preparation and response framework must exist at the local government level for practical disaster risk-sensitive planning involving climate risk-sensitive planning (Disaster Risk Reduction in Sri Lanka, 2018). Further bridging gaps between national DRR initiatives and their implementation at local government levels must also be ensured (Disaster Risk Reduction in Sri Lanka, 2018).

Effective disaster risk reduction relies on regulatory frameworks, policies, and disaster management plans (Sri Lanka National Disaster Management Plan 2013–2017 Disaster Management Centre Ministry of Disaster Management 2nd, 2014). For efficient DRR, it is imperative to align these governance instruments with global agendas (Upadhaya et al., 2021). Siriwardana et al. (2018) noted minor alignments between local disaster management frameworks, policies, and global agendas. Institutional arrangements for disaster risk reduction activities play a pivotal role. Unfortunately, multiple policies and unclear roles and responsibilities within institutions create challenges to successfully implementing SFDRR (Siriwardana et al., 2018).

Technical capacity remains one of the primary hurdles to implementing SFDRR successfully. As an emerging nation, Sri Lanka is actively building technical capabilities to mitigate disaster-induced effects in each phase of disaster management—preparedness, response, recovery, and mitigation. An inadequate disaster risk governance system challenges SFDRR implementation. These deficiencies include inadequate institutional mechanisms for DRR at the local level, ineffective communication of national-level decisions to local governments, insufficient involvement in national-level decision-making processes, and lack of clarity around roles and responsibilities (Malalgoda et al., 2016).

Early warning systems and hazard maps must be integrated into land use planning, area development, and construction projects to enhance disaster preparedness. Furthermore, environmental Impact Assessments (EIA) should be maximized as planning tools. Disaster risk considerations must also be included when planning development projects. Numerous vulnerabilities stem from a lack of resources, insurance, housing opportunities, education, or income streams. As poverty reduction and increased equity should be at the core of government priorities, poverty reduction must remain at the top of policymakers' minds. Addressing poverty through various strategies such as risk transfers, social welfare benefits, and subsidies is necessary for effective poverty reduction, particularly protecting those most susceptible, low-income farmers, and environmentally dependent populations, for instance, from adverse events. Investment in social protection schemes must also increase to reach all who require help (Disaster | UNDRR, 2023).

4 Conclusion

Conclusion This in-depth review provided invaluable insights into the efforts of Sri Lanka and the challenges related to its endeavours to integrate global agendas like the Paris Agreement, SDGs, and Sendai Framework into coastal built environments. The primary goal was to assess the progress of Sri Lanka with these international frameworks by aligning national policies to them while also highlighting areas for improvement, with advice provided for policymakers and stakeholders on advancing these essential objectives within the unique context of Sri Lanka.

The coastal regions in Sri Lanka are vulnerable to climate change impacts such as sea-level rise, extreme weather events, and coastal erosion. While their efforts in adapting measures have certainly made a difference in addressing these threats, which are commendable, our analysis has identified several findings worthy of note.

Sri Lanka has taken significant strides toward aligning its policies with international agreements, but significant challenges remain. The lack of comprehensive climate and socio-economic data poses one barrier, hindering effective planning and decision-making processes and implementing adaptation measures within coastal built environments. Furthermore, the lack of robust climate regulations/ standards creates further difficulties.

Second, it is vitally important that stakeholders work more collaboratively and closely together. The progress of Sri Lanka toward meeting global agendas requires collaboration from government bodies, non-government organizations, private sector entities, and local communities alike. Any gaps between knowledge sharing or coordination must be bridged for a unified coastal climate adaptation approach to be realized.

Finally, localizing climate adaptation initiatives with an eye toward local phenomena and vulnerabilities is critical for Sri Lanka. Educational efforts and awareness-raising campaigns on climate adaptation must take precedence to proactively empower coastal communities to protect their coastal environments.

Integrating the Paris Agreement, SDGs, and Sendai Framework into the coastal built environments in Sri Lanka is a complex yet ongoing endeavour. While challenges persist, Sri Lanka remains dedicated to sustainable development and climate resilience despite some limitations; by diligently addressing data deficiencies, strengthening regulatory frameworks, encouraging collaboration, prioritizing locally tailored adaptation strategies, prioritizing local adaptation plans, and prioritizing locally tailored adaptation strategies, Sri Lanka can overcome them and move confidently toward creating more resilient and sustainable coastal futures.

References

Baseline Study For National Spatial Data Infrastructure Final Report Baseline Study for National Spatial Data Infrastructure Contents. (2017).

Climate Change and Its Impact on Coastal Wetlands of Sri Lanka. (2018).

Demel, M. S. V., et al. (2019). Developing a socio-economic vulnerability index using two agricultural communities exposed to climate change: A case study in Wanathawilluwa in Puttlam District of Sri Lanka. *Applied Economics and Business*, *3*(1), 70–79. https://doi.org/10.4038/AEB.V3I1.57

Disaster | UNDRR. (2023). Available at: https://www.undrr.org/terminology/disaster. Accessed August 28, 2023.

Disaster Management Centre. (2017). Available at: http://www.dmc.gov.lk/index.php?lang=en. Accessed September 18, 2023.

Disaster Risk Reduction in Sri Lanka. (2018).

Disaster Risk Reduction in Sri Lanka. (2019a).

Disaster Risk Reduction in Sri Lanka. (2019b).

Doukas, H., et al. (2018). From integrated to integrative: Delivering on the Paris Agreement. *Sustainability*, *10*(7). https://doi.org/10.3390/SU10072299

Eckstein, D., et al. (2018). *Global climate risk index 2019 who suffers most from extreme weather events? Weather-related loss events in 2017 and 1998 to 2017*.

Executive Board of the United Nations Development Programme, the United Nations Population Fund, and the United Nations Office for Project Services Draft country programme document for Sri Lanka (2018–2022). (2017). Available at: http://megapolis.gov.lk/wp-content/uploads/2016/09/Megapolis-Plan_Jan16_English.pdf

Fernando, N., et al. (2021). *Settling the ripples: An examination of Sri Lanka's approach to addressing cascading impacts of the COVID-19 pandemic* (pp. 554–569). National Science Foundation of Sri Lanka. Available at: https://pure.hud.ac.uk/en/publications/settling-the-ripples-an-examination-of-sri-lankas-approach-to-add. Accessed September 19, 2023.

Final_NDC_IP_-_10072023. (2021).

Global Assessment Report on Disaster Risk Reduction 2019. (2019). United Nations.

Government of the Democratic Socialist Republic of Sri Lanka. (2018).

Gunathilaka, R. P. D., & Samarakoon, P. S. M. K. J. (2022). Adaptation to climate change by vegetable farmers in Sri Lanka. *Climate Change and Community Resilience*, 415–430. https://doi.org/10.1007/978-981-16-0680-9_27/FIGURES/4

Hazard Profiles of Sri Lanka. (2015). Available at: https://www.dmc.gov.lk/hazard/hazard/Report.html. Accessed September 18, 2023.

Highlights of the Year. (2019). Available at: www.energy.gov.lk

Hippola, H. M. S. S., et al. (2020). Gap assessment of warning and dissemination process of early warning system in coastal areas of Sri Lanka. *Lecture Notes in Civil Engineering*, *44*, 36–44. https://doi.org/10.1007/978-981-13-9749-3_4

innovative-financing-for-development-in-ap. (2017).

Implementation of the Sendai Framework for Disaster Risk Reduction 2015–2030 Report of the Secretary-General. (2016). Available at: www.desinventar.net

Institutional Disaster Management Plan. (2019). Available at: www.dmc.gov.lk

IPCC. (2022). *Global Warming of 1.5 °C*. Cambridge University Press. Available at: https://doi.org/10.1017/9781009157940

Jayasekara, R. U., et al. (2021). Identifying gaps in existing early warning mechanisms and evacuation procedures for tsunamis in Sri Lanka with a special focus on the use of social media gaps in early warning mechanisms and evacuation procedures for tsunamis in Sri Lanka, with a special focus on the use of social media. *International Journal of Disaster Resilience in the Built Environment* [Preprint]. https://doi.org/10.1108/IJDRBE-02-2021-0012

Key aspects of the Paris Agreement | UNFCCC. (2019). Available at: https://unfccc.int/most-requested/key-aspects-of-the-paris-agreement. Accessed September 19, 2023.

Kumarasiri, B., & Dissanayake, P. (2021). Barriers to implementing waste-to-energy projects in Sri Lanka: A PESTEL analysis. *Built Environment Project and Asset Management, 11*(4), 544–558. https://doi.org/10.1108/BEPAM-03-2020-0051/FULL/XML

Iizuka, A. (2020). Developing capacity for disaster risk reduction: Lessons learned from a case of Sri Lanka. *Progress in Disaster Science, 6*, 100073. https://doi.org/10.1016/J.PDISAS.2020.100073

Malalgoda, C., Amaratunga, D., & Haigh, R. (2016). Overcoming challenges faced by local governments in creating a resilient built environment in cities. *Disaster Prevention and Management, 25*(5), 628–648. https://doi.org/10.1108/DPM-11-2015-0260

Munasinghe. (2018). *Sustainable Sri Lanka 2030 vision, strategic path and sustainable development goals (SDG)*. Available at: https://www.researchgate.net/publication/336000669

Munasinghe, J. (2019a). *National physical planning policy and the plan 2050 Sri Lanka*. National Physical Planning Policy and Plan 2030 [Preprint]. Available at: https://www.academia.edu/43983783/National_Physical_Planing_Policy_and_the_Plan_2050_Sri_Lanka. Accessed September 18, 2023.

Munasinghe, M. (2019b). *Sustainable Sri Lanka 2030 Vision and Strategic Path (Presidential Expert Committee (PEC))*.

National Adaptation Plan for Climate Change Impacts in Sri Lanka Climate Change Secretariat Ministry of Mahaweli Development and Environment 2016. (2016).

Nianthi, R., Shaw, R., & Nianthi, K. W. G. R. (2015). *Climate change and its impact on coastal economy of Sri Lanka special issue 'second edition of urban disaster resilience and sustainability' view project poster view project climate change and its impact on coastal economy of Sri Lanka*. Available at: https://www.researchgate.net/publication/267419810

(PDF) Sustainable Sri Lanka 2030 Vision and Strategic Path. (2023). Available at: https://www.researchgate.net/publication/327221768_Sustainable_Sri_Lanka_2030_Vision_and_Strategic_Path. Accessed September 18, 2023.

Perera, M. D. N. D., et al. (2018). Risk of seawater intrusion on coastal community of Bentota river basin Sri Lanka. *Procedia Engineering, 212*, 699–706. https://doi.org/10.1016/J.PROENG.2018.01.090

Pitigala Liyana Arachchi, I. S., et al. (2022). Evaluation of societal trust on multi-hazard early warning (MHEW) mechanism: Sri Lankan context. *International Journal of Disaster Resilience in the Built Environment, 13*(5), 533–553. https://doi.org/10.1108/IJDRBE-01-2021-0010

Ranasinghe, R., & Jongejan, R. (2018). *Marine science and engineering editorial climate change, coasts and coastal risk*. https://doi.org/10.3390/jmse6040141

Randil, C., Siriwardana, C., & Hewawasam, K. (2020). Framework to analyze Sri Lanka disaster management mechanism. *International Journal of Disaster Resilience in the Built Environment, 11*(2), 289–307. https://doi.org/10.1108/IJDRBE-07-2019-0044

Secretariat, C. C. (2015). *National adaptation plan for climate change impacts in Sri Lanka 2016 to 2025, Ministry of Mahaweli Development and Environment, Sri Lanka*.

Sendai_Framework_for_Disaster_Risk_Reduction_2015–2030. (2015).

Sri Lanka Updated Nationally Determined Contributions. (2021a).

Sri Lanka Updated Nationally Determined Contributions. (2021b).

Sri Lanka Updated Nationally Determined Contributions. (2021c).

Sri Lanka Comprehensive Disaster Management Programme (SLCDMP). (2014).

Sri Lanka NEXT: A Blue-Green Era—Bio Energy Association of Sri Lanka. (2019). Available at: https://www.bioenergysrilanka.lk/sri-lanka-next-a-blue-green-era/. Accessed September 18, 2023.

Sri Lanka National Disaster Management Plan 2013–2017 Disaster Management Centre Ministry of Disaster Management 2nd. (2014).

Sustainable Development Goal | Sustainable Development Council of Sri Lanka. (2021). Available at: https://sdc.gov.lk/en/sustainable-development-goal/29. Accessed September 18, 2023.

The Paris Agreement | UNFCCC. (2015). Available at: https://unfccc.int/process-and-meetings/the-paris-agreement. Accessed September 15, 2023.

The Sustainable Development Goals Report. (2021).

Tsunami disaster—Sri Lanka Summary situation report 26–31 Dec 2004—Sri Lanka | ReliefWeb (2004). Available at: https://reliefweb.int/report/sri-lanka/tsunami-disaster-sri-lanka-summary-situation-report-26-31-dec-2004. Accessed September 18, 2023.

Upadhaya, B., et al. (2021). *For more information, visit Follow PEFA on social media* www.pefa.org *PEFA partners integrating climate change and livelihood within public investment policies: A cross-country assessment in South Asia (India, Nepal, and Sri Lanka)*. Available at: www.pefa.org.

What is the Sendai Framework for Disaster Risk Reduction? | UNDRR. (2015). Available at: https://www.undrr.org/implementing-sendai-framework/what-sendai-framework. Accessed September 18, 2023.

Wijenayake, V., et al. (2020). Policy gaps and needs analysis for the implementation of NDCs on adaptation and loss and damage in Bangladesh, Nepal, and Sri Lanka. *APN Science Bulletin, 10*(1). https://doi.org/10.30852/sb.2020.1283

Guidance Note on Implementing Local Adaptation Strategies in the Coastal Built Environment

María Merino Gonzalez-Pardo, Ignacio Aguirre Ayerbe, Laura Pérez García, Mo Hamza, Fernando Nishara, Pasindu Jayathilake, Lalith Rajapakse, Chandana Siriwardana, Bawantha Rathnayaka, Shavindree Nissanka, Chamindi Malalgoda, Ruben Paul Borg, and Glorianne Borg Axisa

1 Introduction

Coastal cities and settlements are at the forefront of climate risk. The dynamic interaction between ocean and climate drivers together with varied coastal geographies influences the nature of coastal risks, including many that are unique to cities and settlements located by the sea (Glavovic et al., 2022).

Coastal zones are complex, both from the point of view of natural and human systems and the interaction between them. This, coupled with the uncertainties associated with climate change-related analyses, makes planning and implementing

M. M. Gonzalez-Pardo (✉) · I. A. Ayerbe · L. P. García
IHCantabria—Instituto de Hidráulica Ambiental de la Universidad de Cantabria, Cantabria, Spain
e-mail: maria.merino@unican.es

M. Hamza
Lund University, Lund, Sweden

F. Nishara · P. Jayathilake
University of Colombo, Colombo, Sri Lanka

L. Rajapakse · C. Siriwardana · B. Rathnayaka
University of Moratuwa, Moratuwa, Sri Lanka

S. Nissanka · C. Malalgoda
University of Huddersfield, Huddersfield, England

R. P. Borg · G. B. Axisa
University of Malta, Msida, Malta

S. Nissanka
Department of Quantity Surveying, Faculty of Engineering, Sri Lanka Institute of Information Technology (SLIIT), Malabe, Sri Lanka

C. Malalgoda et al. (eds.), *Climate Change Adaptation in the Built Environment*,
https://doi.org/10.1007/978-3-031-75826-3_4

climate change adaptation strategies a major challenge for managers, administrators and professionals.

In addition to mitigation-oriented strategies to reduce greenhouse gas emissions, adaptation to climate change is necessary to reduce the negative effects that climate change may cause and thus to protect the coastal built environment. Adaptation is considered in this study according to the Intergovernmental Panel on Climate Change (IPCC) definition as the process of adjustment to actual or expected climate and its effects, to moderate harm or exploit beneficial opportunities (IPCC, 2021).

2 Objective

This study constitutes a guidance note to support coastal practitioners and local managers in the implementation of local climate change adaptation measures in coastal built environments. To that end, this guidance note presents:

1. a number of general recommendations to plan and develop coastal adaptation strategies,
2. instruments available to support local adaptation to climate change at the national level in target countries,
3. some examples of climate change adaptation actions undertaken by communities and government agencies that may be useful to inspire others when identifying and implementing adaptation options in their site-specific context.

This piece of work summarizes the results of "Output 5" of the project BEACON (Built Environment leArning for Climate AdaptatiON) funded by the European Union (EU). This three-year collaborative research initiative aimed to develop trans-disciplinary and innovative research-based learning in the built environment to tackle climate change in coastal regions. The participating countries in this project and the target countries of this guidance note are Spain, United Kingdom, Sweden, Malta and Sri Lanka.

3 Methodology

This guidance note has been prepared following both a research analysis and a participatory validation process involving key stakeholders from the five participating countries. The methodology was agreed among the BEACON partners, as well as the set of templates designed to ensure systematic data collection and analysis.

First, an exhaustive bibliographic review was carried out, including scientific and scholarly literature and national and international reports and guidelines. The bibliographic review allowed to identify the following aspects:

1. general recommendations for the formulation and implementation of climate change adaptation measures;

2. country-specific available instruments to support the implementation of climate change adaptation strategies at the national level, including regulatory and financial instruments, strategies, existing plans and guidelines related to climate change adaptation, and available datasets and scientific facilities to support climate change adaptation studies in coastal areas; and
3. case studies in Europe and Sri Lanka addressing different components (physical, economic, social, environmental and governance-classification adapted from UoM, 2023–) and sectors (tourism, finances and insurance, fisheries and agriculture, transport and energy and urban) of the coastal built environment, to elaborate a comprehensive catalogue of case studies.

The analysis of the case studies allowed for the identification of best practices and challenges faced during the implementation of climate change adaptation measures at the local level, which were taken into account in the definition and selection of the most relevant general recommendations. The preliminary identification and selection of general recommendations was discussed and agreed among the project consortium, made up by scientific institutions from the five participating countries.

After the bibliographic review and analysis, a participatory process was carried out with key stakeholders for coastal management and climate change adaptation in the five target countries. The participatory process aimed to:

- validate the preliminary selection of general recommendations;
- validate the list of available instruments at national level;
- present the catalogue of case studies to further analyse the adequacy of different types of adaptation options in each country. This analysis was based on the perception of stakeholders on the suitability of different typologies of adaptation options to the national frameworks and supporting instruments.

The participatory process was carried out through semi-structured interviews that were conducted both face-to-face and via online video calls. 92 stakeholders were interviewed including national and local authorities, civil and professional associations, academia, community associations and private sector actors. The distribution of interviewed stakeholders by category and country is shown in Fig. 1.

The complete guidelines are fully presented in BEACON project output 5 report (IHCantabria, 2023a). The catalogue of cases studies, including more than 30 best practices and experiences (IHCantabria, 2023a) is also available at the University of Malta dedicated BEACON online e-learning platform (BEACON Research UM, 2023) and at the BEACON project website (https://beacon-researchproject.org/outputs/).

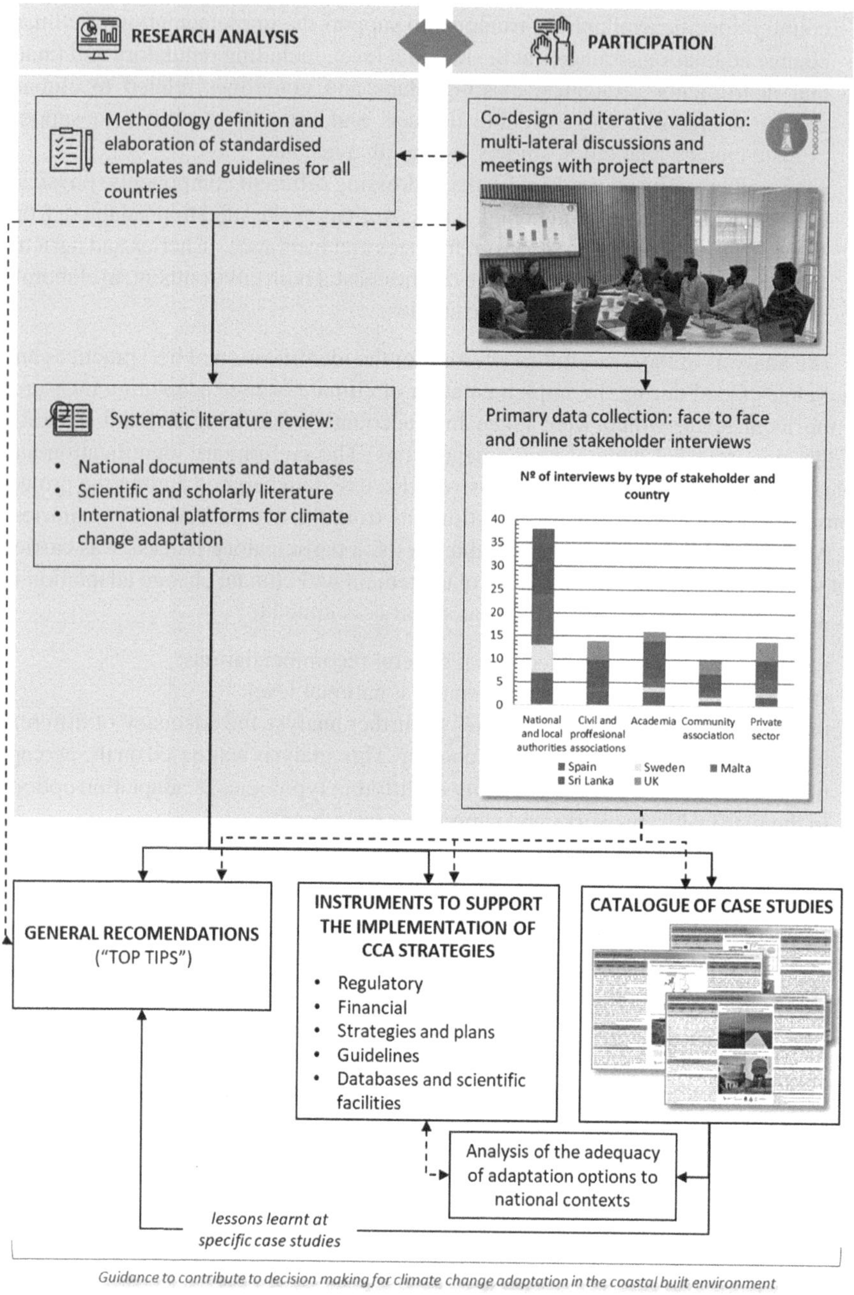

Fig. 1 Methodology flowchart

4 General Recommendations

This section presents a set of general recommendations to guide institutions, managers and professionals on the essential aspects to be considered when planning and implementing climate change adaptation measures in coastal urban areas.

These recommendations have been outlined following international and national guidelines and studies focused on climate change adaptation, coastal management and sustainable urban planning, such as the "*Adapting to climate change: A guide for local councils*" (DEFRA, 2010) or the Spanish "*Guide for the Elaboration of Local Plans for Climate Change Adaptation*" (Feliu et al., 2015) and have considered the lessons learnt from the selected case studies and from the implementation of research and consultancy projects developed worldwide by project partners. In addition, the proposed recommendations were validated through a participatory process with BEACON project partners including online and face to face focus group meetings.

4.1 Quantify Current and Future Climate-Related Risks in Land Use Planning and Urban Development

Develop an in-depth diagnosis of climate-related risks, based on scientific data, as well as their continuous monitoring. This requires:

- the identification of natural hazards in the target study area and their quantification, preferably based on numerical models;
- the assessment of the vulnerability of the different dimensions of the coastal built environment (including social, economic, infrastructure and environment);
- the analysis and mapping of the resulting risk in current and future scenarios, considering cost–benefit analysis and the planning of post-disaster activities.

For this quantification, it is necessary to establish the most appropriate climate change scenario(s) and horizon year to be analysed, which should be in line with other plans and strategies for the target study area. The risk results provide baseline information for designing site-specific climate change adaptation measures and developing a climate change adaptation plan.

4.2 Design Comprehensive Climate Change Adaptation Plans

Based on a deep understanding of climate-related risks, it is recommended to elaborate comprehensive climate change adaptation plans that allow for the definition and selection of climate change adaptation measures, along with a roadmap for their implementation. Climate change adaptation measures include several typologies that may be grouped in general terms as structural/physical, social and institutional. The plan should also define monitoring and evaluation tools to assess the implementation process and the effect of the measures on the coastal built environment.

4.3 Apply an Adaptive Management Approach

Promote flexible and dynamic adaptive management and measures, which can reduce uncertainty in decision-making by monitoring the coastal system and adapting the proposed measures if necessary. It is also necessary to periodically update climate change adaptation plans. Use monitoring and last-generation data and methodologies to update the assessment of climate-related risks and innovative techniques to design climate change adaptation measures.

4.4 Consider a Holistic Approach to Address Coastal Complexities

Apply a holistic approach preferably following comprehensive principles such as Integrated Coastal Zone Management (ICZM) principles for coastal planning, which are fully integrated with climate change adaptation. In this regard, it is necessary to consider the interaction between all components of the coastal built environment: human, physical, economic, social, environmental and governance; the different sectors in place: tourism, financial, fishing, agriculture, transport, and navigation, urban, etc.; and the different levels of administration: neighbourhood, local, national, etc. All these coastal aspects must be addressed in a balanced manner to achieve sustainable development.

4.5 Capitalise on Previous Efforts and Initiatives

Analyse previous initiatives developed in the target study area to build on existing climate change adaptation activities. Incorporate climate change adaptation approaches into those existing initiatives that can be improved. Additionally, explore synergies between climate change adaptation and mitigation measures.

4.6 Work with Local Planning Authorities and Key Agencies

Involve local authorities and create coordination mechanisms for local climate change adaptation. This will contribute to identifying previous efforts and initiatives, identifying potential opportunities from climate change, avoiding conflicts and overlapping competencies, and designing cross-sectoral climate change adaptation measures that benefit components and sectors of the coastal built environment.

4.7 Involve Local Communities

Promote the participation of indigenous knowledge, local community populations and other local stakeholders, such as public entities, civil organizations, local professional organizations, research entities and the private sector, in the decision-making process on climate change adaptation.

4.8 Explore Available Funding for Climate Change Adaptation

Contact local and national climate change adaptation offices to explore funding opportunities available in the target study area. In addition, it is also important to explore available international and regional funding for climate change adaptation, such as the European Commission programs (e.g.: European Regional Development Fund, Horizon Europe, Innovation Fund, InvestEU and the LIFE Programme), the Green Climate Fund, United Nations programmes and development bank funds (e.g.: the World Bank, Asian Development Bank, or the Inter-American Development Bank), among others. Research centres, universities and private innovation companies can support in applying for international and regional funding.

4.9 Collaborate Across Borders

Make sure to collaborate on adaptation measures across borders, both geographical and administrative. To work effectively, it is necessary to strengthen collaboration with all stakeholders related or affected by the climate change adaptation measures that are planned to be implemented.

4.10 Prioritize Nature-Based Solutions

Explore nature-based solutions (NbS) actively and make necessary systemic changes to promote them. Many NbS solutions have multipurpose functions (co-benefits) and thus, they do not only contribute to adaptation, but also to a more resilient ecosystem, promote greater biodiversity, sustainable development, and contribute to mitigation objectives. Nature-based and hybrid approaches combining ecosystems and built infrastructure are becoming a popular option worldwide (IPCC, 2019).

5 National Frameworks and Available Instruments to Support Local Adaptation to Climate Change

In addition to international agendas, such as the Sustainable Development Goals,[1] the Sendai Framework for Disaster Risk Reduction 2015–2030[2] or the Paris Agreement,[3] and the information provided by international platforms such as the IPCC[4] or Climate-ADAPT,[5] national frameworks provide specific instruments to support the implementation of local climate change adaptation measures.

This section presents the available instruments to support local adaptation to climate change at country level, including regulatory and financial instruments, existing national strategies and plans related to climate change adaptation in the coastal built environment, available guidelines, and existing databases and scientific facilities in each target country (Spain, United Kingdom, Sweden, Malta and Sri Lanka). This list can serve as a reference to understand the type of instruments needed for the implementation of climate change adaptation measures.

5.1 Spain

Spain is a predominantly coastal country, with almost 8000 km of coastline housing 46 ports and 18 provincial capitals, reflecting the importance of the built coastal environment in Spain. According to the Spanish Strategy for Coastal Adaptation to Climate Change (MAPAMA, 2016), coastal systems in Spain are particularly sensitive to the effects of sea level rise, as well as to other climate-related factors, such as the increase in sea surface temperature, acidification, and changes in both intensity and direction in storms and waves.

[1] https://www.undp.org/sustainable-development-goals.

[2] https://www.undrr.org/publication/sendai-framework-disaster-risk-reduction-2015-2030.

[3] https://unfccc.int/process-and-meetings/the-paris-agreement.

[4] https://www.ipcc.ch/.

[5] https://climate-adapt.eea.europa.eu/.

Spain has made significant advances in climate change adaptation in the last years with the publication of the National Climate Change Plan Adaptation (PNACC) 2021–2030 (MITECO, 2020a) and the establishment of the Spanish Office for Climate Change (OECC) under the Environmental Secretariat. This office, together with the National Environmental Education Centre, plays an important role in raising awareness and disseminating climate change adaptation initiatives through the AdapteCCa[6] platform. Under this Secretariat, the National Meteorological Agency is responsible for the coordination of the development of regionalized climate scenarios in Spain. Additionally, the Ministry of Transport, Mobility and Urban Agenda collaborates in various lines of action of the PNACC.

The implementation of climate change adaptation measures is mainly addressed in Spain from a sectoral approach. In the built environment of coastal areas, the agencies with competencies in coastal management and land use planning have an important role in mainstreaming climate change adaptation. At the national level, the General Directorate for the Coast and the Sea manages the coastal public domain and develops coastal protection strategies, including interventions in urban areas. At the regional and local level, the Autonomous Communities have the competencies in urban and regional planning and establish strategic lines to be adopted by the municipalities, which have competencies in urban planning.

In addition to the public entities, local organizations play a significant role in implementing site-specific climate change adaptation actions, many linked to environmental conservation initiatives. Other stakeholders, such as the Spanish Network for Climate and the Federation of Municipalities promote and support climate change adaptation interventions. In this regard, there is still a need for communicating climate change adaptation issues and strengthening the governance framework for climate change adaptation, especially at the local level, by the provision of advanced knowledge, data and procedures to mainstream climate change adaptation in their daily activities.

5.1.1 Regulatory Instruments

In addition to the European directives, the following regulatory instruments shall be considered for the development and implementation of climate change adaptation measures:

- Law 7/2021, of May 20, on Climate Change and Energy Transition
- Law 21/2013, of 9 December, on Environmental Assessment
- Royal Legislative Decree 7/2015, of October 30, approves the consolidated text of the Land and Urban Rehabilitation Law
- Law 22/1988, of July 28, on Coasts and its update Law 2/2013, of 29 May, protection and sustainable use of the coast

[6] https://adaptecca.es/.

- Royal Decree 903/2010, of 9 July, on the Assessment and Management of Flood Risks. Transposed from Directive 2007/60/EC of the European Commission
- Royal Decree 314/2006, of March 17, 2006, on Technical Code of Building Construction and its update Royal Decree 450/2022
- Regional and municipal regulation

5.1.2 Financial Instruments

The following entities and funds are available to support climate change adaptation measures in Spain:

- Funds managed through the Biodiversity Foundation (MITECO, 2022)
- Funds managed through the Institute for Energy Diversification and Savings (MITECO, 2020b)
- Funds managed through the Center for Energy, Environmental and Technological Research (MICIN, 2023)
- Program to support actions to improve energy efficiency (MITMA, 2023a)
- Transportation and mobility grants (MITMA, 2023b)
- Common Agricultural Policy (MITMA, 2023c)

5.1.3 National Strategies and Plans

The following strategies and plans guide climate change adaptation in Spain:

- National Climate Change Adaptation Plan (PNACC) 2021–2030 (MITECO, 2020a)
- The Spanish Strategy for Coastal Adaptation to Climate Change (MAPAMA, 2016)
- Regional Strategies for Coastal Protection in Huelva, Maresme, South Valencia and Castellon coasts, Granada, Cádiz, Málaga, and Almeria (MAPAMA 2013, 2014, 2015a, 2015b, 2017, 2021)
- National Strategy for Green Infrastructure, Ecological Connectivity and Restoration (MITECO, 2021)
- Spanish Climate Change and Clean Energy Strategy (MITECO, 2007)
- Spanish Urban Agenda (Ministerio de Fomento, 2019)
- Instrument of Ratification of the European Landscape Convention (MAEC, 2007)
- Regional and local strategies for climate change adaptation and coastal protection

5.1.4 Available Guidelines

The following guidelines provide advice on developing and implementing climate change adaptation in Spain:

- Guide for the elaboration of local plans for adaptation to climate change—Ministry of Ecological Transition (MITECO) (Feliu et al., 2015)

- Guidelines and directives for environmental assessment (MITECO, 2023a)
- Vulnerability of Spanish Ports to Climate Change (Puertos del Estado, 2016)
- Guidance for the development of adaptation plans to climate change in schools (SEO, 2021)
- Citizen Science for Monitoring Climate Change in Ecosystems (Valdor et al., 2021)
- Guía orientativa para la adaptación de las zonas urbanas costeras al cambio climático (IHCantabria, 2023d)

5.1.5 Databases and Scientific Facilities

The following databases and geoviewers support the elaboration of studies related to climate change adaptation. Additionally, Spain has numerous research and experimental facilities with extensive experience in climate change adaptation.

5.1.6 Databases and Viewers

- Climate Change Adaptation platform: AdapteCCa (MITECO, 2023b)
- Climate Change on the Spanish Coast—C3E Viewer (IHCantabria, 2011)
- National Spatial Data Infrastructure (water, climate change, coast and marine, biodiversity and forest, and environmental quality information) (MITECO, 2023c)
- Met-ocean database (wind, wave and sea level conditions): IHData (IHCantabria, 2023b)
- National Port Authority database (MITMA, 2023d)
- Spanish National Research Council database (SICTI, n.d.)
- National Flood Zone Mapping System (MITECO, 2023d)

5.1.7 Scientific and Experimental Facilities

- Balearic Islands Coastal Observation System (SOCIB, 2010)
- Environmental Hydraulics Institute of the University of Cantabria (IHCantabria, 2023c)
- Cantabria Coastal and Ocean Basin (CCOB, 2022)
- BC3 Basque Centre for Climate Change (BC3, 2008)
- Spanish Institute of Oceanography (IEO, 2014)
- Institute of Physics of Cantabria (IFCA, 2023)

5.2 United Kingdom

With a coast length of approximately 12,000 km, the UK has experienced over the last decade several significant natural hazards associated with climate change events, resulting in significant economic and human impacts on communities, properties, and infrastructure networks (Dale, 2021). In addition, climate change will cause sea levels to rise continuously throughout the twenty-first century. Coastal and offshore infrastructure is also vulnerable to changing patterns of storm conditions (Poo et al., 2021). Around £150 billion of assets in the UK are at risk from coastal flooding. Damages to the UK from coastal flooding are estimated to be of the order of £500 million per year (Howard & Palmer, 2020). Due to coastal erosion, long-term morphological change is also evident in coastal areas, including extensive salt marsh loss (Nicholls et al., 2021).

Climate change adaptation in the UK is governed by the climate change second National Adaptation Programme (2018–2023) identified as NAP hereafter. The main thematic areas covered under the NAP are the natural environment, infrastructure, people and built environment, business and industry and local government sectors. According to the review developed, there are several national adaptation strategies and plans, in most cases complemented by national laws and regulations, such as the National Adaptation Program, the Zeroing Strategy, the Ten-Point Plan, the Clean Growth Strategy, the Low Carbon Transport Innovation Strategy, among others. However, it is also found that there seems to be a lack of ground-level guidelines on how to adopt climate change adaptation across the different sectors. Involving local agencies, planning authorities and the local community in the local adaptation process is essential. It is imperative to understand adaptation from the perspective of the people who are integrating climate change risks and adaptation issues into their day-to-day decisions. It is also highly appropriate to explore innovative financing mechanisms that can be used in climate change adaptation programs.

5.2.1 Regulatory Instruments

UK has the following regulations in place to govern national climate adaptation.

- Climate Change Act 2008
- Environment Act 2021
- Energy Act 2016
- Budget 2021; UK's Net Zero target incorporated into the overall economic policy
- Pension Schemes Act 2021: Mandates Climate adaptation for trustees and pension fund managers
- Agriculture Act 2020: Climate friendly land, and livestock management
- Climate Change Agreements: Climate change levy for energy efficiency and carbon targets
- Finance Act 2020: Carbon tax and carbon pricing regimes
- Carbon Budget Order 2016
- Infrastructure Act 2015; promotes off-site carbon abatement measures.
- UK Flood and Water Management Act—2010
- Climate Change and Sustainable Energy Act
- Company Car Tax Reform

- Subsidy Control Act 2022: environmental principles for the issuance of subsidies

5.2.2 Financial Instruments

The following financial instruments are available to support climate change adaptation measures in the UK.

- Green Finance Strategy
- Energy Savings Opportunity Scheme
- Bio-energy Capital Grants Scheme
- Feed-in Tariffs for renewable electricity
- Water Act 2014 (creation of Flood Re scheme)
- Preferential Tax Regimes for Biofuels

5.2.3 National Strategies and Plans

The following national strategies and plans guide the national and local level climate change adaptation in the UK.

- UK Climate Change Risk Assessment
- National Adaptation Programme (last version covering 2018–2023)
- Resilience for the future: The UK's critical minerals strategy
- Net zero strategy
- Boiler Upgrade Scheme
- Energy Security Strategy
- The Limited Liability Partnerships (Climate-related Financial Disclosure) Regulations 2022, made under the Limited Liability Partnerships Act 2000
- The Companies (Strategic Report) (Climate-related Financial Disclosure) Regulations 2022, made under the Companies Act 2006
- CCUS Investor Roadmap
- Hydrogen Investor Roadmap
- Automotive (investor) roadmap
- The Greenhouse Gas Emissions Trading Scheme Order 2020
- Net Zero Strategy: Build Back Greener
- Greening Finance: A Roadmap to Sustainable Investing
- Heat and buildings strategy
- UK Hydrogen Strategy
- Transport Decarbonisation Plan
- National Planning Policy Framework
- England Trees Action Plan 2021 to 2024
- England Peat Action Plan
- North Sea Transition Deal
- Industrial Decarbonisation Strategy
- Bus Back Better: National bus strategy for England
- Net Zero Innovation portfolio

- Motor Fuel (Composition and Content) and the Biofuel (Labelling) (Amendment) Regulations 2021
- Carbon Accounting (Provision for 2019) Regulations 2021
- Financial Conduct Authority Rules on TCFD
- Energy White Paper
- National Infrastructure Strategy
- Ten Point Plan
- Gear Change, a bold vision for cycling and walking (Cycling and walking plan for England)
- A plan for jobs
- Marine Strategy Part One: UK updated assessment and Good Environmental Status.
- Clean Growth Strategy
- Cycling and walking investment strategy
- Carbon Plan
- Low Carbon Transport Innovation Strategy

5.2.4 Available Guidelines

The following are some guidelines available in the UK for advice on developing and implementing climate change adaptation plans.

- Guidance for local government on preparing for a changing climate
- Preparing for Climate Change: A Climate Change Adaptation Strategy
- Code for Sustainable Homes

5.2.5 Databases and Scientific Facilities

The following databases and scientific facilities are available in the UK to extract climate information for studies and local adaptation plans.

- UK climate change statistics portal
- Met Office-UK climate maps and data
- UKCP summaries and headline findings
- Headline findings
- Climate change projections over land
- Marine climate change projections
- UKCP data
- CEDA catalogue
- UKCP data factsheets
- CCWeatherGen: Climate Change Weather File Generator for the UK
- Climate X
- Airex, London—Buildings
- Boxergy, Edinburgh, Scotland—Energy and Electricity
- Circulor, London—Supply Chain

5.3 Sweden

The 3218 km long Swedish coast is largely along the Baltic Sea. In Sweden, municipalities are in charge of dealing with the issue of climate adaptation, but the system in which they operate is influenced by national organizations, regional actors, the business sector, and legal frameworks. Administratively, Sweden is divided into 290 municipalities and 21 counties. Swedish local governments, particularly those who work in city planning or county administrative boards, are crucial to the governance of climate adaptation (Statens Offentliga Utredningar, 2017). The municipal building monopoly, which stipulates that the municipality is the only entity that regulates the construction, use of buildings, and land-use planning, within their boundaries is a crucial component of local adaptation efforts.

Another significant piece of legislation that affects local climate adaption initiatives in Sweden is the Planning and Building Act (PBL), related to the Environmental Code (Miljöbalken), published in 1998. It provides requirements for the preservation of the environment, including steps to mitigate the effects of climate change. Municipalities in Sweden are also tasked with developing oversight plans to manage climate risks in addition to these two legal frameworks (Statens Offentliga Utredningar, 2017). All 21 of Sweden's regions' county administrative boards, as well as Stockholm's and Skåne's regional supervision plans, have an impact on climate adaptation at the regional level (Government proposition, 2017; Region Skåne, 2022). County administrative boards are crucial to the management of climate adaptation, especially when it comes to infrastructure development and land use planning (Olsson, 2018). The region is a political body that exists in addition to the county administrative board. It supervises the creation of the regional oversight plan and shares duties with the county administration board (Government proposition, 2017; Region Skåne, 2022).

Regional actors in Sweden, including enterprises and civil society organisations, may contribute to attempts to adapt to climate change in addition to these official structures. Sweden's climate adaptation is managed at the national level by the Climate Ordinance (2018:1428) as well as various national laws, regulations, and policies. Swedish efforts to adapt to climate change are influenced by national policies and guidelines in addition to the Climate Ordinance. The Housing Authority is another significant national organisation when it comes to climate adaptation in the built environment (Olsson, 2018). More than anything else, it establishes recommendations for how to deal with certain difficulties and supervise the formulation of guidelines for how towns might interpret the Plan and Building Act.

In summary, there is no single organization in Sweden that is in charge of climate change adaptation. All organizations listed in the 2018 Climate Ordinance are subject to certain requirements. They must conduct a climate and vulnerability analysis, create a plan for climate adaptation, and determine which laws have an impact on their line of work and climate adaptation.

5.3.1 Regulatory Instruments

In addition to the European directives, the following regulatory instruments shall be considered for the development and implementation of climate change adaptation measures:

- Plan and Building Act (Plan-och bygglagen, Förordning 2010:900)
- The Environmental Charter (Miljöbalken, Förordning 1998:808)
- Ordinance of governmental agencies climate adaptation work (Förordning om myndigheters klimatanpassningsarbete, Förordning 2018:1428)
- Financial Instruments

One of the main funding sources is MSB's fund 2:2. In 2022, this fund paid out over 235 million SEK, with total applications amounting to around 300–400 million SEK (MSB, 2022). Only municipalities can apply for this fund, and project selection is done in collaboration with SMHI and SGI. The fund prioritizes large-scale projects and technical solutions for preventing disasters resulting from natural hazards, making it challenging for smaller municipalities to access due to the extensive application process and competition from larger municipalities. In 2022, only 15 out of Sweden's 290 municipalities received funding through this mechanism. Despite this, it is considered a key funding source for climate adaptation. There are other funding options available, but the amount of money varies each year, making long-term planning difficult for municipalities.

SMHI's website (klimatanpassning.se) provides information about various funding opportunities for property owners to adapt and protect their land. However, the application process for these funds can be complex and time-consuming, especially for smaller municipalities with limited resources and personnel dedicated to climate adaptation (SMHI, n.d.).

One potential way to fund climate adaptation efforts in Sweden is through taxes or a municipal charge similar to the one for water and sewage. This could provide a sustainable long-term solution for financing climate adaptation. However, there are insurmountable complexities to that option that would render it out of reach or consideration (Andersson & Nilsson, 2021; Bahr and Ivarsson, 2020; Schultze et al., 2022).

5.3.2 National Strategies and Plans

For people working with climate adaptation the broader plans, such as oversight plans or regional development strategies, are key pieces of documentation for involving climate adaptation plans to regional development. In addition, the expert councils' first report on climate adaptation released in February 2022 attempts to guide actors on what to do. They included 168 suggestions for how to increase the resilience of Sweden towards climate change.

5.3.3 Available Guidelines

The following guidelines provide advice on developing and implementing climate change adaptation in Sweden:

- The housing authority's (Boverket, 2011) guidelines for climate adaptation in the planning process, used by most municipal and regional planners as a knowledge bank for how to interpret situations within the planning process.
- Swedish water's guidelines for downpour calculations (Svenskt vatten): inform officers in the field of how to plan for downpours, and by what magnitude future climate effects should be taken into account. Their recommendation of a 'climate factor', multiplying historical downpour accounts by 1.2–1.4, is praxis in the field (Svenskt vatten, 2020).

5.3.4 Databases and Scientific Facilities

The following databases and geoviewers support the elaboration of studies related to climate change adaptation.

- Swedish meteorological institute's (SMHI) climate adaptation portal.
- Flooding portal from the Swedish Civil Contingencies Agency (MSB).
- Map-tool for erosion and landslides developed by a coalition of governmental agencies.

5.4 Malta

Malta, as a small island state, is particularly exposed to the threats and challenges of climate change. As part of the effort to contribute to the global attempt to address and mitigate the effects of climate change, Malta has established various national policies, strategies and plans for a coherent and coordinated governance.

The Climate Action Act (2015) is Malta's main law on climate change. This Act and its subsidiary legislation enshrine international legally binding commitments into national law, including the 1992 United Nations Framework Convention on Climate Change, the 1997 Kyoto Protocol and the 2015 Paris Agreement, besides other commitments as an EU member state. The European Union Climate Policy has been another important factor in driving local policy (MSDEC, 2017).

The Act establishes a Climate Action Board (CAB) that represents government entities, academia, business and civil society as the representation of all sectors of society to facilitate stakeholder dialogue and to instil climate action governance across the public and private sectors.

In Malta, climate change falls under the responsibility of the Ministry for the Environment, Energy and Enterprise. However, different ministries are responsible for key areas related to climate change adaptation. The Climate Emergency Resolution since 2019 is a nationally binding legal declaration that spearheads obligations for coherent and coordinated governance to deal with the climate crisis. The Climate Change Action strategy involves inter-ministerial collaboration, between the private sector and residents (MECP, 2021). The greenhouse gas mitigation policy-making is a combination of bottom-up sectorial adaptation and implementation of measures by stakeholders in the respective sectors (MSDEC, 2017). The top-down policy process is set also to provide a holistic and coordinated perspective (MSDEC, 2017) with the intent to spearhead adaptation measures that are adopted effectively (MECP, 2021).

Malta has recently taken a further step in cementing its commitment to a sustainable, low-carbon future, with the government announcement that carbon neutrality will be one of the five main pillars of government economic policy. Malta's Lower Carbon Development Strategy (LCDS) is consistent with Malta's National Energy Climate Plan, which aims to ensure the achievement of the EU's 2030 objectives and targets in line with the Paris Agreement commitments. The LCDS is also aligned with the Long-Term Renovation Strategy, which supports the renovation of the national building stock into a highly energy-efficient and decarbonised building stock by 2050 and is in line with the EU Energy Performance of Buildings Directive.

5.4.1 Regulatory Instruments

In addition to the European directives, the following regulatory instruments are referred to and considered in the development and implementation of climate change adaptation measures:

- CAP 543—Climate Action Act (2015)
- CAP 623—Building and Construction Authority Act
- CAP 549—Environment Protection Act
- CAP 10—Code of Police Laws
- L.N. 261 of 2008 The Energy Performance of Buildings Regulations
- L.N. 376 of 2012 The Energy Performance of Buildings Regulations
- L.N. 134 of 2020—Energy Performance of Buildings (Amendment) Regulations, 2020
- L.N. 47 of 2018 Building Regulation Act (CAP. 513) Energy Performance of Buildings Regulations (2018)
- L.N. 434 of 2015 Building Regulation Act (CAP. 513) Conservation of Fuel, Energy and Natural Resources (Minimum Requirements on the Energy Performance of Buildings) Regulations
- Technical Document (effective from 01.01.2016): it 1 sets Minimum Energy Performance Requirements for buildings (part 1) and for building services (part 2)
- The EPRDn methodology is used for dwellings while the iSBEMmt methodology is used for Non-Dwellings
- L.N. 538/2010, Promotion of Energy from Renewable Sources Regulations
- L.N.339 of 2009 Infrastructure for Spatial Information Regulations
- L.N. 184 of 2011, Waste Regulations

- L.N. 128 of 1997, Deposit of Waste and Rubble (Fees) Regulations
- L.N. 22 of 2009 on Waste Management from Extractive Industries Regulations
- L.N. 538/2010, Promotion of Energy from Renewable Sources Regulations
- SL 545.11, Promotion of Energy from Renewable Sources Regulations
- SL 545.16, Energy Efficiency and Cogeneration Regulations
- SL 623.01—Energy Performance of Building Regulations
- SL 623.03—Energy Performance of Building (Fees) Regulations
- SL 623.04—Conservation of Fuel, Energy and Natural Resources (Minimum Requirements on the Energy Performance of Buildings Regulations)
- SL 623.05—In Building Physical Infrastructure (Access to Electronic Communication Services) Regulations
- SL 623.06—Avoidance of Damage to Third-Party Property Regulations
- SL 623.07—Energy Efficiency Regulations
- SL 623.08—Construction Site Management Regulations
- SL 549.49—Environmental Impact Assessment Regulations

5.4.2 Financial Instruments

Financing the transition to a circular, carbon–neutral society, is crucial to the success of the national goal of achieving climate neutrality by 2050 in Malta.

The Climate Action Act sets up a Climate Action Fund, which has a separate juridical personality and serves to act as the financial instrument which supports the implementation of the Act. The Malta Development Bank (MDB) was set up in 2017 to support sustainable productive and viable operations suffering from market failures or where the market is unable or unwilling to accommodate such activities. In 2017, Malta Enterprise and the Energy and Water Agency launched an initiative to assist businesses with an annual electricity consumption in the range of 10,000–75,000 kWh to carry out an Energy Audit. Furthermore, the Energy and Water Agency (EWA) and the Malta Business Bureau have launched a scheme encouraging enterprises to enter into a voluntary agreement to implement various energy-efficient measures. The EWA energy audit grant offers financial support for small to medium businesses to carry out energy audits. On the other hand, the Malta Enterprise Smart and Sustainable measure supports business investments that lead to more sustainable and digitalised processes, leading to the enhancement of competitiveness through resource efficiency. The Malta Enterprise Investment Aid in Energy Efficiency measure supports businesses in implementing energy efficiency projects.

The Recovery and Resilience Plan (European Commission) funds for Malta are intended to also facilitate several green transition reforms and investments, including projects related to energy-efficient renovations of private and public buildings, the decarbonization of transport, the renovation and greening of public hospitals and schools, the construction of a carbon–neutral school, enhanced investment in renewable energy in roads and public places, the construction of a new ferry landing to promote alternative modes of transport as well as other initiatives to enhance the uptake of electric vehicles (Central Bank of Malta, 2022). Furthermore, the European Investment Fund (EIF) has reached an agreement with some Maltese Banks to deploy €54 m of loans and €2.82 m of interest rate subsidy to finance projects in renewable energy (EIF, 2020). Green Bonds are exploited as a financial instrument to finance green projects.

5.4.3 National Strategies and Plans

The following strategies, policies and plans can be applied to and guide climate change adaptation in Malta:

- The National Climate Change Adaptation Strategy for Malta
- Malta Low Carbon Development Strategy (2021)
- Malta Long-Term Renovation Strategy (2021)
- Malta's Research and Innovation Smart Specialisation Strategy (RIS3) 2021–2027 (2021)
- Storm Water Action Plan (MRA, 2013)
- The Water Catchment Management Plan (WCMP)
- Energy and Climate Plan (2019)
- National Energy Policy (2012)
- National Energy Efficiency Action Plan (NEEAP) (2014)
- National Renewable Energy Action Plan (NREAP) (2011)
- The National Renewable Energy Action Plan 2015–2020, Energy and Water Agency
- National Nearly Zero-Energy Buildings (NZEB) Plan (2015)
- Malta Environment and Planning Authority—Development Control Design Policy, Guidance and Standards (2015)
- Malta's, National Energy and Climate Plan (2018)
- Malta's Sustainable Development Vision for 2050 (2018)
- Cost Optimality Studies of Building Stock in Malta (2018)
- C&D Waste: Construction and Demolition Waste Management Strategy (2021–2030)
- Long-Term Waste Management Plan (2021–2030)

5.4.4 Available Guidelines

The following guidelines and standards provide advice and support on developing and implementing climate change adaptation in Malta:

- SM 5100:2015 Photovoltaic Installations—Requirements for electrical safety of single phase systems
- SM 5200:2017 Solar Photovoltaic (PV) Installations—General requirements for PV systems installations

- SM 3700:2017 Green Roofs—Criteria for the planning, construction, control and maintenance of Green Roofs
- SM 810:2022 Recycling oriented deconstruction, controlled excavation works and classification of waste
- Kamra tal-Periti—A Modern Building and Construction Regulation Framework for Malta, June 2020.

5.4.5 Databases and Scientific Facilities

The following databases and geoviewers support and can be applied to climate change adaptation research and initiatives. Additionally, there are scientific facilities and training platforms that support climate change adaptation efforts:

5.4.6 Databases and Viewers

- MapServer of the Planning Authority
- Lands Authority Survey Sheets for the Maltese islands
- MARIA/Eta High-Resolution Atmospheric Forecasting System for the Central Mediterranean and Maltese Islands
- MARIA/Wave forecasting system at IOI-Malta
- Malta Shelf Hydrodynamical Model—ROSARIO
- Malta GPS Wave Buoy Interface

5.4.7 Scientific Facilities

- University of Malta (Faculty for the Built Environment, Faculty of Engineering, Faculty of Science, Various Institutes at the University of Malta covering Climate Change, Sustainable Energy and Earth Systems.)
- Malta College for Arts Science and Technology
- Malta Council for Science and Technology
- Research and Training Facility; The Water Tower Research Station in Materials Engineering and Structural Health Monitoring; University of Malta.

5.4.8 Training and eLearning Platforms

- Decarbonisation of the Construction Industry Course: Continuous Development Course for Engineers, Architects, EPC Auditors, Building Industry Stakeholders
- CESBA– Sustainable Med Cities (Common European Sustainable Building Assessment) e-Learning Platform, University of Malta
- BEACON (Built Environment Learning for Climate Adaptation) e-Learning Platform, University of Malta.

5.5 *Sri Lanka*

As a tropical nation with a 1620 km coastline, it is extremely vulnerable to climate change. The IPCC predicts that small islands, low-lying coastal areas, and deltas will be especially exposed to the risks associated with sea level rise (IPCC, 2018). The coastal zone covers 24% of Sri Lanka's total land area and is home to 33% of the country's population. Important sectors of the country are tourism, fisheries, agriculture, food security, and biodiversity (De Costa, 2008; De Silva, 2006; Marambe et al., 2013, 2015), productions related to export crops (Nissanka et al., 2011; Ranasinghe, 2013; Wijeratne et al., 2007), health and human settlements (ME, 2010a, 2010b). Accordingly, every aspect of the built environment is highly susceptible to the impacts of climate change.

Originally, climate change was identified as a meteorological issue in Sri Lanka and as such, the Department of Meteorology was the national focal point for the IPCC in Sri Lanka. Later the issue of climate change was taken under the purview of the Ministry of Mahaweli Development and Environment (Pallawala, 2018). The Climate Change Secretariat (CCS), which is currently the national focal point on climate change was established in 2008. In addition, the National Expert Committees of Climate Change (NACC) and the Inter-Ministerial Coordination Committee were established to address climate-related matters under the CCS.

National Adaptation Plan (NAP) 2016–2025 is the key policy document related to climate change adaptation in Sri Lanka. It emphasized the importance of collaborative efforts in climate change adaptation. In order to implement adaptation measures at the local level, the NAP (2016–2025) has proposed Sectorial Climate Cells, National Working Groups, Regional Climate Cells, and Civil Organizations. CCS acts as the coordinating agency between these organizational bodies (CCS and MoMDE, 2016). The Central Government has an administrative structure that connects with the grassroots level. District Secretariats and Divisional Secretariats are the main representatives of the national-level ministries that reach the local level. Further, the local governments also have a major role in the implementation of climate change adaptation and mitigation measures in collaboration with central government. However, Provincial Councils (PC) and Local Councils (LC) such as Municipal Councils, Urban Councils, and Pradeshiya Sabas do not have a clear local mandate on climate change. Both PCs and LCs carry out actions relevant to climate change under their current mandate though not recognized as climate actions (Pallawala, 2018).

5.5.1 Regulatory Instruments

There is no Act on Climate Change in Sri Lanka for the implementation of climate adaptation activities. According to interviews with representatives of government organizations, it was evident that the Climate Act is presently in the drafting stages. Therefore, several other laws, acts, policies, and guidelines related to the environment and other categories could be used for climate adaptation. The Legal Instruments are as follows:

- The Flood Protection Ordinance (1924)
- The Fauna and Flora Protection Ordinance (1940)
- Wildlife Conservation Act (1947)
- The Soil Conservation Act No. 25 of 1951
- The National Environment Act, No. 47 of 1980
- The Coast Conservation Act, No. 57 of 1981
- The Sri Lanka Land Reclamation and Development Corporation Act No. 52 of 1982
- Urban Development Authority (Amendment) Act, No. 49 of 1987
- The Biodiversity Conservation Act (2002)
- Disaster Management Act, No. 13 of 2005
- The Fauna and Flora Protection Ordinance
- Marine Environment Protection Authority Act (2018)

5.5.2 Financial Instruments

The Government of Sri Lanka faces difficulty in allocating funds for adaptation activities due to the financial crisis in the country. The following funding sources provide financial and technical assistance through different implementation bodies in the local context. The CCS endorses all project proposals when obtaining funds through the following sources:

- World Bank
- Asian Development Bank
- International Finance Cooperation
- Adaptation Fund
- World Food Project
- United Nations Climate Technology Centre and Network funds
- Global Environmental Facility
- Green Climate Fund
- United States Agency for International Development

5.5.3 National Strategies and Plans

The following national strategies, policies, and plans can be applied to climate adaptation.

- National Environmental Action Plan (NEAP) (1992)
- National Forest Policy (1995)
- Forestry Sector Master Plan (1995–2020)
- National Biodiversity Strategic Action Plan (NBSAP) (1998)
- National Air Quality Management Policy (2000)

- National Solid Waste Management Strategy (2000)
- Initial National Communication (developed in 2000)
- National Climate Change Policy (2003)
- National Environmental Policy (2004)
- National Wetlands Policy (2006)
- National Land Use Policy (2007)
- Sector Vulnerability Profiles: Water, Health, Agriculture and Fisheries, Urban Development, Human Settlements and Economic Infrastructure (2007–2016)
- Nationally Appropriate Mitigation Action on Energy Generation and End-Use Sectors (2010)
- Second National Communication
- National Climate Change Adaptation Strategy (NCCAS) (2011)
- National Climate Change Adaptation Strategy (2011–2016)
- Technology Needs Assessment and Technology Action Plans for Climate Change Adaptation (for food, health, and water sectors) (2011–2016)
- National Policy on Urban Climate Resilience (2013)
- Sri Lanka Climate Resilience Strategy and Action Plan (2013)
- National Policy on Biological Diversity (2014)
- Nationally Appropriate Mitigation Actions (NAMA) on Transportation (Draft) (2015)
- National Adaptation Plan for Climate Change Impacts (2016–2025)
- The Long-Term Electricity Generation Expansion Plan (2013–2032)
- National Biodiversity Strategic Action Plan (2016–2022)
- National Climate Change Policy (NCCP) (2017)
- National Environmental Policy (NEP) (2017)
- National Policy on Solid Waste Management (2017)
- Coastal Zone and Coastal Resources Management Plan of 2018
- National Climate Change Adaptation Strategy (2018)
- National Policy on Disaster Risk Reduction (2018)
- National Policy on Water Resources Management (2019)
- National REDD+ Investment Framework and Action Plan
- Sustainable Development Goals 2015–2030
- World Meteorological Organization (WMO) Regulations (WMO, 2021), (WMO, 2023)

5.5.4 Available Guidelines

The following regulations and guidelines can be applied to climate adaptation:

- Environmental Impact Assessment guidelines
- Guidelines of National Building Research Organization

5.5.5 Databases and Scientific Facilities

The following bullet points indicate the existing databases and e-repositories in Sri Lanka that can be utilized for climate adaptation. Majorly, e-repositories share different kinds of publications that can be used in climate change adaptation research and initiatives. Other databases provide statistical and spatial data which can also be used in climate adaptation research and initiatives.

5.5.6 Databases

- Open Data Initiative of the Government of Sri Lanka
- Risk Info database of the Disaster Management Center, Sri Lanka
- Disinventra of the Disaster Management Center, Sri Lanka
- Database of the Department of Meteorology
- Geological records and building material records of the National Building Research Organization (NBRO, 2023)
- GIS database of the Land Use Policy Planning Department (LUPPD)

5.5.7 Repositories

- Sri Lanka Climate Change Knowledge Repository
- National Science Foundation of Sri Lanka, National e-Repository
- Sri Lanka Journals Online
- Coconut Research Institute Sri Lanka Repository
- Central Environmental Authority Repository
- Hector Kobbekaduwa Agrarian Research and Training Institute
- Rubber Research Institute of Sri Lanka Repository
- Tea Research Institute of Sri Lanka Repository

5.5.8 E-Repositories of Sri Lankan Universities

- Repository: Southeastern University of Sri Lanka
- E-Resources: University of Sri Jayawardanapura
- Digital Archive: The Open University of Sri Lanka
- Digital Library: University of Peradeniya
- Institutional Repository: University of Ruhuna
- Institutional Repository: Rajarata University of Sri Lanka
- E–Repository: University of Colombo
- Institutional Repository: University of Moratuwa
- Institutional Repository: Eastern University of Sri Lanka

6 Case Studies and Good Practices

In the last decades, governmental agencies, local authorities and other entities, including NGOs and the private sector have developed various strategies for adaptation to the impact of climate change in coastal regions. This section presents different typologies of adaptation options and the results of the assessment of their adequacy to the target countries, based on the perception of key stakeholders.

6.1 Categories of Adaptation Options and Case Studies

Over the years, several categories of options have been identified and applied in different contexts. These options include a wide range of actions that can be categorized according to different criteria depending on the objective pursued, thus any categorization is unlikely to be universally agreed on (Noble et al., 2014).

For instance, the Sixth Assessment Report (AR6) of the IPCC presents diverse adaptation options and categories in the different thematic chapters and cross-chapter papers related to the coastal built environment, such as those focused on ocean and coastal ecosystems, cities, small islands, or decision-making options for managing risk, among others. Specifically, Cross-Chapter Paper 2, Cities and Settlements by the Sea (Glavovic et al., 2022), presents adaptation strategies for coastal cities and settlements, classified according to the typical classification of adaptation options in coastal areas, in terms of: (i) protection; (ii) accommodation; (iii) advance; and (iv) retreat.

Previously, the Fifth Assessment Report (AR5) of the IPCC (Noble et al., 2014) organized adaptation options into three general categories:

- Structural/Physical
  - Engineered and built environment
  - Technological
  - Ecosystem-based
  - Services
- Social
  - Educational
  - Informational
  - Behavioural
- Institutional
  - Economic
  - Laws and regulations
  - Government policies and programs

This classification aims to include the diversity of adaptation options for different sectors and stakeholders and is commonly used because of its simplicity and versatility to be applied in different environments. These adaptation options should be considered overlapping rather than discrete, and are often pursued simultaneously as part of adaptation plans (Noble et al., 2014).

Following this classification, a catalogue of case studies was developed under the BEACON project (IHCantabria, 2023a). It presents examples of adaptation options in the coastal built environment in European countries, the United Kingdom and Sri Lanka. The catalogue features individual factsheets for each case study, including a brief summary, background, climate impacts addressed, actions taken, stakeholders involved, key outcomes and benefits for coastal communities, barriers encountered and top tips for implementing similar actions in other places.

Additionally, each factsheet indicates which components of the built environment (i.e.: physical, economic, social, environmental and governance-adapted from UoM, 2023–) and sectors (i.e.: tourism, finances and insurance, fisheries and agriculture, transport and energy and urban) are addressed by each case study. This information can help coastal professionals to easily identify inspiring case studies to apply in their work areas.

The "Structural/Physical" category is the most representative of the catalogue, as these measures have been traditionally adopted in many regions. Indeed, engineering, and the multidisciplinary teams that engineers work with (architects, planners, legal experts, etc.), are often at the forefront of adaptation technologies and strategies. The catalogue presents case studies intended to adapt to impacts such as coastal erosion or flooding by using adaptation measures such as engineering 'hard' solutions, sand nourishment or the elaboration of policy instruments for coastal protection. In addition, NbS are increasing in the last few years, partly due to the support of international agencies and donors. Technological solutions are also presented in this catalogue, that focus on improving energy efficiency in buildings and urban areas or efficient drainage. Finally, it should be noted that no case study related to services was identified under this category, which could be a sign of the lack of experience in this area.

Case studies categorized as "Social" include examples of actions to adapt to impacts such as coastal erosion or flooding based on educational, informational and behavioural solutions. The catalogue includes case studies that emphasize community responsibility and coordination between different stakeholders. In addition, examples of the importance of communication with citizens are also included.

The "Institutional" case studies address adaptation through the development of laws, regulations, policies, and government programs, and also the provision of financial support and economic incentives to share the cost of adaptation. One of the case studies focuses on the importance of the existence and accomplishment of laws related to coastal activities or on wise policies. In addition to this, the significance of holding sharing seminars, and expert consultation sessions is also highlighted.

6.2 Adequacy of Adaptation Options to National Contexts

The national context and governance framework sets the conditions that enable or hinder the development and implementation of the different categories of adaptation options. In other words, the existence of a favourable regulatory framework, financing mechanisms or social acceptance, among other factors, facilitate the development, implementation and success of the different adaptation options.

In the participating countries, the adequacy of the different types of adaptation options has been analysed through a participatory process, considering the existence of an enabling governance framework and instruments of each country. Based on their experience, knowledge and perception, stakeholders ranked the ten categories of adaptation options presented in the previous section as follows:

- Very adequate: the existing legal and institutional framework promotes the implementation of the adaptation option category, there are financial and technical resources or other factors in place to support its development and successful implementation, it is accepted by the society, etc.
- Moderately adequate: only some aspects of the enabling framework are in place, but some other aspects need to be improved.
- Unsuitable: the existing framework does not support the implementation of the adaptation option category.

The results of the stakeholder consultation are presented in Fig. 2. It is worth mentioning that the cultural and participatory contexts in each country are quite different. Therefore, the assessment of adequacy aims to provide an overview of stakeholder perceptions and a general comparison among countries, but it is not intended to provide a reference baseline or a detailed analysis.

On average, the "Technological", "Government policies and programs" and "Engineering and built environment" categories are considered the most suitable measures. For all countries, "Technological" options are adequate and supported by a comprehensive governance framework. In the case of Sweden, "Technological" and "Engineering and built environment" options are much better ranked than other categories. In Sweden, there are solutions in place that combine technological and social measures such as building walls to protect from flooding or changing surfaces to permeable to deal with rainwater, such as some projects developed in Malmö and Lund.[7] In Spain, where the "Engineering and built environment" category is also highly ranked, there is a large experience in coastal protection solutions from the 50s and 60s, when there was a strong development of tourism and urban areas in the Mediterranean area. However, engineering and hard infrastructures in coastal areas are becoming less socially accepted and there is an increase in "Government policies and programs", such as the *PIMA Adapta Program*. In Malta, stakeholders support the infrastructure works, such as the upgrading of port facilities or stormwater

[7] All examples provided in this section are part of the catalogue of case studies developed under Output 5 of BEACON project. The catalogue is available at: https://beacon-researchproject.org/outputs/.

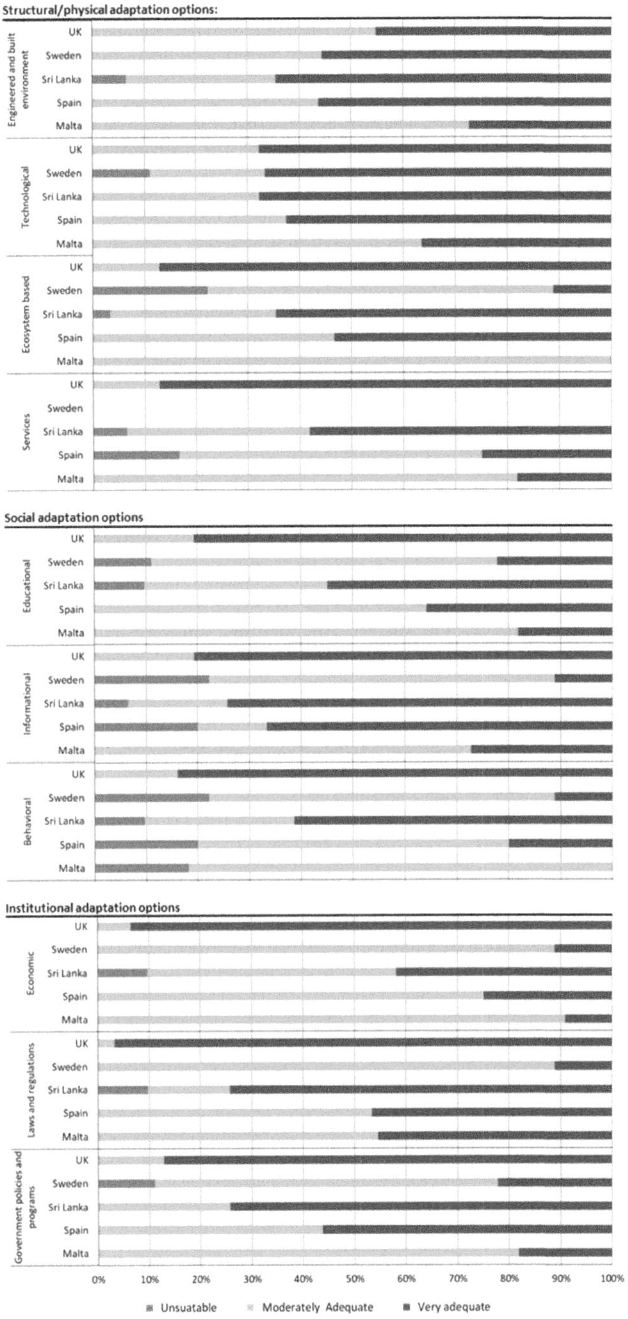

Fig. 2 Results of the stakeholder consultation on the adequacy of the different types of adaptation options in the participating countries

infrastructure that includes underground tunnels to relieve flooding from low-lying flood-prone. In Sri Lanka, stakeholders mentioned that the country has a problem achieving the "Technological" and "Engineering and Built Environment" categories due to the high cost. The main constraint for the implementation of the "Government policies and programs" category is its dependence on political will in the country.

These categories are followed by "Laws and regulation", "Services", "Ecosystem-based" and "Informational" categories. Stakeholder comments mainly focused on the "Ecosystem-based" and "Informational" categories, perhaps due to the increased number of initiatives developed, although comprehensive frameworks for their implementation are not yet in place. In Sweden, NbS, according to stakeholder views, are currently not the most suitable adaptation strategy, because the system and financial regulations favour more engineering 'hard' solutions. However, there are some successful pilot projects. In Spain, some initiatives to restore estuaries are an example of the increasing trend in implementing NbS. Regarding the "Informational" category, UK, Sri Lankan and Spanish stakeholders ranked this category higher. UK stakeholders highlight the need to have more accurate and timely data on future climate information together with the needs to accommodate improved communication of risks and impacts. Sri Lankan stakeholders identify a need in to develop common data-sharing mechanisms and improvements in real-time flood forecasts, hazard mapping, and a platform for storing climate data. In Spain, there are a few examples of informational initiatives.

On average, "Economic", "Educational" and "Behavioural" categories obtained the lowest rank, maybe due to the lack of knowledge and experience in implementing this category of adaptation options. In the case of UK, "Economic" options are better ranked above other categories. For instance, the UK is taking an initiative on implementing local adaptation measures to protect the Humber Estuary coastline in coordination with stakeholders, to promote the pooling of resources and cost-sharing.

Economic initiatives have been more numerous than in other countries, but being at a preliminary stage, a better understanding of all aspects of adaptation investments is needed to make the most of past and current initiatives in these areas. In Malta, training was identified as a major area for action by different stakeholders, even though there are some courses in place in the frame of the construction industry. In Sri Lanka, respondents highlighted the need to integrate climate change adaptation education at the school level, from primary education to tertiary education.

7 Conclusions

Coastal managers and practitioners face numerous challenges in addressing climate change adaptation in the built environment, such as lack of a solid governance framework, lack of understanding climate change impacts on the built environment, lack of funding, or working with uncertainties in a complex environment, among others. The objective of this work is to provide them with general guidelines to facilitate the development and implementation of climate change adaptation measures in the built coastal environment.

To that end, this guidance note presents, firstly, ten key recommendations to implement coastal adaptation strategies, that call for more holistic, participative and scientific-based management of the coastal built environment, merging ICZM, disaster risk management and climate change adaptation approaches. Additionally, this guidance note emphasizes the need to mainstreaming climate change adaptation into sectoral approaches (such as urban planning, construction, and environmental conservation), so that any project or initiative in the coastal built environment intrinsically may involve a climate change adaptation strategy.

Then, this guidance note presents instruments available to support local adaptation to climate change in target countries (Spain, UK, Sweden, Malta and Sri Lanka) that constitute enablers for the development of climate change adaptation measures. These include regulatory and financial mechanisms in place, available strategies, plans and guidelines as well as databases to support climate change adaptation studies. These instruments can be used as a reference for other communities when developing their tools and mechanisms to support climate change adaptation. However, based on the findings of the analysis performed, it is recommended to design instruments focused on the coastal built environment and local needs.

Finally, this guidance note aims to support managers and professionals of coastal built environments in identifying appropriate adaptation measures that can be applied in their site-specific context. To this end, a catalogue of case studies has been developed, providing some examples of actions undertaken by communities and government agencies. The catalogue is structured according to the general classification of climate change adaptation measures proposed by the Fifth Assessment Report (AR5) of the IPCC (Noble et al., 2014). In this sense, it is worth mentioning that climate change adaptation plans shall address different complementary typologies of measures.

The national context and existing instruments, such as a favourable regulatory framework, financing mechanisms or social acceptance, among other factors, facilitate the development, implementation and success of the different adaptation options. Indeed, the catalogue of climate change adaptation measures shows that the different countries have focused their efforts on some specific categories, which may be a consequence of the available instruments but also of the technical capacities in place and the cultural and societal context.

Through the review of this catalogue, the diversity of challenges and solutions faced by different regions to adapt to climate change has been observed. Therefore, and following IPCC AR6 (IPCC, 2022), it is concluded that, to effectively address the problems associated with climate change, it is necessary to adopt a holistic perspective including environmental, social, economic and cultural dimensions.

References

Andersson, A., Nilsson, P. (2021). *Kommunala Avgifter OCH Klimatanpassning En Juridisk Översyn* (Unpublished report). COWI, Göteborg.

Bahr, V. E., & Ivarsson, M. (2020). *Finansieringsmodeller för Klimatanpassnings-Åtgärder*. COWI, Göteborg. Retrieved February 15, 2023, from https://projects.sgi.se/RKS-SH/wp-content/uploads/2020/06/Finansieringsmodeller_for_klimatanpassning_slutversion_20200604.pdf

BC3 Basque Centre for Climate Change. (2008). Retrieved 07, 2023, from https://www.bc3research.org/

BEACON Research UM. (2023). e Learning MOODLE Platform at the University of Malta. https://beacons.research.um.edu.mt/moodle/

Boverket. (2011). *Klimatanpassning i byggande och planering—analys, åtgärder och exempel*. Retrieved November 12, 2022, from https://www.boverket.se/globalassets/publikationer/dokument/2011/klimatanpassning-i-planering-och-byggande-webb.pdf

Cantabria Coastal and Ocean Basin (CCOB). (2022). Retrieved 07, 2023, from https://ccob.ihcantabria.com/

Central Bank of Malta, Policy Note. (2022).

Climate Action Act. (2015). https://environment.gov.mt/en/decc/Pages/climateActionAct.aspx

Climate Change Secretariat, Ministry of Mahaweli Development and Environment (CCS and MoMDE). (2016). *National adaptation plan for climate change impacts of Sri Lanka—2016–2025*. http://www.climatechange.lk/NAP/NAP%20For%20Sri%20Lanka_2016-2025.pdf

Dale, M. (2021). *Managing the effects of extreme sub-daily rainfall and flash floods—A practitioner's perspective* (vol. 379, no. 2195, pp. 20190550).

De Costa, W. A. J. M. (2008). Climate change in Sri Lanka: Myth or reality? Evidence from long-term meteorological data. *Journal of the National Science Foundation of Sri Lanka, 36*, 63–88.

De Silva, C. S. (2006b). *Impacts of climate change on water resources in Sri Lanka*. Paper presented at the 32nd WEDC International Conference, November 13–17, 2006, Colombo, Sri Lanka.

DEFRA (Department for Environment, Food and Rural Affairs). (2010). *Adapting to climate change: A guide for local councils*. DEFRA, Nobel House, 17 Smith Square London SW1P 3JR.

Fauna and Flora Protection (Amendment) Act No. 22 OF 2009 (SL). https://stepbysteptrade.lk/media/ordinanceeng_1.pdf

Feliu, E., García, G., Gutiérrez, L., Abajo, B., Mendizabal, M., Tapia, C., & Alonso, A. (2015). *Guía para la elaboración de Planes Locales de Adaptación al Cambio Climático*. Oficina Española de Cambio Climático. Ministerio de Agricultura, Alimentación y Medio Ambiente. Retrieved 07, 2023. https://www.miteco.gob.es/es/cambio-climatico/publicaciones/publicaciones/guia_local_para_adaptacion_cambio_climatico_en_municipios_espanoles_tcm30-178446.pdf

Fundación Instituto de Hidráulica Ambiental de Cantabria (IHCantabria), Universidad de Cantabria, Ministerio de medio Medio ambiente Ambiente y medio Medio rural Rural y marinoMarino. (2011). *Cambio climático en la costa española*. Proyecto C3E/C3Sim. Retrieved 07, 2023, from https://adaptecca.es/recursos/buscador/cambio-climatico-en-la-costa-espanola-proyecto-c3e-c3sim

Glavovic, B. C., Dawson, R., Chow, W., Garschagen, M., Haasnoot, M., Singh, C., & Thomas, A. (2022). Cross-chapter paper 2: Cities and settlements by the sea. In H.-O. Pörtner, D. C. Roberts, M. Tignor, E. S. Poloczanska, K. Mintenbeck, A. Alegría, M. Craig, S. Langsdorf, S. Löschke, & V. Möller (Eds.), *Climate change 2022: Impacts, adaptation and vulnerability. Contribution of working group II to the sixth assessment report of the intergovernmental panel on climate change*.

Government Proposition. (2017). *Prop. 2017/18:266*. Retrieved February 12, 2023, from https://www.regeringen.se/49c106/contentassets/ff848ae3ebf642ce9680ea5c1bdfa97a/171826600webb.pdf

Howard, T., & Palmer, M. D. J. E. R. C. (2020). *Sea-Level Rise Allowances for the UK, 2*(3), 035003.

Instituto de Física de Cantabria (IFCA). (2023). Retrieved 07, 2023, from https://ifca.unican.es/es-es

Instituto de Hidráulica Ambiental de Cantabria (IHCantabria). (2023a). *Annex I. Output 5. Case studies. Guidance note on implementing local adaptation strategies in the coastal built environment*. BEACON Project.

Instituto de Hidráulica Ambiental de Cantabria (IHCantabria). (2023b). *Met-ocean database (wind, wave and sea level conditions): IHData*. Retrieved 07, 2023, from https://ihcantabria.com/software-y-servicios-tic/ihdata/

Instituto de hidráulica Ambiental de Cantabria (IHCantabria). (2023c). Retrieved 07, 2023, from https://ihcantabria.com/

Instituto de hidráulica Ambiental de Cantabria (IHCantabria). (2023d). *Guía Orientativa para la Aadaptación de las Zonas Urbanas Costeras al Cambio Climático*. Retrieved 07, 2023, from https://ihcantabria.com/beacon-guia/

Instituto Oceanográfico de España (IEO). (2014). Retrieved 07, 2023, from https://www.ieo.es/es/

IPCC. (2018). *Global warming of 1.5 °C. An IPCC Special Report on the impacts of global warming of 1.5°C above pre-industrial levels and related global greenhouse gas emission pathways, in the context of strengthening the global response to the threat of climate change, sustainable development, and efforts to eradicate poverty*. In V. Masson-Delmotte, P. Zhai, H.- O. Portner, D. Roberts, J. Skea, P. R. Shukla, A. Pirani, W. Moufouma-Okia, W. Pean, C. Pidcock, R. Connors, S. Matthews, J. B. R. Chen, Y. Zhou, X. Gomis, M. I. Lonnoy, E. Maycock, T. M. Tignor, & Waterfield T. (Eds.). https://www.ipcc.ch/

IPCC. (2019). Summary for policymakers. In H.- O. Pörtner, D. C. Roberts, V. Masson-Delmotte, P. Zhai, M. Tignor, E. Poloczanska, K. Mintenbeck, A. Alegría, M. Nicolai, A. Okem, J. Petzold, B. Rama, & N. M. Weyer (Eds.), *IPCC special report on the ocean and cryosphere in a changing climate* (pp. 3–35). Cambridge University Press, Cambridge, UK and New York, NY, USA. https://doi.org/10.1017/9781009157964.001

IPCC. (2021). Annex VII: glossary. In J. B. R. Matthews, V. Möller, R. van Diemen, J. S. Fuglestvedt, V. Masson-Delmotte, C. Méndez, S. Semenov, & A. Reisinger (Eds.), *Climate change 2021: The physical science basis. contribution of working group I to the sixth assessment report of the intergovernmental panel on climate change* [V. Masson-Delmotte, P. Zhai, A. Pirani, S. L. Connors, C. Péan, S. Berger, N. Caud, Y. Chen, L. Goldfarb, M. I. Gomis, M. Huang, K. Leitzell, E. Lonnoy, J. B. R. Matthews, T. K. Maycock, T. Waterfield, O. Yelekçi, R. Yu, & B. Zhou (Eds.) (pp. 2215–2256). Cambridge University Press, Cambridge, United Kingdom and New York, NY, USA. https://doi.org/10.1017/9781009157896.022

IPCC. (2022). Climate change 2022: impacts, adaptation and vulnerability. In H.- O. Pörtner, D. C. Roberts, M. Tignor, E. S. Poloczanska, K. Mintenbeck, A. Alegría, M. Craig, S. Langsdorf, S. Löschke, V. Möller, A. Okem, B. Rama (Eds.), *Contribution of working group II to the sixth assessment report of the intergovernmental panel on climate change* (p. 3056). Cambridge University Press, Cambridge, UK and New York, NY, USA. https://doi.org/10.1017/9781009325844

Klimatanpassning.se. (no year). Retrieved November 12, 2022, from https://www.klimatanpassning.se/klimatanpassa

Land Use Policy Planning Division of Sri Lanka. (2023). *National land use policy (2007)*. https://luppd.gov.lk/images/content_image/downloads/pdf/national_land_use_policy.pdf

Malta, EIF (2020). Retrieved October 1, 2023. https://www.eif.org/what_we_do/guarantees/news/2020/malta-eur-54-million-unlocked-with-eu-fund-energy-projects.htm

Malta Low Carbon Development Strategy. (2017). https://meae.gov.mt/en/Public_Consultations/MSDEC/Documents/MSDEC%20LCDS%20Vision.PDF

Malta, Government of Malta, The National Renewable Energy Action Plan (NREAP) (2011).

Malta, Government of Malta, The National Energy Efficiency Action Plan (NEEAP). (2014).

Malta, Ministry for Sustainable Development and Climate Change. (2017). *Malta's low carbon development strategy*. MSDEC.

Malta, Ministry for the Environment, Climate Change and Planning. (2021). *Malta low carbon development strategy public consultation document June 2021*. MECP.

Malta, The National Climate Change Adaptation Strategy for Malta. (2012). https://environment.gov.mt/en/Documents/Downloads/maltaClimateChangeAdaptationStrategy/nationalAdaptationStrategy.pdf

MAPAMA. (2013). Estudio de la dinámica litoral, defensa y propuesta de mejora en las playas con problemas: Estudio de actuación del tramo de costa comprendido entre las desembocaduras de los ríos Guadiana y Guadalquivir. Retrieved 07, 2023, from https://www.miteco.gob.es/es/costas/temas/proteccion-costa/estra-sosten-guadiana-guadalquivir_final_vprensa_tcm30-163208.pdf

MAPAMA. (2014). Estudios de dinámica litoral, defensa y propuestas de mejora en las playas con problemas erosivos, considerando los efectos del cambio climático: estrategia de actuación en el Maresme. Retrieved 07, 2023, from https://www.miteco.gob.es/es/costas/temas/proteccion-costa/Estrategia%20de%20actuacion%20en%20el%20Maresme%20(Final)%20AVANCE%20(1)-optimizado_tcm30-163212.pdf

MAPAMA. (2015a). Estudio de dinámica litoral, defensa y propuestas de mejora en las playas con problemas erosivos, considerando los efectos del cambio climático: estrategia de actuación en la costa sur de valencia (Puerto de Valencia-Puerto de Denia). Retrieved 07, 2023, from https://www.miteco.gob.es/es/costas/temas/proteccion-costa/estrategiaactuacionvalenciafinalmemoriared1b_tcm30-163219.pdf

MAPAMA. (2015b). Estudios de dinámica litoral, defensa y propuestas de mejora en las playas con problemas erosivos, considerando los efectos del cambio climático: Estrategia de actuación del tramo de costa comprendido entre el puerto de Castellón y el puerto de Sagunto (Castellón Sur). Retrieved 07, 2023, from https://www.miteco.gob.es/es/costas/temas/proteccion-costa/estrategiaactuacioncastellonsurfinalmemoriared1b_tcm30-163216.pdf

MAPAMA. (2016). Estrategia de adaptación al cambio climático de la costa española (Strategy for Adaptation to Climate Change on the Spanish Coast). Dirección General de Sostenibilidad de la Costa y del Mar, Ministerio de Agricultura y Pesca, Alimentación y Medio Ambiente (120 pages). Madrid. Retrieved 07, 2023, from https://www.miteco.gob.es/es/costas/temas/proteccion-costa/estrategiaadaptacionccaprobada_tcm30-420088.pdf

MAPAMA. (2017). Estudios de dinámica litoral, defensa y propuestas de mejora en las playas con problemas erosivos, considerando los efectos del cambio climático: Estrategia de actuación en la costa de Granada. Retrieved 07, 2023, from https://www.miteco.gob.es/es/costas/temas/proteccion-costa/estrategia-proteccion-granada_tcm30-422730.pdf

MAPAMA. (2021). Estrategias para la protección de la costa en las provincias de Cádiz, Málaga y Almería considerando los efectos del cambio climático. Retrieved 07, 2023 from https://www.miteco.gob.es/es/costas/temas/proteccion-costa/estrategias-proteccion-costa/cadiz/estrategia-proteccion-cadiz-malaga-almeria.aspx

Marambe, B., Pushpakumara, G., Silva, P., Weerahewa, J., & Punyawardena, R. (2013). Climate change and household food security in homegardens of Sri Lanka. In H. P. M. Gunasena, H. A. J. Gunathilake, J. M. D. T. Everard, C. S. Ranasinghe, & A. D. Nainanayake (Eds.), *Proceedings of the International Conference on Climate Change Impacts and Adaptation for Food and Environmental Security, Colombo* (pp. 87–100).

Marambe, B., Punyawardena, R., Silva, P., Premalal, S., Rathnabharathie, V., Kekulandala, B., Nidumolu, U., & Howden, M. (2015). Climate, climate risk, and food security in Sri Lanka: Need for strengthening adaptation strategies. In: W. Leal (ed.), *Handbook of climate change adaptation* (pp. 1959–1789). Springer-Verlag.

ME. (2010a). *Sector vulnerability profile: Urban development, human settlements, and infrastructure*. Ministry of Environment, Colombo, Sri Lanka.

ME. (2010b). *Sector vulnerability profile: Health*. Ministry of Environment, Colombo, Sri Lanka (Updated Nationally Determined Contributions Under The Paris Agreement on Climate Change Sri Lanka, 2021)

MEPA. (2015). *Strategic environmental assessment environment report*. The Malta Environment and Planning Authority (MEPA) (p. 122).

Ministerio de Ciencia e Innovación (MICIN). (2023). Centro de Investigaciones Energéticas, Medioambientales y Tecnológicas. Retrieved 07, 2023, from https://www.ciemat.es/
Ministerio de Ciencia, Innnovación y Universidades (SICTI). (n.d.). Información y Documentación de la Ciencia en España (ÍnDICEs-CSIC). Retrieved 07 2023, from https://indices.csic.es/
Ministerio de Fomento. (2019). *Agenda Urbana Española*. Retrieved 07, 2023, from https://apps.fomento.gob.es/CVP/handlers/pdfhandler.ashx?idpub=BAW061
Ministerio de Transportes, movilidad y Agencia Urbana (MITMA). (2023a). *Program to support actions to improve energy efficiency*. Retrieved 07, 2023, from https://www.mitma.gob.es/ministerio/proyectos-singulares/prtr/vivienda-y-agenda-urbana/programa-de-ayuda-las-actuaciones-de-mejora-de-la-eficiencia-energetica-en-viviendas
Ministerio de Transportes, Movilidad y Agenda Urbana (MITMA). (2023b). Transportation and mobility grants. Retrieved 07, 2023, from https://www.mitma.gob.es/ministerio/proyectos-singulares/prtr/transporte/convocatorias-ayudas-transporte-movilidad
Ministerio de Transportes, Movilidad y Agenda Urbana (MITMA). (2023c). *Política Agraria Común*. Retrieved 07, 2023, from https://www.mapa.gob.es/es/pac/
Ministerio de Transportes, Movilidad y Agenda Urbana (MITMA). (2023d). *Database of the National Port Authority*. Retrieved 07, 2023, from https://www.puertos.es/es-es/oceanografia/AccesoSimplificado/Paginas/Tablas.aspx
Ministerio para la Transición Ecológica y el Reto Demográfico (MITECO). (2007). Estrategia Española de Cambio Climático y Energía Limpia. Retrieved 07, 2023, from https://www.miteco.gob.es/es/cambio-climatico/legislacion/documentacion/est_cc_energ_limp_tcm30-178762.pdf
Ministerio para la Transición Ecológica y el Reto Demográfico (MITECO). (2020a). Plan Nacional de Adaptación al Cambio Climático 2021–2030. Retrieved 07, 2023, from, https://www.miteco.gob.es/es/cambio-climatico/temas/impactos-vulnerabilidad-y-adaptacion/pnacc-2021-2030_tcm30-512163.pdf
Ministerio para la Transición Ecológica y el Reto Demográfico (MITECO). (2020b). *Institute for Energy Diversification and Savings*. Retrieved 07, 2023 from https://www.idae.es/
Ministerio para la Transición Ecológica y el Reto Demográfico (MITECO). (2021). Estrategia Nacional de Infraestructura Verde y de la Conectividad y Restauración Ecológicas. Retrieved 07, 2023, from https://www.miteco.gob.es/es/biodiversidad/temas/ecosistemas-y-conectividad/eniv_2021_tcm30-515864.pdf
Ministerio para la Transición Ecológica y el Reto Demográfico (MITECO). (2022). *Fundación Biodiversidad*. Retrieved 07, 2023, from https://fundacion-biodiversidad.es/
Ministerio para la Transición Ecológica y el Reto Demográfico (MITECO). (2023a). *Guías y directrices para la evaluación ambiental*. Retrieved 07, 2023, from https://www.miteco.gob.es/es/calidad-y-evaluacion-ambiental/temas/evaluacion-ambiental/guias-directrices/
Ministerio para la Transición Ecológica y el Reto Demográfico (MITECO). (2023b). *Climate Change Adaptation platform: AdapteCCa*. Retrieved 07, 2023, from https://adaptecca.es/
Ministerio para la Transición Ecológica y el Reto Demográfico (MITECO). (2023c). *Infraestructura de datos espaciales*. Retrieved 07, 2023, from https://www.miteco.gob.es/es/cartografia-y-sig/ide/
Ministerio para la Transición Ecológica y el Reto Demográfico (MITECO). (2023d) *Sistema Nacionl de Cartografía de Zonas Inundables*. Retrieved 07, 2023, from https://sig.mapama.gob.es/snczi/index.html?herramienta=DPHZI
Ministro de Asuntos Exteriores y de Cooperación (MAEC). (2007). Instrumento de Ratificación del Convenio Europeo del Paisaje (número 176 del Consejo de Europa. Retrieved 07, 2023, from https://www.boe.es/eli/es/ai/2000/10/20/(1)
MSB—Myndigheten för samhällsskydd och beredskap. Statsbidrag för naturolyckor. (2022). Retrieved 05 Nov, 2022, from https://www.msb.se/sv/amnesomraden/skydd-mot-olyckor-och-farliga-amnen/naturolyckor-och-klimat/statsbidrag-vid-naturolyckor/
MSB—Myndigheten för Samhällsskydd och Beredskap. (no year). Översvämmningsportalen. Retrieved 12 Nov, 2022, from https://gisapp.msb.se/Apps/oversvamningsportal/index.html

Myndigheten för Samhällsskydd och Beredskap. (2022). *Sammanställning av beslutade bidrag 2022*. Retrieved Feb 06, 2023, from https://www.msb.se/siteassets/dokument/amnesomraden/skydd-mot-olyckor-och-farliga-amnen/naturolyckor-och-klimat/informera-allmanheten/sammanstallning-av-statsbidrag-vid-naturolyckor/1sammanstallning-av-beslutade-bidrag-for-2022.pdf

NAMA Design Document for Transport Sector of Sri Lanka DRAFT. (2015). http://www.climatechange.lk/NAMAs/SL%20Transport%20Sector%20NAMA-Semi%20Final.pdf

National Building Research Organization. (2023). Guidelines of National Building Research Organization (NBRO). https://www.nbro.gov.lk/index.php?option=com_content&view=article&id=275&Itemid=471&lang=en

Nicholls, R. J., Hanson, S. E., Lowe, J. A., Slangen, A. B. A., Wahl, T., Hinkel, J., Long, A. J. (2021). Integrating new sea-level scenarios into coastal risk and adaptation assessments: An ongoing process. *Wiley Interdisciplinary Reviews: Climate Change, 12*(3). https://doi.org/10.1002/wcc.706

Nissanka, S. P., Punyawardena, B. V. R, Premalal, K. H. M. S., & Thattil, R. O. (2011). Recent trends in Partners Report 84 | P a g e annual and growing seasons' rainfall of Sri Lanka. In *Proceedings of the International Conference on the Impact of Climate Change in Agriculture. Faculty of Agriculture, University of Ruhuna, Mapalana, Sri Lanka. December 20, 2011* (pp. 249–263)

Noble, I. R., Huq, S., Anokhin, Y. A., Carmin, J., Goudou, D., Lansigan, F. P., Osman-Elasha, B., & Villamizar, A. (2014). Adaptation needs and options. In C. B. Field, V. R. Barros, D. J. Dokken, K. J. Mach, M. D. Mastrandrea, T. E. Bilir, M. Chatterjee, K. L. Ebi, Y. O. Estrada, R. C. Genova, B. Girma, E. S. Kissel, A. N. Levy, S. MacCracken, P. R. Mastrandrea, & L. L. White (Eds.), *Climate Change 2014: Impacts, Adaptation, and Vulnerability. Part A: Global and Sectoral Aspects. Contribution of Working Group II to the Fifth Assessment Report of the Intergovernmental Panel on Climate Change*. Cambridge University Press, pp. 833–868.

OECD. (2023). Report on green, social and sustainability bonds issued by multilateral development banks and its use for infrastructure financing. Organisation for Economic Co-operation and Development. https://one.oecd.org/document/DAF/CMF/AS(2023)3/REV2/en/pdf (1st September 2023).

Olsson, D. (2018). *Conditions of 'Sustainability' The Case of Climate Change Adaptation in Sweden*. Faculty of Arts and Science Karlstad University.

Pallawala, R. (2018). *Mapping of Institutional Arrangement in Sri Lanka for Climate Change Mitigation*. Janathakshan GTE Limited.

Poo, M.C.-P., Yang, Z., Dimitriu, D., Qu, Z., Jin, Z., & Feng, X. (2021). Climate change risk indicators (CCRI) for seaports in the United Kingdom. *Ocean and Coastal Management, 205*, 105580. https://doi.org/10.1016/j.ocecoaman.2021.105580

Puertos del Estado, IMEDEA, AEMET y CEDEX. (2016). Vulnerabilidad de los puertos españoles ante el cambio climático. Retrieved 07, 2023, from https://www.puertos.es/es-es/BibliotecaV2/VULNERABILIDAD_completo_alta.pdf

Ranasinghe, C. S. (2013). Impact of climate change in coconut and identification of adaptation measures with special emphasis on reproductive development. In: H. P. M. Gunasena, H. A. J. Gunathilake, J. M. D. T. Everard, C. S. Ranasinghe, & A. D. Nainanayake (Eds.), *Proceedings of the International Conference on Climate Change Impacts and Adaptations for Food and Environment Security* (pp. 101–113). Hotel Renuka.

Regeringskansliet. (1998). (Förordning 1998:808). Retrieved 04 Nov, 2022, from https://rkrattsbaser.gov.se/sfst?bet=1998:808

Regeringskansliet. (2010). Förordning 2010:900. Retrieved 04 Nov, 2022, from https://rkrattsbaser.gov.se/sfst?bet=2010:900

Regeringskansliet. (2018). Förordning 2018:1428. Retrieved 04 Nov, 2022, from https://rkrattsbaser.gov.se/sfst?bet=2018:1428

Region Skåne. (2022). Regionplan. Retrieved 12 Feb, 2023, from https://geodata.skane.se/regutv/regionplan/2022/granskning/dokument/regionplan-for-skane-2022-2040-granskningshandling.pdf

Schultze, L., Keskitalo, C., Bohman, I., Johanesson, R., Kjellström, E., Larsson, H., Lindgren, E., Storbjörk, S., & Vulturius, G. (2022) Första rapporten från Nationella expertrådet för klimatanpassning.

Sistema de observación y Predicción Costeros de las Illes Balears (SOCIB). (2010). Retrieved 07, 2023, from https://www.socib.es/

Sociedad Española de Ornitología (SEO). (2021). Guía para la elaboración de planes de adaptación al cambio climático en escuelas. Retrieved 07, 2023, from https://seo.org/wp-content/uploads/2021/08/Guia_Adaptacion_CC_Escuelas_SEO_BirdLife.pdf

Statens Offentliga Utredningar. (2017). 'Vem har ansvaret?' Betänkande av Klimatanpassningsutredningen, 42.

Svenskt vatten. (2020). 'Rekommendationer vid val av nederbördsstatistik för dimensionering av dagvattensystem'. Retrieved 12 Nov 2022, from https://www.svensktvatten.se/globalassets/rornat-och-klimat/klimat-och-dagvatten/svensktvatten_smhi_pm-april-2020.pdf

Swedish Metrological and Hydrological Institute, SMHI. (n.d.). 'Finansiering i Sverige'. https://www.klimatanpassning.se/klimatanpassa/underlag-for-klimatanpassning/finansiering/finansiering-i-sverige-1.186869

University of Moratuwa (UoM), Department of Civil Engineering. (2023). *Output 3. A framework on adapting coastal built environment to the effects of climate change during design, construction, and retrofitting—A systematic review*. BEACON Project.

Valdor, P. F., Cloux, I., Rabadán-González, J., Gandía-Navalón, C., Nuero, M., Bastardas, J., Belmonte, J., Blanco-Aguiar, JA., Canepa, A., Carracedo, V., De La Fuente, P., Díaz, S., Fdez-Arroyabe, P., Gago, J., García Martínez, MC., García Medina, N., García Teruel, A., García, D.A., González Mazón, JM., González-Fernández, N., Gracia, A., Hidalgo, M., Hinojosa, MB., Ibarra, I., López-López, L., Manzanas, R., Mazarrasa, I., Meira, P., Mercado, JM., Padrosa, T., Parra de la Torre, A., Polo Sáinz, J., Prat, E., Salazar Guerra, M., Santurtún Zarrabeitia, A., Torres Leal, MA., Valeiras, J., Vega Pozuelo, RF., Velasco, E., Velázquez, J., Vilella, M., Heras, F., & Pelacho, M. (2021). GUÍA RED4C. Ciencia ciudadana para el seguimiento del cambio climático en los ecosistemas. Cantabria. Red Cambera. Retrieved 07, 2023, from https://red4c.es/wp-content/uploads/2021/08/guia_RED4C_digital.pdf

Wijeratne, A., Anandacoomaraswamy, A., Amarathunga, M. K. S. L. D., Ratnasiri, J., Basnayake, B. R. S. B., & Kalra, N. (2007). Assessment of impact of climate change on productivity of tea (Camellia sinensis L.) plantations in Sri Lanka. *Journal of National Science Foundation Sri Lanka, 35*(2).

World Meteorological Organization (WMO). (2021). *Manuals-guides-technical regulations | world meteorological organization*. World Meteorological Organization. Retrieved December 15, 2022, from https://community.wmo.int/manuals-guides-technical-regulations

World Meteorological Organization (WMO). (2023). *World Meteorological Organization (WMO) regulations*. https://community.wmo.int/en/manuals-guides-technical-regulations

Mitigation Strategies for Human-Elephant Conflicts: Two Case Studies on Built Environment Solutions for Human-Elephant Conflicts (HECs) in Sri Lanka

Thushara Kamalrathne and Dhammika Herath

1 Introduction

Long-standing human-elephant conflicts have been a critical environmental and social issue in Sri Lanka. HECs could be regarded as a disaster, considering the range of direct–indirect and cascading challenges within the human–environment interface in the country (Gunawardhana, 2018). Globally, the second highest HEC-induced human mortality and the highest per capita death rate from HECs are reported from Sri Lanka (Prakash et al., 2020). Sri Lanka is believed to be the highest HEC intense country in the world (Anuradha et al., 2019). Therefore, HEC in Sri Lanka has assumed the significance of a national disaster/issue drawing the attention of concerned public, media, civil society, political parties, and the highest policy-making authorities in the country. Sri Lanka loses on average 250 elephants and 70 humans annually, demonstrating the dire need for sustainable and effective methods to mitigate the challenges posed by human-elephant conflicts (NAO, 2019). HEC has largely escalated during the previous three decades resulting in 1734 human deaths, 1053 human injuries and 4837 elephant deaths between 1991 and 2018 (Rathnayake et al., 2022). Ironically, The year 2022 marked the highest annual elephant deaths in Sri Lankan history, with 433 elephant deaths and 145 human lives (LEF, 2023).

The human-elephant conflict involve several losses: human lives and injuries, deaths of elephants and injuries, property damage, crop damage, abandonment of agricultural lands, and dozens of indirect and cascading costs (Dharmaratne & Magedaragamage, 2014). Some indirect impacts and costs seemed to be overlooked,

T. Kamalrathne (✉)
Global Disaster Resilience Centre, University of Huddersfield, Huddersfield, UK
e-mail: Thushara.Kamalrathne@hud.ac.uk

T. Kamalrathne · D. Herath
Department of Sociology, University of Peradeniya, Peradeniya, Sri Lanka

C. Malalgoda et al. (eds.), *Climate Change Adaptation in the Built Environment*,
https://doi.org/10.1007/978-3-031-75826-3_5

such as the lost time of the farmers in their attempt to guard farmlands and properties and compromising of family security and psychological strain. Different mitigating mechanisms developed by the government, farmer communities, and individuals to confront the HECs have their own opportunity costs.

HECs revolve around the issue of land sharing between humans and elephants (and other wild animals) and are related with broader issues of the need for increasing food production, employment generation and consequent deforestation, and unplanned land consumption in Sri Lanka. Most of the large-scale development programmes since 1930s such as Galoya and Mahaweli irrigation development programmes have resulted in massive land acquisition and deforestation in the dry zone of Sri Lanka (Fernando et al., 2005). Forest density has been going down contentiously as a result of expansion of human settlements and legal and illegal land encroachments. Therefore, development activities, which were concentrated in the dry zone, have more or less been the primary factor which engendered and intensified the HECs. Recent mega-development projects in the South of the country such as construction of the Hambantota harbour, Mattala international airport, Mattala cricket stadium and the convention center have not only aggravated the deforestation but also introduced pockets of elephant concentration in places such as Hambantota, Mattala and Demaliya. Even greater scale deforestation is anticipated in the light of proposed large-scale projects in Hambanthota. It is highly like that these development programmes further aggravate the HECs in dry-zone and Southern Sri Lanka. Further, relocation programmes connected with disaster-induced and development-induced displacements also have had detrimental impact on land use patterns in many areas (Kamalrathne & Senanayake, 2023). Therefore, land use patterns have drastically changed with increasing acquisition or encroachment for development purposes, which is a primary driver of HECs (Rathnayake et al., 2022). In addition, land use pattern in dry zone agriculture, including traditional chena cultivation, has had adversely impacted on elephant habitats (Anuradha et al., 2019; Fernando et al., 2005).

Authorities in Sri Lanka have used a variety of strategies to mitigate challenges encountered in human-elephant confrontation, particularly where likelihood of HEC is severely high. Methods used to keep elephants away from the human settlements and agricultural lands can mainly be divided into two categories: (1) combative methods that are used for chasing elephants away before or at the time of confrontations with humans (2) preventive measures that are long-term strategies to minimise human-elephant interactions (Santiapillai, 1996). A primary preventive method that Department of Wildlife Conservation (DWC) has tested, and use is translocation of 'rogue elephants'. However, existing research has highlighted that many of translocated elephants return to their origin habitat after some time while a few die or move to other forest areas due to various reasons (Cabral de Mel et al., 2022; Fernando et al., 2012).

In addition, detaining 'rogue elephants', within a fence protected area, especially those violent elephants having records of killing and attacking humans or habitually damaging properties has also been used and tested as a long term preventive measure (Bandara, 2019). The government of Sri Lanka had allocated funds in 2012 budget proposal to initially to build four elephant holding grounds (EHGs) and had planned to establish two of those elephant holding grounds in Horowpothana and Lunugamwehera national parks within a three-year plan at first round (NAO, 2019). However, there have been many barrier and constraints encountered with the first and so far the only elephant holding ground in Horowpothana, including social unrest in the particular area, several cases of elephants escaping the elephant holding ground and significant number of elephant deaths due to severe starvation and other weaknesses of the elephant holding ground (Bandara, 2019; NAO, 2019; Hettiarachchi, 2022). Within this context, this chapter aims to examine the challenges and weaknesses of the existing EHG in Horowpathana as a built environment solution to the HECs and apply the insights we generate from this examination to explore the potential of the proposed Lunugamwehera elephant rehabilitation centre (PERC).

2 Methodology

Multi-method qualitative approach was used in the data collection with case study as the overall research design focusing one elephant holding ground (EHG) and a proposed elephant rehabilitation centre (PERC). In line with the research design, field investigation was conducted in 2018 based on Horowpothana elephant holding ground in Anuradhapura District and in the general areas surrounding the Lunuganwehera national park where the PERC is to be built. We conducted informal and semi-structured interviews and focus group discussions to collect data from wildlife experts, wildlife officers and public officers, local level politicians and religious leaders. In the second stage of data collection, we interviewed potential beneficiaries or project affected communities in villages of two project areas. We also conducted FGDs in the two communities. The purposive sampling was used to select respondents base on their relevance to the HEHG and PERC. In total fifteen interviews and 9 FGDs were conducted as follows (Tables 1 and 2).

Villages were selected based on the severity of HEC in the area and proximity to elephant holding ground in Horowpothana and proposed elephant rehabilitation centre in Lunugamwehera. Statistics available in regional DWC offices on HEC incidents reported from villages were also used to select villages. Respondents were selected based on the experience and involvement of HEC mitigation activities in selected areas. In addition, researchers carefully selected respondents in terms of their gender, age, and livelihoods to represent socio-cultural variations within respondents.

We undertook several field observations in Horowpothana EHG accompanied by wildlife officers to collect technical data on nature of forest being used, fences of the elephant holding ground, process of elephant detention, and problems faced by detained elephants. We also undertook independent field observations to the adjacent villages without the presence of wildlife officers. Similar field observations were conducted in the Lunugamwehera where the PERC is to be built, in order to gather data on site selection, technical aspects of proposed rehabilitation centre, including

Table 1 List of interviews conducted to gather primary data

Person/official	Area of expert/profession	Place
Park warden	Wildlife conservation	Lunugamwehera National Park
Deputy park warden	Wildlife conservation	Horowpothana
Ex director general and consultant	Wildlife conservation	Department of Wildlife Conservation
Coordinator	Wildlife conservation and project management	Eco System Conservation and Management Projects (ESCAMP)
Buddhist monk	Volunteer and a community leader	Lunugamwehera
Divisional secretary	Public service	Thanamalwila
Divisional secretary	Public service	Lunugamwehera
Divisional secretary	Public service	Katharagama
Range forest officer	Forestry and wildlife	Lunugamwehera
Member, Pradesheeya Saba	Local level politics	Katharagama
District secretary	Public service	Hambanthota
Member, Tourist Guide Association	Tourism	Lunugamwehera
3 Former officers	Wildlife conservation	Horowpothana

Table 2 List of focus group discussions conducted with communities and officials

Place	Participants
Wildlife office, Horowpothana	Wildlife officers
Horowpothana	Community around the elephant holding ground
Wildlife office, Lunugamwehera	Wildlife officers
Sittarama village, Thanamalwila	Community and agricultural officers
Kithulkote village, Tahanamalwila	Community
Bodagama village, Thanamalwila	Community
Karawilya village, Katharagama	Community
Senukkuwa village, Thanamalwila	Community and wildlife officers
Ranawaranawa village, Thanamalwila	Community
Thanamalwila DS office	Government officials including GNs, Samurdhi officers, agriculture extension officers and development officers attached to divisional secretariats
Katharagama DS office	Government officials including GNs, Samurdhi officers, agriculture extension officers and development officers attached to divisional secretariats
Lunugamwehera DS office	Government officials including GNs, Samurdhi officers, agriculture extension officers and development officers attached to divisional secretariats

plan of the fence, elephant monitoring system, habitat development and social support system in the project area.

We also collected unpublished records maintained by regional offices of Department of Wildlife Conservation in study areas. These records contain information on HECs, unauthorised human activities in park areas, and community activities associated with wildlife conservation. Data analysis was conducted using thematic analysis method.

3 Results

Aligning with the broader objective of this research, data analysis was done aiming to understand two distinguished aspects of the selected cases as follows:

(a) identify technical and social implications encountered with Horowpothana elephant holding ground in Sri Lanka as a case point of lessons learnt.
(b) assess potential benefits and drawbacks of the proposed elephant rehabilitation centre in Lunugamwehera national park, which is the first ever experiment of the concept of elephant rehabilitation.

Case 1: Horowpothana Elephant Holding Ground (HEHG)

The first elephant holding ground in Sri Lanka is located within the Horowpothana national park which was established in 2015 in accordance with the Cabinet Memorandum No.12/0151/549/001 dated 17 March 2012. The elephant holding ground here was proposed to mitigate HECs focusing to meet the following objectives.

(a) Minimizing human-elephant conflict by removing the 'rogue elephants' from human habitats.
(b) Receiving socio-economic and socio-cultural benefits by way of removing violent wild elephants from the areas where human-elephant altercations occur.
(c) Using the Elephant holding ground as an income earning venue by opening it as a tourist destination through the ecotourism (NAO, 2019)

The HEHG is a fenced environment, which extends to 997 hectares of dry-mixed evergreen forest and scrub and grasslands. It is believed to be the first elephant holding ground in the world (Fernando et al., 2012). The terms 'elephant detainee centre' or 'elephant conservation park' have been used as alternatives to "elephant holding ground" (Gunawardhana, 2018). Many translocated elephants, who have records of causing human deaths and property damage were also kept in the Horowpothana elephant holding ground (Bandara, 2019).

As seen in the Fig. 1, HEHG is covered by a 16 km long strong fence made of steel cables and concrete posts that were considered to be unbreachable by elephants. Also, concrete posts are studded with sharp steel spikes (Fig. 2) because the experts believed that elephant could push the fence with their forehead with a formidable force. Two electric fences are placed between the main fence in order to distract the elephants which attempts to escape or breach the main fence.

The HEHG was originally designed to accommodate only 30–40 elephants, however, recent data suggests that approximately 50–60 elephants have been kept at the HEHG since 2015 (NAO, 2019; Hettiarachchi, 2022). An unexpected number of elephant deaths have been reported since the inception of the HEHG, sparking criticism over its failure to address several significant aspects of the HEHG in terms

Fig. 1 Layout of the fence-protected area of the HEHG. Adopted from Hettiarachchi (2022)

Fig. 2 Structural view of the main fence and a concrete post of the HEHG

Fig. 3 Recent news articles highlighted the elephant deaths in HEHG[2]

of planning and implementing the project. As a result of those complications, 12[1] elephants were reported to have died, largely caused by starvation, according to the elephant census data in 2019 (NAO, 2019). This scenario fuelled the activism against to HEHG at national and international forums and many journalists and activists revealed the depth of the problem in the HEHG and neglect of the authorities (Fig. 3).

[1] This number seemed to be significantly higher compared with the number of elephant deaths reported by journalists and wildlife conservation activists. See Hettiarachchi (2022).

[2] The Leader, Elephant Holding Ground at Horowpathana: Jumbos Perish due to Malnutrition, Monday 24 January 2022 (https://english.theleader.lk/news/1907-elephant-holding-ground-at-horowpathana-jumbos-perish-due-to-malnutrition).

The morning, Horowpathana Elephant Holding Ground, 1st December 2023 (https://www.themorning.lk/articles/166734).

Ceylon News, Sri Lanka, Elephants at the Horowapathana Elephant Holding Ground in Sri Lanka Starve to Death, 9 March 2022 (https://earthjournalism.net/stories/elephants-at-the-horowapathana-elephant-holding-ground-in-sri-lanka-starve-to-death).

As investigated in the field study, another significant problem reported from the HEHG was the unexpected and surprising escape of a tusker and an elephant over the fence, which is about 7 feet tall. These escapes of elephants from a holding ground considered to be unbreachable caused chaos and fuelled increasing social unrest among the communities around the Horowpothana HEHD area. The incidents damaged the trust the community had on invincibility of security aspect of the HEHG. Many respondents of the communities in this area highlighted the fact that they no longer trust that they are protected from the violent elephants remaining at the HEHG. This distrust has further escalated into a hidden conflict between wildlife officers of the HEHG and famers in surrounding villages.

According to the wildlife experts, this project was based on trial and error as this is known to be the first ever elephant holding ground in the world (Bandara, 2019). Therefore, HEHG is considered to be an experiment of detaining the 'rogue elephants' in Sri Lanka.

Implications and Lessons Learned from the Horowpothana Elephant Holding Ground

Findings discussed in this section were derived from the field study, especially responses from wildlife experts, officials of HEHG and the Horowpothana park, and selected informants including village leaders, farmer societies, religious leaders and key informants representing the communities living proximate to the HEHG.

EHG in Horowpothana, despite some negative publicity, has a solid infrastructure and so far, has not suffered any structural failure in its physical infrastructure, although two elephants ingeniously managed to escape from it to the bewilderment of the wildlife officials. The DWC has monitored its performance and time to time has made certain improvements. However, the escape of a few elephants (including one tusker, as it can use its tusk to break the electric wire as the tusks do not conduct electricity) and subsequent coincidental attack by another wild elephant, which, according to DWC officials, was not one of those escaped, generated some public anxiety and collapse of trust about the HEHG. As mentioned above, after this incident, communities proximate to the HEHG became nervous as they anticipated further threats posed by HEHG, assuming no longer people are safe from unexpected breaching of the fence. The FGDs we had with communities brought out the perceptions maintain by the concerned community that several people had been killed and some farmlands ruined by elephants escaped from the HEHG. Moreover, the community believe that many of elephants find their way out from the HEHG when the scarcity of food is critical. Also, there is a high risk of escaped elephants being killed by anxious farmers as they believe that the escaped elephants are violent as they were captured and retained in the HEHG.

The most distressing fact about the HEHG as we observed is the weak assessments of the site selection and inadequate habitat development before releasing elephants to the protected area. According to the wildlife experts, EHG in Horowpothana is

---

Ground Views, Journalism for Citizen, The Horror of Horowpathana: Starving Elephants to Death, 05 April 2022, (https://groundviews.org/2022/05/04/the-horror-of-horowpathana-starving-elephants-to-death/).

located in a primary and secondary types of a forest in which vegetation is dry-mixed evergreen, scrub and grassland (Bandara, 2019). However, Primary and secondary forests are considered inappropriate as elephant habitat as these forests fail to generate sufficient food for elephants. This has been a major factor behind the significant number of elephant deaths caused by starvation. Also, it was noted that there were no habitat development programmes within the HEHG since its inception. Even in primary and secondary forests, habitat development such as growing grass, could have reduced the elephant deaths caused by lack of food. However, inability of the DWC to undertake habitat development has been caused by the absence of infrastructure development within the HEHG, including access roads. The absence of transport facilities completely restricted any further involvement and activities associated with habitat development inside the fence protected area. Therefore, providing food from outside has become difficult in terms of financial and supplying constraints every year. Lack of quality and quantity of food and a proper mechanism to feed elephants from outside and, inadequate supervision of food quality and feeding have been identified as key issues (NAO, 2019).

It came out from the expert interviews that the scarcity of water supply within the HEHG area is a matter of serious concern and has deprived the elephants of basic needs. This critical infrastructure issue has been a result of weak environmental assessment before the project implementation. It goes without saying that proper assessments would have guided the necessary improvements within the elephant habitat, including natural water reserves and built facilities such as waterholes and ponds. A recent report highlighted that, fetching water to the waterholes and ponds in the HEHG during the drought season is a challenge due to scarcity of water in the park area and lack of resources, such as water bowsers (Fig. 4) (NAO, 2019).

Further, there had no proper assessment about how these 'rogue elephants' will interact with other elephants in this inadequate space where food and water is extremely limited. Further, experts believe that the HEHG completely obscures the natural life of elephants. An elephant consumes 200 L of water per day and water level of a pond should have enough to fill the water to the trunk at a stretch (NAO, 2019). Also, on average, elephants are believed to be roaming about 17 h per day to have meals in order to meet its food requirement, which is challenging within the mixed and dry evergreen forest chosen for the HEHG.

It is important to note that despite its physical resilience, the EHG in Horowpothana suffers from certain infrastructural, operational and management loopholes

Fig. 4 Water scarcity in the HEHG area during the drought period. Adopted from NAO (2019)

such as the absence of interior roads inside the facility for monitoring and effective care of elephants, watch towers. It is important to note that there are deficiencies in human resources such as the absence of services of a veterinary sergeant and support staff, etc. It is observed that the essential information on elephants, such as number of elephants living in the facility, causes of deaths of elephants, and medical treatments etc., was unsatisfactory. Likewise, there was no proper mechanism and resources to reach the elephants needing treatment, including blind, limped and wounded elephants. No proper assessment is done of the health of the elephants before they are released into the HEHG.

WE conducted focus group discussions (FGDs) with selected residents in Horowpothana to ascertain if the establishment of the EHG in Horowpothana has had impact on the security and livelihoods of the people in surrounding communities. Informants we met in focus group discussions (FGDs) in our field study, mentioned that they had never experienced security threats for due to the HEHG. Nevertheless, there is some anxiety about the presence of elephants inhabiting the forest that surrounds the HEHG. Some residents mentioned of occasions in which elephants roaming outside of the EHG caused some damage to farmlands. Officials of the DWC as well as residents interviewed in Horowpothana were of the opinion that the wild elephants habitually walked towards the interior of the forest, but the HEHG is located in the interior and therefore, for elephants outside the HEHG, the access is effectively blocked to interior part of the forest. This has resulted in elephants looking for alternative places during the rainy season and has exposed some farmlands and households for possible risk of attacks by wild elephants living in or migrating to the area surrounding the HEHG.

Two key state officials who had served 13 years in this area confirmed that the construction of the EHG in Horowpothana blocked elephant pathways and corridors. These officials mentioned that because of the EHG, elephants turned towards the villages and started to rampage farmlands. They further pointed out that they had to request some elephant fences to protect lives but were not able to protect farmlands. Therefore, it seems the HEHG considerably affected the livelihoods of the farmers.

It is clear that weak planning, management and monitoring of the HEHG has engendered a variety of serious issues for both the human settlements and wild animals. However, the officials of the DWC mentioned that they have taken serious note of these lapses and informed the research team that these would be rectified in the design and construction of the next elephant holding ground, which is likely to be the PERC, Lunugamwehera.

Case 2: The Proposed Elephant Rehabilitation Centre (PERC) in Lunugamwehera

Department of Wildlife Conservation is the key state organ, which is primarily responsible for both the preservation of wildlife and protection of people against attacks by wild animals. The proposal to establish the PERC emerges in response to the long-drawn-out HECs in the country. The proposal to establish the PERC by the DWC comes in the wake of loss of lives and damage to property in several parts of the country and to extend the capacity of elephant holding grounds, as per the original cabinet proposal made in 2012 to build 4 EHGs in Sri Lanka. It should be noted that what is to be constructed in Lunugamwehera is an elephant rehabilitation center not an elephant holding ground as in Horowpathana. DWC already has previous experience of translocating elephants and detaining them in the elephant holding ground as a "solution" to HECs, but existing data amply confirms that translocation and detaining are not effective. The EHG in Horowpothana provides insightful lessons for the design of the PERC in the Lunugamwehera national park (LNP). The establishment of the PERC within the LNP has different aims; the so-called 'rogue elephants' will be translocated but they will also be rehabilitated unlike in HEHG (Herath & Kamalrathne, 2018).

Ecosystem Conservation and Management Project (ESCAMP) was expected to implement the proposal to build the PERC through a credit line extended by the World Bank under supervision of the Ministry of Mahaweli Development and Environment, through a Project Management Unit (PMU). The Department of Wildlife Conservation is supposed to implement the sub-project activities in order to achieve several broader objectives. As per the project proposal the key objective of the overall project by DWC is mitigating the Human Elephant Conflict by providing a problem elephant rehabilitation centre where the problem elephants can be held as free ranging elephants under a natural wilderness in an area of around 3000 hectares, which is more than three times the HEHG. It has the ultimate aim of rehabilitation and possibly return such elephants to the wild as fully free ranging elephants. The proposed project has the following specific objectives.

1. To undertake an assessment of the situation with regard to elephants already in the existing Elephant Holding Ground in Horowpothana for learning the lessons to be incorporated in the PERC in Lunugamwehera.
2. To design and establish a PERC in Lunugamwehera National Park based on lessons from the existing HEHG.
3. To release 'problem elephants' captured from high HEC areas into PERC and provide basic requirements for the elephants to live as free ranging animals under natural wilderness conditions including creating a suitable habitat for food and water.
4. To study the behaviour and habitat utilization pattern of problem elephants released into the PERC.
5. Monitor and document the behaviour and welfare of the elephants released into the PERC.
6. Explore the possibility of rehabilitation and return of these elephants to the wild as fully free ranging elephants (Herath & Kamalrathne, 2018)

According to DWC data, there were plans for establishing 4 EHGs in the country in Horowpothana, Lunugamwehera, Maduruoya and Galgamuwa (Fig. 5). Initially, the proposal was presented to the Sri Lanka parliament in 2012. As a result of that Horowpothana EHG was established in 2015. However, ESCAMP plans to convert the second EHG as a rehabilitation centre for the problem elephants because the DWC believes that there is the potential for releasing rehabilitated elephants to the natural wilderness once their behaviour becomes nonviolent. The experts believe

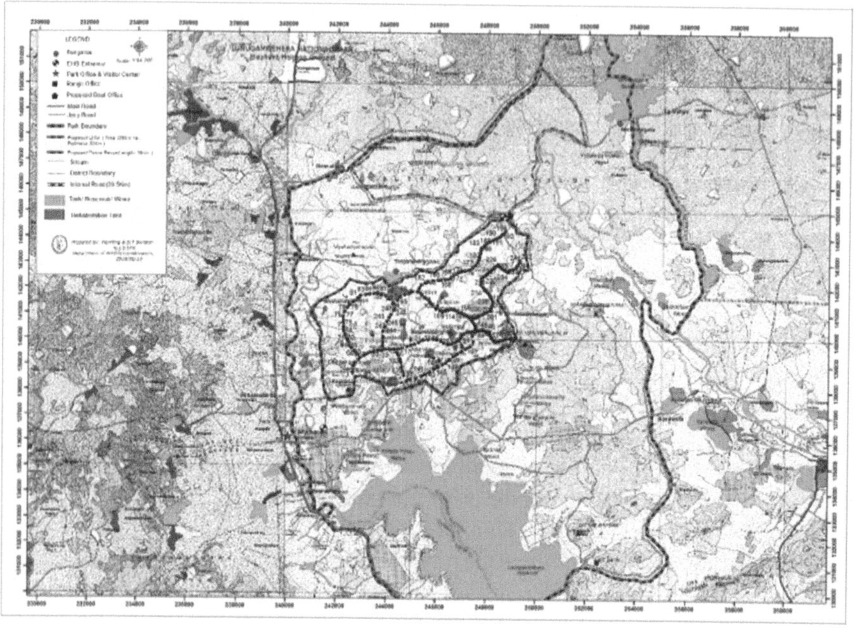

Fig. 5 Map of the proposed PERC within the Lunugamwehera park area. *Source* Herath and Kamalrathne (2018)

that violent elephants, when kept away from conflict situation for period, tends to shed that violence. Therefore, the project in Lunugamwehera will be the first of elephant rehabilitation centre in the country and perhaps in the world, according to local wildlife experts.

Proposed PERC (black dotted line in the map) will be established within the LNP. Understanding the nature of population among the proximate villages and Divisional Secretary divisions (DSDs)[3] is essential in determining the social impact of the project. Thanamalwila, Lunugamwehera, and Katharagama DS divisions in Monaragala and Hambanthota DS division in Hambanthota district form the boundary of the LNP and therefore, are relevant to the project. Thanamalwila DS division has 14 Gramaniladari (GN) divisions and the population is 26,683. Lunugamweherea DS division includes 36 GN divisions with a population of 33,664 while Katharagama DS division has 5 GN divisions with a population of 21,867.

Authors of this chapter conducted a social impact assessment (SIA)[4] to pre-assess the direct and indirect impacts of the PERC to the communities in villages mentioned

[3] Divisional secretariats are the third-level administrative divisions of the country and there are currently 331 divisional secretariats in Sri Lanka. Provinces are divided into districts, which are further divided into divisional secretariats.

[4] Herath and Kamalrathne (2018).

above. It is important to note that there was no such SIA before establishing Horowpothana EHG. Hence, WDC has already shown that it has learnt lessons from the mistakes it did in the past.

Incorporating the Lessons from Horowpothana Elephant Holding Ground to the PERC

Insights from the HEHG as the first EHG in Sri Lanka seemed to have impacted the overall plan of the ERC to be established in Lunugamwehera national park. Officials of DWC and experts involved in the PERC informed the research that they had taken into consideration certain primary factors, which had been ignored in the HEHG.

Although LNP is also a primary and secondary forest, the PERC is to be set up in a forest land, which the wildlife officials consider as far more appropriate as a habitat for elephants. Also, habitat development will be an integral technical aspect of the PERC and proposals were called for experts to involve in habitat development as a specific component of the project.

One of a key concerns among the communities and experts about the HEHG was distracting elephant pathways and corridors when HEHG was constructed. The PERC in Lunugamwehera would, thus, have to undertake an in-depth study as to whether it would interfere with habitual elephant corridors and pathways and devise a monitoring mechanism even after the project completion to monitor elephant movements outside of the PERC.

Hight of the fence in the PERC was proposed to be 11 feet but the fence of HEHG could not avoid escaping elephants. Also, experts were of the view that there must be an appropriate gap between the ground and the bottom cable to make sure smooth movements of other wild animals, such as deer, and small species like rabbits, in and out of the fence covered area in order to protect the biodiversity of the LNP. This means the fence will only barricade the elephants but not many other smaller animals.

A key concern about the HEHG was the detrimental impacts of separating male elephants from their eco-system and herds. Also, experts believe that captivating strong male elephant and detaining them in an EHG disturbs the genetic transmission and decrease the biological and environmental resilience among elephants in the long-term. Therefore, PERC was expected to minimise the issue of restricting 'rogue elephants' interacting with herds in Lunugamwehera forest area as rehabilitated elephants are expected to be released to the wilderness gradually once they show behavioural changes. It was expected that, behavioural movements of elephant would be monitored using radio collars and other movement tracking systems. Key infrastructure missing in HEHG including watch towers, internal roads, transport system, effective veterinary service and efficient emergency protocol are to be placed at the PERC. The most significant aspect of the PERC is the promise of releasing rehabilitated elephants back to their herds and homelands appropriately, after a proper assessment about behavioural changes of elephants by wildlife experts. The PERC cannot benefit from cross-country experiences as there is no adequate evidence from other countries to show how versatile and effective this strategy would be. On a positive note, this will generate novel insights and possibly valuable lessons for countries grappling with 'rogue elephants'.

Apparently, there will be tremendous challenges posed by the ERC to the ecosystem and the elephants and hence these must be considered in policy and practice. The biggest impact is the inevitable disruption likely to happen in the exco-system balance of the Lunugamwehera National Park from which more than 3000 ha would be allocated to construct the PERC. The forest area used by other wild elephants roaming in the forest outside of the PERC and animals will be shrunk due to establishment of the PERC. Therefore, wildlife officers would have to take necessary steps to prevent intensity of HEC in villages adjacent to the park area. Also, it is necessary to assess the threat posed by large trenches and electric fences of the PERC to other wildlife population. In addition, another potential aspect that wildlife officers highlighted in the discussions was the challenge of environmental pollution, largely water pollution that can likely to be caused by drainage system of PERC in the Lunugamwehera national park area. Severe disturbance due to increasing human and machinery involvements in wildlife sensitive area is another consequence cited by wildlife officers during FGDs.

4 Discussion

According to the DWC survey data, there were about 5879 wild elephants living in Sri Lanka by 2011, mostly, in the dry zone (Fernando et al., 2011). Unsustainable levels of deforestation and patterns of land use for agriculture and for other development projects have intensified the HEC in the country during previous two decades. During 2005–2017, among 2832 elephant deaths, 1543 deaths were deliberate killings through shooting, electrocution, poisoning and increasingly through improvised explosive devises. Simultaneously, 937 people were killed by wild animals during this time (DWC, 2017). Evidently, around 200–300 elephants and 60–70 people are killed annually as a direct consequence of HECs in Sri Lanka (DWC, 2017; LEF, 2023; NAO, 2019). Table 3 shows increasing nature of human deaths, elephant deaths and property damages caused by HECs in Sri Lanka during previous decade. Ironically, both elephant and human deaths have doubled within a 9-year-period showing the national significance for immediate and long-lasting policy formulation in mitigating HEC in the country.

Brutal killing of elephants has become a serious concern among wildlife experts, policy makers and practitioners, environment conservationists and general public as an unresolved matter during the previous 25 years which has received international attention on the basis of conserving endangered Asian Elephants (Fernando et al., 2012; Bandara, 2019). Table 4 shows the causes of human deaths from 2005 to 2017. According to the DWC data, most of the elephants killed during this time were of adult male elephants relative to juveniles and adult females.

When comparing with 2022 data, gunshots felled the highest number of elephants (58), overtaking jaw bombs known as '*hakka patas*' (55), which was the leading cause of elephant deaths in 2021 (LEF, 2023).

Table 3 Statistics of human and elephant deaths and damages

Year	Elephant deaths	Human deaths	Property damages	Physical damages
2014	231	67	1279	82
2015	205	63	1120	55
2016	279	88	1593	87
2017	256	87	1302	75
2018	319	96	838	37
2019	364	110	993	55
2022	433	145	–	–

Source Adopted from NAO (2019) and LEF (2023)

When looked at the spatial distribution as illustrated in Table 5, higher human deaths have been reported largely from North Western, Mahaweli project area, Eastern and Southern regions where land use pattern is critical due to large agricultural lands and recent development activities. However, human deaths also have doubled during previous two decades, which was 77 in 2005 and 145 in 2022 (DWC, 2017; NAO, 2019).

Understanding Key Drivers of the Conflict and Mitigation Strategies

Drastic changes of the land use pattern and depletion of the elephant habitats seemed to be the primary driver of the HEC in Sri Lanka (Anuradha et al., 2019; Fernando et al., 2005; Thaufeek et al., 2014). Many of the agriculture-resettlement programmes, including Mahaweli Development Programme, were concentrated in the dry zone in Sri Lanka resulting in significant land separation for large tanks, reservoirs, and farmlands. Unplanned land use and deforestation in the dry zone with the establishment of resettlements and further expansion of human habitats has been critical a factor responsible for increasingly human-elephant confrontations in dry-zone regions. In addition, changes seen in the traditional chena cultivation, including seasonally altering cultivations, which allowed resource sharing among wild animals and people, is also a significant driver of HEC in Sri Lanka (Fernando et al., 2005). A recent study suggested that land cover changes in many areas has a clear connection with increasing trend of HEC in Sri Lanka (Rathnayake et al., 2022).

Further, increase of human involvements in the forest areas, including reserved forest and parks, has detrimental impact on the human-wildlife confrontations. For example, during the field investigation, we found that there are illegal human activities in the Lunugamwehera park area, where the PERC was proposed to be established. The research team investigated existing records on the human elephant conflict in the proximate villages of Lunugamwehera National Park and consulted official reports on the same issue including past attacks by elephants, damages caused, and human deaths caused by elephants in order to understand the gravity of the issue in study area. It is noted that many elephant deaths reported in the Lunugamwehera park area were caused by the human involvement. The official records of the DWC regional offices and data of FGDs conducted with public officers and communities confirmed

Table 4 Causes of elephant deaths, from 2005 to 2017

Year	Gun Shot	Unknown	Electrocuted	Poisoned	Jaw bombs	Accidents	Natural	Train Accident	Other Causes	Total
2005	73	14	9	–	–	14	3	–	10	**123**
2006	93	16	21	3	–	7	4	–	19	**163**
2007	85	40	19	6	–	23	16	–	–	**189**
2008	123	30	20	10	–	18	22	–	1	**224**
2009	113	49	19	4	–	14	8	–	21	**228**
2010	87	30	17	8	11	14	2	–	58	**227**
2011	67	62	18	16	36	22	2	–	32	**255**
2012	45	77	22	7	36	14	11	9	29	**250**
2013	35	54	27	6	32	6	25	7	14	**206**
2014	52	63	12	7	33	9	33	10	12	**231**
2015	44	39	17	5	51	6	11	12	20	**205**
2016	52	54	19	5	47	17	39	12	30	**275**
2017	47	55	25	8	51	28	22	7	13	**256**
Total	**916**	**583**	**245**	**85**	**297**	**192**	**198**	**57**	**259**	**2832**

Source Records of elephant deaths, DWC regional offices

Table 5 Regional distribution of human deaths due to elephant attacks 2005–2017

Region	2005	2006	2007	2008	2009	2010	2011	2012	2013	2014	2015	2016	2017	Total
Northwestern	34	34	32	33	19	35	15	14	25	7	12	6	5	271
Mahaweli/Polonnaruwa	18	17	16	16	14	14	22	10	12	16	10	26	16	207
Eastern	9	17	15	9	10	14	5	15	18	22	15	20	23	192
Southern	12	10	4	8	3	9	5	8	3	9	7	10	10	98
Central	2	1	3	3	3	4	6	3	3	2	4	0	3	37
Yala Bundala	2	–	2	2	1	6	0	0	0	0	–	–	–	13
Uva	–	–	–	–	–	–	–	2	2	4	2	4	7	21
Anuradhapura	–	–	–	–	–	–	–	11	7	5	8	13	11	55
Kilinochchiya	–	–	–	–	–	–	–	–	–	1	0	1	1	3
Vavuniya	–	–	–	–	–	–	–	–	–	1	1	2	3	7
Trincomalee	–	–	–	–	–	–	–	–	–	–	4	4	5	13
Puttalam	–	–	–	–	–	–	–	–	–	–	–	2	3	5
Total	77	79	72	71	50	82	53	63	70	67	63	88	87	922

Source Records of human deaths, DWC regional offices

Table 6 Illegal human involvements reported to the park office, 2015–2018[5]

Nature of the case	2015	2016	2017	2018	Total
Entrance to the park without permission	03	06	14	01	24
Hunting	04	08	15	12	39
Cannabis cultivation	05	11	19	13	48
Illegal fishing	02	02	03	10	17
Use or maintaining illegal weapons	03	05	13	01	22
Sand mining and selling inside the park	06	07	04	01	18
Use forest for grazing cattle	01	01	03	0	05
Others	03	06	08	04	21
Total	27	46	79	42	194

Sources Lunugamwehera National Park Office

that illegal human involvements within the Lunugamwehera Park have detrimental impact on increasing human and elephant deaths. Table 6 shows the number of illegal incidents reported to the park office during 2015–2018.

As illustrated in Table 6, it is clear that human involvement has been increasing annually and that it adversely affects the forest and wildlife. Specially, hunting, cultivation of cannabis in the forest, use of illegal weapons and sand theft have increased during the given period. In addition to that, killing of elephants, illegal logging, digging mines for gems, destroying archaeological sites to find treasures and other types of theft were reported under 'other' category. These activities represent the illegal human activities within the park. For instance, all four cases reported under "other" in first four months in 2018 were involved in killing of elephants. Therefore, it is possible to assume that reported as well as under-reported human involvement have significantly intensified the issue of HEC in the area.

Several combating methods used by people to chase elephants from their farmlands and settlements also seemed a significant driver of the HEC in research area. For instance, we were told by community members that some mass-plantations around the Lunugamwehera use powerful lights mounted to tractors that are supposed to patrol around the farmlands at night to chasing elephants. As a result of that elephants suddenly approach nearby villages exposing human settlements and other cultivated lands to risk. Moreover, according to the field data, a considerable number of people die due to elephant confrontations while trying to chase them away using firecrackers, noisemaking, and shouting.

Resource sharing among elephants and humans in villages close to elephant habitats is also intensify the HEC. Apparently, many dry zone agricultural villages in Sri Lanka are located close to Large Lakes, ponds, and other water streams. Water scarcity in drought periods make the resource-competition acute among wild animals

[5] According to the FGDs had with park officers, public officers and communities, actual number of the incidents have been largely underreported. The table presents only the activities that caught by the park officers and legal actions taken against the incident.

and people, especially in villages proximate to forest lands. During crop-seasons in droughts, water sharing becomes highly critical and human-elephant confrontation is inevitable due to elephant migration towards water sources and proximate forest areas. Risks of conflict between human and elephants caused by water sharing in rural villages have been frequent in other countries as well (Shaffer et al., 2019).

Habitat fragmentation due to development programmes has caused HECs intense in Sri Lanka, which has been largely visible in southern and eastern region. In our FGDs, communities in southern region which are prone to HEC expressed that HECs increased after mass-development activities in the region, including Mattala airport, Hambantota port, southern highway, etc. Intense of HECs has been critical in the Eastern regions with the end of the war in 2009 (Rathnayake et al., 2022). Re-occupying agricultural lands and increasing population movements and development activities have gradually increased in former war-affected areas making HEC intense (Prakash et al., 2020). Further, re-occupying roads and railways in war-affected areas in Sri Lanka after 2009 and developing road networks across forest lands have largely increased elephant deaths caused by train and vehicle accidents.

From the beginning of the HECs, several mitigation strategies have been used by farmers who have been exposed to the HEC-induced risks and relevant authorities. Primarily, use of spotlight, firecrackers, use of infrasound and shooting over the elephants have been used as combative method to chasing away elephants when they confront with humans. In addition, use of repellent, use of buffer-zone, trenches, physical and bio fences, electric fences, and translocation are some of the commonly used preventive strategies that are aimed to provide long-term solutions (Cabral de Mel et al., 2022; Santiapillai, 1996). However, wherever possible, co-living is believed to be the best mitigation strategy that used common sense to deal with elephants, which conservationists most often emphasised (Santiapillai, 1996). Also, co-existence has been the long-lasting method that local communities have used to prevent confrontations with elephants-Understanding elephant behaviour and considering elephants as a part of the eco-system is part and parcel of this strategy (Anuradha et al., 2019).

Combating methods that are aimed to chasing elephants, who confront with human settlements and farmlands, are merely temporary strategies. However, many of those methods seemed ineffective in dealing with rogue elephants who are male, roaming alone without a heard and have records of aggressive behaviours. Among the long-term preventive strategies, translocation has been largely used to mitigate the severe risks of HEC caused by ‘problematic elephants’ in Sri Lanka, including detaining them in the elephant holding ground (Bandara, 2019). Wildlife experts considers translocation of elephants as a management method with regard to ‘rogue elephants’ but, nevertheless, points out that translocated elephants return to their former habitat after some time (Cabral de Mel et al., 2022). From 2007 to 2011, 68 elephants were translocated and 4 among them died due to several physical, environmental and other causes. Fernando et al. (2012) argue that the capture and translocation of elephants do not mitigate HEC as most of translocated elephants continue to cause damage to human beings and farmlands. The capture and translocation operations have an elephant mortality of approximately 6% due to accidents. GPS monitoring via radio-collar units confirm the return of the translocated elephants to the original habitat and makes translocation highly ineffective (Fernando et al., 2012).

As learned by the HEHG, use of elephant holding ground or detaining rogue elephants without a proper assessments and management systems lead in grave danger for starving to death or deliberate imprisonment for life, which is ethically unacceptable. Also, according to the experts, this strategy challenges the principles of natural order of the elephant conservation, especially genetic inheritance from male elephants who have been detained. The serious weakness of the HEHG is the neglection of the science of wildlife and extreme failure to respect general principles behind the care of male adult elephants. For example, selecting primary and secondary forest for the project, shrinking per capita space, which is about 10 km^3 are major considerations. Usually a home-range of an elephant is believed to be a roaming distance of at least about 25 km^3 to find food and water. Yet, currently, there is no elephant monitoring and management system.

The PERC in Lunugamwehera would be an advanced elephant rehabilitation centre which can potentially address many of the lessons learned from HEHG if expert advises are heeded. For instance, PERC must be largely aligned with space for elephants, development and maintenance of habitat, mechanism to manage food and water in drought seasons, treatment plans to ill-elephants and, allowing elephants to mix with home-range herd which will help them to change the behaviour. Furthermore, effective monitoring system and especially, a mechanism for releasing rehabilitated elephants to the natural wildness once appropriate is the real key for success.

5 Conclusion

Increasing human and elephant casualties call for appropriate mitigation methods to protect both human and elephant lives and property. We believe that even though HECs have been a critical concern in the country in terms of its heavy casualties and collateral damage including severe impact on the ecosystem, the issue has been inadequately addressed by national policies and action strategies including budget allocations. Further, preparedness to recognize HEC as a disaster in Sri Lanka by disaster mitigation and management authorities, such as disaster management centre, and academia is apparently inadequate. The HEC is largely seen as a wildlife issue or a rural farmer's issue in HEC-prone areas in the country. Even though root causes of the HECs are largely linked with the unplanned land use and development activities that cannot be undone, most of direct impacts, such as human and elephant deaths, injuries and property damages, are believed to be preventable if appropriate precautions are taken. However, the biggest challenge the country undergone is managing 'problem elephants', known as rogue elephants, and exploring appropriate mitigation strategies to reduce implications encountered with problem elephants. Ironically, rogue elephants seem to be the result of some combating methods, such as using crackers, gun shots, throwing things to chase elephants, and organised chasing, used by humans. Thus, educating communities involved in HEC and changing their attitudes towards elephant confrontation is vital. Also, attitude changing of people living in HEC prone areas towards co-existence of humans and elephants is very much important in terms of behavioural changes. Establishment of the HEHG was be a swift response to manage the public and political pressure caused by an unprecedented increase in HECs and attendant consequences posed by HECs. Therefore, first HEHG became an failure from an elephant conservationist point of view. Lack of pre-assessments, inadequate use of scientifical knowledge, ineffective management and poor monitoring and supervision were the key aspects that had been detrimental in the HEHG. Discourse of the built environment solutions, especially HEHG, for HEC, sparked the need for policy measures, practice and jurisdictions in terms of elephant conservation, management of land acquisition and expansion of human settlements, protecting forest cover and sustainable agricultural practices, and built environment in natural habitats. It is perhaps too early to think about the sustainability of the ERC in Lunugamwehera but, if managed and implemented properly with expert help, it does have the potential to address the whole spectrum of implications encountered in the HEHG.

References

Anuradha, J. M. P. N., Fujimura, M., Inaoka, T., & Sakai, N. (2019). The role of agricultural land use pattern dynamics on elephant habitat depletion and human-elephant conflict in Sri Lanka. *Sustainability, 11*, 2818.

Bandara, S. D. M. N. (2019). Study behaviour of translocated marauding elephants in Horowpothana elephant holding ground, Sri Lanka. *WILDLANKA,* 7, 99–111

CABRAL DE MEL, S. J., SENEWEERA, S., DE MEL, R. K., DANGOLLA, A., WEERAKOON, D. K., MARASENI, T. & ALLEN, B. L. (2022). Current and future approaches to mitigate conflict between humans and Asian elephants: The potential use of aversive geofencing devices. *Animals, 12*, 2965.

Dharmaratne, M. P. J., & Magedaragamage, P. C. (2014). *Sciscitator, 01*, 56–58

DWC. (2017). Regional reports of HEC incidents: 2015–2017. Department of Wildlife Conservation.

Fernando, P., Wikramanayake, E., Weerakoon, D., Jayasinghe, L. K. A., Gunawardene, M., & Janaka, H. K. (2005). Perceptions and patterns of human–elephant conflict in old and new settlements in Sri Lanka: Insights for mitigation and management. *Biodiversity and Conservation, 14*, 2465–2481.

Fernando, P., Jayewardene, J., Prasad, T., Hendavitharana, W., & Pastorini, J. (2011). Current status of Asian elephants in Sri Lanka. *Gajah, 35*, 93–103.

Fernando, P., Leimgruber, P., Prasad, T., & Pastorini, J. (2012). Problem-elephant translocation: Translocating the problem and the elephant? *PLoS ONE, 7*, e50917.

Gunawardhana, L. M. A. P. (2018). An analysis of human-elephant conflict as a disaster: A case study in Anuradhapura district of Sri Lanka. *Journal of Tropical Environment, 1*, 40–55.

Herath, D. & Kamalrathne, T. (2018). Social impact assessment: problem elephant rehabilitation centre (PERC), Lunugamwehera National Park. Colombo: Department of Wildlife Conservation. Available: https://www.researchgate.net/publication/345435906_Social_Impact_Assess

ment_Problem_Elephant_Rehabilitation_Centre_PERC_Lunugamwehera_National_Park_in_Sri_Lanka#fullTextFileContent

Hettiarachchi, R. S. (2022). Elephants at the Horowapathana elephant holding ground in Sri Lanka starve to death. *Ceylon News, 24* [Online]. Available: https://earthjournalism.net/stories/elephants-at-the-horowapathana-elephant-holding-ground-in-sri-lanka-starve-to-death#:~:text=The%20audit%20report%20reveals%20that,between%202014%20and%202019%20alone

Kamalrathne, T., & Senanayake, A. (2023). Relocated or displaced? A social inquiry of tsunami-induced relocation programme in Southern Sri Lanka. In M. Hamza, D. Amaratunga, R. Haigh, C. Malalgoda, C. Jayakody, & A. Senanayake (Eds.), *Rebuilding communities after displacement: Sustainable and resilience approaches.* Cham: Springer International Publishing.

LEF. (2023). *Elephant deaths: SL records highest number In 2022* [Online]. Lanka Environment Fund. Available: https://lankaenvironmentfund.org/elephant-deaths-sl-records-highest-number-in-2022/#:~:text=The%20highest%20number%20of%20elephant,also%20claimed%20145%20human%20lives. Accessed October 10, 2023.

NAO. (2019). Performance of the Horowpothana elephant holding ground. Sri Lanka: National Audit Office.

Prakash, T. G. S. L., Wijeratne, A. W., & Fernando, P. (2020). Human-elephant conflict in Sri Lanka: Patterns and extent. *Gajah, 51*, 16–25.

Rathnayake, C. W. M., Jones, S., Soto-Berelov, M., & Wallace, L. (2022). Human–elephant conflict and land cover change in Sri Lanka. *Applied Geography, 143*, 102685.

Santiapillai, C. (1996). Mitigation of human—elephant conflicts in Sri Lanka. *Gajaii, 15*, 1–7.

Shaffer, L. J., Khadka, K. K., Van Den Hoek, J., & Naithani, K. J. (2019). Human-elephant conflict: A review of current management strategies and future directions. *Frontiers in Ecology and Evolution, 6*.

Thaufeek, U. L., Padmalal, U. K. G. K., & Fernando, P. (2014). Land use and human-elephant conflict in the Sigiriya Sanctuary, Sri Lanka. *Gajah, 40*, 26–30.

Challenges of Built Environment's Stakeholders in Climate Change Adaptation

Shavindree Nissanka, Chamindi Malalgoda, Dilanthi Amaratunga, Richard Haigh, Nesha Ranaweera, María Merino Gonzalez-Pardo, Ignacio Aguirre Ayerbe, Mo Hamza, Fernando Nishara, Lalith Rajapakse, Ruben Paul Borg, and Glorianne Borg Axisa

1 Background

International recognition of the significance of climate change adaptation in the built environment is widespread. Approximately 56% of people on Earth reside in cities, and the rate of urbanisation has been rising globally (Liu and Ebrahimi, 2024). Europe, for instance, has a 75% urbanisation rate (Liu and Ebrahimi, 2024). It shows the growing need for urban areas while many greenfield lands are converted into brownfield areas (Wan et al., 2024). Built environments, or urban areas, are now essential to every economy, and several academics have recognised built environments as economic drivers (Heinzlef et al., 2022). Since the built environment

S. Nissanka (✉) · C. Malalgoda · D. Amaratunga · R. Haigh · N. Ranaweera
University of Huddersfield, Huddersfield, UK
e-mail: chrishani.shavindree@hud.ac.uk

M. M. Gonzalez-Pardo · I. A. Ayerbe
IHCantabria—Instituto de Hidráulica Ambiental de la Universidad de Cantabria, Santander, Spain

M. Hamza
Lund University, Lund, Sweden

F. Nishara
University of Colombo, Colombo, Sri Lanka

L. Rajapakse
University of Moratuwa, Moratuwa, Sri Lanka

R. P. Borg · G. B. Axisa
University of Malta, Msida, Malta

S. Nissanka
Department of Quantity Surveying, Faculty of Engineering, Sri Lanka Institute of Information Technology, Malabe, Sri Lanka

C. Malalgoda et al. (eds.), *Climate Change Adaptation in the Built Environment*,
https://doi.org/10.1007/978-3-031-75826-3_6

consists of physical elements such as industry, housing, transportation, and infrastructure, it has a higher vulnerability to climate change impacts. On the other hand, 62% of greenhouse gas emissions are generated by cities (Pereira et al., 2024). Due to the increasing impacts of extreme weather, rising sea levels, heat waves, and altered precipitation patterns resulting in frequent flooding, climate change has gained significant attention worldwide (Gasper et al., 2011; Rathnayaka et al., 2023).

The possibility of climate change was first recognised in the 1960s. Later, in 1972, with the publication of Edward Goldsmith's book, "A Blueprint for Survival," climate changes and their consequences became a topic of discussion, and with rising pollution rates, acid rains, and severe climate changes, the world began to pay attention to climate change adaptation (Roaf et al., 2009). After conducting the first World Climate Conference in 1979, the first Intergovernmental Panel on Climate Change (IPCC) was established in 1990, and its findings resulted in the establishment of the United Nations Framework Convention on Climate Change in 1992 (Nations, 1992). Different countries use a variety of policies to mitigate the effects of climate change. The UK government was the first to implement a carbon-cost policy for all development projects. Even so, the effectiveness of these rules in large-scale construction and infrastructure development projects remains questionable. The City of Barcelona established the first European solar law allowing new houses to use solar energy, particularly for hot water, in 2005 (Roaf et al., 2009). Japan was the first country to implement carbon labelling on food and products. Similarly, different countries have taken different approaches and sometimes established distinct policies. However, with the increased pace of urbanisation and complications in climate change legislation, implementing effective climate change adaptation has proven difficult. With the global economic, social, and environmental effects of climate change in mind, the 2015 Paris Agreement provided legally binding guidelines for international policy to mitigate those effects (Williams et al., 2020). Following all previous interventions, the Paris Agreement specified that global warming must be kept below 2.0 degrees Celsius to achieve net zero emissions. Without maintaining global warming levels, cities 'coastal city dwellers are more likely to experience coastal flooding. Adaptive capacity must be improved to increase resilience and decrease vulnerability to climate change, particularly in built environments (Masson et al., 2020). Because more people live in the built environment, especially in urban areas, and higher density of economic activities, climate change sensitivity is markedly higher in the built environment (Zuniga-Teran et al., 2019).

However, the built environment has also had a significant contribution in climate change, particularly in the construction sector, which accounts for one-third of greenhouse gas emissions worldwide (Lai et al., 2023). Max Fordham and Partners have further proved this by indicating that almost 50% of fossil fuel usage in developed economies is used for buildings (Roaf et al., 2009). Another major factor in greenhouse gas emissions is transportation. Deforestation and changes in land use patterns have further multiplied these impacts (Kunte & Bhat, 2024). Thus, the core of efforts to adapt to climate change is the built environment (MacGregor et al., 2018). Many urban areas are now at risk of flash floods because there is not enough green space to increase infiltration during periods of extreme precipitation (Agonafir et al., 2023).

Living in flood-prone areas puts over a billion people at risk of social vulnerability (Leta & Adugna, 2023). The social vulnerability of climate change has increased due to urban sprawl and the rapid land use change in developing countries (Lindley et al., 2006). Because developing countries are still in the process of incorporating climate change adaptation methods into their development agendas and because of their higher level of uncontrolled urbanisation, their influence on climate change can be significant (Malalgoda et al., 2013). This has been further proven By Pradosh K. Nath and Bhagirath Behera (Nath & Behera, 2010). Although the built environment is still in the most vulnerable position, it has emerged as one of the key factors contributing to the rising impacts of climate change. Therefore, several studies have emphasised the necessity of climate change adaptation measures in the built environment (Lizarralde et al., 2015, Murtagh et al., 2020). Despite promoting resilient development at the international policy level, implementations of resilient development and climate change adaptation are still limited at the national and local levels (Heinzlef et al., 2022). Consequently, climate change adaptation in the built environment is essential in every context.

The built environment is complex and involves many users and stakeholders (Heinzlef et al., 2022; Zuniga-Teran et al., 2019). The built environment includes also architects, contractors, land surveyors, urban planners, and other stakeholders, including municipal councils, national policymakers and the public (Meyer et al., 2018). Therefore, stakeholders are involved in the built environment from the national to the local levels (Malalgoda et al., 2013). For instance, many countries developed national development goals to achieve national economic growth, and these goals always involve the built environment of that country, including major infrastructure development and industry development. For instance, the National Planning Framework in Scotland, UK and the National Physical Plans in Sri Lanka. These development policies can be critical tools or instruments to implement climate change adaptation strategies. On the other hand, these policy frameworks might encourage higher urbanisation with new development without knowing the real impacts behind the target developments. Thus, to mitigate the impacts of climate change in the built environment, national and local climate change adaptation policies and plans must be established in every country.

The built environment is constantly complex and includes several stakeholders. It is difficult to get various stakeholders to the same table to discuss how to increase their actions related to climate change adaptation in the built environment (Hürlimann et al., 2022). Therefore, it is essential to investigate their challenges in adjusting to climate change in the built environment. Although some countries have committed to implementing measures to adapt to climate change, they are currently only at the policy level and not at the implementation level (MacGregor et al., 2018). The relevance of climate change adaptation strategies at the national and local levels remains limited, even though the built environment has been extensively studied globally regarding climate change adaptation (Malalgoda et al., 2013). Consequently, there is an urgent need to support local community-led climate change adaptation; nevertheless, community-led adaptation may not be well-known in some contexts (Uchiyama et al., 2021). Because of this, it is critical to determine what challenges prevent stakeholders from fully implementing climate change adaptation in the built environment. Therefore, is this research aims to identify the challenges that the built environment's stakeholders face in executing climate change adaptation. In order to determine the challenges faced by built environment stakeholders in climate change adaptation, this study selected five primary stakeholder categories. The national and local government, community, private sector, academia and research organisations, civil organisations and professional bodies are stakeholder groups.

1.1 National and Local Government

To implement strategies and policies for climate change adaptation, national and local governments to enforce the necessary laws and allocate funding (Wellstead & Stedman, 2011). For instance, the Swedish Government has allocated SEK 8 billion to adopt climate-resilient strategies. Sometimes, this funds allocation could be beyond their county boundaries. For example, the UK government has allocated 100 million pounds to vulnerable countries to tackle the consequences of climate change. They oversee the entire process, from the national to the local level. On the other hand, because the national government represents public opinion, it may be utilised to raise public awareness of climate change adaptation (Vij et al., 2023). More crucially, the national government allocates funding and implements climate change adaptation plans on the municipal and regional levels. Simultaneously, the national government may work with international organisations to implement climate change adaptation solutions and policies (Queiroz-Stein et al., 2023). Therefore, decision-making authority at the national level can be necessary in implementing climate change adaptation policies and strategies in the built environment.

The local government is the authority in charge of the built environment at the local level, reporting to the national government (Williams et al., 2020). Depending on the nature of the law, local authorities in various countries may have distinct policies and regulations in place for climate change adaptation. Local governments are primarily responsible for managing the local built environment, monitoring building laws, analysing the effect of new development, and implementing national-level policies into practice on the ground (Malalgoda et al., 2013). Local governments are always considerably more accessible to the community or general public than national governments. Local governments significantly raise public awareness of climate change adaptation (Dupuits et al., 2024). On the other hand, local governments play an important role in allocating funding for climate change adaptation, especially for research and development (Mees et al., 2019). Therefore, local governments have authority in decision-making, public leadership, communication, funding, and collaboration on climate change adaptation implementation with other sectors.

1.2 Community

The community is one of the most significant stakeholders in the built environment; because they are end users, community demand defines the built environment's needs (Strange et al., 2024). Conversely, the community influences climate change by increasing consumer patterns, such as increasing energy demand and waste, directly influencing higher demand for fossil fuel usage. Therefore, the community has been selected as the most effective mechanism to implement climate change adaptation strategies (Mees et al., 2019). Community-led climate change adaptation is one of the concepts that allows communities to make decisions at the grassroots level, and communities may explain the need for climate change adaptation strategies (Strange et al., 2024; Uchiyama et al., 2021). Also, the community might be a great place to start decreasing the consequences of climate change by increasing awareness (Hügel and Davies, 2024). Consumers must manage their consumption patterns to maintain climate resilience, which can substantially impact climate change. Therefore, it is essential to acknowledge the community as an important stakeholder in climate change adaptation in the built environment.

1.3 Academia and Research Organisations

Knowledge is critical for climate change adaptation policies (Fombad, 2024). Climate change research in the built environment, as well as analysis of the impact and future changes and trends of climate change, is constantly required in order to develop successful national and local policies (Díaz Tautiva et al., 2022). Therefore, academic and research institutions are among the most significant players in climate change adaptation in the built environment (Boorse & Jablonski, 2024). Collaboration with national and local governments, conducting climate change research, and providing technical knowledge to all parties involved in decision-making, including the community, are among the primary responsibilities of academia and research organisations (Alejandro et al., 2024; Ebekozien et al., 2022). Therefore, academics and researchers can collaborate with all other stakeholders in the built environment to produce information about climate change adaptation.

1.4 Private Sector

The private sector is the next most significant stakeholder in the built environment, responsible for product and service supply. The private sector can function as users and facilitators of the built environment (Akomea-Frimpong et al., 2022). The private sector encompasses all firms and entrepreneurs; every stakeholder has a monetary interest in the built environment. Many governments rely on the private sector to keep prices stable and to deliver critical infrastructure (Arevalo, 2024). Additionally, the private sector can provide funding for climate change adaptation (Akomea-Frimpong et al., 2022). Many private industries encourage climate resiliency to boost market value and be more sustainable in the business world. However, the private sector, such as construction and industries, greatly influences greenhouse gas emissions (Lai et al., 2023). Therefore, the private sector must be considered an essential stakeholder in climate change adaptation in the built environment.

1.5 Civil Organisations and Professional Bodies

Civil organisations play an important role in climate change adaptation (Delalieux et al., 2023). Civil organisations can serve as a means of communicating the public's demands, as well as representing and organising the community to adopt climate change adaptation strategies (Strange et al., 2024). Professional organisations oversee local, regional, and national developments, monitoring and assessing the effects of regulations and developments (Cole, 2011). When implementing climate change adaptation measures, certain civil organisations, such as non-governmental organisations, serve as a bridge between the professional entities in the built environment and the community (Delalieux et al., 2023). Conversely, they can also provide funds to deliver grassroots community benefits, increasing the community's resilience (Lützkendorf et al., 2011). Therefore, civil organisations and professional bodies must address climate change adaptation in the built environment.

Consequently, it is essential to examine their challenges to develop successful implementation methods that take an integrated approach. To uncover these challenges, this study examined various case studies from across the world, each representing different economic levels, to analyse differences and similarities. The case studies are the United Kingdom, Sweden, Spain, Malta, and Sri Lanka. The study has demonstrated the challenge of identified stakeholders on a national scale and indicated the common and critical level of challenges that must be addressed in implementing climate change adaptation in the built environment.

2 Methodology

The methodology employed in this study is a component of the overarching approach utilised in the BEACON project, which focuses on Built Environment leArning for Climate Adaptation. The primary goal of the BEACON project is to create an analytical framework for examining the stakeholders involved in climate change adaptation within the built environment. As part of the project, a preliminary scoping analysis was carried out at the national level, with project partners from the United Kingdom, Sweden, Spain, Malta, and Sri Lanka. These nations are experiencing significant effects from climate change, including the increase in sea levels, global warming, and alterations in rainfall patterns and heat waves. These implications are also affecting the rapidly developing built environment in these countries. In order to improve the implementation of climate resilient solutions, it is essential to carefully analyse the challenges faced by the stakeholders listed before.

As part of the BEACON project, the country-level studies were conducted concurrently and divided into two sections. In the beginning stage of this study, a contextual analysis was carried out to comprehend the overall circumstance. Following that, comprehensive semi-structured interviews were carried out to pinpoint the precise challenges encountered by each party involved in the execution of climate resilient measures in the built environment. During the initial stage, contextualization was conducted to examine the secondary data of the country. This was done to understand the current policies, climate resilient projects, and key players in each category. The preliminary contextual analysis defines the analytical framework of the research and categorises the stakeholder groups involved in climate change adaptation, including Local and National governments, Community and civil organisations and Professional Bodies, Academia and research organisations and the Private sector. Following the initial phase, the project collaborators in each country carried out semi-structured interviews to pinpoint the issues they encountered as an extra component of the BEACON project. This method involves the involvement of 92 persons. The analytical framework that was developed under the BEACON project considering the stages of the built environment's life cycle, aspects of climate change adaptation, and categories of stakeholders which is presented in Fig. 1.

This chapter only discussed the challenges encountered by each group of stakeholders involved, utilising the established analytical framework depicted in Fig. 1. This research used the analytical framework of the BEACON project to investigate the challenges encountered by stakeholders in five countries. The comparative study is substantiated by the findings in the literature, which further validate the highlighted challenges. The identified challenges were further confirmed by each partner in this

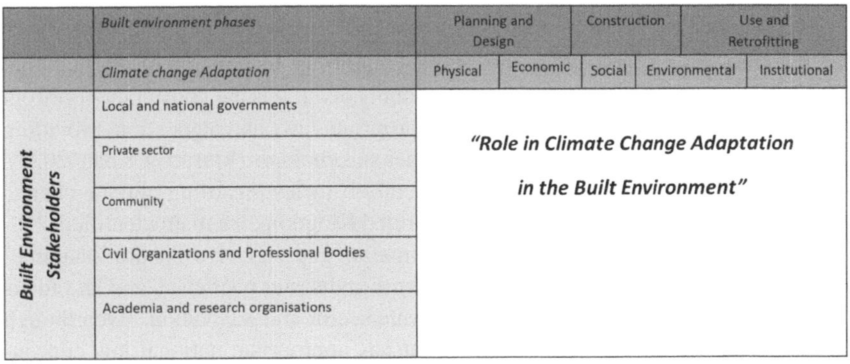

Fig. 1 Analytical Framework used for identifying the challenges Built Environment stakeholders face in climate change adaptation

project before conducting a comparative analysis. The research was carried out using NVivo (Version 20) software for data analysis.

It is important to consider that, the development of the analytical framework has been addressed in a separate paper. However, this study only concentrated on the highlighted challenges. The approach employed in this study is a subset of another study. This technique specifically outlines the methodologies used to identify the challenges encountered by stakeholders in five distinct countries.

3 Results

As described in the methodology, the study examined the challenges faced by local and national governments, the private sector, the community, civil organisations, and professional bodies, as well as academia and research organisations, with a focus on five major case studies: the United Kingdom, Sweden, Spain, Malta, and Sri Lanka. These stakeholder categories were identified as part of the BEACON project objectives. Following the first contextual study, the project partners performed an in-depth scoping review to classify the stakeholders. Typically, a built environment involves several stakeholders such as contractors, industry developers, designers, government organisations, and end customers. The BEACON project aimed to align with the broader context and identify specific stakeholders in the built environment relevant to climate change adaptation. Therefore, the findings are presented for each stakeholder category, focusing on country-specific findings. Furthermore, the highlighted challenges have been examined in the literature to give further scientific proof and critical comprehension.

3.1 National Government

The national government holds the primary authority to implement international climate change adaptation measures at the county level. It is the body at the forefront of decision-making in terms of enacting appropriate laws and legislation, providing funds, and facilitating collaboration with other stakeholders (Masud & Khan, 2023). However, the national government has several obstacles regarding climate change adaptation in the built environment. In the United Kingdom, the main identified challenges for the national government are information gaps, a lack of institutional integration and collaboration, a lack of investments in disaster resilience, and less focus on international research efforts on the private sector and adaptation. Even though the United Kingdom's national government has appropriate national law concentrating on climate change adaptation, Sweden continues to struggle with the clarity of existing legislation in terms of who can/should pay for climate adaptation initiatives. There is a considerable shortage of information among politicians about the need for climate change adaptation in the built environment, and politicians even represent the public's voice. In Spain, there is an urgent need to include climate change adaptation into sectoral activities such as local land use planning, as local managers frequently lack the expertise and resources to incorporate climate change issues and implement adaption adaptation strategies properly. Furthermore, improving coordination across national, autonomous, and local governments, as well as developing local technical capacities, is critical for obtaining European financing and successfully implementing climate change adaptation measures. The main obstacles in Malta's national government are fragmentation between different entities, investment reluctance from certain spheres of the private sector, delays in public procurement procedures and lack of enforcement. In the case of Sri Lanka, the national government is experiencing a lack of information, resources, and coordination at the national level. The other major difficulty in Sri Lanka is the lack of funding and little political support for climate change adaptation. According to the results, national governments face a number of challenges, including severe information gaps (Mohammad et al., 2014), inadequate investment in disaster resilience (Hürlimann et al., 2022), and limited funding for climate adaptation measures (Vij et al., 2023). Raising public awareness of the hazards of climate change is essential, including the corporate sector in adaptation measures (Queiroz-Stein et al., 2023). However, executing national plans in the face of diverse stakeholder interests creates major challenges (Masud & Khan, 2023). In addition, (Madnor et al., 2024) identified that the lack of updated national policies slows the implementation of effective climate-resilient strategies. Overcoming these issues requires effective information sharing (Hürlimann et al., 2022), increased investment in resilience measures (Mohammad et al., 2014), and inclusive decision-making procedures that prioritise long-term sustainability, which this study's findings have also proved.

3.2 *Local Government*

Local governments tackle challenges similar to those faced by national governments. Major challenges include a lack of power in local administration, as well as a lack of institutional integration and coordination. In Sweden, municipalities have long-term planning difficulties, particularly for smaller ones, and climate adaptation will be addressed over several decades. Additionally, land availability for towns to adopt mitigation measures is limited. In Spain, adapting/adjusting national legislation to the local level and inadequate capacities are significant challenges. In Sri Lanka, the local government confronts issues similar to those in the UK, Sweden, and Spain, such as a lack of laws, funding and resources, knowledge, and integration with other sectors. The complexity of government laws and legislation adds to the difficulty of defining who is responsible for adopting climate change adaptation methods in the built environment in Sri Lanka. In Malta, the local governments as small municipalities, given the economies of scale, have limited responsibilities and refer primarily to the national legal and policy frameworks. Local government,the most closely related administrative organisation to the community, is facing significant challenges in climate change adaptation. Lack of power and resources blocks proactive disaster risk reduction initiatives, leaving communities vulnerable to the increasingly severe effects of climate change (Malalgoda et al., 2013). Despite the critical need to raise public awareness about these risks, poor scientific information limits effective communication and planning (Wellstead & Stedman, 2011). Political factors commonly address the need for regional land use plans (Williams et al., 2020). The lack of understanding regarding the consequences of climate change among local government officials slows down decision-making (Uchiyama et al., 2021). Funding restrictions limit the implementation of climate change adaptation measures, such as employing and performing essential studies (Mohammad et al., 2014). Existing land rights legislation fails to address climate change adaptation and environmental preservation (McNaught, 2024). Besides, a lack of local specialists working on climate adaptation plans limits the ability to establish comprehensive and effective policies (Hürlimann et al., 2022). Addressing these difficulties will require coordinated efforts at all levels of government and engagement with stakeholders to develop resilience and adapt to the impacts of climate change successfully.

3.3 Community

The community is another key group of stakeholders involved in climate change adaptation in the built environment. Recent research has highlighted the necessity of creating community-led climate change adaptation and new development strategies to promote participatory planning and development (Uchiyama et al., 2021). However, communities encounter several challenges in the built environment. In the United Kingdom, challenges include an inadequate understanding of resilience measures, the ability to act within the community, the complexity of responding to climate change, and a lack of appreciation and recognition of community contributions. In Sweden, similar community difficulties can be observed, such as a lack of engagement in climate adaptation, a lack of ability, a lack of understanding about how they might benefit or minimise risk, and inadequate collaboration and coordination. A similar community problem may be recognised in Spain, such as a lack of knowledge and less attention on strengthening participatory governance. In Malta, the inadequate understanding of the complexities of climate change limits community engagement and initiatives except for active Environmental NGOs. However, the community still makes use of and benefits from significant national incentives, including retrofit subsidies in existing buildings to address and adapt to climate change. In Sri Lanka, there are Inadequate opportunities for the community to contribute to the drafting and amendment of national policies, a lack of support from governmental organisations to implement climate change adaptation measures locally, and a decline in community participation in programme implementation are other major challenges. The challenges of climate change adaptation exceed government levels and appear inside communities (Mees et al., 2019). One significant barrier is a lack of awareness about resilience mechanisms, which limits communities' capacity to plan for and respond to climate-related hazards effectively (Lindley et al., 2006). The complexities of tackling climate change increase these difficulties, making it difficult for community members to navigate and execute solutions (Heinzlef et al., 2022). Furthermore, lacking recognition of community efforts reduces motivation and involvement (Cole, 2011). Fewer people are willing and able to drive organisational development, involvement, and leadership in community projects. Inadequate financing causes these issues by reducing the resources available to support community activities (Carlson-Ostroff and Ostrander, 2024). Therefore, it is evident from this study that to address climate change at the grassroots level effectively, everyone must work together to increase knowledge, capacity, and resources within communities and foster a culture of collaboration and empowerment.

3.4 Private Sector

The private sector is the primary source of funding and product creation for climate change adaptation in the built environment; their participation is critical in attracting customers to climate-resilient goods and finance (Lizarralde et al., 2015). However, the private sector in the United Kingdom continues to be under-represented in climate change adaptation efforts. In Sweden, the private sector faces the challenge of treating climate adaptation as a whole rather than as a series of isolated incidents in each neighbourhood and collaborating with local government, private sector consultants, and citizens who own land most of the time. In Spain, there is a pressing need to mainstream climate change adaptation (CCA) across various sectors, where professionals often encounter challenges such as a lack of understanding of climate change impacts, insufficient skills to integrate climate considerations into sectoral initiatives, and poor communication skills. Furthermore, there is a substantial lack of understanding and distribution of current climate change research and expertise, demanding greater coordination between national and local governments. In Sri Lanka, the private sector has obstacles due to public conflict produced by individuals, as well as a lack of climate-related data for studies and modelling. In Malta, the private sector is well represented and active in national fora to contribute to climate change adaptation policies. However, there are gaps in the understanding of climate adaptation challenges and investment. The challenge is the private sector's inability to contribute to creating national climate change adaptation policies, which limits the capacity to coordinate local efforts with larger initiatives (Lützkendorf et al., 2011). Private groups' essential contributions to policy formation sometimes go unnoticed, creating collaboration gaps (Masud & Khan, 2023). Besides, a lack of access to modern technology and resources slows progress towards climate change adaptation goals (Madnor et al., 2024). The lack of reliable climate change prediction models limits the market predictions. Rigid law adds another layer of complication, limiting activity even when people are eager to spend (Wallius et al., 2022). Therefore, this study's findings and literature indicate that strategic leadership, improved coordination and communication, investment in technology and resources, and increased private sector participation are required to enable successful climate change adaptation at all levels of government.

3.5 Academia and Research Organisation

Academic and research organisations, particularly those that explore new information, share knowledge, and provide new technologies to the rest of the stakeholders, play a vital role in climate change adaptation (Díaz Tautiva et al., 2022). However, research organisations have a number of obstacles in adapting to climate change. Communication and cooperation are key challenges for research institutions in the United Kingdom, and the same is true in Sweden. Significant research has been conducted in Spain's coastal areas and cities. There are also good CC-related research institutes in Spain. However, there is a need for enhanced efforts to transfer scientific results (theoretical approaches, data, methods, and results) to coastal and urban management. In Malta, there are various research initiatives. The main challenges refer to the coordination and communication between entities involved in the research activity, together with human and financial resources. In Sri Lanka, government authorities provide minimal assistance to researchers in giving permissions, funds, and participation. More importantly, the professional organisations engaged in climate change are hesitant to share data with academics and researchers. Inadequate scholars and researchers with experience in climate change adaptation and rigid academic responsibilities limit their capacity to contribute to climate change adaptation in Sri Lanka.

Addressing climate change successfully requires extensive research to guide evidence-based efforts and policy (Ebekozien et al., 2022). However, government bodies provide inadequate funds to encourage research in this critical field, indicating a lack of understanding about the importance of such efforts (Fombad, 2024). Also, rigid regulations and lengthy permission processes for data sharing hinder research activities, reducing the availability of essential data (Hügel and Davies, 2024). Additionally, language barriers prevent efficient communication among government officials, compounding further problems (Hürlimann et al., 2022). Local research institutions tend to concentrate on teaching over research. On the other hand, communication barriers limit the sharing of scientific knowledge with non-scientific audiences and politicians (Meyer et al., 2018). Given the changing nature of disaster risk management due to climate change, it is important to design strategies that account for existing and expected threats, requiring a comprehensive and accountable approach to disaster risk reduction (MacGregor et al., 2018). To overcome these barriers, additional research funding, quicker approval processes, improved communication methods, and a stronger emphasis on research activities are required to develop evidence-based solutions to climate change concerns (Hürlimann et al., 2022). It is more important to transfer all the scientific knowledge into practice by knowledge sharing rather than keeping all only within academics.

3.6 Civil Organisations and Professional Bodies

Another key sector is the involvement of civil society and professional bodies in climate change adaptation. The UK study found that civil organisations' catastrophe response and recovery are more significant than disaster risk mitigation. In Spain, professionals feel a shortage of awareness of climate change adaptation and the dissemination of existing studies and background knowledge. In Sri Lanka, some challenges include the inadequate opportunities for various non-governmental and international organisations to participate in climate change-related initiatives and the lack of assistance governmental bodies provide to adopt climate change adaptation strategies. Disturbances created by pressure groups such as religious leaders and members of the military forces might be highlighted as challenges for civil organisations. In addition to civil organisations, professional entities in Sri Lanka confront a number of obstacles, including public conflicts generated by people's refusal to alter their conduct, an inadequate understanding of climate change strategies among team members, and a lack of climate-related data.

Civil organisations and professional bodies suffer unequal opportunities to participate in climate change-related activities, restricting the diversity of viewpoints and solutions (Mohammad et al., 2014). Government entities typically fail to appropriately assist in adopting climate change adaptation measures, limiting civil organisations' involvement (Queiroz-Stein et al., 2023). Limited climate change training in professional bodies worsens the problem by leaving decision-makers with insufficient information to handle adaptation and mitigation consequences adequately. At the same time, unclear obligations and financial constraints intensify further challenges (Hürlimann et al., 2022). Bridging the gap between scientific information and practical application in planning procedures remains a key problem within professional bodies (Hügel and Davies, 2024, McNaught, 2024). Therefore, to successfully address the effects of climate change, more cooperation, capacity building, and the development of inclusive decision-making procedures that incorporate multiple viewpoints and prioritise evidence-based solutions are required.

4 Discussion

The results clearly indicate that all stakeholders in the built environment are encountering challenges in relation to climate change adaptation and implementation. Despite the considerable influence held by national governments in making decisions on climate change adaptation, these five countries' governments confront significant policy disagreements and a lack of cohesion across institutions. Particularly, developed economies such as the UK, Sweden, and Spain have implemented adequate policies to address climate change adaptation. The main difficulty is in the collaboration and integration, particularly in underdeveloped nations such as Sri Lanka, where there is a shortage of financing and resources. Therefore, it is clear that the implementation of climate change adaptation in the built environment might be challenging due to the characteristics of national governments and the financial and resource capacities in a certain country. Local governments play a crucial role in implementing climate change adaptation measures at the grassroots level. However, they face challenges such as a lack of regulations, complex implementation processes, insufficient information, limited capacity, and most importantly, inadequate funding. These factors contribute to a lack of motivation to take action on climate change adaptation in built environments. In developed countries, the local government has the authority and legal framework to address climate change adaptation. However, they encounter difficulties in terms of resources, information, and collaborations. On the other hand, countries like Sri Lanka face significant challenges related to legal provisions and complex regulatory policies within their governing systems, in addition to the common challenges. The conflict between the government and private sector is significantly impeding the implementation of climate change adaptation interventions. In countries like the UK, the private sector has not been adequately recognised in climate change adaptation efforts, and their lack of engagement is also significant in terms of knowledge and collaboration. In Sri Lanka the private sector is facing challenges related to complex laws and regulations, as well as a lack of information and knowledge. These factors are restricting their ability to take action. Although the private sector in Malta is actively involved in climate change adaptation, yet targets are still not met. These issues indicate that despite the private sector having ample finances and seeking business possibilities for investment, the lack of collaboration and inflexible regulations restrict their involvement. The assistance they require from government authorities is significantly inadequate. The stakeholders responsible for investigating new innovations and knowledge connected to climate change adaptation in the built environment are also encountering major challenges in research organisations across all countries studied. The main challenge faced by academia is the effective collaboration and communication with responsible implementation agencies. In countries like Sri Lanka, academia encounters issues related to finding sufficient resources, despite having the capacity to conduct significant research and make substantial contributions. On the other hand, civil organisations and professional bodies in every examined country are facing a lack of sufficient possibilities. It demonstrates that civil organisations are motivated to participate in climate change adaptation in the built environment. However, their actions are constrained by the limited opportunities available to them for climate change adaptation. Compared to all the professionals involved in climate change adaptation in the built environment, the community at the bottom, who are the end users of any societal changes, are showing significant motivation for climate change adaptation. However, their motivation to act on climate change adaptation is limited due to the lack of opportunities, lack of appreciation, recognition, and inadequate support from professional bodies. Community participation in climate adaptation has been widely recognised as an effective approach for resilient initiatives (Strange et al., 2024). However, without support from the national and local systems, there is a need to address and develop effective strategies to involve the community in this process. Thus, it is evident from this chapter that all stakeholders, from top to bottom, have significant challenges while participating in the process of adapting to climate change in the built environment.

5 Conclusion

Understanding the challenges stakeholders in the built environment experience that restrict their ability to adapt to climate change is critical for establishing effective, inclusive, and sustainable climate change adaptation policies and plans. This study aimed to identify the challenges of stakeholders in the built environment adapting to climate change in the United Kingdom, Sweden, Spain, Malta, and Sri Lanka. The study analysed the challenges of national and local governments, the private sector, the community, academia and research organisations, civil organisations, and professional bodies that represent all stakeholder categories in the built environment, as their roles are critical in implementing climate change adaptation policies and strategies. This study conducted an analysis of 92 participant interviews in order to identify the common challenges and distinguish the challenges in different contexts, with the aim of facilitating a comparative conversation. As part of the BEACON project, the study utilised the analytical framework developed under the same project to assess the challenges within each category and present them in a comparative manner in this chapter. The key results show that every type of stakeholder, including developed countries, has fund-related challenges. While there are insufficient finances for national and local governments to act on climate change adaptation in countries such as Sri Lanka, the United Kingdom, Malta, Sweden and Spain are dealing with the complexities of legislation, funding, and responsibility allocations. All the country's national and local governments are struggling to execute climate change adaptation strategies due to a lack of coordination and engagement with other stakeholders. More significantly, a lack of enthusiasm and understanding among national and local politicians presents a serious problem. The private sector is struggling to engage with other stakeholders, and their role in climate change adaptation has been restricted due to a lack of data and information, as well as uncertainty. Inadequate collaboration and communication are also an issue for academic and research institutions in charge of delivering knowledge and information about climate change adaptation. More crucially, the strict laws and procedures involved in the research process limit their ability to act on climate change adaptation despite their inevitable academic obligations in the academic environment. While there are numerous challenges at the top level, the community that consumes the built environment is also hesitant to act on climate change adaptation due to a lack of motivation, knowledge, and information, as well as insufficient opportunities for community participation in decision-making processes. Civil organisations and professional bodies that operate as intermediaries between the community and professionals are struggling due to a lack of knowledge, capacity and funding and failing to include them in decision-making processes. Among the various stakeholders, financial constraints remain at the heart of the difficulties, while limited collaboration, insufficient ability, and expertise intensify the challenges. Therefore, it is critical to conduct more research to find effective financing methods for climate change adaptation while encouraging community-led adaptation and supporting inclusive decision-making. At the same time, it is critical to encourage more research on climate change adoption while investing adequate finances to equip decision-makers with the necessary knowledge, information, and tools to aid the decision-making process since these challenges are common in every country studied. Future research can use the findings of this study to develop effective and inclusive climate change adoption implementation strategies. Consequently, the study findings not only contribute to the research related to climate change adaptation but also to policy and decision makers involved in climate change adaptation to identify their implementation challenges to encourage climate change adaptation worldwide, promoting sustainable development.

References

Agonafir, C., Lakhankar, T., Khanbilvardi, R., Krakauer, N., Radell, D., & Devineni, N. (2023). A review of recent advances in urban flood research. *Water Security, 19.*

Akomea-Frimpong, I., Kukah, A. S., Jin, X., Osei-Kyei, R., & Pariafsai, F. (2022). Green finance for green buildings: A systematic review and conceptual foundation. *Journal of Cleaner Production, 356.*

Alejandro, A., Maertens, L., Cheli, Z., Fragnière, A., & Sarrasin, O. (2024). Designing role-play simulations for climate change decision-making: A step-by-step approach to facilitate cooperation between science and policy. *Environmental Science & Policy, 152.*

Arevalo, J. A. T. (2024). Climate change versus economic growth: case of greenhouse apply a study of european union countries and England from 2010 to 2019 using linear regression and neural networks. *Sustainability, 16.*

Boorse, D. F., & Jablonski, L. M. (2024). Strengthening partnerships for the sustainable development goals: engaging faith-based organizations and scientists in youth climate change learning and action. *Sustainable Earth Reviews,* 7.

Carlson-Ostroff, J., & Ostrander, J. (2024). Social Work and emergency management: Climate change, disasters, and why we need to work together more now than ever. *Journal of Policy Practice and Research.*

Cole, R. J. (2011). Motivating stakeholders to deliver environmental change. *Building Research & Information, 39*, 431–435.

Delalieux, G., Kourula, A., & Pezet, E. (2023). Civil Society Roles in CSR Legislation. *Journal of Business Ethics, 190*, 347–370.

Díaz Tautiva, J. A., Huaman, J., & Ponce Oliva, R. D. (2022) Trends in research on climate change and organizations: A bibliometric analysis (1999–2021). *Management Review Quarterly, 74*, 227–261.

Dupuits, E., Garcés, A., Llambí, L. D., & Bustamante, M. (2024). Strategies for monitoring and evaluation of climate change adaptation: localizing global approaches into Andean realities. *npj Climate Action, 3.*

Ebekozien, A., Aigbavboa, C., & Aliu, J. (2022). Built environment academics for 21st-century world of teaching: Stakeholders' perspective. *International Journal of Building Pathology and Adaptation, 41*, 119–138.

Fombad, M. C. (2024). Knowledge management for climate change in South Africa: A proposed strategy. *IFLA Journal.*

Gasper, R., Blohm, A., & Ruth, M. (2011). Social and economic impacts of climate change on the urban environment. *Current Opinion in Environmental Sustainability, 3*, 150–157.

Heinzlef, C., Barroca, B., Leone, M., & Serre, D. (2022). Urban resilience operationalization issues in climate risk management: A review. *International Journal of Disaster Risk Reduction, 75.*

Hügel, S., & Davies, A. R. (2024). Expanding adaptive capacity: Innovations in education for place-based climate change adaptation planning. *Geoforum, 150.*

Hürlimann, A. C., Nielsen, J., Moosavi, S., Bush, J., Warren-Myers, G., & March, A. (2022). Climate change preparedness across sectors of the built environment—a review of literature. *Environmental Science & Policy, 128*, 277–289.

Kunte, G., & Bhat, V. (2024). Deforestation, climate change and the sustainability of agriculture: A review. *Journal of Resources and Ecology, 15.*

Lai, K. E., Abdul Rahiman, N., Othman, N., Ali, K. N., Lim, Y. W., Moayedi, F., & Mat Dzahir, M. A. (2023). Quantification process of carbon emissions in the construction industry. *Energy and Buildings, 289.*

Leta, B. M., & Adugna, D. (2023) Characterizing the level of urban Flood vulnerability using the social-ecological-technological systems framework, the case of Adama city Ethiopia. *Heliyon, 9*, e20723.

Lindley, S. J., Handley, J. F., Theuray, N., Peet, E., & McEvoy, D. (2006). Adaptation strategies for climate change in the urban environment: Assessing climate change related risk in UK Urban Areas. *Journal of Risk Research, 9*, 543–568.

Liu, H.-Y., & Ebrahimi, B. (2024). A scalable method for identifying key indicators to assess urban environmental sustainability: A case study in Norway. *City and Environment Interactions, 22.*

Lizarralde, G., Chmutina, K., Bosher, L., & Dainty, A. (2015). Sustainability and resilience in the built environment: The challenges of establishing a turquoise agenda in the UK. *Sustainable Cities and Society, 15*, 96–104.

Lützkendorf, T., Fan, W., & Lorenz, D. (2011). Engaging financial stakeholders: Opportunities for a sustainable built environment. *Building Research & Information, 39*, 483–503.

Macgregor, C., Dowdell, D., Jaques, R., Bint, L., & Berg, B. (2018). The built environment and climate change: A review of research, challenges and the future. *BRANZ Study Report SR403.* Judgeford, New Zealand.

Madnor, M. S., Harun, A. N., & Ros, F. C. (2024). Integrating adaptation of climate change to strengthen Malaysia's disaster risk governance. *Environment and Social Psychology, 9.*

Malalgoda, C., Amaratunga, D., & Haigh, R. (2013). Creating a disaster resilient built environment in urban cities. *International Journal of Disaster Resilience in the Built Environment, 4*, 72–94.

Masson, V., Lemonsu, A., Hidalgo, J., & Voogt, J. (2020). Urban climates and climate change. *Annual Review of Environment and Resources, 45*, 411–444.

Masud, S., & Khan, A. (2023). Policy implementation barriers in climate change adaptation: The case of Pakistan. *Environmental Policy and Governance, 34*, 42–52.

McNaught, R. (2024). The application of collaborative governance in local level climate and disaster resilient development—A global review. *Environmental Science & Policy, 151.*

Mees, H. L. P., Uittenbroek, C. J., Hegger, D. L. T., & Driessen, P. P. J. (2019). From citizen participation to government participation: An exploration of the roles of local governments in community initiatives for climate change adaptation in the Netherlands. *Environmental Policy and Governance, 29*, 198–208.

Meyer, M., Hendricks, M., Newman, G., Horney, J., Berke, P., Masterson, J., Sansom, G., Cousins, T., van Zandt, S., & Cooper, J. (2018). Participatory action research: Tools for disaster resilience education. *Int J Disaster Resil Built Environ, 9*, 402–419.

Mohammad, H., Mojtahedi, S., & Lan Oo, B. (2014) Stakeholders' approaches to disaster risk reduction in built environment. *Disaster Prevention and Management, 23*, 356–369.

Murtagh, N., Scott, L., & Fan, J. (2020). Sustainable and resilient construction: Current status and future challenges. *Journal of Cleaner Production, 268.*

Nath, P. K., & Behera, B. (2010). A critical review of impact of and adaptation to climate change in developed and developing economies. *Environment, Development and Sustainability, 13*, 141–162.

NATIONS, U. (1992). Climate change: The 1990 and 1992 IPCC Assessments. Canada.

Pereira, P., Wang, F., Inacio, M., Kalinauskas, M., Bogdzevič, K., Bogunovic, I., Zhao, W., & Barcelo, D. (2024). Nature-based solutions for carbon sequestration in urban environments. *Current Opinion in Environmental Science & Health, 37.*

Queiroz-Stein, G. D., Seifert JR, C. A., & Luiz, A. M. M. T. (2023). climate change, denialism, and participatory institutions in Brazil: Effects of the Bolsonaro Government's Environmental Strategy (2019–2022). *Brazilian Political Science Review.*

Rathnayaka, B., Robert, D., Siriwardana, C., Adikariwattage, V. V., Pasindu, H. R., Setunge, S., & Amaratunga, D. (2023). Identifying and prioritizing climate change adaptation measures in the context of electricity, transportation and water infrastructure: A case study. *International Journal of Disaster Risk Reduction, 99.*

Roaf, S., Crichton, D., Nicol, F. (2009). *Adapting Buildings and Cities for Climate Change, A 21st Century Survival Guide*, Elsevier Ltd.

Strange, K. F., Satorras, M. & March, H. 2024. Intersectional climate action: the role of community-based organisations in urban climate justice. *Local Environment*, pp. 1–21.

Uchiyama, C., Ismail, N., & Stevenson, L. A. (2021). Assessing contribution to the Sendai Framework: Case study of climate adaptation and disaster risk reduction projects across sectors in Asia-Pacific (2015–2020). *Progress in Disaster Science, 12.*

Vij, S., Stock, R., Ishtiaque, A., Gardezi, M., & Zia, A. (2023). Power in climate change policy-making process in South Asia. *Climate Policy, 24*, 104–116.

Wallius, V., Kunttu, J., Leskinen, P., van Brusselen, J., & Näyhä, A. (2022). Stakeholder perceptions of wood-based products in the built environment: A literature review. *European Journal of Wood and Wood Products, 81*, 287–299.

Wan, Y., Chen, S., Liu, J., & Jin, L. (2024). Brownfield-related studies in the context of climate change: A comprehensive review and future prospects. *Heliyon, 10*, e25784.

Wellstead, A. M., & Stedman, R. C. (2011). Climate change policy capacity at the sub-national government level. *Journal of Comparative Policy Analysis: Research and Practice, 13*, 461–478.

Williams, D. S., Celliers, L., Unverzagt, K., Videira, N., Máñez Costa, M., & Giordano, R. (2020). A method for enhancing capacity of local governance for climate change adaptation. *Earth's Future, 8.*

Zuniga-Teran, A. A., Staddon, C., de Vito, L., Gerlak, A. K., Ward, S., Schoeman, Y., Hart, A., & Booth, G. (2019). Challenges of mainstreaming green infrastructure in built environment professions. *Journal of Environmental Planning and Management, 63*, 710–732.

Land Use Planning and Adaptation in the Built Environment

Analysis and Prediction of Flood Events in Colombo District, Sri Lanka Using Standard Precipitation Index (SPI) and Rainfall Anomaly Index (RAI)

R. M. A. S. D. Rajakaruna and B. G. N. Sewwandi

1 Introduction

Climate change enhances the occurrence of extreme meteorological disasters (Feng et al., 2020)which have caused global issues such as climate migration that leads to many other environmental, economic and political problems (Cao & Chen, 2012; Liang et al., 2023). Floods are natural hazards triggered by hydro-meteorological conditions and often result in widespread economic damage and loss of human lives (Leskens et al., 2014; Nied et al., 2014). Variety of diseases triggered by floods threaten the public health (Anderson et al., 2019). Heavy rainfall is a major natural factor that causes severe flood damage in addition to anthropogenic factors such as improper land use and deforestation (Acosta-Coll et al., 2018; Tingsanchali, 2012). Hydrological cycle is also negatively affected by rapid urbanization making urbanized areas more vulnerable to floods induced by climate change (Ma et al., 2024; Peng et al., 2024). Global flood risk will accelerate by 187% in 2050, with a strong regional variability inducing more adverse impacts in Asia (Arnell & Gosling, 2016). Furthermore, Merz et al. (2021) have also identified a flooding hotspot in Asia due to changes in climate in socio-economic factors.

Flood monitoring is a critical measure for flood control. Current flood monitoring is conducted by several methods such as conventional gauge methods and image-based methods that rely on continuous manned monitoring (Lo et al., 2015; Sun, 2013). Although monitoring and detecting flood risk is difficult, different indices could be utilized to monitor hydrological conditions in flood-plain areas and predict

R. M. A. S. D. Rajakaruna
Faculty of Silviculture and Forest Engineering, Transilvania University of Brasov, Brasov, Romania

B. G. N. Sewwandi (✉)
Department of Zoology and Environmental Management, Faculty of Science, University of Kelaniya, Kelaniya, Sri Lanka
e-mail: sewwandidh@kln.ac.lk

C. Malalgoda et al. (eds.), *Climate Change Adaptation in the Built Environment*,
https://doi.org/10.1007/978-3-031-75826-3_7

potential flood events posed by heavy rainfall. Standard Precipitation Index (SPI) and Rainfall Anomaly Index (RAI) are such indices that were initially developed for drought monitoring but still could be utilized for the analysis of extreme wet events leading to floods (Aryal et al., 2022; Olanrewaju & Reddy, 2022; Ologunorisa & Tersoo, 2006; Seiler et al., 2002). Both indices rely on rainfall as the single input variable and normalize rainfall data (Hänsel et al., 2016; Tirivarombo et al., 2018). McKee et al. (1993) formed the Standard Precipitation Index (SPI) that is now widely applied in the assessment of rainfall variability on a spatial and a temporal basis (Karavitis et al., 2011; Khan et al., 2008; Livada & Assimakopoulos, 2007). A study conducted in Pearl river basin, China analyzed the wet and dry conditions using SPI (Zhang et al., 2009). A study by Zhao et al. (2020) has found that SPI can reflect the annual and seasonal flood changes. Li et al. (2013) have utilized SPI to study precipitation distribution and droughts and flood occurrence in Fuhe basin, China. Another study in Jilin province, China has identified seasonal characteristics of floods using SPI (Qin et al., 2010). Wang et al. (2022) have used SPI to detect flood prone areas in Taihu lake basin, China. SPI has been identified as a tool for predicting long-term dry and wet periods in eastern China (Bordi et al., 2004). Fauzi et al. (2017) have monitored the occurrence of floods in Terengganu, Malaysia using SPI of 3, 6, 24, and 48-month time scales that revealed the higher probability of floods during wet periods. Rainfall duration and severity are critical factors that affect flood intensity. Rauf and Zeephongsekul (2011) have quantified those factors using SPI. The Rainfall Anomaly Index which has been developed by Van-Rooy (1965), utilizes a ranking procedure for categorizing positive and negative anomalies (Aryal et al., 2022; Raziei, 2021). A study conducted in North-West Oman has observed that extreme wet conditions indicated by both SPI and RAI, induce flash floods (Abushandi & Al Ajmi, 2022). They also have shown the consistency and suitability of using SPI and RAI in flood and drought monitoring. Ologunorisa and Tersoo (2006) have studied an association between extreme wet events and the occurrence of flood events in Northern Nigeria using the Rainfall Anomaly Index. Chaubey et al. (2021) have found that the variation in intense and heavy rainfall is linked to extreme flood events using RAI and SPI. RAI has been utilized to investigate precipitation variability that causes extreme climate events in Danube Plain, Bulgaria (Nikolova & Vassilev, 2005). Nevertheless, many studies in Sri Lanka have utilized SPI and RAI for drought analysis with a focus on agricultural production in dry zone (Alahacoon et al., 2021; Begum & Najim, 2020; Burchfield & Gilligan, 2016; Hemachandra et al., 2021). Therefore, it is important to apply SPI and RAI for flood monitoring in Sri Lanka which has not been conducted before.

Colombo District has the highest population density in Sri Lanka and has experienced frequent floods in the last two decades that destroyed the properties and affected the mental and physical well-being of people. For instance, 172,000 people have been affected by the severe floods in Colombo District in May 2016 (Samarakoon & Abeykoon, 2018). A study by Samantha (2018) observed a decline in the revenues of enterprises in the Colombo District due to the May 2016 flood event. The large-scale development projects in Colombo District are often disrupted by floods caused by heavy rainfall (Weerasinghe et al., 2018). Dahanayake and Wickramasinghe (2022) have found a 1.29% increase in highly vulnerable areas for floods from 1999 to 2018 due to urban expansion in Colombo District. Despite the congested housing conditions and poor drainage management in Colombo, the increased rainfall intensity and frequency make the area highly vulnerable to floods (Hewawasam & Matsui, 2020). Therefore, the present study aims to analyze the effect of rainfall variability on the occurrence of floods in the Colombo District using 1, 3, and 6-month time scales of the Standard Precipitation Index (SPI) and Rainfall Anomaly Index (RAI) in predicting flood events for flood risk management in Colombo District.

2 Methodology

2.1 Study Area and Data Collection

Colombo District, located in the Southwest of Sri Lanka, is the most populated administrative district in Sri Lanka, with a population of 2,324,349 in 2012 within an area of 699 km^2 (Department of Census & Statistics, 2012). It is the commercial and administrative center of Sri Lanka, where frequent flooding occurs due to heavy rainfall and unplanned urbanization (Dammalage & Jayasinghe, 2019). This study focuses on extreme rainfall in Colombo District leading to floods. Climate Hazards Group InfraRed Precipitation with Station data (CHIRPS) is a freely available global dataset with a high resolution of 0.05 and data availability since 1981 (Climate Hazard Center, 2023). Bandara et al. (2022) have found that CHIRPS data can be used in hydrological applications such as flood forecasting and climate change impact assessment in Sri Lanka. Therefore, daily rainfall data of thirty-year period from 1991 to 2020 were collected from CHIRPS dataset for thirteen Divisional Secretariat Divisions (DSD) in Colombo District, Sri Lanka, namely Colombo, Dehiwala, Homagama, Kaduwela, Kesbewa, Kolonnawa, Maharagama, Moratuwa, Padukka, Ratmalana, Seethawaka, Sri Jayawardenepura Kotte, and Thimbirigasyaya by restricting the range of the gridded data using relevant time period and geographic coordinates of each DSD. The map of the study area is shown in Fig. 1. RStudio and Microsoft Excel software were used for the SPI and RAI analysis. The 'Precintcon' package was used for Rstudio and the data was fitted to the Gamma distribution in the SPI calculation. Flood impact data of each DSD of Colombo District was obtained from the Disaster Information Management System (DesInventar) database from 1991 to 2020 to compare the occurrence floods with wet conditions categorized by SPI and RAI (UNDRR, 2020).

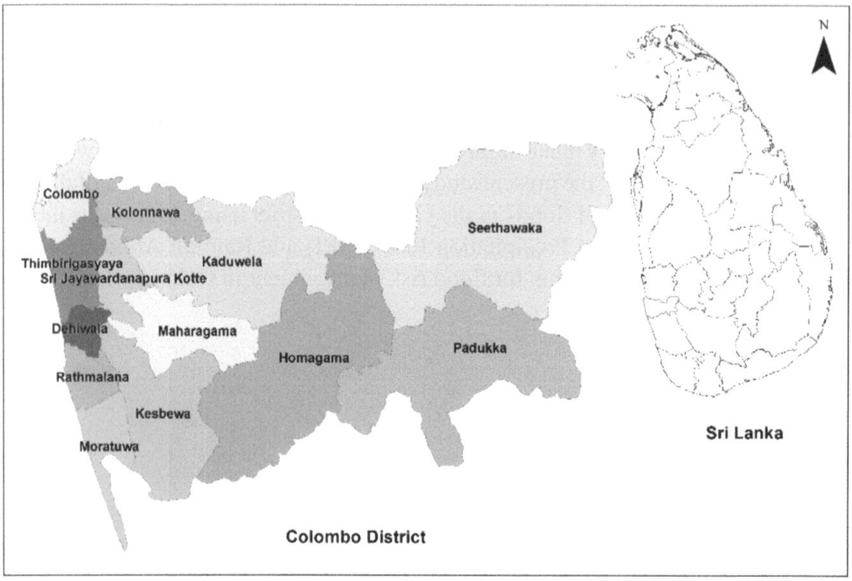

Fig. 1 Map of the study area

2.2 *Standard Precipitation Index (SPI)*

SPI was calculated for the periods of 1, 3 and 6 months using Eq. 1 to get SPI 1, SPI 3 and SPI 6, respectively for all thirteen study locations (Begum & Najim, 2020; McKee et al., 1993). SPI has been widely utilized for the analysis of hydrological conditions leading to flood events (Fauzi et al., 2017; Rauf & Zeephongsekul, 2011; Zhang et al., 2009). SPI ranges from −2 to 2. Positive SPI values indicate that the precipitation exceeds the mean value while negative SPI values indicate the precipitation below the mean value. The categorization of SPI values given in Table 1 was used to classify dry and wet events into different classes (Begum & Najim, 2020).

$$SPI = \frac{(X_i - \overline{X})}{S} \tag{1}$$

where, X_i= Seasonal precipitation, $\overline{X}$ = Long-term seasonal mean precipitation, S = Standard deviation.

2.3 *Rainfall Anomaly Index (RAI)*

Rainfall Anomaly Index (RAI) has been used by many researchers to investigate extreme rainfall events (Chaubey et al., 2021; Nikolova & Vassilev, 2005). RAI was

Table 1 Categorization of SPI values

SPI value	Category
> 2.0	Extremely wet
1.5–1.99	Severely wet
1.0–1.49	Moderately wet
−0.99 to 0.99	Near normal
−1.0 to −1.49	Moderately dry
−1.5 to −1.99	Severely dry
< −2	Extremely dry

Table 2 Categorization of RAI values

RAI value	Category
Above 4	Extremely humid
2–4	Very humid
0–2	Humid
−2 to 0	Dry
−4 to −2	Very dry
Below −4	Extremely dry

calculated for negative and positive anomalies using Eq. 2 and 3, respectively (Aryal et al., 2022). The categorization of RAI values is shown in Table 2.

$$RAI = -3\left[\frac{N - \overline{N}}{\overline{M} - \overline{N}}\right] \tag{2}$$

$$RAI = 3\left[\frac{N - \overline{N}}{\overline{X} - \overline{N}}\right] \tag{3}$$

where N = current monthly/annual rainfall, $\overline{N}$= average monthly/annual rainfall of historical series, $\overline{M}$= average of ten highest monthly/annual rainfall, $\overline{X}$= average of ten lowest monthly/annual rainfall.

2.4 Prediction of Flood Events

Return period and probability of exceedance of wet events linked to occurrence of floods which can predict the severity and likelihood of floods. Eqs. (4) and (5) were used for calculating the Return period (T) and Probability of Exceedance, respectively (Olanrewaju & Reddy, 2022). 30-year SPI data was extrapolated to 100 years using linear extrapolation method.

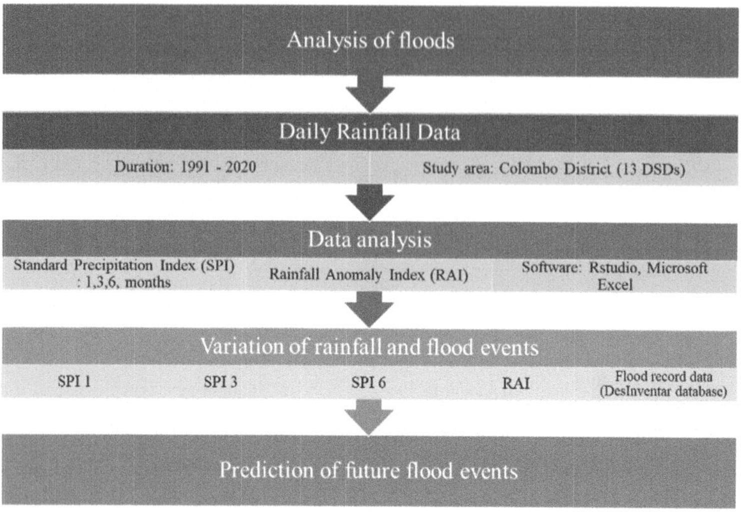

Fig. 2 Methodological framework of the present study

$$T = \frac{n}{f} \tag{4}$$

$$Probability of Exceedance = \frac{1}{T} \tag{5}$$

where, $n =$ number of years (100 years), $f =$ frequency of flood event (30 years data for SPI extrapolated to 100 years).

Figure 2 shows the overall methodology used in this study.

3 Results and Discussion

SPI 1 is considered more reflective of short-term meteorological conditions than SPI 3 and SPI 6 (WMO, 2012). According to SPI 1 analysis, all thirteen study locations in Colombo District have faced extreme wet conditions (SPI > 2) in five periods: October 1997, December 2005, March 2008, December 2014, and May 2016. For example, the variation of SPI 1 from 1991 to 2020 in Colombo DSD is shown in Fig. 3. In addition, at least seven out of thirteen DSDs in the Colombo District experienced extreme wet conditions in February 2011, July 1998, and 2008, August 2014 and 2019, September 2019, and November 2010. Extreme wet periods in March indicate the influence of the First Inter-monsoon season, while extreme wet periods in May, July, August, and September suggest the impact of Southwest-monsoon on the Colombo District. Extreme wet periods experienced in October and November indicate the influence of the Second Inter-monsoon, and extreme

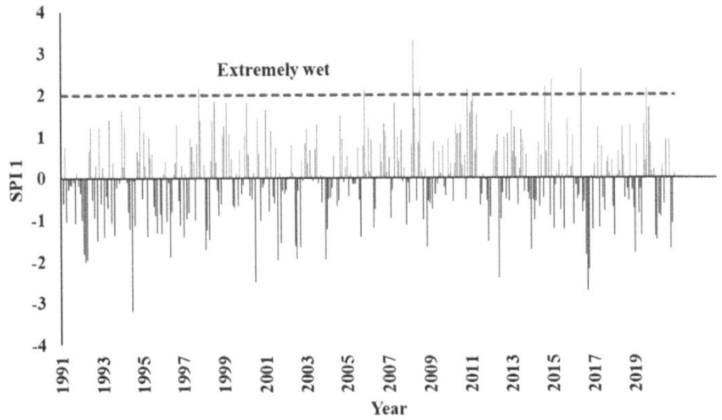

Fig. 3 Variation of SPI 1 in Colombo DSD

wet conditions in December and February show the effect of Northeast-monsoon on the Colombo District. Nevertheless, most extreme conditions occurred between May and September, conveying Southwest-monsoon's higher impact towards extreme wet events in the Colombo District.

SPI 1 analysis also revealed more extreme wet conditions in the last 15 years after 2005 compared to the beginning of the study period. That has also been proved by Fig. 4, which shows the number of people directly and indirectly affected by flood events in each DS Division of Colombo District according to flood impact data obtained from the Disaster Information Management System (DesInventar) database. It shows an increase in the number of people affected by flood events after 2005. For instance, in Colombo DS Division, flood events occurred in April 2008, May 2007, and 2016, June 2008, September 2019, and November 2005 and 2010 with extreme wet conditions. Du et al. (2013) and Olanrewaju and Reddy (2022) also found that short term scale SPI explains the occurrence of required conditions for the development of flood events and its potential to predict flood risk. Therefore, SPI 1 analysis is highly applicable for early warning and assessing the severity of flood events. Understanding extreme rainfall patterns that lead to floods using SPI 1 could be beneficial to reduce the flood damages in heavily urbanized areas like Colombo District where rainfall could trigger the occurrence of flash floods due to abundance of impervious surfaces.

SPI 3 reflects the short and medium term moisture conditions of soil caused by rainfall while SPI 6 indicates the seasonal rainfall trends of an area (WMO, 2012). The variation of SPI 3 and 6 values from 1991 to 2020 in Colombo DSD is shown in Fig. 5. According to SPI 3 and SPI 6 analysis in Colombo District, extreme wet events have been identified in five years, which are 1997, 2008, 2011, 2016, and 2019 during the study period. These extreme wet peaks indicates the saturation of soil that leads to higher storm flow generation compared to a drier soil (Olanrewaju & Reddy, 2022). As shown in Fig. 6, extreme wet conditions have aligned with major flood events (Table 3) in May 2016 in Colombo, Moratuwa, and Sri Jayawardenepura Kotte. A

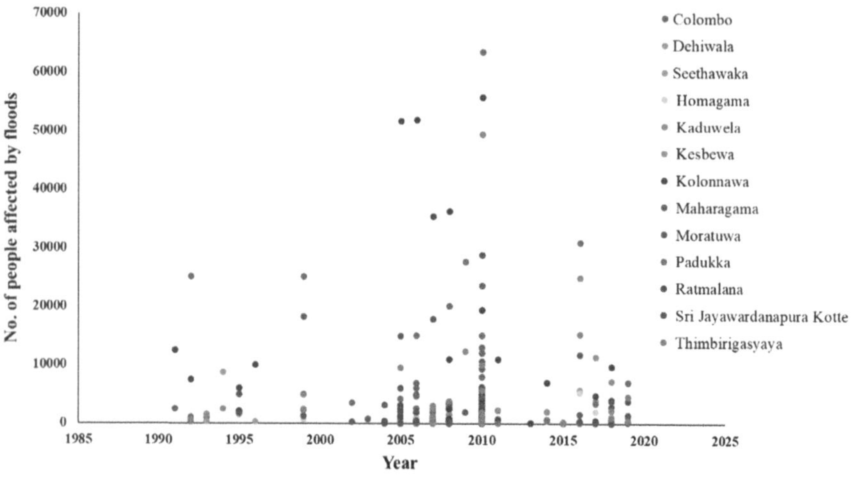

Fig. 4 Number of people affected by flood events in Colombo District from 1991 to 2020

similar variation was observed for Thimbirigasyaya, and Maharagama DSDs, which obtained extreme wet SPI 3 and severely wet SPI 6 values in May 2016. In Colombo DS Division, SPI 3 and SPI 6 analyses well explained the flood events (Table 3) that occurred in May 2016 and April 2008 due to extremely wet and very wet conditions (Fig. 5) in rainfall. Zhai et al. (2010) has also found a high correlation between long term runoff variation and SPI in six rivers in China which indicates that the precipitation is the primary cause of runoff generation.

The flood event occurred in June 2008 in Homagama DSD (Table 3) has been followed by SPI 3 extreme wet periods in March (SPI 2.33), April (SPI 2.44), and May (SPI 2.15) of 2008 and SPI 6 severely wet conditions in April (SPI 1.66) and May (SPI 1.75) of 2008 as shown in Fig. 7. The same SPI 3 and SPI 6 pattern (Fig. 7) was identified in flood events in Thimbirigasyaya, Kolonnawa, Sri Jayawardenepura Kotte, and Seethawaka DSDs in June 2008 (Table 3). It shows that the periods with high SPI 3 and SPI 6 values lead to the occurrence of floods. Furthermore, in Kaduwela, Thimbirigasyaya, Moratuwa, Sri Jayawardenepura Kotte, and Maharagama DSDs, flood events in September 2019 (Table 3) occurred under extreme wet conditions in SPI 3 analysis and severely wet conditions in SPI 6 analysis in each DSD. A study by Yuan and Zhou (2004) also suggested that the combined consideration of different scales of SPI gives a satisfactory explanation for flood events in China. Nevertheless, with the increase in time scale of SPI the response to short-term precipitation variation become slower (Olanrewaju & Reddy, 2022; Yuan & Zhou, 2004). Therefore, it is clear that SPI 3 and SPI 6 could be utilized to identify major wet periods that could be associated with the increased streamflows and rising surface and ground water levels.

According to the RAI analysis, October 1997 and May 2016 were identified as extremely humid periods (RAI > 4) in all the DSDs. For instance, Fig. 8 shows the variation of Rainfall Anomaly Index (RAI) values over time in Colombo DSD.

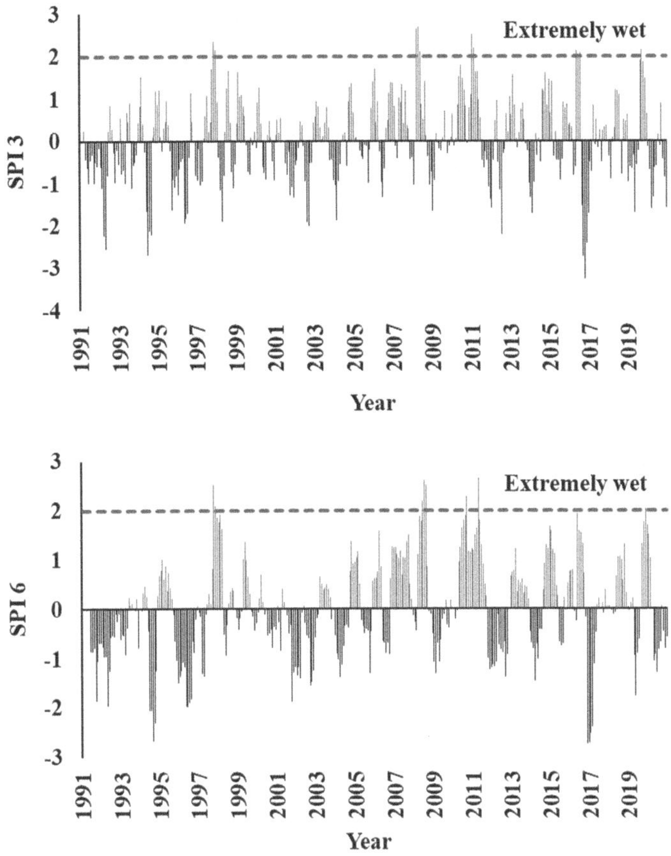

Fig. 5 Variation of SPI 3 and 6 in Colombo DSD from 1991 to 2020

As shown in Fig. 9, the RAI values of Colombo DSD in 2016 (RAI 5.08) were well aligned with the historic floods occurred in May 2016 in Colombo (Table 3). Similarly, extremely humid conditions in other DSDs such as Kesbewa (RAI 4.17), Homagama (RAI 4.54), Kaduwela (RAI 4.67), Moratuwa (RAI 4.68), Padukka (RAI 4.47), Ratmalana (RAI 4.83), Seethawaka (RAI 4.63), Sri Jayawardenepura Kotte (RAI 4.96), Thimbirigasyaya (RAI 5.08), and Maharagama (RAI 4.77) were also followed by floods in May 2016 (Table 3). An extremely humid condition (RAI 4.17) was also observed in Kesbewa DSD in September 2019. Similar results have been observed in Adamawa State, Nigeria by Sadiq et al. (2020) using RAI to identify extreme wet conditions that led to exacerbating flood damages to farmlands. Studies have applied RAI to analyze rainfall patterns causing floods which have shown the association between major flood events and extreme rainfall periods (Abashiya et al., 2017; Udosen, 2012). Therefore, the evaluation of flood events using RAI could be a reliable, effective, and a simple method for climate adaptation.

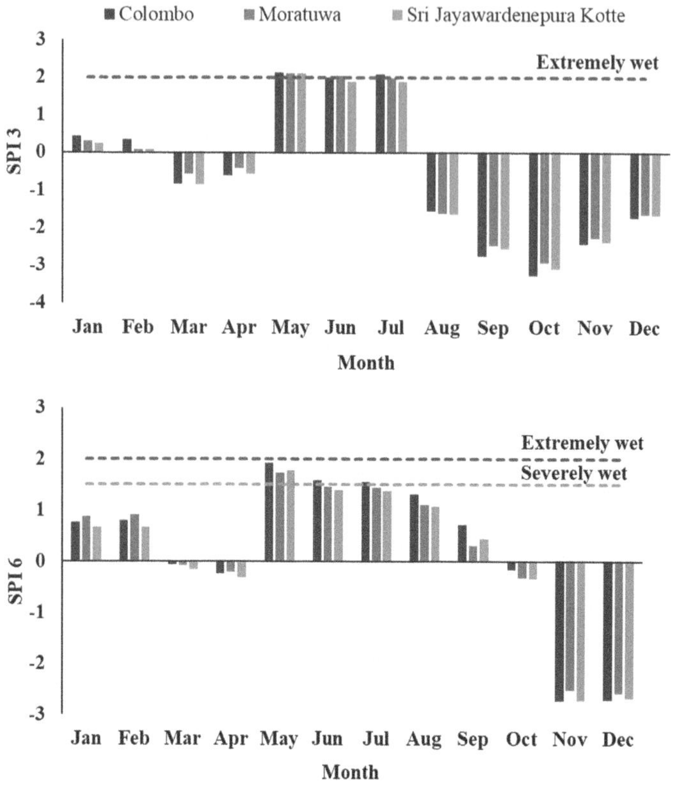

Fig. 6 Major wet events occurred in 2016 at Colombo, Moratuwa, and Sri Jayewardenepura Kotte DSDs

According to the SPI and RAI analysis, it was concluded that the SPI 1 analysis aligns well with the occurrence of floods compared to SPI 3 and SPI 6 analyses, which reveal only the major flood events that occurred in the past. The same results have been obtained by previous studies in Hunan Province, China (Du et al., 2013) and in Spain (Vicente-Serrano & López-Moreno, 2005), where a high correlation between short-time scale SPI and surface flows was observed. Therefore, SPI 1 was used for the prediction of future flood events. Table 4 shows the results of the flood prediction analysis. Extreme wet periods (SPI > 2) under SPI 1 analysis were considered potential flood occurrence periods, also confirmed by historic records according to the above analysis.

Return period is used to determine the rareness of a particular extreme event. Table 3 shows the intensity and severity of extreme wet events that could cause flooding in Colombo District. According to Table 3, Dehiwala and Ratmalana have the highest frequency (11) of extreme wet events during the study period, with the lowest return period (2.72 years) in 100 years among the DSDs of the Colombo District. It means that extreme wet events could occur on average of nearly every

Table 3 Flood events occurred in Colombo District (UNDRR, 2020)

Flood events in DS divisions												
Colombo	Dehiwala	Homagama	Kaduwela	Kesbewa	Kolonnawa	Maharagama	Moratuwa	Padukka	Ratmalana	Seethawaka	Sri Jayawardenepura Kotte	Thimbirigasyaya
1991 Jun	1992 Jun	1992 Jun	2004 Sep	1993 May	1991 Jun	1999 Oct	1991 Jun	1993 May	1992 Jun	1996 Sep	1992 Jun	1991 Jun
1992 Jun	1994 Oct	1999 Apr	2005 Nov	1999 Oct	1992 Jun	2005 Nov	1994 Oct	2010 May	1994 Oct	2005 Now	1995 May	1992 May
1992 Oct	1995 May	2005 Nov	2006 Sep	2004 Sep	1992 Oct	2006 Mar	1995 May	2011 May	1995 May	2006 Jun	1999 Apr	1992 Jun
1992 Dec	1999 Oct	2006 Jun	2007 May	2005 Nov	1993 Jun	2007 May	1999 Oct	2014 Jun	1995 Nov	2007 May	1999 Oct	1992 Oct
1993 May	2007 May	2006 Oct	2007 Nov	2006 Dec	1995 May	2008 Oct	2003 May	2016 May	2004 Sep	2008 Apr	2002 Oct	1993 May
1999 Oct	2010 May	2008 Jun	2008 Apr	2007 May	1995 Nov	2009 Nov	2004 Sep	2017 May	2005 Nov	2008 Jun	2005 Nov	1994 Oct
2002 Oct	2010 Nov	2010 May	2008 Jun	2009 Nov	1996 Sep	2010 May	2004 Nov	2018 Oct	2006 Oct	2008 Sep	2006 Oct	1999 Apr
2005 Nov	2018 May	2011 May	2008 Oct	2010 May	2004 Sep	2010 Nov	2004 Dec		2007 May	2010 May	2006 Nov	1999 Jun
2006 Oct		2016 May	2010 May	2010 Nov	2005 Nov	2015 May	2005 Nov		2010 May	2010 Nov	2007 May	1999 Oct
2007 May		2017 May	2010 Nov	2015 May	2006 Oct	2016 May	2005 Nov		2010 Nov	2011 May	2008 Jun	2002 Oct
2008 Apr		2018 May	2011 May	2016 May	2007 May	2018 Oct	2006 Nov		2015 Sep	2013 Mar	2008 Oct	2005 Nov

(continued)

Table 3 (continued)

Flood events in DS divisions												
Colombo	Dehiwala	Homagama	Kaduwela	Kesbewa	Kolonnawa	Maharagama	Moratuwa	Padukka	Ratmalana	Seethawaka	Sri Jayawardenepura Kotte	Thimbirigasyaya
2008 Jun		2018 Oct	2014 Jun	2018 May	2008 Apr	2019 Sep	2007 May		2016 May	2014 Jun	2009 Nov	2007 May
2008 Oct			2016 May	2018 May	2008 Jun		2008 Oct		2018 Oct	2016 May	2010 May	2008 Jun
2009 Nov			2017 May		2010 May		2010 May			2017 May	2010 May	2009 Nov
2010 May			2018 May		2010 Nov		2010 Nov			2018 May	2015 Sep	2010 May
2010 Nov			2018 Oct		2011 May		2016 May			2018 Oct	2016 May	2010 Nov
2013 Jun			2019 Sep		2013 Jun		2018 Oct				2017 May	2014 Jun
2016 May					2014 Dec		2019 Sep				2018 May	2015 May
2017 May					2015 May						2018 Oct	2016 May
2018 May					2015 Sep						2019 Sep	2018 May
2018 Oct					2017 May							2018 Oct
2019 Sep					2018 May							2019 Sep
					2018 Oct							

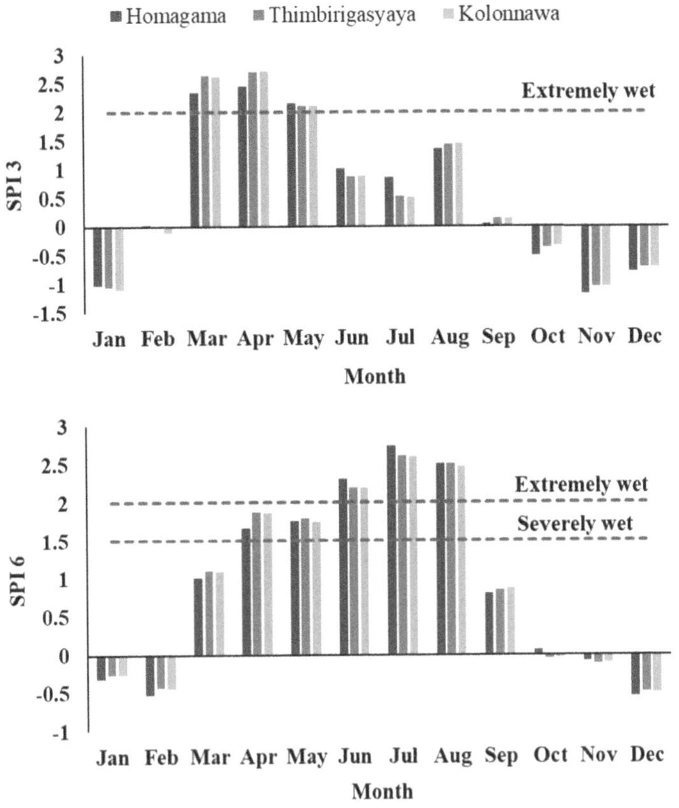

Fig. 7 Variation of SPI 3 and SPI 6 in 2008 at Homagama, Thimbirigasyaya, and Kolonnawa DSDs

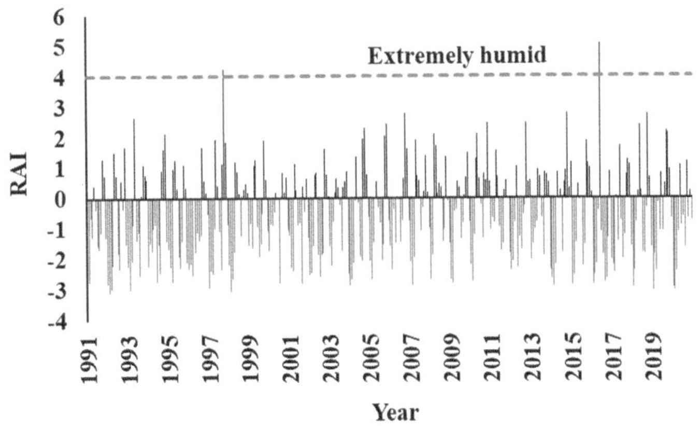

Fig. 8 Variation of RAI in Colombo DSD from 1991 to 2020

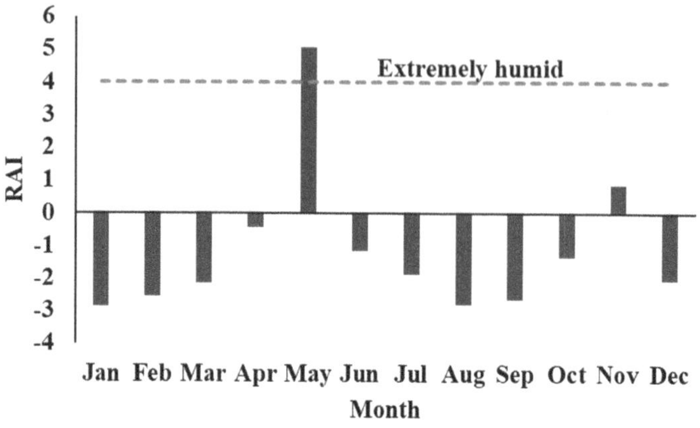

Fig. 9 Variation of RAI in 2016 at Colombo DSD

Table 4 Intensity and severity of extreme wet events

Divisional secretariat division	Frequency in 30 years	Return period (T) in 100 years	Probability of exceedance (P = 1/T)
Colombo	9	3.33	0.30
Dehiwala	11	2.72	0.37
Homagama	9	3.33	0.30
Kaduwela	10	3.00	0.33
Kesbewa	9	3.33	0.30
Kolonnawa	9	3.33	0.30
Maharagama	9	3.33	0.30
Moratuwa	10	3.00	0.33
Padukka	8	3.75	0.27
Ratmalana	11	2.73	0.37
Seethawaka	9	3.33	0.30
Sri Jayawardenepura Kotte	9	3.33	0.30
Thimbirigasyaya	9	3.33	0.30

3 years in those DSDs. The return period represents the probability of occurrence of extreme wet events and indicates the need to protect the people of those areas from adverse consequences of those meteorological conditions (Volpi, 2019). Therefore, the highest exceeding probability of meteorological flood occurrence (37%) due to extreme precipitation in 100 years resulted in Dehiwala and Ratmalana areas compared to others. Kaduwela and Moratuwa DSDs have a return period of 3 years in 100 years, with an exceeding probability of 33% for extreme wet events. In addition, eight out of thirteen DS Divisions, namely Colombo, Homagama, Kesbewa,

Kolonnawa, Maharagama, Seethawaka, Sri Jayawardenepura Kotte, and Thimbirigasyaya have a 30% exceeding probability for flood events posed by extreme wet conditions in those areas in 100 years. Similar results were obtained by De Silva et al. (2012) by using a 2-D inundation model to estimate flooding under extreme rainfall in the Kelani river basin. In that study, DS divisions, including Kaduwela, Kolonnawa, and Colombo, were identified to be inundated under precipitation of a 50-year return period. According to the flood prediction analysis, Padukka is the least affected area, with a return period of 3.75 years in 100 years. It has the lowest exceeding probability (27%) for extreme wet events in 100 years. According to the flood prediction analysis, SPI 1 predicts the severity of flood events occurred by short-term meteorological conditions such as flash floods that allows early warning for mitigation of flood damages.

3.1 Limitations of the Study

Same as all station data, the CHIRPS data used for the study could contain measurements that are interpolated due to fewer rain gauge stations in the study area. Nevertheless, CHIRPS integrates satellite data and fills in such areas with reliable data. Furthermore, the suitability of using CHIRPS data for hydrological applications in Sri Lanka has been proved by previous studies (Bandara et al., 2022).

4 Conclusions

The present study results suggest the potential of using the Standard Precipitation Index (SPI) and Rainfall Anomaly Index (RAI) to analyze and predict the flood events. Application of SPI and RAI will be a beneficial, less demanding, and reliable approach for flood risk management. Both SPI and RAI can be utilized to explain the development of meteorological conditions leading to occurrence of floods, which is essential for proper flood monitoring. The analysis capacity of SPI could be broadened by utilizing several time scales of SPI such as 1-, 3-, and 6-months periods. SPI 1 values were more reflective of short-term meteorological conditions that led to the occurrence of floods. SPI 1 analysis shows an increase in extreme wet events after 2005 in Colombo District. All thirteen DSDs experienced extreme wet events in 1997, 2005, 2008, 2014, and 2016. According to the SPI 3 and SPI 6 analysis, extreme wet events posed by the Southwest monsoon followed the major flood events in June 2008, May 2016, and September 2019. RAI analysis indicated that the severe flood events that occurred in May 2016 were in compliance with the extremely humid conditions in Colombo District. SPI 1 was used for the flood prediction analysis, which suggests an exceeding probability of $\geq 27\%$ for extreme wet conditions leading to flood events in the Colombo District in 100 years. The results of this study show the applicability of SPI and RAI indices for flood monitoring and prediction by explaining the meteorological wet conditions that lead to the occurrence of floods. The variation of flood events can be observed through the variation of extreme wet events. Therefore, it is useful for the decision-makers on future flood disaster preparedness plans towards a disaster-resilient community.

References

Abashiya, M., Abaje, I. B., Iguisi, E. O., Bello, A. L., Sawa, B. A., Amos, B. B., & Musa, I. (2017). Rainfall characteristics and occurrence of floods in Gombe Metropolis Nigeria. *Ethiopian Journal Environmental Studies Management, 10*, 44. https://doi.org/10.4314/ejesm.v10i1.5

Abushandi, E., & Al Ajmi, M. (2022). Assessment of hydrological extremes for arid catchments: A case study in Wadi Al Jizzi. *North-West Oman Sustainability, 14*, 14028. https://doi.org/10.3390/su142114028

Acosta-Coll, M., Ballester-Merelo, F., Martinez-Peiró, M., & De la Hoz-Franco, E. (2018). Real-time early warning system design for pluvial flash floods—A review. *Sensors, 18*, 2255. https://doi.org/10.3390/s18072255

Alahacoon, N., Edirisinghe, M., & Ranagalage, M. (2021). Satellite-based meteorological and agricultural drought monitoring for agricultural sustainability in Sri Lanka. *Sustainability, 13*, 3427. https://doi.org/10.3390/su13063427

Anderson, G. B., Barnes, E. A., Bell, M. L., & Dominici, F. (2019). The future of climate epidemiology: Opportunities for advancing health research in the context of climate change. *American Journal of Epidemiology, 188*, 866–872. https://doi.org/10.1093/aje/kwz034

Arnell, N. W., & Gosling, S. N. (2016). The impacts of climate change on river flood risk at the global scale. *Climate Change, 134*, 387–401. https://doi.org/10.1007/s10584-014-1084-5

Aryal, A., Maharjan, M., Talchabhadel, R., & Thapa, B. R. (2022). Characterizing meteorological droughts in Nepal: A comparative analysis of standardized precipitation index and rainfall anomaly Index. *Earth, 3*, 409–432. https://doi.org/10.3390/earth3010025

Bandara, U., Agarwal, A., Srinivasan, G., Shanmugasundaram, J., & Jayawardena, I. M. S. (2022). Intercomparison of gridded precipitation datasets for prospective hydrological applications in Sri Lanka. *International Journal of Climatology, 42*, 3378–3396. https://doi.org/10.1002/joc.7421

Begum, S. L. R., & Najim, M. M. M. (2020). A ssessment of climate extremes based on SPI in WL1 a agroecological Zone—A case study from Kuruwita Sri Lanka. *Sri Lankan Journal of Technology, 1*, 1–6.

Bordi, I., Fraedrich, K., Jiang, J.-M., & Sutera, A. (2004). Spatio-temporal variability of dry and wet periods in eastern China. *Theoretical and Applied Climatology, 79*, 81–91. https://doi.org/10.1007/s00704-004-0053-8

Burchfield, E. K., & Gilligan, J. (2016). Agricultural adaptation to drought in the Sri Lankan dry zone. *Applied Geography, 77*, 92–100. https://doi.org/10.1016/j.apgeog.2016.10.003

Cao, Z. J., & Chen, S. J. (2012). Migration mechanism of climate migrants and analysis of relevant concepts. *Advances in Materials Research, 518–523*, 5070–5074. https://doi.org/10.4028/www.scientific.net/AMR.518-523.5070

Chaubey, P. K., Srivastava, P. K., Gupta, A., & Mall, R. K. (2021). Integrated assessment of extreme events and hydrological responses of Indo-Nepal Gandak River Basin. *Environment, Development and Sustainability, 23*, 8643–8668. https://doi.org/10.1007/s10668-020-00986-6

Climate Hazard Center, (2023). CHIRPS: Rainfall Estimates from Rain Gauge and Satellite Observations [WWW Document]. Clim. Hazard Center, University of California, Santa, Barbar. URL https://www.chc.ucsb.edu/data/chirps

Dahanayake, H. D., & Wickramasinghe, D. (2022). Impacts of floods on Colombo during two decades: Looking back and thinking forward. *Progress Physical Geography Earth Environment, 46*, 697–715. https://doi.org/10.1177/03091333221097794

Dammalage, T.L. & Jayasinghe, N.T. (2019). Land-use change and its impact on urban flooding: A case study on colombo district flood on May 2016. *EngineeringTechnology Applied Science Research, 9* 3887–3891. https://doi.org/10.48084/etasr.2578

De Silva, M. M. G. T., Weerakoon, S. B., Herath, S., Ratnayake, U. R., & Mahanama, S. (2012). Flood inundation mapping along the lower reach of Kelani River Basin under the impact of climatic change. *Engineer Journal of the Institution of Engineers Sri Lanka, 45*, 23. https://doi.org/10.4038/engineer.v45i2.6938

Department of Census and Statistics, (2012). Census of population and housing of Sri Lanka, 2012 [WWW Document]. URL http://www.statistics.gov.lk/pophousat/cph2011/pages/activities/Reports/District/Colombo/A3.pdf

Du, J., Fang, J., Xu, W., & Shi, P. (2013). Analysis of dry/wet conditions using the standardized precipitation index and its potential usefulness for drought/flood monitoring in Hunan Province China. *Stochastic Environmental Research and Risk Assessment, 27*, 377–387. https://doi.org/10.1007/s00477-012-0589-6

Fauzi, M. F., Abdul Rauf, U. F., Wan Din, W. R., & Hussin, A. G. (2017). Evaluation of wet indices using standard precipitation index: A case study of Terengganu states. *Journal of Physics: Conference Series, 890*, 012159. https://doi.org/10.1088/1742-6596/890/1/012159

Feng, G.-L., Yang, J., Zhi, R., Zhao, J.-H., Gong, Z.-Q., Zheng, Z.-H., Xiong, K.-G., Qiao, S.-B., Yan, Z., Wu, Y.-P., & Sun, G.-Q. (2020). Improved prediction model for flood-season rainfall based on a nonlinear dynamics-statistic combined method. *Chaos, Solitons & Fractals, 140*, 110160. https://doi.org/10.1016/j.chaos.2020.110160

Hänsel, S., Schucknecht, A., & Matschullat, J. (2016). The Modified Rainfall Anomaly Index (mRAI)—is this an alternative to the Standardised Precipitation Index (SPI) in evaluating future extreme precipitation characteristics? *Theoretical and Applied Climatology, 123*, 827–844. https://doi.org/10.1007/s00704-015-1389-y

Hemachandra, E. M. G. P., Dayawansa, N. D. K., & De Silva, R. P. (2021). Application of Standard Precipitation Index (SPI) to assess rainfall variability in a major agricultural area in dry zone of Sri Lanka. *Tropical Agricultural Research, 32*, 276. https://doi.org/10.4038/tar.v32i3.8491

Hewawasam, V., & Matsui, K. (2020). Equitable resilience in flood prone urban areas in Sri Lanka: A case study in Colombo divisional secretariat division. *Global Environmental Change, 62*, 102091. https://doi.org/10.1016/j.gloenvcha.2020.102091

Karavitis, C. A., Alexandris, S., Tsesmelis, D. E., & Athanasopoulos, G. (2011). Application of the Standardized Precipitation Index (SPI) in Greece. *Water, 3*, 787–805. https://doi.org/10.3390/w3030787

Khan, S., Gabriel, H. F., & Rana, T. (2008). Standard precipitation index to track drought and assess impact of rainfall on watertables in irrigation areas. *Irrigation and Drainage Systems, 22*, 159–177. https://doi.org/10.1007/s10795-008-9049-3

Leskens, J. G., Brugnach, M., Hoekstra, A. Y., & Schuurmans, W. (2014). Why are decisions in flood disaster management so poorly supported by information from flood models? *Environmental Modelling and Software, 53*, 53–61. https://doi.org/10.1016/j.envsoft.2013.11.003

Li, R. F., Cheng, L. J., Ding, Y. S., Chen, Y. X., & Khorasani, K. 2013. Spatial and temporal variability analysis in rainfall using standardized precipitation index for the fuhe basin, China. *International Conference on Intelligent Computing for Sustainable Energy and Environment.*

Liang, B., Shi, G., Sun, Z., Babul, H., & Zhou, M. 2023. Evolution trend and hot topic measurement of climate migration research under the influence of climate change. *Frontiers in Ecology and Evolution, 11*. https://doi.org/10.3389/fevo.2023.1118037

Livada, I., & Assimakopoulos, V. D. (2007). Spatial and temporal analysis of drought in Greece using the Standardized Precipitation Index (SPI). *Theoretical and Applied Climatology, 89*, 143–153. https://doi.org/10.1007/s00704-005-0227-z

Lo, S.-W., Wu, J.-H., Lin, F.-P., & Hsu, C.-H. (2015). Visual sensing for urban flood monitoring. *Sensors, 15*, 20006–20029. https://doi.org/10.3390/s150820006

Ma, S., Wang, L.-J., Jiang, J., & Zhao, Y.-G. (2024). Land use/land cover change and soil property variation increased flood risk in the black soil region, China, in the last 40 years. *Environmental Impact Assessment Review, 104*, 107314. https://doi.org/10.1016/j.eiar.2023.107314

McKee, T.B., Doesken, N.J., & Kleist, J. 1993. The relationship of drought frequency and duration to time scales. *Procedings 8th Conference on Applied Climatology, 17*, 179–183.

Merz, B., Blöschl, G., Vorogushyn, S., Dottori, F., Aerts, J. C. J. H., Bates, P., Bertola, M., Kemter, M., Kreibich, H., Lall, U., & Macdonald, E. (2021). Causes, impacts and patterns of disastrous river floods. *Nature Reviews Earth & Environment, 2*, 592–609. https://doi.org/10.1038/s43017-021-00195-3

Nied, M., Pardowitz, T., Nissen, K., Ulbrich, U., Hundecha, Y., & Merz, B. (2014). On the relationship between hydro-meteorological patterns and flood types. *Journal of Hydrology, 519*, 3249–3262. https://doi.org/10.1016/j.jhydrol.2014.09.089

Nikolova, N., & Vassilev, S. (2005). Variability of summer-time precipitation in Danube plain, Bulgaria. *Journal of Geographical Institute "Jovan Cvijic" SASA, 54*, 19–32 WE-Emerging Sources Citation Index (ESCI).

Olanrewaju, C. C., & Reddy, M. (2022). Assessment and prediction of flood hazards using standardized precipitation index—A case study of eThekwini metropolitan area. *Journal of Flood Risk Management, 15*, 1–12. https://doi.org/10.1111/jfr3.12788

Ologunorisa, T.E., & Tersoo, T. 2006. The changing rainfall pattern and its implication for flood frequency in Makurdi, Northern Nigeria. *Journal of Applied Science Environmental Management, 10*. https://doi.org/10.4314/jasem.v10i3.17327

Peng, L., Wang, Y., Yang, L., Garchagen, M., & Deng, X. (2024). A comparative analysis on flood risk assessment and management performances between Beijing and Munich. *Environmental Impact Assessment Review, 104*, 107319. https://doi.org/10.1016/j.eiar.2023.107319

Qin, L.J., Zhang, H., Li, M., & Zong, S.W. (2010). Analysis of drought-flood spatial-temporal characteristics based on Standard Precipitation Index (SPI) in Jilin Province during 1958–2009. Chinese Perspect. Risk Analysic Crises Response.

Rauf, U.F.A., & Zeephongsekul, P. (2011). Modelling rainfall severity and duration in north-eastern victoria using Copulas. In *9th International Congress on Modelling and Simulations 2011*

Raziei, T. (2021). Revisiting the rainfall anomaly index to serve as a simplified standardized precipitation index. *Journal of Hydrology, 602*, 126761. https://doi.org/10.1016/j.jhydrol.2021.126761

Sadiq, A. A., Suleman, M. U., & Mohammed, U. B. (2020). An estimation of rainfall anomaly index and its impact on crop production in Yola and environs. *African Journal Environment Natural Science Research, 3*, 35–53.

Samantha, G. (2018). The impact of natural disasters on Micro, Small and Medium Enterprises (MSMEs): A case study on 2016 flood event in Western Sri Lanka. *Procedia Engineering, 212*, 744–751. https://doi.org/10.1016/j.proeng.2018.01.096

Samarakoon, U., & Abeykoon, W. (2018). Emergence of social cohesion after a disaster. *Procedia Engineering, 212*, 887–893. https://doi.org/10.1016/j.proeng.2018.01.114

Seiler, R. A., Hayes, M., & Bressan, L. (2002). Using the standardized precipitation index for flood risk monitoring. *International Journal of Climatology, 22*, 1365–1376. https://doi.org/10.1002/joc.799

Sun, Q. (2013). Research on quick monitoring and evaluating of flood disaster in Poyang lake area based on remote sensing. *Applied Mechanics and Materials, 303–306*, 740–743. https://doi.org/10.4028/www.scientific.net/AMM.303-306.740

Tingsanchali, T. (2012). Urban flood disaster management. *Procedia Engineering, 32*, 25–37. https://doi.org/10.1016/j.proeng.2012.01.1233

Tirivarombo, S., Osupile, D., & Eliasson, P. (2018). Drought monitoring and analysis: Standardised precipitation evapotranspiration index (SPEI) and standardised precipitation Index (SPI). *Physics and Chemistry Earth Parts a/b/c, 106*, 1–10. https://doi.org/10.1016/j.pce.2018.07.001

Udosen, C. (2012). Rainfall trends in Uyo-Akwa Ibom state and its implication on urban flooding. *JEAS-Journal of Engineering and Applied Science, 7*, 79–85.

UNDRR, 2020. DesInventar Sendai [WWW Document]. Sendai Framew. *Disaster Risk Reduct.* URL https://www.desinventar.net/DesInventar/profiletab.jsp?countrycode=lka

Vicente-Serrano, S. M., & López-Moreno, J. I. (2005). Hydrological response to different time scales of climatological drought: An evaluation of the standardized precipitation index in a mountainous Mediterranean basin. *Hydrology and Earth System Sciences, 9*, 523–533. https://doi.org/10.5194/hess-9-523-2005

Volpi, E. (2019). On return period and probability of failure in hydrology. *Wiley Interdisciplinary Reviews:Water, 6*. https://doi.org/10.1002/wat2.1340

Wang, D., Dong, Z., Jiang, F., Zhu, S., Ling, Z., & Ma, J. (2022). Spatiotemporal variability of drought/flood and its teleconnection with large-scale climate indices based on standard precipitation index: A case study of Taihu Basin China. *Environmental Science Pollution Research, 29*, 50117–50134. https://doi.org/10.1007/s11356-022-19329-1

Weerasinghe, K. M., Gehrels, H., Arambepola, N. M. S. I., Vajja, H. P., Herath, J. M. K., & Atapattu, K. B. (2018). Qualitative flood risk assessment for the Western Province of Sri Lanka. *Procedia Engineering, 212*, 503–510. https://doi.org/10.1016/j.proeng.2018.01.065

WMO. (2012). *Standardized precipitation index user guide.* World Meteorological Organization.

Yuan, W., & Zhou, G. (2004). Comparison between standardized precipitation index and Z-index in China. *Acta Phytoecol Sin, 28*, 523–529.

Zhai, J., Su, B., Krysanova, V., Vetter, T., Gao, C., & Jiang, T. (2010). Spatial variation and trends in PDSI and SPI indices and their relation to streamflow in 10 large regions of China. *Journal of Climate, 23*, 649–663. https://doi.org/10.1175/2009JCLI2968.1

Zhang, Q., Xu, C.-Y., & Zhang, Z. (2009). Observed changes of drought/wetness episodes in the Pearl River basin, China, using the standardized precipitation index and aridity index. *Theoretical and Applied Climatology, 98*, 89–99. https://doi.org/10.1007/s00704-008-0095-4

Zhao, W., Yan, T., Peng, S., Ding, M., & Lai, X. (2020). Drought and flood characteristics evolution in peninsular region of Shandong Province based on standardized precipitation index. *IOP Conference Series Earth Environmental Science, 601*, 012021. https://doi.org/10.1088/1755-1315/601/1/012021

Climate Adaptation in Built Environment: Urban Agriculture and Food Security with Innovative Approaches

Wadduwa Pathirage Thilini Deepashika Perera, Senevirathne Navaratne, and Indira Wickramasinghe

1 Introduction

The living environments made of natural and built environments are called Habitats. The natural environment is the prevalent macro and microclimatic condition at a particular geographic location. The built environment is deliberate, man-made, ordinally derived from local materials with related skills (Aysha & Mani, 2017; Bekchanov et al., 2022; Habash, 2022; Kaklauskas & Gudauskas, 2016), and surrounded by buildings and infrastructural systems. The built environment depends on energy for its operation and maintenance to keep productivity and liveability in the living environment (Aysha & Mani, 2017). Climate change is an alteration in weather patterns and changes in oceans and land surfaces (Nkrumah, 2018). Natural processes (ex, changes in the Sun's radiation or internal variability in the climate system) or human influences (ex, land use or emissions) may trigger climate change (Nkrumah, 2018). Global concerns (ex, climate change) have critically affected food system sustainability and security in recent years (Challinor et al., 2014; Dubbeling & de Zeeuw, 2011; Flörke et al., 2018; Warren et al., 2015; Yan et al., 2022). Hence, ensuring food security is becoming difficult due to the considerable challenges of climate change. The Food and Agriculture Organization of the United Nations (FAO) defined the words "food security" in 1996 at the World Food Summit as 'when all

W. P. T. D. Perera (✉) · S. Navaratne · I. Wickramasinghe
Department of Food Science and Technology, University of Sri Jayewardenepura, Nugegoda, Sri Lanka
e-mail: thilinip@sci.sjp.ac.lk

S. Navaratne
e-mail: sbnava@sci.sjp.ac.lk

I. Wickramasinghe
e-mail: indiraw@sjp.ac.lk

C. Malalgoda et al. (eds.), *Climate Change Adaptation in the Built Environment*,
https://doi.org/10.1007/978-3-031-75826-3_8

people, at all times have physical and economic access to sufficient, safe and nutritious food to meet their dietary needs and food preferences for an active and healthy life'. It is necessary to understand the links connecting food security, food security drivers and climate change to recognize the impacts of climate change on food security (Ziervogel & Ericksen, 2010). Due to the increase in the impact of climate change, it is necessary to establish more sustainable, shorter agricultural production practices that can use resources optimally in the built environment and food supply chains (Maheshwari, 2021; Steenkamp et al., 2021). Urban Agriculture (UA) is an emerging field of research that aims to improve urban resilience to climate change and food security (Dubbeling & de Zeeuw, 2011; Lucertini & Di Giustino, 2021). UA can be described as crops and livestock production within urban areas (ex, towns and cities) (Chatterjee et al., 2020). Previous studies have discussed the topic of UA (Appolloni et al., 2021; Ayambire et al., 2019; Follmann et al., 2021; Pinheiro & Govind, 2020; Yan et al., 2022). However, only a few studies have addressed climate change adaptation and mitigation responses in built environments with UA and food security. Climate changes and rapid urbanization pose challenges to maintaining sustainable agriculture and food security within urban areas while the operational processes of built environments are successfully being carried out. Therefore, it is essential to understand how climate change links with these three main elements: built environment, UA and food security. Hence, this chapter focuses primarily on the impact of climate change on built environment, agriculture, and food security in urban areas, as well as adaptation and mitigation strategies, particularly UA, that minimize those impacts. This study centred on the built environment of cities in which the population is dense. Furthermore, this study suggests innovative approaches to sustainable UA in the built environment.

2 Methodology

Peer-reviewed publications during the past two decades relevant to climate change, built environment, UA and food security were searched in renowned international journals and databases, and information was gathered. Then, the gathered scholarly literature was evaluated. The chapter was organized under six main headings, which are: (1) Characteristics of Buildings/Built Environment, (2) Human influenced Climate Change, (3) Adaptation and Mitigation Responses to Climate Change, (4) Vertical Farming (VF) in Urban Areas, (5) VF and human nutrition, (6) Innovative concept to store leafy vegetables under a self-reliant facility in the Built Environment.

3 Characteristics of Buildings/Built Environment

Building characteristics include geometry concerning the location's solar geometry, material configuration, amenability alteration, cultural performance, etc. (Aysha & Mani, 2017). A Building's primary function is to protect its occupants from elements of nature. As buildings embody life (occupants), their ability to support life (house) decides their adaptability (Aysha & Mani, 2017). In the modern context, a building's basic functional performance includes various aspects such as its functioning regarding structure, acoustics, ability to withstand weathering, lighting, material durability, energy efficiency, thermal comfort, and resilience to hold out against extreme weather conditions such as high winds, rain and snow. All these are adaptation traits that judge the capability of buildings to adapt to local environmental changes, a recent functional requirement anticipated from buildings (Aysha & Mani, 2017). There are three types of buildings: air-conditioned, passive solar, and partially air-conditioned (Aysha & Mani, 2017). In human habitations, buildings are present as a group, and it is called a Cluster.

The various features/aspects of buildings affect the health of their occupants/users. For example, design features of hospitals (ex, better lighting, ventilation and layout) help reduce the stress and fatigue of patients and staff and thus improve overall health (Younger et al., 2008). In addition, neighbourhood design affects mental health, social connectedness, physical activity, chronic disease risk, respiratory and cardiac health, and injury risk (Younger et al., 2008). However, several current community designs in practice have adversely contributed to global climate change (Younger et al., 2008). Buildings are considered the largest consumers of resources in the world and comprise the largest ecological and energy footprints (Aysha & Mani, 2017; Mani & Reddy, 2012). Globally, more than 50% of global emissions are from buildings (Aysha & Mani, 2017). Buildings have been discovered as the major emitters that emit other non-carbon dioxide (non-CO_2) GreenHouse Gases (GHGs), which include halocarbons, hydrochlorofluorocarbons (HCFCs), and chlorofluorocarbons (CFCs). Because these non-CO_2 GHGs are used for building cooling, fire suppression, refrigeration, and halocarbons in insulation materials (Oguntona & Aigbavboa, 2016). Therefore, the way of constructing the built environment is crucial. Many sectors are involved in developing the built environment: urban planning, engineering, architecture, transportation design, land conservation, local and regional governments, and environmental psychology. As the building characteristics and built environment influence health, public health professionals must also be included in the decision-making process regarding land use and transportation (Younger et al., 2008).

4 Human Influenced Climate Change

The increment in GHGs in the atmosphere results in global warming (Etkin & Ho, 2007; Oguntona & Aigbavboa, 2016), and CO_2 is defined as the primary GHG contributing to the recent changes in climate. Its emission occurs due to human activities such as industrial and agricultural activities (Mattos et al., 2014; Oguntona & Aigbavboa, 2016). The CO_2 levels in the atmosphere have rapidly increased than that over the past million years (Mattos et al., 2014). Naturally, atmospheric CO_2 is removed by the photosynthesis of plants and by absorbing CO_2 by oceans to reduce pH levels (Cotgrave & Riley, 2013; Oguntona & Aigbavboa, 2016). Among the various challenges being faced by humankind around the world, climate change and its impact pose more threats.

4.1 Impact on Environment, Human, and Built Environment

Rowlands et al. (2012) have presented that the global average temperature will increase by 1.4-3 °C by 2050. This temperature increment can be anticipated to lead to changes in rainfall patterns and extreme weather conditions more frequently (Maheshwari, 2021; Mattos et al., 2014). In other words, climate disruption impacts the physical state of the planet (Mattos et al., 2014). Climate impacts include sea level rise, intense droughts and food shortages, growing deserts, more frequent floods from heavy rainfall, and changes in storm paths and frequency (Al-Humaiqani & Al-Ghamdi, 2022; Arain, 2011; Misra, 2014; Oguntona & Aigbavboa, 2016; Younger et al., 2008).

Urbanization factors cause the urban heat island effect. These factors include increment in local anthropogenic heat output caused by energy consumption, changes in surface materials (ex, vegetation and free water areas decrease while the proportion of non-evaporative surfaces such as concrete and asphalt increase), etc. (Shimoda, 2003). The excessive air temperature in the daytime of the summer caused by the urban heat island scenario leads to thermal stress (Deng et al., 2023; Shimoda, 2003). Heat stroke is one of the most common acute illnesses caused by thermal stress, and in such an outdoor situation, a significant change in lifestyle and work can be expected (Shimoda, 2003). Therefore, heat stress, compromised air quality, and altered patterns of infectious diseases are some health consequences of climate change (Harvison et al., 2011; Mattos et al., 2014).

The increment in air temperature is accompanied by an increased number of room air-conditioners, air-conditioned space, operating hours, cooling load, peak electricity demand, and a decrease in the Coefficient of Performance (COP) of a refrigerator (Shimoda, 2003). The number of household air-conditioners increases for two reasons: the urban heat island effect/global warming and general improvement in living standards (Shimoda, 2003). Energy consumption for generating hot water decreases with the increment of temperature in tap water, while energy consumption for cooling increases with the increase in air temperature since the cooling load increases. Furthermore, increased energy consumption in summer results in increased air pollutant emissions. When combined with "dust dome" formation, it can potentially increase photochemically and non-photochemically active air pollutant concentration in urban areas (Shimoda, 2003).

In addition, changes in rainfall and snow will significantly affect hydropower generation (Shimoda, 2003). As seawater is used as a coolant in thermal and nuclear power plants, the efficiency of power generation operation is estimated to be impacted by the changes in seawater temperature (Shimoda, 2003). As global warming increases thunder clouds, developing an improved lighting protection technique associated with renewable energy infrastructure and information technology may be required (Shimoda, 2003). Hence, there are different views on the total energy consumption in the building sector, whether it will decrease or increase due to global warming.

4.2 Impact on Agriculture and Food Security

Climate change has become the most remarkable global environmental issue. Therefore, immediate considerations are required to evaluate the impact of climate change on agriculture (Mattos et al., 2014) and food security. Globally, agricultural production has been reduced with the rise in population and the demand for food (Misra, 2014). According to studies, the agricultural yield may be severely affected during the next hundred years due to unusual rates of climatic changes (Jarvis et al., 2010; Misra, 2014; Thornton et al., 2011). Leafy vegetable cultivation is traditionally carried out in open fields in developing countries. Moreover, it can be seen that Asian paddy rice fields are diversified and transformed to upland vegetable production (Kleinhenz et al., 1996). Also, it was reported in China that many paddy fields are being converted for growing vegetables (Qin et al., 2016). For instance, small-scale leafy vegetable cultivators in Sri Lanka mostly use lands left after paddy cultivation or near paddy fields as it gives advantages, such as ease of providing water, holding moisture in the soil, applying fertilizers, controlling weeds, and receiving enough sunlight. However, uncertainties are involved in this practice due to climate change, which can expose plants to different types of disasters, such as droughts, landslides, and floods. Thus, those can crucially affect food security by remarkably lowering production yield.

In summer, tap water consumption increases due to increased air temperature (Shimoda, 2003). With the global temperature increase, there will be a considerable reduction in agricultural yield and freshwater resources at the end of the twenty-first century (Misra, 2014). Factors such as the type of plant tissue, crop growth stage and nature of temperature stress affect the crop's response to temperature stress (Mattos et al., 2014; Porter & Semenov, 2005). Exposure of vegetable crops to high temperatures can lead to physiological disorders and other related external and internal symptoms. During the development of crops, high temperature affects respiration, plant hormone levels, water relation, photosynthesis, membrane stability and primary/secondary metabolites (Magan et al., 2011). Temperature increases mainly affect the loss of leafy vegetables such as spinach and lettuce (Mattos et al., 2014). Adverse effects of climate change on the production of lettuce and Brassica species have been reported (McKeown et al., 2006; Warland et al., 2006). Besides, the harvesting time of vegetable crops varies depending on climate conditions, exposure to direct sunlight, water regime, temperature management, pest control, cultivar, maturity index, and cultural practice (Mattos et al., 2014). Therefore, changes in climate conditions also affect the harvesting process of food crops. Water and food shortages are massive global problems, severely affecting the semi-arid and arid regions/countries. An acute decline in crop yield in semi-arid and arid areas in the world has led to food scarcity and a diverse increase in food inflation (Misra, 2014). Moreover, there will be changes in the movement of crop diseases and pest attacks with climate changes such as temperature and CO_2 rise and precipitation patterns. In addition, ozone formation happens under solar irradiation and high temperatures, usually through the summer season (Mattos et al., 2014). Ozone exposure causes increased leaf yellowing in leafy vegetables, declining biomass production, inhibition of photosynthetic rate, and eventually impaired postharvest qualities concerning colour, overall appearance, and flavour (Mattos et al., 2014).

5 Adaptation and Mitigation Responses to Climate Change

Adaptation and mitigation are responses to climate-changing issues (Oguntona & Aigbavboa, 2016). Mitigation aims at the source of climate change, where adaptation deals with its consequences (Oguntona & Aigbavboa, 2016; Schipper, 2006). Mitigation focuses on declining energy consumption in buildings and limiting GHG emissions. Adaptation mitigates the consequences of natural threats, taking root in prevention and protection concepts (Oguntona & Aigbavboa, 2016). Theoretically, more mitigation will result in less adaptation, and vice versa (Huq & Grubb, 2003; Oguntona & Aigbavboa, 2016). Adaptation to global climate change is one of the crucial processes to ensure the survival of human beings (Mattos et al., 2014; Rockström et al., 2009). Rapid growth has occurred in developing climate change adaptation responses worldwide during the past two decades (Arnott et al., 2016; Moser, 2011; Owen, 2020). The increased understanding of global warming and the documentation of its recognizable impacts have directed the development of climate change adaptation responses worldwide (Owen, 2020).

Several standards, guidebooks, manuals and other references are available from different governments and organizations to assist in adapting to projected climate change impacts and developing adaptive responses (Al-Humaiqani & Al-Ghamdi, 2022; Tyler & Moench, 2012). The effectiveness can be analyzed by routinely evaluating the adaptation processes (Adger et al., 2005; de Coninck et al., 2018; Owen, 2020). Several practitioners and scholars have constructed frameworks for guiding the evaluation design in adaptation (Eriksen et al., 2011; Owen, 2020; Vogel et al., 2017). Though it is necessary to assess the outcome and impact of the adaptation, it is often missing (Owen, 2020). Owen (2020) has also mentioned that evidence is there to say that adaptation activities to climate change frequently do not work as intended. Further, he has mentioned that insufficient resources, conflicting or competing priorities for action, prohibitive policies, and uncertainty about future changes are barriers that consistently hinder adaptation efforts (Biesbroek et al., 2013; Moser & Ekstrom, 2010; Owen, 2020).

5.1 Climate Adaptation and Energy Saving in the Built Environment

In order to address building adaptation, it requires addressing both physical (geometry, size and height of the building, orientation, building envelope, internal layout, durability of the structure and services) and functional (thermal, lighting, acoustical needs, safety, occupant comfort, fire protection, and security) attributes (Aysha & Mani, 2017). Deng et al. (2023) and Brager and de Dear (1998) have reported that an occupant undergoes three components of adaptation, which are physiological (physiological responses to thermal factors by acclimatisation or genetic adaptation), behavioural (personal, environmental/technological, and cultural adjustments by occupants), and psychological (change in perception of sensory information) adaptations, to adjust with the indoor environment. The comfort requirements of occupants differ from one person to the other. For example, occupants in Asian countries have entirely different requirements than those in Western countries. In naturally ventilated buildings, effective and timely use of balconies, doors, windows, and curtains are the adaptation mechanisms (Table 1). Occupants in naturally ventilated buildings have more relaxation and tolerance towards temperature swings, whereas those living in air-conditioned buildings require a more uniform-cool environment (Aysha & Mani, 2017; Brager & de Dear, 1998). A study conducted in India (Chennai) revealed that, during hot and humid climate/thermal discomfort conditions, greater use of windows and balconies as one of the adaptive practices can help occupants to acclimatise up to a temperature of 31.8 °C even though they prefer 27 °C (Aysha & Mani, 2017; Rajasekar & Ramachandraiah, 2010). Although open windows provide comfort at lower outdoor temperatures, they cause discomfort/uneasiness at higher temperatures (Aysha & Mani, 2017). Therefore, architects/designers have to learn to use the natural resources around the building site effectively to provide cold and heat as per the needs of occupants, and they should also construct methods to build passively and hastily (Roaf et al., 2009).

The dependence on mechanical cooling to supply the increased cooling demand strengthens the urban heat island effect by increasing anthropogenic heat release and global warming by increasing CO_2 emissions (Shimoda, 2003). Therefore, passive cooling techniques that maintain thermal comfort in indoor environments (without

Table 1 Adaptive response strategies in a building's unique physical manifestations and comfort requirements (to resist heat gain and promote heat loss) (Aysha & Mani, 2017; Nayak & Prajapati, 2006)

Climate classification	Unique physical manifestations	Comfort requirements	
		To resist heat gain	To promote heat loss
Hot-dry	Insulation of building envelope; airlocks/ lobbies/ verandahs/ balconies/ overhangs trees; a massive structure	Decrease exposed surface area, air exchange rate; increase thermal resistance, surface reflectivity; thermal capacity; shading; and buffer spaces	Increase air exchange rate and humidity levels; ventilation of appliances
Warm-humid	Reflective roof surface; roof and wall insulation; verandahs/ balconies; desiccant cooling/ dehumidifiers	Decrease exposed surface area; increase thermal resistance, buffer spaces, surface reflectivity, and shading	Increase air exchange rate and decrease humidity levels; ventilation of appliances
Moderate	Roof/west and east wall insulation; fins/ trees/ overhangs for east and west wall	Decrease exposed surface area; increase thermal resistance, shading, and surface reflectivity	Increase air exchange rate; ventilation of appliances
Cold-cloudy/ cold-sunny	Roof and wall insulation; sun spaces/greenhouses/ thicker walls; double glazing; darker colors; airlocks/lobbies; Trombe walls; weather stripping	Decrease exposed surface area, air exchange rate; increase thermal resistance, surface absorptivity, thermal capacity, and buffer spaces	Utilize heat from appliances; trap heat; reduce shading
Composite	Thicker walls; roof and wall insulation; exhausts; dehumidifiers/ desiccant cooling	Decrease exposed surface area, air exchange rate; increase thermal resistance, surface reflectivity, thermal capacity, shading, and buffer spaces	Increase air exchange rate; ventilation of appliances; increase humidity levels during summer; decrease humidity levels during monsoon

air-conditioners) or reduce the cooling load are becoming increasingly important (Shimoda, 2003). They include using reflective surfaces, the creation of air tracks for natural ventilation, controlling solar radiation by vegetation, earth tubes, eaves and blinds, thermally massive buildings, and rooftops/ terrace planting (Table 1) (Gill et al., 2007; Shimoda, 2003). Nevertheless, night purging becomes less effective as the urban heat island effect increases the night-time temperature more than in the daytime (Shimoda, 2003). When the outdoor Relative Humidity (RH) value decreases, it might be necessary to do the planting and make fountains to humidify the spaces outside the building from the perspective of health, static electricity and fire prevention (Shimoda, 2003). In some cases, architectural and structural features can negatively affect the thermal adaptation of occupants. For instance, defects in following building regulations, wrong positioning of thermostats, using curtains as they block vertically pivoted windows, wall-mounted furniture as blocking the operation of windows, and using curtain glazing on external walls, etc. (Aysha & Mani, 2017; Indraganti et al., 2014).

The building sector should enforce the usage of renewable energy sources, and a sustainable building should be capable of upholding consistent performance with less material and energy input (Aysha & Mani, 2017). For example, building envelope insulation in a temperature zone can reduce carbon emissions and energy loss while allowing pleasant comfort/interior. Solar collectors can be installed for water heating as they can reduce energy consumption (Aysha & Mani, 2017; Wilkinson et al., 2009). Wind power can be used in regions with enough wind speed. However, variations in wind patterns can acutely affect the output. With the increased urbanization, many cities have implemented low energy/zero carbon initiatives. For instance, Freiburg im Breisgau in Germany has given attention to energy efficiency and renewable energy; Dongtan City in China has executed a zero-emission vision; Wallington City in South London has adopted energy-efficient building designs (Aysha & Mani, 2017). The measures include increased insulation and heat reduction, use of energy-efficient appliances, solar panel installation, orienting buildings to face North–South direction and creating green roofs, gardens and open space with maximum vegetation (Aysha & Mani, 2017). In addition to traditional air conditioning, Helsinki in Finland and Stockholm in Sweden have adopted centralized district cooling systems to reduce the peak power of electricity during summer periods and increase efficiency (5–10 times) (Aysha & Mani, 2017).

The built environment can reduce its effect on long-term climate change by using enhanced design and sustainable materials to influence climatic behaviour and buildings' sustainability significantly (Beddoes & Booth, 2012; Oguntona & Aigbavboa, 2016). Eco-architecture is an effort to adapt to the global environmental problems caused indirectly or directly by the built environment, such as GHG emissions, exhaustion of natural resources, etc. It includes optimized use of material and energy, use of renewable energy, minimal use of water cycles and constructing buildings in terms of their landscape, natural resources, environment, history and society (Aysha & Mani, 2017). A building's ability to emulate natural processes and function like an ecosystem (in its creation/design, use, and maintenance) has been reported to have the potential to be part of a regenerative and resilient built environment (Oguntona & Aigbavboa, 2016; Zari & Storey, 2007). For example, the Lilypad or floating Ecopolis, is a model for a completely self-sufficient floating city designed by architect Vincent Callebaut (Oguntona & Aigbavboa, 2016; Rao, 2014). By this, he intended to provide a solution to the issue of rising seawater levels due to global warming. Suppose the buildings are located in coastal areas (ex: Japan). Then, it is essential to consider subterranean water tables as rising can occur due to global warming. Hence, subterranean water table control will be important in flood protection measures in the short term. Furthermore, strengthening the foundations and buildup of artificially rising ground levels in risk areas will be necessary for long-term protection measures (Shimoda, 2003).

5.2 UA and Food Security in Urban Areas

Developing an effective food security plan and its implementation and maintenance/ upgrading is essential in the built environment. According to the United Nations Report in 2015, the world's population by the year 2050 will be 9.7 billion (Chatterjee et al., 2020; Maheshwari, 2021). The dearth of agricultural lands occurs with the rapid growth of the population worldwide. This scenario increases the demand for more food and land to grow more. Moreover, most crop failures worldwide are related to excess or lack of rainfall (Misra, 2014). For a better agricultural yield, precise climate forecasting can help in pre/post-decision-making processes and reduce the risk of crop failure (Misra, 2014). However, the forecast should be specific and in time to be valuable. Hence, early warning systems and effective development programs may be required to increase resilience in urban areas as climate change increases the risk of Global food insecurity (Maheshwari, 2021).

UA is the process of farming, processing and distributing perishable products such as vegetables and animal products in urban areas (Anwar et al., 2023; Yan et al., 2022; Zulfiqar et al., 2021). UA contributes in various ways (ex, economically, socially, and environmentally) to the sustainability of cities (Chatterjee et al., 2020). Nowadays, UA is in demand to face challenges arising from rapid urbanization. The advantages associated with UA can be mainly identified as fighting environmental challenges, enhancing food nutrition and security, alleviating poverty, and adequately using land (Chatterjee et al., 2020). UA motivates the re-usage of organic waste as it reduces the urban energy footprint. Furthermore, UA improves the city environment and inspires cities to become more resilient by diversifying income opportunities for poor and urban food sources, keeping open spaces covered with the greeneries and making microclimate worth living, forbidding building construction in risky areas, producing fresh commodities which decrease GHG emission and using limited energy during farm-to-plate process in developed countries, and reusing wastewater productively (ex, recycling urban wastewater for fertilization/irrigation of crops) helps to fight freshwater crisis and save water bodies and canals from being polluted (Chatterjee et al., 2020). Unconventional water holds a major role in handling agricultural activities and water resources in semi-arid and arid areas (Misra, 2014). Also, wastewater reuse and recycling can help minimize the impact of climate change on water resources and crop yield (Misra, 2014).

It is estimated that 60% of Earth's population will inhabit cities by 2030. Thus, poverty will move from villages to metro cities by then (Chatterjee et al., 2020). UA enhances the nourishment of the poor class as well as food security. It is reported that families engaged with UA can have a variety of diets (Chatterjee et al., 2020), and these people consume more greens and herbs than others. Hence, it helps for a better mitigation of diseases as they consume medicinal plants with better nutritional and medicinal properties. Rapid urbanization in developing countries has caused population growth in slum areas and, thus, difficulty in creating sufficient employment opportunities. Hence, UA can meet the employment requirements to some extent in the aforementioned cities (Chatterjee et al., 2020). Engaging in UA causes more physical health due to the exercise and enhances their self-esteem as they don't have to depend on food aid and gifts (Chatterjee et al., 2020). Moreover, UA ensures food security during wars and natural calamities.

A major challenge for food production in urban areas, especially in population-dense areas, is the availability of land and access (Chatterjee et al., 2020). Scientists conceptualized the "edible city" and introduced the Continuous Productive Urban Landscape (CPUL) concept, recommending a rational introduction to interlinked productive landscapes in cities as a critical component in sustainable urban infrastructure (Chatterjee et al., 2020). Scientists and practitioners aim to build large-scale food production in and on buildings in high-density urban areas. Low-space or no-space agricultural technologies guide space-confined growth in such areas (Chatterjee et al., 2020). Besides the various advantages, some disadvantages of UA are linked with potential health risks. The improper livestock management and reuse of untreated irrigation water from urban streams lead to health risks. The intensive use of pesticides, fertilizers, and fungicides in UA may cause the presence of agro-chemical residues in crops/groundwater (Chatterjee et al., 2020). This risk mainly presents in commercial urban farming areas. However, proper management procedures and health risk reduction measures are required to reduce the rate of these associated risks (Chatterjee et al., 2020). Therefore, sustainable, innovative and integrated approaches should be developed for the built environment and farming activities (Maheshwari, 2021). Hence, farmers/vegetable growers and entrepreneurs nowadays focus on exercising sustainable vegetable production to overcome climate changes, dearth of agricultural lands, increased demand, and difficulty in traditional cultivation methods. The solution is called VF, a UA technique for growing crops (Chatterjee et al., 2020).

6 VF in Urban Areas

Adapting advanced techniques makes farming more productive and easier. Increased vegetable production for a country is required with the increment in population, per capita income, and tourism and export industry expansion (Weerakkody & Mawalagedera, 2020). The input availability and the natural environment may not provide the optimum condition for a given crop (Panwar et al., 2011). Because the maximum response in crops depends on the extent to which the environmental parameters are balanced. Environmental parameters such as RH, soil moisture content, air and soil temperature, sunlight, and air composition (ex, CO_2 concentration) cannot be controlled in open-field agricultural farming (Panwar et al., 2011). Therefore, indoor farming (soilless) is becoming favoured among plant factories to provide fresh, nutritious, and safe leafy greens and herbs. The VF system was first developed by Dickson

Despommier in 1999, based on conventional hydroponic and aeroponic techniques to produce farm yields at a consistent rate (Maheshwari, 2021). Afterwards, VF became a farming technology that involved large-scale agricultural production. Though VF is a newer concept in Sri Lanka and India, it is well-adopted in Korea, Japan, and the Netherlands (Maheshwari, 2021) as a solution to many challenges. In VF, plants are grown in vertically arranged layers that may reach several feet, meters, or even several stories high. VF facilitates fast crop growth and production by managing nutrient solutions and maintaining environmental conditions with the basis of hydroponics technology (Chatterjee et al., 2020). Hence, VF is much less vulnerable to changes in climate and can provide year-round production. Moreover, this guarantees growers a consistent income and consumers a steady flow of agricultural goods.

This farming technique is rapidly growing and getting more and more attention from entrepreneurs in many cities due to its beneficial role in agriculture (Chatterjee et al., 2020). Hence, VF, as a UA technique, can reduce the cost of transportation as it is adjacent to buyers. Moreover, the production and the growing condition of herb plants can be enhanced by adjusting growing conditions (ex, RH, temperature, lighting, etc.) (Chatterjee et al., 2020). As the growing conditions are controlled and wastewater is recycled in indoor farming, it requires less water than traditional outdoor farming. VF's other advantages over traditional cultivation are reducing crop failures, farm inputs, and restoring farmland (Chatterjee et al., 2020). Hence, these facts have enabled people/scientists to execute VF technology in large-scale production. There are three main types of VF techniques. Those are hydroponics, aquaponics, and aeroponics (Chatterjee et al., 2020).

6.1 Hydroponics

It is the uppermost growing technique used in VF. Hydroponics is a soilless growing technique for plants or vegetables, where a mineral nutrient solution is mixed with water (Ruengittinun et al., 2017; Son et al., 2020). However, root aeration is critical for successful cultivation in soilless culture (Baudoin et al., 2013). Hydroponic cultivation gives numerous advantages, including precise control of nutrients, reduction of fertilizer and water consumption, achieving high yield, absence of soil-borne pathogens, reducing pollution of water bodies by phosphates and nitrates, etc. (Baudoin et al., 2013). However, hydroponic cultivation requires technical skills and has high installation costs. Most fertilizers in nutrient solution preparation are highly soluble inorganic salts with some inorganic acids. For example, Albert's solution, which is used in hydroponic cultivation, contains calcium (Ca), magnesium (Mg), zinc (Zn), iron (Fe) like micronutrients, soluble potash (K_20) and phosphoric acid (P_2O_5). Therefore, it is necessary to control the parameters of this nutrient solution (Ruengittinun et al., 2017). Those parameters include concentration or pH value and Electrical Conductivity (EC). The rate of mineral uptake in crops is affected by environmental factors such as light, RH and temperature, and it varies considerably during the growing period, especially in long-cycle crops (Baudoin et al., 2013). The properties of the growing medium (concentration of nutrients and other elements, pH, moisture, total salinity) also influence nutrient uptake.

In hydroponic cultivation, the optimal pH of 5.5 to 6.5 is maintained in the root zone of most of the crop species. However, pH between 5.0–5.5 and 6.5–7.0 may not be problematic for most crops. Most plants show growth restrictions when exposed to pH > 7 or < 5 (Baudoin et al., 2013). EC measures the total salt concentration in a substrate or nutrient solution. EC of nutrient solution is an important parameter in soilless culture. If the EC is too low in the nutrient solution, it may be due to the inadequate supply of some nutrients. Likewise, if the EC is too high, it exposes plants to salinity (Baudoin et al., 2013). However, the plant's yield response to the nutrient solution's EC may vary widely among species.

Hydroponic growing systems play an essential role in indoor farming plant production, such as in factories equipped with artificial lighting (Son et al., 2020). Among different hydroponic growing techniques, Nutrient Film Technique (NFT) and Deep Flow Technique (DFT) are used commercially using re-circulated nutrient solutions (Son et al., 2020). In addition, Deep Water Culture (DWC) and float hydroponics are also included in hydroponic methods (Baudoin et al., 2013). However, the main hydroponic systems used for leafy vegetable growing are the NFT and DFT systems (Son et al., 2020). In NFT systems, the nutrient solution flows through channels on a 1.2–3.0% slope (Baudoin et al., 2013). Therefore, the nutrient solution flows as a very thin layer (film) in channels wherein the plant's bare roots lie, thus keeping the roots completely wet. However, in DFT, the nutrient solution continuously flows at a 50–150 mm depth (Baudoin et al., 2013). Nowadays, Light-Emitting Diodes (LEDs) technology and controlled-environment agriculture have been linked irrevocably (Bantis et al., 2018). Fruit and vegetable production in controlled environments under LEDs have been shown to enhance crop yield while improving their nutritional properties (D'Souza et al., 2015; Morrow, 2008; Perera et al., 2022a, 2022b; Yeh & Chung, 2009). As LED is reported to be a great light source that affects crops positively, it can be used to precisely manipulate a crop's lighting regime to enhance crop quality and yield (Jones, 2018; Lee et al., 2014).

Urban hydroponics is not a recently invented technique. In the past, the Floating Gardens of Aztecs beautified the cities for an extended period. There, fruits and vegetables were cultivated. Currently, urban hydroponics is used by modern cities for physical and psychological relaxation (Chatterjee et al., 2020). It also improves the urban environment. For example, in Bangkok, the railway overpasses and concrete roads are covered with ornamentals grown in hydroponic systems. Also, the in-house air quality can be improved by constructing hydroponic systems on interior walls. For instance, indoor hydroponics is used to decorate commercial centres for improved indoor air quality (Chatterjee et al., 2020). Hydroponics decreases the temperature and increases the humidity in arid climatic areas. Hence, it prevents global warming to a certain extent, as it helps to reduce the overall discharge of CO_2 (Chatterjee et al., 2020). In addition, VF on the top of traditional buildings performs as a large heat sink. Hydroponic systems can thermoregulate buildings, trapping winter heat and cooling summer buildings (Chatterjee et al., 2020). As the space is limited in cities, hydroponic gardens are generally constructed vertically.

6.2 *Aquaponics*

One step forward from the hydroponic system is called aquaponics. Therein, the plants and fish are combined in the same ecosystem. In other words, it combines hydroponics with aquaculture. The nutrient-rich aquaculture water feeds the plants in vertical farms. Then, plants filter and purify the fed water and recycle it into the aquaculture tank. A Canadian scientist, Savidov, has described that, most likely, the organic components in aquaponic systems make trace elements needed for proper growth of plants readily available to plants, and thus recirculating aquaculture water decreases root diseases in growing crops and resulting increased crop yield compared to conventional hydroponics (Chatterjee et al., 2020). As no fertilizers and pesticides are used in aquaponic systems, the vegetables grown in these systems can easily be qualified for organic product certification (Chatterjee et al., 2020). Some scientists are expected to create "sky farming", which is the construction of VF in skyscrapers (Chatterjee et al., 2020). It may also embody the aquaponics in these buildings to establish a better source of fresh fish. Therefore, aquaponics systems are advantageous as they are easier and cost-effective, no additional expenses are needed to clean fish tanks, plants are grown organically, no pesticides harm fish, and diseases can be reduced (Chatterjee et al., 2020). Thus, aquaponics makes itself a very suitable urban farming technology.

6.3 *Aeroponics*

The National Aeronautics and Space Administration (NASA) was the first to develop this innovative indoor growing technique (Chatterjee et al., 2020). Aeroponics is still in the growth phase of the world of VF. In this technique, plant roots are suspended in air and then sprayed with nutrient-dense mist. Here, up to 90% less water is used than other effective hydroponic systems (Chatterjee et al., 2020). Moreover, plants grown in aeroponics systems absorb more vitamins and minerals and develop into more nutritious and healthier plants (Chatterjee et al., 2020). Therefore, the aeroponic system is an efficient plant-growing technique in the VF world.

7 VF and Human Nutrition

Leafy greens (ex, salads and leafy vegetables), herbs, and microgreens are VF's most commonly used crops. When selecting crops for growing in a VF system, few factors can be considered for successful cultivation. Those are the crop type, size at the harvesting stage, how long it takes to reach the harvesting stage, the quantity needed to grow to avoid failures and losses, and the profit that can be earned from a harvested batch. Regarding leafy greens, herbs, and microgreens, they grow quickly and hold an increased demand for their fresh produce. Leafy vegetables can be grouped as dark green and pale green (ex, lettuce and cabbage) (Suraiya & Mahendran, 2009), and they are enriched in carotene, vitamins (ex, vitamin C, folic acid, and riboflavin), minerals [Ca, Fe, and Phosphorous (P)], and crude fibre (Natesh et al., 2017; Sheela et al., 2004; Suraiya & Mahendran, 2009). Metal ions are essential for humankind's well-being and normal functioning as they maintain protein structures and serve as cofactors for enzymatic reactions (Natesh et al., 2017). Fe deficiency in children and women causes the development of anaemia. Zn deficiency causes impaired immune and gastrointestinal functions. The α-linolenic acid (precursor of omega-3 fatty acid) is found in green leafy vegetables (Natesh et al., 2017). Omega-3 fatty acids are known to be important in normal growth and development. It plays a critical role in preventing and treating diabetes, hypertension, cancer, arthritis, coronary artery diseases, and other autoimmune and inflammatory disorders (Natesh et al., 2017). Therefore, including these non-nutrient and nutrient molecules in the human diet can reduce the risk of chronic diseases (ex, obesity, cardiovascular diseases, diabetes, and certain cancers) (Dias, 2012; Perera et al., 2022a, 2022b; Ülger et al., 2018). In dietary guidelines for Americans, taking five servings of vegetables per day has been recommended based on 2000 Cal intake (Natesh et al., 2017). Also, one of these five servings should be from green leafy vegetables. Thus, leafy green vegetables are necessary for a healthy diet and contain low calories (Enloe, 2018). On that account, VF, as a UA technique, improves the health status of people living in urban areas while ensuring food security. Furthermore, though leafy greens are currently widespread in VF systems, there will be a variety of crops suitable for growing in VF systems in the near future.

In addition to the nutritional benefits, VF gives other benefits as well. Climate change adversely affects the overall function of the built environment. Besides, rapid urbanization causes increased food insecurity, more consumption of resources and energy in the built environment, water and air pollution, etc. As climate changes and rapid urbanization limit the resources available in the built environment, VF in urban areas is suitable not only for improving UA but also for reducing the further depletion and waste of energy and input resources. Most importantly, VF must be operated at maximum efficiency to meet sustainability goals (Delorme & Santini, 2022). As mentioned, VF can provide optimum crop growth conditions with reduced fertilizer and water consumption and year-round production regardless of extreme climate changes. As VF reduces the pollution of water bodies by phosphates and nitrates, it decreases the cost of water treatment. As mentioned in Sect. 6.1, establishing indoor VF on walls can improve the air quality, decrease air temperature and increase indoor RH. On top of traditional buildings, VF can perform as a large heat sink. Moreover, as the indoor VF gives a pleasant view, it may indirectly cause better health in people (occupants) who live in the buildings. Therefore, VF as a UA technique can withstand challenges caused by climate changes and rapid urbanization to maintain sustainable agriculture and food security in urban areas.

8 Innovative Concept to Store Leafy Vegetables Under a Self-reliant Facility in the Built Environment

A storage condition with reduced temperature and high RH (95%) is most suitable for extending the postharvest life of leafy vegetables. Also, postharvest illumination (especially with LEDs) has gained more interest due to its advantages, such as being a highly efficient, residue-free, and non-chemical preservation technique (Perera et al., 2022a, 2022b). A shelf-life of fresh produce is the time until the commodity is no longer acceptable for selling to consumers. As leafy vegetables have high respiration and transpiration rates, their shelf-life decreases unless they are stored under proper storage conditions after harvesting. The postharvest life of leafy vegetables under normal ambient conditions ranges from around 2 days (Garande et al., 2019). However, the postharvest loss of leafy vegetables can be minimized if the harvesting is performed when the customers need it.

Hence, as an innovative concept, the space/area/cabin with controlled conditions (ex, temperature, RH, light) can be developed in buildings to store hydroponically grown leafy vegetables and herbs as it is (without separating from stems) under biologically optimum conditions to extend their shelf-life as well as to reduce postharvest loss and waste. Hence, this concept can contribute to ensuring food security and providing people with enough vegetables. Therein, the nutrients, light and growing conditions can be provided to hydroponic plants artificially, which provides plants with a "self-reliant facility". Here, LEDs can be used for illumination as they are mostly used in VF/hydroponic cultivation. LEDs generally have a long-life expectancy with small mass and volume, minimum heating with relatively cool surfaces, and high energy conversion efficiency (Kokalj et al., 2016; Perera et al., 2022a, 2022b). Most importantly, the spectral composition of light can be controlled by using LEDs (D'Souza et al., 2015; Perera et al., 2022a, 2022b). LED illumination regulates physiological, developmental, anatomical, morphological, metabolic, and photosynthetic parameters in horticulture species (Bantis et al., 2018). It is necessary to understand the spectral combinations and cultivation conditions to produce high-quality horticulture commodities because the lighting requirement among plant species is different in terms of light quality and quantity (D'Souza et al., 2015). The blue (430–450 nm) and/or red (650–670 nm) LED illumination particularly contributes to photosynthesis in plants and thus improves nutrition quality in vegetables (Kim et al., 2016; Olle & Viršile, 2013). With further research, this concept can be made feasible for various crops of leafy vegetables and herbs on a large scale.

9 Conclusion

Climate changes occur due to natural processes or human influences interlinked with the built environment, agriculture and food security. The rise in global average temperature causes changes in rainfall patterns and extreme weather conditions, ozone formation, adverse health effects, increased water and electricity consumption, changes in the movement of crop diseases and pest attacks, declining crop yield, and food scarcity. Adaptive strategies prepare the built environment to withstand the adverse effects of climate change and help to create a comfortable living environment for its occupants. Building's functioning regarding structure, acoustics, ability to withstand weathering, lighting, material durability, energy efficiency, thermal comfort, and resilience to hold out against extreme weather conditions such as high winds, rain and snow are adaptation traits that judge the capability of buildings to adapt to local environmental changes. Passive cooling techniques (ex, the use of reflective surfaces, the creation of air tracks for natural ventilation, control of solar radiation by vegetation, earth tubes, eaves and blinds, thermally massive buildings, and rooftops/ terrace planting) are environmentally friendly to use to maintain thermal comfort in indoor environments than the mechanical cooling which strengthens the urban heat island effect. The building sector should enforce the usage of renewable energy sources, and more attention has to be given to Eco-architecture as it includes optimized use of material and energy, use of renewable energy, minimal use of water cycles and constructing buildings in terms of their landscape, natural resources, environment, history and society. UA, which produces crops and livestock within urban areas, is an effective strategy to mitigate the impacts of climate change. UA supports the sustainability (ex, economically, socially, and environmentally) of built environments as it contributes to food security, livelihood improvement, reduced air pollution, resource conservation, and urban beautification. However, there are some disadvantages of UA, namely, potential health risks due to improper livestock management, reuse of untreated irrigation water from urban streams, and presence of agrochemical residues in crops/groundwater due to intensive use of pesticides, fertilizers, and fungicides. VF is the most suitable and beneficial UA technique, and it can withstand challenges caused by climate change to maintain sustainable agriculture and food security in urban areas. VF systems contribute to human nutrition in the built environment as leafy vegetables are the most commonly grown crop. Furthermore, as an innovative concept, a room/area can be developed in buildings to store hydroponically grown leafy plants with a self-reliant facility while artificially controlling temperature, RH, and light, particularly affixing LEDs to control spectral composition. However, more innovative approaches are needed to strengthen the UA and food production to feed the hunger with the increase in world population.

Acknowledgements This work was supported by the University of Sri Jayewardenepura, Sri Lanka, under Grant (ASP/01/RE/SCI/2018/29).

Declarations of Competing Interest The authors declare no potential conflict of interest.

References

Adger, W. N., Arnell, N. W., & Tompkins, E. L. (2005). Successful adaptation to climate change across scales. *Global Environ Change, 15*, 77–86. https://doi.org/10.1016/j.gloenvcha.2004.12.005

Al-Humaiqani, M. M., & Al-Ghamdi, S. G. (2022). The built environment resilience qualities to climate change impact: Concepts, frameworks, and directions for future research. *Sustainable Cities Soc, 80*, 103797. https://doi.org/10.1016/j.scs.2022.103797

Anwar, M. M., Breuste, J. H., Ahmad, A., Aziz, A., & Aldosari, A. A. (2023). Quantifying the impacts of urbanization on urban agriculture and food security in the megacity Lahore Pakistan. *Sustainability, 15*, 12143. https://doi.org/10.3390/su151612143

Appolloni, E., Orsini, F., Specht, K., Thomaier, S., Sanye-Mengual, E., Pennisi, G., & Gianquinto, G. (2021). The global rise of urban rooftop agriculture: A review of worldwide cases. *Journal of Cleaner Production, 296*, 126556. https://doi.org/10.1016/j.jclepro.2021.126556

Arain, M. A. (2011). Climate change impacts on water quality and quantity. *Global Bioethics, 24*(1–4), 77–78.

Arnott, J. C., Moser, S. C., & Goodrich, K. A. (2016). Evaluation that counts: A review of climate change adaptation indicators & metrics using lessons from effective evaluation and science-practice interaction. *Environmental Science & Policy, 66*, 383–392. https://doi.org/10.1016/j.envsci.2016.06.017

Ayambire, R. A., Amponsah, O., Peprah, C., & Takyi, S. A. (2019). A review of practices for sustaining urban and peri-urban agriculture: Implications for land use planning in rapidly urbanising Ghanaian cities. *Land Use Policy, 84*, 260–277. https://doi.org/10.1016/j.landusepol.2019.03.004

Aysha, S., Mani, M. (2017). Adaptation of buildings to climate change. Elsevier, Netherlands, pp. 331–349. https://doi.org/10.1016/B978-0-12-409548-9.10202-7

Bantis, F., Smirnakou, S., Ouzounis, T., Koukounaras, A., Ntagkas, N., & Radoglou, K. (2018). Current status and recent achievements in the field of horticulture with the use of light-emitting diodes (LEDs). *Scientia Horticulturae, 235*, 437–451. https://doi.org/10.1016/j.scienta.2018.02.058

Baudoin, W., Nono-Womdim, R., Lutaladio, N., Hodder, A., Castilla, N., Leonardi, C., De Pascale, S., Qaryouti, M., Duffy, R. (2013). Good agricultural practices for greenhouse vegetable crops: Principles for mediterranean climate areas. FAO plant production and protection paper (FAO).

Beddoes, D., Booth, C. (2012). Insights and perceptions of sustainable design and construction. In: Booth, C. A., Hammond F, N., Lamond, J.E., Proverbs DG (Eds.) Solutions for climate change challenges in the built environment. Wiley, pp. 127–139. https://doi.org/10.1002/9781444354539.ch11

Bekchanov, M., Wijayasundara, M., de Alwis, A. (2022). Circular economy-A treasure trove of opportunities for enhancing resource efficiency and reducing greenhouse gas emissions. In: Asif, M. (Ed.) Handbook of energy and environmental security, (Chapt. 24), Academic Press, pp. 481–499. https://doi.org/10.1016/B978-0-12-824084-7.00016-3

Biesbroek, G. R., Klostermann, J. E., Termeer, C. J., & Kabat, P. (2013). On the nature of barriers to climate change adaptation. *Regional Environmental Change, 13*, 1119–1129. https://doi.org/10.1007/s10113-013-0421-y

Brager, G. S., & de Dear, R. J. (1998). Thermal adaptation in the built environment: A literature review. *Energy Build, 27*, 83–96. https://doi.org/10.1016/S0378-7788(97)00053-4

Challinor, A. J., Watson, J., Lobell, D. B., Howden, S. M., Smith, D. R., & Chhetri, N. (2014). A meta-analysis of crop yield under climate change and adaptation. *Nature Climate Change, 4*, 287–291. https://doi.org/10.1038/nclimate2153

Chatterjee, A., Debnath, S., Pal, H. (2020). Implication of urban agriculture and vertical farming for future sustainability. In: Urban horticulture-necessity of the future. IntechOpen.

de Coninck, H., Revi, A., Babiker, M., Bertoldi, P., Buckeridge, M., Cartwright, A., Dong, W., Ford, J., et al. (2018). Strengthening and implementing the global response. In: Masson-Delmotte, V.,

Zhai, P., Pörtner, H. O., Roberts, D., Skea, J., Shukla, P. R., Pirani, A., Moufouma-Okia, W., Péan, C., Pidcock, R., Connors, S., Matthews, J. B. R., Chen, Y., Zhou, X., Gomis, M. I., Lonnoy, E., Maycock, T., Tignor, M., Waterfield, T. (Eds.) Global warming of 1.5 °C. Cambridge University Press, (Chapt. 4), Cambridge and New York, pp. 616. https://doi.org/10.1017/9781009157940.006

Cotgrave, A., & Riley, M. (2013). *Total sustainability in the built environment*. Palgrave Macmillan.

D'Souza, C., Yuk, H.-G., Khoo, G. H., & Zhou, W. (2015). Application of light-emitting diodes in food production, postharvest preservation, and microbiological food safety. *Comprehensive Reviews in Food Science and Food Safety, 14*, 719–740. https://doi.org/10.1111/1541-4337.12155

Delorme, M., & Santini, A. (2022). Energy-efficient automated vertical farms. *Omega, 109*, 102611. https://doi.org/10.1016/j.omega.2022.102611

Deng, X., Nie, W., Li, X., Wu, J., Yin, Z., Han, J., Pan, H., & Lam, C. K. C. (2023). Influence of built environment on outdoor thermal comfort: A comparative study of new and old urban blocks in Guangzhou. *Building and Environment, 234*, 110133. https://doi.org/10.1016/j.buildenv.2023.110133

Dias, J. S. (2012). Nutritional quality and health benefits of vegetables: A review. *Food and Nutrition Sciences, 3*, 1354–1374. https://doi.org/10.4236/fns.2012.310179

Dubbeling, M., De Zeeuw, H. (2011). Urban agriculture and climate change adaptation: ensuring food security through adaptation. In: Resilient cities: Cities and adaptation to climate change-proceedings of the global forum 2010. Springer, pp. 441–449.

Enloe, A. (2018). The 13 Healthiest leafy green vegetables. Healthline. https://www.healthline.com/nutrition/leafy-green-vegetables. Accessed 29 Jun 2022

Eriksen, S., Aldunce, P., Bahinipati, C. S., Martins, R. D., Molefe, J. I., Nhemachena, C., O'brien, K., Olorunfemi, F., Park, J., Sygna, L., & Ulsrud, K. (2011). When not every response to climate change is a good one: Identifying principles for sustainable adaptation. *Climate and Development, 3*(1), 7–20. https://doi.org/10.3763/cdev.2010.0060

Etkin, D., & Ho, E. (2007). Climate change: Perceptions and discourses of risk. *Journal of Risk Research, 10*(5), 623–641. https://doi.org/10.1080/13669870701281462

Flörke, M., Schneider, C., & McDonald, R. I. (2018). Water competition between cities and agriculture driven by climate change and urban growth. *Nature Sustainability, 1*, 51–58. https://doi.org/10.1038/s41893-017-0006-8

Follmann, A., Willkomm, M., & Dannenberg, P. (2021). As the city grows, what do farmers do? A systematic review of urban and peri-urban agriculture under rapid urban growth across the global south. *Landscape Urban Planning, 215*, 104186. https://doi.org/10.1016/j.landurbplan.2021.104186

Garande, V. K., Raut, P. D., Shinde, U. S., Dhumal, S. S., Sonawane, P. N., Sarvade, S. A. (2019). Studies on storage behavior of primary processed leafy vegetables under different storage conditions. *International Journal of Current Microbiology and Applied Sciences, 8*(6), 2249–2272. https://doi.org/10.20546/ijcmas.2019.806.268

Gill, S. E., Handley, J. F., Ennos, A. R., & Pauleit, S. (2007). Adapting cities for climate change: The role of the green infrastructure. *Built Environment, 33*(1), 115–133. https://doi.org/10.2148/benv.33.1.115

Habash, R. (2022). 9—Urbanization as an intelligent system. In: Habash, R. (Ed.) Sustainability and health in intelligent buildings, woodhead publishing series in civil and structural engineering. Woodhead Publishing, pp. 239–257. https://doi.org/10.1016/B978-0-323-98826-1.00009-0

Harvison, T., Newman, R., Judd, B. (2011). Ageing, the built environment and adaptation to climate change. In: National climate change adaptation research facility (NCCARF), Australian climate change adaptation research network for settlements and infrastructure (ACCARNSI). UNSW.

Huq, S., Grubb, M. (2003). Scientific assessment of the inter-relationships of mitigation and adaptation. Scoping paper for IPCC Fourth Assessment Report 2003. IPCC, Geneva.

Indraganti, M., Ooka, R., Rijal, H. B., & Brager, G. S. (2014). Adaptive model of thermal comfort for offices in hot and humid climates of India. *Building and Environment, 74*, 39–53. https://doi.org/10.1016/j.buildenv.2014.01.002

Jarvis, A., Ramirez, J., Anderson, B., Leibing, C., Aggarwal, P. (2010). Scenarios of climate change within the context of agriculture. In: Climate change and crop production. CABI Wallingford UK, pp. 9–37.

Jones, M. A. (2018). Using light to improve commercial value. *Hortic Research, 5*, 1–13. https://doi.org/10.1038/s41438-018-0049-7

Kaklauskas, A., Gudauskas, R. (2016). 17—Intelligent decision-support systems and the internet of things for the smart built environment. In: Pacheco-Torgal, F., Rasmussen, E., Granqvist, C. G., Ivanov, V., Kaklauskas, Arturas Makonin, S. (Eds.) Start-up creation. Woodhead Publishing, pp. 413–449. https://doi.org/10.1016/B978-0-08-100546-0.00017-0

Kim, M.-J., Mikš-Krajnik, M., Kumar, A., & Yuk, H.-G. (2016). Inactivation by 405 ± 5 nm light emitting diode on Escherichia coli O157:H7, Salmonella typhimurium, and Shigella sonnei under refrigerated condition might be due to the loss of membrane integrity. *Food Control, 59*, 99–107. https://doi.org/10.1016/j.foodcont.2015.05.012

Kleinhenz, V., Schnitzler, W. H., & Midmore, D. J. (1996). Diversification and transformation of Asian paddy rice fields to upland vegetable production. *Plant Research and Development, 43*, 81–94.

Kokalj, D., Hribar, J., Cigić, B., Zlatić, E., Demšar, L., Sinkovič, L., Šircelj, H., Bizjak, G., & Vidrih, R. (2016). Influence of yellow light-emitting diodes at 590 nm on storage of apple, tomato and bell pepper fruit. *Food Technology and Biotechnology, 54*(2), 228.

Lee, Y. J., Ha, J. Y., Oh, J. E., & Cho, M. S. (2014). The effect of LED irradiation on the quality of cabbage stored at a low temperature. *Food Science and Biotechnology, 23*, 1087–1093. https://doi.org/10.1007/s10068-014-0149-6

Lucertini, G., & Di Giustino, G. (2021). Urban and peri-urban agriculture as a tool for food security and climate change mitigation and adaptation: The case of mestre. *Sustainability, 13*(11), 5999. https://doi.org/10.3390/su13115999

Magan, N., Medina, A., & Aldred, D. (2011). Possible climate-change effects on mycotoxin contamination of food crops pre-and postharvest. *Plant Pathology, 60*(1), 150–163. https://doi.org/10.1111/j.1365-3059.2010.02412.x

Maheshwari, S. (2021). Vertical farming: Resilience towards climate change. In: Kateja, A., Jain, R. (Eds.) Urban growth and environmental issues in India. Springer, Singapore, pp. 207–221. https://doi.org/10.1007/978-981-16-4273-9_13

Mani, M., & Reddy, B. V. (2012). Sustainability in human settlements: Imminent material and energy challenges for buildings in India. *Journal of the Indian Institute of Science, 92*, 145–162.

Mattos, L. M., Moretti, C. L., Jan, S., Sargent, S. A., Lima, C. E. P., Fontenelle, M. R. (2014). Climate changes and potential impacts on quality of fruit and vegetable crops. In: Ahmad P, Rasool S (eds) *Emerging technologies and management of crop stress tolerance*, (Chap. 19), Academic Press, San Diego, pp. 467–486. https://doi.org/10.1016/B978-0-12-800876-8.00019-9

McKeown, A. W., Warland, J., & McDonald, M. R. (2006). Long-term climate and weather patterns in relation to crop yield: A minireview. *Botany, 84*, 1031–1036. https://doi.org/10.1139/b06-080

Misra, A. K. (2014). Climate change and challenges of water and food security. *International Journal of Sustainable Built Environment, 3*, 153–165. https://doi.org/10.1016/j.ijsbe.2014.04.006

Morrow, R. C. (2008). LED Lighting in horticulture. *HortScience, 43*, 1947–1950. https://doi.org/10.21273/HORTSCI.43.7.1947

Moser, S. C. (2011). Entering the period of consequences: The explosive US awakening to the need for adaptation. In: Ford, J., Berrang-Ford, L. (Eds.) Climate change adaptation in developed nations. Springer, Dordrecht, pp. 33–49. https://doi.org/10.1007/978-94-007-0567-8_3

Moser, S. C., & Ekstrom, J. A. (2010). A framework to diagnose barriers to climate change adaptation. *PNAS, 107*(51), 22026–22031. https://doi.org/10.1073/pnas.1007887107

Natesh, H. N., Abbey, L., Asiedu, S. K. (2017). An overview of nutritional and antinutritional factors in green leafy vegetables. *Horticulture International Journal, 1*(2), 58–65. https://doi.org/10.15406/hij.2017.01.00011

Nayak, J. K., Prajapati, J. A. (2006). Handbook on energy conscious buildings. Project Report, IIT Mumbai.

Nkrumah, B. (2018). Edible backyards: Climate change and urban food (in)security in Africa. *Agriculture & Food Security, 7*, 45. https://doi.org/10.1186/s40066-018-0196-y

Oguntona, O. A., Aigbavboa, C. O. (2016). Biomimetic strategies for climate change adaptation in the built environment-a literature review. In: Building the future-sustainable and resilient environments. 9th International conference of faculty of architecture research unit (FARU). University of Moratuwa, Colombo, pp. 87–96.

Olle, M., & Viršile, A. (2013). The effects of light-emitting diode lighting on greenhouse plant growth and quality. *Agricultural and Food Science, 22*, 223–234. https://doi.org/10.23986/afsci.7897

Owen, G. (2020). What makes climate change adaptation effective? A systematic review of the literature. *Global Environmental Change, 62*, 102071. https://doi.org/10.1016/j.gloenvcha.2020.102071

Panwar, N. L., Kaushik, S. C., & Kothari, S. (2011). Solar greenhouse an option for renewable and sustainable farming. *Renewable and Sustainable Energy Reviews, 15*, 3934–3945. https://doi.org/10.1016/j.rser.2011.07.030

Perera, W. P. T. D., Navaratne, S., & Wickramasinghe, I. (2022a). Impact of spectral composition of light from light-emitting diodes (LEDs) on postharvest quality of vegetables: A review. *Postharvest Biology and Technology, 191*, 111955. https://doi.org/10.1016/j.postharvbio.2022.111955

Perera, W. P. T. D., Navaratne, S. B., & Wickramasinghe, I. (2022b). Review on effect of postharvest illumination by fluorescent and ultraviolet light waves on the quality of vegetables. *Journal of Food Process Engineering, 45*, e13960. https://doi.org/10.1111/jfpe.13960

Pinheiro, A., & Govind, M. (2020). Emerging global trends in urban agriculture research: A scientometric analysis of peer-reviewed journals. *Journal of Scientific Research, 9*(2), 163–173.

Porter, J. R., & Semenov, M. A. (2005). Crop responses to climatic variation. *Philosophical Transactions of the Royal Society b: Biological Sciences, 360*, 2021–2035. https://doi.org/10.1098/rstb.2005.1752

Qin, H. L., Zhang, Z. X., Lu, J., Zhu, Y. J., Webster, R., Liu, X. L., Yuan, H. Z., Hou, H. J., Chen, C. L., & Wei, W. X. (2016). Change from paddy rice to vegetable growing changes nitrogen-cycling microbial communities and their variation with depth in the soil. *European Journal of Soil Science, 67*, 650–658. https://doi.org/10.1111/ejss.12365

Rajasekar, E., Ramachandraiah, A. (2010). Adaptive comfort and thermal expectations–a subjective evaluation in hot humid climate. In: Proceedings of the adapting to change: new thinking on comfort, London.

Rao, R. (2014). Biomimicry in architecture. *International Journal of Advanced Research in Civil Structural Environmental and Infrastructure Engineering and Developing, 1*(3), 101–107.

Roaf, S., Crichton, D., & Nicol, F. (2009). *Adapting buildings and cities for climate change: A 21st century survival guide*. Routledge.

Rockström, J., Steffen, W., Noone, K., Persson, Å., Chapin, F. S., Lambin, E. F., Lenton, T. M., Scheffer, M., et al. (2009). A safe operating space for humanity. *Nature, 461*, 472–475. https://doi.org/10.1038/461472a

Rowlands, D. J., Frame, D. J., Ackerley, D., Aina, T., Booth, B. B., Christensen, C., Collins, M., Faull, N., et al. (2012). Broad range of 2050 warming from an observationally constrained large climate model ensemble. *Nature Geoscience, 5*, 256–260. https://doi.org/10.1038/ngeo1430

Ruengittinun, S., Phongsamsuan, S., Sureeratanakorn, P. (2017). Applied internet of thing for smart hydroponic farming ecosystem (HFE). In: 10th International conference on Ubi-Media computing and workshops (Ubi-Media). IEEE, pp. 1–4. https://doi.org/10.1109/UMEDIA.2017.8074148

Schipper, E. L. F. (2006). Conceptual history of adaptation in the UNFCCC process. *Review of European Community & International Environmental Law, 15*(1), 82–92. https://doi.org/10.1111/j.1467-9388.2006.00501.x
Sheela, K., Nath, K. G., Vijayalakshmi, D., Yankanchi, G. M., & Patil, R. B. (2004). Proximate composition of underutilized green leafy vegetables in southern Karnataka. *Journal of Human Ecology, 15*, 227–229. https://doi.org/10.1080/09709274.2004.11905698
Shimoda, Y. (2003). Adaptation measures for climate change and the urban heat island in Japan's built environment. *Building Research & Information, 31*, 222–230. https://doi.org/10.1080/0961321032000097647
Son, J. E., Kim, H. J., Ahn, T. I. (2020). Hydroponic systems. In: Kozai, T., Niu, G., Takagaki, M. (Eds.) Plant factory, (Chapt. 20), 2nd edn. Academic Press, pp. 273–283. https://doi.org/10.1016/B978-0-12-816691-8.00020-0
Steenkamp, J., Cilliers, E. J., Cilliers, S. S., & Lategan, L. (2021). Food for thought: Addressing urban food security risks through urban agriculture. *Sustainability, 13*(3), 1267. https://doi.org/10.3390/su13031267
Suraiya, M. N. M., Mahendran, T. (2009). Comparative nutritional study and minimal processing of dark green leaf vegetables found in Batticaloa District. http://www.digital.lib.esn.ac.lk/handle/123456789/13611. Accessed 6 Apr 2024
Thornton, P. K., Jones, P. G., Ericksen, P. J., & Challinor, A. J. (2011). Agriculture and food systems in sub-Saharan Africa in a 4 °C+ world. *Philosophical Transactions of the Royal Society A, 369*, 117–136. https://doi.org/10.1098/rsta.2010.0246
Tyler, S., & Moench, M. (2012). A framework for urban climate resilience. *Climate and Development, 4*, 311–326. https://doi.org/10.1080/17565529.2012.745389
Ülger, T. G., Songur, A. N., Çırak, O., Çakıroğlu, F. P. (2018). Role of vegetables in human nutrition and disease prevention. In: Asaduzzaman, Md., Asao, T. (Eds.) Vegetables—Importance of quality vegetables to human health. IntechOpen, London, pp. 7–32. https://doi.org/10.5772/intechopen.77038
Vogel, J., Letson, D., & Herrick, C. (2017). A framework for climate services evaluation and its application to the Caribbean agrometeorological initiative. *Climate Services, 6*, 65–76. https://doi.org/10.1016/j.cliser.2017.07.003
Warland, J., McKeown, A. W., & McDonald, M. R. (2006). Impact of high air temperatures on Brassicacae crops in southern Ontario. *Canadian Journal of Plant Science, 86*, 1209–1215. https://doi.org/10.4141/P05-067
Warren, E., Hawkesworth, S., & Knai, C. (2015). Investigating the association between urban agriculture and food security, dietary diversity, and nutritional status: A systematic literature review. *Food Policy, 53*, 54–66. https://doi.org/10.1016/j.foodpol.2015.03.004
Weerakkody, W. A. P., Mawalagedera, S. M. M. R. (2020). Recent developments in vegetable production technologies in Sri Lanka. In: Marambe, B., Weerahewa, J., Dandeniya, W. (Eds.) Agricultural research for sustainable food systems in Sri Lanka. Springer, Singapore, pp. 189–214. https://doi.org/10.1007/978-981-15-2152-2_9
Wilkinson, S. J., James, K., & Reed, R. (2009). Using building adaptation to deliver sustainability in Australia. *Structural Survey, 27*(1), 46–61. https://doi.org/10.1108/02630800910941683
Yan, D., Liu, L., Liu, X., & Zhang, M. (2022). Global trends in urban agriculture research: A pathway toward urban resilience and sustainability. *Land, 11*, 117. https://doi.org/10.3390/land11010117
Yeh, N., & Chung, J.-P. (2009). High-brightness LEDs-energy efficient lighting sources and their potential in indoor plant cultivation. *Renewable and Sustainable Energy Reviews, 13*, 2175–2180. https://doi.org/10.1016/j.rser.2009.01.027
Younger, M., Morrow-Almeida, H. R., Vindigni, S. M., & Dannenberg, A. L. (2008). The built environment, climate change, and health: Opportunities for co-benefits. *American Journal of Preventive Medicine, 35*(5), 517–526. https://doi.org/10.1016/j.amepre.2008.08.017
Zari, M. P., & Storey, J. B. (2007). An ecosystem based biomimetic theory for a regenerative built environment. *Lisbon sustainable building conference (SB07)* (pp. 620–627). IOS Press.

Ziervogel, G., & Ericksen, P. J. (2010). Adapting to climate change to sustain food security. *Wires Climate Change, 1*, 525–540. https://doi.org/10.1002/wcc.56

Zulfiqar, F., Shang, J., Yasmeen, S., Wattoo, M. U., Nasrullah, M., & Alam, Q. (2021). Urban agriculture can transform the sustainable food security for urban dwellers in Pakistan. *GeoJournal, 86*, 2419–2433. https://doi.org/10.1007/s10708-020-10208-1

Analysis of the UK Homeowners' Practices Prompting Housing Retrofits: A Practice Theory Approach

Chamara Panakaduwa, Paul Coates, and Mustapha Munir

1 Introduction

1.1 Housing Retrofits

Housing retrofit has become a critical aspect of energy efficiency and sustainability in the United Kingdom (UK) (Climate Change Committee, 2017). Retrofitting is the process of improving existing buildings to make them more energy-efficient, environmentally friendly, healthy, and comfortable for occupants (BSI, 2019). The UK government has set a target to achieve net-zero emissions by 2050, and retrofitting houses is a crucial step towards achieving this goal (Passivhaus Trust, 2021). There are several benefits to retrofitting houses, including reducing energy consumption and bills, improving the comfort of the living environment, and reducing the carbon footprint of the built environment (HM Government, 2021). The UK government has introduced several policies and initiatives to encourage homeowners to retrofit their properties, such as the Green Homes Grant scheme (HM Government, 2020).

However, retrofitting houses can also be costly, and there are challenges that need to be addressed, such as the lack of skilled workers and a shortage of affordable financing options (BEIS, 2021; TheIET, 2020). Despite these challenges, the benefits of housing retrofit are significant, and the UK government is committed to achieving its net-zero target by 2050 (BEIS, 2021). Retrofitting homes is a critical component of achieving this goal, and there is a need for increased investment, collaboration, and innovation to ensure that homes across the UK are energy-efficient, sustainable, and comfortable for their occupants.

C. Panakaduwa (✉) · P. Coates · M. Munir
University of Salford, Greater Manchester, Salford M5 4WT, UK
e-mail: c.s.panakaduwagamage@edu.salford.ac.uk

C. Malalgoda et al. (eds.), *Climate Change Adaptation in the Built Environment*,
https://doi.org/10.1007/978-3-031-75826-3_9

1.2 Importance of the Homeowner

Homeowners play a crucial role in housing retrofit for several reasons. First and foremost, homeowners are the primary occupants of a property, and as such, they have a vested interest in the comfort and energy efficiency of their homes. Retrofitting homes can significantly improve energy efficiency, reducing energy bills and improving the living environment for occupants. Accordingly, involving occupants in the retrofit process is highly important (Chiu et al., 2013).

Secondly, homeowners can benefit from various government incentives and schemes aimed at encouraging retrofitting. For example, the UK government's Green Homes Grant scheme provides funding to households to make energy-efficient improvements to their homes. Homeowners can take advantage of these incentives to reduce the cost of retrofitting their properties (BEIS, 2021).

Finally, the homeowners are the key decision makers of a house, in the case of owner-occupied and privately rented houses. Without their decision-making for retrofit, the progress of the retrofit will not be contained (Liu et al., 2022). Accordingly Gultekin et al. (2014) also highlighted the importance of involving homeowners in the retrofit decision-making process. When the homeowner decides they need to retrofit, it will create demand and demand will drive the retrofit.

1.3 Why Practice Theory?

Practice theories offer a valuable perspective on the role of everyday activities in shaping social life and social change, and they have been applied to a wide range of social phenomena, from consumption to work and health to environmentalism. Bourdieu's theory of practice, for example, emphasizes the role of habitus, or the set of dispositions and embodied practices that shape individual and collective behaviour (Oerther & Oerther, 2018). Theodore Schatzki's social practice theory emphasizes the importance of understanding social practices as the primary unit of social life, rather than focusing solely on agency or structure. Social practices are seen as the building blocks of social life, as they involve a complex interplay between material, social, and cultural elements (Loscher et al., 2019).

Anthony Giddens' theory of structuration is a sociological framework that emphasizes the interplay between social structures and individual agency in shaping social life. According to Giddens, social structures are both the medium and outcome of social action, and they are constantly being reproduced and transformed through human agency (Chatterjee et al., 2019).

The practice theory approach can be a useful framework for driving housing retrofit because it takes into account the social and cultural factors that influence day-to-day behaviour and decision-making (Bartiaux et al., 2014). The approach emphasizes the importance of understanding how people engage with the built environment and how their actions and practices shape energy use and sustainability. According to Karvonen (2013), the existing technology-oriented or consumer-oriented retrofit approaches are unlikely to achieve the required level of progress in housing retrofit. A more socially oriented retrofit approach is required.

The practice theory approach recognises that housing retrofit is not just about technical solutions but also about social and cultural factors that shape behaviour (Tjørring & Gausset, 2018). For example, retrofitting may require changes in daily routines or household practices, such as adjusting heating and cooling systems or changing energy consumption patterns. Hargreaves (2011) suggests that the practice theory approach can help to understand how these practices are embedded in social norms and cultural values and develop strategies to promote behavioural changes.

Overall, the practice theory approach can provide a useful framework for driving housing retrofit by taking into account the social and cultural factors that shape behaviour and decision-making. By understanding how people engage with the built environment and developing targeted interventions to address barriers and promote behaviour change, the approach can help accelerate the transition to a more sustainable and energy-efficient housing stock.

1.4 Objective

To identify and analyse the practices of homeowners that prompt housing retrofit decision-making.

The study started with the introduction to housing retrofit, homeowners' stake in the retrofit process and the application of practice theories. Secondly, the literature review section studies practice theories and their evolution. Considering the need to influence the decision-making behaviour of the homeowners to promote housing retrofit, Fogg's behaviour model was studied next. After justifying the literature gap, the research methodology was attended next. The data collection and analysis chapter comes next to show the results of the interviews with homeowners. After discussing the findings, the final chapter is for the conclusion of the study.

2 Literature Review

2.1 Practice Theories

2.1.1 Giddens' Theory of Structuration

Anthony Giddens presented how social structures and human behaviours are connected and are constantly reshaping. This is called "Duality of structure". The social life of an individual is constantly revolving between the individual agency and social structures. Social structures provide the rules and norms for guiding human behaviour. Individual agency is the ability of an individual to make their own choices. These choices are constrained by the limitations of the structure (Giddens, 1984). Structures are not just constraints. They can be resources and prompts for action too. It is the perspective of how the structure is viewed.

Rather than the traditional dichotomous view of structure and agency, Giddens rejects the separation of structure and agency. He argues that structure and agency cannot be separated but go together. Human agency is influenced by the structure and social structures are shaped by the collective influence of individual agency (Giddens, 1984). The Giddens' theory of structuration is a valuable concept for understanding the complex phenomena of human behaviour in terms of individual agency and social structures.

Before the Giddens' structuration theory, human behaviour had been interpreted with concepts such as materialism, functionalism, and evolutionism. All of these concepts mainly focused on the structural aspects of human behaviour. The individual agency was considered as a separate actor. However, Anthony Giddens tried to find the combined complex influence of structure and agency towards human behaviour (Bertilsson, 1984).

Baber (1991) proposed that the theory of structuration of Giddens was important to highlight the value of individual agency. Most of the previous concepts have only focused on the structure in describing human behaviour. What Giddens did was emphasise the role of individual agency in influencing human behaviour and coupling it with structure. This argument rejected the previous dichotomy of structure and agency. However, Giddens did not clearly suggest how structure and agency together influence human behaviour.

Although Anthony Giddens has not explicitly talked about practice theories, his concept aligns with the concept of practice theories. Giddens' theory talked about the interplay between structure and agency to shape individual behaviour. This is in line with the practice theory concepts of individual practices are shaped by social structures and vis-a-versa. Considering these factors, Gidden's theory of structuration can be viewed as an early proposal to practice theories.

2.1.2 Bourdieu's Theory of Practice

Pierre Bourdieu was a sociologist, anthropologist and French Philosopher who has greatly contributed to social science. Earlier, human action was not properly theorized. It was often argued that human action is influenced by structure and agency by the structuralists. (Agency is the ability of a person to make their own choices. Structure is how power relations have established the constraints in society to execute one's agency. E.g., A student can exercise his agency to become a doctor by using the education system, however, his agency is limited by the structural constraints of the education system). Bourdieu argued that the action is influenced by habitus but not directly from the structure or agency as others. He presented these ideas through his book "Outline of a theory of practice" (Maggio, 2017).

Habitus is an important term in Bourdieu's theory of practice. Bourdieu's habitus is the context which consists of the person's internal agency and external structure. This is a much broader concept than the agency and structure. A person's habitus is shaped by a number of aspects; society, education, upbringing, family background, life experiences, culture and so on. The habitus explains the behaviours and actions of a person. Accordingly, Bourdieu argued that people make decisions and take actions due to the practices influenced by their habitus (Maggio, 2017).

Habitus is Bourdieu's method of removing dichotomous values such as objective/ subjective or agency/structure. The action of an individual is an outcome arising from the relationship among the habitus, capital, and field. Capital has several forms. The main four capital forms are economic, cultural, social, and symbolic capital. There are fields around each of these fields. E.g., economic capital may have fields such as monetary field, legal field, educational field, etc. (Power, 1999). In general, an individual's action is influenced by its habitus (Oerther & Oerther, 2017).

The practice theory of Bourdieu can be identified as one of the earliest theories of the practice theory paradigm in sociology and anthropology. The main contribution of this theory is the rejection of an individual's agency and the structure in decision-making and proposing a broader concept of habitus, which is a dynamic and evolving context of an individual. The context of the habitus is broader, sometimes beyond the general understandability of a general academic as Bourdieu expected it to be. Individual decision-making results from the influence of this habitus, which is clearly coming out of individual practice, shaped by culture, society, education, life experiences and a number of exhaustive factors.

2.1.3 Schatzki's Theory of Social Practice

Theodore R. Schatzki further developed the concepts of practice theory body of knowledge. He made significant contributions to individual action and behaviour within the social context. Schatzki said the life of a human being consists of a number of interconnected practices. He emphasized that these practices are not isolated. Together, they shape and give meaning to human life. When it comes to practices, there are both explicit and background practices. These background practices are embedded with social and cultural contexts and influence explicit practices (Schatzki, 1996).

Schatzki further talked about a social ontology. The ontology (how the truth is viewed) of society is a collective approach to individual practices. Individual practices interact with each other and form the collective social ontology. It does not exist on its own. In earlier social theories, human agency and social structures were separate. However, Schatzki argues human agency remains embedded within their practices. Accordingly, individual agency and social structures are not mutually exclusive. He also emphasizes the importance of language, norms, and rules in influencing practices. Importantly, Schatzki criticizes the traditional view of individualism. He says individuals cannot be excluded from the social context. They are complimentary to each other (Schatzki, 1996).

In his next book, Schatzki improves the idea of social practice. He reiterates the interconnections between society and human practices. None of them exists without the other. Practices are the coordinated and repetitive actions which give meaning and structure to human life. Practice is not something totally virtual. They have relationships with physical aspects such as places, colours, objects, or anything material. The practices are continuously evolving, and they shape how a society behaves (Schatzki, 2002).

Figure 1 shows the flow of events according to Schatzki's site ontology, which consists of practices and material arrangements. This is an ever-evolving interconnected set of events.

In the social practice theory of Schatzki, the interaction among the activities, practices and materials is important to note. First, activities are the things that people do. These are the outcomes of human behaviour or actions coming out of decision-making. Not all these activities are practices. Practices are the activities that people do on a frequent basis. By studying human activities, these practices can be easily identified. Further, these practices are related to material arrangements. This means the practices are often related to the physical world (Loscher et al., 2019).

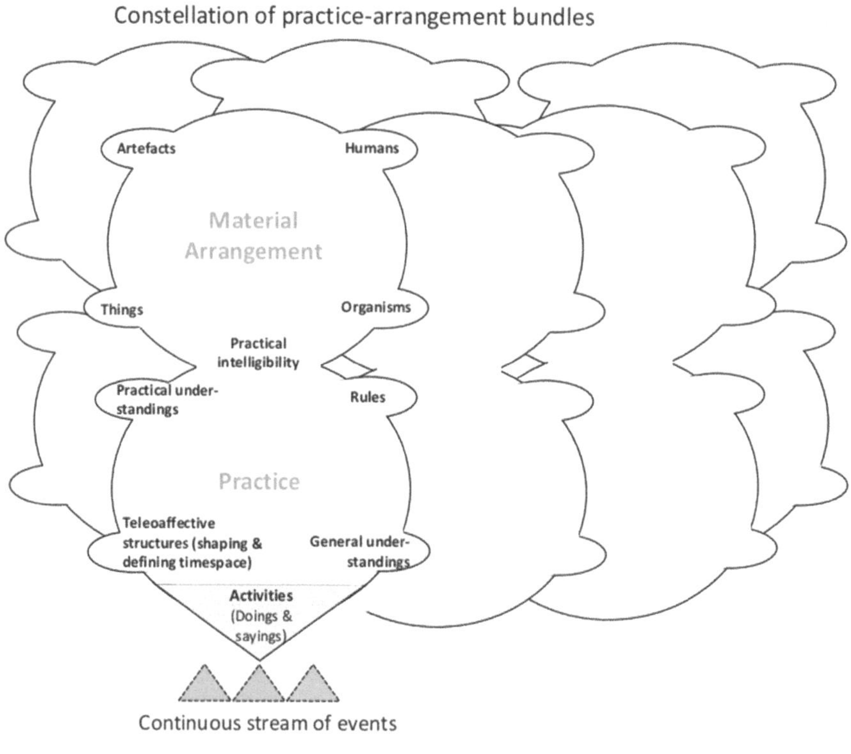

Fig. 1 Overview of the Schatzki's site ontology (Loscher et al., 2019)

2.1.4 Social Practice Concepts by Elizabeth Shove

Elizabeth Shove has also contributed to the body of knowledge of practice theories. In her book titled "The Dynamics of Social Practice", she has highlighted the concept of practice theories in sociology. She explored the importance of social practices which shape the daily activities of the people or vice versa. Finally, these daily practices shall lead to societal change.

Shove identifies practices as the behaviours, actions or values which form a part of the daily routines of the individuals. In agreement with Schatzki (1996), Shove also argues that the daily practices are not isolated. They are interconnected with each other. Practices can lead to creating or modifying social norms and thus leading to societal change. While noting the behavioural aspects, Shove argues that the practices have relationships with the material world. This is also in line with the idea of Schatzki (2002) site ontology (Shove et al., 2012).

Elizabeth Shove goes beyond the philosophical arguments of theories of practice. One of the reasons can be that her work is more contemporary, and these practice theories are already well established. Accordingly, Shove has more focused on bringing societal change with practice theories (Shove & Walker, 2014). Shove proposes three change mechanisms; cultural redefinitions (new meanings), material changes (new technology), and reconfiguration of activities (Shove et al., 2012).

Importantly, she proposes how practice theories can be related to improving sustainability through changing consumption behaviours (Shove, 2012). This is something that needs to be considered for policy decisions related to societal level behavioural changes. Consumption behaviour is an example of that as previously noted.

2.2 Fogg's Behaviour Model

The Fogg's behaviour model was developed by Dr B J Fogg of Stanford University, USA. This model helps to understand the human behaviour in a simple and concise manner. According to this model, there are three pillars that need to be together to create human action. They are the motivation, ability, and triggers (Fogg, 2009). This is graphically shown in Fig. 2.

There are three core motivators according to Dr Fogg. They are sensation, belonging and anticipation. Each of these three motivators has two sides: pleasure/pain, acceptance/rejection, and hope/fear. The second pillar is ability. Ability comes in a chain with five links. They are the time, physical effort, routine, mental effort, and money. The weakest link is the maximum level of ability. Fogg has renamed triggers as prompts later. The subsequent text will use the word "Prompt" accordingly. He introduces three types of prompts as facilitators, signals, and sparks. They should be used according to the available levels of ability and motivation (Fogg, 2022). Figure 3 shows the three types of prompts according to Fogg's model.

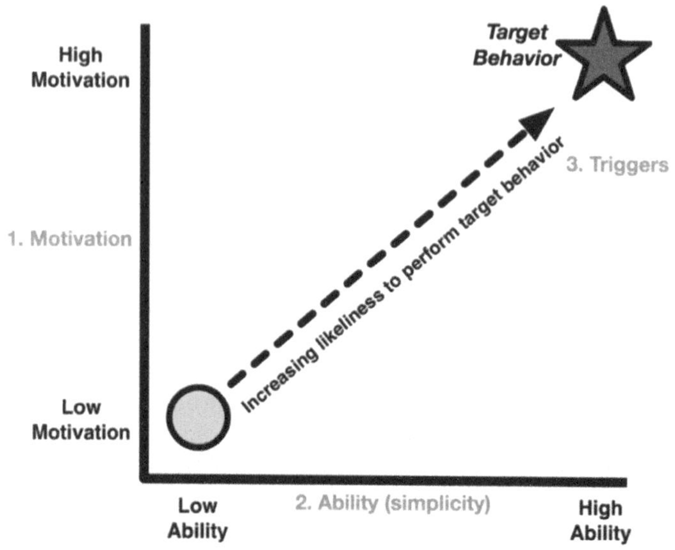

Fig. 2 Fogg behaviour model (Fogg, 2009)

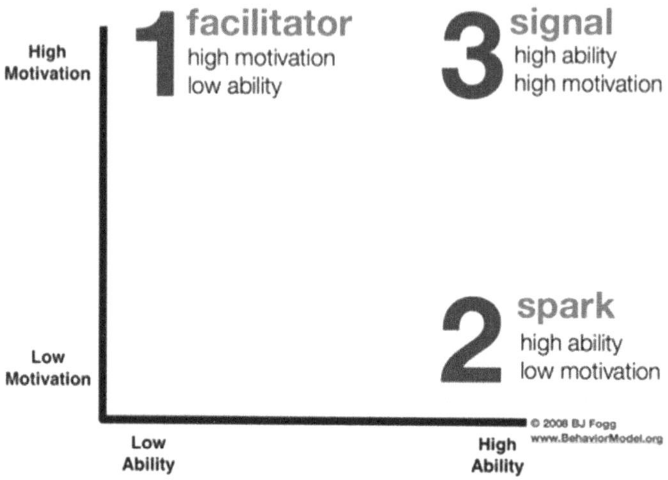

Fig. 3 Three types of prompts (Fogg, 2022)

According to Fogg, the behaviour will not occur without having all the pillars completed. Further, the ability and motivation can be traded off. If the ability is lacking, this can be fulfilled with motivation or vis-a-versa (Fogg, 2009).

Apart from Fogg's behaviour model, there are some other models that explain how human behaviour occurs. Susan Michie introduced COM-B model for behavioural change interventions. In this model, there are three segments required to change the behaviour: capability, opportunity, and motivation (Michie et al., 2011). Another key theory is the self-determination theory. In this theory, basic psychological needs and types of motivation are discussed together with other requirements for behavioural actions. The theory focuses on human behaviour and motivation through psychological needs and self-determination (Ryan & Deci, 2000).

2.3 Literature Gap

As far as the literature is concerned, the studies of practice theory approach to housing retrofit are observed to be low. According to Rickaby (2019), housing retrofit is observed to be product or service driven in the UK, which should be rather process-driven. Further, DellaValle et al. (2018), Gram-Hanssen (2014) highlight that the people factor in housing retrofit should not be neglected. They propose to focus on context-specific behavioural practices and cognitive biases of people in driving housing retrofit. Gram-Hanssen (2014) criticizes the current oversimplified retrofit processes based on technological and economic models, forgetting the importance of engaging people. This has resulted in homeowners' poor engagement in housing retrofits.

Bartiaux et al. (2014) have presented a novel recommendation in this regard. They have endorsed Shove et al. (2012) recommendations for using practice theories for sustainable behavioural interventions. Further, they propose that energy retrofits should be socially contextualised and co-evolved with laws, norms, and procedures. The homeowners cannot be motivated by just providing the right information and putting it in technological frameworks. Finally, Bartiaux et al. (2014) point out the presence of retrofit professionals in homeowners' social networks. Homeowners can be influenced by these social networks.

Karvonen (2013) argues that the existing rule-based retrofit programs are not effective. Accordingly, he proposes to use a practice-based approach through community partnerships. Theoretically, current retrofit approaches focus on the structure, while the author recommends focusing on social practices. In agreement with Gram-Hanssen (2014), Judson and Maller (2014) also criticize the current technical and rational models of promoting housing retrofit, focusing on individual behaviour. They recommend retrofit interventions focusing on the everyday life of the homeowners under social practice theories.

Although the available literature is not rich, it suggests that there is a clear mismatch of focus in driving housing retrofit. Due to the poor progress in housing retrofit with structure and agency-focused methods, the practice theories suggest that retrofit should be driven by focusing on social practices. This research has focused on the homeowner practices to drive housing retrofit under the Fogg behaviour model. What needs to be done can be identified as a behavioural change in the homeowner decision-making. According to Fogg (2009), one of the three factors of behaviour is prompt or trigger, which is expected in homeowners' everyday practices. COM-B model also has the "Opportunity" which is similar to the meaning of "Prompt" in

Fogg's model. Ideally, both models have no significant differences considering the purpose. However, Fogg's model was used considering the successful application of the work by Weeks et al. (2014) regarding the engagement of householders in housing retrofit. After studying the practice theories and Fogg's behaviour model, the research gap is identified here. The next section is to justify the research methodology of the study.

3 Materials and Methods

3.1 Theoretical Aspects

This study is underpinned by the ideas of practice theories such as Bourdieu's theory of practice, Schatzki's theory of social practices and Giddens' theory of structuration. This approach prompted to focus on the homeowners' daily behaviours, which are called "practices" rather than the individual characteristics and traits or society structure (laws, regulations, norms, etc.). According to practice theorists, it is important to focus on the practices in order to drive behavioural change over the society structures or individual behaviour.

Further, Fogg's behaviour model was also considered in this study. As the requirement is to drive housing retrofit, the action of the homeowner is the missing part. Fogg's behaviour model suggests that motivation, ability and prompts lead to action (Fogg, 2009). Assuming motivation and ability a constant, the study tried to find the prompts. By combining Fogg's behaviour model and practice theories, an effort was made to find prompts in homeowners' practices in their daily lives for housing retrofit.

3.2 Data Collection and Analysis

Figure 4 shows the conceptual framework used for this study. Starting with the literature review, semi-structured interviews were conducted with homeowners to identify the triggers for housing retrofit in their everyday practices. This was systematically achieved with the data analysis, thematic analysis, conclusions and recommendations.

The data collection was done by way of semi-structured interviews with five homeowners. The homeowners were conveniently selected. The semi-structured interviews were supported by a questionnaire to motivate the homeowners to further reflect on the subject. A study conducted by Vlasova and Gram-Hanssen (2014) about tenants' everyday practices and housing retrofit was referred. In this study, they have used four or five interviews with tenants for each case. Considering the scope of this study, it was only required to propose an example of the potential of everyday

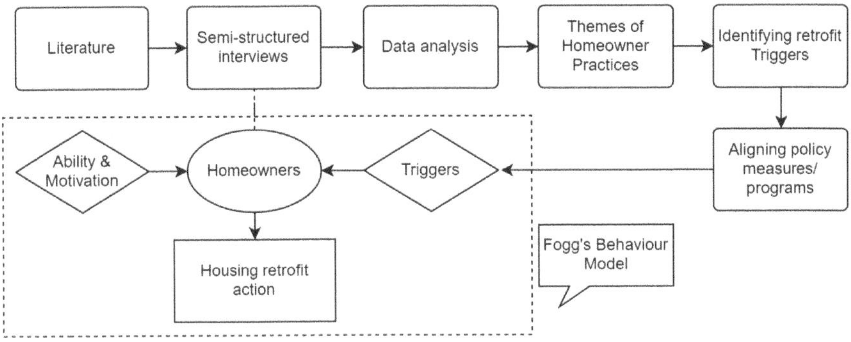

Fig. 4 Conceptual framework (Author)

practices for housing retrofit prompts. Accordingly, the number of interviews was limited to five, while the depth of the interview was extended to identify the practices in detail.

The interviews were recorded and transcribed. Nvivo software was used for coding and thematic analysis of the interview transcripts. A qualitative research approach was used in this study. The homeowner practices were identified from the interviews first. Then, they were separated into key themes for better clarification. At the next step, the prompt points for housing retrofit were identified in their daily practices, if any. Finally, recommendations were made on how the findings can be used in driving housing retrofit.

4 Data Analysis

4.1 Data Collection

The interviews were conducted by using a questionnaire, prepared with pre-coding arrangements for daily, weekly, monthly, and seasonal themes. The interviewees were asked to comment on their routine behaviour under these themes. The background details of the interviewees are given in the mentioned Table 1.

The homeowners were between the ages of year 30–50. They all represent the middle-class income group. Except for one homeowner who is a businessman, others were engaged in employment. Further, they all have two children. All of them had some sort of higher education. The house types represented semi-detached, detached, flat and terraced house types.

Table 1 Background details of the interviewees (Author)

	Participant 1	Participant 2	Participant 3	Participant 4	Participant 5
Age	40–50	30–40	30–40	40–50	40–50
Gender	Male	Male	Male	Female	Male
Education	Masters	Bachelor's	Diploma	Masters	Masters
Income	Clerk	Businessman	Marketer	Teacher	Care worker
Household	4 people	2 people	4 people	4 people	4 people
House	Semi-detached	Detached	Flat	Terraced	Flat

4.2 Presentation of the findings

Table shows the codes of the interview findings. According to Table 2, homeowner practices were identified periodically.

The highest number of practices were identified in the homeowner's daily behaviour. There are 24 practices identified. Weekly practices represent 14 number of practices. In general, the homeowners did not express monthly practices. However, there were contrasting practices observed in the summer and winter seasons. In addition to the practices, they were asked questions about their perceptions of the house. Tables 3 and 4 depict data collection related to these practices in detail.

As there were a large number of practices observed in daily code, they were sub-coded for morning practices, afternoon practices and general practices. The practices identified from all the interviews were consolidated and presented above. In addition to the daily practices, the homeowners talked about weekly practices. However, there is only one practice that could be noted as monthly. The weekly practices are mainly with buying groceries, washing, and drying clothes, cooking meals, and engaging with the children. Table 5 shows the differences in practices between the summer season and winter season.

Table 6 is prepared to identify the perceptions of the house by the homeowners. It has identified nine such perceptions. There are contrasting routine behavioural changes of the homeowners between the summer season and the winter season. This is mainly about heating the house and cooling the house. There is a higher number

Table 2 Data coding frequencies (Author)

Code	Frequency
Daily practices	24
Weekly practices	14
Monthly Practices	1
Summer season practices	4
Winter season practices	8
Perception of the house	9

Table 3 Homeowners' daily practices (Author)

Code	Practices
Daily	Morning 1. Having a daily shower in the morning 2. Preparing meals for the kids in the morning 3. Preparing kids for the school 4. Driving kids to the school 5. Having breakfast 6. Drinking a glass of water in the morning 7. Engaging in religious activities in the morning 8. Having a coffee on the way to work 9. Doing house chores during the morning 10. Taking spouse to the railway station Afternoon 1. Picking up children from the school and nursery 2. Watching TV in the evening 3. Having a small drink in the evening 4. Coming home for the lunch 5. Returning home after work 6. Helping the kids with their studies General 1. Working from home 2. Feeding kids morning and evening 3. Cooking two or three times a day 4. Heating foods with the microwave 5. Counting calory intake of the foods and beverages 6. Cleaning the house daily 7. Using the air fryer instead of the oven 8. Cooking rice daily

Table 4 Weekly and monthly homeowner practices (Author)

Code	Practices
Weekly	1. Going to the supermarket a few times a week 2. Spending some time in the garden 3. Cleaning the garden 4. Using the washing machine during the weekend 5. Busy weekends due to people coming to the house 6. Washing and drying clothes a few times a week 7. Playing with the kids at the playground or park 8. Watching a movie with kids two or three times a week 9. Cooking a special meal every week 10. Taking kids to an extra class 11. Conducting online classes 12. Cooking dishes 2–3 times a week 13. Going out at the weekend with the family 14. Cleaning the house
Monthly	1. Visiting friends usually monthly

Table 5 Seasonal behaviour of the homeowners (Author)

Code	Practices
Summer season	1. Keeping the windows open 2. Using the fan occasionally 3. Using heaters occasionally 4. Drying clothes under the sun in the summer
Winter season	1. Using heaters from November to February 2. Heating the house two times a day 3. Keeping the heating on all the time 4. Wearing winter clothes to make warmer 5. Turning off heaters when nobody is around 6. Using a small electric heater for the room 7. Using timers to control the temperature 8. Turning off heaters during the night

Table 6 Perceptions about the house (Author)

Perceptions about the house
1. Small dampness issue in the ensuite bathroom
2. Small mould patches in the living room and bathroom
3. Satisfied about the comfort levels of the house
4. No need for any upgrades to the house
5. Solar panels may be potential
6. A bigger house may be required in the future
7. Believe paying lesser energy bills as EPC rating A house
8. Believe they pay lesser energy bills as the house is a flat
9. Believes they are paying average energy bills

of practices observed in the winter season as this involves using heaters to manage the temperature.

In addition to the practices of the homeowners, the interview inquired about homeowners' perceptions of their houses. In general, all of them think they have a good house and they do not need any improvements to their houses. Generally, they all are satisfied with their houses. Although the recent energy price increases have affected them, they believe they are paying average or lower energy bills. The next section is to discuss the findings with reference to the literature and theoretical points of departure.

5 Discussion

5.1 Key Themes and Prompts

After analysing the homeowner practices collected under the codes of daily, weekly, monthly, and seasonally, they were analysed for the key themes. The following key themes were summarized from the identified practices. According to Fogg's behaviour model, there are three requirements to result in an action. They are the motivation, ability and prompt (Fogg, 2009). Other than identifying the practices, the key prompts were noted which may lead to housing retrofit action. Figure 5 is prepared to illustrate the key themes of the everyday practices of homeowners.

Cooking

Everybody is engaged in cooking at least once a day. However, some families cook three times a day. They use a microwave to heat the food. Some interviewees said they cook rice daily while preparing a main dish one or two times a week. The cooking patterns are highly diverse from household to household. Prompts for housing retrofit.

- Concerns about the number of cooking times due to energy consumption.

As cooking is one of the main practices in the household and cooking is involved with a high level of energy consumption, this can highly influence the homeowner to go for energy efficiency measures. Usually, most of the British houses use gas for cooking (Energy UK, 2022). In this situation, unless the energy consumption of the house is very low, this can cause homeowners to have a strong motivation to retrofit the house. However, the ability of retrofit is still questionable as there is no single retrofit measure reported to address this problem. Accordingly, this is a

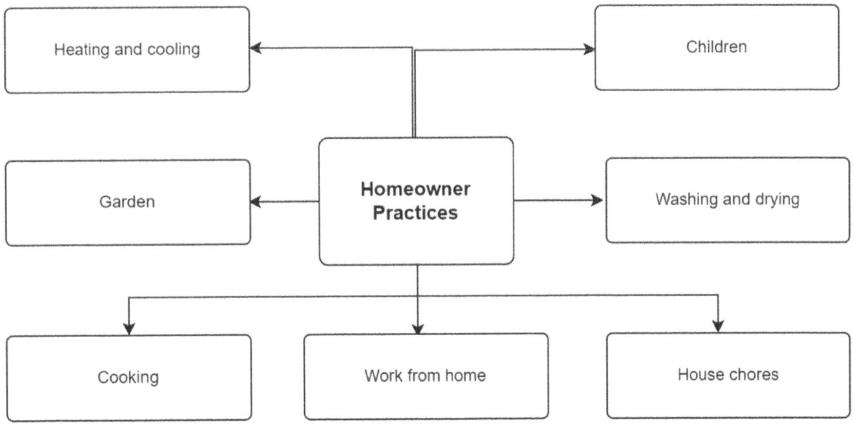

Fig. 5 Key themes of homeowner practices (Author)

high motivation low ability situation. Fogg recommends facilitating the situation to activate the prompt (Fogg, 2009).

Children

Four out of five interviewees have children. The interviewees are in the practice of taking them to school and back home. Further, the morning chores are involved with preparing children for school and preparing meals for them. After school time, these parents engage with the children in different activities. For example, helping them with studies, watching television with them and feeding them. Prompts can be noted as;

- Comfortable levels of the children
- Health and safety concerns of the house
- Upgrading the house for more space. E.g., Loft conversion.

The practices related to children may not be directly involved with energy efficiency. One of the facts is everyone agreed that they will need a bigger house as the children grow. However, as the retrofit is not only about energy, the health, safety, and comfort levels of the residents can be strong motivations for housing retrofit. During the daily practices of the residents (mainly related to children), there will be prompts to think of housing retrofit. Accordingly, this can be a situation of high motivation, low ability. The ability is considered to be lower as the availability of small measures to improve health, safety, and comfort levels is lower in general.

Washing and Drying

The majority of interviewees wash and dry clothes a few times a week. When it comes to drying, they used to dry clothes with a drier during winter seasons. During the summer seasons, they try to dry them outside under the sun. Washing and drying clothes prompts retrofit requirements directly due to the level of energy consumption of these activities. If they had a more energy-efficient house, the less they had to be concerned about the times of washing and drying. Prompts can be noted as,

- Concern for the number of times washing and drying clothes.
- Energy efficiency levels of washers and dryers.

Some residents were observed to be reducing the number of washing and drying cycles to reduce energy consumption. However, washing and drying were not identified as the activities of higher motivation for housing retrofit. The ability can be varied. If the focus is to avoid the total energy consumption, that would be a whole house retrofit. However, some levels of energy efficiency can be achieved with high-efficiency appliances and behavioural changes. According to (Fogg, 2009), when the motivation is lower, sparks are required to stimulate action. A rainy season or the winter season can be sparks where the residents have to use the dryer more frequently.

Heating and Cooling

Due to the cold weather in the UK, heating is an important aspect related to the house in winter months. Everyone talked about using a timer to control heating and heating the house one or two times a day. Further, some interviewees use an electric heater whenever necessary for single rooms. During the summer months, some interviewees said they used an electric fan. In terms of prompts for housing retrofit, heating and cooling can be the main point of prompt. Especially, when there are children around, parents try to keep the house warm. However, the cost of energy bills challenges this requirement. Prompts can be noted as follows,

- The time of receiving an energy bill
- The time of controlling the temperature of the heater

Heating or cooling can motivate residents highly with regard to housing retrofits as they can create a number of prompts. One of the prompts can be the time of receiving the energy bills. Residents were observed trying to find the equilibrium of balancing energy bills and heating comfort. Although the motivation for housing retrofit can be higher in heating and cooling, the ability of energy retrofits may not be higher due to the high cost and disruption of retrofit. According to Fogg (2009), this is a situation of high motivation and low ability. The decision-makers should be facilitated to activate the action in such a scenario. Offering housing retrofit help coupled with the housing retrofit prompts can be a good way of promoting retrofit. E.g., Furnishing information about the help available to housing retrofit together with the energy bill.

Working from Home

According to the findings, one interviewee is involved in a full-time role working from home. Another interviewee conducts online classes on a weekly basis. Another interviewee runs a second-hand vehicle buying and selling business, mainly during the weekends. It is difficult to identify direct prompt points for housing upgrades with working from home. However, this will increase the residents' engagement with the house, which may lead to prompt indirect retrofit requirements. Further, it is difficult to assess the level of ability and motivation related to working from home. It can vary a lot among different practices.

House Chores

Daily house chores are a part of the daily practices of most people. These works include activities such as cleaning the kitchen, sweeping floors, mopping the bathroom, and arranging the house. Further, buying groceries or visiting the beautician can also be grouped under this. Most of them prefer to do these things daily while one said about cleaning the house weekly. During the daily chores of the house, people often find renovation requirements. E.g., a leaky faucet. These can be an ideal prompt to retrofit the house along with renovations. Further, they experience poor health, safety, comfort, and energy efficiency issues. E.g., Mould growth on the ceiling, dampness on window glazing or high energy consumption of the refrigerator. Tradespeople can be useful in spreading the message of housing retrofit in these prompts. Prompts noted under house chores are,

- Noticing renovation requirements of the house

- Experiencing poor health, safety, and comfort
- Identifying poor energy efficiency

According to Bartiaux et al. (2014), tradespeople are part of the social networks of the residents. They often interact with the residents in casual gatherings as well as formal situations such as housing renovations. Even the groceries, supermarkets and gas stations can be useful to spread the message of help available for housing retrofit. However, these house chores cannot be considered as prompts with high motivation or high ability.

Garden and Outdoor

Although not everyone mentioned the practices related to the garden, everyone talked about going outdoors. Especially on sunny days, people use to go out and spend some time outside. The weekend could be a more popular time for people to go outside as the children are not schooling. In addition, the garden and going out can be meeting points with other people. E.g., Meeting friends, neighbours, and family. These can be prompting points of housing retrofit if they talk about the houses (Negative or positive social influence). Some of the prompts can be suggested as,

- Noticing the better comfort of the outdoor atmosphere. E.g., Air quality
- Discussing house performance with others and sharing ideas and experience
- Comparing houses with each other and getting advice. E.g., Energy bills.

One of the important points to note in this section is that these social interactions can impact retrofit decision-making either positively or negatively. Positive interactions will motivate housing retrofit and negative interactions will demotivate retrofit. Accordingly, even if there were prompts during these situations, there would not be a positive use of them. Unfortunately, it can lead to a loss of opportunities to drive a homeowner to retrofit the house. The next section is to recommend how the findings can be applied in promoting housing retrofit.

Synthesis with the Literature

This study has found several practices of the homeowners according to the interviews conducted. However, being the study is not exhaustive but to signpost the possibilities with examples, the identified numbers of practices are limited. Walker et al. (2014) have done a study to find the energy-related practices of the homeowners. However, the purpose of the study was not to identify the retrofit prompts. Tjørring and Gausset (2018) suggest that the most of changes to the houses are triggered not due to technical reasons, but personal and family reasons. In this study, homeowner practices involved with a high level of energy consumption were noted as retrofit prompts. For example, homeowners talked about the time of receiving energy bills and controlling temperature during the winter season. The energy bill factor has been also identified by Bardhan et al. (2014) as a trigger for housing retrofit. These are some of the findings which are related to the literature. However, the study can be generally considered as an extension of the existing studies. The existing studies such as Bartiaux et al., 2014, Judson & Maller, 2014, Karvonen, 2013, Vlasova & Gram-Hanssen, 2014 have made efforts to highlight the importance of focusing practice theories on energy efficiency and housing retrofit. This study focused on taking a further step by studying some examples of such practices to find the prompts and proposing behavioural changes with Fogg's behaviour model.

5.2 Application of the Findings

According to the practice theories, practices lead to action more than awareness. Fogg's behaviour model describes three pillars of behaviour as motivation, ability, and prompt. By combining the practice theories and Fogg's behaviour model, the study has tried to highlight the homeowner practices that prompt housing retrofit. Motivation and ability are not in the scope of this study. However, the finding of the study will be able to support driving housing retrofit by aligning it with the prompts in homeowner practices, subject to the provision of ability and motivation. In general, policymakers, institutional stakeholders and marketing stakeholders will be able to make use of the findings.

Recommendations can be drawn to consider how to link retrofit drives with the homeowners' practices. As per this study, the practices are segregated under the themes of cooking, heating, washing/drying, garden & outdoor, working from home, children, and house chores. According to Tjørring and Gausset (2018), current retrofit programs are mainly technically driven. This has become one of the reasons why these retrofit drives are not successful. The recommendation of this study is for the policymakers to design socially driven retrofit drives by focusing on homeowners' practices and prompts of these practices.

Further, the recommendations can be drawn to the institutional stakeholders such as professional bodies and government related bodies. E.g., Climate Change Committee (CCC). For example, RIBA recommendation report to the government highlights the technical aspects of housing retrofit. They can improve their recommendations more practically with socially contextualized recommendations, although someone can argue that it's beyond their scope. Importantly, it is important to consider the supply chain and contractors in mass scale retrofit delivery. The marketing stakeholders will be able to improve their sales by focusing their sales strategies along with homeowners' everyday practices. Specifically, the promotion campaigns can be tailor-made to the prompts of daily activities. E.g., Advertising heat pumps in energy bills.

According to the findings about homeowners' perceptions about their houses, everybody thinks they have a reasonably comfortable and good house where they do not feel the requirement of any upgrades to the house. However, if objectively compared their houses with the Enerphit or Energiesprong certified house, they are still paying higher energy bills while experiencing lower health, safety, and comfort. Further, these homeowners may not have an idea how a net zero/energy plus house performs. Awareness is important. However, by further motivating them to witness high levels of performance of retrofitted houses through case studies, prompts can be created. Ability will be still lacking according to Fogg's model. While the ability is mainly related to financial capacity and time availability, prompts in homeowners' practices will also play a key role in driving housing retrofit. The next section of the study is to conclude the study. It will acknowledge the limitations and show the potential for further studies.

6 Conclusion

The study has considered the practice theories and Fogg's behaviour model for driving housing retrofit at a scale. The idea of practice leading to action in practice theories and Fogg's behaviour model (ability, motivation and prompt leading to behaviour) are considered as the theoretical points of departure. Accordingly, the objective of the study is to identify and analyse the practices of homeowners that prompt housing retrofit decision-making. However, the study will not provide an exhaustive list of retrofit prompts that lead to action. The idea is to signpost that by focusing on prompts of homeowners' practices, retrofit can be better promoted. Accordingly, this study provides a practical example sourced from several empirical studies.

The study was able to find that the main themes of homeowners' practices are in the themes of cooking, heating, washing/drying, gardening & outdoors, working from home, children, and house chores. The data analysis has further elaborated the identified practices and potential prompts in these themes. The study wishes to conclude that aligning the characteristics of retrofit programs with the practices that prompt housing retrofits can be more productive.

As far as the limitations are concerned, the hypothesis is subject to further validation. The purpose of this study is to study the potential of mapping retrofit initiatives with the daily practices of homeowners through examples. Accordingly, the study is indicative, but not comprehensive. A practical retrofit drive aligned with homeowner practices or broader case studies can be recommended as future studies. Retrofit programs such as government grants, policy-making, institutional campaigns or marketing campaigns by retrofit contractors can make use of these findings for their initiatives.

References

Baber, Z. (1991). Beyond the structure/agency dualism: An evaluation of giddens'theory of structuration. *Sociological Inquiry, 61*(2), 219–230. https://doi.org/10.1111/j.1475-682x.1991.tb00276.x

Bardhan, A., Jaffee, D., Kroll, C., & Wallace, N. (2014). Energy efficiency retrofits for US housing: Removing the bottlenecks. *Regional Science and Urban Economics, 47*, 45–60.

Bartiaux, F., Gram-Hanssen, K., Fonseca, P., Ozoliņa, L., & Christensen, T. H. (2014). A practice–theory approach to homeowners' energy retrofits in four European areas. *Building Research & Information, 42*(4), 525–538. https://doi.org/10.1080/09613218.2014.900253

BEIS. (2021). *Heat and buildings strategy*. GOV.UK. https://www.gov.uk/government/publications/heat-and-buildings-strategy

Bertilsson, M. (1984). Review article: The theory of structuration: Prospects and problems*. *Acta Sociologica, 27*(4), 339–353. https://doi.org/10.1177/000169938402700404

BSI. (2019). *PAS 2035:2019 Retrofi tting dwellings for improved energy effi ciency— Specifi cation and guidance*. The British Standards Institution.

Chatterjee, I., Kunwar, J., & den Hond, F. (2019). Anthony Giddens and structuration theory. *Management, Organizations and Contemporary Social Theory*, 60–79. https://doi.org/10.4324/9780429279591-4

Chiu, L. F., Lowe, R., Altamirano, H., & Raslan, R. (2013). *Post-occupancy interview report: Key findings from a selection of Retrofit for the future projects*. discovery.ucl.ac.uk. https://discovery.ucl.ac.uk/id/eprint/10056896

Climate Change Committee. (2017). *About the committee on climate change—committee on climate change*. Committee on Climate Change. https://www.theccc.org.uk/about/

DellaValle, N., Bisello, A., & Balest, J. (2018). In search of behavioural and social levers for effective social housing retrofit programs. *Energy and Buildings, 172*, 517–524. https://doi.org/10.1016/j.enbuild.2018.05.002

Energy UK. (2022). *Energy in the UK 2022*. www.energy-uk.org.uk. https://www.energy-uk.org.uk/index.php/publication.html?task=file.download&id=8329

Fogg, B. J. (2009). A behavior model for persuasive design. *Proceedings of the 4th International Conference on Persuasive Technology—Persuasive '09*(40), 1–7. https://doi.org/10.1145/1541948.1541999

Fogg, B. J. (2022). Behavior model. *Behaviormodel*. https://behaviormodel.org/

Giddens, A. (1984). *The constitution of society: Outline of the theory of structuration*. Polity Press.

Gram-Hanssen, K. (2014). Retrofitting owner-occupied housing: Remember the people. *Building Research & Information, 42*(4), 393–397. https://doi.org/10.1080/09613218.2014.911572

Gultekin, P., J. Anumba, C., & Leicht, R. M. (2014). Case study of integrated decision-making for deep energy-efficient retrofits. *International Journal of Energy Sector Management, 8*(4), 434–455. https://doi.org/10.1108/ijesm-12-2013-0002

H. M. Government. (2020). *The ten point plan for a green industrial revolution*. GOV.UK. https://www.gov.uk/government/publications/the-ten-point-plan-for-a-green-industrial-revolution

H. M. Government. (2021). *Net zero strategy: Build back greener*. Controller of Her Majesty's Stationery Office.

Hargreaves, T. (2011). Practice-ing behaviour change: Applying social practice theory to pro-environmental behaviour change. *Journal of Consumer Culture, 11*(1), 79–99. https://doi.org/10.1177/1469540510390500

Judson, E. P., & Maller, C. (2014). Housing renovations and energy efficiency: Insights from homeowners' practices. *Building Research & Information, 42*(4), 501–511. https://doi.org/10.1080/09613218.2014.894808

Karvonen, A. (2013). Towards systemic domestic retrofit: A social practices approach. *Building Research & Information, 41*(5), 563–574. https://doi.org/10.1080/09613218.2013.805298

Liu, G., Ye, K., Tan, Y., Huang, Z., & Li, X. (2022). Factors influencing homeowners' housing renovation decision-making: Towards a holistic understanding. *Energy and Buildings, 254*, 111568. https://doi.org/10.1016/j.enbuild.2021.111568

Loscher, G., Splitter, V., & Seidl, D. (2019). Theodore Schatzki's practice theory and its implications for organization studies. In S. Clegg & M. P. Cunha (Eds.), *Management, organizations and contemporary social theory* (pp. 115–134). Routledge.

Maggio, R. (2017). *An analysis of Pierre Bourdieu's Outline of a theory of practice*. Macat International Ltd.

Michie, S., Van Stralen, M. M., & West, R. (2011). The behaviour change wheel: A new method for characterising and designing behaviour change interventions. *Implementation Science, 6*(42). https://doi.org/10.1186/1748-5908-6-42

Oerther, S., & Oerther, D. B. (2018). Pierre Bourdieu's theory of practice offers nurses a framework to uncover embodied knowledge of patients living with disabilities or illnesses: A discussion paper. *Journal of Advanced Nursing, 74*(4), 818–826. https://doi.org/10.1111/jan.13486

Oerther, S., & Oerther, D. B. (2017). Pierre Bourdieu's Theory of practice offers nurses a framework to uncover embodied knowledge of patients living with disabilities or illnesses: A discussion paper. *Journal of Advanced Nursing, 74*(4), 818–826. https://doi.org/10.1111/jan.13486

Power, E. M. (1999). An introduction to Pierre Bourdieu's key theoretical concepts. *Journal for the Study of Food and Society, 3*(1), 48–52. https://doi.org/10.2752/152897999786690753

Rickaby, P. (2019). PAS2035, *Each home counts and a career in energy efficiency—an interview with Dr Peter Rickaby*. www.youtube.com. https://www.youtube.com/watch?v=HpV_-SF4GW8

Ryan, R. M., & Deci, E. L. (2000). Self-determination theory and the facilitation of intrinsic motivation, social development, and well-being. *American Psychologist, 55*(1), 68–78. https://doi.org/10.1037/0003-066X.55.1.68

Schatzki, T. R. (2002). *The site of the social*. Pennsylvania State University Press.

Schatzki, T. R. (1996). *Social practices: A Wittgensteinian approach to human activity and the social*. Cambridge University Press.

Shove, E., Pantzar, M., & Watson, M. (2012). *The dynamics of social practice*. SAGE.

Shove, E. (2012). Putting practice into policy: Reconfiguring questions of consumption and climate change. *Contemporary Social Science, 9*(4), 415–429. https://doi.org/10.1080/21582041.2012.692484

Shove, E., & Walker, G. (2014). What is energy for? Social practice and energy demand. *Theory, Culture & Society, 31*(5), 41–58. https://doi.org/10.1177/0263276414536746

TheIET. (2020). *Scaling up retrofit 2050* (www.theiet.org, Issue. https://www.theiet.org/media/8758/retrofit.pdf

Tjørring, L., & Gausset, Q. (2018). Drivers for retrofit: A sociocultural approach to houses and inhabitants. *Building Research & Information, 47*(4), 394–403. https://doi.org/10.1080/09613218.2018.1423722

Passivhaus Trust. (2021). *Passivhaus retrofit in the UK*.

Vlasova, L., & Gram-Hanssen, K. (2014). Incorporating inhabitants' everyday practices into domestic retrofits. *Building Research & Information, 42*(4), 512–524.

Walker, S., Lowery, D., & Theobald, K. (2014). Low-carbon retrofits in social housing: Interaction with occupant behaviour. *Energy Research & Social Science, 2*, 102–114.

Weeks, C., Delalonde, C., & Preist, C. (2014). *Power law of engagement: Transferring disengaged householders into retrofitting energy savers*. ICT for Sustainability 2014 (ICT4S-14)

The Cultural Landscape of Portovenere, Cinque Terre, and the Islands (Palmaria, Tino and Tinetto) in Northern Italy: How to Tackle Climate Change in the Case of a Coastal UNESCO Site

S. Cecere, A. Di Maggio, G. Melis, and A. Rosso

1 Introduction

1.1 How to Define a Cultural Landscape?

According to the World Heritage Centre, (2016), cultural landscapes are unique assets that constitute heritage resources, blending tangible elements—such as ancient monuments, historic cities, or natural landscapes—with intangible aspects—such as the folklore, traditions, and knowledge (O'Donnell, 2023). Consequently, their conservation often requires the integration of cultural aspects with environmental, social, and economic dimensions (von Droste et al., 1999). Thus, they can play a crucial role in fostering connectivity among communities, improving well-being and quality of life, and encouraging sustainable economic growth (Rössler, 2006).

The concept of "cultural landscape" was initially defined as an area modified by human activity (Potthoff, 2013), and later expanded to encompass natural landscapes modified by cultural groups (Luengo & Ramsay, 2020). Nowadays, it is widely recognized across various disciplines as an illustration of the evolved interactions between a community and its natural environment and resources (Pearson and Gorman 2023).

S. Cecere · A. Di Maggio (✉) · G. Melis · A. Rosso
LINKS Foundation, Torino, Italy
e-mail: andrea.dimaggio@linksfoundation.com

S. Cecere
e-mail: serena.cecere@linksfoundation.com

G. Melis
e-mail: giulia.melis@linksfoundation.com

A. Rosso
e-mail: andrea.rosso@linksfoundation.com

C. Malalgoda et al. (eds.), *Climate Change Adaptation in the Built Environment*,
https://doi.org/10.1007/978-3-031-75826-3_10

In 1990-s, the United Nations Educational, Scientific and Cultural Organization (UNESCO) introduced the "Operational Guidelines for the Implementation of the World Heritage Convention" (1992), which classified Cultural Landscapes according to three dimensions: (i) cultural landscapes created by humans (such as natural parks, cultivated terraces, and gardens); (ii) naturally grown landscapes, relict or still in use with an active social role in contemporary society; and (iii) associative cultural landscapes justifiable by virtue of the powerful religious, artistic or cultural associations of the natural element, rather than material cultural evidence, which may be insignificant or even absent (Ontario Heritage Trust, 2012). Subsequently, the Council of Europe adopted the European Landscape Convention in 2000, describing a (cultural) landscape as any part of the territory "as perceived by people, whose character is the result of the action and interaction of natural and/or human factors" (Council of Europe, 2000). Consequently, UNESCO and the European Landscape Convention apply different conservation approaches to cultural landscapes. While UNESCO looks to conserve only internationally acknowledged heritage areas, the European Landscape Convention aims to conserve landscapes valued at a national level and improve the quality of life of local communities (Dastgerdi et al., 2022).

1.2 How is Climate Change Threatening Coastal Cultural Landscapes?

Cultural landscapes are not static areas trapped in time; rather they are social-ecological systems that continually interact with their surroundings and respond to global changes, shocks, and crises. Recognizing the contemporary challenges confronting cultural heritage is crucial for its preservation and transmission to future generations (Orr et al., 2021).

Among these challenges, the impacts of climate change, primarily driven by anthropological activities, exacerbate pressures by altering the occurrence and intensity of temperature, precipitation patterns, atmospheric moisture, and wind intensity. These alterations pose significant threats to the already fragile UNESCO cultural sites and their intrinsic values (Kapsomenakis et al., 2023; Richards et al., 2020).

One of the most conspicuous manifestations of climate change's impacts upon coastal cultural landscapes is the relentless rise of sea levels, leading to erosion, flooding, and potential inundation, in addition to changes to inland fluvial dynamics. Terraced slopes, cliffs, and ancient archaeological sites are gradually being consumed by the encroaching waters (Sesana et al., 2021). Moreover, climate change amplifies the frequency of extreme weather events, which, when they hit coastal regions, often take the form of storms, hurricanes, and wildfires. These events directly wreak havoc on cultural heritages sites through landslides, as well as corrosion of various materials when combined with air pollution (Grøntoft & Cassar, 2020).

1.3 How to Act?

Ecologists have recently demonstrated that coastal communities possess greater resilience compared to inland counterparts due to a diversified economy, extensive connections, and a diversity of niches (Garmestani et al., 2019).

Indeed, cultural landscapes, characterized by their irreplaceable nature and unique significance, hold immense potential for contributing to several of the United Nations' Sustainable Development Goals (SDGs), particularly when cultural, environmental, and social practices are seamlessly integrated. This concept was first addressed during the 29th session of the World Heritage Committee (WHC) in 2005, marking UNESCO's initial exploration of climate change issues (Nguyen & Baker, 2023). During this session, the Committee urged State Parties to develop management plans to address the impacts of climate change, recognizing the importance of climate adaptation efforts within the realm of cultural landscapes. A noteworthy study conducted in the Netherlands by Fatorić and Egberts, (2020) shed light on the potential benefits of climate change adaptation actions for cultural heritage resources and for their value preservation (Fatorić & Biesbroek, 2020). These actions encompass various approaches. Soft responses, such as ecosystem-based adaptation and nature-based solutions, prioritize non-intrusive methods. Conversely, hard responses, such as the construction of infrastructure like dams, are commonly utilized, particularly in northern Europe. Accommodation responses involve using technologies and innovations to physically modify exposed infrastructures, such as protecting buildings individually or developing floating housing.

Soft protection measures, such as dune rehabilitation and sand nourishment, adopt an integrated approach often referred to as the "building with Nature strategy", aiming to preserve ecosystem services (de Schipper et al., 2021). Hard responses do not provide long-term solutions to coastal hazards and can often alter the coastal ecosystem functioning. Lastly, accommodation responses may involve strengthening monitoring capacity, establishing new rules and policies, producing and disseminating useful information, and promoting safer behaviour. To be successful, all soft, hard, or accommodation techniques should be harmonised with local stakeholders' engagement (Bongarts Lebbe et al., 2021). This alignment is crucial to recognize the inextricable interdependency between nature and communities within heritage sites, a key factor in bolstering resilience to climate change impacts. Indeed, locally informed stakeholders are likely to be pivotal contributors to building holistic adaptation strategies that consider the rich historical and cultural heritage found in UNESCO-designed sites. Support this notion, a report from UNESCO titled "Climate Change Adaptation for Natural World Heritage Sites," highlights that engaging local communities in policymaking and governance is vital for implementing adaptation strategies at cultural heritage sites (Hite & Bray, 2010). This report also emphaticises the importance of local peoples' knowledge in climate change assessment, which that climate scientists may not consider.

Several adaptation strategies are in place to mitigate the impact of climate change in coastal areas. For example, Cuba serves as a global model for natural hazard risk management, demonstrated during the Hurricane Georgia, which resulted in only four casualties, far fewer 600 deaths recorded in other countries in the region. Cuba's strategy for coping with disasters is based on three key pillars: disaster preparedness, prevention, and response (Abella and Wisner, 2019). All these aspects are reinforced by an education curriculum at all levels of schools and workplaces, ensuring every individual is knowledgeable about interpreting the information provided by the Cuban Institute of Meteorology (Abella and Wisner, 2019). Another example is the Amalfi Coast in Italy, which is a UNESCO site threatened by coastal erosion, landslides and increasing tourism. Adaptation efforts here focus on stabilizing cliffs while managing tourist activities undermining the fragile ecosystem. Similarly, the Angkor Archaeological Park in Cambodia showcases best practice in conservation efforts, which involve water management and dry-stone wall maintenance.

2 Methodology

Systemic qualitative literature review serves as a valuable tool for identifying and synthesizing key literatures within a specific field of application. By applying a descriptive analysis approach coupled with systemic methods for categorizing diverse source of information, this article aims to highlight viable directions for researchers and policymakers to tackle Climate change threats to Cultural Heritage site (specifically UNESCO-nominated ones).

The process includes the definition of a specific research question (RQ):

RQ How does the cultural landscape of Portovenere Cinque Terre, and the Islands is tackling climate change threats?

2.1 Data Collection

The collection has incorporated a crawler-based web search engine and a bibliographic database of more than thirty articles, starting from review's year range of ten years. This range was purposefully selected to compare the climate changes in different territories.

Through a systematic extraction of relevant information, this article has followed the following two steps: (i) identification of key concepts emerging from articles focusing on the Cultural Landscapes reaction to climate change; (11) extraction and synthesis of common information through a thematic analysis to identify patterns between UNESCO cultural sites and other landscapes; and (ii) clusterization of key concepts into.

2.2 Data Analysis

Clustered concepts were organized into an Excel spreadsheet to consolidate fundamental information from each publication. Subsequently, a content analysis was undertaken to categorize various articles into five overarching macro themes. Thematic analysis was then employed to identify, scrutinize, and extract values from these macro themes. This methodology offers a flexible approach and research tool for researchers to unearth novel insights and concepts derived from the literature (Braun & Clarke, 2006).

This study follows a methodology developed around the values of the site that can be useful for other cultural heritage sites (both inscribed in the UNESCO WHL or not) that are facing similar challenges.

UNESCO sites have been nominated for a series of values—physical and immaterial—listed in the OUV (Outstanding Universal Value) declaration that management bodies must protect. Climate change threats were examined by UNESCO for the first time in 2005. The World Heritage Committee then encouraged single states to review the nature and scale of climate change risks and devise proper management strategies (Nguyen & Baker, 2023). On these premises, the present article will approach the situation of the UNESCO site of "Portovenere, Cinque Terre and the Island (Palmaria, Tino and Tinetto)" following the steps:

1. Description of the site, context and features
2. List of the values to be protected
3. Description of the governance framework
4. Analysis of specific threats linked to climate change
5. Existing projects and approaches for values protection
6. Review and critical analysis.

3 The Case Study of Portovenere, Cinque Terre, and the Islands (Palmaria, Tino and Tinetto)

3.1 Context and Features

Extended for about 15 km along the eastern Ligurian coast between Levanto and La Spezia, the Italian UNESCO site includes the territories of Portovenere and its three islands (Palmaria, Tino and Tinetto), as well as the renowned Cinque Terre area, named after its five villages: Riomaggiore, Manarola, Corniglia, Vernazza, and Monterosso.

Over centuries, the rugged coastal landscape has been shaped by the construction of traditional dry-stone walls and terraces used for cultivating vines and olive trees. Many of these terraces, some extending up to 2 kms in length, date back to the twelfth century, following the cessation of Saracen incursions from the sea. Nowadays, these dry-stone walls are meticulously built and maintained using traditional methods, employing rough sandstone blocks joined with pebbles sourced from the site. The care of these terraces and the cultivation of olive trees and vines are upheld through a collaborative and cooperative approach between local inhabitants and agricultural maintenance staff, a partnership critical for sustaining agricultural activity in the area.

The maquis and natural garrigue survive intact in the upper parts of the slopes and ridges, providing nourishment and shelter to a vast range of animal species. Much of the UNESCO site area is protected within the boundaries of the Cinque Terre National Park and the Natural Park of Porto Venere, ensuring the preservation of its ecological richness and cultural heritage (Fig. 1).

Human communities have adapted to this unique environment, living in compact settlements and villages, built directly on the rocky terrain, characterized by narrow streets. The prevalent use of natural stone for roof covering gives these villages a distinctive aesthetic. The terraced landscape is also dotted with small, isolated or grouped stone shelters, utilized as temporary shelter during harvesting seasons.

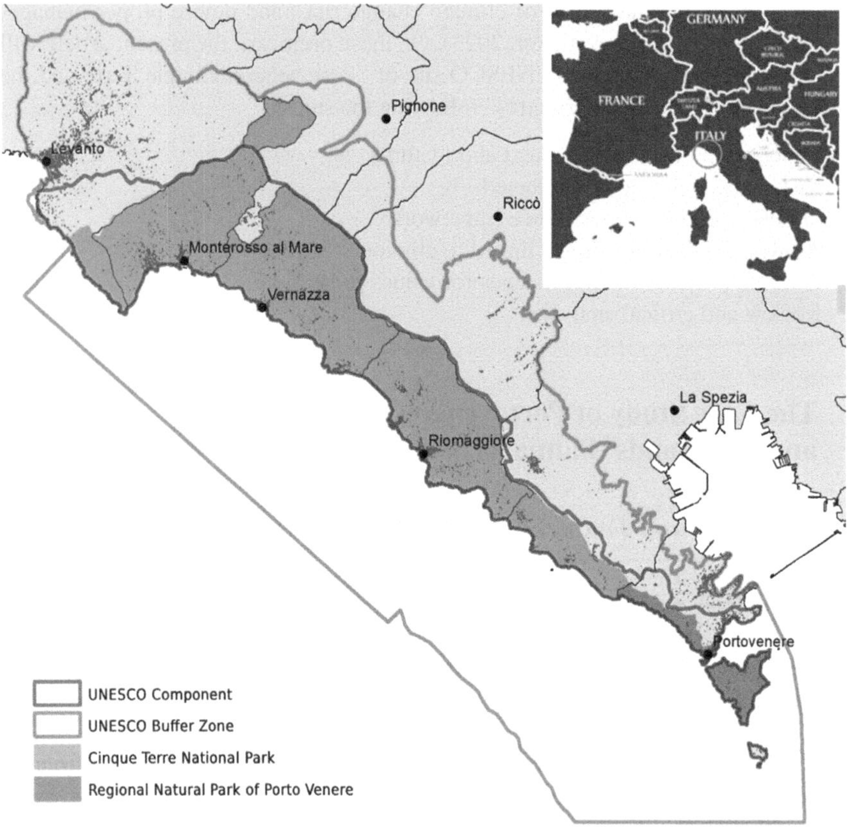

Fig. 1 UNESCO site component and buffer zone

Fig. 2 Grape harvest in Cinque Terre (Vernazza Municipality)

The rugged and visually extraordinary coastal landscape, with its densely packed settlements and spectacular terraces formed over centuries, serve as an exceptional testament to the enduring interaction of traditional communities within a challenging and isolated environment. Recognizing the exceptional nature of this landscape, characterized by its system of villages, terraces, paths, and coastal and mountain connections, along with the 'know-how' of the local culture, UNESCO nominated it as a Cultural Landscape in 1997 (https://whc.unesco.org/en/list/826) (Fig. 2).

However, the inhabitants not only face intrinsic challenges due to the morphology of the territory but also meteorological events that cause landslides on the slopes, floods in the lower valleys, and fires in the summer season. The UNESCO Site, therefore, offers valuable insights into how a cultural landscape deals with the pressing issue of climate change. This site stands apart from other coastal cultural sites due to its distinct physical characteristics and its intangible dimension, a distinction duly acknowledged by UNESCO. This inherent quality is underscored by its classification within the World Heritage List (WHL) as a 'Cultural Landscape', together with other 120 sites of exceptional value around the world. Consequently, tackling the risks of terrace abandonment, village depopulation, and impairment of age-old intangible heritage, exacerbated by adverse weather conditions and climate change impacts, is of primary importance. In response to these multifaceted challenges and the imperative to safeguard its intrinsic values, the site has equipped itself with a comprehensive and multi-actor governance framework. This framework seeks to address the diverse challenges posed by climate change in a holistic, multidisciplinary, and organic manner. In recent years, the various actors within this governance system have set up an organic strategy for the ordinary and extraordinary management of the site's values, with a keen focus on climate change challenges, and have implemented a series of strategies and actions aimed at achieving tangible results.

3.2 *Being a UNESCO Site: The OUV (Outstanding Universal Value) to Preserve*

The Outstanding Universal Value (OUV) (https://whc.unesco.org/en/list/826/) outlines a description of the site, the justification for the inscription on the UNESCO World Heritage List and the fulfilment of the criteria. Specifically, the values are substantiated by physical elements, the attributes, which materially support them, as well as by immaterial aspects (the human capital, the passing-on of oral traditions, the agricultural and dry-stone walls maintenance know-how) (Table 1).

Criterion II underlines an important value with a more immaterial character, the "traditional lifestyle" which has existed for a thousand years, and which continues to play an important socio-economic role in the life of the community. This value is concretized in traditional activities and knowledge that support agriculture, forestry, and fishing activities, such as the maintenance of dry-stone walls, and traditional agricultural and fishing products, therefore in relation to several of the attributes expressed in the table.

Finally, **Criterion IV** highlights the relationship between coastal settlements and the sea: although not included in the definition of the site boundaries, the sea assumes a scenic-perceptive, environmental, and economic role (fishing and tourist-beach activities). The coastal spaces are multifunctional, besides agriculture, serving as hubs for fishing (unloading goods, boat docking, etc.) as well as recreational and leisure areas for residents and visitors (Fig. 3).

3.3 *The Governance of the UNESCO Site*

To ensure the conservation, enhancement, sustainable, and equitable development of the UNESCO Site, based on its values and in compliance with the tangible and intangible attributes that sustain them and make them understandable, a coordinated and shared management strategy is imperative for all institutional players.

On 1 August, 2016, a Program Agreement was signed by all responsible entities for the site and its buffer zone: Liguria Region, Ministry of Cultural Heritage and Activities and Tourism, Cinque Terre National Park, Municipality of Porto Venere—Regional Natural Park of Porto Venere, Municipalities of Levanto, Monterosso al Mare, Pignone, Riomaggiore, La Spezia, Vernazza, Beverino, Riccò del Golfo.

Table 1 Values and attributes of the UNESCO site

Values expressed by the criteria		Physical attributes and characteristics	CRITE-RIA
II	The eastern Ligurian Riviera between Cinque Terre and Portovenere is a cultural site of outstanding value that illustrates a traditional way of life that has existed for a thousand years and continues to play an important socio-economic role in the life of the community	Valleys, basins, and cliffs: in rapid succession and in close contact with the sea, they contribute to the scenic quality of the landscape	IV
IV	The Ligurian coastal region from Cinque Terre to Portovenere is an outstanding example of landscape where the layout and disposition of small towns, historically stratified, in relation to the sea, and the shaping of the surrounding terraces that overcame the disadvantages of a steep, uneven terrain, encapsulates the continuous history of human settlement in this region over the past millennium	Natural habitats: garrigue and Mediterranean maquis environments, with flaps of holm oaks and other wooded formations, in a mosaic juxtaposed with the cultivated areas; presence of fauna species of naturalistic interest	IV
V	Portovenere, Cinque Terre, and the Islands (Palmaria, Tino and Tinetto) is a remarkable cultural landscape created by human endeavour over a millennium in a rugged and dramatic natural environment. It represents the harmonious interaction between people and nature to produce a landscape of exceptional scenic quality	Terraces and dry-stone walls, morphological peculiarities induced by man, suitable for soil conditions. Walls built with local materials and traditional techniques	II, IV, V
		Rural artefacts in terraced areas, supporting agriculture	IV, V
		Network of paths and pedestrian paths in rural areas, for agricultural use and connecting villages, hiking trails, historical and devotional itineraries; several types of paths and ground (dirt track, grassy edges, drywall heads, vie di lizza, stone stairways, etc.); crossings and complementary works	IV, V

(continued)

Table 1 (continued)

Values expressed by the criteria		Physical attributes and characteristics	CRITE-RIA
		Compact and separate settlements, coastal and on the slope, adapted to geomorphology; historically, built, stratified from the archaeological phase to the contemporary era	IV, V
		Mooring system, which substantiates the relationship with the sea from a commercial, economic and tourist point of view	IV, V
		Traditional cultivations, expression of the various characteristics of the soil, the steepness, and the climate (olive tree, vine)	II, IV, V
		Typical agri-food production and related places and methods of processing products (local wines, anchovies, etc.)	II, IV, V

This formal Act delineates the roles and commitments of each party in site management, protection, and enhancement activities.

The Agreement establishes a site management structure, comprising various bodies composed of signatories of the Act:

- Coordination committee—a direction and supervisory body aimed at ensuring the effective management of the Site;
- Community of buffer zone Municipalities responsible for coordinating objectives, problems, and activities relating to the buffer zone;
- Technical-administrative work group responsible for implementing the Management Plan and common activities with operational functions for implementing the coordination committee guidelines;
- UNESCO site office, carries out joint site promotion activities and ensures the smooth running of the technical-administrative board;
- Technical Secretary supports the Coordination Committee, the Working Group, and the UNESCO site office;
- Consulting committee responsible for advisory functions directed to the Coordination Committee regarding the knowledge programmes, general plan guidelines, specific projects, monitoring the implementation of the Agreement and the periodic report.

Fig. 3 Dry-stone Walls maintenance in Cinque Terre

Although non-institutional stakeholders are not officially included in the governance, a continuous series of participation activities are carried out by municipalities, UNESCO office, and Cinque Terre National Park on several topics ranging from tourism management, land management, and agriculture.

3.4 Climatic Context and Identified Vulnerabilities

Liguria region, situated at the intersection of Alpine and Mediterranean climates, is climatologically sensitive to climate change, with most natural hazards occurring between the late summer and autumn. These hazards, such as excessive rainfall events, are exacerbated by warm water and large-scale atmospheric conditions, leading to strong mesoscale convective systems (MCS) (Cassola et al., 2023; IFAB, 2023). As such, Liguria stands out as one of Italy's most hydrologically vulnerable areas, with approximately 40% of its residents inhabiting high-risk zones. It is not uncommon for several Ligurian locations to have up to 500 mm of rainfall (30–40% of the annual average) in a single day (Gallus et al., 2018).

In recent decades the flood events frequency on the regional territory has been practically annual, with catastrophic events recorded approximately every twenty years (1970, 1992, 2011) (SIGEA, 2011).

The 1970 flood on the Ligurian coast was the consequence of exceptional rainfall, which exceeded the critical levels observed until then. Some weather stations recorded more than 400 mm rain fallen in 24 h. The 1992 floods affected Genoa and Savona, while one of the most dramatic events in recent years for the Cinque Terre was the flood that hit the area on 25 October 2011. The precipitation that fell throughout the event was 542 mm, compared to the annual average of around 1500 mm. The event caused 11 victims, 2 missing people and extensive damage. Monterosso al Mare and Vernazza were particularly affected, where the roads turned into overflowing rivers and carried the mud and debris collected on their way to the sea. There was also severe damage to the infrastructure due to landslides and mudslides with 4 bridges collapsed and 20 roads closed, telecommunications interrupted, and several villages and towns unreachable for days.

The province of La Spezia had already been hit in 2009 with the flooding of the Magra river and in 2010, with a sequence of cloudbursts including the event that hit Portovenere on 13 August 2010.

Moreover, the region's extensive coastal areas are under constant threat from sea-level rise, leading to increased risks of flooding and erosion (Copernicus, 2023). These processes gradually wear away shorelines, terraced slopes, cliffs, and the UNESCO archaeological site (UNESCO n.d.). In addition, landslides are becoming more frequent and intense due to extreme weather events, including tornados and thunderstorms (Franceschini, 2022).

As outlined in the Disaster Risk Management Plan, landslides—caused by a combination of factors such as deforestation, anthropological activities, and extreme rainfall—inflict severe damages to cultural heritage site and due to inadequate management of the terraced landscape, underscore the vulnerability of Cinque Terre sites to natural disasters. This condition emphasizes the urgency for proactive risk preparedness measures.

Among the relevant impacts on coastal area of Cinque Terre, it is also important to consider the intrusion of saline waters into rivers and aquifers, which complicates the drainage of water towards the sea during extreme events. Indirect impacts concern changes in the functions of coastal ecosystems and human activities along the coast, driven by transformations in coastal areas and changing climatic conditions in terms of precipitation frequency and temperature variation.

Increasing precipitation intensity and frequency are also accompanied by more days of drought, which, in the more inland area of Cinque Terre, pose a risk of desertification. With the Mediterranean regions recording a warming 20% greater than the global average, and the increase of extreme weather patterns in Cinque Terre, such as heatwaves and windstorms, the onset of forest wildfires over large fire-prone areas is facilitated (Bacciu et al., 2022).

Depopulation is another critical factor contributing to the increased vulnerability of the area and affecting all UNESCO-protected municipalities. This ongoing trend poses a potential threat to the UNESCO site as it gradually diminishes the human influence, which is an integral and essential part of the cultural landscape and contributes to the site's value. Depopulation has consequently led to a decline in agricultural industry, causing degradation in terraced landscapes. Currently, more than 77% of these areas in the Cinque Terre have been abandoned, leading to instability of the terraces (Brancucci, 2007). As demonstrated by Shrestha et al. (2004), the main cause of terrace degradation is related to agricultural activities. As such, current impacts of climate change together with the loss of traditional agricultural practices, represent a challenge to identifying effective mitigation and adaptation strategies.

3.5 The Actions to Contrast Climate Change and Preserve the Cultural Landscape

The activities and the approach carried out by the UNESCO site contribute to a wider strategy at national and regional levels, which includes the Liguria "**Regional strategy for adaptation to climate change**" (approved in January 2023), a tool that enables the planning of actions, interventions, and best practices to manage and limit the effects of climate change on the economy and the territory.

At the local level, the strategy comprises several instruments and approaches implemented by local institutional stakeholders: a loose multi-actor and multisectoral framework aimed at organically targeting various aspects of a complex territory.

The main proponents of this strategy are the UNESCO Office and the National Park of Cinque Terre.

In recent years, the UNESCO office has incorporated Climate Change aspects into the UNESCO site's Management Plan and activated measures and projects to ensure the protection and enhancement of the site's universal value. These actions range from educating younger generations on agricultural and maintenance practices, to mitigating the negative impacts of tourism and implementing physical interventions for slope stabilization (Fig. 4).

The **Disaster Risk Management Plan**, drafted by the UNESCO Office in collaboration with interested stakeholders, is an 'umbrella' tool that identifies potential risks, establishes the governance structure in case of disasters, develops communication strategies for prevention aimed at both tourists and residents, and serves as a roadmap to guide administrators, local stakeholders, and the population in dealing with extreme events. The document describes and analyses the main risk elements associated linked with disasters, divided into two macro categories:

- The main dynamics in progress linked to the theme of disasters, since they contribute to the vulnerabilities of the territory, such as socio-demographic changes (depopulation) abandonment of agriculture and terracing, tourism, climate change, etc.
- Disasters such as floods, landslides and instability, fires, heavy snowfalls, hailstorms, earthquakes.

The analysis proceeds with an identification of the site's vulnerabilities. Vulnerability indicates the "propensity to damage" and measures the fragility of structures,

Fig. 4 Dry-stone walls maintenance manual by Cinque Terre National Park

anthropogenic and social elements, and the environment in the face of events capable of causing damage. Landscape vulnerability depends on factors such as the level of soil degradation, ongoing fragmentation, and the degree of autochthony of the vegetation.

To provide a synthetic framework explicitly referencing the Outstanding Universal Value, the DRMP proposes an assessment of risk factors through a tabular analysis, crossing territorial threats and vulnerabilities with a list of material values and attributes.

Methods of prevention and management of emergencies are examined, setting out the regulatory framework, plans, and procedures for preparation and management during disasters, as well as describing the actors involved and management structures that are activated in the various phases of the emergency, the procedures for assessing and managing post-disaster damage.

Finally, the actions to improve and systematize the prevention and management of risks are illustrated, including initiatives and projects in place for territory protection by responsible bodies (Region, Cinque Terre National Park, etc.), actions undertaken by the UNESCO Site (listed in this Management Plan), and improvement indications for risk prevention and management.

The UNESCO site is participating in the **RescueME** project (Equitable RESilience solutions to strengthen the link between CULtural landscapEs and coMmunities), funded by Horizon Europe, which focuses on a landscape approach to resilience. The project aims to protect cultural heritage and landscapes by mobilizing resources, engaging stakeholders, facilitating decision-making, and enhancing the implementation of co-created and just resilience solutions. Project activities will also be developing, testing, and demonstrating innovative tools and processes for climate adaptation and disaster risk management. It assesses risks and opportunities, co-developing inclusive and just resilience strategies, and equipping local communities with advanced tools to protect European coastal landscapes. RescueME is building a resilience baseline, studying the impact chains related to floods and landslides, activating a co-creation framework to develop and test innovative solutions.

Another initiative, funded through the LIFE Programme of the European Commission, is the project **StoneWallsForLife**. It demonstrates how drystone walls are likely to mitigate climate change. The activities aim at repairing drystone walls and ensuring their long-term maintenance, to protect the territory and its inhabitants against the effects of extreme weather events. Drystone walls are thus environmentally relevant since they effectively improve both the resistance and the resilience of the territory to climate change. Furthermore, the project highlights the social and cultural value of drystone walls and supports their restoration by bringing together experienced builders with migrants and unoccupied people, thus facilitating their job placements and social inclusion. Currently, the park is working on an Interreg proposal that integrates agriculture and climate change management, particularly in viticulture. This project would complement the aforementioned StoneWallsForLife initiative and focus on the agricultural aspect of the study area.

In the context of the EU project, the National Park is also currently (February 2024) drafting the "**Plan for the Adaptation to Climate Change**", which will consider both the protection of biodiversity and traditional agricultural activities as risk mitigation factors.

Additionally, the **Maregot** project (Interreg Programme) focuses on the prevention and management of risks deriving from coastal erosion and intends to contribute to a growth in knowledge and awareness of the main targets with respect to the environmental issue of integrated coastal management. To achieve this aim, a communication plan for the project will be prepared which will contain specific actions aimed at individual targets and provides for an important impact on the territory. The project will define a cross-border communication strategy which will also deepen the knowledge of local contexts to respond to the needs of each individual territory.

The UNESCO Office is currently promoting studies on the **Tourist Carrying Capacity** of the site to prevent natural resource overexploitation, loss of human capital and resident population, as well as negative impacts on the landscape and waste management system due to high (and increasing) tourist numbers.

These activities reflect the site's commitment to landscape maintenance and disaster mitigation. Particularly following the flood on October 25th, 2011, the site has implemented a series of actions aimed at mitigating the damage caused by natural disasters.

An oversight body, known as the **Geological Risk Studies Center,** has been established. It comprises professionals, the Geological Association, the park authority, the University of Federico II, and the University of Genoa's DISTAV—Department of Earth, Environmental, and Life Sciences. This center is responsible for monitoring landslides, analyzing flood events, and tracking all ongoing hydrogeological instability phenomena.

The municipality of Riomaggiore has long adopted its own regulations in the event of an alert, with corresponding self-protection measures for the villages of Riomaggiore, Manarola, and Volastra. Procedures for communication have been developed to maximize the impact of catastrophic events and post-alert actions. These include geological assessments to analyze the state of the terrain and specific interventions such as trail maintenance teams, a dedicated professional figure created by the park for trail upkeep and natural meteoric water flow control. A weather alert has been established by mayoral ordinance, which suspends access to the trails, and access is revoked following verification by the geologist summoned by the park authority.

By the end of 2023, the site is a candidate for inscription in the register of historic rural landscapes. Looking ahead, there is a desire to include the entire UNESCO site in the Globally Important Agricultural Heritage Systems (GIAHS), not just the part within the park. GIAHS are agroecosystems inhabited by communities that have an intricate relationship with their territory. These evolving sites are resilient systems characterized by remarkable agrobiodiversity, traditional knowledge, invaluable cultures, and landscapes, sustainably managed by farmers, herders, fisherfolk, and forest people in ways that contribute to their livelihoods and food security. Through the GIAHS, the Food and Agriculture Organization of the United Nations (FAO) has designated over 60 sites around the world.

4 Discussion

As seen, the UNESCO site's approach to the topic is varied, with several stakeholders engaged in individual aspects, sometimes in an organic and synergistic manner, and sometimes independently. The current response to Climate Change, therefore, reveals a wealth of actions but a partially synergistic approach to the topic.

Firstly, while the most important and recognizable values and attributes can profit from a varied set of measures (dry-stone walls, villages, agriculture traditions, etc.) not all values recognized by UNESCO feature appropriate protection measures, underlying the lack organic take on the Climate Change threats topic.

Furthermore, as seen, the UNESCO site is inherently complex, from a morphological to a governance point of view. The mosaic of administrative and non-formalized stakeholders allows for organic governance to a certain extent. Moreover, efforts are underway to include local non-institutional stakeholders in the governance process, shaping a multi-actor and increasingly multi-sectoral governance structure, as also requested by ICOMOS.

This process has been initiated through projects focused on participation, such as RescueME, which engage local stakeholders in studying and addressing issues related to resilience and responses to climate change. Other projects take a top-down approach, and the mosaic of responses currently appears still fragmented.

Furthermore, there is a lack of an organic monitoring model for actions dedicated to climate change, with monitoring of project progress and effectiveness left to individual proponents.

Despite these challenges, the site seems to be moving towards a more organic approach.

5 Conclusions

The UNESCO site represents an interesting example of how a coastal cultural landscape is likely to mitigate and adapt to the impacts of climate change anticipated in the future. Its recognition on the World Heritage List underscores the imperative to safeguard such a landscape, prompting local authorities to develop tailored-specific measures that could serve as inspiration for many other coastal areas in Southern Europe.

The strategies and actions proposed in the UNESCO management plan, along with other complementary plans and instruments, span from the local to the European level. These encompass agreements with scientific institutions (such as research foundations and universities), collaboration with local communities (information and dissemination), support to the local stakeholders (help to agricultural stakeholders), as well as actions to maintain territorial strength and stability (terraces, rivers, forests, etc.). Methods and actions to manage risks (Disaster Risk Management Plan) have been put in place, and participation in climate change related international projects (such as Horizon Europe, LIFE, Interreg etc.) is feeding the local strategies for climate change adaptation.

References

Abella, E. A. C., & Wisner, B., (2019). Natural hazards governance in Cuba. In *Oxford Research Encyclopedia of Natural Hazard Science.*

Bacciu, V., Mereu, V., Sirca, C., Saura, J. M. C., Ribotta, C., & Spano, D. (2022). A review of recent climate evolution, projected climate change and its impacts on future fire regime and behavior in the Italy-France maritime cooperation area (Sardinia, Corsica, Tuscany, Liguria, and Provence-Alpes-Côte d'Azur). *Environmental Sciences Proceedings, 17*(1), 68.

Bongarts Lebbe, T., Rey-Valette, H., Chaumillon, É., Camus, G., Almar, R., Cazenave, A., & Euzen, A., (2021). Designing coastal adaptation strategies to tackle sea level rise. *Frontiers in Marine Science*, p. 1640.

Brancucci, A. (2007). The terraced landscapes of Cinque Terre: Historical and environmental considerations. In: *Terraced landscapes in Italy: State of the art and future challenges.*

Braun, V., & Clarke, V. (2006). Using thematic analysis in psychology. *Qualitative Research in Psychology, 3*(2), 77–101.

Cassola, F., Iengo, A., & Turato, B. (2023). Extreme convective precipitation in Liguria (Italy): A brief description and analysis of the event occurred on October 4, 2021. *Bulletin of Atmospheric Science and Technology, 4*(1), 4.

Copernicus., (2023). Soil erosion indicators for Italy from 1981 to 2080. Retrieved from: Soil erosion indicators for Italy from 1981 to 2080 (copernicus.eu).

Council of Europe. (2000). *European landscape convention.* Florence.

Dastgerdi, A. S., Sargolini, M., Allred, S. B., Chatrchyan, A. M., Drescher, M., & DeGeer, C. (2022). Climate change risk reduction in cultural landscapes: Insights from Cinque Terre and Waterloo. *Land Use Policy, 123*, 106359.

de Schipper, M. A., Ludka, B. C., Raubenheimer, B., Luijendijk, A. P., & Schlacher, T. A. (2021). Beach nourishment has complex implications for the future of sandy shores. *Nature Reviews Earth & Environment, 2*(1), 70–84.

von Droste, B. M. Rössler and S. Titchen, (1999). *Linking Nature and Culture. Report of the Global Strategy Natural and Cultural Heritage Expert Meeting*, (Eds.) 25–29 March 1998, Amsterdam, The Netherlands.

Fatorić, S., & Biesbroek, R. (2020). Adapting cultural heritage to climate change impacts in the Netherlands: Barriers, interdependencies, and strategies for overcoming them. *Climatic Change, 162*(2), 301–320.

Fatorić, S., & Egberts, L. (2020). Realising the potential of cultural heritage to achieve climate change actions in the Netherlands. *Journal of Environmental Management, 274*, 111107. https://doi.org/10.1016/j.jenvman.2020.111107

Franceschini, R., Rosi, A., Del Soldato, M., Catani, F., & Casagli, N. (2022). Integrating multiple information sources for landslide hazard assessment: The case of Italy. *Scientific Reports, 12*(1), 20724.

Gallus, W. A., Jr., Parodi, A., & Maugeri, M. (2018). Possible impacts of a changing climate on intense Ligurian Sea rainfall events. *International Journal of Climatology, 38*, e323–e329.

Garmestani, A., Craig, R. K., Gilissen, H. K., McDonald, J., Soininen, N., van Doorn-Hoekveld, W. J., & van Rijswick, H. F. (2019). The role of social-ecological resilience in coastal zone management: A comparative law approach to three coastal nations. *Frontiers in Ecology and Evolution, 7*, 410.

Grøntoft, T., & Cassar, J. (2020). An assessment of the contribution of air pollution to the weathering of limestone heritage in Malta. *Environmental Earth Sciences, 79*(12), 288.

Hite, S., & Bray, T. M., (2010). International organizations in education: UNESCO's international institute for educational planning. *International Encyclopedia of Education.*

IFAB, (2023). What is the European Extreme Events Climate Index. Retrieved from: European Extreme Events Climate Index—IFAB International Foundation (ifabfoundation.org).

Kapsomenakis, J., Douvis, C., Poupkou, A., Zerefos, S., Solomos, S., Stavraka, T., & Zerefos, C. (2023). Climate change threats to cultural and natural heritage UNESCO sites in the Mediterranean. *Environment Development and Sustainability, 25*(12), 14519–14544.

Luengo, M., & Ramsay, A. (2020). Cultural landscapes and their transformative potential for social innovation. *Heritage, 6*(12), 7674–7705. https://doi.org/10.3390/heritage6120404.

Nguyen, K. N., & Baker, S. (2023). Climate change and UNESCO world heritage-listed cultural properties: a systematic review, 2008–2021. *Heritage, 6*(3), 2394–2420.

O'Donnell, P. M., (2023). Cultural landscapes: integrating culture and nature to uplift global sustainability through the lenses of the UN SDGs 2030 Agenda. In *Placemaking and Cultural Landscapes* (pp.61–78). Springer.

Ontario Heritage Trust. (2012). *Cultural heritage landscapes – an introduction.*
Orr, S. A., Richards, J., & Fatorić, S. (2021). Climate change and cultural heritage: A systematic literature review (2016–2020). *The Historic Environment Policy & Practice, 12*(3–4), 434–477.
Pearson, D., & Gorman, J. (2023). Acknowledging landscape connection: Using sense of place and cultural and customary landscape management to enhance landscape ecological theoretical frameworks. *Land, 12*(4), 729. https://doi.org/10.3390/land12040729.
Potthoff, K. (2013). The use of 'cultural landscape' in 19th century German geographical literature. *Norwegian Journal of Geography*. https://doi.org/10.1515/norwegian-journal-of-geography.
Richards, J., Orr, S. A., & Viles, H. (2020). Reconceptualising the relationships between heritage and environment within an Earth System Science framework. *Journal of Cultural Heritage Management and Sustainable Development, 10*(2), 122–129.
Rössler, M. (2006). World heritage cultural landscapes: A UNESCO flagship programme 1992–2006. *Landscape Research, 31*(4), 333–353. https://doi.org/10.1080/01426390601004210
Sesana, E., Gagnon, A. S., Ciantelli, C., Cassar, J., & Hughes, J. J. (2021). Climate change impacts on cultural heritage: A literature review. *Wiley Interdisciplinary Reviews Climate Change, 12*(4), e710.
SIGEA, (2011). Italian Society of Environmental Geology. Retrieved from. https://www.sigeaweb.it/documenti/gda-supplemento-dissesti-costa-tirrenica.pdf
Shrestha, D. P., Zinck, J. A., & Van Ranst, E. (2004). Modelling land degradation in the Nepalese Himalaya. *Catena, 57*(2), 135–156. https://doi.org/10.1016/j.catena.2004.01.002.
UNESCO, (n.d.). Portovenere, Cinque Terre, and the Islands (Palmaria, Tino and Tinetto). Retrieved from: Portovenere, Cinque Terre, and the Islands (Palmaria, Tino and Tinetto) —UNESCO World Heritage Centre.
United Nations, (2015). Transforming our world: The 2030 Agenda for Sustainable Development. *New York: United Nations, Department of Economic and Social Affairs, 1*, p. 41.
World Heritage Centre, 2016. The Operational Guidelines for the Implementation of the World Heritage Convention.

Lifelong Learning of Coastal Communities: An Investigation on Nature-Based Solutions to Mitigate Climate Change Effects on Coastal Built Environments

Achini De Silva, K. B. Seenapatabendige, Upul Liyanage, Dhammika Ranathunga, and S. K. Gunathilake

1 Introduction

Southern coastal belt of Sri Lanka is renowned for its rich cultural heritage, a landmass which has tasted the flavors of colonization under Portuguese, Dutch and British and the country's most socially and economically attractive land stretch home for 2.6 million multi ethnic population. Fisheries, agriculture and tourism are the main industries. Coastal belt and its livelihood options; fisheries and tourism are central to the economy of the southern province. Marine fisheries are the oldest and most important coastal livelihood option of the coastal dwellers since ancient times. Southern coastal belt is one of the richest natural resources bases that offer invaluable livelihood opportunities, particularly resource-based economic activities such as fishing, tourism and

A. De Silva (✉)
Department of Agribusiness Management, Faculty of Agricultural Sciences, Sabaragamuwa University of Sri-Lanka, P.O. Box 02, Belihuloya, Sri Lanka
e-mail: desilva.achini@yahoo.co.uk

K. B. Seenapatabendige
Department of Architecture, Faculty of Built Environment and Spatial Science, Southern Campus, General Sir John Kotelawala Defense University, Sooriyawewa, Sri Lanka

U. Liyanage
Regional Research Center, National Aquatic Resources Research and Development Agency, Kapparathota, Weligama, Sri Lanka

D. Ranathunga
Ministry of Fisheries, New Secretariat, Maligawatta, Colombo 10, Sri Lanka

S. K. Gunathilake
Department of Natural Resources, Faculty of Applied Sciences, Sabaragamuwa University of Sri-Lanka, P.O. Box 02, Belihuloya, Sri Lanka

C. Malalgoda et al. (eds.), *Climate Change Adaptation in the Built Environment*,
https://doi.org/10.1007/978-3-031-75826-3_11

recreation. Marine fisheries sector comprised 3 subsectors, coastal, offshore and deep-sea fishing where fishing communities shared the common pool resources. Moreover, small scale fisheries are the most socially and economically important sub sector holding lion's loin.

1.1 Built and Ecological Environment of the South Coast of Sri Lanka

South coast of Sri Lanka has divided into 3 administrative districts; Galle, Matara and Hambantota. Coastal belt is considered as one of the most economically, socially and environmentally vital land area of the country. Built environment of the South coast is structurally designed and developed to cater the needs of tourism, fisheries and residencies. Buildings and other infrastructure facilities are key components of the built environment while natural environment, both coastal and marine environment host the built environment. Traditional fish landing sites (Sabbioni et al., 2006), 12 fishery harbors, auction halls, markets, houses of the fishing communities created the coastal built environment. On the other hand, tourist hotels, guest houses, restaurants, cafes, and tourism infrastructure acquired considerable space already and resource grabbing for tourism reached to alarming levels. Figure 1 present the fish landing sites, both traditional and modern, and the fishing harbors located along the southern coastal belt. Fish landing sites and harbors are central to the coastal life and each location surrounded by the bundle of fishing villages. Tourism, another key industry which places significant demand on coastal resources, especially on construction of star class hotels, restaurants and other infrastructure. Distribution of diverse range of tourist hotels present in Fig. 2.

1.2 Context: Coastal Resources, Competing Demands, Nature-Based Solutions and Lifelong Learning

Lifelong learning and its interventions to manage the coastal disasters and building resilience was the underlying theme of the present research. Research gap analysis facilitates identifying the gaps and the status of the published research. A content analysis of published research used to interpret the content of text data through the systematic classification process of coding and identifying themes and patterns (Hsieh & Shannon, 2005). Results of the content analysis (Fig. 3) of the objectives of the published research on climate change impact on coastal ecosystems and its industrial components. Special emphasis was brought to identify the key focused

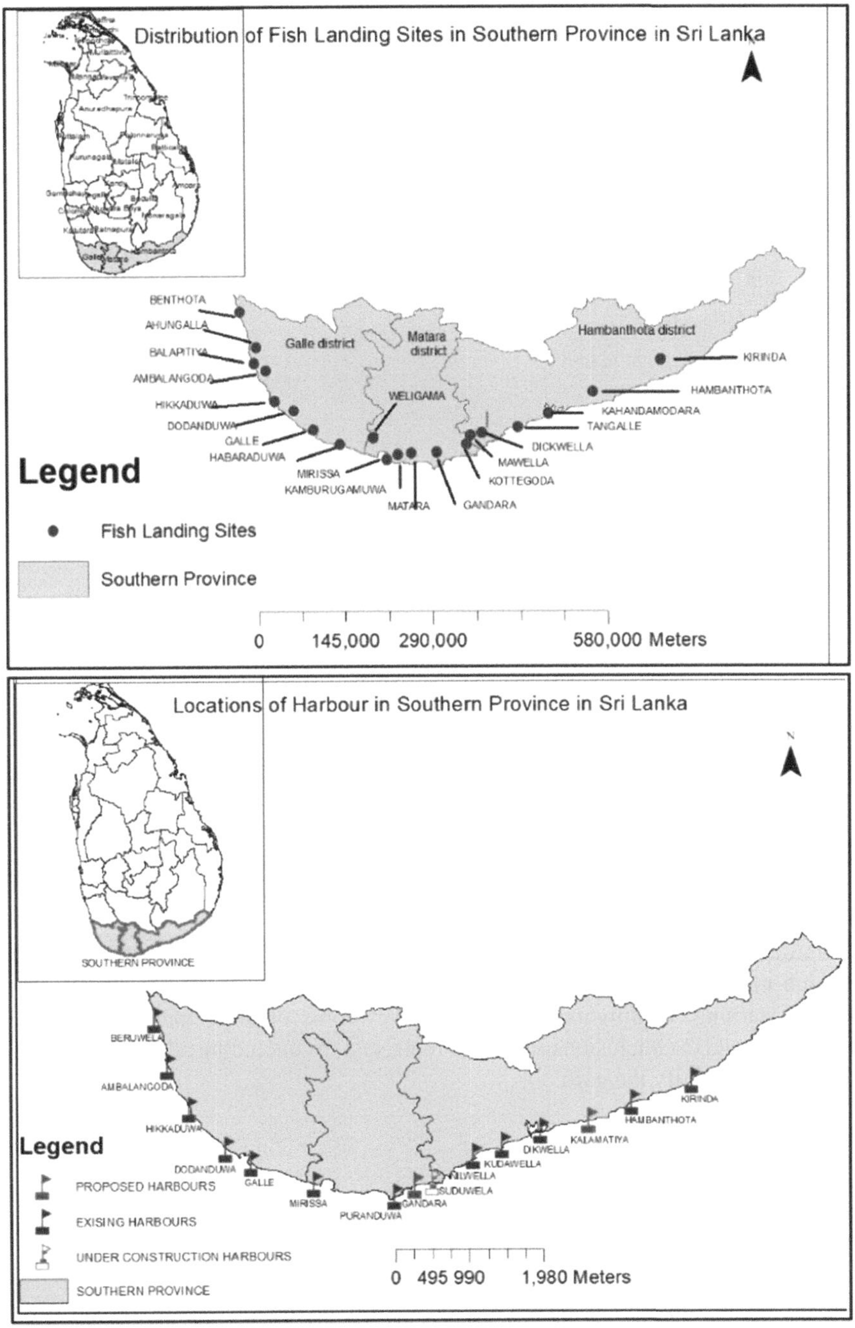

Fig. 1 Fishery harbors and landing sites as key components of the coastal built environment of the southern coast of Sri Lanka

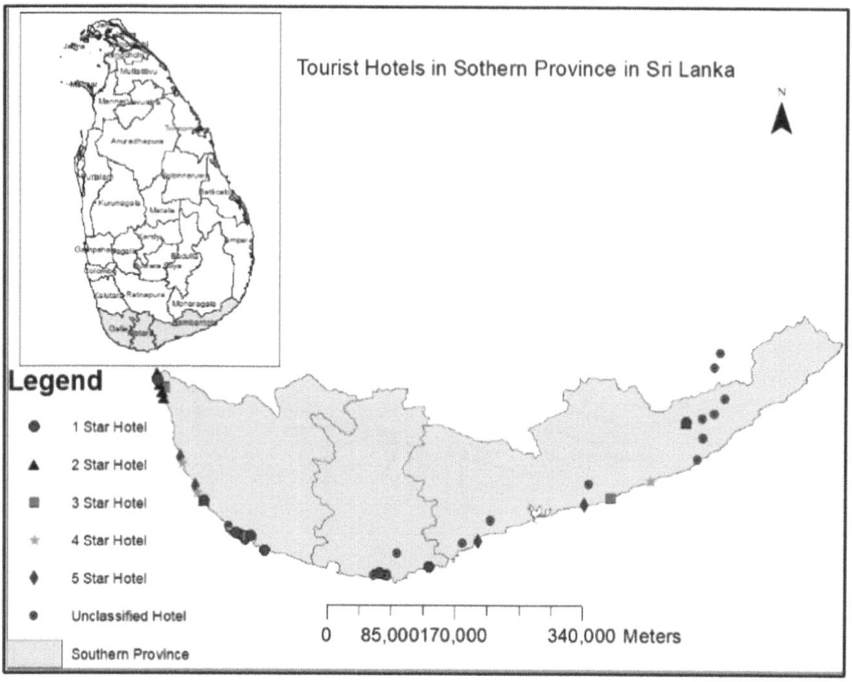

Fig. 2 Map highlighting the typology of the tourist hotels located along the southern coastal belt, a significant influencer to the coastal built environment

areas of the published research. Coastal areas, climate change, strategies, management, adaptation and knowledge were the key themes and lifelong learning, nature-based solutions and community approaches were not highlighted. Content analysis of published research opens up the research gaps.

Figure 4 brought the results of the content analysis of the conclusions of published research on climate impact of coastal zones, hazards, adaptation and mitigation through learning. Highlights of the figure, keywords of large fonts indicate the highly appeared in conclusions and small fonts, less in concluding remarks. Analysis facilitated to identify the gaps.

2 Conceptualization: Climate Change Adaptation, Disaster Risk Reduction and Lifelong Learning

Climate change impacts on coastal ecosystem of island nations reached to alarming levels and expected to rapidly worsen if there are no sustainable solutions along with massive transformation to a low carbon society is implemented (IPCC, 2022; UNDRR, 2020). Coastal built environments are facing two types of climate induced

Fig. 3 Results of the content analysis highlighted the key focused areas of the published research and gaps. *Source* (Bektas & Sakarya, 2023; Ezeah et al., 2014; Hsieh & Shannon, 2005; Karki et al., 2011; Macgregor et al., 2018; Rathnayaka et al., 2021; Rowberry, 2022; Sabbioni et al., 2006; Wang et al., 2022; Wilby, 2007; Zieba et al., 2020)

Fig. 4 Results of the content analysis of the conclusions of published research highlighted the conclusions drawn in published research on climate induced coastal zones hazards and role of knowledge-based approaches. *Source* (Bektas & Sakarya, 2023; Ezeah et al., 2014; Hsieh & Shannon, 2005; Karki et al., 2011; Macgregor et al., 2018; Rathnayaka et al., 2021; Rowberry, 2022; Sabbioni et al., 2006; Wang et al., 2022; Wilby, 2007; Zieba et al., 2020)

incidents; sudden onset or extreme events (floods, storms, etc.) and slow onset events (coastal erosion, sea level rise, etc.) (UNFCCC. CMA., 2018). Climate change impacts adding extra pressure on coastal communities and their livelihood in varying degrees and level of risk attached to the resource richness, exposure, knowledge and skills of the community. Low income, capital deficit developing nations with limited research and development efforts on disaster risk reduction and climate change adaptation are highly vulnerable in one hand. On the other hand, poor coping mechanisms, institutional involvement and infrastructure worsen the situation (IPCC, 2022). Nature-based solutions (NbS) gives much hope for highly vulnerable coastal communities to manage disaster risks (Châles, 2023; UNDRR, 2020). Nature based solutions received global recognition as viable solution to environmental crisis in more integrated manner, role that nature can play in climate mitigation and adaptation (Seddon et al., 2020). NbS plays multidimensional role in disaster risk reduction, sustainably manage and restore natural ecosystems that address the societal challenges effectively and adaptively while providing human well-being (Cohen-Shacham et al., 2016).

Communities as a key player in design and develop disaster risk reduction (DRR) and climate change adaptation (CCA) their exposure, knowledge, skills, research and practices are essential in sustainable eco-system management. The emergence of grassroots CCA and DRR initiatives calls for adult learning strategies that empower and promote the adaptive capacity of communities, and enable communities to develop sustainable solutions (Quang & Wit, 2020). Action-oriented learning, active community engagement, place-based learning, inclusive remote learning and timely responses to local needs of global CCA and DRR agenda. Figure 5 explain the conceptualization process.

3 Methodology

Present study aimed to investigate the role of lifelong leaning in experiencing and applying NbS in climate change adaptation of the coastal communities of the South of Sri Lanka. Socially, economically and naturally diverse southern coastal belt provide unique research location to investigate the life long learning of the communities. Further, lifelong experiences enriched with the knowledge systems of Portuguese, Dutch and British rulers, ethnic and cultural diversity along with the eco-system services of rich resources base. Therefore, research approach aimed to connect with selected respondents from various sub sectors of the local economy and investigate the applications of lifelong learning.

Multidimensional tools were instrumental to collect the data (Table 1). Field survey based on structured a questionnaire (n = 135 fishery value chain actors), 5 focus group interviews with key informants of the coastal communities and tourism and other industry representatives, resource mapping and audit, transect walk, develop seasonal calendar on climate effects, wealth ranking and digital maps

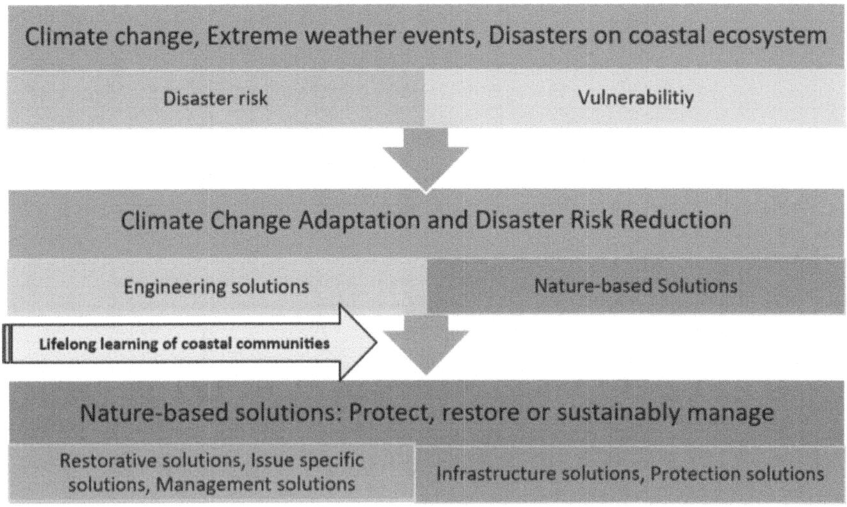

Fig. 5 Conceptualization: connection between climate change, disaster risk reduction through lifelong learning on nature-based solutions. *Source* Adopted from UNDRR (2020)

were instrumental in developing the lifelong learning context of the coastal communities on nature-based solutions. A village transect walk performed in 5 selected small scale fishing communities; Balapitiya, Ambalangoda, Gandara, Kudawella and Godawaya of south coast of Sri Lanka. Systematic walk along a defined path started from the fishing village towards the fish landing site and beach together with selected leaders of the fishing community to explore the status and availability of NbS, by observing, discussing, capturing into photographs and developing transect diagrams. Transect walk was performed in 3 main fishing harbors; Dondra, Kudawella and Tangalle and the main focus was to observe the natural and man-made protective mechanisms.

Further, Ahungalla, Hikkaduwa, Unnawatuna, Mirissa and Tangalle were selected to explore the status of natural and man-made protective mechanisms and special attention has been paid to examine the built environment and available space for natural barriers. In-depth interviews were conducted with selected members of small scale and commercial fishing communities and the management of the tourist hotels. Storytelling exercise was performed with selected community leaders to explore the historical perspectives on natural barriers. Resilience building and coastal disaster risk reduction mechanisms of the south coast along with knowledge sharing and capacity building through learning were obtained from the focus group discussions. Each focus group comprised 8–10 members representing the fisheries, tourism and community categories. Focus group guide, a set of open-ended questions, were instrumental in conducting the focus group discussions. All discussions were recorded and analyzed qualitatively (Zhang and Wildmuth, 2009).

Table 1 A snapshot of the data collection tools, types of respondents and responses

Study methods	Climate change impact	Formal learning	Informal learning	Adaptation and mitigation measures			
	Fisheries (small scale)	Fisheries (commercial)	Tourism	Communities			
Village transect	5 villages	3 harbors	5 tourist hubs	5 villages			3 districts
Ethnography	5 villages	3 harbors	5 tourist hubs	5 villages			3 districts
Field observations	5 villages	3 harbors	5 tourist hubs	5 villages			3 districts
Focus group discussions (impact)	5 villages	3 harbors	5 tourist hubs	5 villages	3 districts	3 districts	3 districts
Focus group discussions (vulnerability)	5 villages	3 harbors	5 tourist hubs	5 villages	3 districts	3 districts	3 districts
Focus group discussions (resilience)	5 villages	3 harbors	5 tourist hubs	5 villages	3 districts	3 districts	3 districts
In-depth interviews	15 small scale fishermen and 12 female fish processors	30 commercial fishers, 10 boat owners	Mangers of 8 five-star hotels, 5 three-star hotels, 10 beach restaurants,	5 community groups	15 members	15 members	15 members
Field survey	115	72	38	12	178	178	237
GIS mapping	3 districts	3 districts	3 districts	3 districts	3 districts	3 districts	3 districts

4 Results and Discussion

Southern coastal belt of Sri Lanka is highly vulnerable to climate induced disasters (De Silva Achini et al., 2023). Economically and socially important landmass, southern coast and its users struggling to manage the available space among competing uses. The Indian Ocean tsunami in 2004 was considered the largest single disaster which crippled the southern coastal belt resulting in huge damage to its built environment, loss of human life and appeal for new coastal zone management policies to face future disasters (De Silva & Yamao, 2007). Natural disasters; heavy rains, floods, coastal inundation, storm surges, coastal erosion, etc., appear from time to time in different severity with impact levels throughout the generations. Coastal communities, especially fishing communities being exposed to prolonged natural disasters, inhabitants developed their own adaptive mechanism. Figure 6 explains the lifelong learning process of the coastal communities.

Learning process comprised of two main subsections; formal and informal. Formal learning process facilitated through the formal education system of both school and universities. Formal education system focuses more on educating public in general on climate change, disasters, policy interventions and disaster management. Informal systems are more focus on educating general public living in coastal belt. Informal system aimed to empower the adaptive capacity of the communities through the lessons learnt from other places. Participatory disaster management and resilient building approaches were commonly practiced with the participation of community

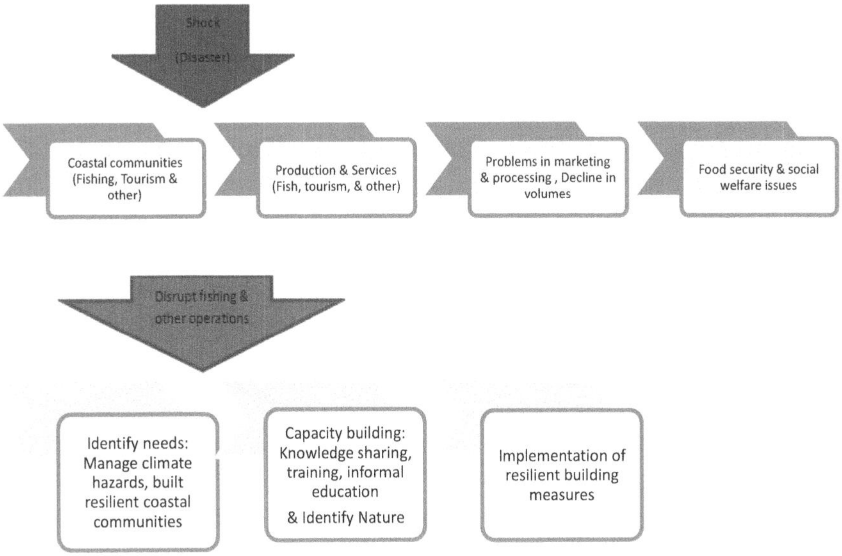

Fig. 6 Lifelong learning of coastal communities—the process

organizations, utilizing the common knowledge and nature-based solutions available locally.

Heavy coastal erosion and inundation is prevalent in the southern province, especially the issue of Galle district reaching an alarming level now, threatening to redesign its main transport system. Figure 7 shows the living example of nature-based solutions in building coastal resilience. Coastal erosion is not prevalent in the areas where natural solutions are available. Especially, coral reefs provide great protection against coastal erosion and inundation and some areas are protected by both mangroves and coral reefs.

Densely populated, economically most important locations of the southern coastal belt badly affected by environmental vandalism; clearing coastal vegetation for construction purposes, coral and sand mining. Further, changes in the coastal built environment; constructing seawalls, roads, buildings and or interfering with natural sand movements made permanent structural changes.

Seenapatabendige et al. (2023) highlighted the cause-and-effect relationship between nature and learning. Converging evidence strongly suggests that experiences of nature boost academic learning, personal development and environmental stewardship. Coastal communities are closely tied with its environment for their livelihood as well as life. Fishing villages are scattered along the coast, allowing

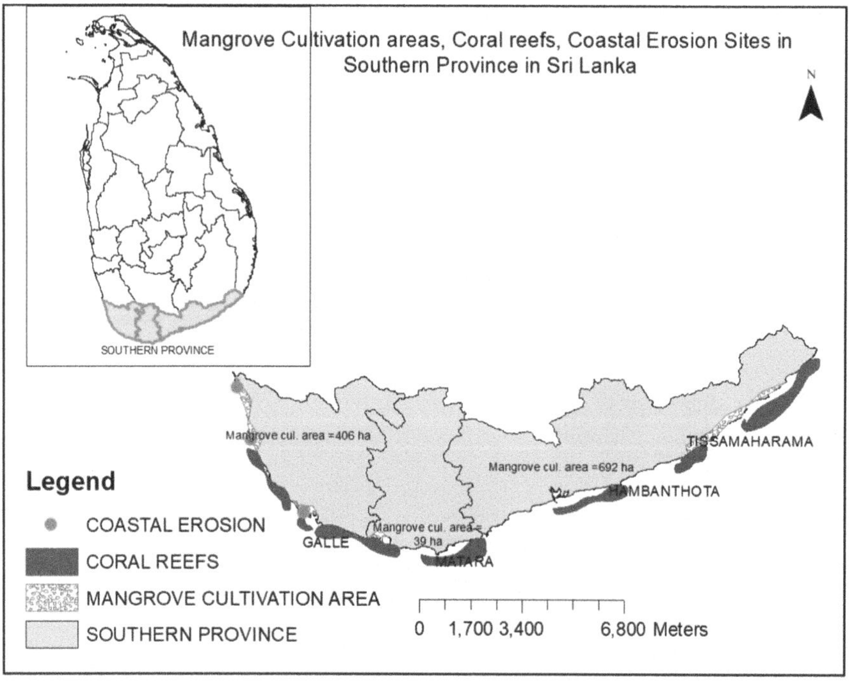

Fig. 7 Graphical analysis of marine nature-based solutions, mangroves, coral reefs and coastal erosion of the southern coastal belt of Sri Lanka

more open spaces among houses, without any permanent fences or boundary walls and open spaces shared among families for fish drying, net mending, playground for kids and adults, etc. Open spaces allow them to interact with the environment, join with the community and learn from adults and nature (Seenapatabendige et al., 2023). Nature and its continuants are playing an important role in the development of pro-environmental behavior of individuals (Kuo et al., 2019). Coastal ecosystems and its dynamic inter relationships promote learning by improving people's attention, self-discipline, interest and enjoyment in learning, physical activity and knowledge sharing. Learning by doing things or experiences with nature enhance learning ability along with skills and empower learners as environmental stewards who are keen on change, adaptation and building resilience.

Figure 8 summarizes the relationship with nature and lifelong learning. The process, experiencing nature and how it enriches the learning process of coastal dwellers was developed through the participatory approach, especially storytelling exercises adopted during the focus group discussions to facilitate the identification of the interactions. With reference to Kuo et al. (2019), the term nature implied not only in wilderness but also within the man-made contexts of the coastal built environment. Both formal and informal engagements with nature; fishing operations, landing, weather forecasting, vessel building and repairing, fish processing, identifying seasonal pattern, ocean current and changes, flora and fauna, hazards, etc. and experiences considered to understand the context of lifelong learning on coastal disasters associated with climate change. Real world settings facilitate effective learning by directly fostering the learner's capacity to learn by providing a more supportive context for learning (Kuo et al., 2019). Moreover, community leaders, seniors, and parents who hold the wealth of experience involved in the process as gatekeepers (Julia et al., 2019; Audefroy and Sanchez, 2017) .

Access to nature as well as availability of natural environment greatly support the learning process. Most of the seniors were worried about the capacity building of youth without experiencing the nature. Findings of the study revealed the lifelong learning outcomes of the diverse coastal communities (Table 2). Natural resources, especially nature-based solutions and strategic interventions of different groups of the coastal communities on disaster management highlighted the importance of learning. Co-management of coastal and marine resources identified as a sustainable approach to utilizing the resource base while conserving for future generations (Elo et al., 2014). Table 3 presents the findings of the lifelong experiences of co-management of coastal and marine resources along with the co-management practice, locations and deliverables. Both coastal and marine resources are managed by local fishers, communities in cooperation with local and central governments while sharing the authority and responsibility.

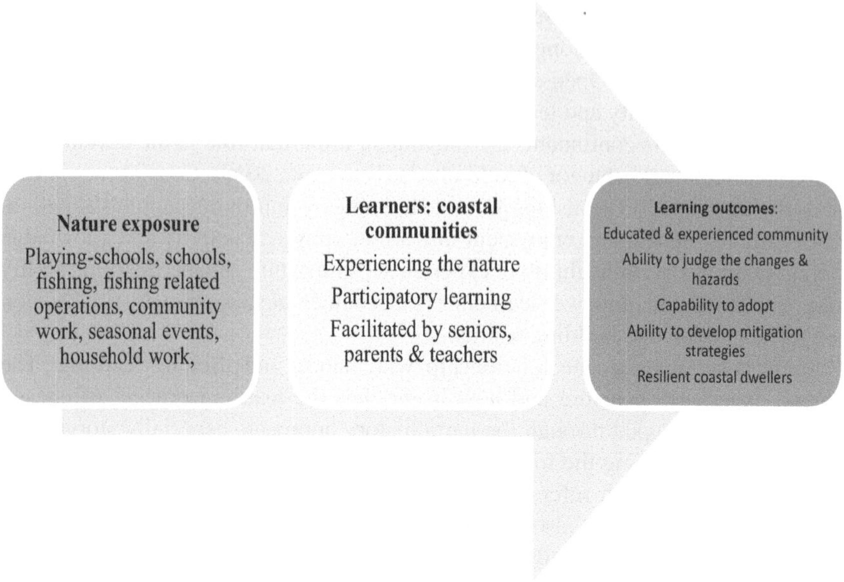

Fig. 8 Nature inspired learning process, exposure methods, learning process with participants and learning outcomes

4.1 Lifelong Experiences of the Coastal Communities

Inventory of climate induced hazards on different sub sectors of the southern coast discussed in Fig. 8. Adaptive responses to risk reduction coupled with climate change. Hazards and lifelong learning outcomes of the small-scale fisheries, commercial fisheries, tourism, communities, cities and coasts described in Fig. 9.

4.2 Nature Based Solutions, A Lifelong Experience of Coastal Communities of South of Sri Lanka

Southern coastal belt is unique in several ways. In terms of rich biodiversity, both coastal and marine ecosystems and exposure and experiences to series of climate effects throughout the centuries. Coastal nature-based solutions are creations of nature to restore ecosystems, manage biodiversity loss, cycling and recycling of both biotic and abiotic resources and tackle the negative effects of climate change on people, society and the built environment. Nature-based solutions play an incomparable role in mitigating resource degradation and vulnerability of coastal environments through restoring the ecosystem functions of the natural coastal environment. Mangroves, coastal vegetation; from gasses to trees, sand dunes, coral reefs, seagrass beds, wetlands, etc. tackling the climate change impact in one and on the other hand

Table 2 Lifelong learning and nature-based solutions of coastal communities

Community type	Nature based solution/ resource	Practice to manage/built resilient community
Small-scale fishermen	Mangrove replanting, avoid clearing Mangrove aim to protect fish breeding grounds, coastal habitat conservation, use of less harm fishing gears and methods for sustainable use of resources, resource sharing, shifting from marine fishery to lagoon fishery	Shifting to other fisheries or fisheries related work, find alternative jobs in tourism sector, work as a multi-day fish workers (net mending), Three wheel driver, grocery store and fish processing;-dry fish, Maldive fish and salted fish, utilized underutilized species, fishing ground management methods, shifting traditional craft to modern craft, speed and powerful engines to reduce travel time and catch more fish, illegal activities, such as drug dealing, smuggling, anti-social activities and disturbing to traditional norms
Commercial fishermen	Organize and participate community activities on conservation of beach and coastal vegetation, seasonal change of fishing gear to match with the target species, follow rules on close seasons	Invest in non-fisheries sectors; marine tourism, hotels and restaurants, agriculture ventures, vehicles, transport and logistics, etc. Fish processing and exports, investments in infrastructure development—ice plants, transport, storage facilities for the industry upgrading, investing in the tourism sector and engaging in illegal activities; drug dealing, smuggling, transport of migrants etc.
Coastal dwellers	Follow traditional practices and conservation efforts of coastal vegetation, beach sand, coral reef, marches and living breakwaters (bivalve shelf)	Attracted to the tourism related work (guides, employment in hotels and restaurants), cottage industries like Batik, needle crafts, fish processing-dry fish, coir rope making, rent out the diving equipment, motor cycles, three wheelers to tourists, manage small boutique hotels. vehicle hiring or work as drivers
Fishing communities	Follow community-based fisheries management practices and traditional norms on closed seasons, tenure rights, traditional fish processing; sun drying, salting, processing seasonally available species	Small scale fish processing: dry fish, Maldive fish, salted fish, work as tourist guides and other service providers, work as domestic helper
Boat owners-substantial fishing	New safe landing sites have been created through sedimentation of the river mouth lagoons and bays	Boats hanging on the sea due to coastal erosion and acclimation of the dead coral rubbles on the beach

(continued)

Table 2 (continued)

Community type	Nature based solution/ resource	Practice to manage/built resilient community
Boat owners-commercial fishing	Boats anchored in lagoons, bays or waterways or hanging in open sea, Find new fishing ground, species etc. Adopting to land good quality fish (follow safety and quality management practices)	Due to an accelerated sedimentation of the existing harbors, there's a need to find or build new harbors or dredging. Boat modification for lengthy fishing trips, adopt smart/ digital technology like potential fish forecasting, strong digital communication facilities, fuel efficient boats, electric boats, use of solar power for lighting, insurance for crafts, gear and crew
Fish traders	Reduce the wastage and optimum use (second grades for dry fish processing). Low value fish, waste generated in harbors and landing sites often use to prepare fish meal, animal feed and fertilizers	Find new export destinations, new methods to store and transport, rigorous promotion, heavy use of social media, minimize post-harvest losses, modern fish retails with clod room facilities, customer services
Beach seine operators	Seeking new beaches, operate seasonally, search for underutilized species	Abundant fisheries operations due to beach erosion, shifting to the multi day or small-scale fisheries operations or tourism, usage of mechanical power, Sea worthy craft than the traditional crafts, modifications of the nets (mesh size, materials used, length)
Fish processors—women/ small-scale	Find new species, processing underutilized species and alternative methods for processing	Adapting to new technology, leaving the industry due to high raw material price and shortage of raw materials, engaging in tourism, restaurants, catering, finding new jobs in hotels, garments etc.
Fish processors—commercial	Find new or underutilized species for processing and export, value addition, new products	Importation and expansion of the local raw material collection network, building more storage capacity, new products, and high end value addition lending loans for the fishermen, manage diving teams, business diversification due to risk
Fisheries coop	Facilitate community-based work on conservation efforts, knowledge sharing and capacity building of the fishery value chain	Most of them are inactive, Joints with the new organizations, seeking govt. or NGO supports

(continued)

Table 2 (continued)

Community type	Nature based solution/ resource	Practice to manage/built resilient community
Community organizations	Capacity building of coastal communities, record traditional knowledge, implement as community participatory work on conserve and protect natural assets	Numbers are increasing, many organizations created to address multiple issues including gender, poverty, rights, infrastructure, environmental issues etc. supporting to capacity building, home businesses, cottage industries, money lending and savings, ecosystem conservation, mangrove replanning, nature-based tourism
Tourist hotels	Used areas where less erosion for building hotels, expanding some hotels above waters, building hotels restaurants on the water, construction of natural pools with elevated sea levels, seeking areas with high biodiversity and less pollution for building hotels, nature-based tourism, promotion for extraction of none consumptive values of degraded resources	Ecosystem conservation, empowerment of rural communities, building impediments for protection property and beach from erosion, beach cleaning programs, treatment of wastewater and garbage before release

Source Filed survey, 2022

facilitate the resilience building mechanism. Figure 10 brings the living examples of nature-based solutions of the southern coastal belt of Sri Lanka.

Nature-based solutions and its ability to tackle the coast and coastal livelihood from natural and man- made disasters were experienced by the coastal communities. Centuries old knowledge and know-how shared from generation to generation to adopt the changes of environment, weather patterns and built environment. Guided by traditional knowledge and know-how, and applications of modern scientific applications enrich the campsites of the communities to better manage the coastal disasters and build resilience. Figure 11 brings the hybrid applications; nature-based solutions of the coastal built environment. A case study on tourist hotels was used to explore the nature-based solution of the coastal built environment.

Table 3 Lifelong experiences of co-management of coastal and marine resources

Co-management practice	Fishery, location	Community practice	Deliverables
Brush Park fishery society	Brush Park fishery, Negombo Lagoon	Sharing of information and implementation of local rules	Strong property rights Forecasting of production variations and weather impacts Knowledge sharing of the effects of climate change
Beach seine fisher societies	Beach seine fishing, Southern province	Community management practices and implementation of local rules	Knowledge sharing Local rules
Fisher societies	Stilt fishing, Southern province	Sharing of traditional knowledge	Information sharing on weather, fishing season, feeding, and migratory behavior of target fishery
Fisheries management committees	Lagoon-based coastal fisheries in Chilaw, Negombo, and Puttalam	Fisher's participation in the decision-making process	Institutional strengthening Engagement with government institutions

5 Conclusions

Coastlines are diverse ecosystems which act as the primary defensive line of many disasters of island nations. Southern coastal belt of Sri Lanka is not a difference from its structure and functions but economically and socially one of the most vital coasts of the island nation. Rich biodiversity, flora and fauna along with diverse geographical locations of the coastline recognized globally as an important place to conserve in one hand. On the other hand, its resource base opens up diverse livelihood options to coastal communities as well as the international community. Nature-based solutions; coastal vegetation, mangroves, coral reefs, seagrass beds, sand dunes, etc. have been actively participating in the process of protecting and conserving the coastal line throughout the generations. Coastal communities and users of coastal and marine resources are well aware of the protective mechanism of nature-based solutions and learn to use the resources while conserving for future generations. Experience, exposure and knowledge sharing of seniors made the continuous knowledge flow from generation to generation.

Most economically viable and socially important southern coastline was an attractive landing site for Portuguese, Dutch and British invasions which has created massive changes to the coastal built environment. Economic diversification and rapid urbanization were the hidden disasters which consumed most of the natural resources, especially the nature-based solutions available on the front defense line of

Fig. 9 Schematic presentation of climate induced hazards on southern coastal zone of Sri Lanka and resilient building through lifelong learning

the southern coastal belt. Early generations who gained the exposure and lived experience on coastal natural protective mechanisms and defenses against the disaster's impacts, especially the climate change impacts were able to create sustainable solutions to the coastal built environment. Adaptation mechanisms of the components of the built environments were derived from the lessons learnt from nature. In contrast, unavailability of nature's solutions due to rapid exploitation made a vacuum for experience, exposure and knowledge transfer. Tackling future disasters and developing adaptation strategies, without or limited touch of nature's solutions is questionable for many generations to come. Therefore, the present study revealed the urgent need of conserving available nature's solutions, scaling up the resource status through applied research, participatory community-based approaches to earn and share the knowledge and capacity building for young generations on knowledge on nature-based solutions. Further, interventions of applied research are vital to design adaptation mechanisms for the future built environment which are supposed to face the consequences of climate change impacts.

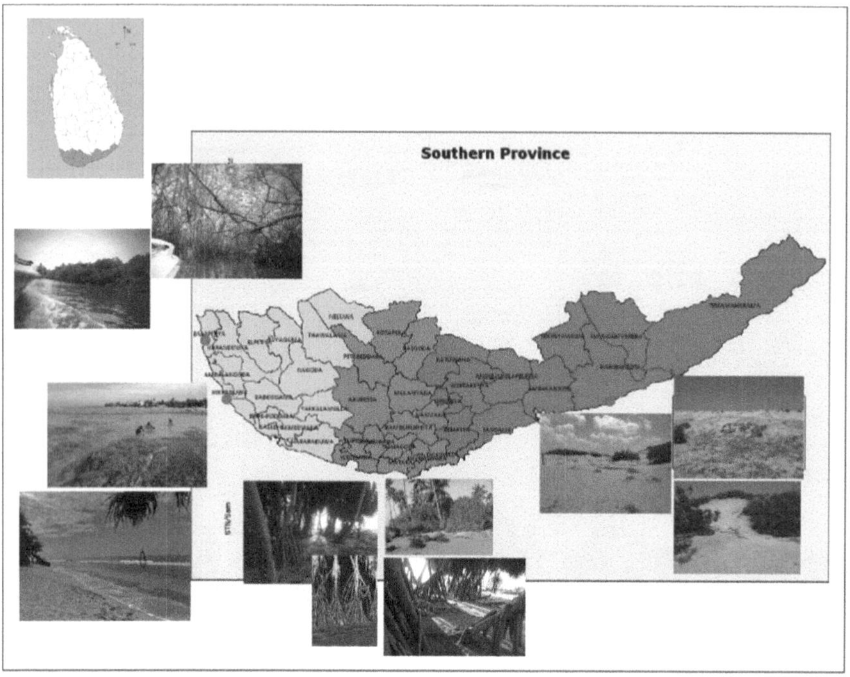

Fig. 10 Nature-based solutions of southern coast of Sri Lanka

Parameters

Courtyards

Open to sky center Courtyards

Semi outdoor courtyards

Courtyards with water ponds

Room separated from the corridor

Operable louvers

Fig. 11 Nature-based solutions of the coastal built environment: a case study on tourist hotels and structural modifications to tackle the climate change impact

Parameters	

Landscaping with surrounding garden	
High ceiling walls	
Overhangs roof Even Shading roof	

Fig. 11 (continued)

References

Audefroy, F., & Cabrera Sánchez, B. (2017). *Integrating local knowledge for climate change adaptation in Yucatan, Mexico.*

Bektas, Y., & Sakarya, A. (2023). *The relationship between the built environment and climate change: The case of Turkish Provinces.*

Châles, F. (2023). *Coastal nature-based solutions for climate adaptation and mitigation in the national policies of Pacific Small Island Developing States.* Economics and Finance. Université de Bretagne occidentale—Brest. Retrieved from coastalNBS.pdf.

Cohen-Shacham, E., Walters, G., Janzen, C., & Maginnis, S. (Eds.). (2016). *Nature-based Solutions to address global societal challenges.* IUCN.

De Silva A. M., Sandamith, M., Wicramarthne, I., Basnayake, B. M. R. L., Ekanayake, J., Seenapatabendige, K. B., & Gunatilake, S. K. (2023). *Displaced fishermen off the coast: Impact of multiple hazards on life above the water, rebuilding communities after displacement.* Springer Nature. Retrieved from https://link.springer.com/chapter/10.1007/978-3-031-21414-1_5

De Silva, D. A. M., & Yamao, M. (2007). Consequences of the tsunami on fisheries and coastal livelihood: A case study of tsunami ravaged southern Sri Lanka. *Journal of Disaster Policy Management, 31*(4), 386–404. https://doi.org/10.1111/j.1467-7717.2007.01015

Elo, S., Kääriäinen, M., Kanste, O., Pölkki, T., Utriainen, K., & Kyngäs, H. (2014). Qualitative content analysis: A focus on trustworthiness. *SAGE Open, 4*(1), 2158244014522633. https://doi.org/10.1177/2158244014522633

Ezeah, C., Ohiaeri, B., & Khatib, J. (2014). *A review of the impact of climate change on the built environment of rural tropical regions.*

Hsieh, H. F., & Shannon, S. E. (2005). Three approaches to qualitative content analysis. *Qualitative Health Research, 15*(9), 1277–1288.

IPCC. (2022). *Climate change 2022: Impacts, adaptation and vulnerability.*

Julia, K., Perdita, P., Viviana, B., & Thorsten, P. (2019). *Encounters between experiences and measurements: The role of local knowledge in climate change research.*

Karki, M., Pokhrel, P., & Adhikari, J. (2011). *Integrating indigenous, local and modern knowledge for sustainable conservation and management of forest ecosystems in Nepal.*

Kuo, M., Barnes, M., & Jordan, C. (2019). Do experiences with nature promote learning? Converging evidence of a cause-and-effect relationship. *Frontiers in Psychology, 10*, 305.

Macgregor, C., Jaques, R., Dowdell, D., Bint, L., & Berg, B. (2018). *The built environment and climate change: A review of research, challenges and the future.*

Quang, N. M., & de Wit, J. (2020). Transformative learning and grassroots climate adaptation: case studies in Vietnam's Mekong delta. *Nature Conservation-Bulgaria* 39.

Rathnayaka, B., Siriwardana, C., Amarathunga, D., Haigh, R., & Robert, D. (2021). *Climate change impacts on built environment: A systematic review.*

Rowberry, R. (2022). *Climate change, coastal built heritage, and critical challenges facing the heritage law frameworks of the United States, United Kingdom, and France.*

Sabbioni, C., Brimblecombe, P., & Drdacky, M. (2006). *Global climate change impact on built heritage and cultural landscapes.*

Seddon, N., Daniels, E., Davis, R., Chausson, A., Harris, R., Hou-Jones, X., Wicander, S., et al. (2020). Global recognition of the importance of nature-based solutions to the impacts of climate change. *Global Sustainability*, 3.

Seenapatabendige, K. B., & De Silva, A. M. (2023). *An architectural analysis of tsunami resettlement villages of South of Sri Lanka.* Springer Nature. Retrieved from https://link.springer.com/chapter/10.1007/978-3-031-21414-1_9

UNDRR. (2020). *Ecosystem-based disaster risk reduction: implementing nature-based solutions for resilience.* United Nations office for disaster risk reduction—Regional Office for Asia and the Pacific, Bangkok, Thailand. Retrieved from https://www.undrr.org/publication/ecosystem-based-disaster-risk-reduction-implementing-nature-based-solutions-resilience

UNFCCC. CMA. (2018). *Report of the Conference of the Parties serving as the meeting of the Parties to the Paris Agreement on the third part of its first session, held in Katowice from 2 to 15 December 2018. Addendum. Part two: Action taken by the COP*. Decisions adopted by the COP serving as the meeting of the Parties to the Agreement. FCCC/PA/CMA/2018/3/Add.2.
Wang, X., Li Lai, X., Sheng, C., & Chi hsiang, W. (2022). *Reflections on coastal inundation, climate change impact, and adaptation in built environment: Progresses and constraints.*
Wilby, R. (2007). *A review of climate change impacts on the built environment.*
Zhang, Y., & Wildemuth, B. M. (2009). In B. Cronin (Ed.), *Qualitative analysis of content.*
Zieba, Z., Dabrowska, J., Marschalko, M., Pinto, J., Mrowczynska, M., & Lesniak, A. (2020). *Built environment challenges due to climate change.*

Climate Change Education

Skill and Knowledge Gaps in Climate Change Adaptation: A Case of Sri Lanka

Nishara Fernando, Pasindu Jayathilake, and Malith De Silva

1 Background

Various organizations have defined climate change in different ways. The Intergovernmental Panel on Climate Change (IPCC) has defined it as "a change in the state of the *climate* that can be identified (e.g., by using statistical tests) by changes in the mean and/or the variability of its properties and that persists for an extended period, typically decades or longer" (IPCC, 2013). The United Nations Framework Convention on Climate Change (UNFCCC) defines climate change as "a change of climate which is attributed directly or indirectly to human activity, that alters the composition of the global atmosphere and which is in addition to natural climate variability observed over comparable time periods" (UNFCCC, 1992).

Climate change has been a natural part of Earth's history, with fluctuations occurring over geological timescales. For example, during the Pleistocene epoch which spanned from approximately 2.6 million to 11,700 years ago, there was a series of glaciations known as ice ages with intermediate warmer interglacial periods (IPCC 2013). However, in the present context, anthropogenic activities alter the average composition of the atmosphere. Greenhouse gas emissions unequivocally caused global warming through human activities, resulting in a global surface temperature increase of 1.1 °C above 1850–1990 levels during the period of 2011–2020. Climate change increasingly impacts water and food production, health and well-being of

N. Fernando (✉)
Department of Sociology, University of Colombo, Colombo, Sri Lanka
e-mail: nishara.fernando@soc.cmb.ac.lk

P. Jayathilake
Department of Geography, University of Kelaniya, Kelaniya, Sri Lanka
e-mail: piumalj@kln.ac.lk

M. De Silva
Federation of Sri Lankan Local Government Authorities, Colombo, Sri Lanka

C. Malalgoda et al. (eds.), *Climate Change Adaptation in the Built Environment*,
https://doi.org/10.1007/978-3-031-75826-3_12

cities, settlements, infrastructure, ecosystems, and species worldwide. It is speculated that the world would face unavoidable, multiple climate hazards with global warming of 1.5 °C (IPCC, 2021) in the next two decades. Further, the Secretary General of the United Nations, Antonio Guterres cautioned that global emissions are expected to increase by 14% in the next decade, resulting in catastrophic events.

The built environment can be defined as a manmade surrounding that encompasses patterns of human activities and comprises land use, urban design, and transportation systems (Sridarran et al., 2016). Further, the RICS School of Built Environment of the Amity University defined the built environment as the man-made surroundings offering a site for human activity-ranging from buildings and parks to vicinities and cities that can often include auxiliary infrastructure such as water supply, or energy networks. It comprises places and spaces created or reformed by people inclusive of buildings, roads, pavements, parks, and transportation systems. It is a structure in which people live, work, and recreate daily. The built environment enriches the growth and welfare of children and young people and supports healthier communities. According to the definitions of the built environment, every aspect associated with human life is included in the built environment.

The built environment of coastal regions is also vulnerable to damage due to flooding, erosion, and intensifying hazards. The impacts of storms are predicted to be amplified and Sea Level Rise (SLR) leads to substantial challenges, especially for cities, deltas, and small island states (IPCC, 2018). The intensity of tropical cyclones across the world's oceans has already been said to have increased, although the frequency remains the same or has decreased (Elsner et al., 2008; Holland & Bruyère, 2014, as cited in IPCC, 2018). Such intensity changes are associated with increases in maximum wind speed, wave height, and inundation levels, leading to increased exposure to related impacts, such as flooding, reduced water quality, and sediment runoff (Brodie et al., 2012; Wong et al., 2014; Anthony, 2016, as cited in IPCC, 2018). These impacts can increase the risks of physical damage to coastal communities and to ecosystems (Long et al., 2016; Primavera et al., 2016; Villamayor et al., 2016; Cheal et al., 2017, as cited in IPCC, 2018).

As an island nation, Sri Lanka is highly vulnerable to the impacts of climate change (CCS and MoMDE, 2016). According to a report by IPCC (2014a, 2014b), sea levels (MSL) have increased globally by 0.19 m during the last hundred years, and it is projected that it would occur faster during this century. Important sectors in the country such as tourism, fisheries, biodiversity, agriculture, and food security (De Costa, 2008; De Silva, 2006; Marambe et al., 2013, 2015), productions related to export crops (Nissanka et al., 2011; Ranasinghe, 2013; Wijeratne et al., 2007), health and human settlements (ME, 2010a, 2010b) could be affected due to the impacts of climate change. Accordingly, every aspect of the built environment is included within the sector and is highly susceptible to the impacts of climate change. Therefore, it is important to streamline climate adaptation plans and all types of mitigation strategies to reduce the cost of climate change impacts.

Several reasons have been identified as the cause for this poor practical implementation and among them, the lack of knowledge and skills of built environment professionals is imperative. The BEACON project proposes a competency framework with three major tiers for all built environment practitioners to address such skill and knowledge gaps.

Considering the climate change adaptation mechanism in Sri Lanka, initially climate change was identified as a meteorological issue and the Meteorology Department of Sri Lanka was known as the national focal point of the Intergovernmental Panel on Climate Change (IPCC). However, later climate change was taken under the purview of the Ministry of Environment (MoE) (Pallawala, 2018). Later the MoE was renamed as the Ministry of Mahaweli Development and Environment (MoMDE). The Ministry of Environment, which is the National Focal Point for the UNFCCC and Kyoto Protocol established the Climate Change Secretariat (CCS) in 2008 under its purview. The CCS is the main body that addresses climate change adaptation, and it is also the national focal point of climate change, addressing cross-sectoral challenges caused by climate change and fulfilling commitments under the United Nations Framework Convention on Climate Change (UNFCCC) & the Kyoto Protocol. The CCS provides a platform to liaise with other key stakeholders and ensures wider consultation for policy implementation (CCS and MoMDE, 2016). Two expert committees have been established under the CCS, namely the National Expert Committee of Climate Change Adaptation (NECCA) and the National Expert Committee on Climate Change Mitigation (NECCCM) (Pallawala, 2018). These expert committees consist of 31 members from key stakeholder organizations such as ministries, departments, authorities, non-governmental organizations, the private sector, and academia. The main role of these expert committees is to bring key stakeholders together and to coordinate all climate related activities (Pallawala, 2018). The Inter-Ministerial Coordination Committee has been implemented to mainstream climate change into the development processes of the country by providing a perfect platform to coordinate national-level climate change-related matters at the government level. In addition, the Nationally Appropriate Mitigation Action Coordinating Body was approved by MoMDE and CCS in partnership with the UNDP. Institutional mechanisms and relevant institutions for mitigation actions were identified through this process. Nationally Determined Contributions (NDCs) have also proposed legislative frameworks such as the "Climate Commission," even though it is not legally empowered to implement capacity building and other functioning activities due to the current institutional mechanism. The National adaptation plan has suggested ways to implement adaptations with the guidance of the CCS (CCS and MoMDE, 2016). The NAP (2016–2025) has proposed Sectorial Climate Cells, National Working Groups and Regional Climate Cells, and civil organizations to implement adaptation measures at the local level. CCS is the coordinating agency of the above-mentioned bodies (CCS and MoMDE, 2016).

In Sri Lanka, the implementation of climate change mitigation measures is structured through a multi-tiered approach, as outlined in the National Adaptation Plan for Climate Change in 2016. At the national level, cross-cutting issues are managed by the National Working Group (NWG), with the CCS serving as the convener and national lead agencies as members. Regional-level actions are executed by Sectoral

Coordination Committees (SCCs), where the CCS plays a facilitating and coordinating role. These committees operate as working groups focusing on sectoral adaptation actions, ensuring collaboration between sectors for joint implementations when necessary. At the community level, the NWG and SCCs include representatives from Civil Society Organizations (CSOs) and community organizations, recognizing the importance of grassroots involvement. Community-level actions benefit from the support of CSOs and community organizations, including national and international (Non-Government Organizations) NGOs, serving as mobilizers and facilitators. The CSO Forum provides an additional platform for coordinating efforts among these organizations, emphasizing the significance of collaborative approaches at all levels for effective climate change mitigation in Sri Lanka.

Although there are several national policies, guidelines, and institutional mechanisms the practical implementation of climate change adaptation is very weak in Sri Lanka due to different factors.

2 Methodology

This study utilized both primary and secondary data sources for qualitative data collection. Built environment stakeholders, their roles and responsibilities, and existing knowledge gaps were identified through secondary data sources. Accordingly, five built environment stakeholders who work for climate adaptation were identified. 40 respondents (built environment professionals) from the said built environment stakeholder categories were purposively selected and were interviewed using a semi-structured interview series to identify their roles and responsibilities, existing knowledge gaps, and key challenges for climate change adaptation (see Table 1). Semi-structured interviews were recorded digitally with the consent of the interviewees. The qualitative data obtained were analyzed using thematic analysis methods.

Table 1 Stakeholders interviewed for the study by category

Stakeholder category	Number of participants
1. Government and local government sector (high, middle, and ground level)	12
2. Academia and research organizations (high, middle, and ground level)	10
3. Community	07
4. Non-government and intergovernmental organizations (INGOs and NGOs)	05
5. Private sector (high, middle, and ground level)	06
Total	40

Source Own data, 2023

Based on the findings derived from interviews conducted with key stakeholders, significant skill and knowledge gaps were identified. Using these findings as a foundation, a set of competencies were crafted and subsequently presented to the interviewed stakeholders for their feedback and additional validation. Further, structured online questionnaires were used to validate the competency framework. The validation process included two questioning rounds and 25 built environment professionals were selected purposively for this process based on the key stakeholders identified in the NAP (2015–2025). Thematic analysis methods were used to analyze data.

3 Results and Discussion

3.1 Key Built Environment Stakeholders in Sri Lanka

As mentioned in the above sections, the coastal built environment is highly impacted by climate change. Therefore, it is important to identify built environment stakeholders that are responsible for climate change adaptation. According to Amaratunga and Haigh (2011), stakeholders were determined according to the nature, objectives, and content of any project. Chandrasekhar (2012), lists stakeholders based on the division of government agencies, non-government organizations, and community organizations. Furthermore, Shafique and Waren (2016) have divided stakeholders into 'principal' and 'primary'. Additionally, Ginige et al. (2010) mention key built environment stakeholders involved in constructions including national and local governments, the international community, the community, civic society, the private sector and corporate sector, academia, and professional associations.

The National Adaptation Plan (2016–2025) of the CCS of Sri Lanka identifies key stakeholders in climate change adaptation, including the state sector (line ministries, line agencies, the central government, and provincial councils), the private sector (corporate sector and SMEs), civil society organizations, academics, researchers, and other knowledge makers, as well as local community-based organizations. Since there are similarities between climate change stakeholders (NAP, 2016) in Sri Lanka and built environment stakeholders (Ginige et al., 2010), the government sector, private sector, Non-Government Organizations (NGOs), academia, and the public were identified as key built environment stakeholders responsible for climate change adaptation in Sri Lanka for the purpose of this work.

3.2 *Roles and Responsibilities of Key Built Environment Stakeholders in Sri Lanka*

State sector

The national government and local governments play major roles in climate change adaptation and mitigation in terms of policy making and policy implementation. The Ministry of Environment, which is the National Focal Point for the UNFCCC and Kyoto Protocol, established the CCS in 2008 under its purview and the CCS is the main body that addresses climate change adaptation. The CCS provides a platform to liaise with other key stakeholders and ensures wider consultation for policy implementation (CCS and MoMDE, 2016). Furthermore, it is the key policy maker and the guide for implementing climate adaptation strategies. This is supported by the statement:

> The Climate Change Secretariat (CCS) of Sri Lanka is the national focal point of climate change. It is responsible for drafting localized, new adaptation and mitigation policies according to international frameworks and coordinating relations among stakeholders. Other government stakeholders are responsible for implementing adaptation and mitigation initiatives introduced by CCS (Semi-structured interviews 2022).

Climate change has been identified as a sectoral and cross-cutting issue (NAP, 2016). Therefore, line ministries and state departments are responsible to execute climate change adaptation measures together with the CCS. In addition, the Disaster Management Center and the Department of Meteorology (The national focal point of IPCC (Pallawala, 2018) have key responsibilities related to climate change for disaster risk reduction and decision-making. Government sector institutions are responsible when providing approvals to stakeholders and assisting them with technical, financial, and professional development. They are also responsible for funding and implementing initiatives that are outlined in the NAP (2016–2025) and NDCs, as well as organizing programs to raise awareness regarding climate change among professionals, students, and the general public. The CCS developed the National Climate Change Policy in 2012 and the National Climate Change Adaptation Strategy (2011–2016) in 2010, which could be identified as a means of executing their roles and responsibilities. Furthermore, local governments play a major role in the implementation of climate change adaptation and mitigation measures in collaboration with the central government. However, Provincial Councils (PCs) and Local Councils (LCs), such as Municipal Councils, Urban Councils, and Pradeshiya Sabas, do not have a clear local mandate on climate change. Both PCs and LCs carry out actions relevant to climate change under their current mandate, though not recognized as climate actions (Pallawala, 2018).

Private sector

Climate change will become the most significant challenge for economies and societies in the forthcoming decades (Cochu et al., 2019). In the context of climate change, the engagement of the private sector is important whenever governments face problems with their public budgets and rising costs to manage climate change to achieve global climate resilience of businesses, financial and other sectors. This would also help to disclose risks and maintain the profitability of their work. Companies, private associations, cooperatives, banks, investors, and insurance companies could be identified as private sector actors in the global context. All in all, such private sector actors play three roles in adaptation including adapting to climate change according to their institutional level, financing climate adaptation of other organizations, and providing support through products and services for resilience (Cochu et al., 2019).

> We support government agencies to implement climate change adaptation and mitigation measures at national and local levels by financing initiatives. Moreover, we implement sustainable energy production methods, waste management methods, and eco-friendly green cooling methods at the organizational level (Semi-structured interviews 2022).

According to the above quotation, private organizations play a key supporting role in climate change adaptation through financing and implementation. They have their own independent funds and thereby can promote climate-resilient services and products. For instance, some private organizations are funded through financial aid that is received through channels of non-government organizations for climate adaptation initiatives. The results further indicate that the private sector plays a key role in assisting customers that are affected by extreme weather as a result of climate change by providing loans and other services, implementing climate adaptation measures at the organizational level, providing products and services that can be used in the context of climate change adaptation, considering the built environment when designing buildings, and implementing climate adaptation measures prescribed by the government.

Community

According to Eriksen et al. (2015), adaptation should be viewed as "part of the dynamics of societies rather than simply a technical adjustment by society to biophysical change". Climate change adaptation measures can sometimes be harmful to the community. Therefore, it is important to obtain the contribution of the community for climate adaptation and mitigation efforts (Bulkeley et al., 2013). The top to down approach adopted by most countries is highly criticized when imposing state policies and engineering solutions as such efforts have been unsuccessful on certain occasions due to a lack of community consultation, considering that the community has the best understanding of their ground situation. Therefore, various authors advocate for more bottom-up approaches in climate adaptation (Simon et al., 2020) in order to address the issue. Community-based adaptation consists of participation, collective action, social capital, access to information, and local knowledge of risk management (Bryan & Behrman, 2013). Using a community-based climate change adaptation entails many benefits such as reducing the risks of policy implementation failure, increasing understanding of the impacts of climate change at the local level, building trust between local government and communities, and developing a wider range of options for adaptation (Simon et al., 2020).

According to the findings of this study, key roles and responsibilities of the community include active participation in the implementation of climate change adaptation measures and contributing to the creation of national policies and regulations. The views and recommendations of the community and community-based organizations were considered during policy formulation in Sri Lanka. For example, the CCS of Sri Lanka followed such a consultation process when developing the National Adaptation Plan (NAP) and other climate-related policies. According to the findings, it was revealed that it was difficult to identify community-based organizations that are dedicated to climate change adaptation. This fact is confirmed through the following quotation.

> To the best of my knowledge, there aren't any community-based organizations that are fully dedicated to tackling climate change. However, some community-based organizations engage in activities such as using and promoting sustainable cultivation methods, flood and drought-resistant seeds, rainwater harvesting etc. They also contribute to climate adaptation measures by providing inputs to draft policies and by acting as a pressure group to agitate for changes in legislation and policies (Semi-structured interviews 2022).

The above statement also emphasizes the involvement of various communities in the implementation of climate adaptation. Other key roles and responsibilities of the community and community-based organizations include protesting and agitating organizations for policy changes and improving the quality of measures employed by the government, non-governmental or private organizations, supporting policy-makers by actively contributing to policy formulation, advocating to update existing policies, guidelines, and national plans, implementing climate change measures at organizational levels through the use of sustainable cultivation methods, awareness raising programs, etc., and following the guidelines provided by the government or local government authorities. At the grassroots level, the communities have the responsibility of implementing adaptation measures in their localities. The report "Coping with Climate Change and Variability: Lessons from Sri Lankans community" discusses certain Sri Lankan initiatives of community-based adaptations by focusing on farmer and coastal communities and adaptations such as raising awareness of slope management and soil conservation, implementing drinking water projects, developing irrigation canals, rehabilitating of old tanks, developing green belts, rainwater harvesting technology, etc.

Academia and research organizations

Undertaking research and innovations is crucial in climate change adaptation and that responsibility is majorly vested with academic and research organizations. These organizations contribute to the development of national policies on climate change through teaching, training, and research on climate change adaptation. Academic and research organizations educate and train undergraduate and postgraduate students, as well as professionals, on climate change and related phenomena in addition to developing and revising the curriculum of courses that explore CCA. Furthermore, they disseminate and exploit research findings to various segments of society, including the public, and participate in the development of guidelines and codes related to climate change. The quotation below confirms these facts:

> In addition to **teaching** climate change to undergraduate and postgraduate students, there is a climate change unit at the university where I **carry out research activities**. I was part of a

curriculum revision recently where a portion on business adaptation to climate change was added to the curriculum of an undergraduate program. I also recently developed an MSC program titled "Climate Change and Environmental Management." This will provide good **training** for students (Semi-structured interviews 2022).

Non-government and inter-government organizations

These organizations have a significant impact on climate change adaptation through knowledge sharing with the CCS during policy planning. They excel in creating awareness and contributing ideas for reforms in laws and policies. Educational, collaborative, and litigation approaches are utilized to complete these tasks. They conduct workshops, conferences, and symposiums to increase community knowledge as well as to share expertise and experiences with government bodies. Social media campaigns, school programs, and community programs are also conducted (Hettiaratchi, 2020). According to the findings, these organizations are responsible for implementing adaptation measures, financing climate change adaptation strategies, and building the capacity of officials and communities involved in climate adaptation. They make significant contributions to climate adaptation at both local and international levels. This is confirmed by the following statement:

We mostly engage in the implementation of climate change adaptation strategies and mitigation measures…. Moreover, NGOs also play a role in policy and practice advocacy through awareness raising and facilitation of training for government officers by organizing local and international training workshops and seminars…..Another key role is co-funding climate change measures implemented by government agencies (Semi-structured interviews 2022).

Accordingly, all built environment stakeholders have their own roles and responsibilities related to climate change adaptation. While the central government plays a key role in climate adaptation, the cooperation of other stakeholders is mandatory; if one stakeholder does not commit to their role or responsibilities, it weakens the climate change adaptation process. According to the findings, although there are many policies and guidelines in the country, the practical implementation of these climate change adaptations is very poor, mainly due to key challenges experienced by these stakeholders. The following section discusses the challenges related to the performance of duties related to climate change adaptation.

3.3 Key Challenges Experienced by Stakeholders

1. **Lack of specialized financial assistance programs to put climate change adaptation measures into action**

One of the primary reasons for this difficulty is the absence of political will to establish state-funded financial assistance programs. The root cause for this phenomenon is the absence of a legally binding body to enforce climate change adaptation measures. Although CCS exists, they do not have the legal capacity to follow up on policy implementation. If there were a legally binding agency under a specific act for climate change, it could govern and create a local funding mechanism to assist the process. As a result of the lack of specialized financial assistance programs, some implementing organizations compromise their solutions due to high costs. This fact is confirmed by the following statement:

> Funding has many problems but I cannot specifically pinpoint one. However, due to the country's situation the present funding is insufficient. The best solution for the coastal area is costly- therefore we must compromise the solution (Semi-structured interviews 2022).

Relevant political authorities and officials in Sri Lanka are not dedicated to implementing climate change adaptation measures. It is also important to note that their knowledge related to climate change is very poor and they fail to understand the importance of CCA. Unfortunately, although CCS has identified risks, climate consciousness is not satisfactory among other officials and the public. As reiterated by the above statement, most climate adaptation measures are construction-based and come with high costs. However, the Sri Lankan government is unable to allocate funds for climate adaptation due to the severe economic crisis in the country. Although there are several funding sources such as Green Climate Fund (GCF) and Global Environmental Facility (GEF) in the CCS, the proposals that have been submitted for grants have not been approved by agencies. The main reason for such rejection is the lack of knowledge in proposal writing. Furthermore, officials' poor understanding of climate change adaptation negatively impacts financial opportunities.

At the community level, many people suffer due to economic problems, with a considerable number living in poverty. Therefore, it is not feasible for individuals facing poverty to implement climate adaptation measures when they have other pressing needs to fulfill. For example, rural farmers in Sri Lanka are advised to use water collection methods for their cultivation and some semi-urban communities have been advised to use rainwater harvesting methods for non-consumption needs. However, most have refused to use these methods due to financial challenges. In addition, although several non-government and private organizations provide financial assistance for climate change adaptation, these opportunities are lost due to a lack of communication and coordination among different stakeholders.

Although research is crucial for the successful implementation of CCA as mentioned in the previous section, the study reveals that researchers and academics do not receive adequate funding to continue their research activities on climate adaptation. In conclusion, almost all stakeholders face financial issues when adaptation measures put into action.

2. **Lack of coordination among government agencies and other stakeholders**

Despite the presence of several state departments and organizations, they do not adhere to guidelines related to climate change adaptation. The main reason for this is the absence of an agency with legal provisions to enforce climate adaptation. Although CCS is the focal point for climate change in Sri Lanka, it does not possess the departmental power or mandate to follow up, enforce and implement adaptation measures. This is confirmed by the following statement:

> Climate change adaptations are just like guidelines and there is no proper way of regulating its implementation, we don't have a proper legal mandate to do that (Semi-structured interviews 2022).

The study identified that some high-profile officials undergo training related to climate change adaptation and other subjects but fail to share the knowledge gained with lower-level officials upon return. Furthermore, officials from local levels and from within the community, lack a clear understanding of international frameworks on climate change. This indicates a problem in the dissemination of knowledge to lower levels. Furthermore, limited recognition is given to climate change matters in working spaces, as depicted by the following statement, thus further aggravating the issue:

> There's no coordination between government officials when it comes to the protection of the wetlands.... (Semi-structured interviews 2022).

This discourages officials from pursuing knowledge on climate change adaptation. The restrictive nature of government offices also contributes to the limited implementation of climate adaptation. This also depicts restrictions in data sharing among government organizations as they are reluctant to share data with students and officials. Such restrictive regulations and policies of data sharing, along with lengthy approval processes have limited the amount of research on climate change adaptation. Moreover, collaboration becomes difficult when government agencies responsible for climate change do not grasp the significance of carrying out research activities to combat climate change through evidence-based initiatives. Additionally, there are difficulties in freely downloading climate-related data due to the absence of an open data-sharing platform in the country. Although there were some data-sharing platforms, there is no input of timely data by relevant authorities due to coordination problems.

3. **Lack of enforcement of laws and regulations**

The findings indicate that there are several rules and regulations that could be utilized for climate change adaptation, both directly and indirectly, though they may not be effectively applied. The neglect of climate change-related legislation and policies by officials and the public, the lack of support extended by governmental organizations for the implementation of CCA measures, and interference by pressure groups such as religious leaders, members of armed forces, politicians, etc., can be identified as challenges related to law enforcement. Interferences to law enforcement made by politicians and pressure groups are revealed by the following statement:

> In a wetland located in the Western Province of Sri Lanka, a signpost was erected. The villagers came and destroyed the signposts and said that it is their private property and how we [you] can say that this is a wetland. The members of the local council also supported them (Semi-structured interviews 2022).

For example, although there are procedures to gain approvals for housing construction, the essential requirements are only considered in planning. The approved guidelines are disregarded in the construction stage. This problem is also caused by the ignorance of various government officials and the lack of importance given to climate adaptation. Furthermore, the absence of a legally empowered body hampers the direct implementation of climate adaptation measures, as there are no specific climate-related laws to provide clear guidance.

4. **Lack of Equitable Opportunities for Community-Based Organizations, Non-Government Organizations, and Private Organizations to Contribute and Provide Inputs for National-Level Policies and Regulations**

The study identified exclusion in policymaking as powerful organizations and institutions obtained chances to submit their inputs in policy formulation stages while small organizations were excluded from doing so. These issues could be identified through the following statement:

> We do not take part in project implementation activities, rather we are engaged in policy-level activities carried out in our own organizations. Nevertheless, when it comes to government policies, we are not involved- only universities and ministries are part of these activities and they dominate the sector. Occasionally, they request us to participate but there is no continuous participation (Semi-structured interviews 2022).

Furthermore, this statement emphasizes that only certain selected organizations could participate. A major reason for this is the lack of political will to introduce government-funded financial assistance programs. A respondent from an organization who was invited to such policy formation events mentioned that they do not have sufficient financial resources to facilitate more organizations and people. However, if well-established funding mechanisms are available in the consultation process, equitable chances could be given. The lack of awareness about the significance of stakeholder collaboration in climate change, ignorance, and lack of education on climate change among state officials were attributed as challenges of exclusion in policy making.

5. **Knowledge and skill gaps of stakeholders**

The findings of the study revealed another key issue as the disparity in knowledge and skills among various stakeholders. It was revealed that stakeholders lacked certain communication, technical, language, and integrated working skills. Furthermore, they were disadvantaged by knowledge gaps in the practical application of theoretical knowledge and vice versa, green building concepts, climate change mitigation and adaptation knowledge, research and investigation, localization of mitigation and adaptation measures, strategies and legislation, policies, and national plans related to climate change. The lack of state-funded financial assistance, the unwillingness of officials and the public to improve their understanding of climate change, and the limited opportunities for officials and the public to improve their knowledge on climate change are causal factors for this challenge. Further, it was revealed that continuous professional development courses on climate change adaptations were lacking among stakeholder organizations. Most respondents studied climate change phenomena during their tertiary education and did not get an opportunity to update their knowledge thereafter, though some professionals, such as academics for instance, in different stakeholder categories received opportunities for further training. However, due to the lack of foreign training opportunities provided only to a select few, most local experts do not receive foreign exposure. According to the study, such knowledge and skill gaps are major reasons for the barriers in the implementation of CCA and other discussed challenges in Sri Lanka. Therefore, it is important to address the skill and knowledge gaps of built environment professionals.

3.4 The BEACON Competency Framework

The BEACON Project developed a competency framework to address the skill and knowledge gaps of built environment professionals in climate change adaptation, by following two validation rounds with built environment professionals from five countries including one Asian country and four European countries. The study identified key competencies such as improved communication, advanced digital literacy, fluency in global languages, integrated working skills, comprehensive knowledge of climate change, and comprehensive knowledge of research and investigation to fill their knowledge and skills gaps. The produced competency framework consists of three tiers: basic competencies, intermediate competencies, and advanced competencies as the findings of the study revealed different levels of competency gaps exist among different stakeholders (see Fig. 1). The basic competency level aims to provide knowledge related to the fundamentals of climate science and climate change while the intermediate competencies focus on professional capabilities that contribute to the successful execution of CCA. The advanced competencies concentrate on providing planning-related skills specifically tailored for the coastal built environment to address climate change challenges. A detailed version of the competencies is given below. The project aims to develop courses based on this framework and share them through a MOOC (Massive Open Online Course) platform.

Detailed Competencies

Tier 1 competencies

1. *Understand the basics of climate science and climate change:*
   - Understand the uncertainties of sea level rise and other climate components.
   - Comprehend climate variability and change.
   - Identify and understand multidisciplinary areas related to climate change, its causes and effects, and their spatial and temporal changes.
   - Understand the compound effects of climate change and the importance of risk and vulnerability assessment.
   - Assess climate-related disasters, with a disaster risk management approach.
   - Identify the inter-relationship between environmental sustainability, climate change, growth, development, and social and economic issues.
   - Understand the impacts of climate change on coastal environments

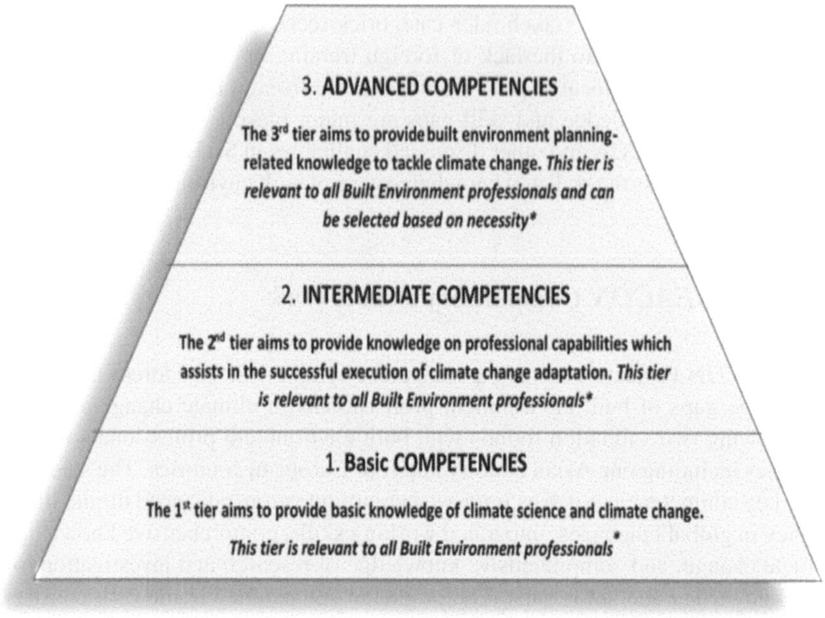

Fig. 1 Structure of the BEACON competency framework. *Source* Own data, 2023

2. *Comprehend new trends and predictions on climate change adaptation.*
   - Understand the importance of quality control, complete and representative data in climate change studies, and decision-making and planning.
   - Understand the uncertainties of climate impact predictions and measures to reduce uncertainties.
   - Incorporate how predictions on sea level rise and coastal impacts can be used in decision-making and planning.
   - Identify global and regional Early Warning (EW) Systems and service providers for EW dissemination, climate change mitigation, and adaptations.
3. *Understand climate change mitigation and adaptation.*
   - Demonstrate the differences between mitigation and adaptation.
   - Identify mitigation measures and adaptation strategies.
   - Identify the importance of technology in climate mitigation and adaptation.
   - Identity nature-based solutions and ecosystem-based solutions versus grey solutions
   - Understand indigenous knowledge and knowledge of local communities on prevention of climate-related impacts (knowledge and experiences of local people on adaptation to Climate Change and their applicability.)

4. *Identify stakeholders (Actors) in climate change adaptation and mitigation and their responsibilities.*

   - Identify stakeholders (Actors) in climate change adaptation and mitigation.
   - Comprehend their responsibilities, challenges, constraints, and suggestions.
   - Understand the importance of an integrated disaster risk management plan for all stakeholders.
   - Realize the thinking and behavioral patterns of people.

5. *Understand the built environment and its relationship to climate change.*

   - Identify built environment impacts on climate change and vice versa (impacts to climate change and how the built environment is impacted by climate change)
   - Comprehend the need for mitigation and adaptation in the context of recent climate change from sectorial and holistic perspectives.
   - Understand indigenous knowledge related to the built environment.

Tier 2 competencies

1. *Use research methods in climate research.*

   - Understand action research in climate change adaptation.
   - Comprehend research techniques.
   - Comprehend location-specific useable research techniques.
   - Understand research proposal writing.

2. *Use technical tools in climate change adaptation.*

   - Understand the uncertainties of climate change modeling.
   - Comprehend the ways of using technical tools, and terminologies in climate adaptation decision-making and planning.
   - Understand different methods used for climate change modeling, climate change impact prediction, and risk assessment.

3. *Comprehend the environmental, social, financial, and governance phenomenon in climate change adaptation.*

   - Application of interdisciplinary and multidisciplinary approaches in climate change adaptation
   - Identify global frameworks and local regulations which address climate change in the planning process.
   - Understand marketing mileage that economic value addition that can be given for tourism-related built environment.
   - Environmental regulations, legislation, and tools
   - Environmental ethics and responsibilities.
   - Comprehend the basics of green financing and climate financing.

4. *Understand the communication strategies in climate change adaptation.*

   - Identify communication approaches.

- Understand negotiation techniques of climate change in different contexts.

5. *Understand the importance of human resource management in CCA.*
    - Understand the basics of human resources management in CCA.
    - Knowledge of the administrative system
    - Environmental Management leadership

Tier 3 competencies

1. *Incorporate the Role of built environment professionals to tackle climate change.*
    - Incorporate policy, legal, and regulatory frameworks in the built environment and local needs to move towards CCA.
    - Identify the responsibilities, challenges, and constraints of built environment professionals and provide suggestions for challenges and constraints.
    - Address sociological, environmental, and economic needs in planning.
    - Understand contextual differences (Industrialized countries versus developing countries)
    - Transdisciplinary approaches, integrating knowledge and skills from different stakeholders in design and construction.
    - Methods related to fulfilling human needs while conserving nature.
    - Understand the importance of bottom-to-top and top to bottom approaches in CCA.
    - Understand the functions and components of the natural environment in a particular area before making decisions.
2. *Understand Environment-friendly planning*
    - Basic understanding of the solar cycle and planning
    - Understand and practical usage of green building concepts.
    - Application of green energies and energy-related adaptations
    - Use of environmentally friendly planning, building materials, sustainable construction, disaster-resilient infrastructure, and techniques
    - Understand the design and technical details to achieve energy efficiency during the design and construction period and the building usage period (post-occupancy).
    - Basic understanding of the importance of Analog Forestry as a nature-friendly environment restoration method (an approach to ecological restoration which uses natural forests as guides to create ecologically stable and socio-economically productive landscapes.)
3. *Incorporate disaster risk reduction in the construction industry*
    - Understand key regulations related to the built environment and CCA.
    - Understand stakeholder engagement in construction.
    - Adopt building codes and safety measurements in construction.
    - Prioritize industrial safety, human rights, and health.
    - Embrace legal backgrounds related to the environment before the construction.

- Understand institutional disaster management planning and business continuity management plan.

4. *Incorporate development planning in different contexts.*
   - Understand the importance of climate resilience planning.
   - Adopt climate change adaptation strategies in city planning, integration of environment and climate priorities into development planning.
   - Understand climate-related DRR and the multi-hazard approach in planning.
   - Application of environmental sustainability in planning
   - Incorporate servitude rights in land use planning together with development activities.
   - Use of new trends and technologies in planning and designing to tackle climate change.
   - Good practices in construction and facilities management to tackle climate change.
   - Understand integrated land-use planning and construction planning. (Interrelationships between environmental sustainability, climate change, growth, development, poverty, social and economic issues)
   - Understand design and resettlement, relocation criteria.
5. *Incorporate household approaches in climate change adaptation.*
   - Application of household plans and designs in climate change adaptation
   - Understand the sociology of housing.
   - Identify types of housing and specific planning and design considerations
   - Understand refurbishment of existing housing
   - Understand the importance of having an insurance scheme.
6. *Incorporate how construction and facilities management address climate change.*
   - Understand the organization and management of construction and maintenance of the built environment.
   - Use of green construction materials and resources (green materials/low co2 emission)
   - Create employment and livelihood opportunities (green job creation)
   - Project management
   - Procurement, contracts, and project delivery
   - Cost estimating, cost control, and contract management
   - Risk financing and risk insurance
   - Housing construction and maintenance
   - Approving plans
   - Stakeholder consultation and participation in community engagement in construction and maintenance
   - Understand the importance of empowering the media sector to promote community participation and engagement in construction and maintenance.
7. *Incorporate Planning in coastal contexts.*

- Understand coastal zone management planning, coastal resource mapping and coastal risk mapping.
- Coastal adaptation strategies
- Identify climate change adaptation strategies in coastal areas.
- Case studies and best practices

4 Discussion

According to the findings of the study, there are several stakeholders in CCA within the built environment. The central government plays a major role among them, and the CCS was identified as the focal point of CCA. However, due to the lack of legal provisions to enforce climate change adaptation, activities are implemented at a slow pace. Additionally, although the National Adaptation Plan (NAP) identified stakeholders for collaboration in CCA, CCS lacks the legal capacity to govern its implementation.

Furthermore, there is evidence of unequal treatment of stakeholders at the policy formation level, resulting in the exclusion of vital stakeholder categories when informing, amending, implementing, and regulating adaptation initiatives. This negatively affects policy formulation. The ideas and the participation of the community are important as they directly are impacted by CCA and are also responsible for grassroots implementation. However, certain adaptation measures have been rejected by the community due to a lack of coordination. This is attributed to inadequate recognition of the needs and capacities of the community before suggesting specific adaptation measures. These challenges and reasons behind the CCA implementation process reveals a cyclic nature of the process (see Fig. 2).

According to the Fig. 2, there is a two-way cyclic relationship among factors. For example, currently, there is no legally empowered body that is active in areas related to climate change adaptation in Sri Lanka and the CCS currently does not have departmental powers to allocate proper funding. In addition, there is a problem of allocating funds to climate adaptation due to the current financial situation of the country. Therefore, respective organizations cannot continue the coordination process caused by insufficient funds. For example, the issue of exclusion (as identified in the previous section) is an underlying cause of funding shortfalls. Furthermore, the lack of coordination among government officials and institutions can cause delays in financial allocations. Corruption at the local level also hinders collaboration with certain stakeholders, as its collaboration can disrupt corrupt mechanisms, and people may refrain from implementing CCA measures due to such corruption. For example, some people build their homes illegally over a canal which would ultimately lead to flooding. As such, measures implemented without consulting the community often end up in failure. Political influence can also discourage the implementation of climate change adaptation. Furthermore, since there is no political influence to provide necessary powers to respective organizations, the bodies that govern the

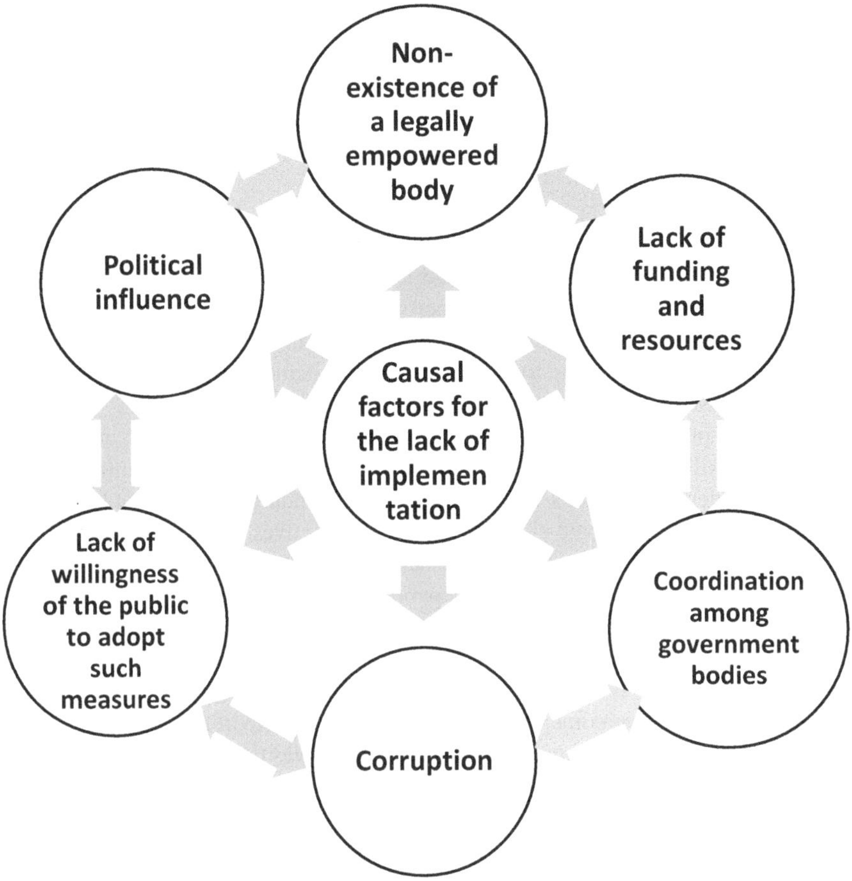

Fig. 2 CCA implementation process. *Source* Own data, 2023

CCA are powerless. In this way, a two-way relationship can be observed among these factors.

In addition to all the above-mentioned factors, there are skill and knowledge gaps related to climate change adaptation. Most people and officials lack sufficient knowledge of climate change and CCA, resulting in a lack of prioritization for climate change adaptation. Additionally, built environment professionals offer solutions based on their expertise, though it is more apt to implement solutions through proper multi-disciplinary consultation. Therefore, addressing these gaps would ultimately help to minimize issues related to climate change adaptation.

In conclusion, it is important to break this cyclic pattern by recognizing the lack of funds as a key reason, while understanding the absence of a legally empowered body to allocate and implement proper funds as the root cause. The funding issue prevents organizations from implementing CCA, and the absence of responsible organizations means that ministries are not legally bound to prioritize CCA, which results in coordination issues.

5 Conclusion

In conclusion, the study illuminates the pivotal role played by various stakeholders within the built environment in climate change adaptation efforts. Despite their significance, these stakeholders encounter numerous challenges that impede the effective execution of their responsibilities. A recurring pattern has been discerned among the causal factors, contributing to the suboptimal implementation of climate change adaptation measures.

One of the prominent challenges highlighted in the study revolves around the existence of skill and knowledge gaps within various categories of built environment stakeholders. These gaps were identified as substantial hurdles that hinder the seamless execution of climate change adaptation initiatives. The recognition of this deficiency underscores the importance of addressing educational and informational needs within the sector to enhance the capabilities of stakeholders.

Furthermore, the study underscores the interconnectedness of these challenges and skill gaps, forming a cyclic pattern that perpetuates the impediments to successful climate change adaptation. This intricate web of factors necessitates a holistic and targeted approach to overcome the barriers faced by built environment stakeholders.

In response to these findings, a proactive initiative was taken to address the identified skill and knowledge gaps. A comprehensive set of competencies was crafted based on the study's insights. This draft of competencies was then presented to the relevant stakeholders for their invaluable input, comments, and further validation. The collaborative engagement with stakeholders played a crucial role in refining the competency framework, ensuring its alignment with the practical needs and challenges faced by those actively involved in climate change adaptation within the built environment.

In essence, the study not only highlights the challenges faced by built environment stakeholders in climate change adaptation but also emphasizes the importance of targeted interventions to bridge knowledge gaps, foster collaboration, and enhance the overall efficacy of climate change adaptation measures within the built environment sector.

References

Amaratunga, D., & Haigh, R. (2011). *Post-disaster reconstruction of the built environment: Rebuilding for resilience*. Wiley.

Bryan, E., & Behrman, J. A. (2013). *Community based adaptation to climate change: Theoretical framework, overview of key issues and discussion of gender differentiated priorities and participation*. International Food Policy Research Institute (IFPRI).

Bulkeley, H., Carmin, J., Broto, V. C., Edwards, G. A., & Fuller, S. (2013). Climate justice and global cities: Mapping the emerging discourses. *Global Environmental Change, 23*(5), 914–925.

Chandrasekhar, D. (2012). Digging deeper: Participation and non-participation in post-disaster community recovery. *Community Development, 43*(5), 614–629.

Climate Change Secretariat, Ministry of Mahaweli Development and Environment (CCS and MoMDE). (2016). *National adaptation plan for climate change impacts of Sri Lanka—2016–2025*. http://www.climatechange.lk/NAP/NAP%20For%20Sri%20Lanka_2016-2025.pdf

Climate change widespread, rapid, and intensifying—IPCC. (2021). IPCC—Intergovernmental Panel on Climate Change. https://www.ipcc.ch/2021/08/09/ar6-wg1-20210809-pr/

Cochu, A., Tobias, H., & Mikael, H. (2019). *The roles of the private sector in climate change adaptation—An Introduction* (p. 26). Extraido el.

De Costa, W. A. J. M. (2008). Climate change in Sri Lanka: Myth or reality? Evidence from long-term meteorological data. *Journal of the National Science Foundation of Sri Lanka, 36*, 63–88.

De Silva, C. S. (2006). *Impacts of climate change on water resources in Sri Lanka*. Paper presented at the 32nd WEDC International Conference, November 13–17, 2006, Colombo, Sri Lanka.

Eriksen, S. H., Nightingale, A. J., & Eakin, H. (2015). Reframing adaptation: The political nature of climate change adaptation. *Global Environmental Change, 35*, 523–533.

Ginige, K., Amaratunga, D., & Haigh, R. (2010). Developing capacities for disaster risk reduction in the built environment: Capacity analysis in Sri Lanka. *International Journal of Strategic Property Management., 14*(4), 287–303. https://doi.org/10.3846/ijspm.2010.22

Hettiaratchi, V. (2020). *Influencing green: climate change, state policy, and civil society | LUP student papers*. Retrieved October 10, 2022, from https://lup.lub.lu.se/luur/download?func=downloadFile&recordOId=9019973&fileOId=9019974

Intergovernmental Panel on Climate Change (IPCC). (2013). Climate change 2013: The physical science basis. In T.F. Stocker, D. Qin, G.-K. Plattner, M. Tignor, S. K. Allen, J. Boschung, A. Nauels, Y. Xia, V. Bex, & P. M. Midgley (Eds.), *Contribution of working group I to the fifth assessment report of the intergovernmental panel on climate change* (1535 pp.). Cambridge University Press.

Intergovernmental Panel on Climate Change (IPCC). (2014a). Climate change 2014: Mitigation of climate change. In O. Edenhofer, R. Pichs-Madruga, Y. Sokona, E. Farahani, S. Kadner, K. Seyboth, A. Adler, I. Baum, S. Brunner, P. Eickemeier, B. Kriemann, J. Savolainen, S. Schlömer, C. von Stechow, T. Zwickel & J. C. Minx (Eds.), *Contribution of working group III to the fifth assessment report of the intergovernmental panel on climate change*. Cambridge University Press.

Intergovernmental Panel on Climate Change (IPCC). (2014b). Climate change 2014: Impacts, adaptation, and vulnerability. Part A: Global and sectoral aspects. In C. B. Field, V. R. Barros, D. J. Dokken, K. J. Mach, M. D. Mastrandrea, T. E. Bilir, M. Chatterjee, K. L. Ebi, Y. O. Estrada, R. C. Genova, B. Girma, E. S. Kissel, A. N. Levy, S. MacCracken, P. R. Mastrandrea, & L. L. White (Eds.), *Contribution of working group II to the fifth assessment report of the intergovernmental panel on climate change* (1132 pp.). Cambridge University Press.

Intergovernmental Panel on Climate Change (IPCC). (2018). Global warming of 1.5°C. In V. Masson-Delmotte, P. Zhai, H.-O. Portner, D. Roberts, J. Skea, P. R. Shukla, A. Pirani, W. Moufouma-Okia, C. Pean, R. Pidcock, S. Connors, J. B. R. Matthews, Y. Chen, X. Zhou, M. I. Gomis, E. Lonnoy, T. Maycock, M. Tignor, & T. Waterfield (Eds.), *An IPCC Special Report on the impacts of global warming of 1.5°C above pre-industrial levels and related global greenhouse gas emission pathways, in the context of strengthening the global response to the threat of climate change, sustainable development, and efforts to eradicate poverty* (616 pp.). Cambridge University Press. https://doi.org/10.1017/9781009157940

Marambe, B., Pushpakumara, G., Silva, P., Weerahewa, J., & Punyawardena, R. (2013). Climate change and household food security in homegardens of Sri Lanka. In H. P. M. Gunasena, H. A. J. Gunathilake, J. M. D. T. Everard, C. S. Ranasinghe, & A. D. Nainanayake (Eds.), *Proceedings of the international conference on climate change impacts and adaptation for food and environmental security, Colombo* (pp. 87–100).

Marambe, B., Punyawardena, R., Silva, P., Premalal, S., Rathnabharathie, V., Kekulandala, B., Nidumolu, U., & Howden, M. (2015). Climate, climate risk, and food security in Sri Lanka: Need for strengthening adaptation strategies. In W. Leal (Ed.), *Handbook of climate change adaptation* (pp 1759–1789). Springer-Verlag Berlin Heidelberg.

Ministry of Environment (ME). (2010a). *Sector vulnerability profile: Health.* Ministry of Environment. Retrieved from http://envmin.nsf.gov.lk/handle/1/378

Ministry of Environment (ME). (2010b). *Sector vulnerability profile: Urban development, human settlements, and infrastructure.* Ministry of Environment, Colombo, Sri Lanka. Retrieved from https://www.climatechange.lk/adaptation/Files/Urban_SVP_Nov-16-2010.pdf

Nissanka, S. P., Punyawardena, B. V. R., Premalal, K. H. M. S., & Thattil, R. O. (2011). Recent trends in annual and growing seasons' rainfall of Sri Lanka. In *Proceedings of the international conference on the impact of climate change in agriculture* (pp. 249–263). Faculty of Agriculture, University of Ruhuna

Pallawala, R. (2018). *Mapping of institutional arrangement in Sri Lanka for climate change mitigation, Colombo, Sri Lanka.* Janathakshan GTE Limited.

Ranasinghe, C. S. (2013). Impact of climate change in coconut and identification of adaptation measures with special emphasis on reproductive development. In H. P. M. Gunasena, H. A. J. Gunathilake, J. M. D. T. Everard, C. S. Ranasinghe, & A. D. Nainanayake (Eds.), *Proceedings of the international conference on climate change impacts and adaptations for food and environment security* (pp. 101–113). Hotel Renuka.

Shafique, K., & Warren, C. M. (2016). Stakeholders and their significance in post natural disaster reconstruction projects: A systematic review of the literature. *Asian Social Science, 12*(10), 1–17. https://doi.org/10.5539/ass.v12n10p1

Simon, K., Diprose, G., Thomas, A. C. (2020). Community-led initiatives for climate adaptation and mitigation. *Kōtuitui: New Zealand Journal of Social Sciences Online, 15*(1), 93–105. https://doi.org/10.1080/1177083X.2019.1652659

Sridarran, P., Keraminiyage, K., & Amaratunga, D. (2016). Community integration and participation to improve the built environment of the Post-Disaster Involuntary Relocations. In *Proceedings of the 12th international conference of the international institute for infrastructure resilience and reconstruction (IIIRR)* (pp. 160–166). IIIRR. ISBN 978-955-589-210-1

United Nations Framework Convention on Climate Change (UNFCCC). (1992). *United Nations framework convention on climate change.* Retrieved from https://unfccc.int/resource/docs/convkp/conveng.pdf

Wijeratne, M. A., Anandacoomaraswamy, A., Amarathunga, M. K. S. L. D., Ratnasiri, J., Basnayake, B. R. S. B., & Kalra, N. (2007). Assessment of impact of climate change on productivity of tea (*Camellia sinensis* L.) plantations in Sri Lanka. *Journal of the National Science Foundation of Sri Lanka, 35*(2).

University Educational Contribution on the Improvement of Perceptions and Behavioral Aspects of Students for Climate Change Mitigation; A Case Study in Sri Lanka

S. A. A. Gunawardana, W. P. Abeyrathna, F. R. Arooz, L. Tupenaite, and R. U. Halwatura

List of Abbreviations

BCM	**Behavior** in **climate** change **mitigation**
BEN	**Behavior** in **energy conservation**
BEC	**Behavior** in **environment** conservation
CC, SD & EC	Climate change, sustainable design and energy conservation
FS	Field of study
GFI	Goodness-of-fit index
PA	**Practical application** during university curricular
PCB	**Perception** of **consumer behavior** on climate change
PEC	**Perception** on **energy conservation**
PI	Personal interest
PMF	**Perception** on climate change **mitigation factors**
PNM	**Perception** on the **need** of climate change **mitigation**
SD	Sustainable development

S. A. A. Gunawardana (✉) · W. P. Abeyrathna · R. U. Halwatura
Department of Civil Engineering, Faculty of Engineering, University of Moratuwa, Moratuwa, Sri Lanka
e-mail: anupamagunawardana77@gmail.com

F. R. Arooz
Department of Architecture, General Sir John Kotelawala Defence University, Southern Campus, Sooriyawewa, Sri Lanka

L. Tupenaite
Department of Construction Management and Real Estate, Faculty of Civil Engineering, Vilnius Gediminas Technical University, Vilnius, Lithuania

C. Malalgoda et al. (eds.), *Climate Change Adaptation in the Built Environment*,
https://doi.org/10.1007/978-3-031-75826-3_13

WCC	**Willingness** to integrate **climate change** mitigations in future professional work
WL	**Weather learned** in curricular
WSD	**Willingness** to integrate **sustainable designs** and energy conservation in future professional work

1 Introduction

Climate change has been determined as the largest global scale hazard of the twenty-first century since the increases in the earth's temperature because due to climate change poses devastating influences on the earth's ecosystems. Enhancing public engagement is much essential to form resilience at individual and societal levels to confine the progression of climate change (Feygina et al., 2019). Scientific understanding of climate change in the community should be enriched and public concern should be increased in order to mitigate climate change and implement actions and strategies to overcome the harmful impacts of climate change (Knutti, 2019). Enhancing education related to climate change, public sentience, training, community-level access to facts, and public contribution is been encouraged to nations by Article 12 of the Paris Agreement (Savaresi, 2016).

The community needs to incorporate climate change mitigation strategies in their professional and day-to-day life while future leaders and policymakers must initiate decisions from an informed position (Mintrom & Luetjens, 2017). Similarly, for the people of the general community to incorporate these concepts within their professional life and showcase the behaviors related to climate change mitigation their cognitive factors such as perceptions, and behaviors should be in line with the climate change mitigation strategies. Perception of an individual can be referred to as a comparatively persistent association of beliefs, emotions, and behavioral propensities headed for the socially important object, person, event, or symbol (Ives et al., 2018). The major factors that influence perception formation are experience, social factors, classical and operant conditioning, and observations. It also involves cognitive processes by the individual's learning, memory, attention, and expectation and connects a person's concepts and expectations by restorative and selective mechanisms. Similarly, the factors that guide perception formation can lead to creating perceptional changes (Fisher & Bonn, 2011). Perceptions of an individual are directly linked with their behavioral contexts. When it comes to human behavior, it can be identified as probable and expressed physical, mental, and social capacities of individuals or groups to react to internal and external stimuli throughout their life. Perceptions and values influenced by thoughts and feelings of individuals have a strong effect on behavior whereas human behavior is prominently shaped by psychological traits as personality types differ from person to person (Clark et al., 2016). The main difference between perception and behavior is that human behavior is the reflection of one's perception towards it. However, it has been identified that perceptions and real behavior aren't constantly aligned (Schlüter et al., 2017).

Even though an individual's perception, and behavior are integrated with a broad spectrum of interconnected factors, there have been identified some major stimuli towards one's overall perceptions and behavior toward climate change (Clayton et al., 2015). Behavior related to climate change mitigations is often associated with knowledge and the personal actions of individuals. For example, if a person has comparatively more knowledge about climate change he or she might give more importance to personal actions such as recycling, emission reductions, buying locally, and actions of environmental justice. Furthermore, psychologists have identified that personal internal influences such as values, and beliefs are also linked to this concept and they have mentioned that awareness, education, persuasion, and guilt are some of the major tools which alter one's perceptions and behaviors (Aminrad et al., 2013). Another important aspect that enhances an individual's perceptions, and behavior toward climate change mitigations is the interest of people. The interest of a person is a psychological state or emotion aroused by specific characteristics of the learning environment. Among these variable interests, the environmental interest of an individual is more related to their view toward climate change mitigations. Environment interest emphasizes the individual's cognitive engagement towards the environment thus it leads to improving the environmental attitude of people toward sustainability. For example, students who are interested in nature or students who have more environmental interests are more likely to be responsible for environmental issues (Uitto et al., 2014). However, it is understood that all these factors are interconnected and collectively work towards enhancing one's perception, and behavior towards climate change mitigations. Climate change education is also another factor that influences on an individual's perception, and behavior toward climate change mitigations (Otto & Pensini, 2017). Applying effective reality and the possible effects of climate change education can also stimulate an individual's ethical perception towards climate change mitigations and therefore, it must be revised and complemented with the new environmental, political, economic, and cultural variability (Boca & Saraçli, 2019).

Countries have become ever more conscious of the prominence of climate change education, as climate change is turning out to be the world's major ecological challenge (Cantell et al., 2019b). In recent years, the concern given for climate change has grown in corresponding to leadership for educational programs and stretched financing that copes with climate change such as including climate change in educational curricular reforms, increasing knowledge of unfamiliar weather patterns, and concerning on probability of economic, social, and environmental changes owing to climate change. Numerous features of climate change have made it a more inspiring topic to explain as it contains multi-disciplinary aspects where the topic of climate change appears to have intensely reverberated with detained morals.

Though climate change education is vital at all stages of education, it is much required at higher education levels for developing a systemic approach (Molthan-Hill et al., 2019). Internationally recognized goals for reducing worldwide mean temperature to 1.5 °C beyond 1990 levels possess a requirement of action gap that will pledge the necessary reductions thus universities are identified as one of the key institutes for diminishing this gap (Molthan-Hill et al., 2019). Youth who are graduating from universities join the drent professions and move towards decision-making policy making levels through their occupational careers. Therefore, it is important to develop critical thinking abilities and assist youth to identify the foundations of the battle regarding climate change and highlight their problem-solving abilities (Leal Filho et al., 2021). Knowledgeable individuals are mindful of the hazards of climate change and well prepared with conversant choices about climate change at local, nationwide and global levels. Therefore, it is of prime importance to identify how universities could better address this challenge.

Since climate change can affect all categories of human activities, major lessons about climate change need to be taught in the form of an interdisciplinary approach (Ayanlade & Jegede, 2016). Approaches for integrating climate change within university education are partial, scattered, and often dedicated to one discipline. A famous approach is the offering of specialized programs and elective modules to students who select them. Universities are capable of integrating new academic programs with a multi-disciplinary approach to ensure that future professionals might acquire a better knowledge of both the challenges of climate change and various strategies for mitigation and adaptation. Both staff and students are needed to be encouraged by the universities to cope with the tasks facing, stimulate research, find solutions, and take the lead in public discourse regarding climate change mitigations (Molthan-Hill et al., 2019). To act effectively with these activities, the knowledge and awareness of climate change and its implications are crucial, and identifying the dynamics of climate change would enable to influence on resulting environmental changes (Ayanlade & Jegede, 2016). Nevertheless, it is identified that every student is required to be aware of mitigation tools for climate change despite their field of studies such as law, chemistry, business, or any other area. Therefore, universities must mainstream climate change education so that every student gets well aware of climate change through their core curriculum (Molthan-Hill et al., 2019).

Similarly, it is also important to determine that whether learning these concepts related to climate change mitigations from the university can actually impact on cognitive and behavioral aspects of individuals. Although, universities have undertaken curricular based approaches to address the issues of climate change and other means of knowledge delivering, still the gap exists whether individual perceptions and behavior towards climate change mitigations are significantly enhanced by university education. Similarly, it should be explored what systemic actions should be carried out to enhance individual's perceptions towards climate change mitigations in terms of university based curricular reforms, because enhancing perceptions towards climate change mitigations can significantly improve one's own behavior towards climate change mitigations.

1.1 Research Gap

Moreover, education is the best mode for delivering these climate change-related concepts to the world since it enhances the knowledge, awareness, and perceptions of the individuals and also it modifies the behavioral patterns of the people. Furthermore, in the avenues of formal education, universities have been determined as one of the best platforms to deliver sustainability to the world by means of curricular integration. Youth who are graduating from the university directly join industry and other sectors for their occupations. The outcome of university education is a professional who can command and lead people and sometimes who ultimately become policy-makers. Therefore, if climate change is integrated in the university curriculum, it provides the learning opportunities for youth to be better equipped with knowledge, skills and practices to deal with sustainable development in all aspects. It also helps to build the interest of students to view climate change holistically, to understand how their actions and decisions affect the environment, and most importantly to promote climate change mitigations in their professional life which ultimately lead to climate change mitigations and sustainable development of the country.

Embedding knowledge of climate change into university education is not enough. It should be measured the effectiveness of this incorporating process. By incorporating these concepts whether it will fill the knowledge gap of the students to use climate change mitigations in their professional life. For that, it is required to know the perceptions of the students towards these concepts. Whether university education has enhanced their perceptions, what is their behavior towards these concepts? Therefore, by doing a perception and behavior analysis there is a possibility of coming into a conclusion as to whether this integration process is successful.

However, individual cognitive variables such as perceptions, behavior and attitudes are complex to measure thus it involves multiple mediations with numerous correlations. In order to handle the researches that involves multiple mediations, SEM examines the structure of interrelationships articulated as a set of equations related to a system of multiple regression equations. SEM is preferably appropriate for analytical models which include multiple mediation variables which are both predicted by other variables and also which predict additional variables.

1.2 Research Objectives

Main intend of the study was to explore how university education has influenced on the development of positive perceptions and behavior towards climate change mitigations. Since factors with multiple mediations are involved in the study, structural equation modeling was used to develop the model. In order to, better understand the underlying conceptions, the variables which are subjected to analysis were categorized. Field of study, which is a main aspect in university education, curricular based learning which involves direct knowledge delivery, practical applications which comes from the curricular based integration and personal interest which is of a personal trait were considered to be determined for the improvement of perceptions of the individuals. Secondly, the individual's perception influence on the behavior related to climate change mitigations were required to be explored.

2 Literature Review

2.1 Scientific Background of Climate Change

Climate change can be identified as an alteration of climate which is endorsed directly or indirectly to anthropogenic activities that modify the global atmospheric compositions in addition to natural climate modifications detected through comparable time periods. Human interventions that change the composition of the atmosphere along with the terrestrial usage can be identified as the prime reason while natural activities such as fluctuations in the internal variability in the climate system, sun's radiation, and volcanoes are also the causes of the climate change (Stachowicz et al., 2002).

Global temperature rise, shrinking ice sheets, warming ocean temperatures, ocean acidification, glacial retreat, sea level increase, diminished snow cover, decreasing arctic sea ice, and are identified as the major consequences of climate change (Wu et al., 2008). Similarly, frequent storms, additional rainfall, frequent and intensified hurricanes and tornadoes mainly throughout severe weather occasions causing floods and other damages, heat waves causing human deaths, endangered habitats homes, and lives, and upsurge in wildfires are also related to climate change (Füssel & Klein, 2006; Heller & Zavaleta, 2009). The impacts of climate change are significantly on the people of developing and undeveloped nations because of their poor ability for adapting to climate change although the majority of the causes of climate change are posed by the developed countries. By lowering susceptibility to the adversative influences of climate change especially by reducing the GHG emissions, impacts of climate change can be mitigated (Stephenson et al., 2010).

2.2 Consumption Pattern's Influence on Climate Change

Consumption patterns of the people have a higher contribution to climate change with direct and indirect GHG emissions across various dimensions of consumption. Among those dimensions, food, transport, and other household consumptions are significant avenues. The global average GHG emissions associated with household consumption is about 6 tCO_2eq/cap. However, due to the high variations in income groups, these are vastly unevenly scattered across nations (Ivanova et al., 2020). Aspects such as everyday life, behavior, and cultural norms are linked with consumption and those are highly influential for embodied emissions with a higher potential for mitigations in various consumption domains. Emissions associated with food can highly vary among different diet categories. The carbon emissions per calorie/kg of the prime invention are significantly lesser for plant foods in comparison to dairy, ruminant, and non-ruminant products with meat generating more releases per unit of production. (Frenette & Bahn, 2020). When it comes to other household consumptions, emissions are linked with features such as prime material, operational energy and embodied energy, use of fuel, thermal insulation, and others. Mitigation strategies for this sector are mainly focused on acquiring renewable power, generating particular renewable current, and choosing heat pumps and renewable-centered heating. Furthermore, consumption linked with transportation are highly based on fossil fuels related to travel. Shifting to non-emission vehicles, carpooling techniques, and using public transport are some of the major changes associated with consumption linked with transport (Wang et al., 2019).

Similarly, growing urbanized populations along with consistent consumption changes have caused GHG emissions to increase. Variations in income levels and spending preferences can cause dissimilarities in per capita consumption which in turn influence GHG emissions (Tian et al., 2017). One of the main drivers of GHG emissions is urbanization because more goods and services are consumed by urban residents than rural, making more direct emissions (Wiedenhofer et al., 2013). As mitigation strategies, consumer demands can be driven toward low-carbon sectors. By providing incentives and highlighting the advantages such as energy savings in long run, customers could be stimulated to provide more priority to green procurement (Zhang et al., 2016).

2.3 Climate Change Impacts in Sri Lanka

Though climate change, adaptation approaches, and modification actions have gained much more attention in public discussions in the developed world, the effects of climate change are further problematic in less developed countries, because the livelihoods are closely tied with the changes in the environment (Barnes et al., 2015). Sri Lanka is a developing country in the South East Asian region that has a higher vulnerability to the consequences of climate change nevertheless, having a minimum contribution to GHG emissions. Changes in precipitation patterns, variations in temperature and high occurrence of extreme events have influenced both directly and indirectly multidimensional sectors such as agriculture, tourism, and society in Sri Lanka (Navaratne et al., 2019).

Around 40% of the total land area is used for the agricultural purposes in the country (Kariyawasam & Kumar, 2019). Around 2.4 million ha of land extent is dedicated for agriculture of which 34% is under irrigation (Esham & Garforth, 2013). The key factors determining the vegetation dynamics are the temperature and precipitation (Kariyawasam & Kumar, 2019). It is revealed that a substantial and systematic warming has happened in every zones of Sri Lanka with the analysis of long-term air temperature data thus the warming rate has surpassed the international regular rate of warming in most of the settings. (Navaratne et al., 2019).

The impacts of droughts and delayed monsoons have stifled the growth of tea which is the other principal crop plantation in Sri Lanka (Esham & Garforth, 2013). Tea grown in Sri Lanka plays a vital role in the improvement of the country's economy and is also greatly impacted by temperature disparities and rainfall (Navaratne et al., 2019).

The impacts of climate change on the coastal environment in Sri Lanka have taken a crucial significance in recent years. Due to the resources existing in coastal areas, the community around the regions mainly derive and depend on fishing and tourism (Lin, 2017; Senevirathna et al., 2018). The sea level increase in Sri Lanka is projected to rise by 1.0 m in 2070. Direct and obvious consequences are felt by low-lying areas with a higher risk of floods and coastal erosion (Senevirathna et al., 2018). Similarly, these hazards have threatened tourism, one of the major incomes of the country, and are likely to affect the socio-economic well-being of the people relying on these eco-system services (Mehvar et al., 2019).

Health risks imposed by climate change are also crucial in Sri Lankan scenario. Sri Lanka is a country with a warm humid climate. People with different fitness levels will be susceptible to dehydration and fatigue as temperature increases and also the frequency of high blood pressure, coronary diseases will be increased among the elderly population. Droughts and floods will result in various sanitation-related illnesses such as cholera, hepatitis, and typhoid among people, especially the low-income community. Higher rainfall conditions will results in water collection in rock pools that will encourage the breeding of Dengue mosquitos (Baba, 2010). Further, extreme events such as floods, storms, and landslides will dislocate heaps of human settlements creating a nationwide emergency while epidemic conditions threaten the food security of the country (De Zoysa & Inoue, 2014).

2.4 Importance of Integrating Climate Change into University Education

Countries have become ever more conscious of the prominence of climate change education, as climate change is turning out to be the world's major ecological challenge. (Cantell et al., 2019a). In order to combat the scientific, environmental, and political issues, the role of climate change education is highly important. (Monroe et al., 2017). The public needs to incorporate climate change mitigation strategies in their professional and day to day life while future leaders and policymakers must initiate decisions from an informed position. Therefore, it is of prime importance to identify how universities could better address this challenge. Paris Agreement addresses new academic programs with a multi-disciplinary approach to ensure that future professionals might acquire a better knowledge of both the challenges of climate change and various strategies for mitigation and adaptation. Both staff and students are needed to be encouraged by the universities to cope with the tasks facing, stimulate research, find solutions and take the lead in public discourse regarding climate change mitigations (Molthan-Hill et al., 2019).

It is important to develop critical thinking abilities and assist youth to identify the foundations of the battle regarding climate change and highlight their problem-solving abilities (Monroe et al., 2017). Knowledgeable individuals are mindful of the hazards of climate change and well prepared with conversant choices about climate change at local, nationwide and global levels. Though climate change education is more vital at all stages of education, it is much required at higher education levels for developing a systemic approach (Molthan-Hill et al., 2019). Approaches for integrating climate change within university education are partial, scattered, and often dedicated to one discipline. A famous approach is the offering of specialized programs and elective modules to students who select them. Nevertheless, it is identified that every student is required to be aware of mitigation tools for climate change despite their field of studies such as law, chemistry, business, or any other area. Therefore, universities must mainstream climate change education so that every student gets well aware of climate change through their core curriculum (Molthan-Hill et al., 2019).

2.5 Individual's Perception, Attitude and Behavior

According to psychology, perception can be defined as the process of identifying, organizing, and interpreting sensory information with the intention of representing and understanding the presented information. It also involves cognitive processes by the individual's learning, memory, attention, and expectation and connects a person's concepts and expectations by restorative and selective mechanisms. Furthermore, the formation of perception seems typically effortless because it happens outside the conscious awareness but it depends on the functions of the nervous system (Pickens, 1998).

Probable and expressed physical, mental, and social capacities of individuals or groups to react to internal and external stimuli throughout their life is termed human behavior. Perceptions and values influenced by thoughts and feelings of individuals have a strong effect on behavior where human behavior is prominently shaped by psychological traits as personality types differ from person to person. Human behavior falls upon a spectrum with limits of acceptable to unacceptable and common to unusual (Fischer, 2017). The acceptability aspect of a behavior highly depended upon the social norms and society's control whereby humans are forced to follow some definite rules and showcase certain behaviors depending on the given society or culture. The main difference between perception and behavior is that human behavior is the reflection of one's perception towards it. However, it has been identified that perceptions and real behavior aren't constantly aligned (Sadusky, 2014).

2.6 *Factors Affecting Individual's Perception, Attitude and Behavior Towards Sustainability*

Even though an individual's perception, attitudes, and behavior are integrated with a broad spectrum of interconnected factors, there have been identified some major stimuli towards one's overall perceptions, attitudes, and behavior toward sustainability. According to the majority of literature sources, pro-environmental behavior can be identified as one of the major influential factors (Sadusky, 2014). Pro-environmental behavior is often associated with environmental knowledge and the personal actions of individuals. For example, if a person has comparatively more knowledge about sustainability, he or she might give more importance to personal actions such as recycling, emission reductions, buying locally, and actions of environmental justice. Similarly, their concerns about sustainability are reflected by their pro-environmental behavior (Aminrad et al., 2013). Environmental education is also another factor that influence on an individual's perception, attitudes, and behavior toward sustainability. Environmental education highlights that knowledge, perceptions, and abilities are correlated to the environment, and through the developed education, the fundamentals of a sustainable environment are recognized and are highly influential on perceptions and behavior. It has been identified that three major concepts of environment, sustainable development, and education coexists, and environment education is based on sustainable environmental focuses on environment knowledge, environmental perceptions, and environmental skills. (Boca & Saraçli, 2019).

Another important aspect that enhances an individual's perceptions, attitudes, and behavior toward sustainability is the interest of people. The interest of a person is a psychological state or emotion aroused by specific characteristics of the learning environment. Among these variable interests, the environment interest of an individual is more related to their view toward sustainability. Environment interest emphasizes the individual's cognitive engagement towards the environment thus it leads to improving the environmental attitude of people towards sustainability. For example, students who are interested in nature or students who have more environmental interests are more likely to be responsible for environmental issues (Uitto et al., 2014).

The ecological values of an individual are also such an important aspect that improves perceptions, attitudes, and behavior towards sustainability. The ecological values of an individual reflect the strength of biocentric values to protect and conserve the environment. Ecological values have the main role in shaping the positive behavior of individuals in terms of environmental growth, and attitudes towards ecological behavior and provide the foundation of close ties with nature. Similarly, it is closely linked with ecological education which enhances the environmental knowledge of individuals (Uitto et al., 2014).

Apart from above-mentioned factors, psycho-social factors such as moral obligations, symbolism, social status, and demographic factors such as gender, and age are also influential on individual's perceptions towards sustainability. All of the above-mentioned factors can be taken under three important lays such as environmental values, psychological variables, and situational variables (Ives et al., 2018). However, it much is understood that all these factors are interconnected and collectively work towards enhancing and one's perception, attitudes, and behavior towards sustainability.

3 Methodology

A process framework was designed and all relevant experiments were carried out according to the following basic framework (Fig. 1). A questionnaire survey was designed and Structural equation modelling which is of a multivariate analysis technique which explores relationships with multiple mediations was used for developing models to determine the correlations among considered variables and to compare the results at considered two country's settings.

3.1 Survey Instrument

The questionnaire survey was designed with 3 major sections and consisted of 32 questions. The questionnaire was structured in the online platform through Limesurvey software https://www.limesurvey.org/. The first section included 04 questions to explore the demographic background of the respondents; gender, age,

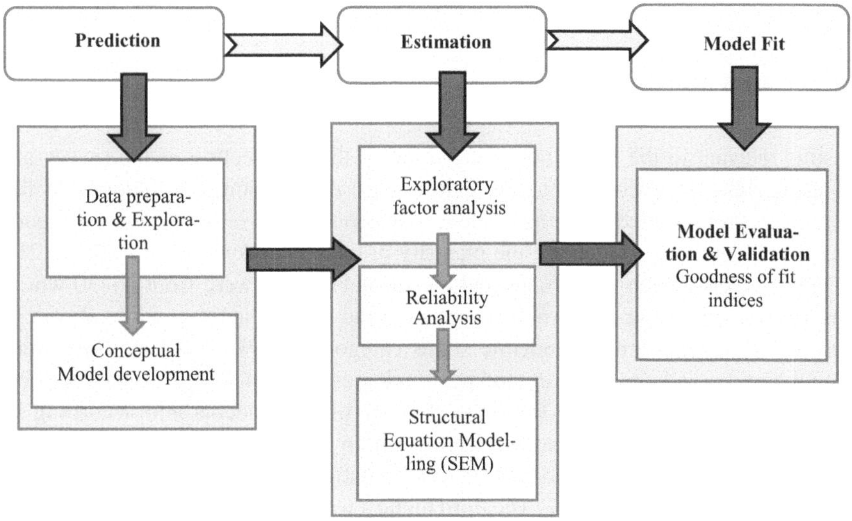

Fig. 1 Process framework

current academic status and field of study. The second section included 17 questions directly targeting climate change. It comprised of questions to reveal the perception of the students regarding the occurrence of climate change, consumer behavior on climate change, their experience with consequences of climate change, their personal interest towards the concept of climate change, and their behavior on climate change mitigation. Also, a set of questions were included to identify the curricular-wise university's contribution to climate change, from a student's point of view. The third section included 11 questions directly targeting sustainable designs and energy conservation. It included questions to identify students' perceptions of the energy crisis, sustainable designs, and energy conservation, their behavior on energy conservation, and a set of questions to explore the curricular wise university's contribution to sustainable designs and energy conservation from a student's point of view.

3.2 Survey Participants

The questionnaire was targeted at final year undergraduates and postgraduates and was distributed among three government universities in Sri Lanka. In order to make comparisons among students of different fields of study, five different fields of study were selected in this research as agriculture, engineering, science, management/arts and law. Altogether, the questionnaire was distributed among more than 1000 students via university e-mails. The survey was conducted from February to June 2020 and a total of 348 responses were obtained. All responses were imported to SPSS data file and relevant statistical analyses were conducted via IBM SPSS 23 version.

4 Results and Discussion

Results relevant to the frequency distribution of the first section of the survey are given in Table 1. According to the gender-wise distribution, the majority of the respondents represented the female category which was 60.1% and the male category was 39.9%. In the age category, the majority of the respondents were from the 21–25 age category which was 75.3% and the second highest were from 26–30 which was 19%. 4.3% of the proportion represented the 31–35 category while above 35 were 1.4%. In the current academic status category, 75.6% of the majority were from the undergraduate category and postgraduates were 24.1%. The final question of the first section identified the field of study of the students and according to the results majority of the students were from an engineering background with a percentage of 44%. The second highest representation was from the agriculture field and their proportion was 36.2%. The third highest was from science background with 12.4% while the fourth highest was from the Arts and Management field with 4.3%. However, the least percentage was from Law category with 3.2%.

Table 1 Results of demographic information

Variable	Group/item	Frequency	% of the respondents
Gender	Female	209	60.1
	Male	139	39.9
Age	21–25	262	75.3
	26–30	66	19
	31–35	15	4.3
	Above 35	5	1.4
Current academic status	Postgraduate	84	24.1
	Undergraduate	263	75.6
Field of study	Agriculture	126	36.2
	Engineering	153	44
	Science	43	12.4
	Management/arts	15	4.3
	Law	11	3.2

4.1 Conceptual Model

Based on the research objectives, following basic conceptual model was developed (Fig. 2). Accordingly, this model can be used to identify how the university education and the personal interest of the students influences on the perceptions of the students and how the perceptions interact to modify the behavior of the students towards climate change mitigations mainly through energy conservation and sustainable designs. Similarly, how their perceptions influences on their willingness to integrate in future professional work.

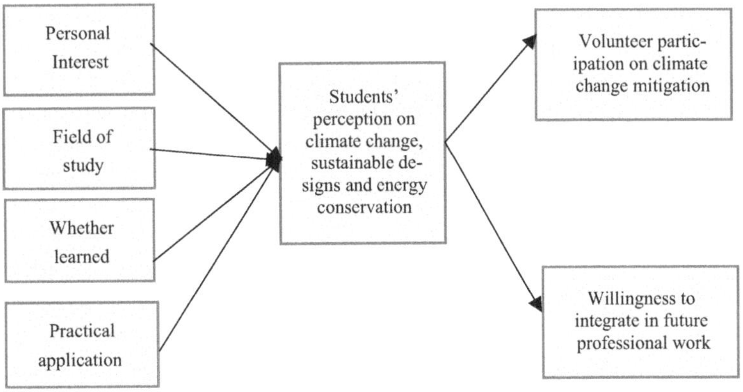

Fig. 2 Basic conceptual model

4.2 Explanatory Factor Analysis (EFA)

In order to explore more about the factors affecting students' perceptions, determine the underlying latent factors of the perceptions, and identify the phenomenon that how perceptions influence individual behavior related to mitigation and willingness to incorporate in future professional work, Exploratory Factor analysis (EFA) was conducted. Before EFA, Kaiser–Meyer–Olkin (KMO) test was conducted to measure the sample adequacy for each variable in the model and the complete modeling. Furthermore, Bartlett's test of sphericity was also conducted to test for equality of variances across groups against the alternative, that variances are unequal for at least two groups.

The formula for the KMO test is:

$$\mathrm{KMO} = \frac{\sum \neq r_{ij}^2}{\sum \neq r_{ij}^2 + \sum \neq u_{ij}^2} \tag{1}$$

where,

$r = [r_{ij}]$ is the correlation matrix,
$U = [u_{ij}]$ is the partial covariance matrix,
Σ = summation notation.

According to the results of the KMO and Bartlett's test, the KMO value was 0.807, and based on Bartlett's test of sphericity results, the significance level was 0.000 (Table 2). Since the KMO value is above 0.5, and the value is between 0.8 and 0.9, the relative appropriateness can be explained as meritorious according to Dragan and Topolsek (2014) which confirmed the sampling adequacy of the data set. Similarly, the obtained significance level confirmed that at least two variables in the data set are significantly correlated therefore, the values of both these tests verified the suitability of applying EFA further.

In EFA, in order to decide the number of factors, a scree plot was generated. Similarly, Principal Component Analysis (PCA) was used to produce the eigenvalues to determine the variance accounted for, by each underlying factor. To get the loading of each of the factors, the Principal axis factor extraction method was used. After these, loadings are rotated and the Varimax rotation method which is an orthogonal rotation was used.

Table 2 Results of KMO and Bartlett's test

Kaiser–Meyer–Olkin measure of sampling adequacy		0.807
Bartlett's test of sphericity	Approx. chi-square	2995.330
	df	325
	Sig.	0.000

Based on the results of the communalities, the extraction values of the items which indicated the proportion of variance in each of the variable ranged between 0.495 and 0.819. Since these values were above 0.3 which is the lowest required level (Howard, 2016), all items were suitable to be applied for further processes. Results of the communalities are given in Appendix. Factor analysis extracted eight factors which had eigen values above one. Generated scree plot also suggested extracting eight factors that accounted for a considerable proportion of total variance (Fig. 3). According to Table 3, it was identified that altogether 55.498% of the total variance was explained by eight factors. The first factor had an eigenvalue of 4.76 which explained 18.33% of the variance and the second factor had an eigenvalue of 2.14 with 8.26% of the variance explained. The third and fourth factors indicated eigenvalues of 1.64 and 1.32 respectively whereas their variances explained were 6.33% and 5.09% respectively. The fifth factor obtained an eigenvalue of 1.234 which explained 4.74% of the variance and the sixth factor had an eigenvalue of 1.19 which supported 4.60% of the variance. The seventh and eighth factors had eigenvalues of 1.09 and 1.01 respectively which described variances of 4.22% and 3.9% respectively.

By the results of the rotated component matrix latent factors were derived and, it was identified that three items were located under the first factor with factor loading ranging from 0.883 to 0.831. Under the second factor, five items were located and their factor loadings were ranging from 0.711 to 0.578. Three items were located under factor three and their factor loadings were ranging from 0.709 to 0.545. Similarly, another three items were located under factor four where factor loadings ranged between 0.755 and 0.490. Two items had come under the fifth factor and their factor loadings were 0.786 and 0. 731. Four items had come under the sixth factor with

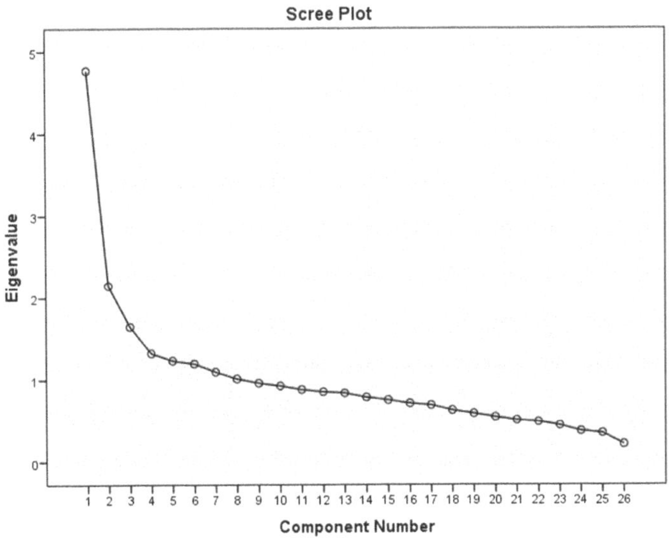

Fig. 3 Scree plot

Table 3 Total variance explained

Component	Initial eigenvalues			Extraction sums of squared loadings		
	Total	% of variance	Cumulative %	Total	% of variance	Cumulative %
1	4.767	18.333	18.333	4.767	18.333	18.333
2	2.148	8.261	26.595	2.148	8.261	26.595
3	1.647	6.336	32.931	1.647	6.336	32.931
4	1.325	5.095	38.026	1.325	5.095	38.026
5	1.234	4.744	42.770	1.234	4.744	42.770
6	1.197	4.603	47.373	1.197	4.603	47.373
7	1.098	4.224	51.597	1.098	4.224	51.597
8	1.014	3.901	55.498	1.014	3.901	55.498
9	0.963	3.702	59.200			
10	0.928	3.571	62.771			
11	0.882	3.391	66.162			
12	0.856	3.292	69.454			
13	0.840	3.232	72.686			
14	0.788	3.032	75.718			
15	0.755	2.905	78.623			
16	0.713	2.743	81.366			
17	0.692	2.662	84.028			
18	0.631	2.426	86.455			
19	0.590	2.270	88.724			
20	0.546	2.102	90.826			
21	0.509	1.958	92.784			
22	0.488	1.878	94.662			
23	0.446	1.714	96.375			
24	0.376	1.445	97.820			
25	0.351	1.352	99.172			
26	0.215	0.828	100.000			

Extraction method: principal component analysis

factor loadings ranging from 0.706 to 0.475. Under the seventh factor, three items were located which had factor loadings ranging from 0.742 to 0.459. Furthermore, another two items were located under the eighth factor and their factor loadings were 0.805 and 0.597. Therefore, based on the results of the EFA, eight latent factors with indicator variables underneath them were identified. Results of the rotated component matrix are given in Appendix.

4.3 Reliability and Validity of Scales

After deriving the latent factors and indicator variables, Cronbach's alpha (α) reliability analysis was used to measure the reliability and the validity of the latent factors used in the study. Cronbach's alpha coefficient measured the internal consistency, of the observed variables which came under each of the latent factor.

The formula for Cronbach's alpha coefficient is:

$$\alpha = \frac{N\bar{c}}{\bar{v} + (N - 1)\bar{c}} \tag{2}$$

where:

N = number of items,
$\bar{c}$ = the average of inter-item covariance among the items,
$\bar{v}$ = the average variance.

The cronbach's α values for the factors 1, 2, 3 and 4 were 0.857, 0.688, 0.640 and 0.722 respectively. The α value for the fifth factor was 0.632 and for the sixth factor was 0.562. Since the sixth factor had a value less than 0.6 which is the minimum required level (Yap et al., 2018), the Cronbach's alpha (α) if an item was deleted was calculated. When the CC13C item is deleted, the α value increases up to 0.635. Therefore, the CC13C item was eliminated from the sixth factor. Furthermore, α values for the seventh and eighth items were obtained as 0.736 and 0.693.

Based on the results of EFA and the developed conceptual model, latent factors were defined as follows:

Factor 1	Perception of climate change mitigation factors
Factor 2	Perception of the need for climate change mitigation
Factor 3	Behavior in environment conservation
Factor 4	Perception of energy conservation
Factor 5	Willingness to integrate sustainable designs and energy conservation in future professional work
Factor 6	Behavior in energy conservation
Factor 7	Behavior in climate change mitigation
Factor 8	Perception of consumer behavior on climate change

Therefore, based on the above derived latent factors, the basic conceptual model of the study was modified and a newfangled model with a higher number of variables

(Fig. 4). The newfangled conceptual model shows the direct and indirect relationship between the research variables that were considered in the study and was further proceed with the newfangled conceptual model.

Similarly, newly derived latent factors and indicator variables were re-coded for the convenience of conducting further analysis (Table 4). Apart from those eight latent factors identified by the EFA, all other considered variables were re-coded as given in Table 5 before conducting the further analyses. With those results, the study continued to justify the newfangled conceptual model by developing the structural equation model (SEM).

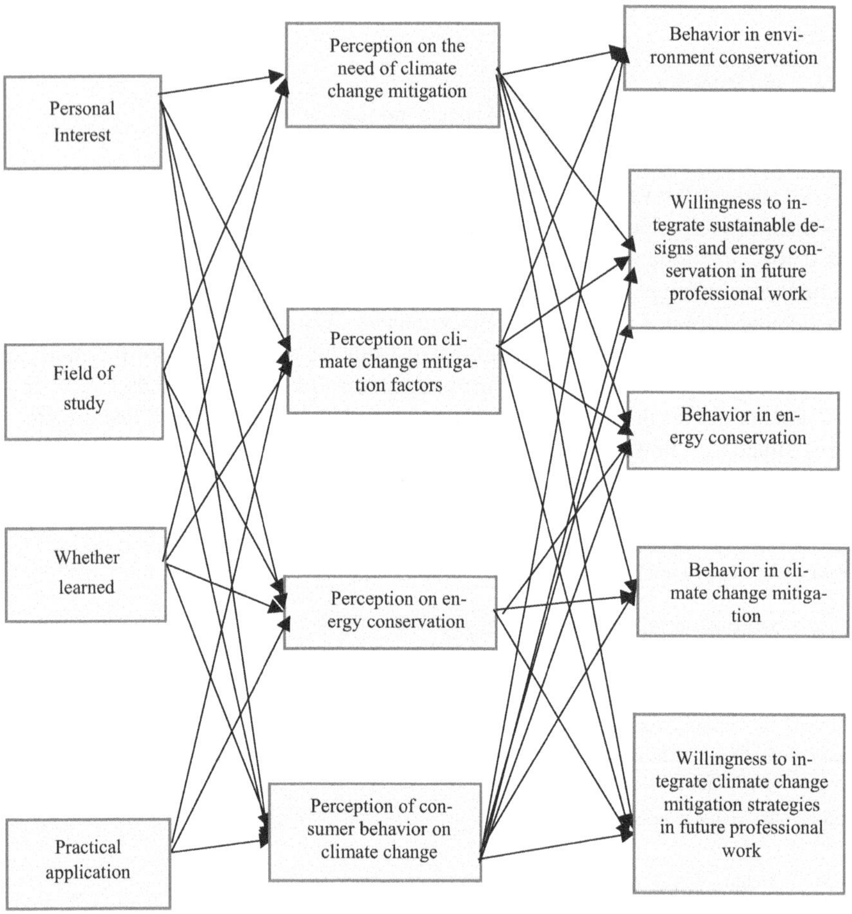

Fig. 4 Newfangled conceptual model

Table 4 Variables and renamed symbols

Factor	Renamed symbol for latent factors	Item number	Renamed symbol for items
Factor 1	PMF	CC14A	PMF1
		CC14B	PMF2
		CC14C	PMF3
Factor 2	PNM	CC1	PNM1
		CC3	PNM2
		SE16	PNM3
		SE22	PNM4
		SE23	PNM5
Factor 3	BEN	CC13G	BEN1
		CC13H	BEN2
		CC13I	BEN3
Factor 4	PEC	SE20B	PEC1
		SE21C	PEC2
		SE21E	PEC3
Factor 5	WSD	SE24	WSD1
		SE25	WSD2
Factor 6	BEC	CC13A	BEC1
		CC13D	BEC2
		CC13F	BEC3
Factor 7	BCM	CC13B	BCM1
		CC13E	BCM2
		CC13J	BCM3
Factor 8	PCB	CC7	PCB1
		CC12	PCB2

Table 5 Renamed variables apart from identified factors

Variable type	Variable	Renamed symbol	Item code	Renamed symbol for items
Independent	Field of study	FS	D5	FS
Independent	Personal interest	PI	CC6	PI
Independent latent	Weather learned in curricular	WL	CC8	WL1
			SE17	WL2
Independent latent	Practical application during university curricular	PA	CC9	PA1
			SE19	PA2
Dependent	Willingness to integrate climate change mitigations in future professional work	WCC	CC15	WCC

4.4 Structural Equation Model (SEM)

As the final step of the analysis, structural equation model was developed using SPSS AMOS 26 version. Standard estimates, regression weights, and correlations, were calculated. Furthermore, the developed structural model was assessed using the goodness of fit indices.

The developed structural equation model is given in Fig. 5. By considering the obtained p values, significance of correlations were determined. Similarly, by observing the r value, the strength of the correlation was identified. P values and r values are given in Table 6. Results revealed that the personal interest (PI) of the students has significantly influenced their perceptions of students. Accordingly, PI has become significant with the perception of the need for climate change mitigation where the strength of the association is considered to be a weak correlation since the r value is below 0.2 ($p = 0.004$, r = 0.227). PI has become significant with the perception of climate change mitigation factors where the relationship has become moderate because the r value is above 0.4 ($p = 0.000$, r = 0.433). Similarly, PI has significantly influenced the perception of energy conservation while the association is considered moderate since the r value is above 0.4 ($p = 0.000$, r = 0.524). Furthermore, PI has become significant with the perception of consumer behavior on climate change and the relationship was determined as moderate since the r value is above 0.4 ($p = 0.000$, r = 0.409). Moreover, it was revealed that other independent variables; field of study (FS), learning from university curricula (WL), and practical application during university curricula (PA) have not influenced the perceptions of the students.

When considering the impact of the perceptions on the behavior related to climate change, sustainable designs, and energy conservation it was identified that perception of energy conservation (PEC) has significantly influenced the behavior in energy conservation where the relationship is considered to be weak since the r value is below 0.4 ($p = 0.002$, r = 0.201). PEC has become significant with behavior in climate change mitigation and that association has happened to be strong because the r value is above 0.6 ($p = 0.017$, r = 0.660). Similarly, it has become significant with behavior in environment conservation, and the association is considered as strong because the r value is above 0.6 ($p = 0.000$, r = 0.726). Furthermore, it has influenced the willingness to integrate sustainable designs and energy conservation in future professional work where that association has been identified as strong since the r value is above 0.6($p = 0.014$, r = 0.635). However, it was noticed that PEC has not become significant with the willingness to integrate climate change mitigations in future professional work ($p > 0.05$).

When considering the influence of perception on climate change mitigation factors (PMF) on behavior, it was revealed that it has become significant with behavior in environment conservation where the relationship is identified as moderate since the r value is above 0.4 ($p = 0.000$, r = 0.457). It has also become significant with the willingness to integrate sustainable designs and energy conservation in future professional work where the association is considered to be moderate as the

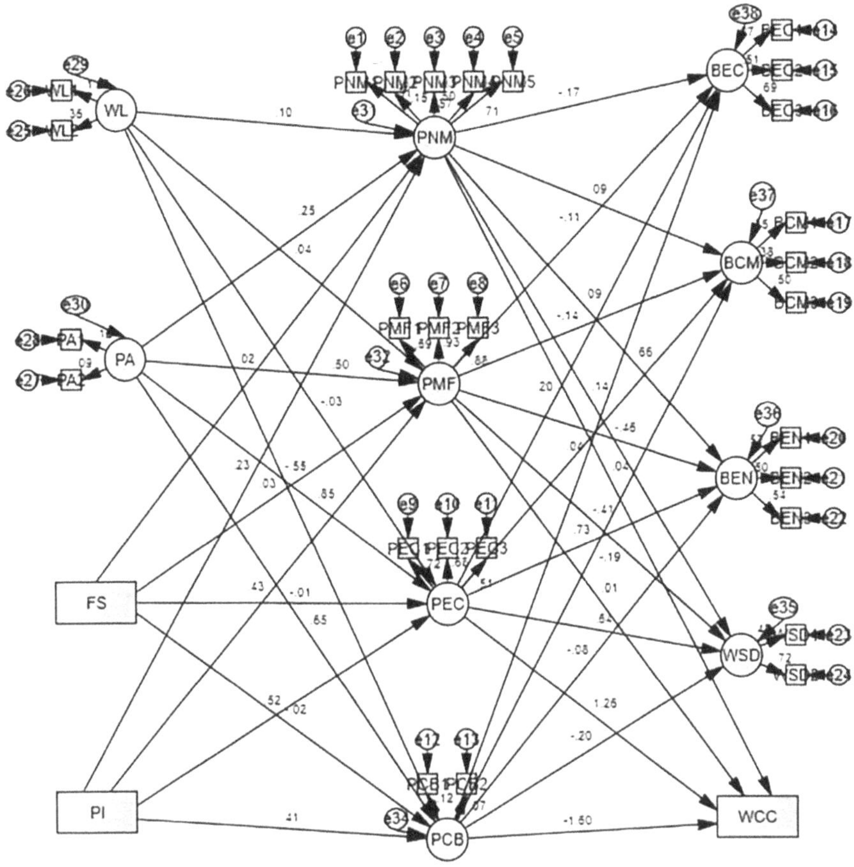

Fig. 5 Structural equation model constructed via IBM AMOS

r value is above 0.4 ($p = 0.007$, $r = 0.408$). However, it has not become significant with behavior in climate change mitigation, behavior in energy conservation, and willingness to integrate climate change mitigations in future professional work ($p > 0.05$).

Nevertheless, from the results, it was revealed that, apart from the above-mentioned, none of the other latent factors have become significantly correlated with other latent factors related to behavior and willingness to integrate with future professional work.

Table 6 Probability and r values of considered correlations

Correlation			*P*	Correlation coefficient (r)	Correlation			*P*	Correlation coefficient (r)
PNM	←	WL	0.142	0.098	BEN	←	PNM	0.349	0.093
PMF	←	WL	0.457	0.037	WSD	←	PNM	0.173	0.141
PEC	←	WL	0.572	− 0.031	BEC	←	PMF	0.359	− 0.105
PCB	←	WL	0.121	− 0.549	BCM	←	PMF	0.300	− 0.141
PNM	←	PA	0.210	0.249	BEN	←	PMF	***	0.457
PMF	←	PA	0.163	0.504	WSD	←	PMF	0.007	0.408
PEC	←	PA	0.159	0.851	BEC	←	PEC	0.002	0.201
PCB	←	PA	0.282	0.648	BCM	←	PEC	0.017	0.660
PNM	←	FS	0.761	0.020	BEN	←	PEC	***	0.726
PMF	←	FS	0.613	0.026	WSD	←	PEC	0.014	0.635
PEC	←	FS	0.816	− 0.013	BEC	←	PCB	0.725	0.042
PCB	←	FS	0.660	− 0.020	BCM	←	PCB	0.313	− 0.195
PNM	←	PI	0.004	0.227	BEN	←	PCB	0.550	− 0.078
PMF	←	PI	***	0.433	WSD	←	PCB	0.272	− 0.197
PEC	←	PI	***	0.524	WCC	←	PNM	0.251	0.043
PCB	←	PI	0.009	0.409	WCC	←	PMF	0.863	0.007
BEC	←	PNM	0.140	− 0.167	WCC	←	PEC	0.216	1.259
BCM	←	PNM	0.440	0.090	WCC	←	PCB	0.125	− 1.600

***Represents probability of 0.000

4.5 *Model Validation*

According to the results of the goodness of fit indices of the measurement model, it was confirmed that the model has met with the superior fit (Table 8). Accordingly, absolute fit indices such as probability level (*P* value < 0.05) and chi-square, RMSEA (RMSEA = 0.039, required level < 0.08), have met the fit levels (Husain, 2019). GFI (GFI = 0.854, required level > 0.9) and AGFI (AGFI = 0.820, required level > 0.9) values are also closer to the required level and therefore, can be considered as met with the fit. Furthermore, the model has met with the minimum discrepancy since the X2/DF value is 1.846 (required level < 5). Incremental fit indices such as NFI, TLI, and CFI have also come closer to the required levels. Even though these fit indices have not achieved exactly the required levels, the model is considered to be validated since the important fit indices such as minimum discrepancy, RMSEA, and probability levels have satisfied the fit values (NFI = 0.665, TLI = 0.776, CFI = 0.806, required levels > 0.9) (Shi et al., 2019). Since model modifications have not been conducted, it is considered to be reasonable to obtain such fit values for the initial structural equation model (Table 7).

Table 7 Goodness of fit indices

	P value	X^2	DF	X^2/DF	RMSEA	GFI	AGFI	NFI	TLI	CFI
Values	0.000	1487.629	806	1.846	0.039	0.854	0.820	0.665	0.776	0.806

4.6 Discussion of Results Gained from Structural Equation Model

When considering the obtained results, it was revealed that out of four variables that were taken into consideration, personal interest (PI) is the only factor that has influenced the perception of the students towards climate change mitigations. All other considered variables such as Field of Study (FS), learning from university curricula (WL), and Practical application (PA) during university curricula have not significantly influenced the perceptions of the students. PI has become influential on all considered variables related to perceptions; perception of climate change mitigation factors, perception of the need for climate change mitigation, perception of energy conservation, and perception of consumer behavior on climate change. Personal interest which can be identified as a psychological variable has gained considerable attention in recent research related to pro-environmental behavior. In a research conducted by Izagirre-Olaizola et al. (2015) they have used 'motivation' instead of personal interest which are both similar in context, to identify the influence of pro-environmental behavior. They have identified this as a cognitive factor apart from being a psychological factor which has become an influential factor for predictions related to environmental behavior. Furthermore, it has been identified that factors such as personal interest can develop from objective environmental knowledge. Universities as centers that generate knowledge should try to fill the gap between objective knowledge and subjective knowledge, and therefore, it will permit students to make better-grounded decisions and also enhance their perceptions (Izagirre-Olaizola et al., 2015). Apart from personal interest, Gifford has identified some other personal factors such as personality and self-construal, cognitive biases, political and world views, goals, childhood experience, sense of control, values, felt responsibility, place attachment, age, and gender as influential factors for pro-environmental concerns and behavior (Gifford & Nilsson, 2014).

There can be several reasons for factors such as FS, WL, and PA to not become influential on the perceptions of the students. Sri Lanka being a lower middle-income country in terms of economy and being a developing country in the Asian continent still face some immense challenges. Therefore, in terms of teaching techniques, scientific literature foundations, practical applications in universities, the teaching and learning culture and services provided, subjective and objective knowledge gained by university students can significantly vary based on the country's setting. So all, these factors might be the reasons for obtaining these sought results. Furthermore, these results were also aligning with research conducted by Boca and Saraçlı, where they revealed that there was no difference in the level of perceptions among the students regarding the importance of environmental education who have specialized in different fields such as engineering, mechanical, electrical and economic (Boca & Saraçli, 2019). When it comes to learning from the university (WL) and practical application (PA) of the students, it can be observed that the subjective knowledge which is taught at the university might not always become influential on the perceptions of the students towards CC, SD & EC. Rather than subjective knowledge, the informal knowledge they obtain from various sources such as newspapers, articles on the internet, family or friends, and social media can be more useful for shaping the perceptions of the students. This argument is also supported by Brito, where they have mentioned in their research that, values acquired at home and practiced at school, can become valuable to identify issues related to sustainability rather than the knowledge obtained from basic education (Brito et al., 2018). However, in some researches that have been previously conducted, some contradictory results appeared. In a study conducted by Bakaç, it was determined that engineering students were highly aware of climate change, and also they were equipped with knowledge regarding environmental effects and familiarity with its causes and these can be useful for them to apply in relevant situations in the future (Bakaç, 2018). Students in such fields get the opportunity to acquire an improved understanding of scientific literature and to develop new skills in the subject through practical applications. It also in return affects the attitudes and perspectives which leads to a growing personal interest in the subject matter. It also allows them to view the problem holistically and to develop a belief that it's an important problem in today's world. Another research conducted by Heyl and the team has determined that the diploma of specialization has significantly influenced the environmental attitudes and behaviors of the students whereas engineering students have shown positive environmental attitudes on average. Similarly, they have identified, that perceived effort (similar to a personal interest in our case) has also impacted the pro-environmental behavior in students (Heyl et al., 2014).

Similar research conducted by Molina and the team to determined the influence of environmental knowledge and other variables on pro-environmental behavior in university students in four countries; the USA, Spain, Mexico, and Brazil which they described as emerging and advanced countries. From the results, it has been identified that type of degree affects only the case of advanced countries while the objective and subjective knowledge of the environment affects both advanced and emerging countries. Similarly, they have identified, motivation and perceived effectiveness as two of the variables which have the highest influence on pro-environmental behavior in all four countries. These two variables; motivation and perceived effectiveness can be interpreted as personal interest when it comes to our case and also be considered psychological factors (Paschalidis et al., 2013). Another similar research has been conducted by Chuvieco and the team, to identify the factors affecting on environmental sustainability habits of university students in three countries; Spain, Brazil, and UAE. In that research, it has been found that self-perceived environmental commitment and field of study have become influential variables for the sustainability habits of the students and there have not been identified any significant differences among country wise comparison where for the students of all three countries, abovementioned factors have become influential (Chuvieco et al., 2018). By considering all these research outcomes, it can be identified that the main challenge for higher educational institutes is to increase both the subjective and objective levels of knowledge related to climate change, sustainable designs, and energy conservation and at the same time, to reduce the gap between these two avenues of knowledge. The results obtained on the insignificance of the WL variable imply that rather than focusing on technical knowledge it's more advisable to teach students about real-time issues with actual experiences which in turn enhances students' cognitive capacity to cope with issues that might arise in the future. However, it has been determined that the relationship between knowledge and perceptions is complex, and based on our study, it requires future research.

When considering the relationship between the student's perceptions of their behavior related to climate change mitigations and their willingness to integrate in the future, a considerable number of significant relationships were determined. Behavior in environmental conservation (BEN) has been influenced by the perception of climate change mitigation factors, the perception of the need for climate change mitigation, and the perception of energy conservation. This might be due to the fact that students have perceived that by contributing toward environment conservation, climate change mitigations can be achieved. This observation was also supported by research conducted by Heyl et al., to identify the factors impacting the environmental behavior of university students. They have identified perceived positive consequences as influential factors for predicting the frequency of students' pro-environmental behavior (Heyl et al., 2014). Similarly, students determine responsible consumption and responsible participation as the key ways of contributing toward environment conservation, therefore, they might have imagined the big picture of protecting the environment as a measure of climate change mitigation.

When considering the variable, Willingness to integrate sustainable designs and energy conservation in future professional work (WSD), it has been influenced by the perception of climate change mitigation factors and by the perception of energy conservation. The reason for this observation might be due to the fact that students have perceived sustainable designs and energy conservation mechanisms as climate change mitigation strategies. The way how perceptions work in people's minds is complex to understand. Sometimes, although they have positive perceptions, they always do not necessarily respond in a manner aligning with their perceptions wise versa. Similar research conducted by Greenville has identified that on some occasions, although students who have a rich background and a strong foundation of knowledge on sustainability, do not want to prioritize sustainability throughout their professional life. But sometimes, students having positive sustainability attitudes such as sustainable purchasing behavior would intend to prioritize sustainability through their professions (Shi & By-nc-sa, 2017).

Similarly, it was also determined that Behavior in climate change mitigation (BCM) has been influenced by the perception of energy conservation. This observation also confirms the student's identification of the connection between climate change mitigation and energy conservation. As per environment conservation, students have perceived energy conservation as a means of climate change mitigation strategy. When energy conservation is considered in a broader aspect, knowledge of several key areas such as GHG emissions, and global warming which are the key factors for climate change, come into consideration. Similarly, it has been identified that there is a positive correlation between environment conservation and energy conservation. Research conducted by Pothitou and the team has identified that positive environmental values, knowledge, and respondents' perception about energy savings are most likely to demonstrate positive energy behaviors, habits, and attitudes that lead to energy savings. Similarly, they have identified that socio-economic factors such as educational levels, income, and employment status also impact the energy-saving behavior of the people (Pothitou et al., 2016).

As a whole, it was determined that, although students have positive perceptions towards CC, SD & EC, it is not always reflected through their corresponding behavior. Sometimes, even though it's represented through their behavior, the proportional frequency can be uncertain. Psychological factors such as social norms and moral obligations can significantly influence decision-making in such behaviors (Esham & Garforth, 2013; Wiedenhofer et al., 2013). Consequently, a message that was derived from both our research and literature was that there do exist some other aspects which are hard to be identified and weighted by researching. So it's difficult to separate such aspects because these variables work together in decision-making processes. Furthermore, it was determined that behavioral aspects do have lots of inter-relationships which are rather complex and challenging to be measured reliably.

5 Conclusions and Recommendations

This study was intended to explore the university curricular contribution on the cognitive levels variables such as perceptions, interests, behaviors, and intentions, and their hidden relationships were assessed on the basis of climate change mitigation strategies. University students are the most significant associates of the university system and enhancing their knowledge, skills, and awareness are much crucial for the sustainable development of the country. They join with both industry and government sector professions after their degree and ultimately, they become the professionals who do important decisions making which is imperative at both individual organizations levels as well as nationwide. Perceptions which are one of the important cognitive factors are vital in decision making similarly, their overall behavior toward the concepts of climate change implies how well they might address the issues related to these conceptions in the future.

5.1 Conclusions

- Based on the study, it was revealed that out of four variables that were taken into consideration, personal interest (PI) is the only factor that has influenced the perception of the students towards climate change. Other considered factors such as Field of Study (FS), learning from university curricula (WL), and Practical application (PA) during university curricula have not significantly influenced the perceptions of the students in the local context.

- Personal interest which is in the category of cognitive also falls on the subset of psychological motive and it significantly influences the perceptions of students towards climate change, sustainable design, and energy conservation. External factors such as socioeconomic status, cultures, and traditions have a significant influence on the perceptions and decision-making processes.
- The insignificance of learning from the university curricular variable implied that along with the technical knowledge, it's more advisable to teach students about real-time issues with actual experiences which in turn enhances students' cognitive capacity to cope with issues that might arise in the future. The study fields such as management, social sciences, and the law should incorporate these concepts because when it comes to decision-making, most of the top ranks in the professional field are comprised of a collective bunch of people. So it should not be limited to science and technology-related fields.
- Perceptions on climate change, sustainable design, and energy conservation not only have implications on tertiary educational level involvement but also many other internal and external factors. Apart from the subjective knowledge gained from the university curriculum, it was determined that objective knowledge has become more influential in the decisions taken by students regarding climate change, sustainable design, and energy conservation. Universities as centers that generate knowledge should try to fill the gap between objective knowledge and subjective knowledge, and therefore, it will permit students to make better-grounded decisions and also enhance their perceptions.
- When considering the relationship between the student's perceptions of their behavior related to climate change mitigations, energy conservation and their willingness to integrate in the future, in the Sri Lankan context, Behavior in environmental conservation (BEN) has been influenced by the perception of climate change mitigation factors, perception on the need of climate change mitigation and perception on energy conservation. Willingness to integrate sustainable designs and energy conservation in future professional work (WSD), has been influenced by the perception of climate change mitigation factors and by the perception of energy conservation. Behavior in climate change mitigation (BCM) has been influenced by the perception of energy conservation.
- Most of the time, students have perceived environmental protection as the strategy for mitigating climate change. Responsible consumption and responsible participation are the common ways of attaining this and, therefore, they are lacking specific strategies for it. However, they might have imagined the big picture of protecting the environment as a means of climate change mitigation.
- Although concepts have been integrated with the tertiary educational levels, perceptions have not always been positively influenced. Therefore, university educational reforms must be designed in a way that improves attitudes, and beliefs more realistically. Sometimes, students might have positive perceptions towards

climate change mitigations, which are not always reflected in their corresponding behavior.

- Although students who have a rich background and a strong foundation of knowledge do not want to prioritize their perceptions throughout their professional life. Sometimes, even though it's represented through their behavior, the proportional frequency can be doubtful. Psychological factors such as social norms and moral obligations can significantly influence decision-making in such behaviors. The way how perceptions work in peoples' minds is complex to understand. Sometimes, although they have positive perceptions, they always do not necessarily respond in a manner aligning with their perceptions wise versa.
- Based on the research outcomes, it was determined that going beyond the level of tertiary education can also significantly enhance the cognitive values of students toward climate change mitigation. Therefore, it might be a good initiative to integrate these concepts even at primary and secondary educational levels, to enhance consciousness of the aforementioned concepts.

5.2 *Limitations of the Study*

- Cognitive factors essentially rely on other unobservable variables and in a research study, it is required to keep these factors as constant as possible for better results outcomes.
- Cognitive aspects such as perceptions and behaviour are assessed in the student's point of view which can be subjective.
- The results of a survey conducted with participants of five fields of study cannot be generalized in a holistic context.

5.3 *Future Studies*

- Future research requires more homogenous samples representing equal proportions from the considered populations.
- The impact of personal factors on the cognitive variables related to climate change, sustainable designs, and energy conservation is a good research area to be explored.
- The factors which develop the personal interest in a certain subject area can be significantly dependent on certain other describable variables and exploring them would be a great initiative for identifying which level of education; (primary, secondary, or tertiary) to incorporate concepts of climate change, sustainable designs, and energy conservation.
- Behavioral aspects do not necessarily depend on the perceived qualities and what influences such variability must be explored.

Table 8 Results of the communalities-extraction values for considered components

Communalities		
	Initial	Extraction
CC1	1.000	0.422
CC3	1.000	0.525
SE16	1.000	0.577
SE22	1.000	0.578
SE23	1.000	0.581
CC14A	1.000	0.712
CC14B	1.000	0.810
CC14C	1.000	0.786
CC7	1.000	0.687
CC12	1.000	0.393
SE24	1.000	0.644
SE25	1.000	0.659
CC15	1.000	0.329
CC13A	1.000	0.385
CC13B	1.000	0.608
CC13C	1.000	0.428
CC13D	1.000	0.549
CC13E	1.000	0.445
CC13F	1.000	0.550
CC13G	1.000	0.533
CC13H	1.000	0.431
CC13I	1.000	0.558
CC13J	1.000	0.354
SE21C	1.000	0.641
SE21E	1.000	0.684
SE20B	1.000	0.561

Extraction method: principal component analysis

Appendix

See Table 8.

References

Aminrad, Z., Zarina, S., Sayed, B., Hadi, A. S., & Sakari, M. (2013). Relationship between awareness, knowledge and attitudes towards environmental education among secondary school students in Malaysia. *World Applied Sciences Journal, 22*(9), 1326–1333. https://doi.org/10.5829/idosi.wasj.2013.22.09.275

Ayanlade, A., & Jegede, M. O. (2016). Climate change education and knowledge among Nigerian University graduates. *Weather, Climate, and Society, 8*(4), 465–473. https://doi.org/10.1175/WCAS-D-15-0071.1

Baba, N. (2010). Sinking the pearl of the Indian ocean: climate change in Sri Lanka. *Global Majority E-Journal, 1*(1), 1–56.

Bakaç, E. (2018). Engineering faculty students' perceptions on climate change. *Environment and Ecology Research, 6*(4), 240–247. https://doi.org/10.13189/eer.2018.060404.

Barnes, H., Carrico, A. R., & Thabrew, L. (2015). A socio-psychological model for analyzing climate change adaptation: A case study of Sri Lankan paddy farmers. *Global Environmental Change, 31*, 85–97. https://doi.org/10.1016/j.gloenvcha.2014.12.010

Boca, G. D., & Saraçli, S. (2019). Environmental education and student's perception, for sustainability. *Sustainability, 11*(6), 1–18. https://doi.org/10.3390/su11061553

Brito, R. M., Rodríguez, C., Aparicio, J. L., Paolacci, J., Sampedro, M. L., & Beltrán, J. (2018). Indicators of sustainability in educational practice: Perception of teachers and students of UAGro-Mexico. *Sustainability, 10*(10), 3733. https://doi.org/10.3390/su10103733

Cantell, H., Tolppanen, S., & Aarnio-linnanvuori, E. (2019a). Bicycle model on climate change education: presenting and evaluating a model. *Environmental Education Research, 0*(0), 1–15. https://doi.org/10.1080/13504622.2019.1570487

Cantell, H., Tolppanen, S., Aarnio-Linnanvuori, E., & Lehtonen, A. (2019b). Bicycle model on climate change education: Presenting and evaluating a model. *Environmental Education Research, 25*(5), 717–731. https://doi.org/10.1080/13504622.2019.1570487

Chuvieco, E., Burgui-Burgui, M., Da Silva, E. V., Hussein, K., & Alkaabi, K. (2018). Factors affecting environmental sustainability habits of university students: Intercomparison analysis in three countries (Spain, Brazil and UAE). *Journal of Cleaner Production, 198*, 1372–1380. https://doi.org/10.1016/j.jclepro.2018.07.121

Clark, B., Stewart, G. B., Panzone, L. A., Kyriazakis, I., & Frewer, L. J. (2016). A systematic review of public attitudes, perceptions and behaviours towards production diseases associated with farm animal welfare. *Journal of Agricultural and Environmental Ethics, 29*(3), 455–478. https://doi.org/10.1007/s10806-016-9615-x

Clayton, S., et al. (2015). Psychological research and global climate change. *Nature Clinical Practice Endocrinology & Metabolism, 5*(7), 640–646. https://doi.org/10.1038/nclimate2622

De Zoysa, M., Inoue, M. (2014). Climate change impacts, agroforestry adaptation and policy environment in Sri Lanka. *Open Journal of Forestry*, 439–456.

Dragan, D., Topolšek, D. (2014). *Introduction to structural equation modeling: review, methodology and practical applications* (pp. 19–21).

Esham, M., Garforth, C. (2013) *Agricultural adaptation to climate change: Insights from a farming community in Sri Lanka* (pp. 535–549). https://doi.org/10.1007/s11027-012-9374-6

Feygina, I. et al. (2019). *Localized climate reporting by TV weathercasters enhances public understanding of climate change as a local problem* (pp. 1092–1100).

Fischer, A. R. H. (2017). *Perception, attitudes, intentions, decisions and actual behavior* (pp. 303–317). https://doi.org/10.1007/978-3-319-50530-5

Fisher, J., & Bonn, I. (2011). *Business sustainability and undergraduate management education: An Australian study* (pp. 563–571). https://doi.org/10.1007/s10734-010-9405-8

Frenette, E., & Bahn, O. (2020). Meat, dairy and climate change: Assessing the long-term mitigation potential of alternative agri-food consumption patterns in Canada. *Environmental Modeling and Assessment*. https://doi.org/10.1007/s10666-016-9522-6

Füssel, H. M., & Klein, R. J. T. (2006). Climate change vulnerability assessments: An evolution of conceptual thinking. *Climate Change, 75*(3), 301–329. https://doi.org/10.1007/s10584-006-0329-3

Gifford, R., & Nilsson, A. (2014). Personal and social factors that influence pro-environmental concern and behaviour: A review. *International Journal of Psychology, 49*(3), 141–157. https://doi.org/10.1002/ijop.12034

Heller, N. E., & Zavaleta, E. S. (2009). Biodiversity management in the face of climate change: A review of 22 years of recommendations. *Biological Conservation, 142*(1), 14–32. https://doi.org/10.1016/j.biocon.2008.10.006

Heyl, M., Díaz, E. M., & Cifuentes, L. (2014). Environmental attitudes and behaviors of college students: A case study conducted at a Chilean University. *Revista Latinoamericana de Psicologia, 45*(3), 487. https://doi.org/10.14349/rlp.v45i3.1489

Howard, M. C. (2016). A review of exploratory factor analysis decisions and overview of current practices: What we are doing and how can we improve? A review of exploratory factor analysis decisions and overview of current practices: What we are doing and how can we improve. *International Journal of Human-Computer Interaction, 7318*. https://doi.org/10.1080/10447318.2015.1087664

Husain, T. (2019). An analysis of modeling audit quality measurement based on decision support systems (DSS). *European Journal of Scientific Exploration.*

Ivanova, D., Barrett, J., Wiedenhofer, D., Macura, B., & Callaghan, M. (2020). Quantifying the potential for climate change mitigation of consumption options OPEN ACCESS quantifying the potential for climate change mitigation of consumption options. *Environmental Research Letters, 15*(9), 093001.

Ives, C. D., Abson, D. J., von Wehrden, H., Dorninger, C., Klaniecki, K., & Fischer, J. (2018). Reconnecting with nature for sustainability. *Sustainability Science, 13*(5), 1389–1397. https://doi.org/10.1007/s11625-018-0542-9

Izagirre-Olaizola, J., Fernández-Sainz, A., & Vicente-Molina, M. A. (2015). Internal determinants of recycling behaviour by university students: A cross-country comparative analysis. *International Journal of Consumer Studies, 39*(1), 25–34. https://doi.org/10.1111/ijcs.12147

Kariyawasam, C. S., & Kumar, L. (2019). *Invasive plant species establishment and range* (pp. 1–24).

Knutti, R. (2019). Commentary closing the knowledge-action gap in climate change. *One Earth, 1*(1), 21–23. https://doi.org/10.1016/j.oneear.2019.09.001

Leal Filho, W., et al. (2021). Handling climate change education at universities: An overview. *Environmental Sciences Europe, 33*(1). https://doi.org/10.1186/s12302-021-00552-5

Lin, L. (2017). *Assessment of vulnerability for coastal erosion with GIS and AHP techniques case study: Southern coastline of Sri Lanka* (pp. 1–12). https://doi.org/10.1111/nrm.12146

Mehvar, S., Dastgheib, A., & Bamunawala, J. (2019). Climate risk management quantitative assessment of the environmental risk due to climate change-driven coastline recession: A case study in Trincomalee coastal area, Sri Lanka. *Climate Risk Management, 25*, 100192. https://doi.org/10.1016/j.crm.2019.100192.

Mintrom, M., & Luetjens, J. (2017). Policy entrepreneurs and problem framing: The case of climate change. *Environment and Planning C: Politics and Space, 35*(8), 1362–1377. https://doi.org/10.1177/2399654417708440

Molthan-Hill, P., Worsfold, N., Nagy, G., Leal Filho, W., & Mifsud, M. (2019). Climate change education for universities: A conceptual framework from an international study. *Journal of Cleaner Production, 226*, 1092–1101. https://doi.org/10.1016/j.jclepro.2019.04.053. ISSN 0959-6526. Downloaded from: https://e-space.mmu.ac.uk/623693/. Version: Accepted Version Publisher, *Journal of Cleaner Production*, 1092–1101.

Monroe, M. C., et al. (2017). Identifying effective climate change education strategies: a systematic review of the research. *Environmental Education Research, 4622*. https://doi.org/10.1080/13504622.2017.1360842

Navaratne, N. W. M. G. S., Wanigasundera, W. A. D. P., & Alahakoon, P. C. B. (2019). Perceptions of climate change and adaptation of climate smart technology by the tea smallholders: A case study of Ratnapura District in Sri Lanka. *Asian Journal of Agricultural Extension, Economics & Sociology*.

Otto, S., & Pensini, P. (2017). Nature-based environmental education of children: Environmental knowledge and connectedness to nature, together, are related to ecological behaviour. *Global Environmental Change, 47*, 88–94. https://doi.org/10.1016/j.gloenvcha.2017.09.009

Paschalidis, V., et al. (2013). *This manuscript was previously submitted to Physical Review Letters and is now being considered for possible publication in Physical Review D as a Rapid Communication.* We If publishable, does this meet our criteria for Rapid Communications as the author. https://doi.org/10.1016/j.jclepro.2013.05.015

Pickens, J. (1998). *Attitudes and perceptions.*

Pothitou, M., Hanna, R. F., & Chalvatzis, K. J. (2016). Environmental knowledge, pro-environmental behaviour and energy savings in households: An empirical study. *Applied Energy, 184*, 1217–1229. https://doi.org/10.1016/j.apenergy.2016.06.017

Sadusky, H. (2014). *Student perception and behavior towards sustainability: Results of a campus survey*. Florida State University Libraries College.

Savaresi, A. (2016). The Paris agreement: A new beginning? *Journal of Energy & Natural Resources Law, 34*(1), 16–26. https://doi.org/10.1080/02646811.2016.1133983

Schlüter, M., et al. (2017). A framework for mapping and comparing behavioural theories in models of social-ecological systems. *Ecological Economics, 131*, 21–35. https://doi.org/10.1016/j.ecolecon.2016.08.008

Senevirathna, E. M. T. K., Edirisooriya, K. V. D., Uluwaduge, S. P., & Wijerathna, K. B. C. A. (2018). Analysis of causes and effects of coastal erosion and environmental degradation in Southern coastal belt of Sri Lanka special reference to Unawatuna Coastal Area. *Procedia Engineering, 212*, 1010–1017. https://doi.org/10.1016/j.proeng.2018.01.130

Shi, D., Lee, T., & Maydeu-Olivares, A. (2019). Understanding the model size effect on SEM fit indices. *Educational and Psychological Measurement, 79*(2), 310–334. https://doi.org/10.1177/0013164418783530

Shi, T., & By-nc-sa, C. C. (2017). *Sustainability attitudes of college students as future business leaders* (pp. 1–23).

Stachowicz, J. J., Terwin, J. R., Whitlatch, R. B., & Osman, R. W. (2002). Linking climate change and biological invasions: Ocean warming facilitates nonindigenous species invasions. *Proceedings of the National Academy of Sciences, 99*(24), 15497–15500. https://doi.org/10.1073/pnas.242437499

Stephenson, J., Newman, K., & Mayhew, S. (2010). Population dynamics and climate change: What are the links? *Journal of Public Health (Bangkok), 32*(2), 150–156. https://doi.org/10.1093/pubmed/fdq038

Tian, X., Chang, M., Shi, F., & Tanikawa, H. (2017). Decoding the effect of socioeconomic transitions on carbon dioxide emissions: Analysis framework and application in megacity Chongqing from inland China. *Journal of Cleaner Production, 142*, 2114–2124. https://doi.org/10.1016/j.jclepro.2016.11.072

Uitto, A., Juuti, K., Lavonen, J., & Byman, R. (2014). Secondary school students' interests, attitudes and values concerning school science related to environmental issues in Finland. *Environmental Education Research*, 37–41. https://doi.org/10.1080/13504622.2010.522703

Wang, Z., Cui, C., & Peng, S. (2019). How do urbanization and consumption patterns affect carbon emissions in China? A decomposition analysis. *Journal of Cleaner Production, 211*, 1201–1208. https://doi.org/10.1016/j.jclepro.2018.11.272

Wiedenhofer, D., Lenzen, M., & Steinberger, J. K. (2013). Energy requirements of consumption: Urban form, climatic and socio-economic factors, rebounds and their policy implications. *Energy Policy*, 1–12. https://doi.org/10.1016/j.enpol.2013.07.035

Wu, S., Bates, B., Zbigniew Kundzewicz, A. W., & Palutikof, J. (2018). *Climate change and water*. Accessed: October 09, 2018. [Online]. Available: http://acikerisim.bahcesehir.edu.tr:8080/xmlui/bitstream/handle/123456789/1056/AE0089.pdf?sequence=1

Yap, R. J. C., Komalasari, F., & Hadiansah, I. (2018). The effect of financial literacy and attitude on financial management behavior and satisfaction. *Bisnis and Birokrasi Journal, 23*(3), 3–5. https://doi.org/10.20476/jbb.v23i3.9175

Zhang, H., Lahr, M. L., & Bi, J. (2016). *Challenges of green consumption in China: A household energy use perspective*, vol. 5314. https://doi.org/10.1080/09535314.2016.1144563.

Developing a Joint Master's Degree Programme in Agroecosystem Resilience for Sri Lanka

A. Carmichael, C. Liyanage, A. De Silva, M. Esham, J. Guttila, and C. Nawaratne

1 Introduction

Collaborative educational programmes are a rapidly growing offering among higher education institutions yet compared with traditional models of delivery they remain relatively rare. Transnational ones are rarer still. This, despite such programmes being encouraged by those involved in setting the educational agenda over the last two decades. The Bologna Process, in establishing the European Higher Education Area in 1999, sought to increase access to higher education, aimed to provide quality assurance of qualifications across national borders, and promoted co-operation between providers. Intercontinental partnerships between universities can share high impact research in European Union countries with institutions expanding their knowledge base in the Global South, helping drive sustainable development. Technology and the opportunities for communication it has afforded has facilitated new possibilities in

A. Carmichael (✉) · C. Liyanage
University of Central Lancashire, Preston, UK
e-mail: acarmichael2@uclan.ac.uk

C. Liyanage
e-mail: clliyanage@uclan.ac.uk

A. De Silva · M. Esham
Sabaragamuwa University of Sri Lanka, Belihuloya, Sri Lanka
e-mail: achini@agri.sab.ac.lk

M. Esham
e-mail: esham@agri.sab.ac.lk

J. Guttila · C. Nawaratne
University of Ruhuna, Matara, Sri Lanka
e-mail: jayasinghe@ageng.ruh.ac.lk

C. Nawaratne
e-mail: champa@agricc.ruh.ac.lk

C. Malalgoda et al. (eds.), *Climate Change Adaptation in the Built Environment*,
https://doi.org/10.1007/978-3-031-75826-3_14

learning. Notwithstanding the potential benefits for learners, academic staff, the institutions, and their host nations, the route to higher education qualifications remains largely within the confines of a single provider and typically classroom based. Clearly there are challenges to developing and implementing an internationally collaborative programme. This chapter details and analyses the efforts of a consortium of higher education partners in creating and effecting one such programme that would have a significant positive impact on national approaches to responding to climate change.

2 The BRITAE Project

Building Resilience in Tropical Agroecosystems (BRITAE) is a project funded under the grant scheme of the European Union Education, Audiovisual, and Culture Executive Agency (Erasmus+: Higher Education—International Capacity Building—Joint Projects). Led by a university based in Sri Lanka, it comprises nine partner universities, five from Sri Lanka, two from the United Kingdom and one each from Estonia and Lithuania. One of the key objectives of the BRITAE project is to develop a joint innovative and adaptive master's degree programme on tropical agroecosystem-based resilience aimed at food security and climate change impacts. The programme was created by the efforts of all partners in accordance with the University Grants Commission (UGC) Sri Lanka Qualifications Framework (UGC, 2015). This satisfies policy 36 of the "National Policy Proposals on Higher Education" (National Education Commission of Sri Lanka, 2019) that states that all post-graduate programs should comply with Sri Lanka Qualification Framework. The programme will be delivered by Sri Lankan universities.

2.1 The Need for the BRITAE Master's Degree Programme

The need for a collaborative postgraduate programme focusing on building resilience in tropical agroecosystems is clear. Tropical agroecosystems have the potential to provide a wide range of food sources amidst optimal growing conditions (Hemachandra et al., 2021). However, the delicate balance in such ecosystems is under threat, and Sri Lanka and its agricultural sector is particularly at risk from climate change (FAO, 2021; IFAD, 2019). There is a necessity to reduce the nation's vulnerability to such change and ensure food security (Tsuchida & Takeda, 2021), as encapsulated in the Sendai Framework for Disaster Risk Reduction (2015–30). These vulnerabilities, and the responses to them, have at times been exacerbated by poorly considered and badly implemented government policies. In April 2021, for example, restrictions on artificial fertilizer and chemicals, and removal of subsidies that encouraged their use in the growing of rice, had major impacts on the production of food, despite the swift reversal of the policy in November of that year (Reuters, 2022). The United Nations Food and Agriculture Organisation identified a fall in overall yield of between 40 and 50% in the latest Sri Lankan harvest over the previous one (SPGlobal, 2022). These shortages have in turn fueled the need for imports, price inflation, and led to protests which ultimately saw the resignation of both Prime Minister and President. And yet the pursuit of an organic agriculture strategy may have been popular with farmers had it been introduced gradually and with appropriate safeguards. Former President Rajapaksa's election campaign in 2019 had promised a ten-year transition (FP Analysis, 2022), instead the change was introduced with very little time to adapt crop management methods and had immediate repercussions.

Clearly a greater appreciation for the outcome of such sweeping changes to complicated agricultural systems would have been an asset in making these decisions. Yet there has been a decline in interest among younger generations to study agriculture, leading to an ageing demographic involved in the sector, and a lack of understanding of new scientific techniques (Amarasekera, 2022). Agriculture graduates typically earn salaries 25% less than the average per hour rate of all graduates (Dunusinghe, 2022). Sustainable agribusinesses utilising well-trained and professional employees with internationally recognized qualifications would be of value in arresting the problem (Amarasekera, 2022). Developing a greater capacity for research and innovation in resilience to climate change and disaster within Sri Lankan higher education institutions would be welcome (Fernando et al., 2021). Beyond the creation of a knowledgeable managerial class, farmers at all levels would gain from promotion of best practice activities that could improve not just yield but how to utilize any surplus and create economic activity. If agricultural policy change can contribute to the fall of a government, then the subject, and its teaching, clearly remains of vital importance to the nation. The BRITAE collaborative master's degree programme is a step towards rebuilding the sector.

3 Collaborative Degree Programmes

Collaborative degree programmes are not new, but at the outset of this project the BRITAE partners were unable to identify an existing example of the same scope and ambition in a similar subject area. Although there were programmes that involved multiple partners across different continents, or provided for digital learning towards their award, covering the subjects of climate change and agriculture at a postgraduate level, there were none that did all these things and no established template upon which to base the work. Higher Education Institutions have initiated more of these programmes in recent, globally competitive times (Hasanen, 2020; Stevens, 2020), they are the logical outcome of the rapid rise in online learning and distance-based qualifications (Bannier, 2016). Yet collaborative programmes appear little defined, and vary according to; the type of programme, the power relationships between partners, institutions' reasons for being in the partnership, and the elements they each deliver (de Róiste et al., 2015). Nor are they the only example of cross border collaboration; joint degree programmes can be found alongside international campus hubs, offering study thousands of miles away from the host university, and online course provision, accelerated by the impact of the Coronavirus pandemic, has made new markets and modes of engagement attainable. While institutions herald such initiatives as a necessary development in providing quality graduates with the skills and world view that our economies now demand (Hasanen, 2020), there appears flexibility in how they are provided when compared with traditional models of education delivery.

Those who have created collaborative programmes highlight their benefits, which can be "many and diverse" (Stevens, 2020, p. 601). For the individual partners, relationships between institutions are developed, students and academics benefit from greater collaboration and mobility, best practice is shared, and there is a consequent increase in the quality of provision in teaching and research (Almansour, 2015; de Róiste et al., 2015; Stevens, 2020). Where institutions currently offer similar singly administered programmes, collaboration can remove the duplication of effort, which in turn is likely to reduce costs, and offer a safeguard against low enrolment numbers, staff recruitment and retention problems (de Róiste et al., 2015). Particularly relevant to BRITAE's geographical partnership, and the trials faced inhabiting some of those locations, is that interruption to study by natural disaster is made less likely (de Róiste et al., 2015). Moreover, if the partnering institutions' nations are at different points in their economic development entrepreneurship, a desirable feature in building resilience in agroecosystems, can receive a boost (della Volpe, 2020).

The challenges that have been suggested in attaining these benefits are equally numerous (van Swet et al., 2012). From a purely practical perspective, different systems of education can create additional considerations in the admissions to, paying for, and awarding of, qualifications (Meehan & Radice, 2016; Stevens, 2020). Institutions not only have potentially different native languages to contend with, each possess subject experts who may come from different research paradigms and fundamentally disagree on aspects of their specialty (van Dam-Mieras et al., 2008). Access to technology and learning materials, particularly library resources, may differ across nations (Vicente et al., 2021). Something as simple as the calendar can be problematic; different observances, holidays, and academic term start times all contributing to bureaucratic challenges (Knobel et al., 2013). Additionally, the legitimacy of their claim to be meeting a need is sometimes questionable (Hasanen, 2020). The literature on transnational higher education is, according to Hasanen (2020) (in citing Chou et al., 2016; Knight, 2006), divided around the idea that it either represents an opportunity for cultural development, or exemplifies a crisis of single culture domination and profit as the prime motivation. These concerns can, however, be overcome by appropriate support and agreed common standards, with a notable lessening of the issues observed when collaboration has centered on the implementation of a joint e-learning platform (Stevens, 2020). This approach has parallels with the intended delivery model of the BRITAE master's degree programme and offered encouragement that success could be repeated.

This chapter consequently details, analyses, and reflects upon the research that supported the BRITAE master's degree programme design, and the programme that was created.

4 Methodology

The BRITAE master's programme development process consisted of interlinked and alternating phases of design, review, and refinement of curriculum content and delivery. These phases embraced a mixed research methods approach. Mixed methods as a form of research improves both the validity of results and level of confidence in the work (Alexander et al., 2008). Quantitative data methods can provide a review of the subject being studied, identifying its overall scope (Kelle, 2006). This is highly relevant for the creation of a new education programme with a global reach. The advantages of qualitative data collection include its "source of well grounded, rich descriptions and explanations of processes in identifiable local contexts" (Miles & Huberman, 1994, p. 1). Use of mixed methods can compensate for any perceived limitations in either quantitative or qualitative methods individually (Caruth, 2013). Moreover, there was confidence from knowing mixed methods had been successfully used in meeting objectives similar to those of the BRITAE project; creating collaborative education programmes between multiple partners in different countries (Vicente et al., 2021), analysing the success of such collaborations between institutions (Jung et al., 2022), and investigating the capacity for strengthening research within an international consortium of agriculture-focused academics (Duda et al., 2023). The mixed methods employed for this project were; literature review, expert consultation, stakeholder workshop, and survey (Fig. 1).

The data generated by these methods come from within the existing academic body of work in agroecosystems, resilience, and related concepts, the expertise of partners in the BRITAE project, and highly qualified stakeholders in the work. Leydens et al. (2004) suggest twelve ways of ensuring that data, and interpretations of it, are trustworthy, with a recommendation that a minimum of two are used. This study used three; peer examination, purposeful sampling, and triangulation. Each stage of development, its purpose, methods employed, and output is discussed in sequence in the remainder of the chapter.

4.1 Method 1—Literature Review and Survey for the BRITAE Framework

The first stage of creating a relevant, and resilience building, master's degree programme in tropical agroecosystems was to examine the state of agriculture in Sri Lanka, the country's existing higher education provision in the subject, and the potential for its development.

Two literature reviews were undertaken. Firstly, a team from one of the Sri Lankan partners undertook a review primarily focused on the local aspects of tropical agroecosystems resilience. The aim was to understand the theoretical background and concepts relating to tropical agroecosystem resilience, and to develop a core of knowledge for further investigation. The findings of these literature reviews, discussed in

Fig. 1 Methods of BRITAE master's programme development

the results section, informed a questionnaire survey to benchmark the provision of agroecosystem resilience learning in Sri Lankan higher education. The aim of the survey was to understand the awareness, knowledge capacities, training and skills, research, and development capabilities among higher education institutions in the country.

4.2 Method 2—Expert Consultation on Programme Objectives/Content

The expert consultation was designed to translate the features identified in the literature reviews, and relevant to the programme, into the content to be included in the master's degree. Effectively, this became a question of what a learner should know and be able to do. For effective digital learning Laurillard and Kennedy (2017) identify a need for the skill of the teacher to be represented in the learning experience. The way to ensure this was for the relevant teaching staff from the institutions of the BRITAE project team to be central to the creation of the programme objectives and content.

While exact criteria are the subject of debate, 'experts' are generally deemed to be worthy of this description based on either their level of knowledge of the subject being researched, their status within a community relevant to the research, or their position in that community (Döringer, 2020). The size, nature, and aim of the BRITAE project provided the ability to call on many people who qualified as experts from within the partners, and a panel was created. This initiative was encouraged by the example provided by Waltz et al. (2015) who describe a comparable task that required 73 strategies to be organised into a much smaller number of categories, something they achieved using a similar expert group process. Programme content was discussed at length by a BRITAE project panel, with comment, refinement, and review over numerous meetings. The proposed modules were then presented, reviewed, and discussed at a full project consortium meeting.

4.3 Method 3—Stakeholder Workshop to Review Objectives/Content

To gauge the suitability of the proposed modules and their associated objectives, an online workshop with invited stakeholders was held. The objectives of the workshop were: To introduce the BRITAE framework and explain the importance of the proposed master's Degree Programme in Building Resilience in Tropical Agroecosystems (in the context of Sri Lanka with specific focus on the Agricultural sector); To discuss and refine the proposed course modules, and their associated Programme Objectives (POs) and Individual Learning Objectives (ILOs); And to identify recommendations from stakeholders about the design of the programme.

4.4 Method 4—Survey of Capabilities/Capacities of BRITAE Partners

Following the refinements to the programme, the capacity and capability of the partners to develop and deliver the proposed modules was measured. This was necessary to ensure that the programme advertised to prospective candidates was achievable.

Two surveys were created using the Qualtrics software platform, one to capture an understanding of how to meet the administrative needs of the programme, the other the teaching demands. The survey questions were created by one of the partners with reference to the programme. These were then discussed with other partners and revisions made. There were 23 questions in the administrative survey, 14 in the teaching one, however they were multiple part questions, so the responses required were significantly higher. The five Sri Lankan partners were encouraged to distribute the surveys to colleagues who would be involved in delivering the programme, with the aim of receiving ten responses per partner.

5 Results and Discussion

The results of each of the four methodological stages are presented and discussed here in turn.

5.1 Results and Discussion of the Literature Review and Survey for BRITAE Framework

The key outcomes of the first literature review were classified under the headings of; ecosystem functions, agroecosystems, the role of higher education institutes in agroecosystems resilience, impact on agroecosystems, resilience in agroecosystems, the necessity of agroecosystem resilience, agricultural vulnerability, existing practices, and challenges. This was followed by an extended literature review by one of the European partners that added global context and perspective to the regional findings. This resulted in the inclusion of; climate change, disaster risk reduction, wider resilience, the role of higher education, international policies and international challenges to considerations. Use of two separate teams from different institutions, geographical regions, academic backgrounds, and positions provided for both breadth and depth to the reviews, this investigative triangulation lending both credibility and validity to the work (Denzin, 1978).

There are 17 government universities in Sri Lanka, and one more campus under the purview of the UGC. However, not all of them offer courses or conduct research relevant to the BRITAE programme. Consequently, the questionnaire survey was provided to the 12 universities that were deemed to be providing applicable education with a request for them to distribute it to staff in the relevant departments. This was necessarily self-selecting, however not only was this a practical consideration in a country that was experiencing challenging circumstances, but additional benefits were to be gained from the method; working through universities raised the profile of the research and began the dissemination of the project aims and it likely generated more responses than had distribution been undertaken independently given the typically low return rate of surveys (Simmons, 2008). There were 113 valid responses to the survey. Based on these, each of the invited universities compiled a report that reflected upon their curriculum, its potential, and any gaps they noted in agroecosystem resilience development.

At the time of research, there were 47 different MSc programmes offered by the institutions taking part in the survey (for example, Agricultural Microbiology, Crop Science, and Tropical Soil Management). The results showed that while there is substantial provision for agroecosystem education in Sri Lanka, there are opportunities for enhancing content and practices when compared with global education and the future needs of the nation. Some gaps identified related to a lack of integration of international policies and frameworks, market orientation, governance in agroecosystems, ecosystems-based approaches for resilience, socioecological perspectives, environmentally friendly farming practices, and sustainable agricultural practices. The knowledge and professionalism of teaching staff was clear, but there is under-capacity in training, soft skills, and entrepreneurial initiatives in the sector. Existing programmes and their staff are already operating at a high level of subject knowledge. What could be expanded is networking, shared experience among institutions, agricultural transformations, industry collaboration, intellectual property, and stakeholder partnerships, both internationally and locally. It is important to respect existing knowledge in this, particularly.

indigenous practices of agroecosystems resilience. What the literature review and surveys effectively identified is that a lack of agriculture-based courses is not the problem, it is the need to create a particular style and focus of course that emphasises areas that have become highly valuable in the modern era and the risk from disasters that is now experienced. These results were discussed by the BRITAE project partners, individuals with wide-ranging specialisms and perspectives, resulting in the creation of the BRITAE framework.

The BRITAE framework (Fig. 2) identifies the key concepts in developing resilience in tropical agroecosystems, how these concepts relate to each other, and where the curriculum of a new master's degree programme can address needs. The framework includes three levels. Initially there are five major areas: resilience, agroecosystems, environment, policy and governance, and higher education. These five areas then cascade to 21 broad subjects in the second layer of the framework, for example within the agroecosystems major area are urban agroecosystems, crops and livestock, and land use patterns. Finally, the third layer comprises 125 specialist subjects, for example within the urban agroecosystems subject area are urban wetlands, conservation practices, and green infrastructure.

Having established an understanding of the resilience of agroecosystems in Sri Lanka, its position and potential development in the nation's higher education provision, there was a foundation from which to develop the collaborative BRITAE master's degree programme content. The process of creating a new study programme in Sri Lanka requires approval from the UGC and involves completing a highly detailed template including rationale for the proposed program, analysis of stakeholder expectations and needs, a curriculum structure detailing courses, intended learning outcomes, course content, teaching, and learning. The template undergoes many review stages with the applicant institutions complying with all SLQF guidelines. At the forefront of these guidelines are outcome-based education and student-centred learning (Bandara, 2020). A curriculum framework that would help regulate and review the content to keep it in line with these requirements once it had been determined and implemented was separately developed and reported. For the curriculum content the process was begun with expert consultation.

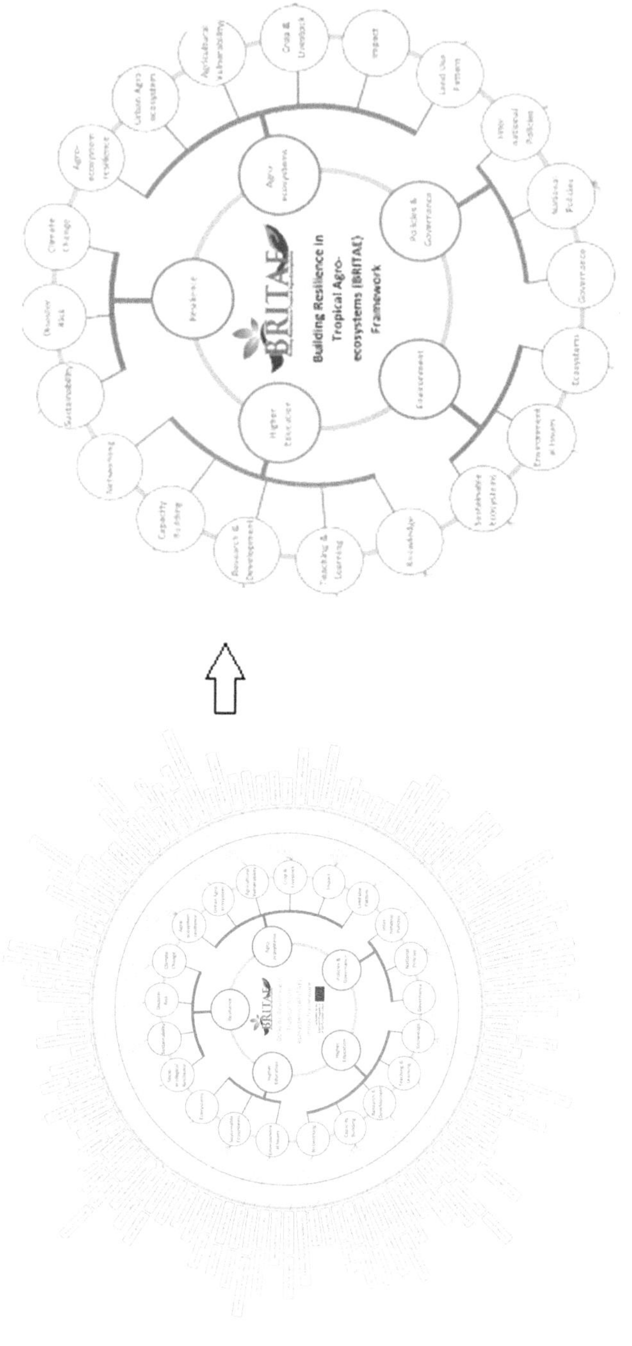

Fig. 2 BRITAE framework

5.2 Results and Discussion of Expert Consultation on Programme Objectives/Content

Initially ten programme objectives were created by expert consultation. These can also be termed 'graduate profiles' as they were a statement of what the learner should be capable of to successfully graduate from the programme. For example, 'To describe the multi-faceted and interconnected linkages between agriculture, biodiversity and ecosystem management'. Accompanying these were ten individual learning objectives that identified the skills that would be developed, allowing completion of the programme objectives. For example, 'On successful completion of this programme learners should be able to determine the social, environmental and economic relationships that contribute to sustainable development'. Adopting this approach utilised James and Brown's (2005) definition of learning outcomes in terms of attainments, understanding, and using. While the content of these objectives was based on the BRITAE framework, the wording was aided by reference to the revised Bloom's taxonomy (Anderson & Krathwohl, 2001).

The BRITAE Framework contained 108 subjects that were relevant to the teaching content of the master's programme. The initial panel discussions resulted in a gradual process of reducing the number by agreement. Mostly it was possible to combine subjects that would not singularly warrant a course module (the standard learning block), for example 'hydroponics' and 'aquaculture' could be taught within one module. This reduced the number to 57 modules, still clearly too many for a programme with a maximum intended two-year duration. and it was noted that some subjects, although important, could be accounted for by other provisions, for example various regulatory subjects and legislative considerations would fall within a module on 'policy development'. These were then grouped to meet the depth and likely timescale of a module, one or two terms, and were put forward to the full consortium meeting. Through continued discussion, and revision, 25 modules were produced that matched the ten programme objectives and ten individual learning objectives.

5.3 Results and Discussion of the Stakeholder Workshop to Review Objectives/Content

Stakeholders from industry, academia, and policy making within Sri Lanka were invited to attend the workshop. They had been identified by Sri Lankan colleagues in the BRITAE partnership as being experts in the subjects covered by the BRITAE framework, possessing an interest in the outcome of the work, and potentially likely to sponsor candidates for the programme. This purposive sampling is an established technique when wanting information-rich data (Patton, 1990) and introduced another layer of expertise and review to the process. In addition, engaging with stakeholders started the dissemination and promotion of the BRITAE project master's degree programme and provided an important safeguard against the suggestions of imbalanced relationships and cultural imposition that can occur in collaborative programmes.

The date chosen for the workshop was a national holiday in Sri Lanka, a factor that could have had both positive and negative consequences for the number of attendees. In the event, 37 senior people from organisations as diverse as the Ministry of Agriculture, Central Bank of Sri Lanka, and the National Science Foundation attended. Under normal circumstances this would have been an in-person event held over the course of a day at a suitable venue, but COVID pandemic restrictions meant it was organised through the online platform Zoom. There are similar competing technologies, but Zoom was familiar to participants and provided for, and integrated well with, innovative methods of data collection. There can be implications for any data collected in a group situation, given how personal views may change in the presence of others (Pain & Francis, 2003), and dissenting views may fail to be voiced in such a forum (Bergold & Thomas, 2012). However, there is also value in attendees learning from fellow participants (Tobin et al., 2020) and the online format provided direct communication channels between individuals and the researchers should anyone wish to express discomfort. Invitees were informed of, and agreed to, the format before the event.

The workshop was divided into three rounds, each round beginning with a short presentation on: the project team's list of proposed programme objectives, proposed individual learning objectives, and the 25 modules that would convey them. After each presentation, the stakeholders were provided with time to complete an online survey relating to each proposal. The survey contained requests for both quantitative and qualitative data, the importance of both forms of data collection in stakeholder engagement research having been established (Tobin et al., 2020). Participants were asked to examine the keywords of the programme objectives, learning objectives, and modules and their relevancy to the aim of the BRITAE project and each other aspect of the programme. They were then invited to score them on a four-point scale ('not relevant', 'less relevant', 'relevant', 'very relevant'). If a stakeholder felt a programme objective, learning objective, or module, was not relevant they were given the opportunity to propose changes or suggest new options. Additionally, the participants were asked which of the proposed modules they felt should be compulsory for everyone taking the master's degree, and which should be elective.

The technology facilitating this proved very robust and allowed for results to be displayed in real-time. This meant any queries could be addressed immediately, potentially leading to better informed discussion and comments. This discussion was conducted via breakout rooms set up within Zoom and named for each of the three areas of expertise of participants: industry ($n = 9$ stakeholders), academia ($n = 12$), and policy making ($n = 16$). A workshop facilitator and a note-taker were assigned to each room with attendees having previously agreed that the discussions could be recorded. A list of the proposed objectives, learning objectives and modules was displayed on the breakout room home screen to remind delegates of the subject, along with the still live links to the online surveys to maximise the completion rate. A time limit was the only restriction placed on breakout room debate, with participants free to express their thoughts as to what the programme should be aiming to provide, how it should achieve this, and the reception such suggestions would receive.

After the three rounds of presentation, survey, and breakout room discussion were completed, an additional presentation on the most effective student-centered teaching methodology and techniques to deliver the programme was provided, followed by a facilitated whole group discussion. A particular strength of the workshop was the planning and timing of each stage, with presentations kept to relatively short, informative summaries, surveys being completed inside five minutes, and the breakout room discussions close to the agreed fifteen minutes. To implement this took a great deal of pre-workshop consultation and planning, and the reward was an audience that stayed throughout the event, remaining engaged with the subject. Insight was gained on all sides during the day, and it was felt this event will have helped strengthen relationships for future dissemination activities.

From the 37 possible responses there were 32 stakeholder votes for each programme objective. Every programme objective was identified as either "relevant" or "very relevant" by a large majority (smallest was n = 24 vote margin) of survey respondents. Six objectives scored as "less relevant" with some stakeholders, but with only n = 3 votes maximum for any one of these. Only two programme objectives were considered "not relevant", these descriptions received n = 1 and n = 2 votes respectively. These results were displayed to participants immediately after the voting closed and are shown in Fig. 3.

The results were further analysed according to the stakeholder group. Academics ranked nine programme objectives as "very relevant" by more votes than any of the other options, the tenth was considered "relevant" by the highest number of votes. Only two objectives received any votes for "not relevant" by academic stakeholders, objective 3 (n = 1) and objective 5 (n = 2). Policy maker respondents did not identify any programme objectives as being "less relevant" or "not relevant". One of the objectives was scored "relevant" by most policy makers, three were scored equally as "very relevant" and "relevant", and the remaining six objectives were considered "very relevant" by most votes. Industry stakeholder respondents were similarly positive in their voting: all programme objectives scored highest in either

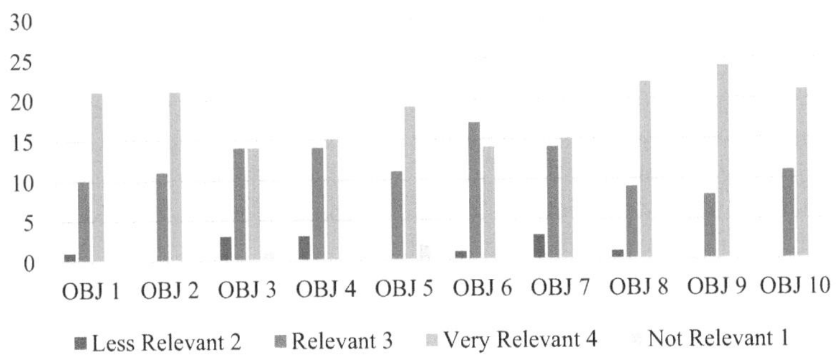

Fig. 3 Programme objectives ranked by all the stakeholders

the "relevant" or "very relevant" category, with six objectives rated "very relevant" by more people, and four objectives "relevant" by more people. Only one vote was recorded as "less relevant" for five of the objectives, and not at all for the other objectives. No-one considered any objective "not relevant," and the conclusion was that there was no divide of opinion based on the stakeholder group background. These results validated the programme objectives that had been created, but in keeping with the mixed methods approach, and employment of multiple ways of ensuring trustworthiness of data, the stakeholder group discussion was reviewed after each proposal.

While the data provided from 15 min discussions was by no means large, it was analysed with descriptive and holistic coding (Saldana, 2013) with the aim of identifying themes. Through this, suggestions relating to the programme objectives were identified. This demonstrated the importance of using mixed research methods as the survey results suggested the proposals were sufficiently acceptable to proceed unchanged. Yet, while the survey only identified $n = 4$ votes for one programme objective being either "not relevant" or "less relevant" analysis revealed it was quite contentious in discussions, and the subject of some disagreement. The objective, it was said, was too broad and not specific. There were also comments that the wording of some objectives was too similar, and one objective, for which the voting had been relatively evenly divided between "relevant" and "very relevant", was queried as to its suitability for the course. It suggested that people were more inclined to vote positively in the survey than they felt, and the opportunity to speak in a trusted environment encouraged different views. These were all keenly noted.

The review of individual learning objectives was similar with a large majority of votes considering them "relevant" and "very relevant" (minimum of $n = 19$ votes for each objective over "less relevant"). No learning objective was considered "not relevant" by any stakeholder. All three stakeholder groups were in broad agreement through the voting. Discussion, however, again raised questions of how the voting related to actual thought. Of note was the possible overuse of terminology in objective descriptions and the need to make them more accessible. Another comment suggested a need to better align a learning objective with the programme. Particularly useful was the remark that the summary of learning relating to research was perhaps more relevant to laboratory experiments, rather than applied to social scientists and agronomists. As with the programme objectives there was encouragement to be more specific in the language used and to include unique appeal for students. It was also thought that the proposals could be better aligned with the needs of the job market and provide the greatest opportunity for postgraduates to find employment after programme completion. While it was necessary to be aware that some of these were isolated comments it was felt if changes could be made that did not impact other opinions, then they should be enacted.

The survey and discussion of the proposed modules had the potential to be the most disparate of all, given that specialists in different subjects had been invited, each potentially ready to defend their interest's inclusion in the programme. In the analysis, the results were relatively clear in terms of which modules stakeholders felt should be compulsory, which elective, and those that were either not considered necessary, or unsure, of inclusion in the programme. Only one module had an even vote between "compulsory" and "elective", and one further module was equally rated between "elective" and "not needed". These two modules needed further consultation, but otherwise the grouping was resolved quite quickly. The discussion provided the main comment that certain proposed modules were widely available in other postgraduate degree programmes, and rather than aiming to cover a lot of diverse subject matter, the content should focus on producing specialists in building resilience in tropical agroecosystems. Towards this, stakeholders suggested additional modules to consider for inclusion in the programme rather than replacement of modules. This offered plenty of consideration because it was not desirable for the programme to expand beyond the 25 modules already proposed for review. However, it was possible to potentially alter parts of the content, and even combine certain modules if the survey and discussion led us to believe that the proposed mention of that subject was excessive.

The immediate outcome of the stakeholder workshop was that the course programme was refined to 23 modules, removing the proposed module that participants had voted equally between "not needed" and "elective", and a module that received some "not needed" votes. Additionally, the single programme objective and single learning objective that attracted the fewest votes for relevancy to the aims of BRITAE were removed, but only after refinement of the remaining objectives in line with stakeholder suggestions. These were the initial actions we felt able to take, while the stakeholders' responses were considered in greater detail. While this did not represent significant change to the programme the results were encouraging and suggested most of the programme had been endorsed by the participants. The workshop confirmed that the processes to date had produced appropriate outcomes and the scope of the master's programme matched the ambitions of the BRITAE project.

5.4 Results and Discussion of the Survey of Capabilities/ Capacities of BRITAE Partners

Despite multiple requests from partners to their colleagues only 17 teaching surveys were completed, a further seven were started but not finished. Even fewer (n = 8) administration surveys were completed, with 14 partial ones submitted. With hindsight, the length of the surveys and the effort required to complete them was probably excessive. The responses received were, however, insightful. The overall attitude towards the proposed master's degree programme was positive, and some colleagues identified they were already utilizing technology to deliver remote learning. However, there were concerns about the stability of the technology and infrastructure within their institutions to support it. Teaching staff also highlighted their existing workload and how they would be able to include additional duties in their schedule. Whilst relevant, important, and in need of addressing, these comments were not aspects that suggested the master's programme was not needed or deliverable, just that their institutions would need to provide resources to support it. One striking feature was that colleagues in the same institution answered questions that should have been a matter of record differently. This was particularly notable when asked if their institution provided online programmes. That both positive and negative answers were received suggested we had potential information deficiency to contend with, that needed to be both factored into the results and emphasized the need to disseminate details of the programme keenly. These results also encouraged the development of a training programme to better prepare partner staff for the implementation of the master's degree.

6 Conclusion

Once the research stages were completed the partners reviewed the proposed master's degree programme prior to its submission for accreditation. It was felt that the programme and individual learning objectives were appropriate and should remain unchanged. When considering the writing of course lectures for each module it was believed that it was possible to further successfully combine some modules, retaining the proposed content of the programme but reducing the administrative needs, something the survey of capabilities and capacities had shown to be of concern. This had the added benefit of making the options for learners simplified. The result was to refine the programme to 16 modules, some now of longer duration, with a clear sequence of learning, and easily understood credit allocation. Across four possible qualification outcomes (Fig. 4) there would be ten compulsory modules and six elective modules. The programme was subsequently approved by the UGC and began accepting students from October 2023.

Qualification	*Credits Required*	*Compulsory Modules*	*Elective Modules*
MSc by Dissertation	60	10	6
MSc by Coursework	37	8	5
PGDip	26	6	5
PGCert	20	5	0

Fig. 4 The BRITAE master's degree programme qualification routes

7 Reflections and Recommendations

There were numerous challenges in the development of the programme, not least of which was the worldwide Covid-19 pandemic which forced all meetings and discussions to be held online rather than at in-person events as planned. Working across different time zones, some of the technology not proving as robust as anticipated, and the potential for miscommunication and misunderstanding that heralded were evident. This, however, provided additional thoughts towards the work and the experience learners might face, being asked as they were to study in an online environment. Such insights might not have been gained in more usual circumstances and there was an opportunity to test likely technology platforms in work situations.

There was much about the process that was positive and to be recommended to anyone attempting similar. The nine partners of the BRITAE project represent different aspects of the academic experience and staff involved in the work encompass a truly multidisciplinary focus. Strengths in some universities were able to compensate for limited exposure to subjects in others. The resources of one could be shared with another. Use of different teams for the literature reviews for example gave added breadth and depth to the studies and provided verification of results. A series of training events helped disseminate knowledge among partners and modules for learners could be developed across a wide range of relevant topics with no compromise on the expertise of the authors. Including consultation and contributions from external experts not only offered valuable insights relating to the content of the programme but gave it an additional channel of awareness and promotion. This consultation included academics from institutions outside of the programme, in some circumstances such a move might be seen as inviting challenge from potential competitors, but the sense was that all institutions would benefit from identifying areas where staff and programmes could be supported, for example in developing soft skills and more entrepreneurial thinking. The inclusion of people in the stakeholder event who not only feel the immediate impacts of policy responses to climate change and disaster risk, but often shape it, was an ambitious step, yet the response rate to invitations was high, the consideration of the objectives and modules in depth, and the comments offered significant. Additionally, it allowed the programme to begin dissemination activities and promotion early, effectively marketing itself to potential candidates or their employers.

By comparison to the usual method of course development familiar to the institutions involved, the BRITAE master's programme benefitted from increased input, scrutiny, and examination. Importantly, many of the research methods described in this paper were originated and managed by the Sri Lankan partners and utilised experts from that country. This was crucial to avoiding suggestion of cultural imposition and an imbalance towards Global North academic experiences. Yet including European institutions potentially reassured learners and their funders to invest in a programme within a Sri Lankan educational system that has been described as evolving and not yet designed towards postgraduate study (National Education Commission of Sri Lanka, 2019). There is the potential for future research to study the experiences of learners on the BRITAE Master's degree programme and the impact they have on decision-making in Sri Lanka, the agriculture sector, and beyond as we contend with the changing climate and the implications that has for tropical agroecosystems.

References

Alexander, V. D., Thomas, H., Cronin, A., Fielding, J., & Moran-Ellis, J. (2008). Mixed methods. In N. Gilbert (Ed.), *Researching social life* (pp. 125–144). Sage.

Almansour, S. (2015). The challenges of international collaboration: Perspectives from Princess Nourah Bint Abdulrahman University. *Cogent Education, 2*(1), 1–13. https://doi.org/10.1080/2331186X.2015.1118201

Amarasekera, K. A. I. M. (2022). Factors affecting entrepreneurship development in agribusiness, among agricultural graduates. In *11th International Conference on Management and Economics at Faculty of Management and Finance*. University of Ruhuna, August 4, 2022.

Anderson, L. W., & Krathwohl, D. R. (Eds.). (2001). *A taxonomy for learning, teaching, and assessing: A revision of bloom's taxonomy of educational objectives*. Allyn & Bacon.

Bandara, D. C. (2020). The Sri Lankan higher education journey. In P. Sarangapani & R. Pappu (Eds.), *Handbook of education systems in South Asia. Global education systems*. Springer. https://doi.org/10.1007/978-981-13-3309-5_65-1

Bannier, B. J. (2016). Global trends in transnational education. *International Journal of Information and Education Technology, 6*(1), 80–84.

Bergold, J., & Thomas, S. (2012). Participatory research methods: A methodological approach in motion. *Forum: Qualitative Social Research, 13*(1), Article 30.

Caruth, G. D. (2013). Demystifying mixed methods research design: A review of the literature. *Mevlana International Journal of Education, 3*(2), 112–122.

Chou, M.-H., Kamola, I., & Pietsch, T. (2016). *The transnational politics of higher education: Contesting the global/transforming the local*. Routledge.

de Róiste, M., Breetzke, G., & Reitsma, F. (2015). Opportunities across boundaries: Lessons from a collaboratively delivered cross-institution master's programme. *Journal of Geography in Higher Education, 39*(3), 470–477.

della Volpe, M. (2020). Entrepreneurial innovation management: the joint Italy–Colombia master's degree. *International Journal of Educational Management, 34*(1), 70-80.

Denzin, N. (1978). *Sociological methods: A sourcebook*. McGraw Hill.

Döringer, S. (2021). 'The problem-centred expert interview' combining qualitative interviewing approaches for investigating implicit expert knowledge. *International Journal of Social Research Methodology, 24*(3), 265–278. https://doi.org/10.1080/13645579.2020.1766777

Duda, K., D'Artibale, A., Moombe, M., Lark, R. M., & Pulford, J. (2023). A mixed-methods evaluation of capacity strengthening within an international conservation agriculture research consortium. *F1000Research*, 12.

Dunusinghe, P. (2022). Analyzing wage differentials by fields of study: Evidence from Sri Lanka. *University of Colombo Review (New Series III), 3*(1), 32–51.

FAO (Food and Agriculture Organisation of the United Nations). (2021). *Climate change impact on crop production in Sri Lanka: Challenges and adaptation options*. Available at: https://www.fao.org/publications/card/en/c/CB3696EN/. Accessed March 30, 2023.

Fernando, N., Hettige, S. T., Weerasinghe, K. D. N., Navaratne, C. M., Siriwardana, C. S. A., Amaratunga, D., Haigh, R., & Liyanage, C. L. (2021). Research and innovation in the context of disaster resilience in the Sri Lankan higher education sector. In *Multi-hazard early warning and disaster risks* (pp. 53–67). Springer International Publishing.

FP Analysis. (2022). *In Sri Lanka, organic farming went catastrophically wrong*. Available at https://foreignpolicy.com/2022/03/05/sri-lanka-organic-farming-crisis/. Accessed March 30, 2023.
Hasanen, K. (2020). "Glocal" transnational higher education: A case study of a finnish-vietnamese collaboration. *International Journal of Learning, Teaching and Educational Research, 19*(2), 305–317.
Hemachandra, K., De Silva, A., Amaratunga, D., Haigh, R., Dissanayaka, D. M. N. A., Hettiarachchi, D., & Navaratne, C. M. (2021). The importance of building resilience in tropical agro-ecosystems for a sustainable future. In *Proceeding of the International Conference on Geography and Global Sustainability (ICGGS), December 9, 2021*. University of Colombo, Sri Lanka.
IFAD. (2019). *Sri Lanka smallholder agribusiness and resilience project design report*. Available at www.ifad.org
James, M., & Brown, S. (2005). Grasping the TLRP nettle: Preliminary analysis and some enduring issues surrounding the improvement of learning outcomes. *The Curriculum Journal, 16*(1), 7–30.
Jung, D., De Gagne, J. C., Choi, E., & Lee, K. (2022). An online international collaborative learning program during the COVID-19 pandemic for nursing students: Mixed methods study. *JMIR Medical Education, 8*(1), e34171.
Kelle, U. (2006). Combining qualitative and quantitative methods in research practice: Purposes and advantages. *Qualitative Research in Psychology, 3*(4), 293–311. https://doi.org/10.1177/1478088706070839
Knight, J. (2006). *High education crossing borders: A guide to the implications of the general agreement on trade in services (GATS) for cross-border education*. Commonwealth of Learning. Available at https://oasis.col.org/handle/11599/208
Knobel, M., Simões, T. P., & de Brito, C. H. (2013). North-South research partnerships between research universities: Experiences and best practices. *Studies in Higher Education, 38*(3), 405–424.
Laurillard, D., & Kennedy, E. (2017). *The potential of MOOCs for learning at scale in the Global South*. Centre for Global Higher Education, UCL Institute of Education.
Leydens, J. A., Moskal, B. M., & Pavelich, M. J. (2004). Qualitative methods used in the assessment of engineering education. *Journal of Engineering Education, 93*(1), 65–72.
Meehan, K., & Radice, G. (2016). Models for international collaborative undergraduate engineering programmes. In *2016 ASEE Annual Conference & Exposition*. https://doi.org/10.18260/p.25745
Miles, M., & Huberman, A. M. (1994). *Qualitative data analysis* (2nd ed.). Sage.
National Education Commission of Sri Lanka. (2019). *National policy proposals on higher education*. Available at http://nec.gov.lk/national-policy-proposals-on-higher-education-2019/. Accessed March 30, 2023.
Pain, R., & Francis, P. (2003). Reflections on participatory research. *Area, 35*(1), 46–54.
Patton, M. (1990). *Qualitative evaluation and research methods*. Sage.
Reuters. (2022). *Fertiliser ban decimates Sri Lankan crops as government popularity ebbs*. Available at https://www.reuters.com/markets/commodities/fertiliser-ban-decimates-sri-lankan-crops-government-popularity-ebbs-2022-03-03/. Accessed March 30, 23.
Saldana, J. (2013). *Coding manual for qualitative analysis*. Sage.
Simmons, R. (2008). Questionnaires. In N. Gilbert (Ed.), *Researching social life* (pp. 182–205). Sage.
SPGlobal. (2022). *Sri Lanka waking up to reality of food shortages after failed organic farming dream*. Available at https://www.spglobal.com/commodityinsights/en/market-insights/blogs/agriculture/081022-sri-lanka-crisis-food-organic-farming. Accessed March 30, 2023.
Stevens, M. G. H. (2020). Collaboration spanning two continents: An online master's degree in tropical animal health. *Journal of Veterinary Medical Education, 47*(5), 594–606.
Tobin, C., Mavrommati, G., & Urban-Rich, J. (2020). Responding to social distancing in conducting stakeholder workshops in COVID-19 era. *Societies, 10*(4), 98. https://doi.org/10.3390/soc10040098

Tsuchida, R., & Takeda, S. (2021). Is resilience socially emerging or embedded?: A review of "resilience" under climate change in Sri Lanka. *Journal of Safety Science and Resilience, 2*(4), 258–266.

UGC. (2015). *Sri Lanka qualifications framework*. The World Bank funded Higher Education for Twenty First Century (HETC) Project of the Ministry of Higher Education, Sri Lanka.

van Dam-Mieras, R., Lansu, A., Rieckmann, M., & Michelsen, G. (2008). Development of an interdisciplinary, intercultural master's program on sustainability: Learning from the richness of diversity. *Innovative Higher Education, 32*(5), 251–264.

Van Swet, J., Armstrong, A. C., & Lloyd, C. (2012). International collaboration as a patchwork quilt: Experiences of developing collaborative practice and research in an international master's programme. *Professional Development in Education, 38*(4), 647–661.

Vicente, C. R., Jacobs, F., de Carvalho, D. S., Chhaganlal, K., de Carvalho, R. B., Raboni, S. M., Qosaj, F. A., & Tanaka, L. F. (2021). Creating a platform to enable collaborative learning in one health: The joint initiative for teaching and learning on global health challenges and one health experience. *One Health, 12*, 100245. https://doi.org/10.1016/j.onehlt.2021.100245

Waltz, T. J., Powell, B. J., Matthieu, M. M., Chinman, M. J., Smith, J. L., Proctor, E. K., Damschroder, L. J., & Kirchner, J. E. (2015). Innovative methods for using expert panels in identifying implementation strategies and obtaining recommendations for their use. *In Implementation Science, 10*(1), 1–3.

STEM Teaching Tools on Climate Education: A Situational Audit on Climate Pedagogy

A. M. S. M. R. S. G. Bandara, D. A. M. De Silva, H. K. N. Udayangani, S. P. Sellapperuma, and H. Galahitigama

1 Introduction

The long-term pattern of the weather in a particular place over a considerable time period is known as the climate. Though the climate has been changing according to the natural variations since the prehistoric times, the rate of changing this climate factors; temperature, humidity, and rainfall has increased than ever before. Along with the recent exponential growth of industrial sectors of many countries of the world, the concept "Climate change" has come to the front table for discussion. Climate change is the dramatic alteration of average weather conditions over several decades or longer, such as growing warmer, wetter, or drier (Climate Change Knowledge Portal, n.d.). According to Climate Change Knowledge Portal (n.d.) the main cause for this variation is uncontrolled human activities.

Climate education is a concept widely in discussion under the umbrella term "sustainable development". Climate education targets using education as a change agent in upbringing awareness, encouragement of public on altering their behaviors and attitudes on climate change. As revealed by Otto et al. (2020) complete halt on climate change requires lowering human intervened carbon emissions to zero

A. M. S. M. R. S. G. Bandara (✉) · D. A. M. De Silva
Department of Agribusiness Management, Faculty of Agricultural Sciences, Sabaragamuwa University of Sri Lanka, Belihuloya, Sri Lanka
e-mail: ruwinibandara@agri.sab.ac.lk

H. K. N. Udayangani
Centre for Computer Studies, Sabaragamuwa University of Sri Lanka, Belihuloya, Sri Lanka

S. P. Sellapperuma
Faculty of Agricultural Sciences, Computer Centre, Sabaragamuwa University of Sri Lanka, Belihuloya, Sri Lanka

H. Galahitigama
Department of Export Agriculture, Faculty of Agricultural Sciences, Sabaragamuwa University of Sri Lanka, Belihuloya, Sri Lanka

C. Malalgoda et al. (eds.), *Climate Change Adaptation in the Built Environment*,
https://doi.org/10.1007/978-3-031-75826-3_15

by 2050. To achieve this goal, it is necessary for the individuals to change their lives and economies. According to Otto et al. (2020) the climate education is vital since climate sustainability cannot be imposed, but it needs to be learned. Moreover, the United Nations Framework Convention on Climate Change (UNFCCC) advocates considering education, training and community awareness as key rejoinders in overcoming climate change (Sustainable Development, n.d.). Further, this enhanced intellectual capacities on climate change will boost individual awareness on the point of view as consumers and it will empower them in arriving at rather ethical and rational decisions on consumption patterns (United Nations, n.d.). A systematic analysis of the literature on climate change education (30 studies conducted during the period 2005–2023) highlighted the availability of the following thematic areas in climate change education (Fig. 1) (Allen & Crowley, 2017; Amanchukwu et al., 2015; Anabaraonye et al., 2018; Apollo & Mbah, 2021; Atta-Ur-Rahman & Shaw, 2015; Ayanlade & Jegede, 2016; Cordero et al., 2008; Davison et al., 2013; Ferrari-Lagos et al., 2019; Filho, 2010; Filho et al., 2023; Jeong et al., 2021; Lal et al., 2021; Leal Filho et al., 2018, 2021; Leichenko & O'Brien, 2020; Li & Liu, 2022; McCowan, 2020; McCright et al., 2013; Mochizuki & Bryan, 2015; Monroe et al., 2019; Nazir et al., 2009; Nwankwoala, 2015; Pérez Salgado et al., 2012; Reimers, 2021; Rousell & Cutter-Mackenzie-Knowles, 2020; Schreiner et al., 2005; Wachholz et al., 2014; Yli-Panula et al., 2022). Though climate change education has been extensively researched, there is an identified gap in mainstreaming climate change education efforts towards the higher education domain.

Thus, the situational audit on climate pedagogies of STEM faculties aimed to (i) identify STEM teaching tools on climate literacy, climate change mitigation

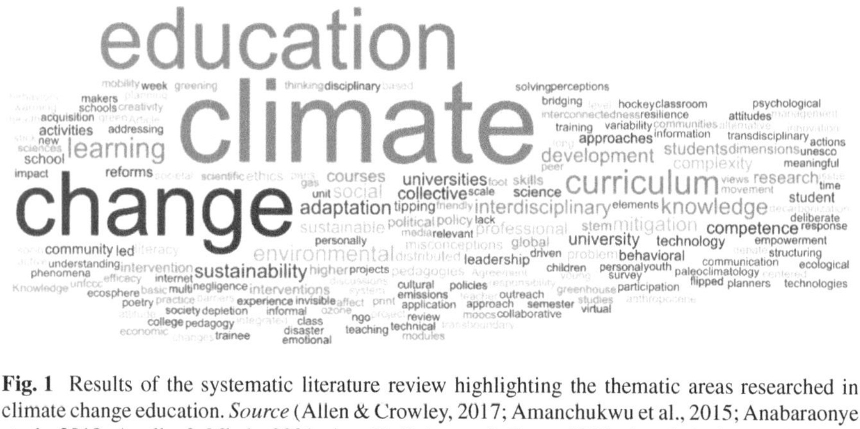

Fig. 1 Results of the systematic literature review highlighting the thematic areas researched in climate change education. *Source* (Allen & Crowley, 2017; Amanchukwu et al., 2015; Anabaraonye et al., 2018; Apollo & Mbah, 2021; Atta-Ur-Rahman & Shaw, 2015; Ayanlade & Jegede, 2016; Cordero et al., 2008; Davison et al., 2013; Ferrari-Lagos et al., 2019; Filho, 2010; Filho et al., 2023; Jeong et al., 2021; Lal et al., 2021; Leal Filho et al., 2018, 2021; Leichenko & O'Brien, 2020; Li & Liu, 2022; McCowan, 2020; McCright et al., 2013; Mochizuki & Bryan, 2015; Monroe et al., 2019; Nazir et al., 2009; Nwankwoala, 2015; Pérez Salgado et al., 2012; Reimers, 2021; Rousell & Cutter-Mackenzie-Knowles, 2020; Schreiner et al., 2005; Wachholz et al., 2014; Yli-Panula et al., 2022)

and adaptation measures and resilience building, (ii) to find out the pedagogies with climate justice and local anchor phenomena and (iii) to find out the pedagogies focusing on climate change learning on solutions along with (iv) finding the pedagogies on building resilient communities.

Considering the Sri Lankan context, the country has been witnessing the adverse impacts of climate change more frequently during the recent past not limiting to increasing temperature, landslides, droughts and floods. As a strategic decision to tackle and develop sustainable mitigation and adaptation strategies for climate change, the National Climate Change Policy was established in 2012. Moreover, Sri Lanka has adopted Intended Nationally Determined Contributions at the U.N. Framework Convention on Climate Change, 2022 to be implemented by 2030 (CFC Sri Lanka, 2023). Accordingly, National Adaptation Plan on climate change has one objective to achieve; "to increase the skills and knowledge on successful practices of adaptation through well designed education, training and awareness programmes" (NAP, 2023).

Hence, it is vital to understand the existing status of climate change education offered by higher education institutions of Sri Lanka, especially embedded in the university curricular. The scholars, graduates would be able to serve as change agents in developing climate literacy in the country. Therefore, an understanding of the current status will enable designing or reforming the university curricular to orient the degree programs to cover climate change mitigation and adaptation approaches in curriculum frameworks. According to Rousell and Cutter-Mackenzie-Knowles (2020), STEM education dominates climate education as climate science and environmental education being a part of it. An analysis of 220 studies on climate change education carried out between 1993 and 2014 found that the majority of them addressed the topic as STEM education (Rousell & Cutter-Mackenzie-Knowles, 2020). Hence it is timely to conduct a situational audit on STEM teaching tools in Sri Lankan Higher education domain.

1.1 Role of the Higher Education Institutes in Climate Education

Climate change education targets on enhance the knowledge on climate, awareness on climate protection means and ultimately the individuals should have enhanced capacity to support creating individual resilience on climate. As embedded in United Nations Sustainable Development Goals 4 and 13 (goal 4: Ensure inclusive and equitable quality education and promote lifelong learning opportunities for all; goal 13: Take urgent action to combat climate change and its impacts), climate education needs to be a part of the national education agenda (United Nations, 2023). Accordingly, some developing countries have included climate change education as a part of their national education system. As revealed by Atta-Ur-Rahman and Shaw (2015), the government of Pakistan has developed national policy on climate change in 2012 that includes climate change education and training as a compulsory module for the forest education system. Further, it points out that higher education institutes (i.e.: universities) should direct young scientists and students for foreign education so that they will be exposed to novel platforms of disaster and climate change science. Though climate change education has been a topic on concern, some studies highlight the negligence climate and environmental education in many developing countries' education systems especially in African continent (Apollo & Mbah, 2021; Leal Filho et al., 2018; Nwankwoala, 2015; Rousell & Cutter-Mackenzie-Knowles, 2020).

The schools and higher education institutes both can target delivering climate change education but the higher education institutes inevitably bear more capacity and strengths in this regard. This is owing to the impracticality of covering a wide spectrum of climate change education at school level. Hence it again consolidates the need of building an accurate climate change education at the university level (Monroe et al., 2019; Reimers, 2021; Wachholz et al., 2014).

The higher education institutes including universities have higher capacities in delivering climate change education to the students so that they can gain universal knowledge, exposure, and finally aids in developing climate competencies in climate change mitigation and adaptation (Reimers, 2021; Schreiner et al., 2005). Higher Education Institutes are crucial in climate education, especially in STEM faculties. They use innovative teaching pedagogies, such as experiential learning, hands-on research, and interdisciplinary collaboration, to foster a deep under-standing of climate-related challenges with incorporate real-world data and case studies, encouraging students to apply scientific principles. They also emphasize the use of advanced technologies and modeling techniques to simulate climate scenarios and their impacts. This multidisciplinary approach equips students with knowledge, skills, critical thinking, and problem-solving, preparing them to be-come future leaders in climate science and sustainable technology development. Moreover, the university students are more exposed to engage in climate related works so that they can design approaches to mitigate climate change, and also to engage in problem-based climate education scenarios. These approaches will help them in conducting community-based outreach activities and design innovations to tackle climate change (Davison et al., 2013; Jorgenson et al., 2019; McCowan, 2020; Reimers, 2021).

Whilst the researches highlight the importance of incorporating climate change education in university curricular, it is essential to explore innovative and more student engaging forms of climate/sustainability education. (Mochizuki & Bryan, 2015) highlights the need of approaching climate education from an interdisciplinary and systems perspective allowing learners to think critically and productively in climate sciences. Further, building theoretical knowledge on climate change won't be satisfactory unless the theoretical foundations are strengthened with three interventions; participation, relevance and interconnectedness in climate change education (Allen & Crowley, 2017). Climate change education should establish the idea of humans being the active and reflexive change agents of climate sustainability that incorporate economic, political, and cultural, psychological, and emotional dimensions of the issue (Leichenko & O'Brien, 2020).

According to the results of a world-wide survey carried out to measure the level of awareness and needs of university students on climate change revealed that the main sources available for them to learn climate change are media (TV, newspapers, radio etc.), university courses, internet, family, friends, clubs and associations where the media recording highest % of responses worldwide surpassing the university level courses (Filho, 2010). This key role of media in climate change education is further confirmed through a study conducted with the Nigerian university graduates in 2016 (Ayanlade & Jegede, 2016). It was also identified that Latin American and North American students had gathered more climate change education from university courses (67% and 71% respectively). Only 22% of sampled students had gathered climate education from universities in Asia which is a comparatively a low value. Thus, it points out the need of focusing more on university level climate change education in Asia so that university level resources and knowledge could be fully utilized to generate sustainable climate education approaches (Filho, 2010). Leal Filho et al., (2018) identifies a list of potential and active barriers that could hinder the climate change education at higher education institute level. These barriers include technological barriers where universities do not have access to updated infrastructure and technologies to provide real-time climate knowledge, perception of climate change education barriers, and publish and perish mind set barriers). Hence, to enhance the climate change education worldwide, it is crucial to overcome these barriers.

2 Methodology

A mixed method approach utilizing both primary and secondary data were applied to explore the teaching pedagogies of STEM faculties. Secondary data were sourced through online search on STEM teaching including the institutional 10 review reports (IR reports between 2017 and 2021) and 15 STEM-program review reports (PR) of the two years 2019 and 2020. A list of STEM related degree programs (that offer climate change education) of Sri Lankan universities was developed and each of the faculty websites were accessed to download the most recently updated handbooks of the relevant STEM programs. Then the handbooks were analyzed under two stages to identify the climate education-related courses offered by the recognized STEM program. The initial search included search for the terms such as "climate change", "disaster risk" and "resilience". The courses that had the key words were analyzed for their content. In the second round of the handbook analyzing, each individual course modules and their contents were analyzed by two researchers and modules that cover climate related content were extracted and listed. Along with the above search for secondary data on the situational audit of climate pedagogies, a detailed literature review on climate change education was conducted. Google scholar, JSTOR, Elsevier and Springer were utilized to download literature related to climate change education. Key words such as "climate change education", "climate literacy", "climate education" and "environmental education", "sustainability education", "university climate education" and "climate pedagogies" were used to retrieve the publications on the above topic. The search resulted in 50 related publications and after removing the repetitions and papers that could not be narrowed under the searched topic, 40 publications during the period 2005–2023 were selected. The articles were reviewed of the title, abstract, objectives, unit of analysis, key results, and key words.

With the background knowledge into climate change education in Sri Lanka, the study intended to collect primary data for the study. The primary data collection tools included focus group discussions, in-depth interviews with key informants, and questionnaires. Focus group discussions were held with 05 groups of academic staff members of the five STEM faculties of Sabaragamuwa university of Sri Lanka (Agricultural Sciences, Applied sciences, Geomatics and Technology). In addition, 05 in-depth interviews were also held with key informants selected from each STEM faculty in the university who were delivering courses related to climate change education. Initial rounds of focus group discussions targeted on exploring the approaches utilized by each of the faculties to provide climate change education, discovered the teaching tools, and discussed the community outreach activities that helps building climate literacy. Semi structured interview forms and additional probe questions were used in this stage.

Second stage of primary data collection included identification of key academics at each of the five faculties who delivered climate change education related courses and they were individually interviewed to find more information on the same criteria. In addition, an online questionnaire had been administered through the google forms with the academics of 5 STEM university faculties; University of Moratuwa, University of Ruhuna, University of Sri Jayewardenepura, University of Colombo and Sabaragamuwa University of Sri Lanka. The questionnaire gathered data on teaching tools, experience with climate change education and pedagogies used for building community resilience and climate justice.

All these varied types of data collection methods helped in data triangulation to verify the accuracy of the data collected from reviewing IR reports, PR reports and program handbooks. Figure 2 depicts the summary of the methodology applied in the study.

3 Results and Discussion

3.1 Audit on the STEM Universities in Sri Lanka

Sri Lanka has 17 national universities established under the university Act No. 16 of 1978. Out of the 17 universities, 16 has STEM faculties. Accordingly, 38 undergraduate program handbooks of 26 STEM faculties were analysed to identify climate related courses offered at the university level. Figure 3 summarizes the universities and STEM faculties where the data has been gathered on the climate related module availability.

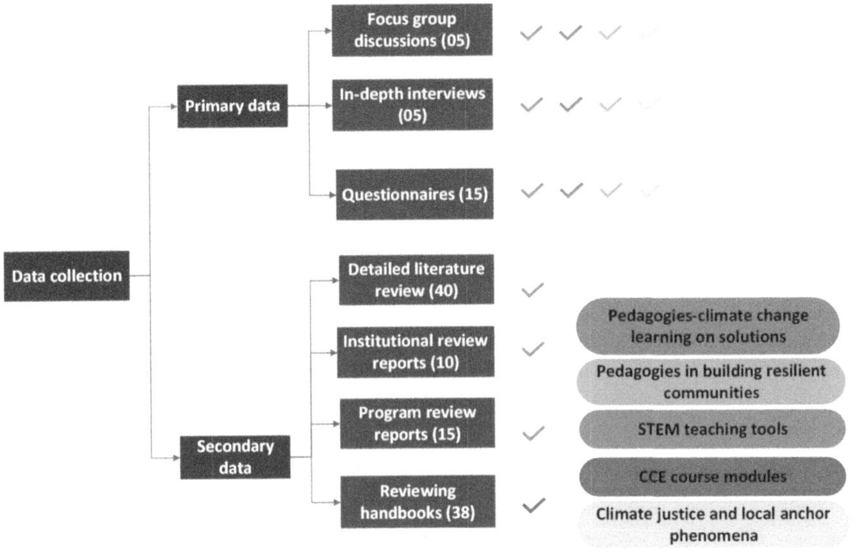

Fig. 2 Data collection approaches employed in the situational audit on climate change education

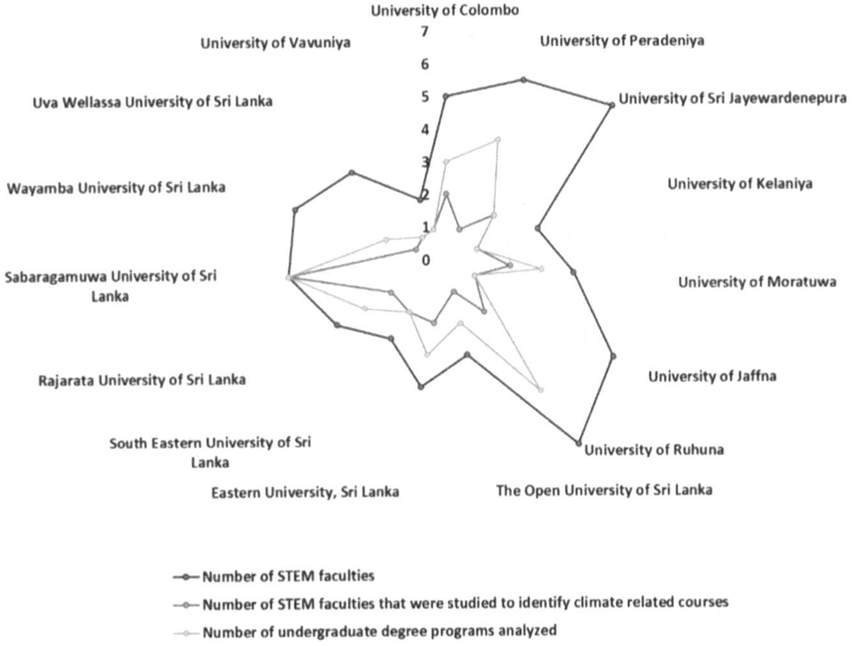

Fig. 3 Composition of the STEM faculties in the state university system. *Source* Results generated from authors' search on STEM universities of Sri Lanka

Accordingly, the degree programs targeting climate change education, mitigation and adaptation were classified under the STEM categories. The degree programs identified are depicted in Fig. 4. Accordingly highest number of STEM related courses were found in agricultural sciences while environmental sciences recorded 07 related courses. Engineering, natural sciences, technology and other applied sciences recorded low number of courses concentrating in climate change education.

The courses that directly and indirectly targets climate change education under each of the STEM degree programs are identified in Figs. 5, 6, 7 and 8. As revealed through Figs. 4 and 5 environmental sciences and agricultural sciences records the highest number of courses targeting the climate change education while engineering discipline recorded 15 climate changed related courses and technology stream found to be delivering 11 courses on climate change education.

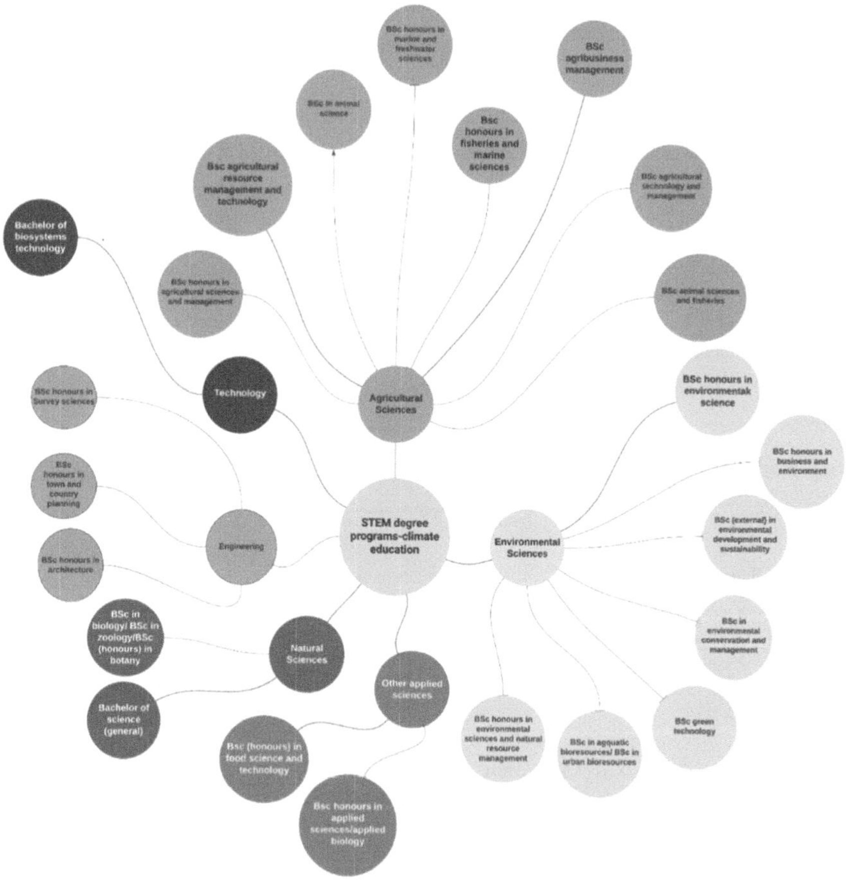

Fig. 4 Undergraduate degree programs nested under climate education domain in national university system

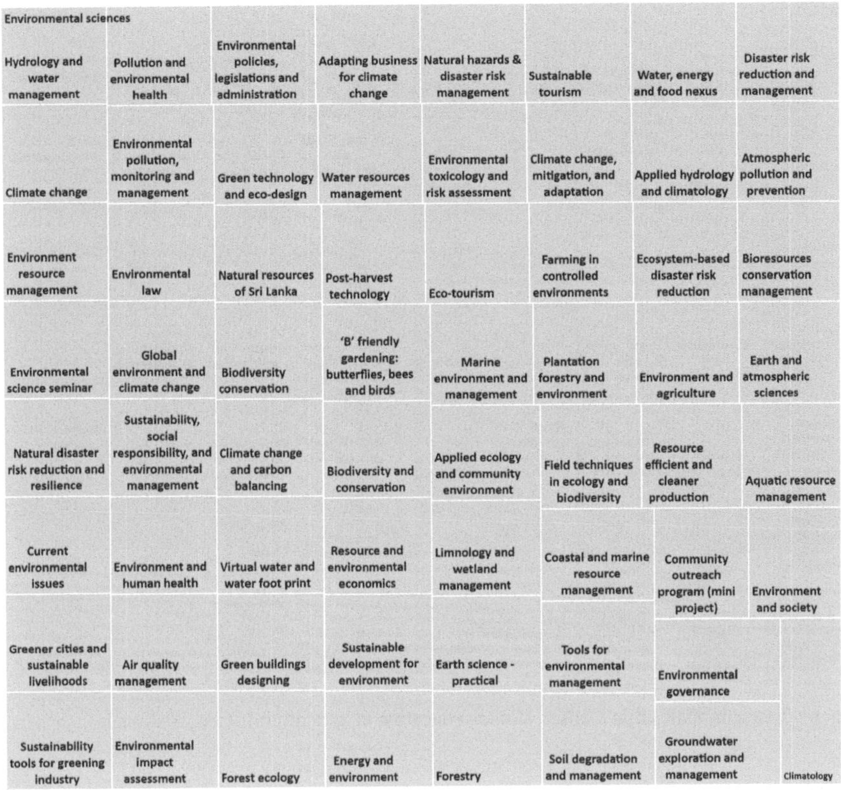

Fig. 5 Course modules that focused climate education in environmental sciences stream

Environmental sciences field is found to be delivering courses covering the thematic areas such as climatology, hydrology, and agronomy, soil sciences, forestry, fisheries, and tourism, energy, green technology, policy and governance, and pollution (Fig. 5).

Agricultural sciences offered courses on climate change education covering wide range of areas such as biodiversity, climatology, livestock management, and crop physiology and production, hydrology, pollution and environmental conservation, indigenous knowledge systems, food systems, and tourism, economics, risk and energy management. Hence the undergraduates of agricultural sciences discipline expect to receive a more holistic view on climate change adaptation and mitigation aspects (Fig. 6).

Engineering discipline offered 15 climate change related course modules of which majority of them are offered under the degree programs B.Sc. (honours) in town and country planning and B.Sc. (honours) in architecture. The climate related modules offered found to be specific to the discipline the students are majoring. Technology

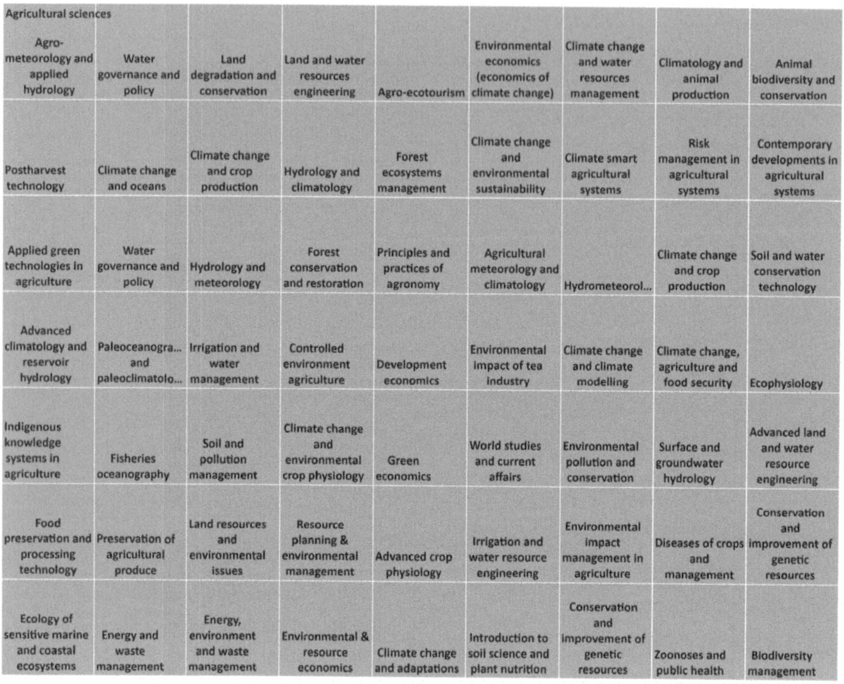

Fig. 6 Course modules that focused climate education in agricultural sciences stream

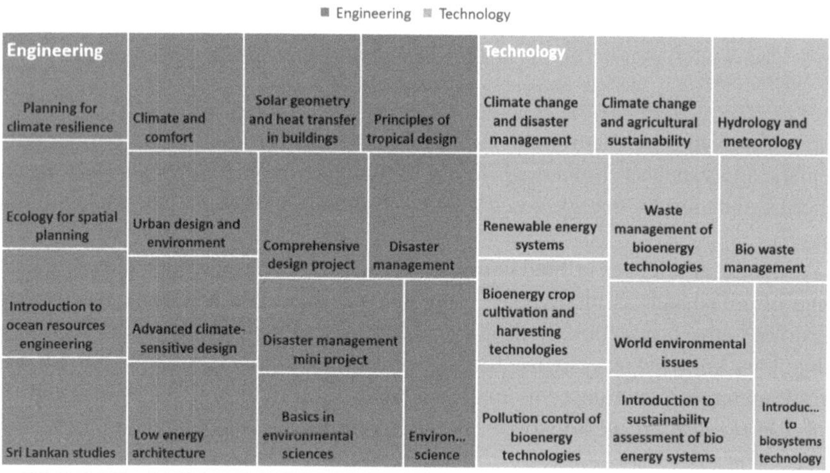

Fig. 7 Course modules that focused climate education in engineering and technology stream

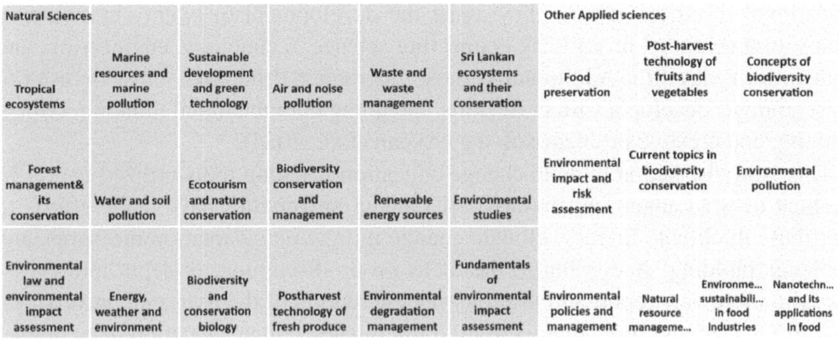

Fig. 8 Course modules that focused climate education in natural and other applied sciences streams

stream has offered courses with higher weightage on bio energy systems, pollution, sustainable waste management and renewable energy sources (Fig. 7).

Natural sciences focused on delivering climate education targeting the thematic areas such as ecosystems, pollution, waste management, tourism and energy conservation. Other applied sciences targeted biodiversity, policy, food industry and technology in delivering climate change education (Fig. 8).

According to the findings, only a few courses contained climate change/resilience keyword in the course modules (eg: climate change and climate modelling, climate change, agriculture and food security, ecophysiology, applied green technology in agriculture etc.) but the content delivered were related to building climate change competencies. Accordingly, the undergraduate courses delivered in the STEM faculties covered the main areas;

- Carbon foot print and direct climate related aspects
- Predictions of future climate change and its impact
- Measures of global climate change including adaptation and mitigation strategies
- Impact of climate change to environment, livelihoods, agriculture and other livelihood means
- Economics of climate change
- Ways of overcoming adverse impacts of climate change
- Legal aspects and policies of climate change
- Climate smart practices
- Climate governance
- Sustainability and innovative practices.

3.2 STEM Teaching Tools on Climate Change Education

Given the present global trend toward STEM education, it is critical to consider teaching methodologies carefully. One crucial aspect of STEM education and its educational resources is how they assist the development of twenty-first century twenty-first century life skills by connecting science, technology, engineering, and math to everyday life. Additionally, those instructional methods must be used to assist students develop a variety of skills, including computational thinking, critical thinking, and creative problem-solving (Akcan et al., 2023).

This study looked at climate change education teaching tools utilized in STEM faculties of Sri Lankan universities. Sri Lankan universities have made efforts to contribute to climate literacy, climate change mitigation, adaptation measures, and resilience building. According to the focus group discussion, in-depth interviews, institutional review report and program review reports of the five state universities, the various teaching tools that were used in climate change education or to deliver climate relevant subjects were identified. The observations are as follows;

- **Interactive websites and simulations**: These tools allow students to explore climate change concepts in a hands-on experience. Students can track greenhouse gas emissions globally using the website "Climate Trace", and they can make climate policy decisions and observe the effects of those decisions using the "simulation choose our future" platform.
- **Videos and documentaries**: These can be excellent tools for involving students and introducing them to the subject of climate change. A few well-known examples are Cowspiracy, Chasing Ice, and An Inconvenient Truth.
- **Books and articles**: Students can learn about climate change from numerous excellent books and articles. The Sixth Extinction, A Crack in Creation, and The People's Climate Manual are a few well-known examples.
- **Field trips and Outdoor activities**: Students can learn a lot about the subject by going on field trips to locations where climate change is an issue. Students might, for instance, travel to a beach that is being eroded by sea level rise or a forest that is experiencing wildfires. Moreover, outdoor activities like tree planting, environmental clean-ups, or monitoring local ecosystems can also engage students in hands-on learning.
- **Projects and activities**: Students can engage in a variety of excellent projects and activities to learn about climate change. For instance, students can make a comic book about climate change or poll members in their neighbourhood to find out what knowledge they have about the issue.
- **Multimedia Presentations**: To illustrate the causes, implications, and mitigation techniques of climate change, it can be employed visually appealing presentations that include photos, films, infographics, and animations. It is possible to use programs like Google Slides, Prezi, or PowerPoint.
- **Educational Games and Apps**: Online platforms and mobile applications offer interactive resources and simulations to enhance understanding. Climate-focused games and apps, like NASA's Climate Kids website or the Climate Challenge game by the BBC, engage students and make learning interactive.
- **Data Analysis and Visualization Software**: Utilize data visualization tools like interactive charts, graphs, and maps to illustrate climate-related data and trends.

Tools like Microsoft Excel, Tableau, or Google Sheets can be used to analyze and visualize climate data, promoting data literacy and critical thinking.

- **Case Studies**: Present real-world case studies to illustrate the impacts of climate change on different regions, ecosystems, or communities. This approach allows students to examine specific examples and understand the localized effects of climate change.
- **Virtual Reality (VR) and Augmented Reality (AR)**: Immersive technologies like VR and AR can provide students with virtual experiences, such as exploring coral reefs, forests, or melting glaciers. These experiences can help foster empathy and a deeper understanding of the consequences of climate change.
- **Role-Playing Games and Simulations**: Engage students in interactive simulations or role-playing activities to simulate the complexities of decision-making related to climate change.
- **Collaborative Projects**: Encourage students to work on group projects that involve researching, analyzing, and proposing solutions to climate change issues. This fosters teamwork, critical thinking, and problem-solving skills while addressing real-world challenges.
- **Guest Speakers and Expert Presentations**: Invite experts, scientists, or professionals working in climate-related fields to share their knowledge and experiences with students. Guest speakers can provide valuable insights and inspire students to take action.
- **Online Databases and Visualizations**: Websites such as NASA's Climate Change: Vital Signs of the Planet or NOAA's Climate.gov offer climate data, maps, and visualizations for analysis and study.
- **Climate Simulations and Models**: Interactive simulations and models, such as Climate Interactive's En-ROADS model or NASA's Climate Interactive Data Explorer, help students understand climate systems.

3.3 STEM Teaching Tools on Climate Literacy, Climate Change Mitigation, Adaptation Measures and Resilience Building

According to the online resources as secondary data, here are some STEM teaching tools suitable for undergraduate level climate education on climate literacy, climate change mitigation, adaptation measures, and resilience building. Figure 9 illustrates the tools that can be used for climate literacy with brief descriptions (Climate Interactive, 2023; Climate Reanalyzer, 2023; NASA, 2023; National Centers for Environmental Information (NCEI), 2023).

- **Climate Interactive's World Climate Simulation**: Group activity simulates international climate talks, involving countries' roles and designing policies, highlighting challenges and promoting understanding of global climate accords.

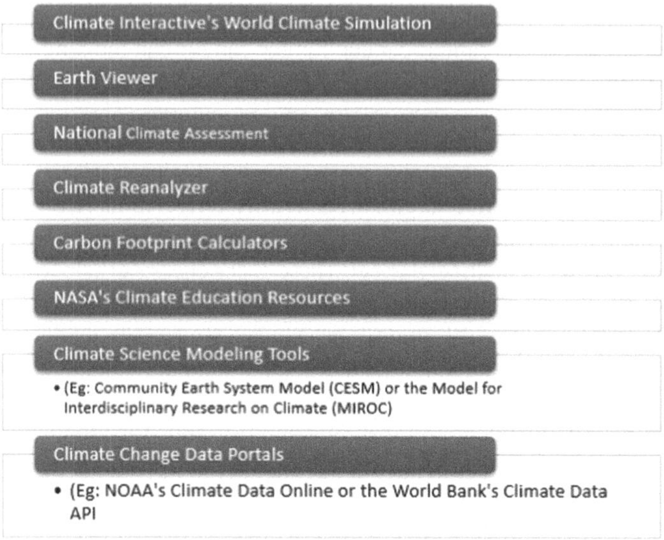

Fig. 9 STEM teaching tools on climate literacy

- **Earth Viewer**: This offers interactive maps, animations, and simulations to explore Earth's history, evolution, and dynamic nature.
- **National Climate Assessment**: The National Climate Assessment evaluates climate change impacts on US sectors, providing scientific findings, risks, and adaptation strategies for policymakers and the public.
- **Climate Reanalyzer**: Climate Reanalyzer is an online tool for real-time climate data analysis, featuring maps, graphs, and data on temperature, precipitation, and sea ice, helping understand current trends and patterns.
- **Carbon footprint calculators**: Carbon footprint calculators estimate greenhouse gas emissions based on energy usage, transportation, and lifestyle choices, raising environmental awareness and encouraging responsible conduct.
- **NASA's Climate Education tools**: NASA offers diverse climate education resources, including interactive websites, videos, lesson plans, and datasets, to teach and learn about climate science, Earth observations, and space-based research.
- **Climate Science Modelling Tools**: Complex climate models simulate Earth's climate systems, aiding researchers and educators in investigating scenarios, examining repercussions, and understanding intricate interactions within the system.
- **Climate Change Data Portals**: Climate change data portals offer access to historical records, trends, and projections, enabling informed decision-making and research on climate change.

The tools listed Fig. 10 are valuable resources for teaching climate change and its solutions, providing a brief overview of their usage. Teachers may assist students

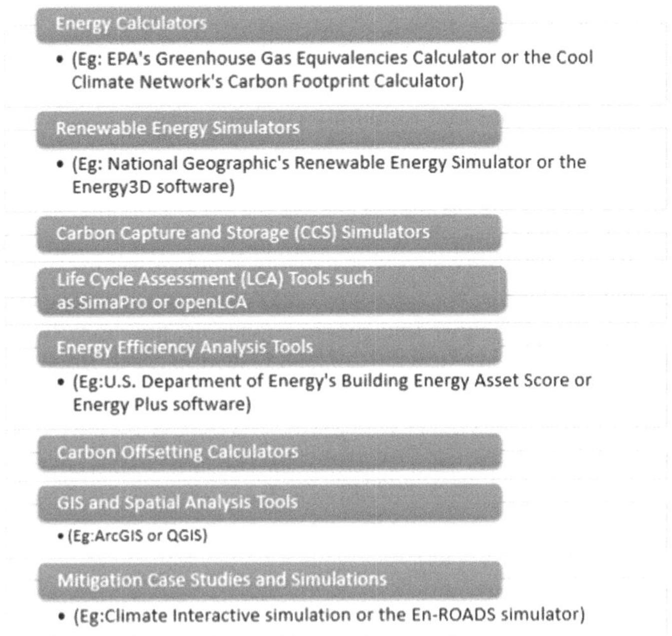

Fig. 10 STEM teaching tools on climate change mitigation

become aware and empowered advocates for climate change mitigation by using these STEM teaching resources (ABB Ability Energy Manager, 2023; Breakthrough Energy, 2023; Climate Change, 2023; Ecometrica, 2023; EnergyPlus 2023; Esri UK, 2023; OpenLCA.org, 2006; SimaPro LCA Software UK, 2023).

- **Energy Calculators**: These tools help students understand the impact of energy usage decisions on greenhouse gas emissions, comparing carbon footprints of different sources and activities.
- **Renewable Energy Kits**: these tool kits offer hands-on experience with renewable energy technologies like solar panels, wind turbines, and hydroelectric generators, educating students about clean energy sources and reducing greenhouse gas emissions.
- **Carbon Capture and Storage (CCS) Simulators**: CCS simulations aid students in understanding carbon emissions capture from industrial processes and power plants, highlighting limitations and potential benefits.
- **Life Cycle Assessment (LCA) Tools**: LCA tools help students evaluate environmental impact of products and processes, enabling broader choices understanding.
- **Energy Efficiency Analysis Tools**: These tools help to evaluate building energy efficiency, optimize usage, and reduce emissions for students.
- **Carbon Offsetting Calculators**: Carbon offsetting tools educate students on actions to reduce emissions, like tree planting and renewable energy support.

- **GIS and Spatial Analysis Tools**: GIS software enables students to visualize and analyse climate change data, including temperature variations, sea level rise, and deforestation, for a comprehensive understanding.
- **Mitigation Case Studies and Simulations**: Case studies and simulations offer students practical understanding of climate change mitigation solutions, enabling investigation and evaluation of their impact on carbon emissions.

STEM teaching techniques enhance students' understanding of climate adaptation and resilience building by fostering critical thinking, problem-solving, and awareness of climate change challenges. These resources aid educators and students in planning for climate resilience, providing frameworks, standards, and tools for enhancing systems and communities' resilience to climate-related impacts. Brief descriptions of STEM teaching tools connected to climate adaptation and resilience building are summarized below, and Fig. 11 depicts the overview of pertinent tools (Asian Development Bank, 2020; Climate Adaptation, 2023; En-ROADS, 2023; Palutikof et al., 2019; Resilience Dialogues, 2023).

- **Climate Resilience Planning Tools**: Resources aid educators and students in understanding and planning climate resilience, offering frameworks, standards, and tools for recovery.

Fig. 11 STEM teaching tools on climate adaptation measures and resilience building

- **Risk Assessment and Mapping Tools**: Risk assessment tools analyse climate change hazards, vulnerabilities, and high-risk areas, aiding decision-making and illustrating impacts.
- **Decision Support Systems**: Decision support systems guide complex climate adaptation planning by integrating data, models, and analysis.
- **Climate Change Scenarios and Simulations**: Climate change scenarios and simulations provide interactive learning tools for students to understand future climate outcomes.
- **Community Resilience Assessment Tools**: Community resilience assessment tools assess preparedness and adaptive capacity for climate impacts, enabling targeted interventions.
- **GIS and Spatial Analysis Tools**: GIS and spatial analysis tools enable climate impacts data analysis, mapping, pattern analysis, and vulnerability assessment.
- **Climate Communication and Stakeholder Engagement Tools**: Climate communication tools enhance diverse audience understanding and engagement, promoting inclusive discussions and adaptation strategies.
- **Case Studies and Real-world Example**: Case studies showcase successful climate adaptation measures, enabling students to understand challenges, solutions, and outcomes.

3.4 *Importance of Pedagogical Strategies for STEM Teaching*

Universities all around the world are becoming more aware of their obligation to provide society and students with the skills necessary to actively contribute to climate change adaptation and mitigation. Universities play this role by adopting and promoting carbon neutral policies and goals (Cordero et al., 2008; Helmers et al., 2021). Modern higher education (HE) providers are increasingly pursuing a dual strategy within this larger environment. In this regard, universities are creating curricula and pedagogical strategies to inform students about the importance of carbon neutrality as well as the mitigation and adaptation to climate change (Leal Filho et al., 2021). Meanwhile, several universities already initiated "carbon neutral strategies" by adopting low carbon operation practices (Leal Filho et al., 2021). Thus, focusing on mentioned twin strategies, the universities target to reduce their own "carbon footprint" through encouraging zero-emission of greenhouse gases (Gardner et al., 2021). Even though, universities are thus becoming more aware of their obligation to prioritize carbon neutrality and incorporate climate change education into their curricula, but sometimes organizational inertia, operational complexity, and a plethora of regulatory requirements that have an impact on governance in the higher education sector can stymie efforts (Leal Filho et al., 2021).

In universities, climate change education (CCE) can take the shape of formal, informal, and non-formal learning and teaching methods, such as higher degree research (HDR), international case studies, and nature-immersive field projects (Molthan-Hill et al., 2021; Walid & Luetz, 2018). However, over the past ten years, there has been a progressive global shift toward larger objectives and innovative pedagogical methodologies in university curricula away from a strict focus on environmental conservation (Leal Filho et al., 2021). In addition, they take the form of sustainable energy, recyclable materials, campus greening, embodied pedagogies, and outdoor instruction that immerses students in the natural world (Buxton et al., 2021; Luetz & Walid, 2019).

A number of past studies suggested that climate change education could be delivered to undergraduates through different pedagogical techniques and approaches including cognitive thinking (Karim et al., 2022). Various methods were implemented by the experts to teach undergraduates relevant to climate change and the environment. Among these strategies, student-centered approaches are highly effective rather than textbook-oriented teaching (Newsome et al., 2023). The diversified methods implemented by educators to teach climate change and environment-related topics such as problem-based learning, inquiry-based learning, experimental studies, arranging of field visits close to nature, action projects, role plays, group discussions, case studies etc. Through these methods, teachers would stimulate student's active engagement to provide a better understanding of climate projection (Karpudewan et al., 2015).

3.5 Pedagogies with Climate Justice and Local Anchor Phenomena

Higher education institutions like colleges and universities all across the world could realign and redefine their goals in teaching, research, and community engagement to build a more equitable, stable, and healthy future by committing to furthering transformative climate justice. Higher education has a variety of new and growing opportunities to combat, reverse, and address climate injustices and associated socioeconomic and health inequities as inequitable climate vulnerabilities rise (Kinol et al., 2023). Climate change education that emphasizes justice is difficult but necessary. Hence, it is needed to develop a collection of resources with a variety of topics to assist educators in developing their ability for this work. Learning about climate science must take place in formal and informal educational settings, outdoors, and in community-based learning environments for students from their childhood to higher education.

Pedagogy on climate justice and local anchor phenomena is a teaching/educational approach that aims to address the disproportionate impacts of climate change on marginalized communities. This approach attends to empower students to understand and address the challenges posed by climate change while also connecting their learning to their immediate environment and community. Climate justice focuses on the intersection of social justice and environmental issues, recognizing that the impacts of climate change disproportionately affect marginalized communities and vulnerable populations (Bartlett et al., 2022). In addition to addressing the causes and effects of climate change, it highlights the need for fair solutions that involve the people most impacted in decision-making.

Local anchor phenomena are events or phenomena that are specific to a particular place. Students can use them in the classroom to better understand the regional effects of climate change. They provide as a beginning point for learning and investigation, allowing students to look into scientific ideas, social problems, and other topics in a setting that has meaning and interest for them personally. For instance, a teacher in a coastal area might use a recent hurricane to instruct children on the dangers of floods and sea level rise.

According to data collected from the academic staff of STEM faculties of Sabaragamuwa University of Sri Lanka and literature review, several ways in incorporating climate justice and local anchor phenomena into pedagogy were identified. Focus group discussion and in-depth interview helped to summarize the benefits of each pedagogy.

3.5.1 Inquiry-Based Learning (IBL)

Inquiry-based learning is a student-cantered approach; hence it is essential to create a supportive and inclusive learning environment that values diverse perspectives and encourages collaboration. Students can gain a deeper grasp of the challenges and gain the skills necessary to act as change agents in the creation of a more just and sustainable future by participating in inquiry-based learning for climate justice (Melville, 2015).

A good way to encourage students' involvement, critical thinking, and problem-solving abilities is through inquiry-based learning. It enable the students to examine the complexities of environmental challenges, comprehend the social and economic components, and build a sense of action to address them when applied to the theme of climate justice. In the context of significant scopes of climate change where the potential for STEM curriculum integration is significant, the ability of an IBL approach to promote connections within and across curriculum disciplines offers a powerful argument for the use of the approach (Fielding, 2015). In addition, high levels of involvement are a well-established important determinant of academic performance. Researchers have found that well-designed inquiry can promote student engagement, problem-solving aptitude, higher-order thinking, and socioemotional skills by integrating direct instruction in response to learner requirements (Darling-Hammond et al., 2020).

Benefits

- Encourage students to come up with sub-questions that will aid them in analysing the main problem or inquiry.
- Motivates students to take ownership of their learning, fostering a deeper understanding of the subject matter.

Fig. 12 Undergraduates questioning and discussing climate smart aquaculture unit development as means of IBL

- Promotes the development of sustainable and innovative solutions due to inquiry-based learning pushes students to study data, assess evidence, and think critically about the social, economic, and environmental components of climate justice.
- Students learn the self-assurance they need to take meaningful action and advocate for change.
- Improves the collaboration and communication skills of students. Because, climate justice is a global issue that affects both individuals and entire communities. By encouraging students to collaborate on projects, share findings, participate in conversations, and present their ideas to others will aids in the growth of student's communication abilities and understand the importance of collective action.

Among the STEM faculties, science faculties revealed employing IBL approaches to teach undergraduate courses related to applied and natural sciences. Figure 12, illustrates students brainstorming the conceptualization of climate resilient aquaculture systems at the faculty farm of Sabaragamuwa university of Sri Lanka. This activity was followed by a short-duration interactive session with the particular lecturer explaining about the discussed content.

> This approach is very effective to students to grasp in-depth knowledge relevant to climate smart aquaculture systems and its' efficiency of application under water-scarcity conditions.

3.5.2 Place-Based Education

This strategy links education to the neighbourhood, its people, and its culture. Students are encouraged to explore how climate change is affecting their community and to come up with localized solutions rather than just typical classroom settings (Catholic Theological Union (CTU), n.d.). It is rooted in the idea that learning should be meaningful, relevant, and connected to the places where students live and learn. Thus, it recognizes the importance of understanding and appreciating one's own community before exploring broader global issues (Johnson et al., 2020). Moreover, it can be integrated into existing curricula or specialized programs that prioritize community and environmental connections in the learning process.

Benefits

- Increase student engagement, exposure and collaborative efforts with community
- Improve academic performance of undergraduates
- Develop critical thinking and problem-solving abilities
- Stronger feeling of community connection.

3.5.3 Community Partnerships

In order to fight climate change and achieve climate justice, community partnerships are essential. Communities, organizations, and governmental bodies can pool their resources, expertise, and efforts by cooperating in order to bring about significant and long-lasting change. Hence, students may get the chance to work on actual climate justice-related problems in the real world by collaborating with regional groups, authorities, and community people. This could entail conducting interviews, contributing to neighbourhood initiatives, or planning neighbourhood events.

Benefits

- Encourages discussion, the sharing of knowledge, and the co-development of solutions to the problems posed by climate change.
- Local communities must be actively involved in the educational process. The information, viewpoints, and experiences of the community are recognized and included into the learning activities.
- Encourages solutions that are suited to the demands of the community.
- It promotes analytical thinking, empathetic engagement, and proactive involvement in creating a more just and sustainable world.

Zero Plastic Community—Sabaragamuwa University of Sri Lanka had organized an event called "ZeroPlastic Trail Walk" and it was implemented with the community participation. Figure 13 illustrates its successfulness.

3.5.4 Interdisciplinary Learning

Interdisciplinary learning plays a crucial role in understanding and addressing climate justice issues. This is because, its mitigation should be distributed fairly and equitably among all people, regardless of their socio-economic background, nationality, or geographic location. It involves recognizing the unequal impacts of climate change on vulnerable communities and taking action to address these disparities. Interdisciplinary learning refers to the integration of knowledge and methods from different disciplines to gain a more comprehensive understanding of complex issues. When applied to climate justice, interdisciplinary learning involves bringing together

Fig. 13 Community involvement on SABRA NATURE'S WALK. *Source* Zero Plastics-SABRA fb page

diverse fields such as environmental science, sociology, economics, political science, law, and ethics.

Benefits

- Enable to study and integrate knowledge from diverse fields, which helps them understand how social justice and climate change are interconnected.
- Encourage collaboration and the exchange of ideas across disciplines for deeper understanding of diverse perspectives and facilitating innovative solutions.
- It helps raise awareness about the intersectional nature of climate change and social justice, empowering them to become informed and engaged global citizens.

3.5.5 Project Based Approach (PjBL)

Project-based learning or project-based instruction is a student-centred teaching method that encourages learning through engaging, real-world, curriculum-related questions or challenges (Krajcik & Shin, 2014). A driving question, the emphasis on learning objectives, engagement in educational activities, student cooperation, the use of scaffolding technology, and the production of tangible artifacts are six characteristics that were identified (Krajcik & Shin, 2014). However, majority of the studies that observed it in terms of the practice were restricted to course descriptions detailing the course scope, instructor requirements, and team size (Guo et al., 2020).

As a collaborative work with other institutes, undergraduates of Sabaragamuwa University of Sri Lanka participated for an environment protection programme at "Hirikatu-oya river basin" to remove polythene/plastic from the river basin and catchment area, s under the project, "zero-plastic". Undergraduates from faculty of Applied Science and Agricultural Sciences had participated for the activity (Fig. 14).

Fig. 14 Undergraduates' active engagement in cleaning the river basin under PjBL

Students were well-instructed before the activity and allowed students to identified numerous plastic types which directly involved for environmental pollution. According to the personal views of project leaders, they explained the project success as follows.

> We have explained the objectives of the current project before the initiation. The guidance was given clearly on the process. We noticed that students actively participated in the project. Furthermore, they were curious to know the possible methods for plastic accumulation in river basin.

3.5.6 Problem-Based Learning (PBL) Approach

In order to promote equality in learning and outcomes, learner-centred instruction aligns with how students learn deeply and how instructors can teach inclusively (National Academies of Sciences, Engineering, and Medicine, 2018; Theobald et al., 2020). As a result, evaluations and suggestions on improving undergraduate STEM education place a strong emphasis on this type of learning. PBL's ability to encourage deep learning is convincingly demonstrated. PBL calls for students to work cooperatively to find and gain the knowledge or skills necessary for a successful solution and challenge welcomes several viable solutions. The procedure is carefully scaffolded, and the teacher sets up the environment so that students can build their comprehension and hone their abilities. PBL is supposed to be predicated on the teacher asking students questions that cause them to reach a deadlock or experience cognitive disequilibrium. The deadlock is broken by applying causal reasoning, formulating predictions or conceptual comparisons, absorbing new information, and abstracting and applying existing knowledge. As a result, managing student confusion is a second requirement for problem-learning (D'Mello & Graesser, 2012). In PBL as in other complex-learning environments, the responsive instructor continuously makes diagnoses of students' thinking, motivations, and emotions and provides

Fig. 15 Field practical on greenhouse cultivation technique targeting PBL

appropriate feedback to promote positive outcomes (Lepper & Woolverton, 2002; Woolf et al., 2009).

Figure 15 shows the undergraduates of faculty of agricultural sciences engaged in field practical's regarding greenhouse cultivation technique. In this regard, students would identify problems related to lower crop production open field cultivation such as drought, salinity, pest and diseases, higher winds etc., and they could suggest possible reasons for this lower production. Simultaneously, students learn to mitigate climate changed related issues through learning novel cultivation (e.g.: greenhouse cultivation). In addition to knowledge relevant to climate change resilience techniques, students have received hands-on experience in greenhouse cultivation techniques.

3.6 Focusing Climate Change Learning on Solutions

Focusing climate change learning on solutions involves directing educational efforts towards understanding and implementing strategies to mitigate and adapt to the impacts of climate change. It goes beyond simply studying the problem and its causes, and instead emphasizes practical approaches to address the challenges confronted.

Following are some crucial components of concentrating climate change learning on solutions:

- **Renewable Energy**: Teaching about renewable energy sources including solar, wind, hydropower, and geothermal power can be a priority in education. Understanding their advantages, advancing technology, and encouraging their widespread adoption are all part of this effort to cut greenhouse gas emissions. For an example, the undergraduates were taught and practically exposed on the bio gas production at the faculty farm as means of sustainable energy utilization (Fig. 16).

Fig. 16 Bio gas unit construction at the *Agrifac* faculty farm- Sabaragamuwa University of Sri Lanka as a climate smart solution. *Source* LP Talks, 2023

- **Energy Efficiency**: It is essential to educate students about energy efficiency and conservation techniques. This entails teaching people, communities, and businesses on how to use less energy, putting energy-efficient technology to use, and implementing sustainable practices to reduce carbon footprints.
- **Sustainable Practices**: Environmental impact is reduced and a more sustainable lifestyle is promoted through educating people and communities about sustainable activities such trash reduction, recycling, composting, sustainable agriculture, and responsible consumerism. The undergraduate students are exposed to develop sustainable packaging materials, and environmentally friendly marketing practices to enrich their lifelong learning and innovation in climate change education (Fig. 17). Moreover, industrial visits on sustainable agribusinesses were also performed to build the awareness and to showcase how the climate change knowledge is used practically in the industry.
- **Adaptation Strategies**: Education about climate change should include developing resiliency and adaptation tactics. This involves educating people about climate hazards, preparing for severe weather, advocating the design of sustainable infrastructure, and assisting local communities in adjusting to the effects of climate change. Beyond giving the knowledge to the undergraduates, they should be given opportunities to incorporate their learning on solving real world scenarios. For an example, group of students can innovate a new food product

Fig. 17 Students innovating eco-friendly means of innovative packaging, product portfolios and marketing their products through agri food stalls

dehydrator as means of proposing a solution to the increased post-harvest losses on fruits and vegetables due to increased rainfall intensity.

- **Policy and Advocacy**: Meaningful change can result from educating people about climate policies and motivating them to participate in advocacy activities. Education about climate change should give people the knowledge and abilities to shape public policy, advance sustainable lifestyles, and support initiatives that address climate change at the local, national, and international levels. This could be achieved via students receiving an opportunity to research into the climate change policies and allowing them to study the existing system and building their capacities in proposing new policies etc. (Fig. 18).
- **Collaboration and Innovation**: Collaboration and innovative thinking should be encouraged through education on climate change. Innovative ideas and positive

Fig. 18 Student ambassador programs that enable the undergraduates to be good decision makers and policy influencers in future on climate change education. *Source* G17 official fb page—Sri Lanka

change can be sparked by putting an emphasis on multidisciplinary approaches, stimulating research, and fostering creative problem-solving.
- **Global Perspective**: An international approach is necessary to address the global challenge of climate change. Education should emphasize how interdependent nations are, stress the value of international cooperation, and foster awareness of the various implications of climate change on different geographic locations.

By focusing climate change education on solutions, the community could be empowered with knowledge, skills, and motivation required to contribute to a sustainable and a resilient future. It encourages group activities to address the pressing issue of climate change, stimulates the development of innovative alternatives, and gives people the power to take action.

3.7 Pedagogies on Building Resilient Communities

Most of countries have already made emergency declarations and developed climate adaptation plans while accepting resilience as a goal or process (Moser et al., 2019). Therefore, academics have identified that beyond government actions, grassroots responses to climate change impacts which can be understood as community-led resilience initiatives (Berkes & Ross, 2013). Community resilience to climate change scenarios would be achieved through increasing the awareness and action over the climate emergency (Bródy et al., 2018). Furthermore, environmental education can contribute to such grassroots mobilization through projects or events nurturing youth agency and empowerment to take climate action within their local communities (Trott, 2020). With an effective education on presence climate crisis leads to individuals and communities' active integration on mitigation of climate change.

Building resilient communities for climate change basically relying on community resources that their members mobilize to cope or adapt to climatic changes while navigating uncertainty (Magis, 2010). Environmental education actions for community could be initiated through providing of training courses, arranging field trips and awareness programmes and other formal and non-formal learning experiences. Further the awareness sessions on potential climate induced hazards floods, landslides food and nutrition security through food preservation enhances the community resilience and buffering capacity against the disasters.

3.7.1 Organizing of Community-Based Training Programmes

Through organizing of community-based training programmes related to climate change aspects would provide a clear image about the way of individual's behavior. Moreover, through these training programmes we could empower and mobilize youth as well as elderly population to understand climate change and environment related topics. Among these youth empowerments is crucial to raise their awareness about the mitigation of climate change and environmental degradation. Very recently, the Department of Export Agriculture, Faculty of Agricultural Sciences has organized a ten days training programme for selected officials from "*Maga-neguma* authority-Sri Lanka" relevant to compost and organic fertilizer production for small scale. The age of the participants ranged between 24 and 40. Throughout the programme we have noticed that participants were actively engaged for the programme. According to the training programme in-charge views on particular activity, he was really pleased with the participants' enthusiasm on given tasks. Moreover, 90% of participants mentioned positive views on their hands-on experience and knowledge about waste management in real scenario in their feedback.

> "We have received great experience on compost and organic fertilizer production. We realized importance of this process as it mitigates the synthetic-polymer accumulation into the ecosystem."

Apart from this several training programmes for school students have been organized by the department of export agriculture, faculty of agricultural sciences to provide adequate knowledge on climate change adaptation strategies. The academic members of the department conducted several class-room lectures and on-farm training sessions relevant to climate resilience activities (Figs. 19 and 20).

According to students' feedbacks on these programmes, it was proved that students actively participated in the training programmes and they enjoyed engaging in relevant activities during the workshop.

Fig. 19 School students and undergraduate training programmes on greenhouse agriculture

Fig. 20 Organic fertilizer production through waste (a workshop conducted for officials of "Maga-neguma" authority, Sri Lanka by the Department of Export Agriculture)

3.7.2 Print Media: Leaflets, Brochures and Magazines

To provide essential knowledge regarding the climate change mitigation strategies it would be important to distribute general articles related to climate change; so that the community could take more information related to this topic. This would be more beneficial for the general community since they do not have in-depth knowledge about climate change. For an instance, the department of agribusiness management, faculty of agricultural sciences quarterly release a magazine under various themes related to agriculture and current situations of the country. Each magazine contains several articles related to climate change. Moreover, the policy briefs are published as means of disseminating research generated knowledge among the general public. Through these publications, academics are able to communicate with the society as well as they can acknowledge people regarding actual circumstances under climate change (Fig. 21).

Fig. 21 AFQ (Agri Food Quarterly) magazines and policy insights issued by the department of agribusiness management

3.7.3 Group Discussions with Community

Universities as the center of higher education has the responsibility to create awareness on the climate change and its adverse impacts. However, this should be done as a cumulative process. Social awareness progressively pervades and influences the practices of the individuals to mitigate the climate change issues (Leal Filho et al., 2021). Therefore, organizing of community awareness programmes collaborate with different parties in the society such as community stakeholders, local/national governments, and cooperate sector would be essential.

The department of agribusiness management has organized a farmer conference very recently which shared recent problems related to climate change such as severe drought and extreme weather conditions. The academia acknowledged the reasons for occurring of such climate risks and consequences and possible mitigation strategies that could be employed mitigating the identified vulnerabilities. Furthermore, this platform allowed several farmers in expressing their experiences of challenging and overcoming climate barriers. Through this, the farmer community were motivated to learn about climate change via learner-centered adult education. The method is also effective in penetrating marginalized communities with climate change education (Fig. 22).

Fig. 22 INNOVATION + 2023 Farmers Mini conference and a farmer led innovation to control the water quality in aquaculture

4 Conclusion

According to the situational audit, it was revealed that STEM faculties of Sri Lankan universities offer climate change related knowledge to undergraduates in different levels through different approaches. One main approach identified was through the compulsory and optional course modules. Though majority of course modules did not directly target the climate change concept, the content indirectly built-up students' knowledge on climate change pertaining to the individual's discipline. All studied STEM courses provided climate education while agricultural sciences, environmental sciences and applied sciences recorded the highest number of course modules oriented towards climate education. It was evident through the module analysis, a sound theoretical foundation was created through the modules, but holistically, it lacked novel approaches in engaging students in innovating solutions to climate change related issues. On the contrary, some universities revealed to be implementing innovative teaching tools on climate education such as IT enabled applications, projects, and case studies. Climate justice and local anchor phenomena which were incorporated in the climate pedagogy (through inquiry-based learning, place based education, and community partnerships, interdisciplinary learning, and project-based learning, and problem-based learning) could empower the undergraduates and the community in climate change adaptation, mitigation and resilience building endeavors. Further, attempts at climate change learning on solutions such as fostering research, innovation, and development of sustainable energy sources, environmentally healthy practices and focusing on climate change policy, advocacy, and governance while exposing undergraduates to global perspectives could broaden the capacities of the climate change education domain. The pedagogies sub setting climate resilience communities, i.e., community-based training programs, awareness dissemination through print media, and community discussions carries the potential to ripple the community's buffering capacity to climate risk and vulnerability.

5 Recommendations

Considering as a single entity, the national higher education system needs to strengthen their existing modules on climate change education with approaches on active problem solving, through curricular reforms. In addition, universities in Sri Lanka can improve the accessibility and efficacy of STEM tools for climate education, equipping students with the abilities and information needed to handle the problems caused by climate change. More emphasis should be placed on partnership with industry and research organizations as well as encouraging students to take part in citizen science programs connected to climate change. Additionally, it is crucial to provide STEM educators with professional development programs and seminars to advance their knowledge and abilities in using climate education resources successfully. Focusing the pedagogy on climate justice, it is necessary to include chances for undergraduates to reflect on the ethical implications of climate change and consider their responsibilities as global citizens. They should be encouraged to think critically about the ethical implications of climate change and consider their own responsibilities and potential actions in promoting climate justice. More focus should be given on practical and innovative teaching tools so that students are inspired to become advocates for climate action and social justice.

References

ABB Ability Energy Manager. (2023). [Internet]. Cited September 19, 2023. Available from: https://new.abb.com/about/our-businesses/electrification/abb-ability/energy-and-asset-manager/abb-ability-energy-manager

Akcan, A. T., Yıldırım, B., Karataş, A. R., & Yılmaz, M. (2023). Teachers' views on the effect of STEM education on the labor market. *Frontiers in Psychology, 14*, 1184730. https://doi.org/10.3389/fpsyg.2023.1184730

Allen, L. B., & Crowley, K. (2017). Moving beyond scientific knowledge: Leveraging participation, relevance, and interconnectedness for climate education. *International Journal of Global Warming, 12*, 299–312. https://doi.org/10.1504/IJGW.2017.084781

Amanchukwu, R. N., Amadi-Ali, T. G., & Ololube, N. P. (2015). Climate change education in Nigeria: The role of curriculum review. *Education, 5*, 71–79.

Anabaraonye, B., Nji, I., & Hope, J. (2018). Poetry as a valuable tool for climate change education for global sustainability. *International Journal of Scientific and Engineering Research, 9*, 81–85.

Apollo, A., & Mbah, M. F. (2021). Challenges and opportunities for climate change education (CCE) in East Africa: A critical review. *Climate, 9*, 93. https://doi.org/10.3390/cli9060093

Asian Development Bank. (2020). *Urban climate change resilience trust fund*. [Internet]. Cited September 19, 2023. Available from: https://www.adb.org/what-we-do/funds/urban-climate-change-resilience-trust-fund

Atta-Ur-Rahman, & Shaw, R. (2015). Disaster and climate change education in Pakistan. In Atta-Ur-Rahman, A. N. Khan, & R. Shaw (Eds.), *Disaster risk reduction approaches in Pakistan, disaster risk reduction* (pp. 315–335). Springer Japan. https://doi.org/10.1007/978-4-431-55369-4_17

Ayanlade, A., & Jegede, M. O. (2016). Climate change education and knowledge among Nigerian University Graduates. *Weather, Climate, and Society, 8*, 465–473. https://doi.org/10.1175/WCAS-D-15-0071.1

Bartlett, M., Larson, J., & Lee, S. (2022). Environmental justice pedagogies and self-efficacy for climate action. *Sustainability, 14*, 15086. https://doi.org/10.3390/su142215086

Berkes, F., & Ross, H. (2013). Community resilience: Toward an integrated approach. *Society & Natural Resources, 26*, 5–20. https://doi.org/10.1080/08941920.2012.736605

Breakthrough Energy. (2023). *Breakthrough energy accelerates climate innovation*. [Internet]. Cited September 19, 2023. Available from: https://breakthroughenergy.org/climate-innovation/?gclid=CjwKCAjwjaWoBhAmEiwAXz8DBYbVaEq4Fx8es5uoN9yRQonfVq_KfOlhdNKq8fLbbwGLep97UShbURoCYxwQAvD_BwE

Bródy, L. S., Chellcri, L., Baró, F., & Ruiz-Mallen, I. (2018). *Enhancing community resilience in Barcelona: Addressing climate change and social justice through spaces of co-management*. In A. Galderisi & A. Colucci (Eds.), *Smart, resilient and transition cities: Emerging approaches and tools for a climate-sensitive urban development* (pp. 203–208). Elsevier.

Buxton, G., Luetz, J. M., & Shaw, S. (2021). *Towards an embodied pedagogy in educating for creation care*. Springer. [WWW Document]. https://link.springer.com/chapter/10.1007/978-981-15-8856-3_20. Accessed July 28, 2023.

Catholic Theological Union (CTU). (n.d). *The "I-Thou" relationship, place-based education, and Aldo Leopold*. [WWW Document]. https://i-share-ctu.primo.exlibrisgroup.com. Accessed July 7, 2023.

CFC Sri Lanka. (2023). *International climate change university (ICCU) in Sri Lanka?* Climate Fact Checks (blog), May 19, 2023. https://climatefactchecks.org/sri-lanka-planning-to-establish-an-international-university-to-study-climate-change/

Climate Adaptation. (2023). *ResilienceLinks*. [Internet]. Cited September 19, 2023. Available from: https://www.resiliencelinks.org/building-resilience/climate-adaptation

Climate Change. (2023). [Internet]. Cited September 19, 2023. Available from: https://education.nationalgeographic.org/resource/climate-change

Climate Change Knowledge Portal. (n.d.). *What is climate change?* [WWW Document]. https://climateknowledgeportal.worldbank.org/overview. Accessed July 28, 2023.

Climate Interactive. (2023). [Internet]. Cited September 19, 2023. Available from: https://www.climateinteractive.org/

Climate Reanalyzer. (2023). [Internet]. Cited September 19, 2023. Available from: https://climatereanalyzer.org/

Cordero, E. C., Todd, A. M., & Abellera, D. (2008). Climate change education and the ecological footprint. *Bulletin of the American Meteorological Society., 89*, 865–872. https://doi.org/10.1175/2007BAMS2432.1

Darling-Hammond, L., Flook, L., Cook-Harvey, C., Barron, B., & Osher, D. (2020). Implications for educational practice of the science of learning and development. *Applied Developmental Science, 24*, 97–140. https://doi.org/10.1080/10888691.2018.1537791

Davison, A., Brown, P., Pharo, E., Warr, K., McGregor, H., Terkes, S., Boyd, D., & Abuodha, P. (2013). Distributed leadership: Building capacity for interdisciplinary climate change teaching at four universities. *International Journal of Sustainability in Higher Education, 15*, 98–110. https://doi.org/10.1108/IJSHE-10-2012-0091

D'Mello, S., & Graesser, A. (2012). Dynamics of affective states during complex learning. *Learning and Instruction, 22*, 145–157. https://doi.org/10.1016/j.learninstruc.2011.10.001

Ecometrica. (2023). *Measuring & reporting greenhouse gas emissions (GHG)*. [Internet]. Cited September 19, 2023. Available from: https://ecometrica.com/solutions/measuring/

EnergyPlus. (2023). [Internet]. Cited September 19, 2023. Available from: https://energyplus.net/

En-ROADS. (2023). [Internet]. Cited September 19, 2023. Available from: https://www.climateinteractive.org/en-roads/

Esri UK. (2023). *What is GIS?—Geographic information system mapping technology*. [Internet]. Cited September 19, 2023. Available from: https://www.esriuk.com/en-gb/what-is-gis/overview

Ferrari-Lagos, E., Martínez-Abad, F., & Ruíz, C. (2019). Education to mobilize society for climate change action: The climate competence in education. In *Proceedings of the Seventh International Conference on Technological Ecosystems for Enhancing Multiculturality. Presented at the TEEM'19: Technological Ecosystems for Enhancing Multiculturality* (pp. 966–970). ACM. https://doi.org/10.1145/3362789.3362853

Fielding, J. (2015). *Identifying core elements of argument-based inquiry in primary mathematics learning*.

Filho, W. L. (2010). Climate change at Universities: Results of a world survey. In W. Leal Filho (Ed.), *Universities and climate change: Introducing climate change to University programmes, climate change management* (pp. 1–19). Springer. https://doi.org/10.1007/978-3-642-10751-1_1

Filho, W. L., Yayeh Ayal, D., Wall, T., Shiel, C., Paco, A., Pace, P., Mifsud, M., Lange Salvia, A., Skouloudis, A., Moggi, S., LeVasseur, T., Vinuesa Antonio, G., Azeiteiro, U. M., Ioannis, N., & Kovaleva, M. (2023). An assessment of attitudes and perceptions of international university students on climate change. *Climate Risk Management, 39*, 100486. https://doi.org/10.1016/j.crm.2023.100486

Gardner, C. J., Thierry, A., Rowlandson, W., & Steinberger, J. K. (2021). From publications to public actions: The role of Universities in facilitating academic advocacy and activism in the climate and ecological emergency. *Frontiers in Sustainability, 2.*

Guo, P., Saab, N., Post, L. S., & Admiraal, W. (2020). A review of project-based learning in higher education: Student outcomes and measures. *International Journal of Educational Research, 102*, 101586. https://doi.org/10.1016/j.ijer.2020.101586

Helmers, E., Chang, C. C., & Dauwels, J. (2021). Carbon footprinting of universities worldwide: Part I—Objective comparison by standardized metrics. *Environmental Sciences Europe, 33*, 30. https://doi.org/10.1186/s12302-021-00454-6

Jeong, J. S., González-Gómez, D., Conde-Núñez, M. C., Sánchez-Cepeda, J. S., & Yllana-Prieto, F. (2021). Improving climate change awareness of preservice teachers (PSTs) through a University science learning environment. *Education Sciences, 11*, 78. https://doi.org/10.3390/educsci11 020078

Johnson, M. D., Sprowles, A. E., Goldenberg, K. R., Margell, S. T., & Castellino, L. (2020). Effect of a place-based learning community on belonging, persistence, and equity gaps for first-year STEM students. *Innovative Higher Education, 45*, 509–531. https://doi.org/10.1007/s10755-020-09519-5

Jorgenson, S. N., Stephens, J. C., & White, B. (2019). Environmental education in transition: A critical review of recent research on climate change and energy education. *Journal of Environmental Education, 50*, 160–171. https://doi.org/10.1080/00958964.2019.1604478

Karim, N., Othman, H., Zaini, Z.-I. I., Rosli, Y., Wahab, M. I. A., Al Kanta, A. M., Omar, S., & Sahani, M. (2022). Climate change and environmental education: Stance from science teachers. *Sustainability, 14*, 16618. https://doi.org/10.3390/su142416618

Karpudewan, M., Roth, W.-M., & Abdullah, M. N. S. B. (2015). Enhancing primary school students' knowledge about global warming and environmental attitude using climate change activities. *International Journal of Science Education, 37*, 31–54. https://doi.org/10.1080/09500693.2014.958600

Kinol, A., Miller, E., Axtell, H., Hirschfeld, I., Leggett, S., Si, Y., & Stephens, J. C. (2023). Climate justice in higher education: A proposed paradigm shift towards a transformative role for colleges and universities. *Climatic Change, 176*, 15. https://doi.org/10.1007/s10584-023-03486-4

Krajcik, J. S., & Shin, N. (2014). Project-based learning. In: R. K. Sawyer (Ed.), *The Cambridge handbook of the learning sciences* (pp. 275–297). Cambridge University Press. https://doi.org/10.1017/CBO9781139519526.018

Lal, A., Walsh, E., Weatherell, A., & Slimings, C. (2021). Climate change education in public health and medical curricula in Australian and New Zealand Universities: A mixed methods study of barriers and areas for further action (preprint). *Epidemiology*. https://doi.org/10.1101/2021.06.11.21258793

Leal Filho, W., Morgan, E. A., Godoy, E. S., Azeiteiro, U. M., Bacelar-Nicolau, P., Veiga Ávila, L., Mac-Lean, C., & Hugé, J. (2018). Implementing climate change research at universities: Barriers, potential and actions. *Journal of Cleaner Production, 170*, 269–277. https://doi.org/10.1016/j.jclepro.2017.09.105

Leal Filho, W., Sima, M., Sharifi, A., Luetz, J. M., Salvia, A. L., Mifsud, M., Olooto, F. M., Djekic, I., Anholon, R., Rampasso, I., Kwabena Donkor, F., Dinis, M. A. P., Klavins, M., Finnveden, G., Chari, M. M., Molthan-Hill, P., Mifsud, A., Sen, S. K., & Lokupitiya, E. (2021). Handling climate change education at universities: An overview. *Environmental Sciences Europe, 33*, 109. https://doi.org/10.1186/s12302-021-00552-5

Leichenko, R., & O'Brien, K. (2020). Teaching climate change in the anthropocene: An integrative approach. *Anthropocene, 30*, 100241. https://doi.org/10.1016/j.ancene.2020.100241

Lepper, M. R., & Woolverton, M. (2002). The wisdom of practice: Lessons learned from the study of highly effective tutors. In J. Aronson (Ed.), *Improving academic achievement, educational psychology* (pp. 135–158). Academic Press (Chapter 7). https://doi.org/10.1016/B978-012064455-1/50010-5

Li, Y.-Y., & Liu, S.-C. (2022). Examining Taiwanese students' views on climate change and the teaching of climate change in the context of higher education. *Research in Science & Technological Education, 40*, 515–528. https://doi.org/10.1080/02635143.2020.1830268

Luetz, J. M., & Walid, M. (2019). Social responsibility versus sustainable development in United Nations policy documents: A meta-analytical review of key terms in human development reports. In W. Leal Filho (Ed.), *Social responsibility and sustainability: How businesses and organizations can operate in a sustainable and socially responsible way, world sustainability series* (pp. 301–334). Springer International Publishing. https://doi.org/10.1007/978-3-030-03562-4_16

Magis, K. (2010). Community resilience: An indicator of social sustainability. *Society & Natural Resources, 23*, 401–416. https://doi.org/10.1080/08941920903305674

McCowan, T. (2020). The impact of universities on climate change: a theoretical framework. In *Transforming universities for a changing climate, working paper series* (Vol. 1). Climate-U, UCL Institute of Education. [WWW Document]. https://www.climate-uni.com/resources. Accessed July 28, 2023.

McCright, A. M., O'Shea, B. W., Sweeder, R. D., Urquhart, G. R., & Zeleke, A. (2013). Promoting interdisciplinarity through climate change education. *Nature Clim Change, 3*, 713–716. https://doi.org/10.1038/nclimate1844

Melville, W. (2015). Inquiry as a teaching strategy. In R. Gunstone (Ed.), *Encyclopedia of science education* (pp. 507–510). Springer. https://doi.org/10.1007/978-94-007-2150-0_191

Mochizuki, Y., & Bryan, A. (2015). Climate change education in the context of education for sustainable development: Rationale and principles. *Journal of Education for Sustainable Development, 9*, 4–26. https://doi.org/10.1177/0973408215569109

Molthan-Hill, P., Blaj-Ward, L., Mbah, M., & Ledley, T. S. (2021). Climate change education at universities: Relevance and strategies for every discipline. In *Handbook of climate change mitigation and adaptation* (pp. 3395–3458). Springer Nature. https://doi.org/10.1007/978-1-4614-6431-0_153-1

Monroe, M. C., Plate, R. R., Oxarart, A., Bowers, A., & Chaves, W. A. (2019). Identifying effective climate change education strategies: A systematic review of the research. *Environmental Education Research, 25*, 791–812. https://doi.org/10.1080/13504622.2017.1360842

Moser, S., Meerow, S., Arnott, J., & Jack-Scott, E. (2019). The turbulent world of resilience: Interpretations and themes for transdisciplinary dialogue. *Climatic Change, 153*, 21–40. https://doi.org/10.1007/s10584-018-2358-0

NAP. (2023). Accessed June 22, 2023. http://www.climatechange.lk/Index_NAP.html

NASA. (2023). *Home—Climate change: Vital signs of the planet*. [Internet]. Cited September 19, 2023. Available from: https://climate.nasa.gov/

National Academies of Sciences, Engineering, and Medicine. (2018). *How people learn II: Learners, contexts, and cultures*. National Academies Press. https://doi.org/10.17226/24783

National Centers for Environmental Information (NCEI). (2023). [Internet]. Cited September 19, 2023. Available from: https://www.ncei.noaa.gov/

Nazir, J., Pedretti, E., Wallace, J., Montemurro, D., & Inwood, H. (2009). *Climate change and sustainable development: The response from education.*

Newsome, D., Newsome, K. B., & Miller, S. A. (2023). Teaching, learning, and climate change: Anticipated impacts and mitigation strategies for educators. *Behavior and Social Issues*. https://doi.org/10.1007/s42822-023-00129-2

Nwankwoala, H. N. L. (2015). Causes of climate and environmental changes: the need for environmental-friendly education policy in Nigeria. *Journal of Education and Practice, 6*, 224–234.

OpenLCA.org. (2006). *OpenLCA is a free, professional Life Cycle Assessment (LCA) and footprint software with a broad range of features and many available databases, created by GreenDelta since 2006*. [Internet]. Cited September 19, 2023. Available from: https://www.openlca.org/

Otto, I. M., Donges, J. F., Cremades, R., Bhowmik, A., Hewitt, R. J., Lucht, W., Rockström, J., Allerberger, F., McCaffrey, M., Doe, S. S. P., Lenferna, A., Morán, N., van Vuuren, D.

P., & Schellnhuber, H. J. (2020). Social tipping dynamics for stabilizing Earth's climate by 2050. *Proceedings of the National Academy of Sciences of the United States of America, 117*, 2354–2365. https://doi.org/10.1073/pnas.1900577117
Palutikof, J. P., Street, R. B., & Gardiner, E. P. (2019). Decision support platforms for climate change adaptation: An overview and introduction. *Climatic Change, 153*, 459–476. https://doi.org/10.1007/s10584-019-02445-2
Pérez Salgado, F., de Kraker, J., Boon, J., & van der Klink, M. (2012). Competences for climate change education in a virtual mobility setting. *International Journal of Innovation and Sustainable Development, 6*, 53–65. https://doi.org/10.1504/IJISD.2012.046053
Resilience Dialogues. (2023). *Connecting experts & communities for resilience.* [Internet]. Cited September 19, 2023. Available from: https://www.resiliencedialogues.org/overview/
Reimers, F. M. (2021). The role of universities building an ecosystem of climate change education. Springer. [WWW Document]. https://link.springer.com/chapter/10.1007/978-3-030-57927-2_1. Accessed July 28, 2023.
Rousell, D., & Cutter-Mackenzie-Knowles, A. (2020). A systematic review of climate change education: Giving children and young people a 'voice' and a 'hand' in redressing climate change. *Children's Geographies, 18*, 191–208. https://doi.org/10.1080/14733285.2019.1614532
Schreiner, C., Henriksen, E. K., & Kirkeby Hansen, P. J. (2005). Climate education: Empowering today's youth to meet tomorrow's challenges. *Studies in Science Education, 41*, 3–49. https://doi.org/10.1080/03057260508560213
SimaPro LCA Software UK. (2023). *Sim-org.* [Internet]. Cited September 19, 2023. Available from: https://www.simapro.co.uk
Sustainable Development. (n.d.). *The 17 goals.* [WWW Document]. https://sdgs.un.org/goals. Accessed July 28, 2023.
Theobald, E. J., Hill, M. J., Tran, E., Agrawal, S., Arroyo, E. N., Behling, S., Chambwe, N., Cintrón, D. L., Cooper, J. D., Dunster, G., Grummer, J. A., Hennessey, K., Hsiao, J., Iranon, N., Jones, L., Jordt, H., Keller, M., Lacey, M. E., Littlefield, C. E., … Freeman, S. (2020). Active learning narrows achievement gaps for underrepresented students in undergraduate science, technology, engineering, and math. *Proceedings of the National Academy of Sciences of the United States of America, 117*, 6476–6483. https://doi.org/10.1073/pnas.1916903117
Trott, C. D. (2020). Children's constructive climate change engagement: Empowering awareness, agency, and action. *Environmental Education Research, 26*, 532–554. https://doi.org/10.1080/13504622.2019.1675594
United Nations. (2023). *The 17 goals.* [Online]. Available at: https://sdgs.un.org/goals. Accessed June 18, 2023.
United Nations. (n.d.). *Education is key to addressing climate change.* [WWW Document]. https://www.un.org/en/climatechange/climate-solutions/education-key-addressing-climate-change. Accessed July 28, 2023.
Wachholz, S., Artz, N., & Chene, D. (2014). Warming to the idea: University students' knowledge and attitudes about climate change. *International Journal of Sustainability in Higher Education, 15*, 128–141. https://doi.org/10.1108/IJSHE-03-2012-0025
Walid, M., & Luetz, J. M. (2018). From education for sustainable development to education for environmental sustainability: Reconnecting the disconnected SDGs. In W. Leal Filho (Ed.), *Handbook of sustainability science and research, world sustainability series* (pp. 803–826). Springer International Publishing. https://doi.org/10.1007/978-3-319-63007-6_49
Woolf, B., Burleson, W., Arroyo, I., Dragon, T., Cooper, D., & Picard, R. (2009). Affect-aware tutors: Recognising and responding to student affect. *IJLT, 4*, 129. https://doi.org/10.1504/IJLT.2009.028804
Yli-Panula, E., Jeronen, E., Koskinen, S., & Mäki, S. (2022). Finnish University students' views on climate change education and their own ability to act as climate educators. *Education Sciences, 12*, 169. https://doi.org/10.3390/educsci12030169

Spatial–Temporal Trend Analysis of Seasonal and Annual Rainfall (1997–2022) Using Nonparametric Analysis with Climatic Zone Changes in Sri Lanka

V. P. A. Weerasinghe and W. S. B. Wickramasingha

1 Introduction

Rainfall is a primary sustenance for ecosystems and plays a pivotal role not only in supporting life forms but also in influencing tropical human and natural systems, which operate dynamically (Pham et al., 2021). Variations in its frequency, quantity, and intensity can have significant repercussions (Baig et al., 2022). Moreover, rainfall stands as a crucial variable in identifying and documenting climate patterns and, subsequently, in facilitating predictions and analyses pertaining to climate change in the future as well as preparedness for natural disasters (Dong et al., 2021; Hackenbruch et al., 2017; Kirsta & Loucka, 2020; Trenberth, 2011).

Tropical regions are particularly susceptible to the impacts of climate change and variability due to the intricate interactions among land, oceans, and the atmosphere. Key climate indicators, such as precipitation and temperature, are frequently utilized to assess global climate alterations. As reported by the Intergovernmental Panel on Climate Change (IPCC) in its fifth assessment report in 2014, global temperatures have risen within the range of 0.65–1.06 °C during the period from 1880 to 2012 (Adopted, 2014). Across Southeast Asia, temperatures have exhibited an upward trend of 0.14–0.20 per °C decade since the 1960s. This trend has been accompanied by a growing number of hot days and warm nights, alongside a decrease in cooler weather occurrences. Importantly, the escalating sea surface temperatures attributed to climate change have brought about significant alterations in precipitation patterns and extremes (Caesar et al., 2011; Manton et al., 2001; Tangang et al., 2007). Furthermore, it is anticipated that future shifts in rainfall will continue to intensify, leading to more frequent occurrences of extreme rainfall events. Excessive

V. P. A. Weerasinghe (✉) · W. S. B. Wickramasingha
Department of Zoology and Environmental Management, Faculty of Science, University of Kelaniya, Colombo, Sri Lanka
e-mail: primali@kln.ac.lk

C. Malalgoda et al. (eds.), *Climate Change Adaptation in the Built Environment*,
https://doi.org/10.1007/978-3-031-75826-3_16

or insufficient amounts of rainfall within specific regions and timeframes can lead to significant harm to the natural environment, economy, and society (Hunt & Menon, 2020; Nillama, 2021; Rahman & Islam, 2019; Zhang & Zhou, 2020). Events such as floods and droughts are occurring with increasing frequency, often influenced by regional precipitation patterns (Hoeppe, 2016; Mujumdar et al., 2020). The planet's rising temperatures are altering global climate conditions, resulting in heightened rainfall levels in certain areas. Notably, North and Central Asia, the eastern regions of North and South America, and northern Europe are experiencing an increase in rainfall (Baig et al., 2022; Gajbhiye et al., 2016; Hamilton et al., 2001). In North America, some regions are witnessing pronounced fluctuations in rainfall, leading to extreme events like floods and droughts (Brimelow et al., 2017; Selin & VanDeveer, 2011). Furthermore, some research suggests a growing prevalence of droughts due to shifting rainfall trends across the globe. For instance, northern and southern Africa, the Mediterranean, and various other regions are receiving decreased rainfall (Gajbhiye et al., 2016; Hamilton et al., 2001).

In recent decades, the scientific community has been increasingly concerned about analyzing the patterns of rainfall changes on various spatial and temporal scales, mainly due to the heightened awareness of global climate change (IPCC, 1996). Research in this area has revealed a slight overall increase in global rainfall trends, although substantial regions are experiencing decreases instead (Kumar et al., 2017). While many studies have examined changes in rainfall patterns at both global and regional levels, more recent research has suggested that it is essential to conduct analyses of hydro-climatic variables at the local level rather than on a larger or global scale (Barua et al., 2013). This shift in focus is driven by the recognition that trends and their impacts can vary significantly from one location to another. Moreover, the scientific research article highlights that Southeast Asia is regarded as highly susceptible to the impacts of climate change, as indicated by the Multiple Climate Hazard Index (Yusuf & Francisco, 2009). It is noted that the same scenario has shown a notable shift in the rainfall patterns in Sri Lanka in recent decades. This shift is attributed to the geographical location of Sri Lanka, where the tropical depressions that are specially generated from the Bay of Bengal, El Niño Southern Oscillation (ENSO), and Indian Ocean Dipole (IOD) systems play a crucial role in governing the seasonal movement of rainfall across different regions of the country throughout the year (Naveendrakumar et al., 2018). The mean annual cycle of rainfall in Sri Lanka is bimodal with a major mode from October to December and a subsidiary mode from April to June (Zubair et al., 2008). Rainfall in Sri Lanka is primarily influenced by the seasonal monsoon system and the corresponding air masses associated with the larger atmospheric wind patterns in South Asia (Burt & Weerasinghe, 2014; Perera et al., 2002). Consequently, Sri Lanka's climate can be described as a tropical monsoon climate, with the country being affected by two distinct monsoon seasons and the meteorological phenomena within the intertropical convergence zone (Karunathilaka et al., 2017; Wickramagamage, 2016). The western, central, and southern regions of Sri Lanka experience wet and humid conditions throughout the year. The two monsoon seasons are as follows: the South-West Monsoon (SWM), also known as the summer monsoon, occurs from May to September and brings heavy rainfall to the low-lying areas in the southwestern part of Sri Lanka as well as the western slopes of the hill country (Jayawardene et al., 2005). Conversely, the North-East Monsoon (NEM), or winter monsoon, occurs from December to February and predominantly affects the northern and eastern parts of the country with its rainfall (Herath & Ratnayake, 2004). In addition to these monsoons, there are two Inter Monsoon (IM) seasons: the First Inter Monsoon season (FIM) from March to April and the Second Inter Monsoon season (SIM) from October to November. During these periods, weather systems such as convergence, convection, and depressions can lead to substantial rainfall (Alahacoon & Edirisinghe, 2021). Furthermore, Sri Lanka is susceptible to tropical cyclones, depressions, and thunderstorms, often associated with the movement of the intertropical convergence zone (ITCZ) during the inter-monsoon seasons (Abeysingha, 2022; Wickramagamage, 2010; Zubair, 2003). Given its tropical location, Sri Lanka experiences relatively stable air temperatures, with the primary climatic variations being associated with rainfall patterns (Zubair, 2003). Sri Lanka can be categorized into four primary climatic regions, namely wet zone, intermediate zone, dry zone, and arid zone (Fig. 1). The wet zone, encompassing the southwestern, central, and eastern provinces, receives over 2500 mm of annual rainfall. Notable cities like Colombo, Kandy, and Nuwara Eliya are situated here. The wet zone experiences heavy rainfall year-round, primarily driven by the South-West monsoon, which is associated with the cultivation season called 'Yala', and North-East monsoon which is associated cultivation season called 'Maha'. It is characterized by lush rainforests, extensive tea plantations, and a wealth of biodiversity, making it ideal for rice cultivation. In contrast, the dry zone (Fig. 1) comprises areas in the northern, north-central, and eastern provinces, receiving annual rainfall from 1250 to 1750 mm. Cities like Anuradhapura and Polonnaruwa are located here. The dry zone experiences a distinct dry season from May to September, followed by a wetter period during the North-East monsoon from November to February. This region grapples with recurring droughts and water scarcity, relying heavily on ancient irrigation systems such as tanks and reservoirs to support agriculture. The intermediate zone (Fig. 1) acts as a transitional area, falling between the wet and dry zones, with annual rainfall ranging from 1750 to 2500 mm. Parts of the north-central and uva provinces belong to this zone. The intermediate zone experiences a balance between the two monsoon seasons, supporting moderate agricultural practices and hosting a mix of semi-evergreen forests and savannas. Lastly, Sri Lanka's southeastern part, including the Hambantota and Monaragala districts, constitutes the arid zone (Fig. 1), receiving less than 1250 mm of annual rainfall. This zone endures prolonged dry seasons, limited rainfall throughout the year, and severe water scarcity. Thorny shrublands and dry grasslands dominate the landscape, presenting significant challenges for both agriculture and human settlement. (Abeysekera et al., 2021; Alahacoon & Edirisinghe, 2021; Baig et al., 2022; Burt & Weerasinghe, 2014; Malmgren et al., 2003; Nisansala et al., 2019; (Herath & Ratnayake, 2004); Somasundaram et al., 2020; Nakagawa et al., 2017).

Rainfall trend analysis shows paramount significance for multiple compelling reasons. Firstly, it operates as a pivotal indicator of climate change. By delving into extensive datasets of historical rainfall, researchers have the capacity to discern

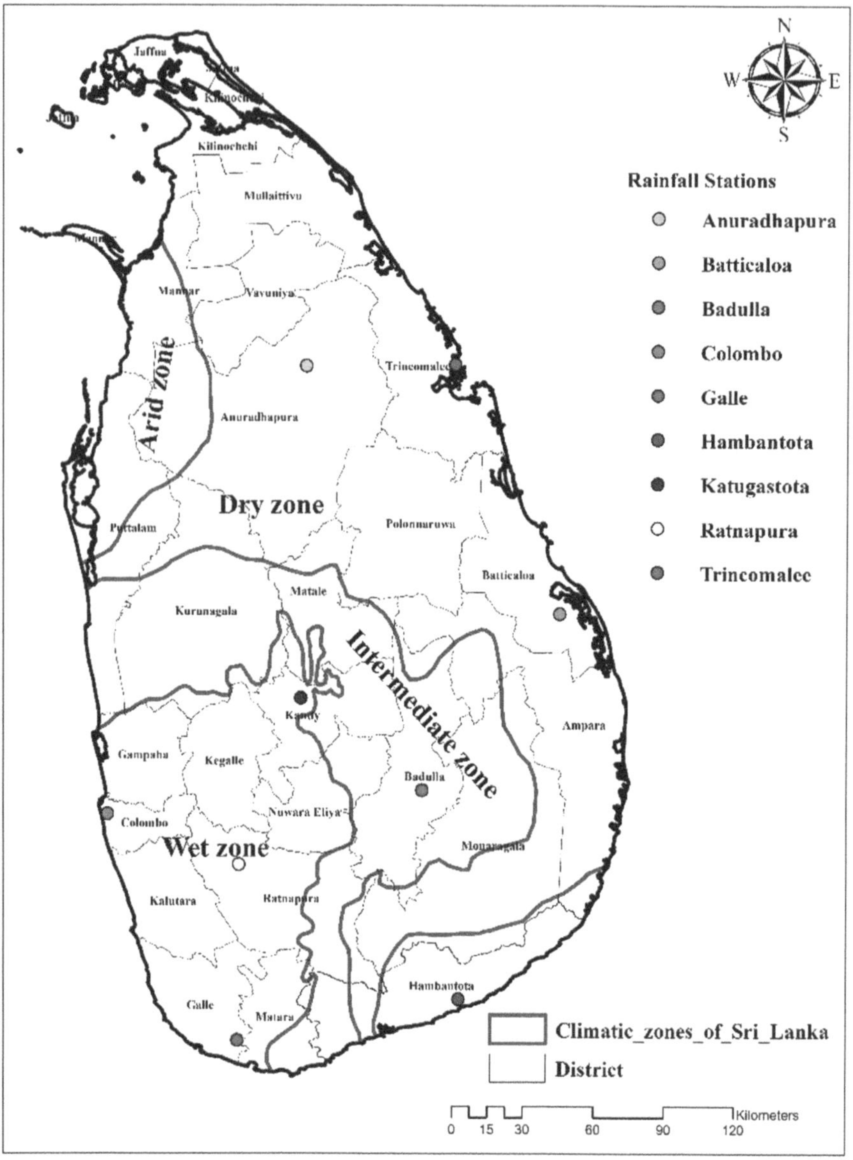

Fig. 1 Districts, climatic zones, and rain gauge stations of Sri Lanka. (*Source* Harischandra et al., 2016; Alahacoon & Edirisinghe, 2021)

discernible patterns such as the amplification or diminishment of precipitation, alterations in seasonal rhythms, and shifts in extreme meteorological occurrences. These profound insights play a pivotal role in the comprehensive assessment of the broader implications of global climate change, thus serving as a cornerstone for the formulation of strategies aimed at both adapting to and mitigating the impacts of climatic shifts (Alahacoon & Edirisinghe, 2021; Kumar et al., 2017; Somasundaram et al., 2020).

Furthermore, the analysis of rainfall trends stands as the linchpin of judicious water resource management. Access to a robust repository of reliable and comprehensive rainfall data holds indispensable value in the meticulous administration of water reservoirs, aquifers, and river systems. This data facilitates the equitable allocation of water resources across diverse sectors including agriculture, industry, and domestic usage. In addition, it proves invaluable in the predictive capacity for impending droughts and floods, enabling timely and efficacious interventions to curtail their far-reaching ramifications. In the realm of agriculture, rainfall trend analysis provides invaluable support for crop scheduling and irrigation oversight. Armed with this data, farmers are empowered to make judicious decisions regarding the optimal timing of planting and harvesting activities, thus optimizing yields and mitigating the risks of crop failures stemming from unpredictable rainfall patterns. This, in turn, enhances food security by underpinning the foundation of a stable and reliable food supply (Aditya et al., 2021; Brema & Anie, 2018; Isioma, 2018).

The prediction and management of natural disasters are profoundly contingent on historical rainfall data. Many catastrophic weather-related events, including floods, landslides, and other calamities, find their origins in precipitation anomalies. A comprehensive comprehension of rainfall trends equips meteorologists and emergency responders with the essential tools to devise early warning systems and emergency evacuation plans, thereby safeguarding lives and property. The domain of resilient infrastructure design is yet another pivotal application of rainfall trend analysis. Engineers and urban planners heavily rely on the insights gleaned from rainfall data to design and construct infrastructure capable of withstanding the rigours of extreme weather events. This encompasses the construction of flood-resistant edifices, the development of efficient stormwater drainage systems, and the formulation of urban blueprints that account for the heightened risk of heavy rainfall in specific geographical areas (Agarwal et al., 2021; Ali et al., 2019).

Safeguarding the health and equilibrium of ecosystems assumes equal importance. Ecosystems rely inexorably on consistent rainfall patterns to preserve biodiversity and ecological stability. The scrutiny of rainfall trends empowers researchers and conservationists to pinpoint regions vulnerable to shifts in precipitation and to devise conservation strategies aimed at shielding fragile species and habitats from adverse impacts. Beyond these immediate applications, rainfall trend analysis exerts a far-reaching influence in domains such as public health, insurance, and risk assessment, an international collaboration concerning climate-related challenges, and the mitigation of crises stemming from droughts and floods (Basistha et al., 2009; Gadgil & Dhorde, 2005; Partal & Kahya, 2006).

Therefore, this study aims to model and visualize the spatiotemporal rainfall current trend using ground-based rain gauging station data in Sri Lanka from 1997 to 2022.

2 Methodology

2.1 Study Area

Sri Lanka, situated at the southern end of the Indian subcontinent, spans from approximately 5°55′–9°51′ N latitude and 79°42′–81°53′E longitude, covering a total land area of 65,610 km^2 (Burt & Weerasinghe, 2014a). The country's geography includes a maximum North–South length of 435 km and an East–West width of 240 km (Pushpakumara, et al., 2014). Daily rainfall data were collected from the Department of Meteorology in Sri Lanka from nine rainfall gauging stations for 26 years, from 1997 to 2022. The rainfall gauging stations were Anuradhapura, Badulla, Batticaloa, Colombo, Galle, Hambantota, Katugastota, Ratnapura, and Trincomalee (Fig. 1).

Figure 1 shows the major climatic zones, namely wet zone, intermediate zone, dry zone, and arid zone, 25 districts, and rainfall stations.

2.2 Seasonal Rainfall Analysis and Climatic Zones Classification

The time series rainfall graph from 1997 to 2022 was plotted using average annual rainfall values. First Inter-Monsoon (FIM) from March to April, South-West Monsoon (SWM) from May to September, Second Inter-Monsoon (SIM) from October to November, and North-East Monsoon (NEM) from December to February were analyzed to identify temporal variations (Jayawardene et al. 2005; Karunathilaka et al. 2017; Wickramagamage 2016) in Sri Lanka. Seasonal rainfall maps were generated by interpolating point average rainfall values using the Inverse Distance Weighting (IDW) method for the specified time periods of rainfall seasons from 1997 to 2022. The average Sri Lankan rainfall map was derived by interpolating average point data using the IDW method. The climate zones map for Sri Lanka was derived using the following threshold values: the wet zone is defined by annual rainfall exceeding 2500 mm, the intermediate zone by annual rainfall ranging from 1750 mm to 2500 mm, and the dry zone by annual rainfall ranging from 1250 mm to 1750 mm and the arid zone by annual rainfall ranging below 1250 mm (Abeysekera et al. 2021; Alahacoon & Edirisinghe 2021; Baig et al. 2022; Burt & Weerasinghe 2014; Malmgren et al. 2003) for the average rainfall map of Sri Lanka. Additionally, the land areas of each climatic zone have evolved over the specified 26-year period and the land area change was analyzed at five-year intervals (1997, 2002, 2007, 2012,

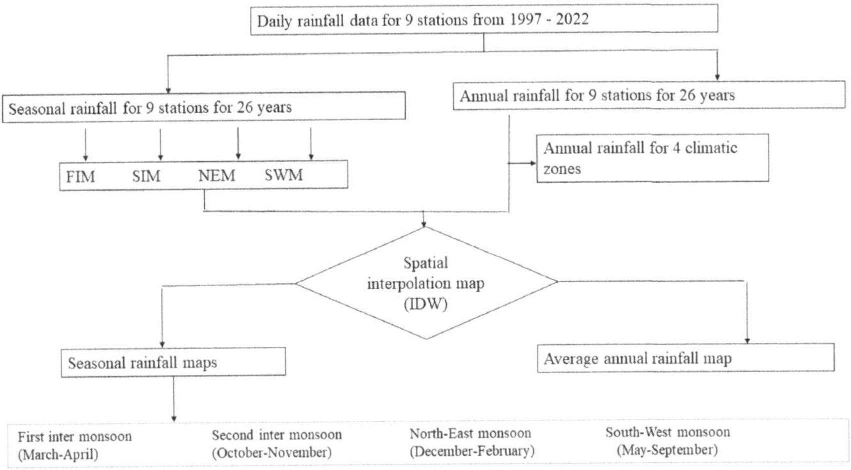

Fig. 2 Descriptive statistics generated using daily rainfall data for spatial and temporal variations from 1997 to 2022

2017 and 2022) in average rainfall maps of those specific six years were subjected to analyse the change detection using ArcGIS 10.8 version software.

2.3 Rainfall Trend Analysis using Mann-Kendall Method

Rainfall was assessed for the direction (whether increasing, denoted as positive, or decreasing, indicated as negative) and magnitude of rainfall trends within seasonal and climatic zones, employing Mann-Kendall's test for direction and Sen's slope estimator for magnitude calculations (Partal & Kahya, 2006). The results of these trend analyses were visualized through spatial maps, which highlighted areas experiencing significant increases or decreases in rainfall patterns. The Mann–Kendall test is a widely used statistical method in scientific research articles for detecting trends in various types of time series data, such as meteorological, hydrological, and agro-meteorological data (Shishutosh et al., 2013 ; Partal & Kahya, 2006). It is a non-parametric test that can determine the significance of a detected trend at different significance levels, typically set at 5%. One of the key advantages of the Mann–Kendall test is its ability to handle data with skewed distributions and missing values, making it robust in various situations (Brown et al., 1992; Chaouche et al., 2010).

Researchers often rely on the Mann–Kendall test, especially in studies related to rainfall trends, as it has a track record of producing reliable results. Its primary objective is to ascertain whether a random variable generally exhibits an increasing or decreasing trend over a specific period, without requiring the assumption of a normal distribution (Basistha et al., 2009; Mishra et al., 2009; Partal & Kahya,

2006). Mann–Kendall test can detect non-linear trends, and it is less sensitive to the presence of outliers because it relies on the sign (sgn) of differences between data points rather than their actual values (Gadgil & Dhorde, 2005; Partal & Kahya, 2006). Mann–Kendall's statistics "S" can be calculated by Eqs. (1) and (2).

$$S = \sum_{i=1}^{n-1} \sum_{j=i+1}^{n} sgn(x_j - x_i) \tag{1}$$

$$sgn(x_j - x_i) = \begin{cases} +1, > (x_j - x_i), \\ 0, = (x_j - x_i), \\ -1, < (x_j - x_i), \end{cases} \tag{2}$$

where x_j and x_i are sequential data values and n is the number of time-series data. The positive and negative values of S represent a "high tendency" and "low tendency", respectively.

If the number of records used for the test is greater than 10 ($N > 10$), then the average distribution statistic is approximately equal to zero and the variance of S is calculated using Eq. (3). Usually, more than ten records are used to analyze the trend of the time series.

$$Var(s) = \frac{n(n-)(2n+5) - \sum_{i=1}^{m} t_i(i)(i-1)(2i+5)}{18} \tag{3}$$

where t_i is the number of bonds specified in sample i. Then, Mann–Kendall's statistics Zc (standard average deviation) for $N > 10$ is calculated using Eq. (4).

$$Zc = \begin{cases} \frac{S-1}{\sqrt{Var(s)}}, S > 0, \\ 0, S = 0, \\ \frac{S+1}{\sqrt{Var(s)}}, S < 0, \end{cases} \tag{4}$$

Here Zc follows the standard normal distribution. *A* positive Zc value shows an increasing trend, while the negative Zc data gives a decreasing trend for the period.

Sen's Slope Estimation

In the context of scientific research articles, the Mann–Kendall test can be complemented by Sen's slope estimator to determine the magnitude of the trend. This estimator, proposed by Sen in 1968, is particularly useful for capturing the varying slope or trend in time series data, rather than providing a single straight-line trend value. Researchers frequently apply Sen's slope estimator when they need to quantify the extent of trends in rainfall patterns over time. It involves calculating slope pairs for all the data points using a specific Eq. (5).

$$T_i = \frac{x_j - x_k}{j - k} f \text{ or } i = 1, 2, 3, \ldots\ldots. n, j > k, \tag{5}$$

where T_i is the slope and x_j and x_k are the data values at time j and k, respectively.

The mean of the n values of T_i is encoded as Sen's slope estimator (Q_i) and is calculated using Eq. (6);

$$Q_i = \begin{cases} \left(T_{(n+1/2)}\right), n \text{ is odd}, \\ , \\ \frac{1}{2}\left(T_{(\frac{n}{2})} + T_{\frac{(n+2}{2})}\right), n \text{ is even} \end{cases} \quad (6)$$

3 Results and Discussion

3.1 Spatial and Temporal Rainfall Analysis in Sri Lanka (1997–2022)

Figure 3 illustrates Sri Lanka's average annual rainfall and changes in climatic zones. It becomes apparent that the highest levels were recorded in 2010, 2011, 2014, and 2021 coinciding with significant flood events in those years. Specifically, the 2010 and 2014 floods occurred in the wet zone, while the 2011 flood event took place in the dry zone. Surprisingly, despite the heavy rainfall accompanying the catastrophic floods in 2016 and 2017, Sri Lanka experienced an overall decline in rainfall during those years. This phenomenon can be attributed to reduced rainfall in the dry zone throughout the 2016–2018 period. A noteworthy observation is the decline in total rainfall across Sri Lanka during the period spanning from 2015 to 2020. However, it's worth highlighting that in 2017 and 2019, the wet zone experienced the most substantial rainfall ever recorded in the past 26 years. Conversely, when examining the dry zone, there was a notable decrease in rainfall during the years 2015–2020, leading to drought and arid conditions during the 'Maha' season from 2016 to 2020. Thus, it is observed that the results of the study conducted by Alahacoon and Edirisinghe (2021a) using Climate Hazards Group InfraRed Precipitation corrected with Stations (CHIRPS) remote sensing data are aligned with the results of the current study conducted using ground-based rain gauge station data.

Figure 4 and Table 1 present key descriptive statistics, including the mean, maximum, minimum, and standard deviation of annual rainfall at the district level. Notably, during the period from 1997 to 2022, Ratnapura and Kalutara districts recorded the highest average annual rainfall, with values of 2891.78 mm and 2868.70 mm, respectively, while the lowest average annual rainfall was documented in Hambantota district at 1565.60 mm. Figure 5 illustrates the spatial distribution of annual and seasonal rainfall in Sri Lanka over the period from 1997 to 2022. The data reveals a broad range of annual rainfall values, from a minimum of 1044.4 mm/year to a maximum of 3745.3 mm/year. In terms of seasonal rainfall, the First Inter-Monsoon (FIM) shows a range from 117.2 mm/year to 591.6 mm/year, the Second

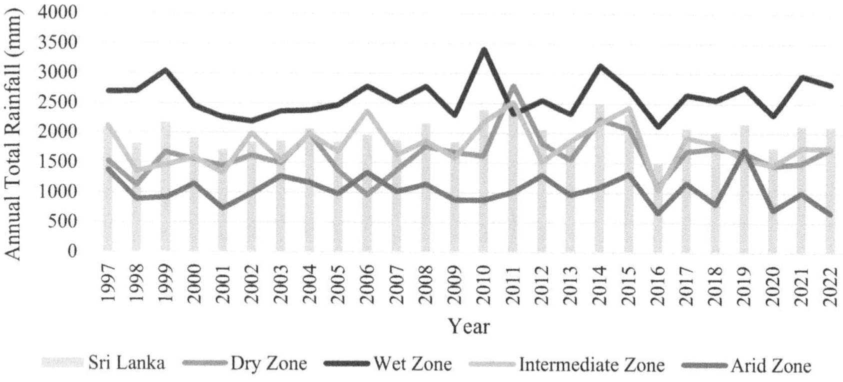

Fig. 3 Annual total rainfall distribution in wet, dry, intermediate, arid zones, and annual average rainfall in Sri Lanka from 1997 to 2022

Inter-Monsoon (SIM) spans from 351.8 mm/year to 841.6 mm/year, and the South-West Monsoon (SWM) exhibits a range from 237.6 mm/year to 1824.0 mm/year. Additionally, the North-East Monsoon (NEM) ranges from 242.5 mm/year to 808.5 mm/year. Hence, there is a noticeable variations in spatial and temporal patterns of the rainfall (Fig 5).

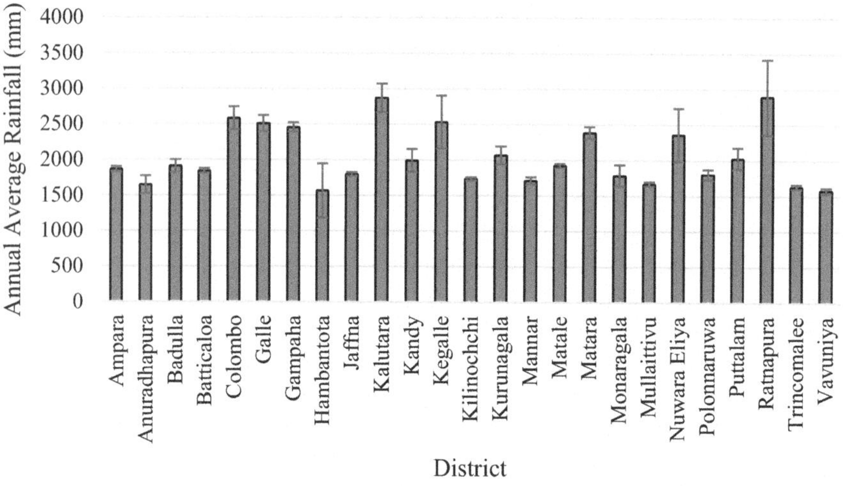

Fig. 4 Descriptive statistics for district annual average rainfall (mm) from 1997 to 2022

Table 1 Descriptive statistics for district annual rainfall (mm) from 1997 to 2022

District	Annual rainfall (mm)	Max	Min	Std.dev
Ampara	1,862.7	1,925.7	1,687.2	38.5
Anuradhapura	1,644.5	1,946.9	1,445.1	126.2
Badulla	1,907.4	2,308.3	1,787.9	89.5
Batticaloa	1,840.6	1,871.2	1,709.3	30.9
Colombo	2,578.8	3,083.6	2,415.2	1,60.0
Galle	2,507.0	2,881.3	2,376.1	111.9
Gampaha	2,453.6	2,704.4	2,308.4	65.4
Hambantota	1,565.6	2,357.0	1,044.4	380.3
Jaffna	1,797.9	1,865.9	1,715.5	24.0
Kalutara	2,868.7	3,431.8	2,460.4	205.3
Kandy	1,992.2	2,697.7	1,864.3	160.7
Kegalle	2,531.7	3,423.7	1,976.6	369.4
Kilinochchi	1,737.5	1,790.9	1,698.8	19.6
Kurunagala	2,066.9	2,374.5	1,826.0	126.0
Mannar	1,710.3	1,848.5	1,582.9	48.6
Matale	1,922.4	1,967.6	1,813.4	29.1
Matara	2,384.7	2,648.8	2,161.8	83.7
Monaragala	1,785.6	2,181.1	1,188.2	148.8
Mullaittivu	1,668.7	1,741.5	1,590.8	28.8
Nuwara Eliya	2,358.6	3,420.6	1,864.9	373.9
Polonnaruwa	1,803.0	1,935.0	1,636.6	71.6
Puttalam	2,027.7	2,386.1	1,758.2	149.5
Ratnapura	2,891.7	3,745.3	1,611.2	529.0
Trincomalee	1,633.1	1,744.6	1,564.4	31.3
Vavuniya	1,578.5	1,659.2	1,478.5	39.3

3.2 *Annual Rainfall Trend in Sri Lanka*

The analysis encompassed the computation of rainfall trends at all measurement stations for both annual and seasonal rainfall data. First, the Mann–Kendall test was applied to identify trends in annual rainfall values. Subsequently, the cumulative seasonal rainfall for the four distinct seasons and climate zone levels was calculated. Sens's slope and Kendall's tau values were computed at a significance level of 0.05 using the Mann–Kendall trend test for annual rainfall data spanning from 1997 to 2022 across 25 districts in Sri Lanka.

One of the scientific research articles examines rainfall trends in Sri Lanka over the century leading up to 1998. The study relies on monthly precipitation data

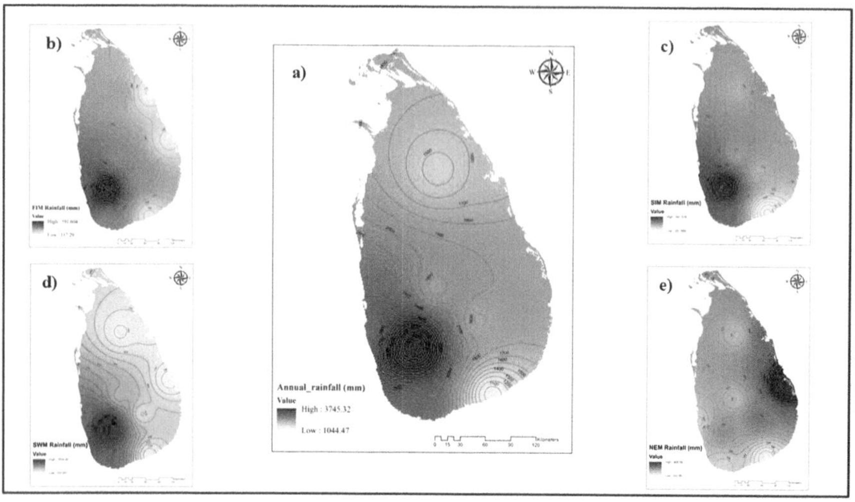

Fig. 5 Rainfall maps for **a** annual rainfall, **b** FIM, **c** SIM, **d** SWM, and **e** NEM of Sri Lanka from 1997 to 2022

collected from 15 measurement stations (Jayawardene et al., 2005). It employs statistical methods, including the Mann–Kendall statistic, Spearman statistic, and least-square regression, to identify these trends. The data were segmented into two time periods: a long-range analysis spanning 98–130 years and a shorter-range investigation covering 36–50 years. The findings indicate a significant long-term increase in rainfall in Colombo, with a 95% confidence level, at a rate of 3.15 mm per year. Conversely, long-term declines in rainfall are observed at two locations: Nuwara Eliya, experiencing a reduction of 4.87 mm per year, and Kandy, with a decrease of 2.88 mm per year. However, recent data records starting from 1949 onwards reveal a decreasing trend at 13 out of the 15 monitoring stations. Notably, in these recent records, the declines in rainfall over recent decades are more pronounced than the long-term variations. Among the recent data, the most substantial downward trend of 11.16 mm per year is observed at Batticaloa.

Another scientific research article has examined the rainfall patterns in Sri Lanka over the past 130 years, starting from 1870, and their correlation with the El Niño-Southern Oscillation (ENSO). The study specifically focuses on analyzing precipitation variations in four different seasons at 15 distinct locations within Sri Lanka. The research provides a comprehensive overview of how rainfall has changed in these 15 locations throughout various seasons (Malmgren et al., 2003).

Figure 6 illustrates the variations in Sen's slope and Kendall's tau values for each district. It's important to note that a two-tailed Mann–Kendall trend test was employed with a 5% significance level for Kendall's tau and 95% confidence for Sen's slope. The outcomes of this trend analysis on annual rainfall reveal a consistent pattern of increasing rainfall in all districts. Notably, except for eight districts, all others exhibit a significant trend towards increased rainfall. The districts that do not exhibit

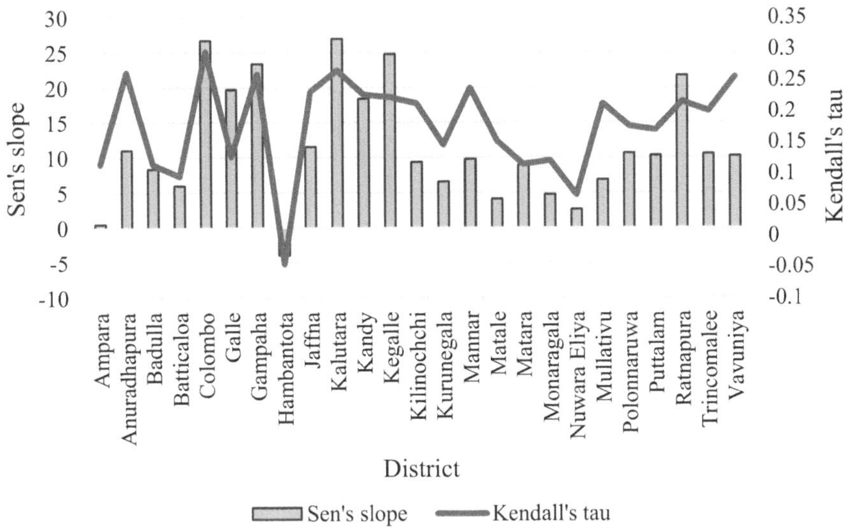

Fig. 6 Kendall's tau value and Sen's slope derived for each district for annual rainfall from 1997 to 2022

a significant trend in rainfall are Ampara, Kalutara, Kegalle, Kilinochchi, Matale, Matara, Nuwara Eliya, and Puttalam (Appendix A: Table A1).

It is noteworthy to mention that over the past 26 years, Hambantota district has exhibited a significant negative trend in rainfall, while all other districts in Sri Lanka have experienced a notably positive rainfall trend. Among the districts with positive trends, Kalutara district stands out with the highest increase at 26.94 mm/year, followed by Colombo (26.69 mm/year), Kegalle (24.77 mm/year), Gampaha (23.37 mm/year), Ratnapura (21.70 mm/year), and Galle (19.72 mm/year). Conversely, the districts with the lowest positive rainfall trends are Monaragala (4.73 mm/year), Batticaloa (6.03 mm/year), and Kurunegala (6.578 mm/year). In general, on an annual scale, the eastern region of the country has shown an increasing trend. Map of annual rainfall trends of the district is shown in Fig. 7.

3.3 *Seasonal Rainfall Trend in Sri Lanka*

The two-tailed Mann Kendall trend test was also utilized to detect seasonal rainfall trends, with significance set at 5% and a confidence level of 95% (Appendix A: Table A4). The graphical illustration of the values determined by the Mann Kendall test for the four seasons of rainfall can be seen in Fig. 8. An important point that emerges from the analysis of Mann Kendall values in FIM, SIM, NEM, and SWM is that no station shows a statistically significant trend in rainfall (p-value > 0.05).

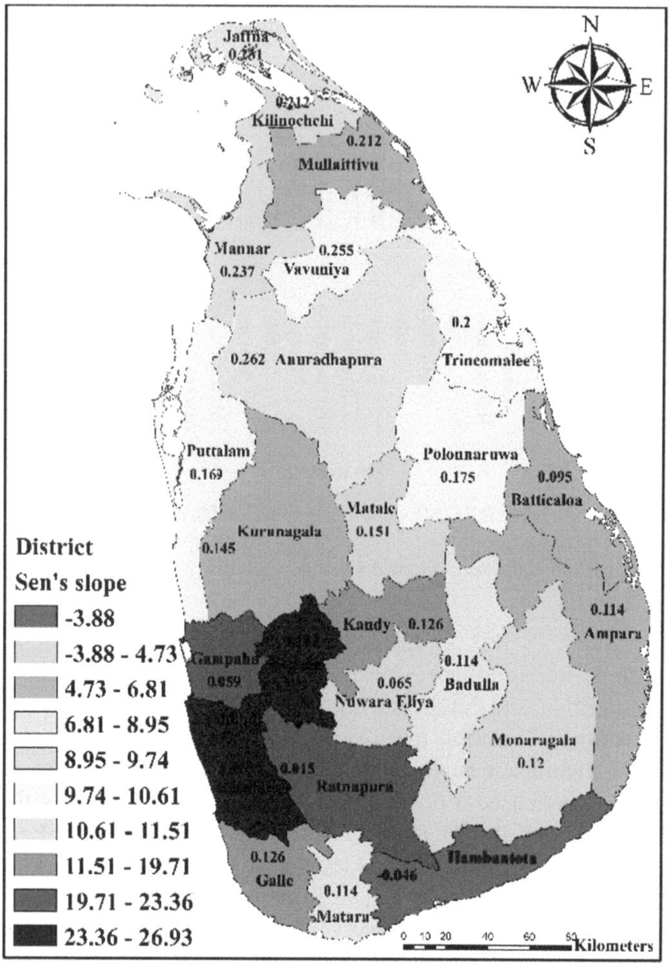

Fig. 7 Representation of Sen's slope (color gradient) and Man Kendall's tau (number on the district map) on changes in annual district rainfall

During the First-Inter Monsoon (FIM) season, the rainfalls at 7 stations show increasing trends while other Batticaloa and Anuradhapura stations show decreasing trends. The highest trend was recorded in FIM from the Ratnapura (6.671 mm/year) station and the lowest from the Anuradhapura (– 1.10 mm/year) station (Appendix A: Table A4). In Fig. 8a, it is evident that there are rising rainfall patterns in the western and south-eastern regions in FIM, whereas the northern regions exhibit declining rainfall trends. During the Second Inter Monsoon (SIM) season, only the station at Hambantota shows a negative trend (– 4.200 mm/year) while others are positive. Figure 8b depicts rainfall patterns for the Second-Inter Monsoon season (October–November). While small areas in the southern and central hill regions of

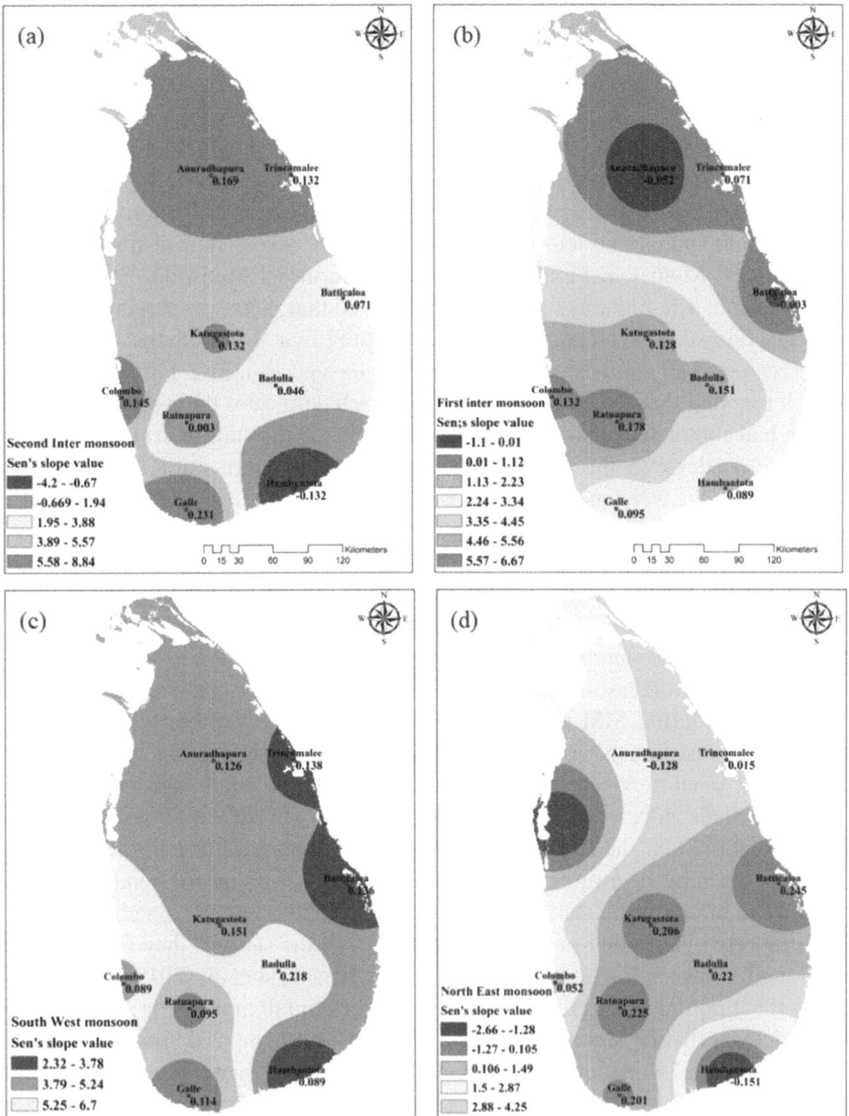

Fig. 8 Representation of Sen's slope (colour gradient) and MannKendall's tau (number on the map) on changes in (**a**) FIM (March–April), (**b**) SIM (October–November), (**c**) SWM (May–September), and (**d**) NEM (December-February) rainfall in nine meteorological stations

the country exhibit a decline in rainfall trends, all other parts of the country have experienced an uptick in rainfall during the analysis period. During this period, the increase in rainfall can primarily be attributed to the presence and movement of low-pressure systems, depressions, and cyclones near or over the country (Alahacoon & Edirisinghe, 2021; Wickramagamage, 2016). Scientists examined changes in Sri Lanka's climate patterns by analyzing daily rainfall data collected from 22 monitoring stations spanning the years 1961–2002. This investigation focused on assessing both spatial and temporal trends in rainfall across four distinct seasons: the Southwest Monsoon, First-inter Monsoon, Northeast Monsoon, and Second-inter Monsoon. The findings of the study indicated that there was a decrease in the number of rainy days at all but one station, Nuwara Eliya. Additionally, annual rainfall decreased at all stations except for the one located in Jaffna. Furthermore, the research reported that the 2000 mm isohyet, which delineates the wet region of the country, had contracted in size. This suggests a noteworthy shift in Sri Lanka's rainfall characteristics over the studied period, particularly in terms of reduced rainfall and changes in the wet zone's boundaries. Similar to the findings in this study, other researchers (Alahacoon & Edirisinghe, 2021; Karunathilaka et al., 2017; Naveendrakumar et al., 2018; Nisansala et al., 2019; Wickramaarachchi et al., 2020) have also noted an increasing trend in FIM (First Inter-Monsoon) at a higher percentage of rainfall monitoring stations. For instance, Nisansala et al. (2019) identified a rising pattern at 76% of the stations during FIM, with 5% of these stations showing statistically significant increases. Additionally, they observed an increasing trend at 51% of the stations during SIM (Second Inter-Monsoon). Furthermore, In the context of Sri Lanka, researchers primarily employ various statistical techniques to identify the directional trends in rainfall. These techniques include Least-squares linear regression, Sen's slope estimator, Mann–Kendall test and its variations, Spearman rank correlation, and the Innovative trend analysis approach. For instance, in the study conducted by Jayawardena et al. (2018) observed an upward trend in rainfall at 80% of the monitoring stations they examined. Conversely, Karunathilaka et al. (2017) reported increasing rainfall trends at 66% of the stations they investigated. Additionally, Naveendrakumar et al. (2018) and Nisansala et al. (2019) both found rising annual rainfall trends in 63% and 65% of the rainfall monitoring stations they analyzed, respectively. Moreover, Alahacoon and Edirisinghe (2021) examined rainfall trends across 25 districts identified a noteworthy pattern and found that annual rainfall had significantly increased in a substantial 84% of these districts, which translates to 21 out of the 25 districts showing statistically significant increases at the 5% significance level. Furthermore, a separate investigation conducted by Amarasinghe (2020) divided the island into four distinct geographical regions and observed a significant annual uptrend rainfall across all four regions.

Nearly all the studies reviewed indicate that Sri Lanka experiences higher rainfall during the inter-monsoon period, encompassing both FIM and SIM seasons.

Interestingly, throughout the SWM, all rain gauging stations show an increasing trend. The highest trend was recorded in SWM from Galle (9.624 mm/year) station and the lowest from the Hambantota (2.315 mm/year) station (Appendix A: Table A4). In Fig. 8c, the variation in rainfall trends during the South-West Monsoon season (May–September) is depicted. An increasing trend in rainfall is observed only in the south-western coastal areas and their adjoining regions, although these trends are notably modest. A decreasing trend in rainfall during the season is seen in the majority of the country. Specifically, a decreasing trend in rainfall is noted in the western, north-eastern, and southern provinces, as well as on the western side of the central hills, when compared to other parts of the country. A study conducted by Samarakoon et al. in 2021 examined the rainfall patterns during the South-West Monsoon in the wet zone of Sri Lanka. They reported that 77% of the monitored weather stations indicated a declining trend in SWM rainfall. However, in a similar study conducted by Alahacoon and Edirisinghe in 2021, using the CHIRPS dataset, they found an increase in SWM rainfall across all districts within the wet zone, and these increases were statistically significant. The RMSE value for the each interpolated map was range from 2.68 to 4.57.

During the NEM season, two of the stations have negative trends. They are the stations at Anuradhapura with − 2.55 mm/yr, and Hambantota with − 2.65 mm/yr (Appendix A: Table A4). The rates of rainfall increase are observed to be high during the North-East Monsoon season. The same method applied to analyze the rainfall patterns across the entire country revealed a noticeable uptrend in rainfall during the North-East Monsoon season. Specifically, Karunathilaka et al. in 2017 observed an 84% increase of rainfall trend, while Naveendrakumar et al. (2018) and Nisansala et al. (2019) reported increases of 80% and 86%, respectively. Additionally, Karunathilaka et al. (2017) and Nisansala et al. (2019) identified significant positive trends in rainfall at 16% and 19% of the monitoring stations in their respective studies. Rainfall trends during the North-East Monsoon season (December-February) are illustrated in Fig. 8d. With the exception of some highly inconsequential areas in the north-west and south, the entire country exhibits an upward trend in rainfall during this season. Furthermore, the spatio-temporal analysis of rainfall trends in this study aligns with the findings of Karunathilaka et al. (2017), Naveendrakumar et al. (2018), and Nisansala et al. (2019). Wickramaarachchi et al. (2020) also highlighted an increase in rainfall at rain gauge stations in the eastern coastal areas during the North-East Monsoon season, with 27% of stations indicating a significant rise. Additionally, there was a substantial 50% increase in NEM rainfall observed in the northwestern coastal areas. Hence, the results of this research indicate a growing trend in North-East Monsoon rainfall for Sri Lanka, with a significant influence observed in the eastern regions of the nation. This suggests that the NEM is likely to make a substantial contribution to the overall increase in annual rainfall.

3.4 *Spatial Variation of Rainfall Trends with the Climatic Zone*

An important finding to highlight is the greater variability in average annual rainfall observed in the wet zone compared to the dry zone. When categorizing Sri Lanka

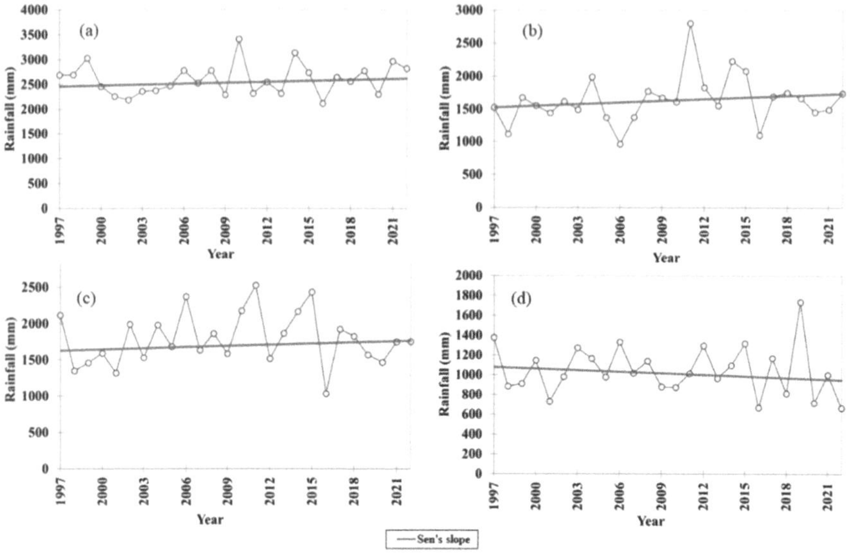

Fig. 9 Rainfall trend for climate zones of Sri Lanka: (**a**) Wet-zone; (**b**) Dry-zone, (**c**) Intermediate-zone, and (**d**) Arid zone

into its four distinct climatic zones, the wet zone exhibits the highest annual average rainfall at 2,600.7 mm, followed by the intermediate zone with 1,787.9 mm, the dry zone with 1,637.9 mm, and finally, the semi-arid zone with 1,044.3 mm. These values are detailed in (Appendix A: Table A2).

Figure 9 illustrates the outcomes of the non-parametric Mann Kendall test and Sen's slope analysis conducted at a significance level of 0.05 for the wet, dry, intermediate, and semi-arid climate zones in Sri Lanka. It's important to note that all of these climatic zones exhibit statistically significant trends except for the arid zone. The wet zone exhibits the highest trend at 25.38 mm/year, followed by the intermediate zone at 17.48 mm/year, and the dry zone at 14.45 mm/year (Appendix A: Table A3). Map of annual rainfall trends of the climatic zones is shown in Fig. 10.

3.5 *Evolution of Sri Lanka's Climatic Zone Land Areas (1997–2022): A 26-Year Analysis*

Table 2 provides a view of how the land areas of the three distinct climatic zones (dry zone, intermediate zone, and wet zone) in Sri Lanka have evolved over a 26-year period from 1997 to 2022. The dry zone experienced the most pronounced changes in land area over the study period. It began with an area of 20,622.37 km^2 in 1997 and underwent rapid expansion in 1998, reaching 45,324.15 km^2. However, it subsequently displayed significant fluctuations, culminating in a peak of 53,195.34 km^2 in

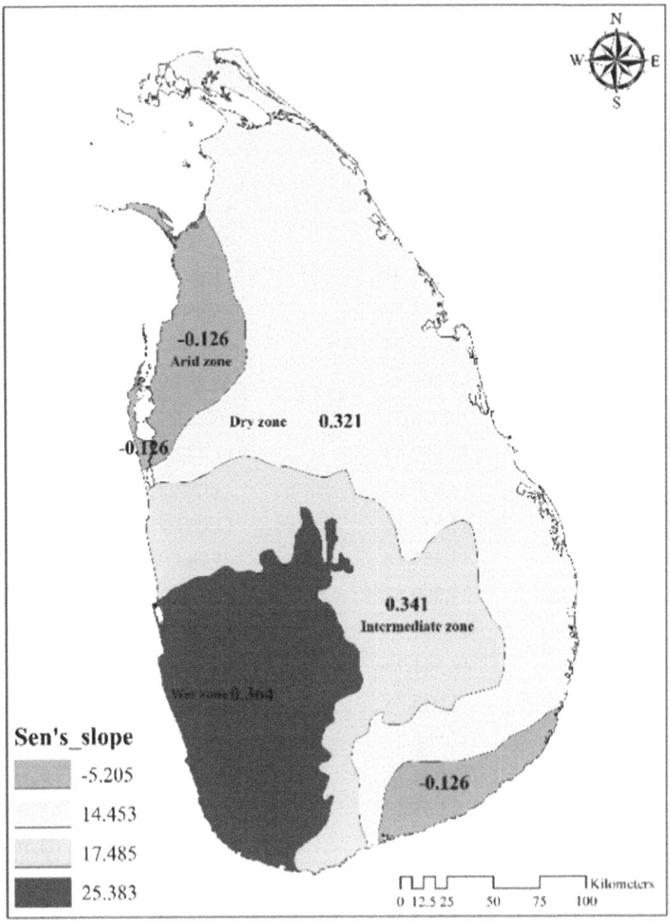

Fig. 10 Representation of Sen's slope (colour gradient) and Kendall's tau (number) on changes in annual rainfall at the each climatic zone

2016. Notably, by 2022, the area had dramatically decreased to 10,470.86 km^2, marking a substantial 79.26% decrease from its peak. The intermediate zone exhibited relative stability in land area during the study period. Beginning with 35,694.33 km^2 in 1997, it experienced steady growth, peaking at 48,558.63 km^2 in 2018. By 2022, it had slightly decreased to 42,294.67 km^2, representing an 18.58% increase over the entire period. The wet zone, initially the smallest in terms of land area at 9,252.50 km^2 in 1997, experienced consistent expansion over the years. By 2022, it covered 12,803.67 km^2, reflecting a substantial 38.33% increase during the study period.

Figure 11 illustrates changes in land area within Sri Lanka's three distinct climatic zones (dry zone, intermediate zone, and wet zone) from 1997 to 2022, with corresponding linear regression models based on precipitation data. In the dry zone, there

Table 2 Land area distribution in Sri Lanka's climatic zones (1997–2022)

Year	Area (km^2)/climatic zone			
	Dry zone	Intermediate zone	Wet zone	Total
1997	20,622.37	35,694.33	9,252.505	65,569.2
1998	45,324.15	12,813.52	7,431.533	65,569.2
1999	20,689.08	32,467.98	12,412.14	65,569.2
2000	30,909.49	28,321.85	6,337.856	65,569.2
2001	43,313.97	18,071.91	4,183.313	65,569.2
2002	43,313.97	19,174.34	3,080.887	65,569.2
2003	39,864.52	20,609.24	5,095.44	65,569.2
2004	12,511.66	46,788.58	6,268.955	65,569.2
2005	39,311.12	20,026.31	6,231.77	65,569.2
2006	28,154.52	25,972.63	11,442.05	65,569.2
2007	39,296.9	18,190.03	8,082.27	65,569.2
2008	9,837.622	44,647.16	11,084.41	65,569.2
2009	2,118.05	33,357.14	3,094.011	65,569.2
2010	8,305.381	35,860.57	21,403.25	65,569.2
2011	2,251.88	45,741.93	17,575.38	65,569.2
2012	6,840.948	48,541.75	10,186.5	65,569.2
2013	25,989.04	34,025.38	5,554.784	65,569.2
2014	1,978.461	37,588.57	26,002.16	65,569.2
2015	1,534.428	51,582.17	1,2452.6	65,569.2
2016	53,195.34	10,033.39	2,340.468	65,569.2
2017	10,544.14	43,550.21	11,474.86	65,569.2
2018	12,413.23	43,858.63	9,297.346	65,569.2
2019	20,743.77	33,653.53	11,171.91	65,569.2
2020	48,205.99	12,257.93	5,105.28	65,569.2
2021	20,348.95	33,144.97	12,075.29	65,569.2
2022	10,470.86	42,294.67	12,803.67	65,569.2

are evident fluctuations and a substantial decline in land area, potentially linked to water scarcity and land degradation, although the linear model captures only a fraction of the variability ($R^2 = 0.1042$). Conversely, the intermediate zone exhibits relative stability in land area with slow growth, suggesting its suitability for agriculture and land use ($R^2 = 0.088$). In the wet zone, a consistent expansion in land area aligns with increased precipitation, offering opportunities for sustainable agriculture and ecological benefits, though the linear model explains a limited portion of the variability ($R^2 = 0.069$). These findings underscore the complexity of land-use

Table 3 Mann–Kendall (M–K) test results for annual rainfall at the district level

District	Kendall's tau	*P* value	Sen's slope	District	Kendall's tau	*P* value	Sen's slope
Ampara	0.114	0.428	5.589	Kilinochchi	0.212	0.134	9.351
Anuradhapura	0.262	0.044	11.082	Kurunegala	0.145	0.001	6.578
Badulla	0.114	0.042	8.364	Mannar	0.237	0.024	9.749
Batticaloa	0.095	0.009	6.031	Matale	0.151	0.290	4.1
Colombo	0.295	0.010	26.69	Matara	0.114	0.428	8.955
Galle	0.126	0.008	19.715	Monaragala	0.12	0.002	4.733
Gampaha	0.259	0.012	23.367	Nuwara Eliya	0.065	0.659	2.654
Hambantota	– 0.046	0.048	– 3.88	Mullativu	0.212	0.014	6.811
Jaffna	0.231	0.103	11.515	Polonnaruwa	0.175	0.017	10.613
Kalutara	0.265	0.059	26.939	Puttalam	0.169	0.234	10.27
Kandy	0.226	0.008	18.418	Ratnapura	0.215	0.030	21.702
Kegalle	0.222	0.041	24.774	Trincomalee	0.2	0.015	10.547
				Vavuniya	0.255	0.041	10.183

** As the computed *p*-value is greater than the significance level alpha = 0.05, there is no trend in the series

dynamics influenced by climate and emphasize the need for region-specific strategies to address environmental challenges and harness opportunities for sustainable development in each climatic zone.

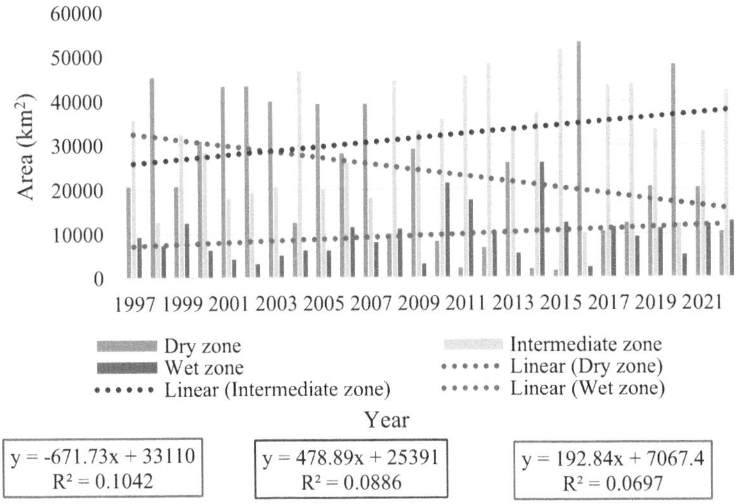

Fig. 11 Temporal analysis of land area changes of each climatic zone in Sri Lanka (1997–2022)

Rainfall pattern fluctuation reinforces the crucial relationship between climate and land use in Sri Lanka. As rainfall patterns fluctuate, the country's climatic zones respond in distinct ways, resulting in varied land area changes. Understanding these dynamics is pivotal for informed land management, agricultural planning, and environmental conservation, particularly in an era of climate change, where sustainable land use practices are imperative for the nation's future well-being. Figure 12 shows the land area changes within Sri Lanka's diverse climatic zones over five-year intervals, spanning from 1997 to 2022, all in the context of shifting rainfall patterns. It reveals valuable insights into how these zones have adapted and evolved in response to the fluctuations in rainfall.

The dry zone exhibited a compelling narrative of expansion during the first two intervals (1997–2002 and 2002–2007), with substantial gains of 20,597.62 km^2 and 39,298.90 km^2, respectively (Appendix A: Table A5). However, this expansion slowed in the subsequent periods (2007–2012, 2012–2017, and 2017–2022), with significantly reduced gains. This trend underscores the sensitivity of the dry zone to rainfall variations and its capacity to adapt gradually, with the initial rapid expansion possibly attributed to favorable precipitation conditions during those years (Fig. 13).

Conversely, the intermediate zone witnessed a more dynamic pattern of land area change. It expanded steadily into the dry zone between 1997 and 2002, gaining

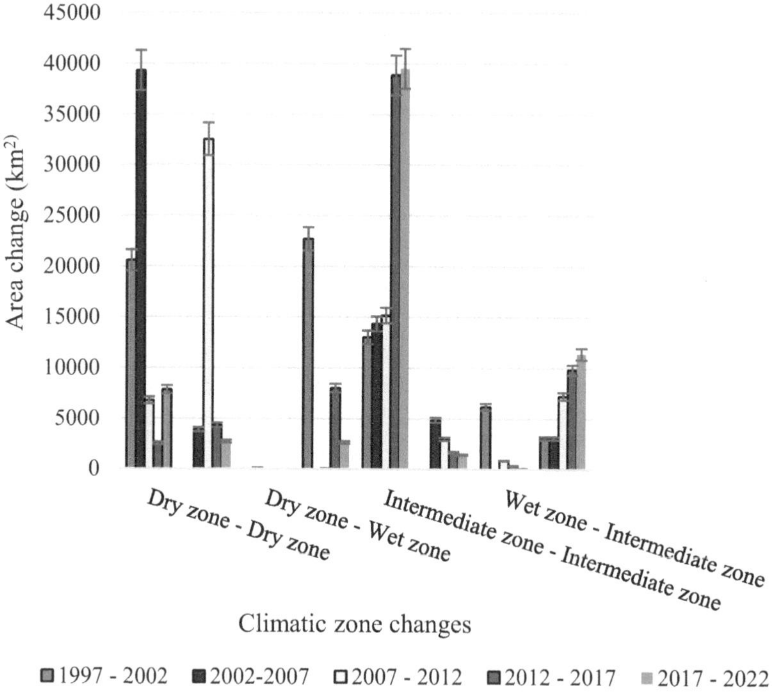

Fig. 12 Land area changes (km^2) in Sri Lankan climatic zones from 1997 to 2022 by five-year intervals

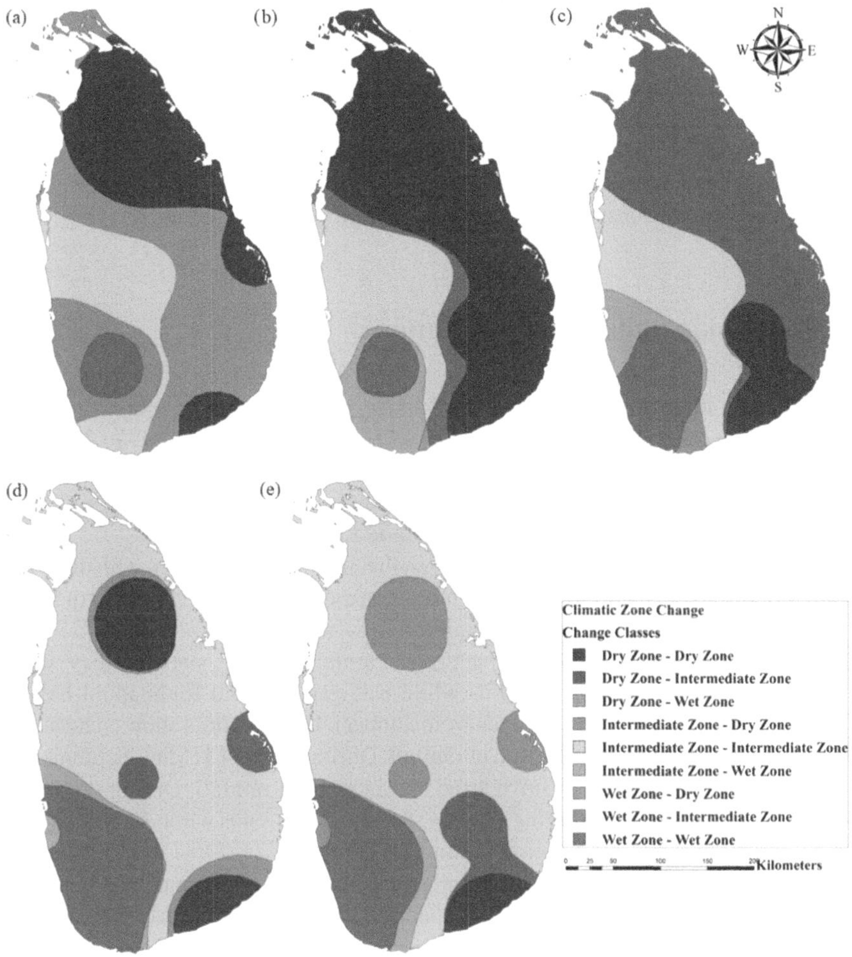

Fig. 13 Changes in land area within Sri Lanka's climatic zones from 1997 through 2022 in five-year intervals (**a**) 1997–2002 (**b**) 2002–2007 (**c**) 2007–2012 (**d**) 2012–2017 (**e**) 2017–2022

22,706.16 km^2, indicative of its capacity to encroach upon the arid region during periods of increased rainfall. The Intermediate zone's most significant growth spurt occurred between 2007 and 2012, when it expanded by a remarkable 32,524.60 km^2, a phenomenon that may reflect shifts in land-use practices in response to changing climate conditions. This zone's dynamic interaction with the dry zone highlights the complex interrelationship between these two climatic regions (Fig. 13a).

In contrast, the wet zone, with its consistently abundant rainfall, exhibited remarkable stability in land area across all five-year intervals. This remarkable resilience suggests that the Wet zone, with its favorably wet conditions, may be less susceptible to environmental shifts, emphasizing the importance of this region for biodiversity and conservation efforts in Sri Lanka (Fig. 13b).

The transitions between climatic zones also unveil fascinating insights. While the dry zone predominantly expanded into the intermediate zone, it had a limited impact on the wet zone, reflecting a clear climatic boundary separating these regions. The intermediate zone, however, showed a consistent trend of growth, particularly into the wet zone, underscoring its adaptability to varying precipitation levels (Fig. 11).

4 Conclusion and Recommendations

In conclusion, the comprehensive analysis of Sri Lanka's rainfall patterns and their impact on the country's climatic zones over a 26-year period from 1997 to 2022 has revealed a wealth of insights.

The climatic zones of wet, dry and intermediate show a positive trend of rainfall and the arid zone shows a negative trend of rainfall.

From 1997 to 2007, rainfall analysis shows that more areas prevailed with dry zone characteristics while from 2007 to 2022 more dry zone areas changed to intermediate zone characteristics due to increasing rainfall.

When considering the seasonal pattern of rainfall gauging stations, they show positive trends in FIM (March–April) while Katugasthota and Rathnapura have a reduction of rainfall. In SIM (October—November), Hambanthota shows a negative trend of rainfall as well as a reduction of rainfall. During the SWM (May–September), all gauging stations show a positive trend and increasing rainfall. During the NEM (November–February), all gauging stations show positive trends while Anuradhapura and Hambanthota show a reduction of rainfall.

These findings shed light on the complex interplay between climate and land use, highlighting the critical need for region-specific strategies to address environmental challenges and harness opportunities for sustainable development and disaster risk reduction in each climatic zone.

One of the most significant observations is the variability in annual rainfall across different regions of Sri Lanka. While some areas have experienced a consistent increase in rainfall, others have witnessed declining trends. Notably, the wetter regions of Ratnapura and Kalutara districts have recorded the highest average annual rainfall, whereas the drier Hambantota district has shown a significant negative trend in rainfall. These variations have direct implications for agriculture, water resource management, natural disaster risk management and overall environmental sustainability. The analysis of seasonal rainfall trends further emphasizes the importance of the inter-monsoon periods, particularly the FIM and SIM seasons, which have exhibited consistent positive trends. Furthermore, the evolution of land areas within Sri Lanka's distinct climatic zones has highlighted the dynamic response of these regions to changing precipitation patterns. The dry zone's expansion during favorable rainfall years underscores its sensitivity to climate fluctuations, while the intermediate zone has demonstrated the ability to encroach upon arid areas during periods

Table 4 Descriptive statistics for annual rainfall (in mm) at climatic zones generated from 1997 to 2022

Climate zone	Average rainfall (mm)	Max	Min	Std.dev
Dry zone	1,637.9	2,802.5	960.2	369.4
Wet zone	2,600.7	3,411.4	2,118.2	313.6
Intermediate zone	1,787.9	2,525.1	1,034.7	363.3
Arid zone	1,044.3	1,734.1	665.7	250.6

Table 5 M–K test results for annual rainfall at the climate-zone level

Climate zone	Kendall's tau	*p*-value	Sen's slope
Dry zone	0.321	0.007	14.453
Wet zone	0.364	0.004	25.383
Intermediate zone	0.341	0.006	17.485
Arid zone	− 0.126	0.378	− 5.205

** As the computed *p*-value is greater than the significance level alpha = 0.05, there is no trend in the series

of increased rainfall. In contrast, the wet zone has maintained remarkable stability, showcasing its resilience in the face of environmental shifts.

Overall, this study underscores the critical importance of understanding the intricate relationship between climate and land use in Sri Lanka. As rainfall patterns continue to fluctuate, land management practices, agricultural planning, disaster risk management and conservation efforts become increasingly vital. These findings provide valuable insights into the challenges and opportunities presented by climate change in Sri Lanka, guiding policymakers, researchers, and stakeholders in developing strategies for a sustainable and resilient future.

Appendix

See Tables 3, 4, 5, 6 and 7

Table 6 M–K test results for SWM, NEM, FIM, and SIM rainfall at the station level

Stations	FIM			SIM			NEM			SWM		
	Kendall's tau	*p*-value	Sen's slope	Kendall's tau	*p*-value	Sen's slope	Kendall's tau	*p*-value	Sen's slope	Kendall's tau	*p*-value	Sen's slope
Anuradhapura	0.052	0.724	– 1.100	0.169	0.234	6.653	− 0.128	0.860	– 2.550	0.126	0.378	5.100
Badulla	0.151	0.290	4.817	0.046	0.758	2.144	0.220	0.402	5.000	0.218	0.123	6.483
Batticaloa	0.003	1.000	– 0.080	0.071	0.628	3.100	0.245	0.930	7.008	0.136	0.343	2.572
Colombo	0.132	0.355	5.786	0.145	0.311	6.850	0.052	0.724	1.300	0.089	0.537	4.963
Galle	0.095	0.508	2.311	0.231	0.103	8.845	0.201	0.103	5.827	0.114	0.427	9.624
Hambantota	0.089	0.537	1.695	− 0.132	0.355	− 4.200	− 0.151	0.290	– 2.657	0.089	0.537	2.315
Katugastota	− 0.028	0.860	4.788	0.132	0.355	5.871	0.206	0.146	6.733	0.151	0.290	4.271
Ratnapura	− 0.058	0.692	6.671	0.003	1.000	0.282	0.275	0.217	6.762	0.095	0.508	8.650
Trincomalee	0.071	0.628	0.675	0.132	0.355	7.233	0.015	0.930	3.500	0.138	0.332	3.230

Table 7 Land area changes (km^2) in Sri Lankan climatic zones from 1997 to 2022 by five-year intervals

Change classes	Area change (km^2)				
	1997–2002	2002–2007	2007–2012	2012–2017	2017–2022
Dry zone—dry zone	20,597.65	39,298.90	6,773.12	2,534.41	7,836.91
Dry zone—intermediate zone	0	3,878.48	32,524.60	4,325.09	2,708.89
Dry zone—wet zone	0	145.92	0	0	0
Intermediate zone—dry zone	22,706.15	0	86.97219	8,011.38	2,630.64
Intermediate zone—intermediate zone	13,012.32	14,331.68	15,147.08	38,863.03	39,500.76
Intermediate zone—wet zone	0	4,874.97	2,976.31	1,666.78	1,432.91
Wet zone—intermediate zone	6,189.99	0	866.38	381.49	113.41
Wet zone—wet zone	3,082.83	3,082.83	7,238.55	9,828.13	11,381.75

References

Abeysekera, A., B. Punyawardena, B. Marambe, I. Jayawardena, T. Sivananthawerl & V. Wickramasinghe (2021). Effect of Indian Ocean Dipole (IOD) Events on the Second Inter-monsoonal Rainfall in the Wet Zone of Sri Lanka. *Tropical Agricultural Research, 32*(3).

Abeysingha, N. S. (2022). A review of recent changes in rainfall trend in Sri Lanka. *Tropical Agricultural Research and Extension, 25*(1), 1. https://doi.org/10.4038/tare.v25i1.5584

Aditya, F., E. Gusmayanti & J. Sudrajat (2021). Rainfall trend analysis using Mann-Kendall and Sen's slope estimator test in West Kalimantan. *IOP conference series: earth and environmental science*, IOP Publishing.

Adopted, I. (2014). Climate change 2014 synthesis report. *IPCC: Geneva, Szwitzerland*, 1059–1072.

Agarwal, S., A. Suchithra & S. P. Singh (2021). Analysis and interpretation of rainfall trend using Mann-Kendall's and Sen's slope method. *Indian Journal of Ecology, 48*(2), 453–457.

Alahacoon, N., & Edirisinghe, M. (2021). Spatial variability of rainfall trends in sri lanka from 1989 to 2019 as an indication of climate change. ISPRS International Journal of Geo-Information, 10(2). https://doi.org/10.3390/ijgi10020084

Ali, S., H.-I. Eum, J. Cho, L. Dan, F. Khan, K. Dairaku, M. L. Shrestha, S. Hwang, W. Nasim & I. A. Khan (2019). Assessment of climate extremes in future projections downscaled by multiple statistical downscaling methods over Pakistan. *Atmospheric Research, 222*, 114–133.

Amarasinghe, A. (2020). Analysis of long-term rainfall trends in Sri Lanka Using CHRIPS estimates. *American Journal of Multidisciplinary Research & Development (AJMRD), 2*(11): 34–44.

Baig, M. R., Shahfahad, N. M. W., Ansari, A. H., Ahmad, S., & Rahman, A. (2022). Spatio-temporal analysis of precipitation pattern and trend using standardized precipitation index and Mann–Kendall test in coastal Andhra Pradesh. *Modeling Earth Systems and Environment, 8*(2), 2733–2752. https://doi.org/10.1007/s40808-021-01262-w

Barua, N., Ng, A. W. M., Perera, B. J. C., Barua, S., & Muttil, N. (2013). Rainfall trend and its implications for water resource management within the Yarra River catchment, Australia)

Rainfall trend and its implications for water resource management within the Yarra River catchment Rainfall trend and its implications for water resource management within the Yarra River catchment. *Australia., 27*(12), 1727–1738. https://doi.org/10.1002/hyp.9311/pdf

Basistha, A., Arya, D. S., & Goel, N. K. (2009). Analysis of historical changes in rainfall in the Indian Himalayas. *International Journal of Climatology, 29*(4), 555–572. https://doi.org/10.1002/joc.1706

Brema, J. & J. Anie (2018). Rainfall trend analysis by Mann–Kendall test for Vamanapuram river basin, Kerala. *International Journal of Civil Engineering and Technology, 9*(13): 1549–1556.

Brimelow, J., Burrows, W., & Hanesiak, J. (2017). The changing hail threat over North America in response to anthropogenic climate change. *Nature Climate Change, 7*(7), 516–522. https://doi.org/10.1038/nclimate3321

Brown, T. B., Barry, R. G., Doesken, N. J. et al. (1992). An exploratory study of temperature trends for Colorado paired mountain—high plains stations. American Meteorol Society 6th Conference on Mountain Metrology, Portland, pp. 181—184.

Burt, T. P., & Weerasinghe, K. D. N. (2014). Rainfall distributions in Sri Lanka in time and space: An analysis based on daily rainfall data. *Climate, 2*(4), 242–263. https://doi.org/10.3390/cli2040242

Caesar, J., et al. (2011). Changes in temperature and precipitation extremes over the Indo-Pacific region from 1971 to 2005. *International Journal of Climatology, 31*, 791–801. https://doi.org/10.1002/joc.2118

Chaouche, K., et al. (2010). Analyses of precipitation, temperature and evapotranspiration in a French Mediterranean region in the context of climate change. *Comptes Rendus Geoscience, 342*, 234–243.

Dong, S., Sun, Y., Li, C., Zhang, X., Min, S., & Kim, Y. (2021). Attribution of extreme precipitation with updated observations and CMIP6 simulations. *Journal of Climate, 34*(3), 871–881. https://doi.org/10.1175/jcli-d-19-1017.1

Gadgil, A., & Dhorde, A. (2005). Temperature trends in twentieth century at Pune India. *Atmospheric Environment, 39*(35), 6550–6556. https://doi.org/10.1016/j.atmosenv.2005.07.032

Gajbhiye, S., Meshram, C., Mirabbasi, R., & Sharma, S. K. (2016). Trend analysis of rainfall time series for Sindh river basin in India. *Theoretical and Applied Climatology, 125*(3–4), 593–608. https://doi.org/10.1007/s00704-015-1529-4

Hackenbruch, J., Kunz-Plapp, T., Müller, S., & Schipper, J. W. (2017). Tailoring climate parameters to information needs for local adaptation to climate change. *Climate, 5*(2), 25.

Harischandra, I. N., Dassanayake, R. S., & De Silva, B. G. D. N. K., (2016). Three sympatric clusters of the malaria vector Anopheles culicifacies E (Diptera: Culicidae) detected in Sri Lanka. Parasites & Vectors, 9(1), 1–10. https://doi.org/10.1186/s13071-015-1286-3

Hamilton, J. P., Whitelaw, G. S., & Fenech, A. (2001). Mean annual temperature and total annual precipitation trends at Canadian biosphere reserves.

Helsel, D. R., & Hirsch, R. M. (n.d.). Techniques of Water-Resources Investigations of the United States Geological Survey Book 4, Hydrologic Analysis and Interpretation Statistical Methods in Water Resources. http://water.usgs.gov/pubs/twri/twri4a3/

Herath, S., & Ratnayake, U. (2004). Monitoring rainfall trends to predict adverse impacts—a case study from Sri Lanka (1964–1993). *Global Environmental Change, 14*, 71–79. https://doi.org/10.1016/j.gloenvcha.2003.11.009

Hoeppe, P. (2016). Trends in weather-related disasters—consequences for insurers and society. *Weather and Climate Extremes, 11*, 70–79.

Hunt, K. M. R., & Menon, A. (2020). The 2018 Kerala floods: A climate change perspective. *Climate Dynamics, 54*, 2433–2446.

IPCC, (1996). "Climate change", The IPCC second assessment report, Houghton JT, Meira Filho LG, Callander BA, Harris N, Kattenberg A, Maskell K (Eds.). Cambridge University Press: New York.

Isioma, I. N., I. I. Rudolph & A. L. Omena (2018). Non-parametric Mann–Kendall test statistics for rainfall trend analysis in some selected states within the coastal region of Nigeria. *Journal of Civil, Construction and Environmental Engineering, 3*(1): 17–28.

Jayawardena, I. M. S. P., Darshika, D. W. T. T., & Herath, H. M. R. C. (2018). Recent trends in climate extreme indices over Sri Lanka. *American Journal of Climate Change, 7*(4), 586–599.

Jayawardene, H. K., Sonnadara, D. U., & Jayewardene, D. R. (2005). Trends of rainfall in Sri Lanka over the Last Century. *Sri Lankan Journal of Physics, 15*(6), 7–17.

Karunathilaka, K. L. A. A., Dabare, H. K. V., & Nandalal, K. D. W. (2017). Changes in rainfall in Sri Lanka during 1966–2015. *Engineer Journal of the Institution of Engineers, Sri Lanka, 50*(2), 39. https://doi.org/10.4038/engineer.v50i2.7251

Kirsta, Y., & Loucka, O. (2020). Annual range of temperature and precipitation forecast for Altai-Sayan Mountain country. *Environmental Dynamics and Global Climate Change, 11*(1), 16–23.

Kumar, N., Panchal, C. C., Chandrawanshi, S. K., & Thanki, J. D. (2017). Analysis of rainfall by using Mann–Kendall trend, Sen's slope and variability at five districts of south Gujarat, India. In MAUSAM, Vol. 68, Issue 2.

Malmgren, B. A., Hulugalla, R., Hayashi, Y., & Mikami, T. (2003). Precipitatioon trends in Sri Lanka since the 1870s and relationships to El Niño-southern oscillation. *International Journal of Climatology, 23*(10), 1235–1252. https://doi.org/10.1002/joc.921

Manton, M. J., Della-Marta, P. M., Haylock, M. R., Hennessy, K. J., Nicholls, N., Chambers, L. E., et al. (2001). Trends in extreme daily rainfall and temperature in Southeast Asia and the South Pacific: 1961–1998. *International Journal of Climatology, 21*(3), 269–284. https://doi.org/10.1002/joc.610

Mishra, A. K., Özger, M., & Singh, V. P. (2009). Trend and persistence of precipitation under climate change scenarios for Kansabati basin India. *Hydrological Processes, 23*(16), 2345–2357. https://doi.org/10.1002/hyp.7342

Mujumdar, M., Bhaskar, P., Ramarao, M. V. S., Uppara, U., Goswami, M., Borgaonkar, H., Chakraborty, S., Ram, S., Mishra, V., Rajeevan, M., Niyogi, D. (2020). Droughts and floods. In: Krishnan, R., Sanjay, J., Gnanaseelan, C., Mujumdar, M., Kulkarni, A., Chakraborty, S. (Eds.) Assessment of climate change over the Indian Region. Springer, Singapore. https://doi.org/10.1007/978-981-15-4327-2_6

Nakagawa, M., KV, W. K., Manabe, T., Ranaweera, L., Nasu, A. (2017). Kaolin deposit at Meetiyagoda, southwestern Sri Lanka. Clay Science. 21(2):29-34. https://doi.org/10.11362/jcssjclayscience.21.2_29

Naveendrakumar, G., Vithanage, M., Kwon, H. H., Iqbal, M. C. M., Pathmarajah, S., & Obeysekera, J. (2018). Five decadal trends in averages and extremes of rainfall and temperature in Sri Lanka. *Advances in Meteorology*. https://doi.org/10.1155/2018/4217917

Nillama, G. V. P. (2021). Modeling rainfall influence to soil sediment yield in corn monocropping systems in Claveria, Southern Philippines through Water Erosion Prediction Project (WEPP). *Model Earth Syst Environ, 7*, 853–869.

Nisansala, W. D. S., Abeysingha, N. S., Islam, A., & Bandara, A. M. K. R. (2019). Recent rainfall trend over Sri Lanka (1987–2017). *International Journal of Climatology, 40*(7), 3417–3435.

Partal, T., & Kahya, E. (2006). Trend analysis in Turkish precipitation data. *Hydrological Processes, 20*(9), 2011–2026. https://doi.org/10.1002/hyp.5993

Perera, H. K., W. I., Sonnadara, D. U. J., & Jayewardene, D. R. (2002). Forecasting the occurrence of rainfall in selected weather stations in the wet and dry zones of Sri Lanka. In Sri Lankan Journal of Physics, Vol. 3.

Pham, H. X., Shamseldin, A. Y., & Melville, B. W. (2021). Projection of future extreme precipitation: A robust assessment of downscaled daily precipitation. *Natural Hazards, 107*(1), 311–329.

Prasanna, R. (2018). Economic costs of drought and farmers' adaptation strategies: evidence from Sri Lanka. *Sri Lanka Journal of Economic Research, 5*(2).

Pushpakumara, G., Weerakoon, D., Silva, P., & Marambe, B. (Eds.). (2014). Climate change adaptation action plan (2015–2018): Western province of Sri Lanka. *Ministry of Agriculture, Agrarian*

Development, Minor Irrigation, Industries, Environment, Arts, and Cultural Affairs (Western Province). ISBN: 978-955-7819-01-3.

Rahman, M. S., & Islam, A. R. M. T. (2019). Are precipitation concentration and intensity changing in Bangladesh overtimes? Analysis of the possible causes of changes in precipitation systems. *Science of the Total Environment, 690*, 370–387.

Selin, H., & VanDeveer, S. (2011). Climate change regionalism in North America. *Review of Policy Research, 28*(3), 295–304. https://doi.org/10.1111/j.1541-1338.2011.00496.x

Shishutosh, B., Nitin, M., Ng, A. W. M., & Perera, B. J. C. (2013). Rainfall trend and its implications for water resource management within the Yarra River catchment Australia. *Hydrological Processes, 27*(12), 1727–1738. https://doi.org/10.1002/hyp.9311

Somasundaram, D., F. Zhang, S. Ediriweera, S. Wang, J. Li & B. Zhang (2020). Spatial and temporal changes in surface water area of Sri Lanka over a 30-year period. *Remote Sensing, 12*(22), 3701.

Tangang, F. T., Juneng, L., & Ahmad, S. (2007). Trend and interannual variability of temperature in Malaysia: 1961–2002. *Theoretical and Applied Climatology, 89*, 127–141. https://doi.org/10.1007/s00704-006-0263-3

Trenberth, K. (2011). Changes in precipitation with climate change. *Climate Research, 47*(1), 123–138. https://doi.org/10.3354/cr00953

Wickramaarachchi, W. W. U. I., Peiris, T. U. S., & Samita, S. (2020). Rainfall trends in the North-Western and Eastern coastal lines of Sri Lanka using non—Non-parametric analysis. *Tropical Agricultural Research, 31*, 41–54.

Wickramagamage, P. (2010). Seasonality and spatial pattern of rainfall of Sri Lanka: Exploratory factor analysis. *International Journal of Climatology, 30*(8), 1235–1245. https://doi.org/10.1002/joc.1977

Wickramagamage, P. (2016). Spatial and temporal variation of rainfall trends of Sri Lanka. *Theoretical and Applied Climatology, 125*(3–4), 427–438. https://doi.org/10.1007/s00704-015-1492-0

Yusuf, A. A., Francisco, H. (2009). Climate change vulnerability mapping for Southeast Asia. In: Economy and environment program for Southeast Asia (EEPSEA), Singapore, pp. 10–15.

Zhang, W. X., & Zhou, T. J. (2020). Increasing impacts from extreme precipitation on population over China with global warming. *Science Bulletin, 65*, 243–252.

Zubair, L. (2003). El Niño-southern oscillation influences on the Mahaweli streamflow in Sri Lanka. *International Journal of Climatology, 23*(1), 91–102. https://doi.org/10.1002/joc.865

Zubair, L., Siriwardhana, M., Chandimala, J., & Yahiya, Z. (2008). Predictability of Sri Lankan rainfall based on ENSO. *International Journal of Climatology, 28*, 91–101.

Enhancing University-Enterprise Collaborations: A Framework for Disaster Resilience in Asia

Kinkini Hemachandra, U. T. G. Perera, Dilanthi Amaratunga, and Richard Haigh

1 Introduction

Collaborations between universities and industries or enterprises have been established for decades. Such collaborations have been formed for several purposes and across several disciplines—for example, skills development, knowledge creation and transfer. In addition, collaborations also support entrepreneurship development by providing solutions to business problems, research and innovation agendas, and developing new products and patents (Guimón, 2013). These collaborations benefit universities, industries and society (Arant et al., 2019; Brundenius et al., 2009; Guimón, 2013; Randil et al., 2018).

Since the 1990s, universities have performed either as teaching, research, or entrepreneurial universities. Nevertheless, most universities promote entrepreneurial qualities and activities to support growth and development (Guimón, 2013). Despite recognising university-enterprise collaborations, the challenge is to establish and sustain these collaborations (Awasthy et al., 2020; Randil et al., 2018).

During the COVID-19 period, the necessity of such collaborations was reemphasised in history. They have collaborated on material transfers and knowledge transfers as well (Puślecki et al., 2021). For example, over one-third of the vaccine candidates are formed via university-enterprise collaborations (Druedahl et al., 2021). Hence, the COVID-19 pandemic could be considered a classic example where collaborative

K. Hemachandra (✉)
Huddersfield Business School, University of Huddersfield, Huddersfield HD1 3DH, United Kingdom
e-mail: K.Hemachandra2@hud.ac.uk

U. T. G. Perera · D. Amaratunga · R. Haigh
University of Huddersfield, Huddersfield HD1 3DH, United Kingdom

U. T. G. Perera
Department of Geography, University of Colombo, Colombo, Sri Lanka

C. Malalgoda et al. (eds.), *Climate Change Adaptation in the Built Environment*,
https://doi.org/10.1007/978-3-031-75826-3_17

work could help to address global disaster risks at the local, national, and global levels.

Asia is the most vulnerable continent, especially to natural hazards (CRED-UNISDR, 2020). Despite this, the present pandemic and many other previous pandemics originated in Asia representing its susceptibility to biological hazards (Fu & Zhu, 2020). Similarly, the region is also affected by many human-induced hazards over the decades. An explosion was recorded in Beirut, Lebanon due to ammonium nitrate storage in 2020 (HRW, 2021), building collapses such as the Mumbai building collapse in India in 2022 (Times of India, 2024) and the Rana Plaza building collapse in Dhaka, Bangladesh in 2013 (ILO, 2023) and further several transport disasters including aviation sector in Malaysia in 2014 (NYTimes, 2014) are few such human-induced disasters recorded in Asia during the past decade. Hence, disaster risk in the region can be considered extremely high (CRED-UNISDR, 2020; Pulhin et al., 2021). In addition, with their low coping capacities and some governance issues, the level of vulnerabilities becomes significant in the region. For example, Asia reports the largest share of poor people living on the earth, making further complexities to these hazards. This is furthermore worsened in small islands and tropical developing countries where the coastal communities represent the poorest group in society (Sjöstedt & Povitkina, 2017).

Having understood this level of disaster risks and vulnerabilities, specifically in Asia, a group of academics have proposed a project to develop and strengthen the collaborative mechanisms between universities and enterprises for strengthening disaster resilience in the region. This is because university-enterprise relationships have not been adequately researched in Asia, though the region is highly vulnerable to disasters (Chung, 2014; Khan & Park, 2012). According to Chung (2014), the Triple Helix (TH) model was adequately examined in Asia and hence, their study examined the characteristics of the TH scholars based on their publications. Nevertheless, Chung (2014) did not study the full coverage of the TR model and emphasized the necessity of further studies for a longer period in Asia. Similarly, Morris et al. (2018) conducted a study in Central Asia explored university enterprise collaboration within the institutional context and revealed how the economic cultures in the Central Area affect the HEIs and their collaboration. Nevertheless, all those previous studies were conducted within diverse ranges rather than focusing on developing UEC within the disaster resilience area. Hence, the paper was written based on a literature review conducted to develop a framework for strengthening UEC to achieve disaster resilience in Asia. The objectives of the paper are limited to:

1. Highlight the importance of university-enterprise collaborations for disaster resilience.
2. Identify the factors affecting university-enterprise collaborations.

These study findings will be used for developing a multilevel university-enterprise collaborative framework for strengthening disaster resilience in Asia. This framework would be helpful for policymakers and practitioners to initiate and strengthen university-enterprise collaborations (UECs) to make the region disaster-resilient.

There are several studies conducted to develop frameworks to support UECs. For example, Awasthy et al. (2020) presented a framework to improve the effectiveness of these collaborations. However, their framework was limited to general UEC, not specific to UEC for disaster resilience. In this context, the paper adds a novel element to UEC literature by introducing a framework for strengthening UEC for disaster resilience.

The paper consists of four sections. Section 2 provides the research methods adopted in the study. Section 3 provides the study findings. Section 4 provides a discussion and conclusions.

2 Research Methods

The expected UEC framework for disaster resilience will be developed by methodologically identifying the challenges and enablers for UECs. Sound research methodology is key to robust findings that direct to valid and reliable solutions (Creswell & Creswell, 2017; Opoku et al., 2016). Research methods are the investigation or data collection tools/techniques (Opoku et al., 2016).

Accordingly, the study's data collection was based on published literature and expert opinions. The study followed a narrative literature review at the global level. The documents were selected based on the research objectives and research scope. Another aspect considered during the data collection was existing papers relevant to the Asian context. Research papers were selected from the Google Scholar and University of Huddersfield Library websites. Several search terms were used for documentary search: university-industry collaborations, university-industry partnerships and university-industry Asia. After considering the topic, the number of citations and the year of publication, the papers were initially selected. After reading the abstracts of initially selected papers, the full papers were selected based on their applicability and relatedness. In addition, several global reports were referred to identify their applicability and concerns about university-enterprise collaborations, specifically within disaster resilience.

The term industry has been interchangeably used throughout the literature with the term enterprise, while collaboration has been swapped with the term partnerships. Hence, the study used both terms to identify relevant documents. However, throughout the paper, the authors use the term university-enterprise collaborations (UECs), which means any partnerships, collaborations, or cooperations between universities and industries.

As the data analysis technique, a thematic analysis was conducted based on the literature review. Accordingly, the findings are presented by research objectives. The literature review will benefit in identifying and understanding the state of the UEC in disaster resilience to identify the factors affecting the UEC. The study findings were further validated by conducting several rounds of discussions with the participating universities to confirm the applicability and suitability of the identified factors.

The identified factors were classified into cultural, structural, relational and material categories.

The outcome of this literature survey will be used to conduct the planned survey across the member countries in Asia and to identify further and validate the findings. Accordingly, the study will finalise its proposed framework based on the outcome of the literature survey and online surveys' outcomes.

3 Findings

3.1 Importance of University-Enterprise Collaborations

3.1.1 Benefits to Universities

Universities benefit from such collaborations. For example, Universities have access to additional funds for teaching, research and collaborations; opportunities to develop knowledge and capacity on the current- state of the art; improved interactions and training opportunities; enhancement of HEI's image as a contributor to the economy are some of such benefits (Azman et al., 2019; Barnes et al., 2002; Kaymaz & Eryiğit, 2011). In addition, universities will benefit from getting opportunities for their students to be involved in the process (Seturidze & Topuria, 2021). Universities' income from academic engagement is usually a high multiple of the income derived from intellectual property (Perkmann et al., 2013). Most importantly, universities will receive more opportunities to access additional public and private funding and increase income through the licensing and patenting process (Barnes et al., 2002).

3.1.2 Benefits to the Industry

As described in the introduction section, UECs are perceived and gained attention across many fields. Some consider UEC as a vehicle to ***enhance innovation*** through knowledge exchange (Ankrah & Omar, 2015; Arant et al., 2019; Giones, 2019). Through these collaborations, both parties are encouraged to develop UECs to ***enhance innovation*** and ***economic competitiveness at institutional levels*** through knowledge exchange (Perkmann et al., 2013). Specifically within the dynamic business world, knowledge creation through universities would be a competitive advantage for industries (Berbegal-Mirabent et al., 2015) since collaborations could provide the path towards gaining a ***competitive advantage*** (Awasthy et al., 2020). As stated above, one of the key instruments for succeeding business opportunities is a ***competitive advantage***. In doing so, universities are the main source which provides new knowledge to most business problems (Berbegal-Mirabent et al., 2015).

These UECs also provide opportunities for ***collaborative research, leading to new patents and intellectual properties*** and ***reorienting the Research and development***

(R&D) agenda (Azman et al., 2019; Kaymaz & Eryiğit, 2011). University-industry collaborations have encouraged ***knowledge and technology exchange*** in the United States, Japan, Singapore, and European Union Countries (Ankrah & Omar, 2015). Others also identify their capacities in enhancing scientific and commercial capabilities (Ergun Demirel & Bayer, 2015). For example, the government of Spain has initiated some programmes to encourage these partnerships to strengthen research and development and knowledge creation. As a result, their ***economic contribution to the industry*** accounted for 632 million Euros in 2010 in Spain (Berbegal-Mirabent et al., 2015). A similar example can be identified in another European country, Italy (Muscio, 2010).

Universities are traditionally assigned the role of teaching and research. However, a new element of universities has been identified. Accordingly, some universities are regarded as ***entrepreneurial universities to commercialise*** their research output to enhance UE collaborations (Muscio, 2010). Through effective UECs, industries have tried to recruit ***more university graduates*** (Azman et al., 2019). Their appointments will facilitate two-way communication and collaboration between HEIs and the learners. Maintaining a continuous engagement with students will benefit the learners and help the institutions ***be more responsive to the market needs*** (Thayaparan et al., 2014). Such UECs are also helpful in transforming economies into ***knowledge and innovation-based economies***. For example, Malaysia has shown how its UEC has been used in this economic transition into knowledge and innovation-based economies (Azman et al., 2019).

In addition to the growth in the business sector and the overall economic growth, universities ***provide benefits to the society***. For example, Philippine universities perform as agents in society for poverty reduction. Universities equip the youth in society through knowledge and through employment generation. Their partnerships with industries prevent them from producing graduates who do not match with labour market requirements. Through solid collaborations between universities and industry, the graduates they produce fulfil industry requirements, and hence, skilled labour will contribute to social development (Orale, 2013).

3.1.3 Importance of UEC in Disaster Resilience

The intensity and frequency of natural hazards have increased unprecedentedly, resulting in devastating impacts on lives and economies (CRED-UNISDR, 2020). The present challenge for practitioners and policymakers is to reduce such impacts and related risks with innovative measures. Climate change has unprecedentedly increased during the last century, causing many deaths and economic losses worldwide (CRED-UNISDR, 2020). Progress in risk reduction and resilience building is uneven worldwide, with some high-risk, low-capacity countries falling behind (UN, 2015; UNISDR, 2015). Decision-making in complex contexts such as disaster risk management requires collaborative approaches to knowledge production (Barton et al., 2020). Hence, this section accounts for the UEC's contribution to supporting disaster resilience worldwide.

UECs are specifically important for addressing disaster risks and making disaster resilient. One way to support disaster risk reduction is through ***knowledge production*** related to disaster risk management. Many universities in developed countries have ***facilitated the development of international curricula and joint degrees, fostered international innovation projects, and supported the exchange of students, staff, and knowledge as a strategy for*** addressing complex challenges, specifically handling disaster risks (Amaratunga et al., 2018). The close partnership between universities and industry related to DRR is essential to reduce community vulnerability (Randil et al., 2018).

In other instances, such collaborations have extended their contribution directly ***to reduce disaster risks*** (Berbegal-Mirabent et al., 2015; Seturidze & Topuria, 2021; Shaw et al., 2020). For example, the Centre for Disaster Management and Public Safety of the University of Melbourne and Volunteer Fire Brigades Victoria (VFBV) have collaborated to support the project Surge Capacity Analysis and Visualisation Tool for the Fire Incident Report System (FIRS). Their collaboration developed a decision support tool to visualise volunteer firefighting capacity across Victoria. A similar example can be highlighted from Sri Lanka, representing developing countries. For example, Dialog Axiata collaborates with several universities in Sri Lanka to improve early warning systems, predicting and disseminating rainfall data (Randil et al., 2018).

Several other examples can be highlighted during the recent COVID-19 pandemic. Many collaborative initiatives were formed to address the global challenge posed by the pandemic (Puślecki et al., 2021; Seturidze & Topuria, 2021). For example, more than 1/3 of ***the vaccines were developed*** through collaborations with universities and enterprises. The Astra Zeneca vaccine was developed collaboratively with the University of Oxford and the Astra Zeneca Company (Mahase, 2021). Another example was found in China. The Fudan University, the Shanghai City Government and the Centre for Disease Control collaboratively developed an artificial intelligence system to speed up their respiratory scanning system. More than 93% of the residents in China used this method.

In addition to the scholarly evidence, some ***leading global policies and agendas*** also emphasise the importance of UEC. For example, ***the Sendai Framework for Disaster Risk Reduction (SFDRR)*** 2015–2030 (Abdeen et al., 2021; UNISDR, 2015) provides the basis for driving towards a risk-informed and resilient development agenda. The SFDRR emphasises that increased international cooperation between countries and organisations is considered essential (Aitsi-Selmi et al., 2016). Prior to this framework, the Hyogo Framework for Action-HFA highlighted knowledge management and education as key priorities for action (UNISDR, 2015). Collaborations are important and can provide education and training as an integral part of capacity building in the disaster management discipline (Thayaparan et al., 2014).

Even though these collaborations provide many benefits to universities, industry and the wider society within different contexts, their existence is not fully operationalised in practice both in developed and developing countries (Amaratunga et al., 2018; Azman et al., 2019). Universities have limited opportunities to work

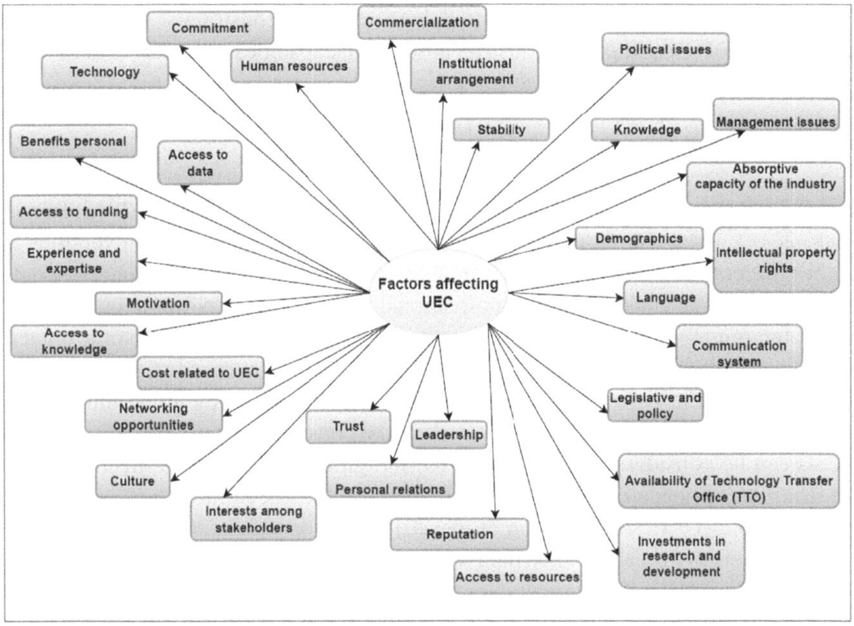

Fig. 1 Factors affecting university-enterprise collaborations (UECs)

collaboratively with related stakeholders. Hence, the following section will present the factors determining the UEC based on the literature review.

3.2 *Factors Affecting UEC: Challenges, Enablers and Good Practices*

After presenting the importance of UEC in general and within the disaster risk reduction context, this section presents the factors that affect either as an enabler or challenge towards UEC, based on the literature. Figure 1 presents the summary of 31 factors found in the literature review. The detailed explanation of each factor is followed by Fig. 1.

3.2.1 Absorptive Capacity of the Industry

Universities produce valuable pieces of evidence for policymakers to address the issues faced by industries and society out of their research. Nevertheless, the usage of such information in the industry is limited due to absorption capacities in industries. For example, Randil et al. (2018) demonstrate that using such university findings in the Sri Lankan industrial sector is limited due to capacity constraints. This highlights the necessity of strengthening industrial absorption capacities (Awasthy et al., 2020).

3.2.2 Access to Data

Industrial data availability is critical for universities' research (Guimón, 2013; Randil et al., 2018). Establishing collaboration with industry is helpful for universities in enabling access to industrial data (Azman et al., 2019). Accordingly, industry data availability can be considered as a motivational factor for universities to collaborate. A counterargument is made by Perkmann et al. (2013), highlighting that some universities are reluctant to reveal their data to industries. This is because of some misconception within the university administration assuming that they are commercially sensitive and with the fear of underestimating the consultancy activities. Hence, data access can be considered a factor towards strengthening UEC. Specifically, within the disaster resilience context, accessibility to data has been identified as a barrier to establishing UEC (Amaratunga et al., 2018). Randil et al. (2018) revealed how difficult to access disaster-related data within the Sri Lankan context. Specially. Their study further explained that most data are not freely available, and authorities hold the ownership and issue data only on request, sometimes with a price (Randil et al., 2018).

3.2.3 Access to Funding

Several studies have identified funding as a factor in strengthening UECs (Azman et al., 2019; Guimón, 2013; Randil et al., 2018). Azman et al. (2019) and Davey et al. (2011) identified that most universities face funding limitations specifically for research and equipment. For example, the allocation of Gross Domestic Product to research and development was around 0.1% in Sri Lanka in 2013. The amount is highly insignificant compared to regional and global averages (Randil et al., 2018). With the continuous decline of public funding for universities, universities are forced to search for different sources of private funding for survival (Azman et al., 2019). Within the context of limited budgetary provisions for universities, specifically state-funded universities, UEC provides opportunities to access funds from industries (Ankrah & Omar, 2015).

3.2.4 Access to Knowledge

Creating scientific knowledge is one of the key functions of universities. Through these collaborations, industries gain access to technological knowledge (Azman et al., 2019; Guimón, 2013). Nevertheless, a minimal number of academically qualified people work in the industry in Sri Lanka (Randil et al., 2018). Randil et al. (2018)

also stated that academia and universities also find difficulties in identifying disaster-related research problems from the industry. Hence, access to knowledge for both parties can be achieved through building UEC. Hence, this can be considered as a key factor for their collaborations.

3.2.5 Access to Resources

Another determinant for UEC is access to resources. Industries will benefit from gaining access to the university's facilities and equipment. Such collaborations can substantially reduce these expensive facilities (Ankrah & Omar, 2015; Guimón, 2013).

Specifically, for small and medium enterprises, the resource constraints can be addressed through UEC (Ankrah & Omar, 2015; O'Reilly & Cunningham, 2017). Similarly, universities will also benefit through such collaborations. Specifically, when there is no adequate allocation for university infrastructure through state funds, access to industrial resources would be an advantage. Hence, the Government of Thailand has set their national policies to support such collaborations allocating more resources by the state. (Tantanee et al., 2018). Hence, resource availability can be considered as a factor for building UEC.

3.2.6 Availability of Technology Transfer Office (TTO)

Technology Transfer Offices help both universities and industry. They support the production of patents and innovations and generating some revenues. They ultimately transfer knowledge from universities to society (Clayton et al., 2018; Giones, 2019). Similar establishments are formed by universities as science parks and incubators to commercialise university outputs (O'Reilly & Cunningham, 2017; Perkmann et al., 2013). Hence, some argue that TTOs are the leading institutions responsible for establishing UEC (Kaklauskas et al., 2018). However, there are instances where TTO limit their service to support university researchers with the administration of licensing, consultancy contracts, patenting and spin-off creation rather than supporting UECs in the Netherlands (Bodas Freitas & Verspagen, 2017). At the time of 2003, there were more than 20 science parks in Sweden to support UECs (Lindelöf & Löfsten, 2003). Apart from the Europe, there are evidence from Asia, for example, in Japan more than 110 science parks operate where as in China more than 100 science parks (Siegel & Phan, 2005).

3.2.7 Benefits-Personal

As stated above, many UECs benefit the university, industry and community. However, individuals who engage in UEC expect some personal benefits through collaborations. Particularly, when working with high-scale academics will be challenged since they will only work with the industry if there is a personal benefit (Perkmann et al., 2013). A counterargument was raised highlighting how the reputation can be achieved for academics by engaging with industry and is regarded as a personal benefit from such collaborations (Barnes et al., 2002; Guimón, 2013). Within this context, academics have been motivated to engage more with industry engagements for personal benefit such as more funding for research and development activities which benefit in the long run for academic promotions (Ankrah & Omar, 2015).

3.2.8 Commitment

Strong commitment by all stakeholders is another factor that affects successful UEC. Through their commitments, projects can deliver outputs effectively (Barnes et al., 2002; Nielsen, 2017). Besides individual commitments towards successful UECs, institutional and/or top management commitment has been identified as a key factor (Ankrah & Omar, 2015).

3.2.9 Commercialization

Commercialisation of the university outputs, academic knowledge, and academic entrepreneurship are some recent developments of UECs (O'Reilly & Cunningham, 2017; Perkmann et al., 2013). Through such commercialisation, universities generate income (Orale, 2013). Some governments have mandated commercialisation to maximise the total contribution of universities to job creation, competitiveness and national economic well-being (O'Reilly & Cunningham, 2017). For example, the recent initiative of the University-Industry Linkages (UILs) programme by the Thailand Government, aimed at the commercialisation of universities' research and development findings specifically among developing countries. In addition, they aim to commercialise technological commercialization specifically among entrepreneurial universities in Thailand.

3.2.10 Communication System

Effective Communication is critical towards successful UECs (Barnes et al., 2002; Plewa et al., 2013; Seturidze & Topuria, 2021). For example, clear communication helps to initiate professional dialogue during the engagement phase (Seturidze & Topuria, 2021). This is because communication culture among universities and industry differs (Azman et al., 2019). Such differences may lead to some negative perceptions about universities and industry. For example, most academic writings are full of technical jargon, which makes them difficult to understand by industrial users (Azman et al., 2019; Thayaparan et al., 2014). Effective communication also helps build personal relationships in which trust develops (Azman et al., 2019). For example, Amaratunga et al. (2018) revealed that the language barrier is a major factor in building university enterprise collaborations in Asia by conducting their study in Bangladesh, Sri Lanka and Asia. This is because language affects the quality of communication and the choice of communication media.

3.2.11 Cost Related to UEC

Some consider the UEC to involve large costs especially during the initial stage of building collaborations (Perkmann et al., 2013). However, through such collaborations, costs can be shared between universities and industries (Ankrah & Omar, 2015).

3.2.12 Culture

Another factor affecting UEC is culture (Bodas Freitas & Verspagen, 2017). According to Amaratunga et al. (2018), cultural norms affect how UECs are operated. For example, the messages are misinterpreted due to some cultural differences among partners. Azman et al. (2019) highlighted other cultural differences between the two sectors. Industries are more outcome-oriented, while universities are not. Ankrah and Omar (2015) also identified sociocultural issues as barriers to successful university-industry collaborations. For example, academics do not admit the importance of industry-based research (Randil et al., 2018).

3.2.13 Demographics

Demographics is considered a factor that determines the engagement of academics in UECs (Link et al., 2007). A similar finding was revealed by Perkmann et al. (2013) stating that male academics engage more industry collaborations and in the commercialisation of their research compared to female academics. Amaratunga et al. (2018)'s study identified a similar pattern across three countries in Asia. They = ir study revelated that male academics' involvement in international research projects and publication with international partners is higher than female academics. They explained the possible reason as the existing cultural contexts in Asia.

3.2.14 Experience and Expertise

Experience among academics is also considered as a factor affecting effective UEC. Guimón (2013) highlights that in most developing countries, such experiences are limited and hence considered challenging. Azman et al. (2019) also identify a lack of academic expertise and reputation as a challenge for building UEC in Malaysia.

Similarly, prior experience in UICs also facilitates preparing and organising such activities and initiatives (Tartari et al., 2012). Established and senior academics can collaborate more individually (Perkmann et al., 2013).

3.2.15 Human Resources

Human resource availability is another key factor for successful collaborations, from university and industry points of view. For example, liaison officers' availability is considered a supportive factor for building effective UECs (Bruneel et al., 2010; Guimón, 2013). Nevertheless, many universities find difficulties in appointing and retaining a skilled workforce (Ankrah & Omar, 2015).

3.2.16 Institutional Arrangement

Many scholars have identified institutional arrangements as vital for successful UECs (Amaratunga et al., 2018; Ankrah & Omar, 2015). In most instances, the institutional setting works as a barrier instead of a supporting UEC (Amaratunga et al., 2018). Their study provides evidence from Bangladesh, Sri Lanka and Thailand. According to their study, institutional arrangements have been identified as the most critical factor among Asian universities. For example, universities' bureaucratic structures, rigid rules and administrative hierarchies are considered challenges for effective collaborations by industries (Azman et al., 2019; Davey et al., 2011).

3.2.17 Intellectual Property Rights (IPR)

Intellectual property rights are a key determinant for successful UEC (Azman et al., 2019; O'Reilly & Cunningham, 2017). Some consider such IPRs and developing university-patented knowledge as the basis for new collaborative contacts (Bodas Freitas & Verspagen, 2017). According to industry representatives, IPRs are often tedious and inflexible. They further highlight the reluctance of universities to collaborate and cooperate regarding IPRs. For example, most IPRs have not been successful and have raised many issues related to their cost involvement and disclosure methods (Azman et al., 2019; Randil et al., 2018). Another aspect is the misalignment of their expectations relating to IPRs. The suggestions are to establish clear agreements in a timely and economical manner to initiate successful IPRs (Bruneel et al., 2010; Guimón, 2013).

3.2.18 Interests Among Stakeholders

Another common factor affecting the successful UEC is the interest among stakeholders towards UEC (Liyanage et al., 2020). Many argue that universities and industries have divergent interests, which disturb collaboration towards common goals (Amaratunga et al., 2018; Bruneel et al., 2010). For example, universities focus on basic research, while industries focus on fast commercial results—similarly, the conflicts between industries' short-term orientation and the long-term perspective of academic researchers. Another aspect of divergent interests among the two parties is that academics are not ready to change their autonomy to setting research agendas, while industry partners are reluctant to allocate resources to long-term research projects (Giones, 2019). Guimón (2013) provides another aspect, stating that industries are interested in delaying publications to avoid disclosing information, while universities are interested in publishing immediately.

3.2.19 Investments in Research and Development

Studies found that limited opportunities to invest in R&D works by the practitioners. For example, Randil et al. (2018) stated that most disaster management agencies in Sri Lanka tend to collaborate with academics through consultative work rather than investing in research and development with their support or participation. This limited investment in R&D can be identified as a challenge for strengthening more collaborations between universities and industry.

3.2.20 Knowledge

Knowledge gaps are challenging for successful UEC (Amaratunga et al., 2018; Azman et al., 2019). Knowledge gaps have been identified within universities in developing countries (Guimón, 2013). For example, in Sri Lanka, most of the postgraduates in disaster risk management do not originate from Sri Lanka. The postgraduates, either master's or PhD candidates, acquire qualifications from abroad (Randil et al., 2018; Siriwardena et al., 2013). This made industries lose their confidence in the knowledge among academics. Hence, as solutions, universities and academics should be able to update their knowledge on the subject (Burritt et al., 2017).

3.2.21 Language

Language become a key contributor to successful UEC. Language effect is a key factor, especially when collaborating with the industry and with international universities or funding agencies. Most academics use technical jargon when communicating with practitioners or industries (Nielsen, 2017). Language also affects the clarity of communication as well. The study by Amaratunga et al. (2018) emphasized the importance of language for building a successful UEC in Asia. Even though it was not significant as institutional support, it still plays a significant role.

3.2.22 Leadership

Leadership and the lead investigator or researcher are other critical factors for successful UECs (Barnes et al., 2002; Randil et al., 2018). From the university's point of view, identifying suitable and capable academics, specifically young academics, is encouraged, while identifying capable managers or leaders from the industry is recommended for supporting effective collaborations (Awasthy et al., 2020). A similar recommendation was made by Randil et al. (2018) within the Sri Lankan university system. Takanashi and Lee (2019) explained how limited attention was given to identifying critical abilities and quality of successful project leaders in establishing UEC within the Japanese university system.

3.2.23 Legislative and Policy Background

Rules and regulations define organisational policies and strategies. Most academics face difficulties in engaging in UECs due to existing institutional rules and regulations (Azman et al., 2019; Liyanage et al., 2018). Such rules and regulations are also related to IPRs. In addition to institutional policies and procedures, some national policies, rules, and regulations also affect the building of UEC (Bruneel et al., 2010). For example, strict rules are applicable within universities to assure transparency. Specifically, more stringent rules will be applied when a project generates some income through collaborations (O'Reilly & Cunningham, 2017). Due to some available regulations, academic consultation charges do not receive monetary benefits (Randil et al., 2018). Another example was given by Perkmann et al. (2013), highlighting that national policies and legislation are applicable when accessing limited funding through governments. In contrast, some government policies promote UEC. For example, since the Lambert Review (2003) was introduced in the UK, there has been a strong trend for promoting UEC in the country. It motivates academics, universities and industries to collaborate through a world-class research base for supporting industries with innovations (Wilson, 2012). In contrast, more than 73% of respondents in a study conducted in Bangladesh, Thailand and Sri Lanka have agreed that institutional and legislative barriers are the strongest barrier for UEC (Amaratunga et al., 2018).

3.2.24 Management Practices

Evidence highlights the importance of institutional settings as a key factor for establishing and supporting UECs (Bruneel et al., 2010; Guimón, 2013). These practices often hinder building UEC (Seturidze & Topuria, 2021). In contrast, effective project monitoring, planning, and clearly defined objectives and responsibilities are critical for successful UECs (Barnes et al., 2002). Hence, institutions and organisational settings are considered as incentives or enablers for successful UECs (Ankrah & Omar, 2015; Bodas Freitas & Verspagen, 2017).

3.2.25 Motivation

Motivation affects both industry and universities (Perkmann et al., 2013). Most university experts and academics do not find adequate motivation to engage in UECs (Azman et al., 2019) due to several reasons. No value or recognition for building such collaborations is found within many universities specifically within their promotion procedures. Hence, motivation is regarded as a significant barrier among academics for successful UECs. Furthermore, their lack of motivation is affected by academic workload. Hence, studies suggest introducing a reward system in the form of recognizing such effects as part of promotion criteria (Azman et al., 2019). Similarly, Nielsen (2017) highlights to incentivise academics to engage with more collaborations, which benefits universities and industries.

3.2.26 Networking Opportunities

Networking is another key determinant affecting UECs (Amaratunga et al., 2018; Guimón, 2013; Randil et al., 2018; Wilson, 2012). Of the reviewed literature, most studies identified networking as a barrier for building UECs. Similar evidences are found from in most Asian universities (Amaratunga et al., 2018) find limited opportunities to build networks with other HEIs, the industry, professional bodies and communities (Amaratunga et al., 2018). Networking is also related to the university's reputation. Solid long-term reputation facilitates the creation of networks (Bruneel et al., 2010; Guimón, 2013). Nevertheless, finding the correct partners is challenging for building successful collaboration (Guimón, 2013). According to Randil et al. (2018), except for personal contacts, limited avenues are to establish networks between industry and university in Sri Lanka. This limits the collaborations to the same set of experts from universities.

3.2.27 Personal Continuity or Personal/Good Personal Relations

Continuity of personnel contact is a contributor. This is also related to another factor of trust between persons and institutions. Because trust tends to develop on a person-specific basis. For example, frequent changes in the contact person could affect UECs (Barnes et al., 2002; Plewa et al., 2013). Prior experience of working together with industry is considered necessary (O'Reilly & Cunningham, 2017).

3.2.28 Political Background

A supportive political atmosphere will facilitate thriving universities and industries for more innovation. Innovation has been identified as a path towards development. Most countries have recognised the importance of innovations through UEC as their development strategy (Nielsen, 2017). Nevertheless, limited political and state support has always been recognized as a challenge within many countries for building UECs (Ankrah & Omar, 2015).

3.2.29 Reputation

Universities are interested in building reputation, especially when fee-paying students are accepted. Among their reputation-building strategies, engagement in UECs become a popular strategy. Building strong bonds with the industry enhances their reputation. Hence, reputation can be considered an enabler for universities to promote UEC (Nielsen, 2017). From the industry point of view, such collaborations are encouraged by industries for their reputation claiming that they have links with reputed universities. Furthermore, through UECs, they can demonstrate their engagement in social responsibility projects and economic benefits to society (Ankrah & Omar, 2015).

3.2.30 Stability

Well-established cooperate partners are identified as enablers of strong UECs. Instead, unstable partners have affected weakening UECs (Barnes et al., 2002). This is because, unstable and not well-established cooperate partners' attention and priorities are more towards stabilising their business rather being collaborative. This is mainly evidenced by small- and medium-sized enterprises (SMEs) (Ankrah & Omar, 2015).

3.2.31 Trust

Trust significantly affects effective communication (Bruneel et al., 2010; Nielsen, 2017). According to Bruneel et al. (2010), mutual understanding of expectations and working practices leads to trust among keyholders. Hence, trust affects or supports parties exchanging information, knowledge, and materials considered confidential (Giones, 2019).and hence identified as an enabler for building UEC collaborations (Amaratunga et al., 2018). In contrast, a lack of trust among partners has been identified as a key issue affecting the success of UEC in many countries (Azman et al., 2019; Plewa et al., 2013; Randil et al., 2018; Siriwardena et al., 2013). For example, in Malaysia, universities are interested in cooperating with industries, while industries are skeptical about sharing ideas and technology with academics. Hence, lack of mutual trust between academics and industrial partners has been identified as a key barrier within Malaysian universities (Azman et al., 2019).

4 Discussion

Several researchers have classified these challenges based on a different basis. Barnes et al. (2002) presented a classification of factors based on a study conducted in Germany. They classified the factors into five categories: universal factors such as trust and commitment, choice of partners, project manager/management, equality and environmental factors, outcomes and cultural issues.

Bruneel et al. (2010) classified the factors they identified into orientation-related, transaction-related, and operation-related categories. Different orientations of collaborating partners cause orientation-related barriers, while transaction-related barriers were identified based on conflicts that arose. Operational barriers are referred to as barriers related to insufficient knowledge regarding collaboration processes and partners and problems in managing collaboration projects.

Perkmann et al. (2013) also grouped factors affecting UEC under three levels of classification: individual level, organisational level, and institutional level. Accordingly, demographics, career trajectories and productivity, attitudes, and motivations are categorised under individual factors. TT support, formal incentives, university quality, leadership, and department culture are categorised under organisational factors. Scientific discipline, regulation and public policy are classified under institutional factors.

Another classification was presented by Azman et al. (2019) by classifying these factors into orientation-related barriers, such as cultural differences, communication styles and differing priorities and transaction-related barriers, such as contractual disputes concerning intellectual property and project deliverables. Personnel-related barriers include bureaucratic inflexibility and lack of communication.

5 Conclusions

The study was conducted as part of an EU-funded project aiming to strengthen UEC for disaster resilience in three Asian countries; the Philippines, Sri Lanka, and Thailand. As an initial task, the development of a comprehensive framework was targeted. The framework development is based on the factors identified through literature and follow-up interviews with the identified stakeholders.

This paper was written based on the initial literature review and consultancy work with partner countries' experts. Accordingly, the study presented the importance of strengthening UECs for universities and industries separately. Furthermore, the paper presented the importance of UECs for building disaster resilience, as part of objective one.

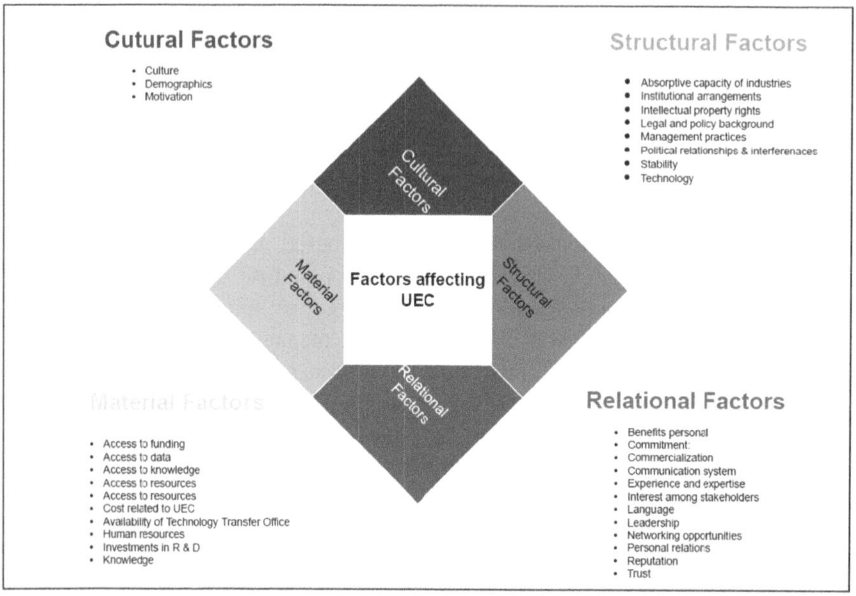

Fig. 2 Classification of factors affecting university-enterprise collaborations

The second objective of the paper was to identify the factors that support or challenge/hinder UECs. Accordingly, the study identified and presented thirty-one factors through the literature review and presented in Fig. 1. The identified challenges were grouped into four categories with the support of the experts, following a thematic analysis method, and presented in Fig. 2. The experts confirmed the identified groups and related factors through a consultative process. Based on their relatedness, these factors were categorised under four different categories as material, structural, relative, and cultural categories.

The above classification of factors presented in the previous studies is somewhat diverse and overlapping. Hence, relatively little help is given to managing them. Hence, this study categorised the challenges into four categories: material, structural, relational, and cultural categories. The reason for introducing such a new categorisation would be more meaningful and more practical than the previous categorisation. Furthermore, the study identifies some factors being considered as a challenge and an enabler. Accordingly, the study refers to material factors as factors that physically affect the establishment and strengthening of UECs. Structural factors are those that are influenced by operational structures and practices. The third category is relational factors. Relational factors refer to factors that affect the relationship between stakeholders. Hence, factors such as trust, language, leadership and so on were categorised under the relational category. Finally, culture, motivation and demographics were categorised under the cultural category.

With this new classification and identification of a list of factors, the study will continue to the next stage to test these identified factors within partner countries in

Asia. The factors identified will guide data collection mechanisms such as developing data collection instruments and deciding the key personals and agencies.

The study findings will be an important contribution to the literature with another new classification. Moreover, the identified themes will benefit from discovering new perspectives and comparing similarities and differences across the selected countries. The study provides inputs for relevant stakeholders when planning and initiating necessary actions to strengthen UECs. In conclusion, the study will be a significant milestone in disaster resilience within Asia by building collaborations between academics and industries.

Acknowledgements The European Commission supports the production of this publication as part of Project SECRA (619022-EPP-1-2020-1-SE-EPPKA2-CBHE-JP) and does not constitute an endorsement of the contents which reflect the views only of the authors, and the Commission cannot be held responsible for any use which may be made of the information contained therein.

References

Abdeen, F., Fernando, T., Kulatunga, U., Hettige, S., & Ranasinghe, K. A. (2021). Challenges in multi-agency collaboration in disaster management: A Sri Lankan perspective. *International Journal of Disaster Risk Reduction, 62*, 102399.

Aitsi-Selmi, A., Murray, V., Wannous, C., Dickinson, C., Johnston, D., Kawasaki, A., Stevance, A.-S., & Yeung, T. (2016). Reflections on a science and technology agenda for 21st century disaster risk reduction. *International Journal of Disaster Risk Science, 7*(1), 1–29.

Amaratunga, D., Liyanage, C., & Haigh, R. (2018). A study into the role of international collaborations in higher education to enhance research capacity for disaster resilience. *Procedia Engineering, 212*, 1233–1240.

Ankrah, S., & Omar, A.-T. (2015). Universities-industry collaboration: A systematic review. *Scandinavian Journal of Management, 31*(3), 387–408.

Arant, W., Fornahl, D., Grashof, N., Hesse, K., & Söllner, C. (2019). University-industry collaborations—The key to radical innovations? *Review of Regional Research, 39*(2), 119–141.

Awasthy, R., Flint, S., Sankarnarayana, R., & Jones, R. L. (2020). A framework to improve university-industry collaboration. *Journal of Industry—University Collaboration.*

Azman, N., Sirat, M., Pang, V., Lai, Y. M., Govindasamy, A. R., & Din, W. A. (2019). Promoting university-industry collaboration in Malaysia: Stakeholders' perspectives on expectations and impediments. *Journal of Higher Education Policy and Management, 41*(1), 86–103.

Barnes, T., Pashby, I., & Gibbons, A. (2002). Effective university-industry interaction: A multi-case evaluation of collaborative R&D projects. *European Management Journal, 20*(3), 272–285.

Barton, T., Beaven, S., Cradock-Henry, N., & Wilson, T. (2020). Knowledge sharing in interdisciplinary disaster risk management initiatives: Cocreation insights and experience from New Zealand. *Ecology and Society, 25*(4).

Berbegal-Mirabent, J., García, J. L. S., & Ribeiro-Soriano, D. E. (2015). University–industry partnerships for the provision of R&D services. *Journal of Business Research, 68*(7), 1407–1413.

Bodas Freitas, I. M., & Verspagen, B. (2017). The motivations, institutions and organization of university-industry collaborations in the Netherlands. *Journal of Evolutionary Economics, 27*(3), 379–412.

Brundenius, C., Lundvall, B.-Å., & Sutz, J. (2009). *The role of universities in innovation systems in developing countries: Developmental university systems-empirical, analytical and normative perspectives*. Edward Elgar Publishing.

Bruneel, J., d'Este, P., & Salter, A. (2010). Investigating the factors that diminish the barriers to university-industry collaboration. *Research Policy, 39*(7), 858–868.

Burritt, R., Guthrie, J., Evans, E., & Christ, K. (2017). Expanding collaboration between industry and business faculties in Australia. Improving collaboration and innovation between industry and business schools in Australia.

Chung, C. J. (2014). An analysis of the status of the triple helix and university-industry-government relationships in Asia. *Scientometrics, 99*(1), 139–149.

Clayton, P., Feldman, M., & Lowe, N. (2018). Behind the scenes: Intermediary organizations that facilitate science commercialization through entrepreneurship. *Academy of Management Perspectives, 32*(1), 104–124.

CRED-UNISDR. (2020). *Human cost of disasters: An overview of the last 20 years 2000–2019.* Centre for Research on the Epidemeology of Disasters. Retrieved December 15, 2020 from https://www.cred.be/publications

Creswell, J. W., & Creswell, J. D. (2017). *Research design: Qualitative, quantitative, and mixed methods approaches.* Sage Publications.

Davey, T., Baaken, T., Muros, V. G., & Meerman, A. (2011). *The State of European university—Business cooperation.* European Commission. Retrieved July 22 2011 from http://ec.europa.eu/dgs/education_culture/repository/education/tools/docs/uni-business-cooperation_en.pdf

Druedahl, L. C., Minssen, T., & Price, W. N. (2021). Collaboration in times of crisis: A study on COVID-19 vaccine R&D partnerships. *Vaccine, 39*(42), 6291–6295.

Ergun Demirel, D., & Bayer, D. (2015). Establishment of cooperation and collaboration platforms between universities and industry to improve education quality. *The Online Journal of Quality in Higher Education, 2*(3), 59.

Fu, K.-W., & Zhu, Y. (2020). Did the world overlook the media's early warning of COVID-19? *Journal of Risk Research*, 1–5.

Giones, F. (2019). *University-industry collaborations: An industry perspective.* Management Decision.

Guimón, J. (2013). Promoting university-industry collaboration in developing countries. *World Bank, 3*, 12–48.

HRW. (2021). *"They Killed Us from the Inside": An Investigation into the August 4 Beirut Blast.* Human Rights Watch. Retrieved March 12, 2024 from https://www.hrw.org/report/2021/08/03/they-killed-us-inside/investigation-august-4-beirut-blast

ILO. (2023). *The Rana plaza disaster ten years on: What has changed?* International Labour Organisation. Retrieved March 12 2023 from https://www.ilo.org/infostories/en-GB/Stories/Country-Focus/rana-plaza#intro

Kaklauskas, A., Amaratunga, D., Haigh, R., Binkyte, A., Lepkova, N., Survila, A., Lill, I., Tantaee, S., & Banaitis, A. (2018). A model and system for an integrated analysis of the iterative life cycle of university-industry partnerships. *Procedia Engineering, 212*, 270–277.

Kaymaz, K., & Eryiğit, K. Y. (2011). Determining factors hindering university-industry collaboration: An analysis from the perspective of academicians in the context of entrepreneurial science paradigm. *International Journal of Social Inquiry, 4*(1).

Khan, G. F., & Park, H. W. (2012). Triple Helix and innovation in Asia using scientometrics, webometrics, and informetrics. *Scientometrics, 90*(1), 1–7.

Lindelöf, P., & Löfsten, H. (2003). Science park location and new technology-based firms in Sweden-implications for strategy and performance. *Small Business Economics, 20*, 245–258.

Link, A. N., Siegel, D. S., & Bozeman, B. (2007). An empirical analysis of the propensity of academics to engage in informal university technology transfer. *Industrial and Corporate Change, 16*(4), 641–655.

Liyanage, C., Thakore, R., Amartunga, D., Mustapha, A., & Haigh, R. (2018). The barriers to research and innovation in disaster resilience in higher education institutions in Asia. *Procedia Engineering, 212*, 1225–1232.

Liyanage, C., McDonald, R., Amaratunga, D., & Haigh, R. (2020). *University-Industry Partnerships (UIPs) for research and innovation in disaster resilience*. Theme Seminar of the Sri Lanka Association for the Advancement of Science: Science for Quality Life.

Mahase, E. (2021). How the Oxford-AstraZeneca covid-19 vaccine was made. *BMJ*, 372.

Morris, D., Leal, X., & Göbel, H. (2018). Promoting university-business cooperation in developing countries: Evidence from Central Asia. *International Journal of Innovation and Regional Development, 8*(3), 214–231.

Muscio, A. (2010). What drives the university use of technology transfer offices? Evidence from Italy. *The Journal of Technology Transfer, 35*(2), 181–202.

Nielsen, C. (2017). A personal reflection: European experiences on value exchange in university-industry collaborations. In *Improving collaboration and innovation between industry and business schools in Australia* (pp. 53–61). RMIT University.

NYTimes. (2014). 2014 Proves a Grim Year in Malaysian Aviation. *The New York Times*. Retrieved March 12 2014 from https://www.nytimes.com/2014/12/29/world/asia/2014-proves-a-grim-year-in-malaysian-aviation.html

O'Reilly, P., & Cunningham, J. A. (2017). Enablers and barriers to university technology transfer engagements with small-and medium-sized enterprises: Perspectives of principal investigators. *Small Enterprise Research, 24*(3), 274–289.

Opoku, A., Ahmed, V., & Akotia, J. (2016). Choosing an appropriate research methodology and method. In *Research methodology in the built environment* (pp. 32–49). Routledge.

Orale, R. L. (2013). Redirecting academe-industry partnership for countryside development. *Countryside Development Research Journal, 1*(01), 9–20.

Perkmann, M., Tartari, V., McKelvey, M., Autio, E., Broström, A., D'este, P., Fini, R., Geuna, A., Grimaldi, R., & Hughes, A. (2013). Academic engagement and commercialisation: A review of the literature on university-industry relations. *Research Policy, 42*(2), 423–442.

Plewa, C., Korff, N., Baaken, T., & Macpherson, G. (2013). University-industry linkage evolution: An empirical investigation of relational success factors. *R&D Management, 43*(4), 365–380.

Pulhin, J. M., Inoue, M., Shaw, R., Pangilinan, M. J. Q., & Catudio, M. L. R. O. (2021). Climate change and disaster risks in an unsecured world. In *Climate change, disaster risks, and human security* (pp. 1–19). Springer.

Puślecki, Ł, Dąbrowski, M., & Puślecki, M. (2021). Development of innovation cooperation in the time of COVID-19 pandemic. *European Research Studies, 24*, 1049–1073.

Randil, C., Jayasiri, G., Perera, C., Siriwardana, C., Liyanage, C., Hettiarachchi, S., & Haigh, R. (2018). Need for strong university-industry partnerships: a case study in Sri Lanka. In *MATEC Web of Conferences*.

Seturidze, R., & Topuria, N. (2021). A way of developing collaboration between universities and businesses in a time of COVID-19. *Kybernetes*.

Shaw, R., Kim, Y.-K., & Hua, J. (2020). Governance, technology and citizen behavior in pandemic: Lessons from COVID-19 in East Asia. *Progress in Disaster Science*, 100090.

Siegel, D. S., & Phan, P. H. (2005). Analyzing the effectiveness of university technology transfer: Implications for entrepreneurship education. In *University entrepreneurship and technology transfer* (pp. 1–38). Emerald Group Publishing Limited.

Siriwardena, M., Malalgoda, C., Thayaparan, M., Amaratunga, D., & Keraminiyage, K. (2013). Disaster resilient built environment: Role of lifelong learning and the implications for higher education. *International Journal of Strategic Property Management, 17*(2), 174–187.

Sjöstedt, M., & Povitkina, M. (2017). Vulnerability of small island developing states to natural disasters: How much difference can effective governments make? *The Journal of Environment & Development, 26*(1), 82–105.

Takanashi, C., & Lee, K.-J. (2019). Boundary spanning leadership, resource mobilisation, and performance of university-industry R&D projects: A study in a Japanese university. *Technology Analysis & Strategic Management, 31*(2), 140–154.

Tantanee, S., Buranajarukorn, P., & Apichayakul, P. (2018). University-industry linkages in the disaster resilience sector: A case study of Thailand. *Procedia Engineering, 212*, 519–526.

Tartari, V., Salter, A., & D'Este, P. (2012). Crossing the Rubicon: exploring the factors that shape academics' perceptions of the barriers to working with industry. *Cambridge Journal of Economics, 36*(3), 655–677.

Thayaparan, M., Malalgoda, C., Keraminiyage, K., & Amaratunga, D. (2014). Disaster management education through higher education-industry collaboration in the built environment. *Procedia Economics and Finance, 18*, 651–658.

Times of India. (2024). *3 killed after building under construction collapses in Mumbai*. Times of India. Retrieved March 12, 2024 from http://timesofindia.indiatimes.com/articleshow/108430998.cms?utm_source=contentofinterest&utm_medium=text&utm_campaign=cppst

UN. (2015). *Women's leadership in risk-resilient development: good practices and lessons learned*. UN Office for Disaster Risk Reduction (UNISDR).

UNISDR. (2015). *Sendai framework for disaster risk reduction 2015–2030*. http://www.preventionweb.net/files/43291_sendaiframeworkfordrren.pdf

Wilson, T. (2012). *A review of business—University collaboration*.

Climate Change Education in Universities: A Sri Lankan Perspective

M. Esham, M. M. S. C. Senevirathne, D. A. M. De Silva, H. S. R. Rosairo, A. S. Karunaratne, E. M. Wimalasiri, M. Korkeakoski, and O. Haapasaari

1 Background and Significance of Climate Change Education

Human activities such as the combustion of fossil fuels and deforestation have significantly contributed to global warming. Greenhouse gas emissions have made the earth warmer by 1.1 °C above the average levels observed between 1850 and 1900 (Masson-Delmotte et al., 2021). If greenhouse gas emissions persist, global warming will continue to increase, and it is estimated that it may reach 1.5 °C within the next two decades (Masson-Delmotte et al., 2021). Each increment of global warming will amplify the occurrence of multiple and simultaneous hazards. Exceeding the threshold of 1.5 °C global warming could trigger multiple climate tipping points associated with tipping elements including ecosystem changes, shifts in the frequency and intensity of extreme events, and the significant release of carbon stocks to the atmosphere (Armstrong McKay et al., 2022; Lenton et al., 2008).

In the South Asian region, it is projected that the average annual maximum temperature could rise by 1.4–1.8 °C by 2030 and 2.1–2.6 °C by 2050. Consequently, areas experiencing heat stress in the region may witness a 12% increase by 2030 and a 21% increase by 2050 (Tesfaye et al., 2017). South Asia is ranked as one of the hotspot regions prone to climate extremes, including droughts, floods, and heatwaves,

M. Esham · M. M. S. C. Senevirathne (✉) · D. A. M. De Silva · H. S. R. Rosairo
Department of Agribusiness Management, Faculty of Agricultural Sciences, Sabaragamuwa University of Sri Lanka, Belihuloya 70140, Sri Lanka
e-mail: sulochana@agri.sab.ac.lk

A. S. Karunaratne · E. M. Wimalasiri
Department of Export Agriculture, Faculty of Agricultural Sciences, Sabaragamuwa University of Sri Lanka, Belihuloya 70140, Sri Lanka

M. Korkeakoski · O. Haapasaari
Finland Futures Research Centre, Turku School of Economics, University of Turku, 20014 Turku, Finland

C. Malalgoda et al. (eds.), *Climate Change Adaptation in the Built Environment*,
https://doi.org/10.1007/978-3-031-75826-3_18

making it a prominent hotspot for such events (Mie Sein et al., 2021; Ullah et al., 2021). The agriculture sector in South Asia, which is the mainstay of the economy will be severely affected by climate change posing a significant threat to food security in the region (Schellnhuber et al., 2013; Yan & Alvi, 2022). As a developing island nation in South Asia, Sri Lanka is not an exemption from these events. The temperature in Sri Lanka under the highest emissions pathway (RCP8.5) is projected to increase by 2.9–3.5 °C by the 2090s compared to the baseline of 1986–2005. It is noteworthy that minimum temperatures are projected to increase at a faster rate than average temperatures. In Sri Lanka, the number of days exceeding 35 °C is potentially increasing from a baseline of 20 days to over 100 days by the 2090s under the RCP8.5 emissions pathway showing signs of threat from extreme heat. Additionally, there is a high degree of variability in projected rainfall trends for Sri Lanka with more rainfall extremes.

Over the years, there has been considerable attention given to education as a tool to combat climate change. The United Nations (UN) initiatives on climate change education (CCE) can be traced back to the Decade of Education for Sustainable Development (DESD) that was observed from 2005 to 2014 under the patronage of UNESCO. In 2009, UNESCO endeavored to integrate CCE into the education policy and curricula. In the 16th session of the COP of UNFCC, Article 6 seeking to reduce the impact of climate change by enabling society to be a part of the solution was unanimously adopted in which education, training, and awareness raising were recognized as integral parts of the effort underscoring the importance of CCE. At the 21st session of the UNFCC held in 2015, the Paris Agreement was adopted, and Article 12 of the Paris Agreement calls on member states to enhance CCE, training, public awareness, public participation, and public access to relevant information. During the same period, the Sendai Framework for Disaster Risk Reduction was adopted highlighting the importance of promoting education. The most recent UN initiative related to Sustainable Development Goal 13 was introduced for urgent action to mitigate climate change and its impacts. Target 13.3 highlights the need to improve education, awareness-raising, and human and institutional capacity on climate change mitigation, adaptation, impact reduction, and early warning. Beyond doubt, it has been realized that education can play a vital role in promoting prompt and fair action to mitigate and adapt to the impacts of climate change to achieve sustainable development goals.

With the increased concern about the consequences of climate change, there is a growing interest in CCE throughout the world (Leal Filho et al., 2021; Monroe et al., 2019). Even though CCE has advanced to a higher level in several countries, the CCE in Sri Lanka is at the infant stage and has not been assessed adequately to understand the current status and provide directions for learning from evolving best practices in other parts of the world.

According to a bibliometric analysis conducted by authors in 2023 using the records available in the Scopus database for the period of 1990–2022, there are 11,724 publications relevant to CCE reported from 159 countries. The highest number of publications were reported from the United States of America (3826) followed by the United Kingdom (1492) and Australia (986). More than 100 publications were reported from 34 countries. However, only 14 publications were published in Sri Lanka. Since a higher number of CCE-related publications are available in a limited number of developed countries, an opportunity exists to popularize CCE and identify and replicate them in other countries including Sri Lanka.

1.1 Purpose and Scope of the Chapter

Currently, the higher education sector in Sri Lanka has experienced a gap in climate change curricula. This chapter highlights the importance of developing a curriculum intended to enable educators to teach about climate change; nature leaders who can influence policy analysis and civic action related to the issue of climate change; create a platform to become a climate intrapreneur; and learn what it takes to influence and lead critical initiatives inside of businesses.

The Climate Change Education (CCE) landscape in Sri Lanka is characterized by numerous fragmented elements. This chapter underscores the significance of adopting an integrated approach that connects key dimensions of climate change.

This chapter is focused on the identification of the gaps in climate education through a structured survey conducted (refer to Sect. 3) with state universities in Sri Lanka, considerations of curriculum design and development, program learning outcomes that comply with Sri Lanka Qualifications Framework (SLQF), innovative teaching and learning pedagogies, assessments methods and recommendations for university academics and policymakers for developing curriculum for undergraduate and postgraduate programs considering global best practices. The Climate University in Finland can be considered as a model for developing courses and adapting innovative teaching and assessment methods beyond traditional university education in Sri Lanka.

2 Climate Change Education

2.1 Global Best Practices in Climate Change Education: Lessons for Sri Lanka

It was reported that education on climate change started with the awareness of societies on adaptation to climate change, which was later developed into curricula of universities and colleges (Perkins et al., 2018). Since climate change is not solely a physical science problem and contains social aspects, CCE is complicated due to site/region-specific social interactions (Perkins et al., 2018). Good CCE should focus on cognitive (learning to know), socio-emotional (to be), action-oriented (to do), and justice-focused (to live together) approaches within the learning process (Gibb, 2016; Kristen & Marcia, 2020). Therefore, the whole institution approach is necessary for

CCE which should be implemented at all educational levels (Gibb, 2016; Kristen & Marcia, 2020). Some of the best practices in CCE at the whole institution level can be categorized as follows (Dunlop & Rushton, 2022; Gibb, 2016; Kristen & Marcia, 2020).

Governance—Best practices in CCE associated with the governance of the institution.

- Establishing a sustainability portfolio and hiring sustainability staff is the basic step. This is important because the whole CCE process depends on the sustainability staff.
- Environmental committees to assist climate-friendly and sustainable initiatives play an important role in CCE.
- Climate action plans with measurable targets are essential at different institutional levels. Due to the variation in social and geographic aspects of CC, it is better to develop targets at different institutional levels.
- Promoting climate change-related research is another important factor that is being practiced across the world as a best global practice in CCE.

Teaching and Learning—Best practices in CCE associated with teaching and learning at the institution.

- Support environmental, climate change, and sustainability education within all subjects is an important teaching and learning-related best practice across the world.
- Provide education resources, programs, and workshops.
- Providing professional development opportunities related to CCE is another good practice in many countries.
- Develop eco-challenges for staff and students.

Community Partnerships—Best practices in CCE associated with the community partnerships of the institution.

- Since the CC is different among locations and societies, learning extended outside the classroom through community partnerships is essential.
- Partner with the different stakeholders to co-develop climate solutions that are being practiced throughout the world.
- Involve farmers and local communities to discuss local effects of climate change.
- Partner on CCE projects is another best practice in CCE associated with the community partnerships of the institution.

Facilities and Operations—Best practices in CCE associated with facilities and operations of the institution.

- Environmental improvements (efficient use of energy, tree planting, composting, etc.).

There are some other best practices in CCE used in different countries as follows.

- Pervasive/Cross-curricular integration—Incorporation of CCE with Education for Sustainable Development (ESD) within a wide range of subjects is one of the best practices. For example, CCE can be used in Mathematics (climate change mathematics-related problems), History (history of climate/resource extraction), Geography (global carbon footprint), Biology (biological aspects of climate change), English (climate-related speeches and activities), Arts (posters, protest art, art installations) and many other subjects as practiced globally (Gibb, 2016; Kristen & Marcia, 2020; NCSE, 2016; Newsome et al., 2023).
- Establish/join networks—Previous studies suggested the importance of networking in climate action students/institutes can share ideas, problems, and resources. These organizations include ministries, local, national, or international organizations (Hargis et al., 2018).
- Make climate change local and relevant. The impact of climate change varies across geographical locations. Evaluation and link to the local meaning can make CCE more engaging (Kuntze, 2023; Monroe et al., 2019).
- Creative technology—Use of advanced technology such as communication platforms and interactive websites in new and exciting ways in CCE (Hargis et al., 2018).
- Connect to Indigenous knowledge—Engagement with local/Indigenous communities can make CCE more relevant and interesting (Kristen & Marcia, 2020).
- Engage diverse climate action actors (Hargis et al., 2018; Kristen & Marcia, 2020)
- Critical pedagogy—critical thinking among learners needs to be encouraged where they may be able to make informed judgments by critically analyzing, evaluating, and interpreting the problems they face due to climate change (Mbah et al., 2022).
- Make it hopeful—since the CCE can trigger anger, guilt, fear, or hopelessness, it is useful to integrate science with solutions (NCSE, 2016).
- Incorporating conservation practices and disaster risk management knowledge in formal and informal education (Mbah et al., 2022).
- Experiential learning (internships, residencies) (Mbah et al., 2022).
- Courses offered in undergraduate and graduate programs and research (various post-graduate theses, research centers) (Leal Filho et al., 2021; Mbah et al., 2022; UBC, 2023).
- Many training events focused on staff and/or students (Leal Filho et al., 2021; Mbah et al., 2022).
- Supporting and promoting disaster risk research in academics (Mbah et al., 2022)
- Mobilizing youth to engage in CC understanding (Mbah et al., 2022)
- CC adaptation knowledge is incorporated in formal and non-formal curricula (Mbah et al., 2022)
- Action-based learning of CCE and promotion of environmentally friendly practices (Mbah et al., 2022)

2.2 Climate Change Education in Sri Lanka: From Primary School to University Level

In his book 'Sustainable Education,' Stephen Sterling describes whole systems thinking as a cognitive and existential attribute essential for transcending prevailing modes of thought, which are analytical, linear, and reductionist in nature (Sterling, 2001). In this context, educating with systems thinking needs not only being able to work with the broader perspective but also to think in circularity and fullness. Systems thinking helps examine not just the relations that connect key components and people but also the blocks, groups of people in the climate movement are beginning to connect the blockages to the current economic system. In the climate education system, some blocks such as many adults have not received CCE and many do not feel equipped to understand the science, the sense of urgency, or how to deal with it. Climate change, disasters, and resilience building essentially need the knowledgeable community to prepare for the future. This highlights the importance of climate education, which should start early, be comprehensively mainstreamed across the education system, and result in a skilled and educated workforce as well as a climate-aware and -conscious society (Mombauer, 2023). Further, teachers at the school level are often not empowered to impart skills to young people in relation to climate literacy, CCE, and response action in emergencies. Therefore, climate-related continuing professional development (CPD) is necessary for school staff and not just STEM teachers and university staff.

Primary education including nursery and pre-schools focuses on climate literacy or introducing basic concepts in the form of general knowledge. Subjects taught at the secondary school level are focused on delivering information on climate change. However, required prominence is only given to some aspects of climate change, especially when it comes to innovative thinking, which allows for combating climate change impact. In the final stage of school education at the senior secondary level, some aspects of climate change are included in the curriculum as taught courses only. Practical exposure is not included. Most importantly, the youth component is ignored in mainstreaming efforts. Countable numbers of youth pay attention to climate change aspects only if they have a deep passion for the subject. With reference to the youth perspectives on climate change and education in Sri Lanka (UNICEF, 2021) 34% can explain what climate change is and 48% highlighted that they have learned about climate change and its consequences in school. Gender and age differences were not observed in the level of climate literacy. Further, the majority of the students (67%) learned the climate change in Geography class while 18% learned in science class (UNICEF, 2021).

In contrast, the state universities of Sri Lanka offer various graduate and postgraduate courses and degrees based on environmental studies including climate change. Unfortunately, only a countable number of students get the opportunity to study climate change subjects deeply, which should be included in the knowledge base of all citizens. Formal education must make arrangements to educate people or give an adequate understanding of climate change and related issues, management strategies, and overall sustainable solutions. About 56% of the students are willing to learn all aspects of climate change while 18% are willing to explore more local actions of climate change (UNICEF, 2021). Further, CCE indirectly enhances socio-economic status and social capital. Educated citizens have a better awareness of climate change and it gives them the knowledge and skills to adapt to flexibility. With regard to learning methods, climate education could benefit from active and experiential learning, such as project-based learning, field trips to relevant projects and facilities, citizen science projects, problem-based learning, or role-playing exercises, potentially tying in with Model UN initiatives or the National Youth Parliament (Mombauer, 2023). Figure 1 explains the CCE system of Sri Lanka along with its institutions.

The informal sector, especially, NGOs are playing a significant role in educating people who left the formal system but playing an important role in operations. Climate literacy programs for both young and adults help to empower field-level operators and prepare them to manage situations. Most of the programs are equipped with problem-based, practice-oriented, and community-level interventions.

2.3 *Framework for Climate Change Education Qualifications in Sri Lanka*

2.3.1 The Sri Lanka Qualifications Framework (SLQF)

Overview

The SLQF serves as a nationally standardized framework encompassing all higher education qualifications offered by both public and private higher education institutions (HEIs) in Sri Lanka, focusing on post-secondary education. It places significant emphasis on two key aspects: the extent of learning undertaken by students and the program learning outcomes expected of qualification holders. This framework consists of twelve levels, each accompanied by level descriptors that provide a detailed description of the expectations at each level. Additionally, it clarifies the number of credits students must earn, aligning with the volume of learning involved. To promote consistency in qualification naming, the SLQF includes designators and qualifiers for each qualification, represented by abbreviations. Furthermore, the SLQF outlines minimum admission requirements and outlines potential pathways for academic progression.

The updated version of the SLQF published by the Ministry of Higher Education was in 2015 with three revisions (the First Edition in June 2012, the Second Edition in October 2012, and the Third Edition in January 2013), and later versions superseded all previous versions (Framework, 2015).

Levels of Qualifications

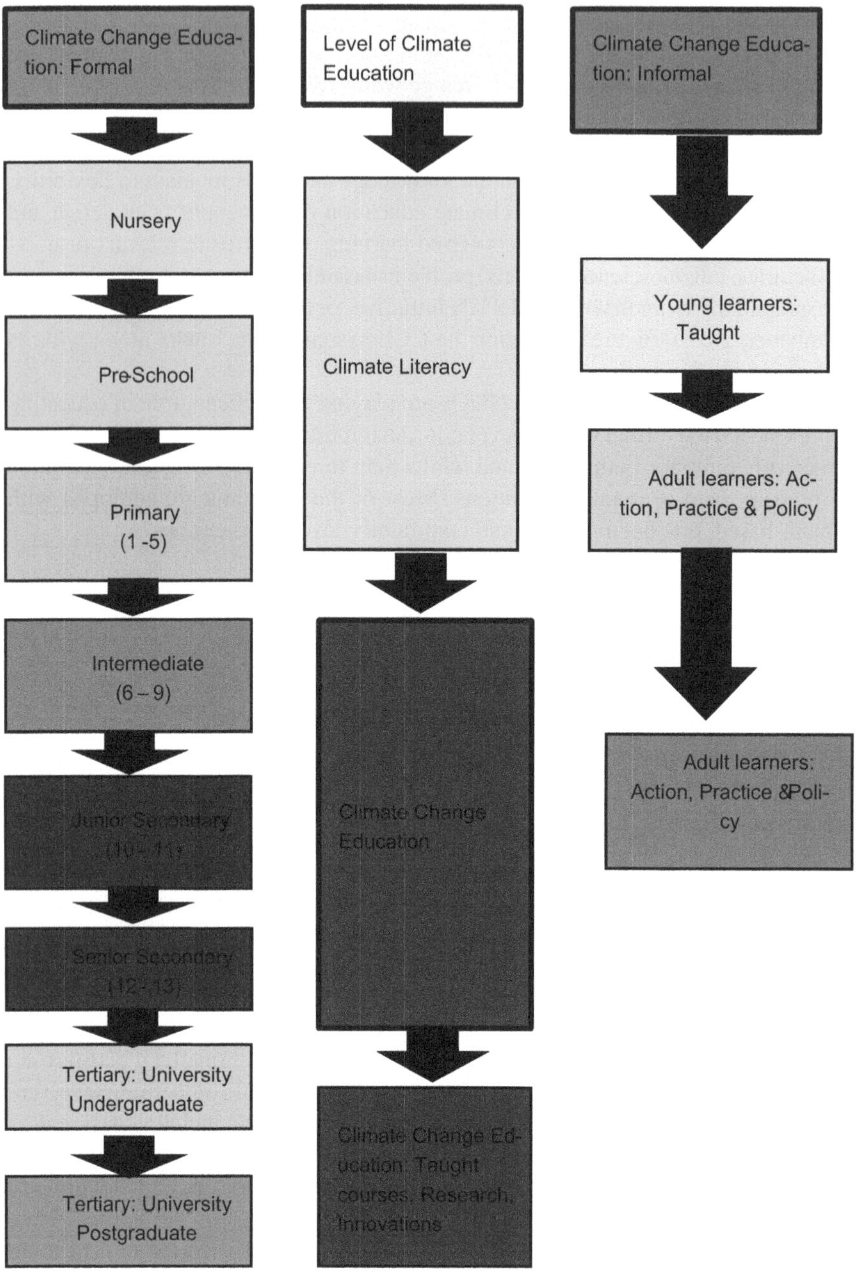

Fig. 1 Climate change education in Sri Lanka—levels and institutional engagement

The SLQF comprises twelve levels. As one progresses through the levels, there is a growing need for more advanced learning outcomes, and the complexity of learning also rises accordingly. The first two levels (levels 1–2) are senior secondary level education qualifications, and the next four levels (levels 3–6) are undergraduate qualifications. The other six levels (levels 7–12) are postgraduate qualifications. The levels are not necessarily directly related to the years of study.

Table 1 provides a summary of the SLQF levels, the corresponding qualifications conferred at each level, and the minimum credit requirements for each level.

Naming of Qualifications

Table 1 Minimum volume of learning required for each level of SLQ

SLFQ level	Qualification awarded	Minimum volume of learning for the award
12	Doctor of Philosophy/M.D. with Board Certification/Doctor of Letters/Doctor of Science	Minimum 3 years of full-time or equivalent time of original research after SLQL 6 or above
11	Master of Philosophy	Minimum 2 years of full-time or equivalent time of original research after SLQL 6 or above
10	Masters with course work and research component	60 credits after SLQL 5 or SLQL 6 including a research component of a minimum of 15 credits
9	Masters by coursework*	30 credits after SLQL5 or SLQL6
8	Postgraduate diploma*	25 credits after SLQL5 or SLQL6
7	Postgraduate certificate*	20 credits after SLQL5 or SLQL6
6	Bachelors honours	120 credits after SLQL 2, Of which 90 credits after SLQL3, Of which 60 credits after SLQL4, Of which 30 credits after SLQL5
5	Bachelors	90 credits after SLQL2, Of which 60 credits after SLQL3, Of which 30 credits after SLQL4
4	Higher diploma	60 credits after SLQL 2
3	Diploma	Of which 30 credits after SLQL 3
2	Advanced certificate (General Certificate of Examination Advanced Level (GCE A/L) or equivalent)	30 credits after SLQL2
1	Certificate (GCE O/L or equivalent)	

Source Sri Lanka Qualification Framework (SLQF)

*Only the extension category study programmes that lead to Postgraduate Certificate, Postgraduate Diploma and Masters by course work are placed at SLQF Levels 7, 8 and 9 respectively. Conversion category study programmes that award Postgraduate Certificate, Postgraduate Diploma or Masters by course work are placed at SLQF Level 6

The first identifier assigned to a qualification is its qualification type. The SLQF includes the following qualification types:

- Senior Secondary Level:—Certificate, and Advanced Certificate
- Undergraduate Level:—Diploma, Higher Diploma, Bachelors, and Bachelors Honours
- Postgraduate Level:—Postgraduate Certificate, Postgraduate Diploma, Masters, and Doctorate.

The second identifier assigned to a qualification is referred to as the designator. This specifies the broad area of study or discipline. All degrees, i.e., Bachelor, Master, and Doctoral degrees have designators. The third identifier assigned to a qualification is known as the qualifier. This is used to specify the field of specialization of a qualification. The qualifier can be found in most qualification types. The term 'in' serves as the connecting word between the qualifier and either the qualification type or its designator, depending on the context. Some examples are Bachelor of Science Honours in Botany. Every qualification type of Levels 1–12 on the SLQF has a credit value allocated to each of its parts and the whole qualification (Table 1).

2.4 *Job Scope in Sri Lanka*

It has been stressed by the United Nations Environmental Program (UNEP) and International Labor Organization (ILO) that sectors such as energy, construction, transportation, basic industries and recycling, food and agriculture, and forestry could complement the creation of green jobs in the future. There are strong sustainability-led opportunities in Sri Lanka, in the agricultural, renewable energy, waste management, and tourism sectors. Sri Lanka has recognized adaptation as a sustainable strategy for addressing the issue of climate change. Further, green accountancy, green energy, green business, green industries, green agriculture, green cities and buildings, green transport, green trading, green infrastructure, green service, and green jobs were identified by the Government of Sri Lanka as national approaches for moving towards a green economy as means of climate change adaptions.

According to Sivarajah et al. (2017), the specialized skills required for sustainability-driven projects in Sri Lanka are limited. For example, most people handling sustainability management in various institutions are not experts with formal education in sustainable development but are persons who have developed their skills over time and with experience. Therefore, there is a need to recognize the importance of education and training to address climate change and its impacts. Universities and other higher education institutes (HEI) in Sri Lanka are becoming more aware of their responsibility to provide students and the public with the skills necessary to actively contribute to the mitigation and adaptation of climate change.

Job opportunities in the green economic sectors in Sri Lanka seem to link with education and training. Several state universities are offering undergraduate and postgraduate level education in the areas of climate change and sustainability. However, study programs directly labeled as climate change or sustainability are rare in Sri Lanka. There are programs with a few related course modules included. Therefore, those who have an interest in making a career in climate change and sustainability find it challenging to enhance their knowledge and skills.

Career opportunities in Sri Lanka within the areas of climate change, natural environment, sustainability, and other allied areas are becoming popular. Mitigation, conservation, research, and social policy analysis are some of the popular areas where related careers exist. Diverse and rewarding careers are available in government ministries, non-governmental organizations, environmental and business consultancies, public sector organizations, and manufacturing and service industries in the private sector. There are numerous entrepreneurial opportunities also available in this sector.

3 Methodology of the Survey

The survey was carried out on climate change and sustainability education (CCSE) relating to curricula implemented in Sri Lankan HEIs. The sample consisted of 33 academics from state universities in Sri Lanka (University of Colombo, University of Ruhuna, Sabaragamuwa University of Sri Lanka, University of Peradeniya, University of Kelaniya, Uva Wellassa University of Sri Lanka, South Eastern University of Sri Lanka, University of Sri Jayewardenepura, Eastern University Sri Lanka, Ocean University of Sri Lanka, Open University of Sri Lanka) that have already integrated climate change-related aspects in the course curricula of degree programs in their respective HEIs. These respondents were in the ranks of senior lecturers, course or subject coordinators, or the heads of relevant departments. The dominant knowledge areas were natural and environmental sciences, environmental management, agriculture, geography, and civil engineering. Other areas included sociology, tourism management, biosystems and engineering technology, and coastal and marine resources management.

To collect data, a semi-structured questionnaire survey was administered using Google Forms. The questionnaire consisted of sections focusing on the subject related backgrounds of respondents, their perceptions of the adequacy of climate change and sustainability education within the university system, the integration methods of Climate Change Education (CCE) into curricula, specific competences, capabilities, and dispositions intended to be fostered in students through CCE, student-centered learning and teaching approaches, techniques employed in student-centered teaching, and the assessment methods utilized to evaluate student learning outcomes. Respondents were selected based on a scoping study done to identify climate change-related courses at the undergraduate and postgraduate levels offered by Sri Lankan universities.

4 Results and Discussion

4.1 Climate Change-Related Courses Offered by Higher Education Institutions in Sri Lanka

Faculties of science, agriculture, engineering, and arts; centers; and several postgraduate institutes offer relevant programs and course modules. Some specific departments of study that offer these programs and courses include environmental studies, natural resource management, geography, civil engineering, agriculture and allied areas, botany, and zoology. Postgraduate institutes of agriculture and science are prominent in offering these programs and courses at postgraduate level.

There are few certificate and short courses, taught undergraduate and postgraduate programs, and research programs available in climate change and sustainability, environmental management, environmental science, disaster risk management, and allied disciplines. Objectives of these programs are to provide sufficient theoretical knowledge to evaluate perspectives on the ecological and economic value of the environmental assets and products; to evaluate the overall issue caused by anthropogenic activity and the impact of climate change as a major global environmental concern; and to enable learning about the strategies such as national efforts, international cooperation in mitigation and adaptation, and the application of other remedial measures in addressing the climate change and the adaptation related to sustainable development with special reference to Sri Lanka. They also attempt to train professionals to undertake these tasks. They also use specific tools such as cost–benefit analysis, environmental impact assessment, Geographic Information Systems (GIS), and remote sensing applications in environmental management.

There are no undergraduate degree programs in Sri Lanka dedicated to climate change. However, there are subjects offered related to climate change. Higher educational institutes offer a few undergraduate degree programs with courses aimed at climate change and a few allied areas. Bachelor's degree programs in B.Sc. in Environmental Science; Environmental Development and Sustainability; Eco-business and Environment; Agriculture, Agricultural Resource Management and Technology; Environmental Management and Forestry; Green Technology; Applied Sciences; Civil and Environmental Engineering; and BA honors in Geography and Environmental Management are some undergraduate degree programs offered by universities that offer subjects related to climate change. Some of the degree programs also focus on Environmental Conservation and Management and a Bachelor of Arts Honours degree in Geography with subjects related to climate change. There are also short courses and diploma programs available in the areas of climate change and its impact on agriculture ecosystems, environmental awareness, tank rehabilitation, and cascade development.

Postgraduate programs such as Master of Science (M.Sc.) in Environmental and Water Engineering, Environmental Science, Development Practices, Agricultural and Biosystems Engineering, Crop Sciences, Plantation Crop Management, Environmental Forestry, and Master of Arts (M.A.) in Geology offer subjects related to climate change and sustainability. There are some composite programs such as M.Sc./P.G. Diploma in Environmental Engineering and Management and M.Sc./ P.G. Diploma in Environmental Management also offers course modules directly related to climate change as Meteorology Climate Change, and Natural Resource Management.

Postgraduate courses designed directly in line with the climate change issue are quite rare in Sri Lanka. However, postgraduate programs are designed to enhance the capacity of the students to cope with climate change challenges through adaptation and mitigation. Every course target student with a relevant academic or professional background. Some course modules such as Climate Change Impact and Adaptation in Water Sector are components of programs such as Sustainable Built Environment offered at the postgraduate level. In addition, few universities offer research-based Master of Philosophy and doctoral programs. As such, these programs will open pathways to diverse and rewarding careers in government ministries and public sector organizations; non-governmental organizations; environmental and business consultancies; and manufacturing and service industries in the private sector.

4.2 Program Learning Outcomes (PLO) of Climate Change Education

Program Learning Outcomes are measurable statements that describe knowledge or skills that students attain upon successful completion of their academic program. A comprehensive and well-developed list of PLOs provides information about student learning, curriculum, and teaching. Learning outcomes are measured at different stages during the educational experience. The degree to which outcomes are attained will play a pivotal role in shaping curricular planning and resource decision-making.

Within the SLQF, the learning outcomes are comprised of two distinct parts. The first part, referred to as attributes of the qualification holders, is a set of general statements of the wider abilities that the typical student is expected to have developed by the end of the course or study program. The second part, referred to as the level descriptor, comprises a series of precise outcome statements. These outcomes are evaluated to determine whether a student can demonstrate them, and they represent the criteria that a graduate must meet to be eligible for the qualification award. The courses that are related to Climate Change (CC) in existing undergraduate and postgraduate study programs in Sri Lanka seem to have the following learning outcomes focused on knowledge and skills.

1. Understanding the multidimensional and interconnected nature of CC.
2. Knowledge of Climate Science and local environmental conditions.
3. Knowledge of mitigation of CC effects, CC adaptation, and DRR.
4. Knowledge of sustainable production and consumption.
5. Managing information related to CC.
6. Thinking critically about CC.

7. Coping with the emotional realities of CC.
8. Preparing for sudden and slow-onset disasters.
9. Preparing for the green jobs.
10. Understanding the offenses committed by various parties within the climate and sustainability aspects.

The perception of these attributes was captured using a structured questionnaire survey conducted with 33 university academics (five out of thirty-three were post-graduate administrators). 50% of the respondents reported that they strongly agreed with the fact that the CCSE intended to cultivate and develop the "understanding of the multidimensional and interconnected nature of the causes and consequences of CC". Around 39% of respondents agreed with the statement and only 7% disagreed with the statement. Considering the respondents responded as "disagreed", they specialized in agriculture engineering and biosystems engineering. Most respondents (39% strongly agreed and 46% agreed) believed that the CCSE was intended to cultivate and develop knowledge of climate science among undergraduates. All 33 respondents believed that the CCSE intended to cultivate and develop the knowledge of CC effects. Around 96% of the respondents believed that CCSE intended to cultivate and develop the knowledge of sustainable production and consumption while 4% disagreed with the statement.

The majority (89% of respondents) believed that the CCSE was intended to cultivate and develop knowledge of CC adaptation and DRR. Almost all the respondents (96%) believed that the CCSE intended to cultivate and develop knowledge of local environmental conditions, associated risks, and management strategies. Results indicated that 36% strongly agreed and 50% agreed with the statement that "CCSE intended to cultivate and develop managing information and thinking critically about CC". About 29% strongly agreed and 39% agreed with the fact that CCSE intended to cultivate and develop knowledge in coping with the emotional realities of CC. 10% of the respondents did not believe that the CCSE was intended to cultivate and develop the knowledge and skills needed for preparing for sudden and slow-onset needs. The majority (72%) agree with the statement while 18% neither agree nor disagree. The biggest share of the respondents (82%) believed that the existing CCSE intended to cultivate and develop the knowledge and skills needed for preparing for green jobs. About 39% of the respondents agreed and 36% strongly agreed with the fact that "CCSE intended to cultivate and develop the knowledge to enhance the understanding various offenses committed by various parties within the climate and sustainability aspects". Meanwhile, 18% reported they neither agree nor disagree (Fig. 2).

Five respondents from postgraduate programs strongly believe that the CCSE intended to cultivate and develop Knowledge of climate science, coping with emotional realities of CC, CCE, and adaptation and DRR, and understanding various offenses related to CC committed by various parties.

Based on the current course learning outcomes and the perceptions of 33-course coordinators/heads of departments in higher educational institutes in Sri Lanka, it is important to integrate these findings along with twenty-first-century skills in defining program learning outcomes for future programs in CCE.

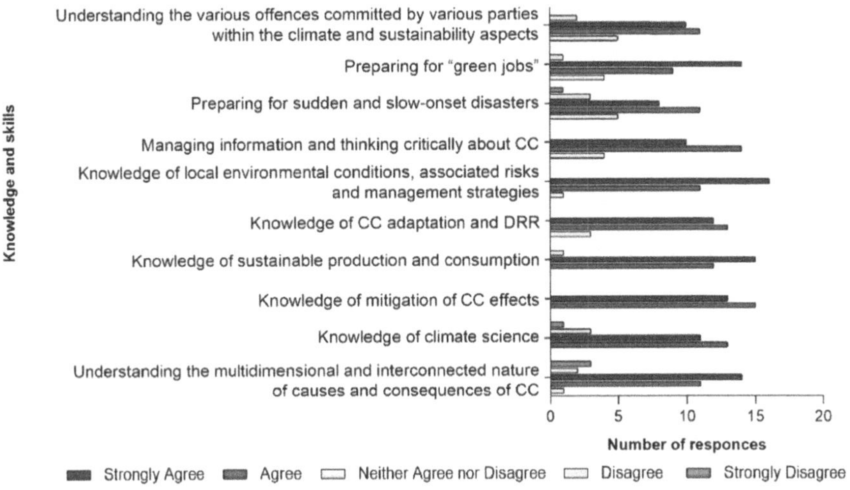

Fig. 2 Areas of knowledge and skills that respective CCSE intended to cultivate and develop in undergraduate students

4.3 Curriculum Design Considerations and Course Contents

The survey revealed that the curricular designers were aware of existing policies implemented in Sri Lanka related to climate change and their impacts. These are compulsory courses or programs at undergraduate and postgraduate levels, they are of a weightage of two to three credits offered by the academic departments and institutes relating to the above main subject categories.

A majority (64%) of the respondents agreed that the number of existing undergraduate degree programs is not sufficient to enhance climate change and sustainability knowledge (Fig. 3).

However, 89% of the respondents believed that the students were able to comprehend the multidimensional and interrelated nature of the causes and consequences of climate change through the existing modules. Therefore, intervention may be necessary. However, 67% of the respondents expressed that existing programs at the postgraduate level are sufficient with an appropriate range of subjects to enhance climate change and sustainability knowledge among students.

According to the responses received, environmental management and environmental sciences (or natural sciences) were identified as the major disciplines where climate change and sustainability education are integrated into the curriculum at the undergraduate level. Other than that, geography and agriculture also offer subjects/courses related to climate change. The field of engineering, civil engineering, environmental engineering, and engineering technology also offers subjects related to climate change and sustainability at the undergraduate level. Environmental sciences, environmental management, green buildings, climate change, and agriculture were

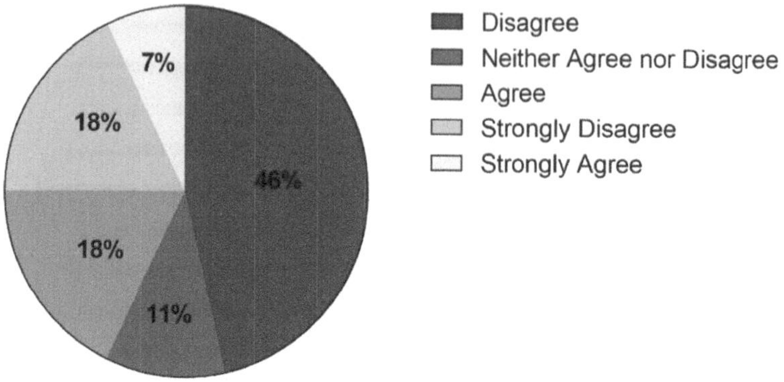

Fig. 3 Responses about the adequateness of knowledge provided to students by the respective undergraduate programs

identified as the major disciplines where the CCSE is integrated into the curriculum at the postgraduate level.

4.4 *Pedagogical Practices and Teaching Methodologies*

Climate change education in the local context is a complex topic and blended with parallel disciplines and the focus was more on awareness building rather than skill development. Countries being challenged by various climate change-induced disasters and highlighting their high vulnerability levels, society needs to learn how to develop their adaptive capacities. To address this challenge, climate change awareness modules were embedded into school and university curricula without emphasis on specific teaching pedagogies (Perkins et al., 2018). The fundamental role of CCE is to help students become better thinkers and life-long learners. This undertaking is not one activity per se, but a collective of practices like the ability to recognize facts, practice reflective skepticism, and adopt different perspectives when evaluating new ideas and issues (Perkins et al., 2018).

Figure 4 explains the outcomes of the survey conducted among selected Sri Lankan universities on teaching and learning pedagogies. Results highlighted that blended learning and e-learning were common delivery approaches while the majority of the respondents never used differentiated instruction methods and inverted learning, massive open courses, and the flipped classroom in their teaching. Only a limited number of educators engage in the delivery of climate-related modules using problem-based learning, especially case studies. Community outreach programs were instrumental in Outward Bound Training (OBT) where students work as groups and address real-time problems. Outreach activities embedded with mini-research, case analysis, seminars, task-based learning, and awareness sessions play an important role.

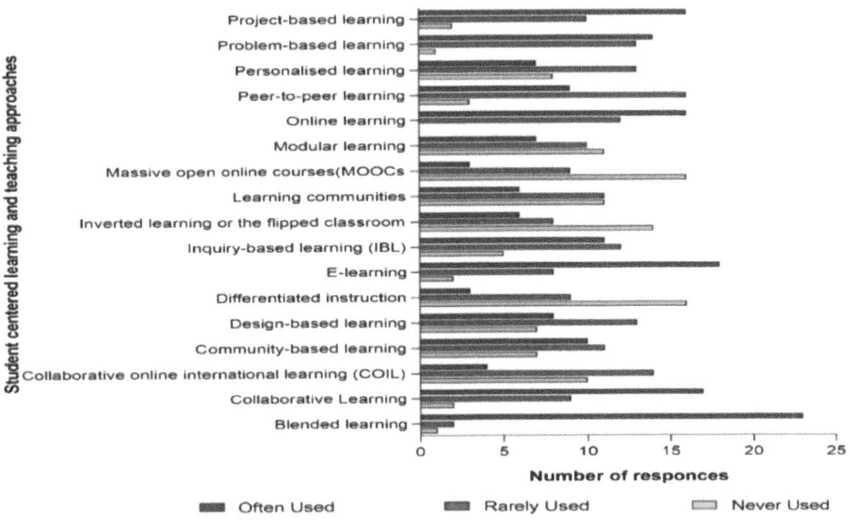

Fig. 4 Student-centered learning and teaching approaches at the undergraduate level

Student-centered learning and teaching techniques were adopted at the undergraduate level, presentations, question and answer, and quizzes were identified as the commonly used techniques (Fig. 5). Role play and the mental model were the least apparent techniques. Production of video programs on identified cases was highlighted as a blended approach where learners were able to analyze the case and were oriented toward skill development.

4.5 Assessment Methods and Evaluation in Climate Change Education

Assessment is one of the most important aspects of the educational process. It is a feedback mechanism that provides feedback to learners on their progress in achieving intended learning outcomes. This information is vital for teachers to make decisions about future actions, and it serves as evidence relating to the effectiveness of course offerings. A good assessment will ensure alignment of the contents and scope of the assessment with the learning outcome of the curricula.

In Sri Lankan universities at postgraduate and undergraduate levels, both summative and formative assessment methods are adopted. However, summative assessments that evaluate a student's achievement at the end of the course are popular at the undergraduate level.

The survey results revealed that written examinations were the most often used assessment method. Among other popular methods coursework, take-home assignments, class presentations, and research reports were commonly used (Fig. 6).

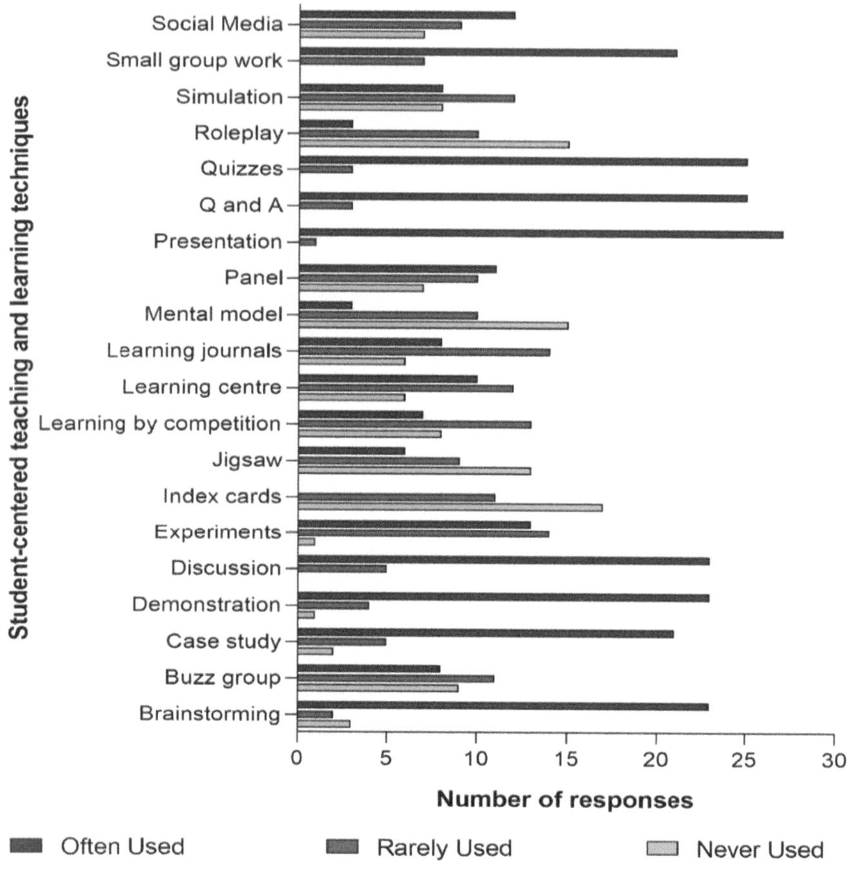

Fig. 5 Student-centered teaching and learning techniques at the undergraduate level

Comparatively at the postgraduate level formative assessment tools were used more often. Peer evaluations are very rarely used in undergraduate-level assessment, nevertheless, at the postgraduate level peer evaluation was used for some course modules.

The final year research component is considered a critical component in most degree programs and evaluation criteria are set to measure the student performance in awareness, learning, practice, and skill improvement. Unfortunately, the assessment methods recorded were unable to measure the student skill development aspects. Undergraduate and postgraduate research work is mainly assessed by evaluating the thesis and conducting oral examinations. Patents, copyrights, and licensing were not used as methods to assess undergraduate work however, some respondents reportedly used these to assess postgraduate work. The publication of research articles was another tool used to assess postgraduate work.

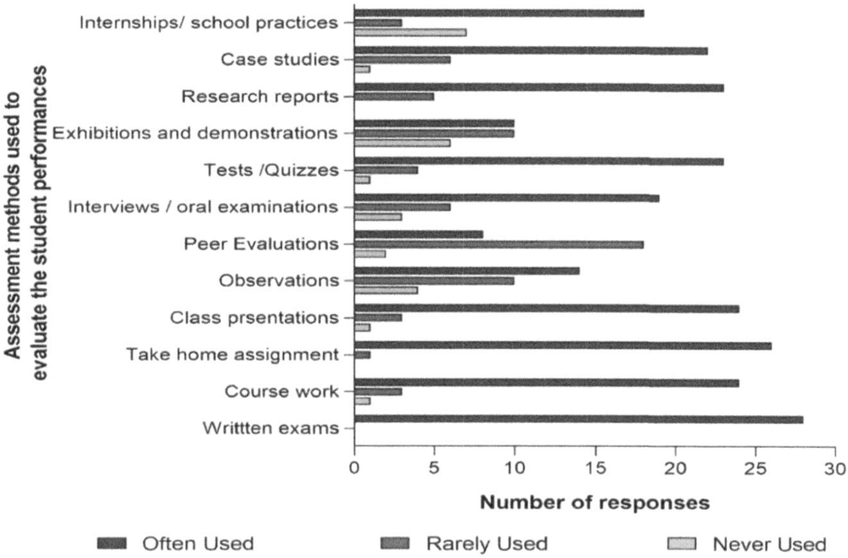

Fig. 6 Assessment methods used to evaluate the student performances in climate change education at the undergraduate level

4.6 Benchmarking in Postgraduate and Undergraduate Climate Change Education

Quality improvement is one of the key requirements of higher education institutions. Especially given the crucial role of CCE in combating the CC threat. Higher education providers need to continuously improve their offerings considering the evolving approaches to provide impactful educational experiences. In this respect, it is important to understand the role of benchmarking to continually improve educational offerings.

Benchmarking is defined as the process of self-evaluation and self-improvement through the systematic and collaborative comparison of practice and performance with similar organizations to identify strengths and weaknesses, learn to adapt, and set new targets to improve performance (Blackstock et al., 2012). Benchmarking in CCE is a systematic process of comparing and evaluating different CCE programs, strategies, and outcomes against predefined criteria and standards. The purpose of benchmarking is to identify best practices, areas for improvement, and successful approaches in CCE adopted by higher educational institutions.

In this section, some of the existing good practices in CCE in Sri Lanka at both undergraduate and postgraduate levels are highlighted, and possible interventions to enhance CCE offerings are discussed.

Four approaches were identified in embedding CCE into curricula (Molthan-Hill et al., 2019). The main approaches are piggybacking and mainstreaming. The other two approaches involve creating a new structure to embed CCE such as specializing where entirely new modules are introduced, and connecting or transdisciplinary approaches where CCE is embedded across disciplines. In undergraduate education in Sri Lanka, piggybacking (integration of CCE within existing course modules) is the most commonly used method (54% of the respondents) followed by mainstreaming (integrating CCE within the existing structure with a broader cross-curricular perspective) (21%) and specialization (21%). Only 4% of respondents reported attempting to embed CCE across disciplines. In postgraduate education, specialization was the most common approach (40% of the respondents) followed by mainstreaming (20%), connecting (20%), and piggybacking (20%). The results show that in postgraduate education reasonable steps have been taken to embed CCE across disciplinary areas as connecting is thought to be the most innovative approach (Molthan-Hill et al., 2019). However, at the undergraduate level, more efforts are needed to take CCE education to a broader segment of learners. The findings from the survey reveal a substantial gap between the priority given to incorporating CCE into curricula at the individual and institutional levels, highlighting the necessity to inculcate the value of CCE within organizational policies.

Pedagogy plays a significant role in influencing the learning outcomes of students. Well-planned pedagogical approaches can promote critical thinking, creativity and innovation, and problem-solving skills among students. Regarding student-centered learning and teaching approaches, it is clear that blended learning and e-learning have become popular approaches in undergraduate and postgraduate education in Sri Lanka. In undergraduate education, some innovative approaches such as community-based learning can be found. However, most of the respondents reported never using differentiated instruction, inverted learning, massive open courses, or the flipped classroom in their teaching of undergraduates. In postgraduate teaching approaches such as problem-based learning, modular learning, flipped classrooms, and community-based collaborative international learning were reported. According to Mintz and Tal (2018), participatory learning was found to be effective in promoting skill development, changes in awareness, and attitudes in courses that addressed sustainable development.

Considering the student-centered learning and teaching techniques, presentations, question and answer, and quizzes were identified as the commonly used techniques used by the majority of the teachers at the undergraduate level. In postgraduate education, teaching techniques frequently used include social media, presentations, learning journals, brainstorming, experiments, demonstrations, and case studies. Both in undergraduate and postgraduate education the least used techniques include simulation, roleplay, mental model, and learning by competition. International experience demonstrates that case studies and the adoption of software tools are found to be effective methods to enhance application and analysis competencies while project-based learning (PBL) and project-based service-learning (PBSL) help students to reach the synthesis level and develop effective outcomes related to sustainability (Bielefeldt, 2013).

Assessment methods regularly adopted at undergraduate and postgraduate for thought courses are similar and include written exams, take-home assignments, coursework and class presentations, tests, and quizzes. Theses and viva voce are the common assessment methods used for the research. It is rare to find peer evaluation, exhibitions, demonstrations, patents, copyrights, and licensing agreements being used as means of assessment at both undergraduate and postgraduate levels.

The survey results revealed that there is a significant gap in the importance of integrating CCE between teachers' interests and organizational priorities. This highlights the need for comprehensive policies that promote and support the integration of CCE into the education system at universities in Sri Lanka.

Necessary policy interventions and infrastructure facilities for climate change education (minimize carbon use/green buildings and other infrastructure contents to blend with climate education) are needed.

5 Conclusions and Recommendations

5.1 Summary of Key Findings

A summary of key findings is as follows.

1. Climate change education appeared in both formal and informal education systems in Sri Lanka beginning with climate literacy and going towards research and policy development agenda at the postgraduate level.
2. Analysis of the existing formal education system emphasized a sequential progression through climate literacy, education, skill, and capacity building. Research and policy intervention should be improved.
3. Undergraduate and postgraduate level CCE modules were prepared to fit into the program but merely cater to a common need.
4. Informal CCE is managed and offered by the NGOs, community organizations, and some government ministries, in a less organized manner. Overlapping programs, objectives and expectations may hinder the overall aim of capacity building.
5. Climate change is undoubtedly this century's most challenging environmental issue. In South Asia, temperatures are projected to rise substantially with potential impacts including heightened heat stress and extreme weather events to create inevitable consequences for all economic sectors including agriculture and food security. Amid these challenges, education has gained attention as a tool to combat CC, with United Nations initiatives stressing the importance of embedding CCE into policies, curricula, and public awareness efforts.
6. It was revealed that Sri Lankan HEIs offer programs and courses related to climate change, sustainability, and environmental management within the main disciplines of environmental sciences, agriculture, engineering, and geography. Standalone undergraduate climate change programs are rare.

7. The courses offered by HEIs at undergraduate and postgraduate levels at varying degrees focus on understanding multidimensional aspects, climate science, mitigation, sustainable practices, adaptation, critical thinking on CC, emotional coping, disaster readiness, green jobs, and legal aspects. The study revealed strong agreement among academics for the need for cultivating these focus areas and integrating them with twenty-first-century skills in designing program learning outcomes for future programs in climate education.
8. Results showed mixed opinions on the adequacy of CCE. A majority indicated CCE is inadequate at the undergraduate level while a majority were satisfied with the coverage of CCE at the postgraduate level.
9. Among pedagogies, blended and e-learning were common with limited use of innovative methods such as community outreach and problem-based learning to enhance real-world engagement. The majority of respondents have never used differentiated instruction methods such as inverted learning, massive open courses, and flipped classroom in their teaching.
10. Sri Lankan universities use both summative and formative assessment methods at undergraduate and postgraduate levels, with written exams being the most common. At the postgraduate level formative assessment tools are used more often. However, assessment methods apparently failed to measure the student skill development aspects.

5.2 Recommendations and Strategies for Enhancing Climate Change Education in Sri Lankan Universities

The scope of CCE should be broadened from basic cognitive skills to higher-order cognitive abilities, as well as personal and interpersonal skills. There was strong agreement among surveyed academics that CCE should be streamlined to harmonize with twenty-first century skills encompassing learning and innovation skills; life and career skills; and information media and technology skills to effectively foster adaptation and mitigation to address climate change. Educators should be provided with opportunities to enhance their abilities to engage students in hands-on experiences related to CC.

Further, the pedagogic approaches employed to attain CCE outcomes should not be confined to overly didactic and teacher-centered individual subjects. Survey findings revealed limited use of innovative pedagogies in Sri Lankan universities such as community outreach and problem-based learning to enhance real-world engagement. Therefore, there is a need to rethink the pedagogic approaches in practice. Proper analysis of formal and informal CCE systems should be done to identify the existing gaps.

National CCE road maps should be introduced with the participation of all stakeholders. In addition, strategic interventions are essential to remove system-wide bottlenecks, understand the ground realities, and propose practical approaches for climate literacy, education, research, and policy development.

Policies of higher education institutions should be geared to accommodate CCE and strong leadership is needed to inculcate a culture to embed CCE into curricula. A systems view should be adopted to ensure alignment between curriculum, assessment, instructional resources, teacher support, and student engagement.

It is important to seek international assistance to streamline CCE to address higher-order cognitive skills and to upgrade physical and human resources. In this regard, the establishment of an international climate university in the country can play a pivotal role in advancing CCE and strengthening the capabilities of educators. The International Climate University in Finland serves as an excellent example for developing courses and adopting innovative teaching and assessment methods beyond the traditional university education system in Sri Lanka.

Sri Lankan researchers have an opportunity to write more on CC and sustainability aspects to popularize them while identifying practices from developed countries and replicating them in Sri Lanka.

Disclaimer Funded by the European Union. Views and opinions expressed are however those of the author(s) only and do not necessarily reflect those of the European Union or the European Education and Culture Executive Agency (EACEA). Neither the European Union nor EACEA can be held responsible for them.

References

Armstrong McKay, D. I., Staal, A., Abrams, J. F., Winkelmann, R., Sakschewski, B., Loriani, S., ... & Lenton, T. M. (2022). Exceeding 1.5 C global warming could trigger multiple climate tipping points. *Science, 377*(6611), eabn7950.

Bielefeldt, A. R. (2013). Pedagogies to achieve sustainability learning outcomes in civil and environmental engineering students. *Sustainability, 5*(10), 4479–4501.

Blackstock, D., Burquel, N., Comet, N., Kajaste, M., Dos Santos, S. M., Marcos, S., ... & Sixto, L. C. V. (2012). *Internal quality assurance benchmarking*. ENQA Workshop Report 20. ENQA (European Association for Quality Assurance in Higher Education).

Dunlop, L., & Rushton, E. A. C. (2022). Putting climate change at the heart of education: Is England's strategy a placebo for policy? *British Educational Research Journal, 48*(6), 1083–1101. https://doi.org/10.1002/berj.3816

Gibb, N. (2016). *Getting climate-ready: A guide for schools on climate action—UNESCO digital library*. United Nations Educational, Scientific and Cultural Organization. https://unesdoc.unesco.org/ark:/48223/pf0000246740

Hargis, K., Chopin, N., & McKenzie, M. (2018). *Ten Canadian schools' stories of climate action: Promising practices from the Getting-Climate Ready project*. Sustainability and Education Policy Network. University of Saskatchewan. https://en.ccunesco.ca/-/media/Files/Unesco/Resources/2018/12/PromisingPracticesCanadianUNESCOSchools.pdf

Kristen, H., & Marcia, M. (2020). *Responding to climate change education: A primer for K–12 education* (pp. 1–18). The Sustainability and Education Policy Network. https://sepn.ca/wp-content/uploads/2021/01/SEPN-CCEd-Primer-January-11-2021.pdf

Kuntze, L. (2023). *Climate education must be locally rooted and globally informed*. Teach For All. https://teachforall.org/blog/climate-education-must-be-locally-rooted-and-globally-informed

Leal Filho, W., Sima, M., Sharifi, A., Luetz, J. M., Salvia, A. L., Mifsud, M., Olooto, F. M., Djekic, I., Anholon, R., Rampasso, I., Kwabena Donkor, F., Dinis, M. A. P., Klavins, M., Finnveden,

G., Chari, M. M., Molthan-Hill, P., Mifsud, A., Sen, S. K., & Lokupitiya, E. (2021). Handling climate change education at universities: An over-view. *Environmental Sciences Europe, 33*(1), 109. https://doi.org/10.1186/s12302-021-00552-5

Lenton, T. M., Held, H., Kriegler, E., Hall, J. W., Lucht, W., Rahmstorf, S., & Schellnhuber, H. J. (2008). Tipping elements in the Earth's climate system. *Proceedings of the National Academy of Sciences, 105*(6), 1786–1793.

Masson-Delmotte, V., Zhai, P., Pirani, A., Connors, S. L., Péan, C., Berger, S., Caud, N., Chen, Y., Goldfarb, L., Gomis, M. I., & Huang, M. (2021). Climate change 2021: The physical science basis. In *Contribution of working group I to the sixth assessment report of the Intergovernmental Panel on climate change* (Vol. 2).

Mbah, M. F., Shingruf, A., & Molthan-Hill, P. (2022). Policies and practices of climate change education in South Asia: Towards a support framework for an impactful climate change adaptation. *Climate Action, 1*(1), Article 1. https://doi.org/10.1007/s44168-022-00028-z

Mie Sein, Z. M., Ullah, I., Saleem, F., Zhi, X., Syed, S., & Azam, K. (2021). Interdecadal variability in myanmar rainfall in the monsoon season (May–October) using eigen methods. *Water, 13*(5), 729.

Mintz, K., & Tal, T. (2018). The place of content and pedagogy in shaping sustainability learning outcomes in higher education. *Environmental Education Research, 24*(2), 207–229.

Molthan-Hill, P., Worsfold, N., Nagy, G. J., Leal Filho, W., & Mifsud, M. (2019). Climate change education for universities: A conceptual framework from an international study. *Journal of Cleaner Production, 226*, 1092–1101.

Mombauer, D. (2023). *Mainstreaming climate change into Sri Lanka's education system.* Daily Financial Times. Accessed March 05, 2022. https://www.ft.lk/columns/Mainstreaming-climate-change-into-Sri-Lanka-s-education-system/4-751415

Monroe, M. C., Plate, R. R., Oxarart, A., Bowers, A., & Chaves, W. A. (2019). Identifying effective climate change education strategies: A systematic review of the research. *Environmental Education Research, 25*(6), 791–812. https://doi.org/10.1080/13504622.2017.1360842

NCSE. (2016). *Teaching climate change: best practices.* National Center for Science Education. https://ncse.ngo/teaching-climate-change-best-practices

Newsome, D., Newsome, K. B., & Miller, S. A. (2023). Teaching, learning, and climate change: Anticipated impacts and mitigation strategies for educators. *Behavior and Social Issues.* https://doi.org/10.1007/s42822-023-00129-2

Perkins, K. M., Munguia, N., Moure-Eraso, R., Delakowitz, B., Giannetti, B. F., Liu, G., Nurunnabi, M., Will, M., & Velazquez, L. (2018). International perspectives on the pedagogy of climate change. *Journal of Cleaner Production, 200*, 1043–1052.

Schellnhuber, H. J., Hare, B., Serdeczny, O., Schaeffer, M., Adams, S., Baarsch, F., ... & Rocha, M. (2013). Turn down the heat: Climate extremes, regional impacts, and the case for resilience. In *Turn down the heat: Climate extremes, regional impacts, and the case for resilience.*

Sivarajah, A., Khasim, M. T. A., & Thushari, N. M. Y. (2017). *Sectoral approaches to skills for green jobs in Sri Lanka, ILO/regional workshop on sectoral approaches to skills for green jobs.* Bangkok, Thailand, January 24–25, 2017.

Sri Lanka Qualification Framework. (2015, September). *Sri Lanka qualification framework.* Retrieved June 3, 2023, from University Grant Commission Sri Lanka: http://www.ugc.ac.lk/attachments/1156_SLQF.pdf

Sterling, S. (2001). *Sustainable education: Re-visioning learning and change.*

Tesfaye, K., Zaidi, P. H., Gbegbelegbe, S., Boeber, C., Rahut, D. B., Getaneh, F., ... & Stirling, C. (2017). Climate change impacts and potential benefits of heat-tolerant maize in South Asia. *Theoretical and Applied Climatology, 130*, 959–970.

UBC. (2023). *Open collections—UBC library open collections* [Open collections]. https://open.library.ubc.ca/

Ullah, F., Qayyum, S., Thaheem, M. J., Al-Turjman, F., & Sepasgozar, S. M. (2021). Risk management in sustainable smart cities governance: A TOE framework. *Technological Forecasting and Social Change, 167*, 120743.

UNICEF. (2021). Rising to the challenge. In *Youth perspectives on climate change and education in Sri Lanka*. UNICEF, Regional Office for South Asia. Accessed February 15, 2022. https://www.unicef.org/rosa/media/14946/file/Rising%20to%20the%20Challenge%20-%20Sri%20Lanka.pdf

Yan, S., & Alvi, S. (2022). Food security in South Asia under climate change and economic policies. *International Journal of Climate Change Strategies and Management, 14*(3), 237–251.

An Innovative Pedagogy in Climate Change Education: A Case of Problem-Based Teaching and Learning

A. S. Hettige, D. A. M. De Silva, and M. A. E. K. Jayasinghe

1 Introduction—Climate Change Education and Problem-Based Learning

The interdisciplinary nature of climate change makes it one of the most severe problems people have to deal with (McCright et al., 2013). Although climate change remains a top priority on the worldwide agenda, educators even those who are engaging in the STEM disciplines (science, technology, engineering, and mathematics) still have limited knowledge of how to address climate change through the education system (Mochizuki & Bryan, 2015). Simply climate change education entails learning how to deal with risk, uncertainty, and fast climate changes (Stevenson et al., 2017), however, it is extremely difficult to impart the science behind climate change through the teaching of content. As climate change is largely the result of interactions between humans and the environment, STEM disciplines can have a substantial impact on mitigating its impact with the adaptation of sustainable practices and developing technologies that change the factors that cause climate change (Reimers, 2021).

People need to be prepared to adopt sustainable behaviors to lessen their impact on climate change and its effects on their lives which can be done through educating them about more sustainable ways to interact with their environment. In an effort to promote "climate literacy" among students, UNESCO (2010) introduced the Climate Change Education for Sustainable Development initiative which connected

A. S. Hettige (✉) · D. A. M. De Silva
Department of Agribusiness Management, Faculty of Agricultural Sciences, Sabaragamuwa University of Sri Lanka, P.O. Box 02, Belihuloya, Sri Lanka

D. A. M. De Silva
e-mail: desilva.achini@yahoo.co.uk

M. A. E. K. Jayasinghe
Department of Export Agriculture, Faculty of Animal Science and Export Agriculture, Uva Wellassa University of Sri Lanka, Badulla, Sri Lanka

C. Malalgoda et al. (eds.), *Climate Change Adaptation in the Built Environment*,
https://doi.org/10.1007/978-3-031-75826-3_19

four programs; (1) climate change, ethics, social and human sciences elements, (2) climate change education, (3) climate change, cultural and biological diversity, and (4) cultural heritage. Apart from that, STEM educators introduced several teaching and learning settings and methodologies to boost STEM learning at the university level, and climate change education provides a suitable platform to put these insights into practice (Handelsman et al., 2004). A majority of STEM education research concentrates on identifying the pedagogies and approaches that are most successful in fostering this change to systems thinking (McCright et al., 2013). Moreover, the higher education system has more flexibility and capacity to re-orient the learning and teaching approaches by upgrading the existing curriculum with knowledge from climate and sustainable development perspectives, compared to primary and secondary education, to achieve the aforementioned level of climate literacy. Universities may choose to educate about climate change using both formal and informal methods, such as fieldwork, case studies, participatory observations, higher degree research, and others (Molthan-Hill et al., 2019, 2022).

In the present scenario, the improved understanding of the ways individuals learn has been steered to the adoption of more student-centered pedagogies and active learning at the university level (Herreid, 1998). The Brookings Institution's Center for Universal Education also attempted in 2011 to discover the most effective strategies for formal and non-formal climate change education material by examining evidence-based research by which they found out that through persistent, active learning activities and the use of integrated, cross-discipline curriculum, climate change literacy can be increased (Anderson, 2012).

1.1 Problem-Based Learning (PBL)

Problem-based learning (PBL), a form of active learning that encourages students to analyze problems, apply their knowledge, and collaborate on pertinent concerns, is an ideal match to educate learners on climate change (Schmidt, 1994). PBL fosters intellectual learners who can evaluate circumstances and come up with meaningful solutions through teamwork and also can gain powerful, effective communication abilities (Major & Mulvihill, 2018). Emerging more than 35 years ago, PBL was first used to improve professional instruction methods (Steinemann, 2003). As per the literature on cognition and meta-cognition (Barrows & Bennett, 1971; Bransford et al., 1986; Steinemann, 2003), PBL allows the learners to work with a real-world problem and engage in active learning at the cognitive level and to assess and improve their problem-solving techniques at the meta-cognitive level. For all these reasons, PBL has significant potential to encourage exploration as an educational strategy in STEM disciplines and can be employed to mediate the gap between concept and practice (Chin & Chia, 2004). According to Brundiers and Wiek (2013), to successfully implement PBL, the learning goals should specifically focus on helping students develop sustainability capabilities, environments should be more interactive, allowing students to collaborate and, more focus should be placed on encouraging independent learning, collaborative problem-solving, and research. To accomplish this stipulation, this chapter mainly focused on mainstreaming PBL on climate change education into undergraduate curricula through innovative pedagogies associated with STEM disciplines.

1.2 *Significance of Mainstream Problem Based Learning on Climate Education*

Mainstreaming climate change education into a formal education system is one of the most crucial and effective ways to build capacity for tackling the climate catastrophe (Mochizuki & Bryan, 2015). Climate change education involves a focus on learning, critical and creative thinking, and capacity building for the learners to understand and determine the appropriate actions to respond to climate change (Stevenson et al., 2017). Higher education institutions have a specific responsibility to create potential future leaders and decision-makers who can comprehend complicated global challenges and offer answers. To comply with these requirements, students engaging in higher education should acquire the much-needed competencies which are described as learnable not teachable to be proactive and deal with the uncertainties associated with climate change (Burandt & Barth, 2010; Green et al., 2009). Therefore, learning environments that foster personal growth and provide them with the tools to deal with difficult circumstances, act after reflection, and make judgments are necessary for the development of competencies (Dochy et al., 2003). This is where PBL learning comes into action which stresses learning by doing. Several PBL courses have been implemented for a time in various regions of the world and the engineering-related studies are those in which the most experience has been documented (Carrió Llach & Llerena Bastida, 2023; Guerra, 2017; Mann et al., 2021). Numerous studies demonstrate that PBL enhances critical thinking skills. Learners are encouraged to identify, analyze, and evaluate problems independently, leading to a deeper understanding of concepts and issues. PBL fosters critical learners who collaboratively analyze situations and solve meaningful problems, acquiring information-seeking methods and organizational skills (Duch et al., 2001; Murray-Harvey & Slee, 2000). Engaging in PBL also cultivates strong communication abilities (Allen et al., 2011). According to Aron et al. (2013), PBL approaches positively impact students' knowledge retention, synthesis, and interest in the subject. However, this method is still not widely applied in most disciplines as it is extremely challenging to implement curriculum-wide, integrated PBL in the present curriculum (Carrió Llach & Llerena Bastida, 2023).

Understanding the modalities of teaching and learning is essential to achieve the targeted learning outcomes in climate change education and the long-term sustainability, along with changes to curriculum in higher education. Current evidence-based research findings on climate change education primarily highlight findings from studies conducted in Europe, the United States (US), Canada, and Australia whereas for developing nations, very little evidence-based research was discovered

(Anderson, 2012). Agriculture, a socially, economically, and environmentally important sector in Sri Lanka is extremely exposed to climate variability and future changes, including an increase in the occurrence of extreme weather events. Over the following several decades, the most significant and immediate effects of climate change will be felt in the agricultural and food sectors (Esham & Garforth, 2013). All agricultural output sub-sectors have high post-harvest losses along the supply chain as a result of Sri Lanka's inadequate post-harvest infrastructure, which is characteristic of most developing nations (Esham et al., 2018). Overall, weather conditions that are necessary for early pre and post-harvest handling are severely affected by unpredictable rainfall patterns and climate variation.

However, the higher education curriculum in Sri Lanka does not cater to the required practical awareness and exposure to adaptation and mitigation of climate vulnerabilities associated with the agricultural sector. PBL is a practical-based pedagogy as it covers all the stages of updated Bloom's Taxonomy of cognitive learning including remembering, understanding, applying, analyzing, evaluating, and creating (Amer, 2006) which can be used to cater to the climate change education related to agriculture. As per the authors' knowledge is concerned, when it comes to the Sri Lanka context, there is no research conducted on PBL in association with climate change education. By focusing on these aspects, this study significantly adds to the body of knowledge on climate change education based on PBL in Sri Lanka and has substantial implications for the current curriculum in STEM disciplines with special reference to Sabaragamuwa University of Sri Lanka. This study mainly aimed at mainstreaming problem-based learning into climate change education in science-based undergraduate curricula with special emphasis on performing a resource audit on learning materials and teaching pedagogies and highlighting the feedback of learners regarding PBL. A smart student-centered learning framework is proposed using the reviewed literature and analyses of the case studies conducted.

2 Methodology

The study employed a mixed approach, integrating both qualitative and quantitative methods. A comprehensive literature review was conducted, focusing on climate problem-based learning and change education. To access relevant literature, various academic databases such as Google Scholar, ResearchGate, and Springer were utilized. Specific keywords, including "Climate change education," "Problem-based Learning," "Innovative pedagogies," "Sustainability education," and "University climate education," were employed to retrieve publications on the topic. A total of 60 related publications were initially identified, and after eliminating duplicates and unrelated papers, 30 publications from 2003 to 2023 were selected. The exclusion criteria encompassed articles that were not original studies, were published in languages other than English, were unrelated to climate education, and fell outside the specified year interval, with the exception mentioned above. The selected articles were thoroughly reviewed based on their title, keywords, objectives, and key findings.

2.1 Research Design

Primary data collection was strengthened with a resource audit on physical resources of the STEM faculties, participatory observation of PBL of undergraduates, focus group discussions with farming communities on adaptation strategies, and feedback survey on PBL. The sample comprised 19 undergraduates from the specialization module, agribusiness management, and 10 paddy farmers of the Uvathenna village situated in closer vicinity to the university. A focus group discussion was carried out with 19 students based on a semi-structured questionnaire which mainly comprised of questions to get an understanding of "What teaching approaches do they mostly prefer when engaging with climate change education", "What is the impact of PBL to their learning process" and "How to improve the learning through innovative pedagogies". The participatory appraisal which is a community-based research method (White & Taket, 1997) was used to get a broader picture of the thoughts and opinions of the community and promote their involvement. Involving the public actively in decision-making processes is encouraged by the participatory method when its application in agriculture is taken into account.

The participatory approach was designed to investigate the impact of climate change on paddy farmers and to understand the strategies they have developed to cope with these changes. The discussion was carried out using a storytelling technique where farmers were guided to express their views freely on the impact of climate change on their livelihoods. The discussion aimed at gathering common constraints faced by the respondents and the common coping strategies they had developed to mitigate those issues.

A resource audit on physical resources and human resources was carried out to identify the learning materials and teaching pedagogies in the Faculty of Agricultural Sciences, Faculty of Applied Sciences, and Faculty of Technology of Sabaragamuwa University of Sri Lanka (Fig. 1). The data obtained were presented as a comparative graph between the three faculties.

The feedback survey comprised 211 undergraduates (120—year I, B.Sc Food Business Management, 19—year III, B.Sc Agricultural Sciences & Management, and 72—year III, B.Sc Food Business Management). The data were collected using an online questionnaire. The questionnaire was designed to get the responses of the students on the pedagogies used to achieve the learning objectives. It mainly comprised five components namely knowledge sharing, interaction, experience, communication, and reflection and outcome for which the answers were collected using the Likert Scale. Objective knowledge on PBL in climate education was measured via 21 items, grouped to represent the five components namely knowledge sharing, interaction, experience, communication, and reflection. Student responses to

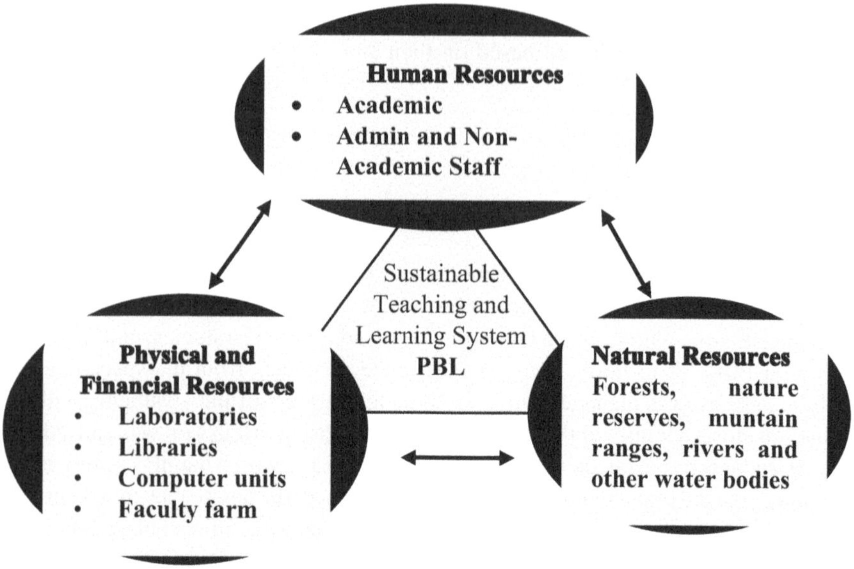

Fig. 1 Profile of the resource audit

these items were collected using the Likert Scale (5—Strongly agree, 1—Strongly disagree). To keep participants' identities confidential, their names or registration numbers were not collected. In the quantitative stage, participants' answers obtained were processed using IBM SPSS Statistics version 21. Both descriptive and inferential analyses were performed on the data. Results were grouped into 3 case studies.

2.2 *Results and Discussion*

2.2.1 Case Study 01: Integrating PBL into the Teaching & Learning System of the Agribusiness Management Curricula: A Case Study on Climate Change Adaptation Through Value Addition

Problem-based learning (PBL) is an instructional approach that encourages students to learn by actively engaging in solving real-world problems. It involves presenting students with authentic, open-ended problems that require critical thinking, collaboration, and self-directed learning. Applying the PBL concept to reduce post-harvest losses in locally found fruits and vegetables can be an effective strategy and combining value addition to products and PBL in climate change education can be a powerful approach to engage students and empower them to tackle environmental challenges.

This practical was conducted as a continuous evaluation assignment for the marketing management module of the agribusiness specialization students. It aimed to generate ideas and develop prototypes for adding value to commonly available fruits and vegetables, as an adaptive measure to minimize climate risks.

Through PBL, students can actively explore innovative solutions for farmers and communities to reduce post-harvest losses and increase the self-life of raw products. Each student group was assigned a locally available fruit or vegetable type, and a generous allocation of 3 hours was provided for their work. The idea generation process took place within the basic product development facilities. The evaluation of the groups' efforts was conducted based on several criteria. Firstly, the number of products developed by each group was taken into account. Additionally, the recipe and taste of the food products created were carefully evaluated. Furthermore, the packaging material used for the products played a role in the assessment. Lastly, the ability of the student groups to effectively answer questions about their creations was considered during the evaluation process and at the end of the process.

This approach allowed us to witness the potential of collaborative thinking and teamwork. Students from different backgrounds and skill sets brought their unique perspectives to the table, resulting in a dynamic and exciting environment. During the assessment, students were presented with various scenarios and challenges related to product development by using commonly found vegetables, fruits, and meat and quality assurance. They were encouraged to utilize their knowledge, and analyze past experiences, both successful and unsuccessful, to identify patterns and insights that can be applied to solve problems and face challenges. The activity was based on several products developed, packing materials, taste, and quality of the developed products.

Furthermore, students are tasked with evaluating the packaging materials used for these products. They explored the significance of sustainable and eco-friendly packaging choices, considering factors such as environmental impact, cost-effectiveness, and aesthetics. This exercise encouraged students to think critically about the broader implications of their choices and the importance of making informed decisions. Students received feedback from panel members and was assessed by the impact of taste, texture, and overall quality on product success. Through this analysis, students gained a deeper understanding of the significance of consumer satisfaction and the role it plays in building a successful product. To gather comprehensive feedback from students regarding the PBL teaching and learning method, an online questionnaire was used. This method allowed the students to provide their opinions and suggestions conveniently.

The feedback obtained from students participating in the PBL initiative focusing on product development and climate change education indicates a positive perception of the pedagogical approach. Figure 2 depicts the level of student interest in the assessment related to PBL. Students' responses to the assessment related to PBL reveal that a significant majority of participants expressed a strong preference for this approach. The findings collectively indicate a positive reception of the PBL approach to teaching and learning among the students.

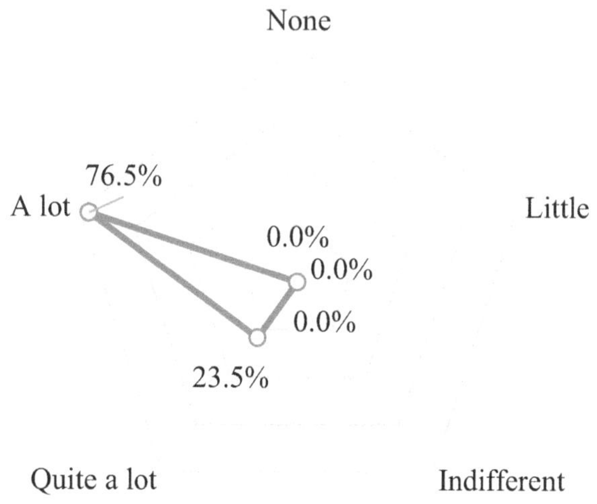

Fig. 2 Feedback assessment of case study 1—students' interest in PBL

Students' viewpoints on enhancing the learning process with PBL in agribusiness management curricula showed a positive perspective on the potential of PBL to improve their learning experience. Further, a majority (94%) of the participating students expressed a clear preference for problem-based teaching and learning over the traditional lecture method. It is evident from these findings that students perceive problem-based teaching and learning as an effective means to improve the learning process, as it enables them to actively participate, collaborate, and share knowledge and experiences.

Figure 3 presents the idea-generation methods utilized by participants during the development of prototypes on climate change adaptation through value addition. The graph illustrates that the participants predominantly employed focus group discussions as their primary method for generating ideas. The participants likely engaged in open discussions, sharing their perspectives, knowledge, and experiences, which fostered creativity and innovation in developing the prototypes (Fig. 3).

Participants were also asked to provide their responses to the evaluation factors using a five-point Likert scale. This scale allowed them to indicate their level of agreement or disagreement with the problem-based learning method and its significance. By utilizing this scale, participants were able to express their opinions and perceptions regarding the effectiveness and importance of the problem-based learning approach.

Several factors, including communication, interaction, knowledge sharing, and outcomes, were considered to measure the effectiveness of the assessment (Table 1) Based on the mean score values, the results obtained from the Likert scale indicate that the respondents strongly agree with the statements, suggesting a significant level of effectiveness in the implementation of assessment with PBL. It revealed that the activity had successfully sparked their creativity and encouraged them to think outside the box. Additionally, students expressed an improved understanding

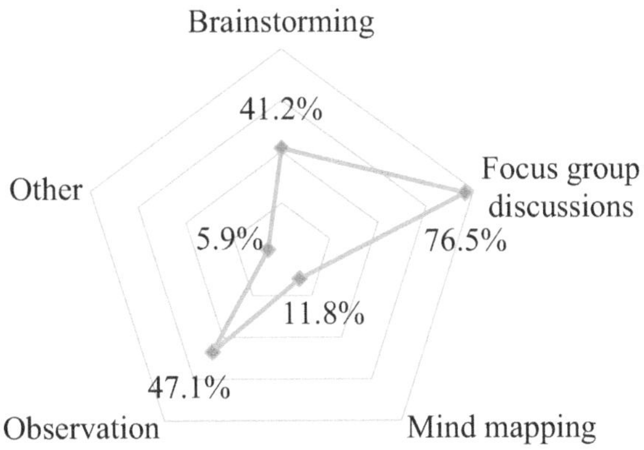

Fig. 3 Feedback from the students on the idea-generation method used on case study 1

of climate change issues and the importance of sustainable practices, emphasizing the relevance of PBL in addressing complex global challenges.

In conclusion, this PBL assessment provided students with valuable hands-on experience in product development, packaging evaluation, and quality assessment. Students had to work together, share ideas, and make collective decisions to bring their concepts to life (Fig. 4). PBL offers an engaging and effective approach to equip students with the necessary skills and knowledge to tackle real-world challenges associated with climate change (Fig. 5).

2.2.2 Case Study 02: Integrating PBL and Co-creation into Food Business Management Curricula: A Case Study on Climate Change Education Through Nature-Inspired Agri-food Solutions

Application of PBL in climate change education through the development of nature-inspired solutions is a strategy that links active learning and climate adaptation successfully. This practical was conducted as a learning assessment for the students specializing in Food Business Management. As agricultural products are mostly perishable and highly vulnerable to climate change, this activity aimed to enhance the active learning and creative thinking ability of the participants as they have to find solutions to adapt to climate change through sustainable and climate-smart innovative solutions. There, students were instructed to develop solutions that are inspired by nature to adapt to climate change. Each group was given the freedom to apply the theory learned into practice and develop student-led project-based ideas. The evaluation was done based on the relevance to the topic, applicability, level of innovation, and sustainability.

The feedback obtained from students on applying PBL to climate change education through the development of nature-inspired solutions indicates a positive perception of the pedagogical approach. Figure 6 describes the level of student interest in assessment-related PBL. The responses indicated that 50% of the participants

Table 1 Mean scores of the evaluated factors—Case study 1

Factor	Question	Mean score
Knowledge sharing	The class welcomed my ideas and suggestions for learning content	4.41
	Sufficient explanations and information were provided in the class for effective learning	
	I willingly invested my time and effort in sharing my learning ideas and suggestions with the class	
	The class provided a conducive environment and ample opportunities for sharing my ideas and suggestions	
	I successfully applied the acquired knowledge to my work	
	I am enthusiastic about generating new ideas for improvement and sharing them with others	
Interaction	I was able to conveniently express my needs during the learning process	4.35
	Sufficient information is provided in the class for learning	
	Adequate interaction between me and the class during the learning process	
	Active role played in interacting with the class to maximize knowledge acquisition	
Experience	The method used made the subject interesting for me	4.51
	The method used was more engaging than traditional lectures	
	The method used enhanced my creativity and problem-solving skills	
	I found the learning process valuable as it allowed me to explore new challenges	
	This was a memorable experience for me	
Communication	The instructions were communicated clearly	4.22
	I found it convenient to express my specific requirements while learning	
	Adequate feedback was received for the work	
Reflecting and Outcome	The case study enabled me to apply my learning to real-world problems	4.35
	Through the case study, I assessed my views on an issue, identifying my strengths and weaknesses	
	The case study facilitated a change in my thinking about an issue based on what I learned	

Source Field survey, 2023

Fig. 4 PBL on climate change adaptation through value addition

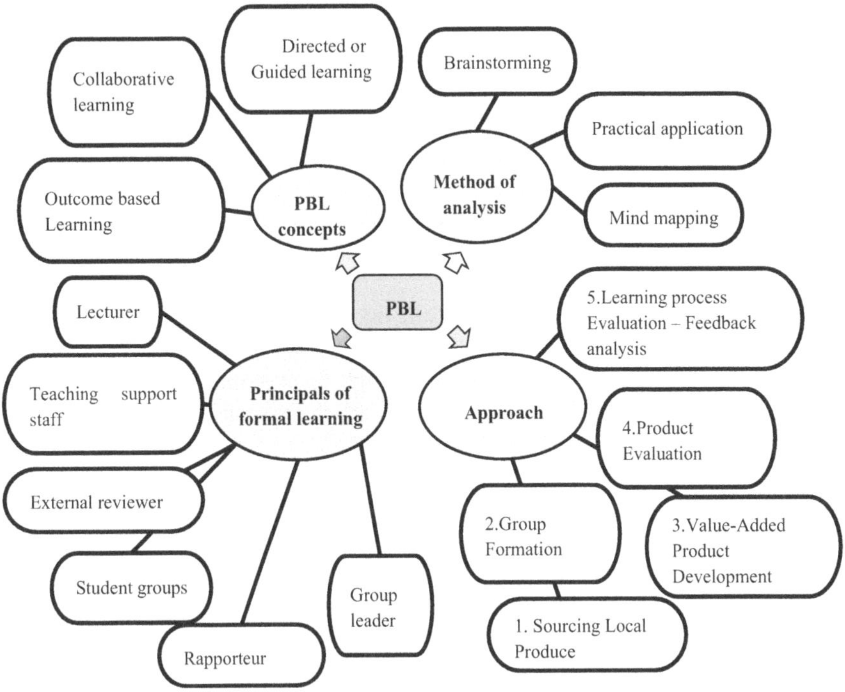

Fig. 5 Summary of the key elements of PBL activity 01

expressed a significant preference ("quite a lot") while 20.8% indicated a strong preference ("a lot") for this method. These findings jointly indicate a clear preference towards the utilization of the PBL as an innovative pedagogy for assessment purposes.

As per the results on students' perspectives on the possibility of improving the learning process through empirical means, a significant proportion of the respondents (85%) answered agreeably, where, 15% of the respondents showed some

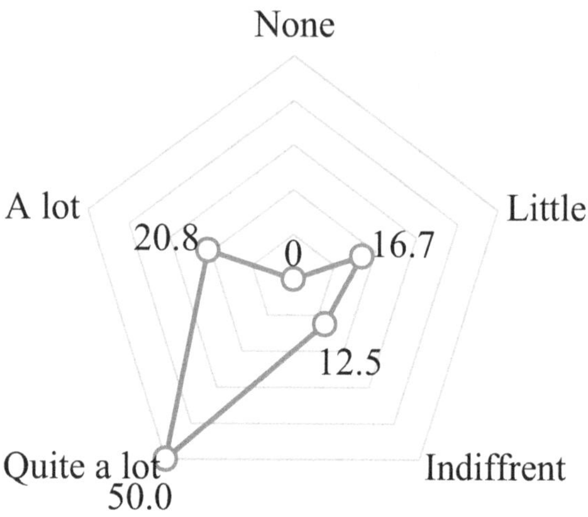

Fig. 6 Students' interest in the assessment

uncertainty regarding the impact of the empirical learning process and answered with "Maybe". This result stresses the prospective benefits of incorporating PBL into the curriculum while highlighting the positive impact it creates on the students. Furthermore, students stated PBL as the most preferred over traditional one-side lecturing.

The idea generation methods employed by the respondents when developing nature-inspired solutions are shown in Fig. 7. The results revealed that the majority of the respondents generated their ideas through focus group discussions, which accounts for 68%. It was followed by 12.8% of the respondents who employed observation as their idea-generation method.

The mean score values obtained from the Likert scale questions which were used to assess the effectiveness of PBL were presented in Table 2. The respondents agreed with all the five factors and the respective statements provided with them as they

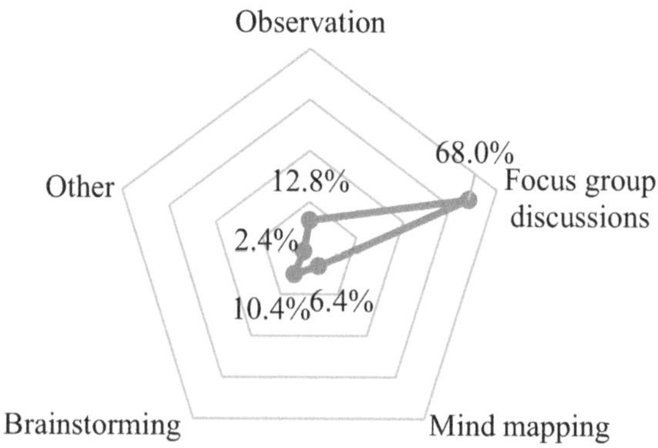

Fig. 7 Feedback from the students on the idea-generation method used in case study 2

ranged from 3.41 to 4.20. It can be highlighted that the respondents were considering PBL as an effective pedagogy to use in climate change education.

Table 2 Mean scores of the evaluated factors—Case study 2

Factor	Question	Mean score
Knowledge sharing	The class welcomed my ideas and suggestions for learning content	3.83
	Sufficient explanations and information were provided in the class for effective learning	
	I willingly invested my time and effort in sharing my learning ideas and suggestions with the class	
	The class provided a conducive environment and ample opportunities for sharing my ideas and suggestions	
	I successfully applied the acquired knowledge to my work	
	I am enthusiastic about generating new ideas for improvement and sharing them with others	
Interaction	I was able to conveniently express my needs during the learning process	3.90
	Sufficient information is provided in the class for learning	
	Adequate interaction between me and the class during the learning process	
	Active role played in interacting with the class to maximize knowledge acquisition	
Experience	The method used made the subject interesting for me	3.94
	The method used was more engaging than traditional lectures	
	The method used enhanced my creativity and problem-solving skills	
	I found the learning process valuable as it allowed me to explore new challenges	
	This was a memorable experience for me	
Communication	The instructions were communicated clearly	4.00
	I found it convenient to express my specific requirements while learning	
	Adequate feedback was received for the work	
Reflecting and Outcome	The case study enabled me to apply my learning to real-world problems	3.96
	Through the case study, I assessed my views on an issue, identifying my strengths and weaknesses	
	The case study facilitated a change in my thinking about an issue based on what I learned	

Source Field survey, 2023

In conclusion, this assessment delivered students with real-world experience in nature-inspired agri-food solutions development (Fig. 8). The PBL method can be employed in the process of advancing climate adaptive innovations in different market segments as it will reduce post-harvest losses, create employment opportunities, and support the economy to grow amid rapid climate change.

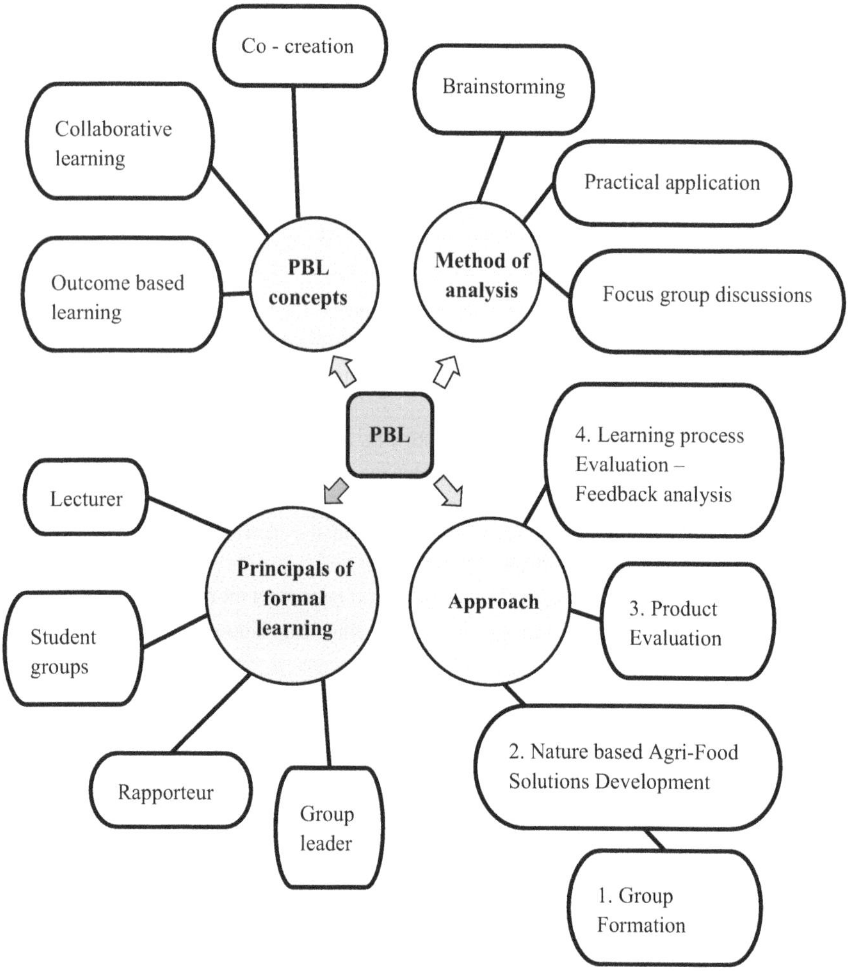

Fig. 8 Summary of the key elements of PBL activity 02

2.2.3 Case Study 3: Integrating Problem-Based Learning into the Food Business Management Curriculum: A Case Study on Climate Change Adaptation through Value Chain Analysis and Mapping

This case study presents an application of PBL, specifically focusing on the subject of food value chain management. The objective of the study was to assess climate change adaptation within the food value chain and provide students with practical exposure to its operations. By actively engaging in problem-solving activities, students were expected to develop a comprehensive understanding of the challenges and resilience mechanisms involved in managing the food value chain amidst the impacts of climate change. Group activity allows students to select local food value chains and explore the actors and functions. Assessment based on the depth of value chain analysis, mapping, identification of climate effects on the selected value chain, proposed climate resilient mechanisms, and ability of student groups to effectively answer questions. An online feedback survey was conducted to evaluate the effectiveness of the PBL.

The students' responses indicated a high level of satisfaction with the PBL approach, emphasizing its practicality and its ability to develop their analytical and problem-solving abilities in the context of the food value chain and climate change education.

Figure 9 reveals that a majority of the students displayed a notable interest in this particular assessment method. This finding implies that the students perceive PBL as a valuable and engaging method for their learning and assessment process (Fig. 9).

A majority of participants expressed a positive belief in the possibility of improving their learning through this approach. These findings highlight the overall positive reception of PBL and its potential for fostering effective learning experiences in the context of climate change education. The results in Fig. 10 reveal that

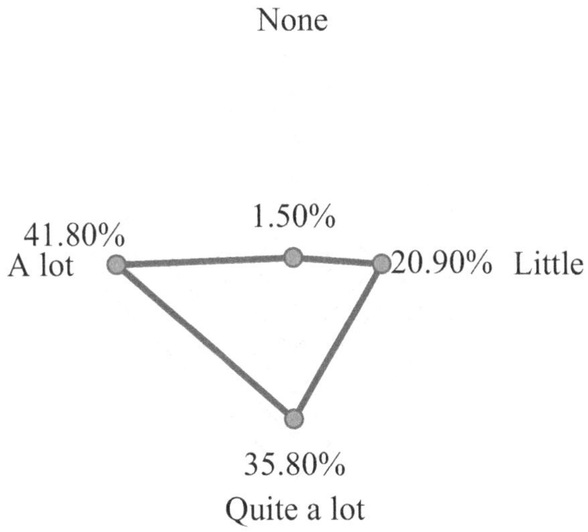

Fig. 9 Feedback assessment of case study 3—students' interest in PBL

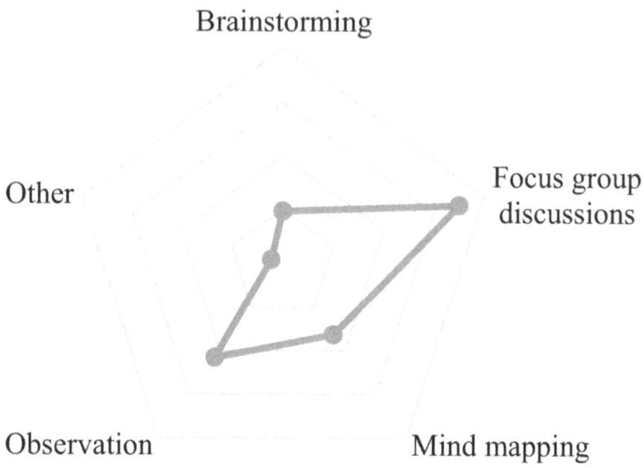

Fig. 10 Feedback from the students on the idea-generation method used in case study 3

students predominantly utilized focus group discussions as their primary method for generating ideas.

In conclusion, the PBL assessment has proven to be an effective tool in climate change education. By integrating climate change education into the PBL framework, students were able to explore the complexities of value chains, assess their climate impacts, and propose innovative solutions. The hands-on nature of the PBL approach provided students with valuable experiential learning opportunities, allowing them to actively engage with real-world challenges associated with climate change. This approach not only equipped students with the necessary skills and knowledge to address these challenges but also fostered collaboration, critical thinking, and decision-making skills (Fig. 11).

2.2.4 Case Study 4: Empowering Farmers Through Problem-Based Learning on Climate Change: A Case Study in Uvathenna Area

The case study delved into the impact of climate change on farmers and the importance of providing them with education on climate change. A combination of focus group discussions and field observations facilitates obtaining data from the farming community in the Uvathenna village. Focus group discussions allowed for in-depth insights into farmers' experiences, perspectives, and strategies in dealing with climate change.

The findings revealed that climate change had diverse and detrimental effects on the farmers in the region. Climate change has led to unpredictable weather patterns, including extreme heat, prolonged droughts, and erratic rainfall. Farmers reported experiencing reduced crop yields and frequent crop failures, impacting their income and food security. As a result of rising temperatures, water sources have become scarce or unreliable, affecting irrigation practices. With diminishing natural habitats and resources due to climate change, wild animals, such as elephants, are encroaching upon agricultural lands in search of food and water (Fig. 12). To cope with the

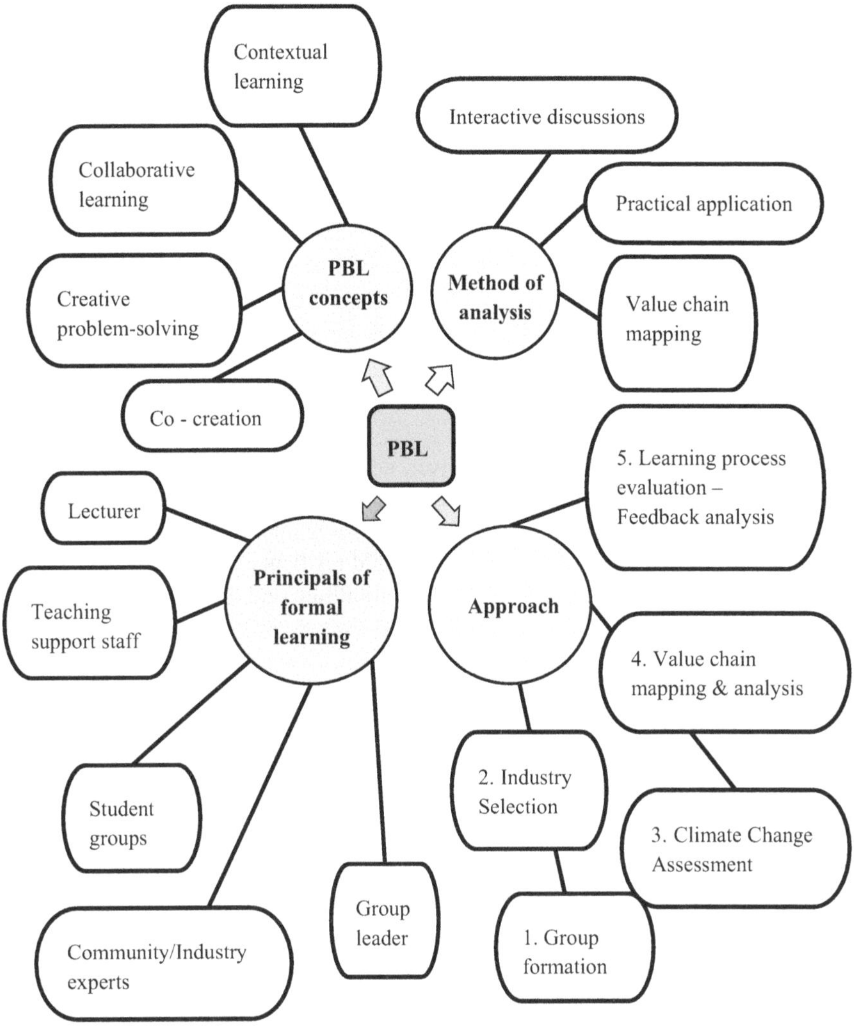

Fig. 11 Summary of the key elements of PBL activity 03

adverse effects of human-wildlife conflicts, farmers have developed various resilient mechanisms such as burning firewood to deter wild animals, although such practices may have negative ecological implications.

Findings highlighted the eager need to provide education on climate change to farmers in the region. By raising awareness about the causes and consequences of climate change, farmers can better understand the importance of sustainable agricultural practices. Promoting climate-smart techniques, water conservation methods, and wildlife conservation efforts can help mitigate the negative effects of climate

Fig. 12 Wild elephant footprints on paddy fields

change on farming communities. Moreover, educating farmers about the implications of their actions on the environment can foster a sense of responsibility, leading to more environmentally friendly practices. This study underlines the crucial role of climate education in empowering farmers to build resilience against climate change and to work towards a sustainable future for agriculture and the environment.

2.2.5 Case Study 5: Focus Group Discussion on Integrating PBL into Agribusiness Management Curricula with Special Reference to Climate Change Education

This case study explored the importance of PBL on undergraduates, especially referring to climate change education. The discussion revealed that the participants are more likely to engage with PBL pedagogies in their curricula. They highlighted that they have improved their capacity to recall and apply what they have learned. They were also better able to link theory to practice. Students have been reported to be able to analyze and synthesize contextual information, absorb fresh content, and integrate it into their existing knowledge base by reflecting on previous experiences. PBL allows students to apply problem-solving techniques to a comparable problem on a relevant topic, which helps students strengthen their problem-solving abilities. Moreover, students actively participated in the group discussion and shared their thoughts and opinions very engagingly. Throughout this case study, it is stressed that realistic issues like climate change can be easily addressed through active learning. The positive feedback of the students regarding PBL proved the effectiveness of integrating innovative pedagogies in climate change education.

Table 3 Human and physical resource comparison between faculties

Resource	Faculty of Agricultural Sciences	Faculty of Applied Sciences	Faculty of Technology	Total
Academic	44	70	23	137
Non-academic	35	52	12	99
Laboratories	11	22	3	36
Libraries	1	0	0	01
Computer units	1	4	1	06
Faculty farm	1	0	0	01
Number of course units containing climate change education	5	3	2	10

Source Survey data, 2023

2.3 Resource Audit

The available resources in three STEM faculties at Sabaragamuwa University of Sri Lanka are mentioned in Table 3. It provides an insight into the feasibility of applying PBL in those faculties.

Figure 13 provides a visual representation of the number of academic and non-academic staff in the STEM faculties.

According to the results on the number of course units containing climate change education in the STEM faculties 50% of the currently available course units are offered by the Faculty of Agricultural Sciences whereas 30% and 20% are offered by the Faculty of Applied Sciences and Faculty of Technology respectively.

3 Smart Student-Centered Learning Framework

Student-centered learning is a learning strategy that is widely connected to and supported by constructivist theories of learning. It also symbolizes the mindset and the culture inside a higher education institution. It is characterized by innovative teaching strategies that encourage communication between students and teachers, deal with these individuals seriously as active participants in their learning, and promote applicable abilities like problem-solving, critical thinking, and reflective thinking (Kurowska & Lewicki, 2017). The four pillars of student-centered learning can be summarized as follows (Fig. 14): learning is personalized, competency-based, and can occur anytime, anywhere. Students take ownership of their learning.

This conceptual framework is driven by what cognitive scientists have learned about how individuals acquire knowledge, comprehend new ideas, and build themselves into experts. Optimizing learning for each student is the fundamental objective of student-centered learning. Thus, the system's functions mainly focus on enabling

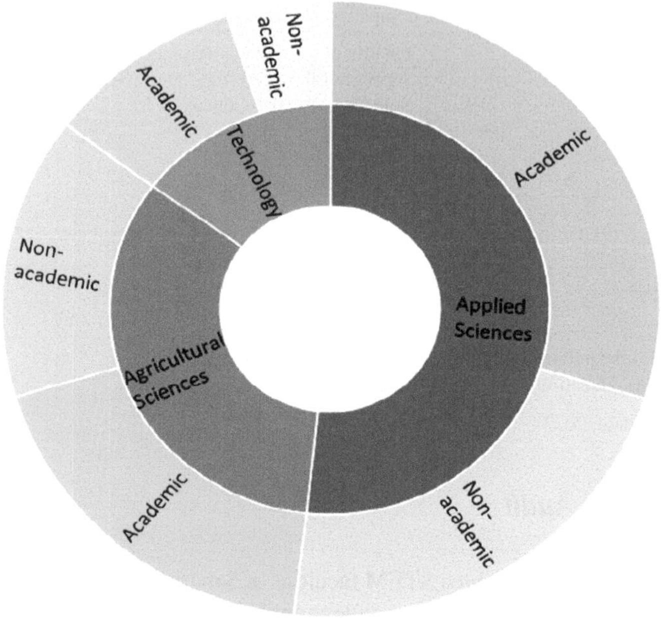

Fig. 13 Comparative analysis of academic and non-academic staff count in STEM faculties at Sabaragamuwa University of Sri Lanka

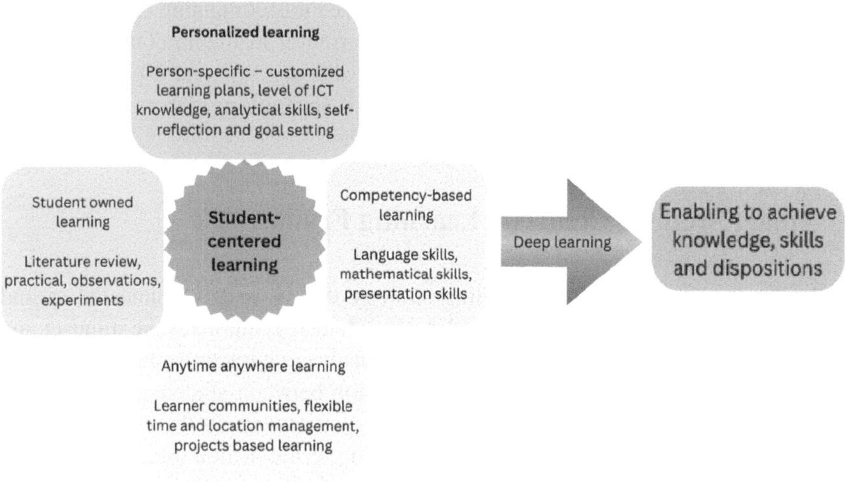

Fig. 14 Framework for student-centered learning *Source* Kurowska and Krasniewski (2015)

personalized learning experiences, which may occur as a result of the learner's direct interaction with an educator as a result of a variety of online and offline experiences. The framework must enable the provision of learning experiences based on unique student strengths, needs, interests, and motivations in addition to relevant, timely feedback offered through a student-centric interface.

The activities that served as the basis for the above case studies were all designed within the framework for student-centered learning. These PBL activities provided a student-owned learning experience for the students as they incorporated their skills and knowledge on decision-making and executed the learning process. Furthermore, those learnings were not limited to the classroom as students visited the farmers and the community to generate value additions and nature-inspired solutions. Competency-based learning was used to evaluate the components of the curriculum and allowed them to demonstrate the knowledge and experience gathered through practical exposure. Students were provided with the objectives to achieve through every activity to measure their competencies and they were gone through performance-based evaluations. It also provided an opportunity for personalized learning as the students were engaged in various activities in which both the student and the instructor could understand their current position and take each student's requirements and preferences. This framework was noticeably successful in implementing PBL and though it is not commonly used within the curriculum, these activities were aimed at mainstreaming and circulating PBL in a more detailed and organized manner to achieve the learning objectives.

Within this framework, students get an opportunity to take an active role in their learning, choose what and how they will learn, and focus on their educational progress as they develop learning strategies and collaborate with other students. Here, the instructors also have responsibilities to acknowledge and make arrangements for various learning styles, establish structure without becoming excessively restrictive, pay attention to the opinions of every student, and support collaborative decision-making among students to yield the maximum advantages from student-centered learning. This framework creates shifts within a classroom as interactions between instructor and students become more flexible, the curriculum is more practical and conceptual, accommodates a wide range of learning preferences, builds on abilities, desires, and experiences and evaluation takes into account multiple intelligences, uses real-world evaluations, and encourages self-reflection.

In contrast to teacher-centric education, student-centered learning places more emphasis on the individual student and the instructional procedures that enable a student-centric learning cycle. The individual learner's learning experiences, resources, and interactions with peers, educators, and other education stakeholders must all be taken into account when designing this framework. Implementing any one of these tenets individually may be advantageous, but for learning to change, all concepts must be accepted and implemented systemically (Glowa & Goodell, 2016). There is no standardized method for adopting student-centered learning, and depending on the goals and resources, implementation will also differ for each higher education institute.

4 Conclusion

Problem-based curriculum offer a learning environment where competency is where competency is created by promoting knowledge-seeking rather than teaching primarily to transfer information. Students engaged with PBL developed abilities related to reasoning and making conclusions, creative thinking that combines intuition and exploration, problem-solving, and communication. They were also better able to apply basic science knowledge to the solution of climate change as PBL increased students' intrinsic interest in the subject. PBL facilitated the growth of beneficial competencies that can be used for lifetime learning, mainly leadership and cooperation. Students-centered learning, a PBL method, is based on principles related to students' prior knowledge, organization of knowledge, motivation, goal-directed practice, and self-directed learning. Being able to study within a smart student-centric environment requires most students to advance their learning skills along with institutional support in developing their learning skills. It includes learning how to learn independently, developing communication skills, taking more control of their learning, teaching others, developing life-long learning skills, and developing their metacognitive skills.

PBL also has several drawbacks. Students expect the lecturers to assist and facilitate their learning rather than to directly impart their knowledge, which might be considered inefficient at some point in time. Knowledge assimilated through PBL can be less organized for students than knowledge acquired through traditional learning if they cannot systematize that information. On the other hand, the skills required by the PBL facilitators are different than the skills required for traditional teaching. The time and cost associated with PBL can be challenging to some faculties with the students who have to learn within an increasingly comprehensive curriculum.

In conclusion, using PBL as a potentially useful strategy for bridging the gap of climate change education in present STEM curricula is essential. However, it is of utmost importance to analyze the most suitable ways to incorporate PBL into climate change education to reap the most out of it while reducing its shortcomings.

References

Allen, D. E., Donham, R. S., Bernhardt, S. A. (2011). Problem-based learning. New directions for teaching and learning. *New Directions for Teaching and Learning, 128* [Preprint].

Amer, A. (2006). Reflections on Bloom's revised taxonomy. *Electronic Journal of Research in Educational Psychology, 4*(1), 213–230.

Anderson, A. (2012). Climate change education for mitigation and adaptation. *Journal of Education for Sustainable Development, 6*(2), 191–206. https://doi.org/10.1177/0973408212475199

Aron, A., Lewandowski Jr, G. W., Mashek, D., & Aron, E. N. (2013) The self-expansion model of motivation and cognition in close relationships. In *The Oxford handbook of close relationships* (pp. 90–115).

Barrows, H. S., & Bennett, K. (1971). Experimental studies on the diagnostic (problem solving) skills of the neurologist and their implications for neurological training. *Transactions of the American Neurological Association, 96*, 51–54.

Bransford, J., Sherwood, R., Vye, N., & Rieser, J. (1986). Teaching thinking and problem solving: Research foundations. *American Psychologist, 41*(10), 1078.

Brundiers, K., & Wiek, A. (2013). Do we teach what we preach? An international comparison of problem-and project-based learning courses in sustainability. *Sustainability, 5*(4), 1725–1746.

Burandt, S., & Barth, M. (2010). Learning settings to face climate change. *Journal of Cleaner Production, 18*(7), 659–665.

Carrió Llach, M., & Llerena Bastida, M. (2023). Exploring innovative strategies in problem based learning to contribute to sustainable development: A case study. *International Journal of Sustainability in Higher Education, 24*(9), 159–177.

Chin, C., & Chia, L. G. (2004). Problem-based learning: Using students' questions to drive knowledge construction. *Science Education, 88*(5), 707–727.

Dochy, F., Segers, M., Van den Bossche, P., & Gijbels, D. (2003). Effects of problem-based learning: A meta-analysis. *Learning and Instruction, 13*(5), 533–568. https://doi.org/10.1016/S0959-4752(02)00025-7

Duch, B. J., Groh, S. E., & Allen, D. E. (2001). *The power of problem-based learning: A practical "how to" for teaching undergraduate courses in any discipline*. Stylus Publishing, LLC.

Esham, M., & Garforth, C. (2013). Climate change and agricultural adaptation in Sri Lanka: A review. *Climate and Development, 5*(1), 66–76.

Esham, M., Jacobs, B., Rosairo, H. S. R., & Siddighi, B. B. (2018). Climate change and food security: A Sri Lankan perspective. *Environment, Development and Sustainability, 20*, 1017–1036.

Glowa, L., & Goodell, J. (2016). Student-centered learning: Functional requirements for integrated systems to optimize learning. *International Association for K-12 Online Learning* [Preprint].

Green, E. I. H., Blashki, G., Berry, H. L., Harley, D., Horton, G., & Hall, G. (2009). Preparing Australian medical students for climate change. *Australian Family Physician, 38*(9), 726–729.

Guerra, A. (2017) Integration of sustainability in engineering education: Why is PBL an answer? *International Journal of Sustainability in Higher Education* [Preprint].

Handelsman, J., Ebert-May, D., Beichner, R., Bruns, P., Chang, A., DeHaan, R., Gentile, J., Lauffer, S., Stewart, J., Tilghman, S. M., & Wood, W. B. (2004). Scientific teaching. *Science, 304*(5670), 521–522.

Herreid, C. F. (1998). Why isn't cooperative learning used to teach science? *BioScience, 48*(7), 553–559.

Kurowska, K., & Krasniewski, A. (2015). Student-Centred Learning in learning outcomes based curricula. In *ICERI2015 Proceedings* (pp. 1629–1638). IATED.

Kurowska, K., & Lewicki, J. (2017). Does the practical profile of studies require a new organization of curricula and innovative methods of education? *Forum Oświatowe, 29*, 203–213.

Major, T., & Mulvihill, T. M. (2018). Problem-based learning pedagogies in teacher education: The case of Botswana. *Interdisciplinary Journal of Problem-Based Learning, 12*(1), 1.

Mann, L., Chang, R., Chandrasekaran, S., Coddington, A., Daniel, S., Cook, E., Crossin, E., Cosson, B., Turner, J., Mazzurco, A., & Dohaney, J. (2021). From problem-based learning to practice-based education: A framework for shaping future engineers. *European Journal of Engineering Education, 46*(1), 27–47.

McCright, A. M., O'shea, B. W., Sweeder, R. D., Urquhart, G. R., & Zeleke, A. (2013). Promoting interdisciplinarity through climate change education. *Nature Climate Change, 3*(8), 713–716. https://doi.org/10.1038/NCLIMATE1844

Mochizuki, Y., & Bryan, A. (2015). Climate change education in the context of education for sustainable development: Rationale and principles. *Journal of Education for Sustainable Development, 9*(1), 4–26. https://doi.org/10.1177/0973408215569109

Molthan-Hill, P., Worsfold, N., Nagy, G. J., Leal Filho, W., & Mifsud, M. (2019). Climate change education for universities: A conceptual framework from an international study. *Journal of Cleaner Production, 226*, 1092–1101. https://doi.org/10.1016/j.jclepro.2019.04.053

Molthan-Hill, P., Blaj-Ward, L., Mbah, M. F., & Ledley, T. S. (2022) Climate change education at universities: Relevance and strategies for every discipline. In *Handbook of climate change mitigation and adaptation* (pp. 3395–3457). Springer.
Murray-Harvey, R., & Slee, P. (2000). *Problem based learning in teacher education: Just the beginning* (pp. 4–6). Australian Association for Research in Education.
Reimers, F. M. (2021). The role of universities building an ecosystem of climate change education. In *Education and climate change: The role of universities* (pp. 1–44).
Schmidt, H. G. (1994). Problem-based learning: An introduction. *Instructional science* (pp. 2.47–250).
Steinemann, A. (2003). Implementing sustainable development through problem-based learning: Pedagogy and practice. *Journal of Professional Issues in Engineering Education and Practice, 129*(4), 216–224. https://doi.org/10.1061/(ASCE)1052-3928(2003)129:4(216)
Stevenson, R., Nicholls, J., & Whitehouse, H. (2017). What is climate change education? *Curriculum Perspectives, 37*. https://doi.org/10.1007/s41297-017-0015-9
UNESCO. (2010). *Climate change education for sustainable development: The UNESCO climate change initiative.* https://unesdoc.unesco.org/ark:/48223/pf0000190101
White, L., & Taket, A. (1997). Beyond appraisal: Participatory appraisal of needs and the development of action (PANDA). *Omega, 25*(5), 523–534.

Digital Competencies, Disaster Risk Reduction, and Education: An Introduction to the Digital Competency Framework to Develop Digital Pedagogical Competencies of Educators in Disaster Risk Reduction

Anuradha C. Senanayake, Aravindi Samarakkody, Chamindi Malalgoda, Dilanthi Amaratunga, Richard Haigh, Champika Liyanage, Ruchira Yapa, Mo Hamza, Artūras Kaklauskas, and Rajib Shaw

1 Introduction

Disasters in the contemporary world are frequently occurring phenomena that affect the whole infrastructure of a given society and its inhabitants. In the year 2022, a total of 387 reported extreme weather events have resulted in 30,704 deaths and have affected 185 million people (Centre for Research on the Epidemiology of Disasters (CRED), 2023). Further, the year has also recorded 32.6 million internal displacements by extreme weather events (The Internal Displacement Monitoring Centre (IDMC) & Norwegian Refugee Council (NRC), 2023). Out of those extreme weather events, floods and wind storms have the highest number of occurrences (Centre for Research on the Epidemiology of Disasters (CRED), 2023).

A. C. Senanayake (✉) · A. Samarakkody · C. Malalgoda · D. Amaratunga · R. Haigh
Global Disaster Resilience Centre, University of Huddersfield, Huddersfield, UK
e-mail: anuradha.senanayake@hud.ac.uk

C. Liyanage · R. Yapa
University of Central Lancashire, Preston, UK

M. Hamza
Lund University, Lund, Sweden

A. Kaklauskas
Vilniaus Gedimino Technikos Universitetas, Vilnius, Lithuania

R. Shaw
Keio University, Minato, Japan

C. Malalgoda et al. (eds.), *Climate Change Adaptation in the Built Environment*,
https://doi.org/10.1007/978-3-031-75826-3_20

In such a background of disasters that significantly impact individuals and societies, this study focuses on education in the field of disasters. It has been recorded that there is an increased interest in the field of education within disaster contexts (Kagawa, 2005). While it has been acknowledged that the conceptualisation of disaster education is yet to come to a consensus, it has been mentioned that disaster education could be conceptualised based on the time context the education is conducted (whether it is regular or irregular), the type of the education (formal or informal) and locating disaster education as a sub-subject within the subject of education (Septima et al., 2019). Further, the dimensions (1) knowledge—the understanding of the disaster contexts, (2) action—motivating students to be active in disaster mitigating activities, (3) response—responding to disasters during and post-disaster events, (4) participation—involvement of the students to build disaster resilience among communities, and (5) integration—structural element of disaster education have been recognised to be vital for disaster education (Herdiansyah et al., 2020). Further, disaster education has also been defined as education in disaster risk, disaster prevention and Disaster Risk Reduction (DRR) (Shaw et al., 2011).

Even though the studies have emphasised the importance of disaster education (Kitagawa, 2021), it has been pointed out that research in the field of disaster education is still in its infancy and has yet to achieve its potential (Mangione et al., 2013). Studies have added that the emergency shift of education towards the online space has significantly impacted disaster education (Senanayake et al., 2023). To address this timely need, this study will be based on the field of disaster education and will follow the footpaths of Shaw et al. (2011), referring to it as DRR education. This study will investigate the competencies that a DRR educator should possess in their digital teaching process. The notion of competency has been defined from several points of view yet is mainly centred on the individual. When the results are known as the worthy outcomes of an individual's behaviour, the state of being competent could be defined as that individual's ability to produce results constantly (Teodorescu, 2006). However, the concept of competencies is multi-faceted, and the rationale for using competencies determines its definition (Hoffmann, 1999). The typology of competencies is dominantly represented under three approaches. Firstly, the traditional American approach is focused on individual characteristics and understands behavioural competency as a means to develop superior performance. In contrast, the UK approach deals with functional competencies and how they are implemented in the workplace (Le Deist & Winterton, 2005). highlighted Lastly, the French and German approach highlights the potential of a more analytical and multidimensional explanation of competencies and argues the need for a more holistic framework that complements the KSA (Knowledge, Skills, and Attitudes) worldview, which considers knowledge, skills, and behaviours as dimensions of competencies (Le Deist & Winterton, 2005).

Digital competency is often intertwined and interchangeably used with terms such as internet competence, digital literacy, e-competence, technology literacy, e-literacy, e-skills, media and information literacy (Ferrari et al., 2012). Mota and Cilento (2020) defined internet competencies as a collection of skills, knowledge, and attitudes towards internet use. Likewise, digital competencies are frequently defined as using knowledge, skills, and attitudes that help an individual effectively use digital technologies to achieve goals in various life contexts (Baartman & de Bruijn, 2011).

Technological transformation challenges the teaching profession at two levels: firstly, it necessitates the educators to develop their digital competencies, and secondly, it expects the educators to design instructional activities that endow their students with the competencies essential to succeed in the digitalised education environment (Hämäläinen et al., 2021). According to ElSayary (2023), in the context of education, digital competency is about teaching to facilitate students' active engagement with digital technologies in their work, lives, and careers. Literature mainly highlights the pedagogical digital competencies relating to skills, knowledge and attitudes. According to From (2017), educators possessing pedagogical digital competencies can better support their students in achieving expected learning outcomes, understand how the learning process works, and relate it to regulating principles.

While studies have been conducted to address digital competencies, no existing competency frameworks or studies address the competencies a DRR educator should hold. Hence, this study will be based on the main objective of developing a competency framework to address the digital competencies a DRR educator should behold. This was the final Output of the two-year research project INCLUDE (INCLUsive Disaster Education), a collaborative research project co-funded by the EU Erasmus + programme. The main objective of this project was to reimagine online distance learning education to better support the diverse DRR community. The project INCLUDE was led by the University of Huddersfield's Global Disaster Resilience Centre, based in the UK. The project consortium also comprises partners from the UK, Sweden, Lithuania and Japan.

2 Methodology

The methodology that was adopted in developing the main competency framework could be explained in three main stages as follows:

2.1 Stage 1—Literature Review to Recognise the Main Categories of Digital Pedagogical Competencies

A detailed literature review was conducted to trace the main types of digital pedagogical competencies recognised in related studies. The review was conducted based on journal articles and already developed competency frameworks that have been developed for the contexts of digital teaching. As principle reference points, the following two competency frameworks were utilised:

1. Educators' Digital Competency Framework—UNICEF Regional Office for Europe and Central Asia, 2022
2. DIGCOMP: A Framework for Developing and Understanding Digital Competence in Europe—Joint Research Centre of the European Commission, 2013

2.2 Stage 2—Reviewing Existing Project Output Reports to Trace Digital Competencies Required for the DRR Educators

As mentioned before, developing the competency framework was the final output of the two-year-long INCLUDE project. It was aimed to reflect the overall findings of the project throughout the various outputs that were conducted, which were dedicated towards online DRR education. The overall output structure could be summarised as follows (Fig. 1).

Hence, in the second stage of developing the initial competency framework, the existing output reports were evaluated to investigate the relevant digital competencies the DRR educators require.

The reviewed reports were:

1. Output 1—A survey of online distance learning strategies used in DRR education and their effectiveness to identify their success factors and associated issues and problems
2. Output 2—A framework to reimagine online distance learning education
3. Output 3—An inclusive University-Industry digital learning platform
4. Output 4—Case studies with the use of disruptive technologies for DRR

The content analysis helped realise the pedagogical and technological aspects relevant to online DRR education. The above reports' content analysis was conducted using Nvivo Plus (V.12). The derived themes were coded, and further refinement and redefinition were performed within the process.

2.3 Stage 3—Development of the Competency Framework and the Validation

Following the previous two stages, the competency framework was developed by mapping out the findings of the first two stages. Following the development of the initial competency framework, the final stage was the validation. The validation was conducted in the following phases:

1. Internal validation—as the first validation phase, the partners' feedback validated the initial framework.

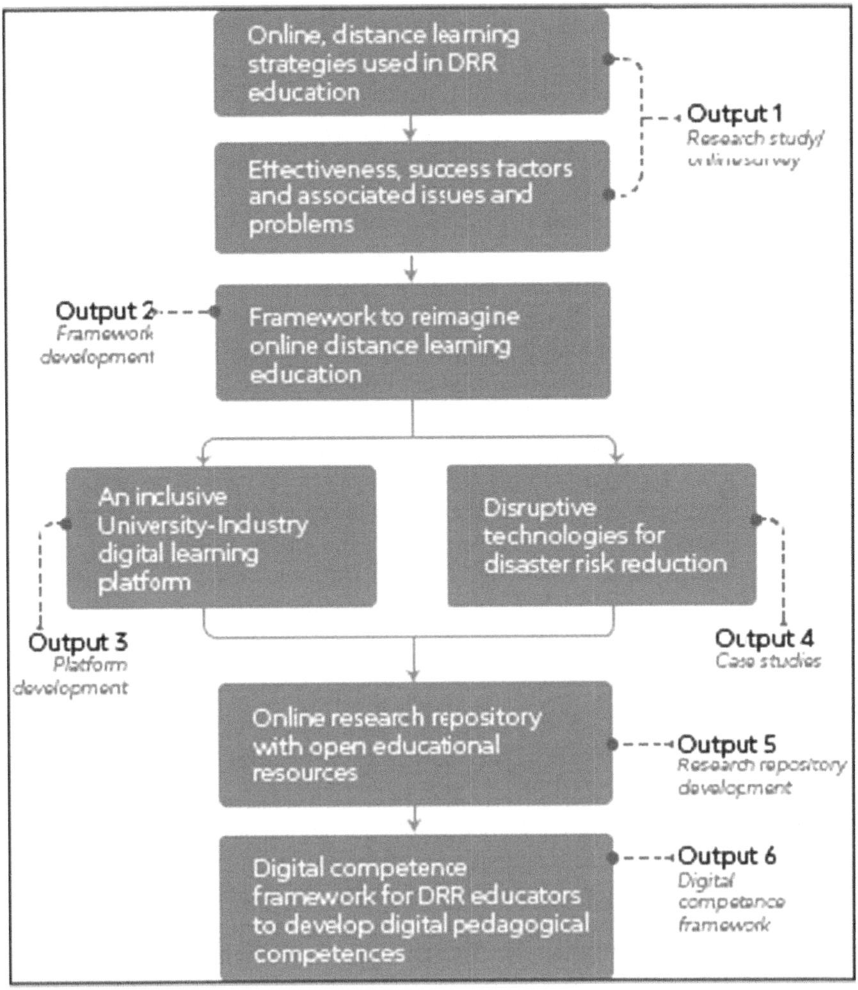

Fig. 1 An overview of the INCLUDE project process and the outputs (*Source* Prepared by the authors)

2. External validation—Following the feedback from the partners, the refined framework was presented to selected experts who are experienced educators in DRR.

3 Findings and Discussion

The findings of the present chapter will be presented based on the stages explained in the previous methodology section.

3.1 *Main Categories of Digital Pedagogical Competencies*

As per the literature that was conducted, the following three main categories were recognised as vital categories of digital pedagogical competencies.

3.1.1 Knowledge

Competency involves putting conceptual and procedural knowledge into action (Perrenoud, 2005). The framework widely used to understand the knowledge educators need to integrate technology into the usual teaching practice effectively is the TPACK (Technological Pedagogical Content Knowledge) framework (Voogt et al., 2013). It describes the teacher knowledge as a holistic complex interaction among the three bodies of knowledge, namely pedagogy (knowledge about teaching/ learning processes, educational theories, instructional design, etc.), content (knowledge about the taught subject matter) and technology (knowledge about technologies and their use in the educational settings) (Koehler & Mishra, 2009; Koehler et al., 2013). A significant dimension of knowledge highlights the educators' ability to design/ develop courses in a way they are mobilised to support students' learning with effective ICT support (From, 2017).

3.1.2 Skills

While digital competency is sometimes interchangeably used as digital skills, Tuamsuk and Subramaniam (2017) recognise the definition of digital literacy entails three primary skills: technical skills (to handle digital devices and applications), cognitive skills (to discern and evaluate data and distinguish between accurate, false and biased data) and emotional-social skills (to understand the impact of the data). The literature further highlights 21st-century digital skills, focusing on the skills for the knowledge-based workforce and for employees to be in charge of their learning (van Laar et al., 2017). The skills that integrate technology into teaching are referred to as techno-pedagogical skills, and they include sub-skills such as basic technological skills, technology usage skills for personal development and knowledge acquisition, and technology usage skills for planning and preparation with lesson plans (Lyonga et al., 2021). Moreover, the literature discourses on skills are in line with the following thematic areas of digital literacy (Hall et al., 2014);

- Being safe in the digital environment.
- Finding, evaluating and applying information.
- Using digital tools/hardware/software.
- Understanding the social responsibility.
- Showcasing achievements.
- Awareness of digital identity.
- Collaborating with community education and work life.

3.1.3 Attitudes

Naturally, educators may develop negative and/or positive beliefs towards the use of technology and their self-efficacy in their changed role in the digital teaching environment, especially if the shift is unprecedented, as it was in the COVID-19 era. An educator's attitude towards integrating technology into teaching will be based on those positive or negative beliefs (Instefjord & Munthe, 2017). As one dimension of digital competency, attitudes influence educators' actions and eventually play an important role in learning (Funkhouser & Mouza, 2013). The research identified technology attitudes under three core areas: general attitudes towards ICT, attitudes towards ICT in education, and ease of educational use of technology (Scherer et al., 2018). At the individual level, the intention to use technology in due course can be determined by levels of educators' self-efficacy, perceived usefulness of technology and perceived ease of use (Joo et al., 2018). Studies have shown that educators' positive attitude towards using technology to support collaboration, learning, and productivity develops their digital competencies and their students' digital competencies (ElSayary, 2023).

3.2 Mapping the Main Categories of Digital Pedagogical Competencies with the Digital Competencies Required for the DRR Educators

As already discussed under Sect. 3.1, knowledge, skills and attitudes are critical to developing a digital competency framework (Ferrari, 2013). Further, the importance of knowledge has been emphasised in the aspects of development, application, sharing and communication (Siina, 2022). In the context of skills, various related studies have emphasised the importance of developing skills in educators (Funkhouser & Mouza, 2013). Hence, this section will present the key insights of the content analysis pattern matched with the findings of the literature review under the themes of knowledge, skills, and attitudes.

3.2.1 Knowledge

This section includes the views of experts captured through the Output 1 (please refer to Fig. 1 presented under the Methodology) expert interviews. Interviews were conducted with 48 educators in the higher education sector in the four countries where the partner universities are located (United Kingdom, Japan, Sweden, and Lithuania) and further extended to explore the perspectives of educators from other countries such as Sri Lanka, Thailand, Australia, Bangladesh, China, India, Indonesia, Japan, and the Philippines. As per the above-discussed TPACK (Technological Pedagogical Content Knowledge) framework, interview findings can be organised into three dimensions.

Pedagogical Knowledge

The online environment allows educators to explore and practice different teaching techniques that could help them build a positive connection with varying types of students, especially those overwhelmed by meeting people in person or who perform better in an anonymous learning environment. One of the highlighted struggles for educators during online teaching was understanding their student cohort without seeing them. This demands them to develop innovative pedagogical styles that help them appreciate and respect students' differences. The innovativeness of such pedagogical styles ranges in a broad spectrum; for instance, they should overcome or minimise the challenges of teaching lab-based/equipment-focused sessions. Educators who taught DRR modules using equipment highlighted the challenge of accessing critical equipment during online learning and the difficulty explaining their use. It has also been observed that students from different cultures respond differently. Especially in DRR programmes, educators should refine their pedagogical knowledge to perform better and be more inclusive in a culturally diverse student body. It is also essential for educators to self-evaluate their competencies to identify their own training/learning needs (Baran et al., 2019), which helps them enhance their pedagogical knowledge in the online teaching environment.

Content Knowledge

While the DRR educators should deeply understand the DRR theories, concepts, and practices, the interview findings reflected on several areas in which DRR educators should also be updated and which are sometimes underrepresented. These could be included in developing future DRR courses, and educators should possess the knowledge to do so. A repeated aspect that needs to be reflected in the DRR curriculum and hence through the educators' content knowledge is the technological development (for example, industry 5.0, society 5.0, smart cities, etc.). Moreover, having a complete understanding means they are also aware of the limitations of technology use. Technology and educators need to be updated in general with all new cases, legislation, standards, and best practices, related to DRR. This also includes the knowledge about progress made towards different development agendas related to DRR. While the DRR curriculum usually covers several development agendas, including the Sendai framework for DRR, the state-of-the-art developments are not sufficiently communicated to the DRR learners through a taught series. Educators need to ensure that the information they share with students is current and that they are using recent materials, although they should still reflect on the past. Positioning disasters in the systematic equilibrium and reflecting on all the cascading and ripple effects they could have on different components in the urban ecosystems is another aspect that needs to be taught to the DRR learners yet is underrepresented. Moreover, DRR is a team effort involving stakeholders from different disciplines. Therefore, DRR learners need to understand the interdisciplinarity in DRR. Hence, educators should understand the interconnected thinking, be willing to reach out to other ways of working and be knowledgeable not only about straightforward relationships but also about extended value-adding networks/interdisciplinarity. DRR discipline deals a lot with data, so understanding data includes analysing and forecasting and discerning their limitations. Therefore, DRR educators must thoroughly understand the data DRR professionals handle to make students explain the whole picture.

Technological Knowledge

Online teaching helps educators to teach better (and present) using different online teaching materials. This includes resources like videos, podcasts, documentaries, etc., and data visualisation methods and representations. The interviewee ID UK_HUD_9 explained this as *"presenting data in various ways and catering to the different learning styles, learning needs and learning plans."* It is also essential to acquire knowledge about handling online resources and preparation strategies, for instance, uploading or creating high-quality videos, demonstrations, and video editing, as there have been observations regarding improved student attention and engagement when the online resources are high-quality and have user-friendly interfaces. Similarly, some educators find accommodating changes to online content a complex task for which they should sharpen their technological know-how. Technology may bring advantages to create inclusivity in classrooms, especially for students with disabilities. A critical limitation for such DRR students with mobility issues would be field trips. With educators' knowledge of technology, those students can still have an immersive experience, which is more convincing than a video of that field visit. It is also essential to have sufficient database management and analytics for collecting and processing vast amounts of data to develop the different types of DRR scenarios in teaching and related research.

3.2.2 Skills

The following thematic orientations could be used to present the aspects of skills that evolved during the content analysis and the literature review.

The Pedagogical and Technological Categorisation

As per Fig. 1, the project's first output was dedicated to recognising distance learning strategies used in DRR education and their effectiveness in identifying their success factors and associated issues and problems. Based on the recognised strategies and challenges, a framework was developed to re-imagine online DRR education. On the other hand, outputs 3 and 4 are dedicated to the technological aspect of online DRR education. Output 3 was dedicated to developing an inclusive University-Industry digital learning platform, and Output 4 evaluated the use of disruptive technology in DRR. Based on the content of these reports, themes were derived that are dedicated towards educational strategies, challenges, and technological aspects. When connecting these themes with digital educator competencies, it was abided by the related studies that have recognised that online teaching competencies fall into the more prominent umbrella themes of pedagogical and technical skills (Sopegina et al., 2016).

The Role of Pedagogy in the Online Space—Digital Pedagogical Skills

Regarding competencies for online education, pedagogy has been considered one of the primary skills required (Baran & Correia, 2014; Dobbin et al., 2009; Habibi, 2021; Koehler et al., 2013). Regarding pedagogy, the Output 2 report relies on the constructivist and cognitive learning theories. The constructivist theory perceives knowledge as something constructed through the learners' previous experience and the collaboration among the learners and the educators (Koohang et al., 2009; Reid-Martinez & Grooms, 2018). On the other hand, the cognitive theory emphasises the interactions between the learner and the content (Malik, 2021). Output 3 lays the foundation for developing an inclusive University-Industry digital learning platform; it relies on the importance of open education and collaboration between higher education and industry.

Using appropriate online pedagogical strategies is another viral aspect of digital skills in online education (Bigatel et al., 2012; Kattoua et al., 2016). While a Learning Management System (LMS) is one of the most popular online learning strategies (Bigatel et al., 2012; Farmer & Ramsdale, 2016), the findings of the Output 1 claim that blended learning, only synchronous learning, only asynchronous learning, flipped classroom and live lectures were also recognised as common online learning strategies. It is also essential to set up protocols for managing the online classroom regarding rules, students' progress and time management (Craddock & Gunzelman, 2013; Munoz Carril et al., 2013).

The content analysis also raised the importance of handling distractions (via chat, unauthorised log-ins, unauthorised removal, and not switching on videos) and conflicts that may occur during the online class. This was recognised as an online educator skill in related studies (Carril et al., 2013). Effectiveness in teaching online was another theme that emerged, and students' performance in online assessments is an essential factor to be considered (Habibi, 2021). Therefore, the ability to develop online rubrics, journals and logs is vital (Kattoua et al., 2016). The use of active learning strategies to enhance participation within an online classroom is another central aspect that evolved from the content analysis, which is widely recognised in the available literature (Bigatel et al., 2012). The content analysis further highlighted the need to adapt active learning strategies to accommodate diverse types of learners, such as less vocal students. The Output 1 and 2 findings claim that educators use various technological strategies such as chat, Zoom polls, and Zoom break-out rooms to keep students engaged within the online space. Further, as educators are faced with the challenge of keeping students who are not very vocally active in online classes, they believe that anonymity options, such as the private chat option and polls, can be used as a benefit to get them engaged.

However, the content analysis further highlights that the development of online pedagogy is based on several factors, such as the level of students (undergraduate/post-graduate), the experience of the students, motivation of the students based on their interactions within the online space, and other commitments of the students (professional/personal). In terms of inclusivity within the classroom, the content analysis further emphasised the importance of having content available within the learners' local language, presenting data that is accessible to anyone, and promoting equal participation within the class.

The Designing of the Online Teaching Content—Design Skills and Content Skills

The themes derived from the content analysis also addressed the importance of adequately designing online teaching content. It is a competency that has been vastly recognised in studies related to online teaching competencies (Baran & Correia, 2014; Guasch et al., 2010; Habibi, 2021; Palloff & Pratt, 2011). The literature also highlights the importance of adequately designing suitable online assessments (Dobbin et al., 2009; Habibi, 2021).

In terms of designing the course content, the content must be of high quality and up-to-date (Koehler et al., 2013). The content analysis emphasises explicitly the content of a DRR course and should consider its interdisciplinary and practical aspects of the subject. Further, it was pointed out that the latest developments in the DRR field should be reflected in the content. In addition, the reliability and authenticity of the content are also vital factors. The findings also emphasise the importance of the educator exploring the under-presented areas in DRR and promoting interconnected thinking with the use of quality content. The results of Output 1 and the validation of Output 2 further claim the importance of constructive alignment in designing online courses, considering the intended learning outcomes. The findings further claim the importance of educators prioritising to assess the workload of both educators and learners and determining the time frame for designing and planning the teaching content to ensure that they are not over-burdening either party. The diversification of the teaching strategies was another critical highlight revealed in the Output 1 report, where the educators believe that such diversification will assist in the flexibility and adaptability of the course content based on the course context. Further, keeping in mind the future unpredictable risks the world is yet to face, the content analysis emphasises the importance of having content available and prepared to be converted into various forms.

On the other hand, educators who specifically teach undergraduate or early career learners must develop content that considers the future career prospects of the learners.

Online classes accommodate more learners than physical classes (Jena, 2020). Online learning tools expand the scalability of courses, making personalisation of education easy (Mäkelä et al., 2020). Educators must consider the broader context in designing courses in this context. The content analysis claims the importance of considering geographical contexts, language, and disabilities and other learning needs.

The Social and Communicational Setups in Online Classrooms—Social and Communication Skills and Classroom Management Skills

An online educator should have social and communication skills (Bigatel et al., 2012; Salam, Mohamad, & Besar, 2011). The interaction during online lessons is a vital point highlighted in the content analysis (Bigatel et al., 2012). Feedback, on the other hand, is another essential point. There are two types of feedback: getting feedback from the students for further development (Bigatel et al., 2012) and giving feedback to the students (Albrahim, 2020). The content analysis further emphasised the importance of enhancing the inclusivity within the class and having a more excellent knowledge of the backgrounds of the students, such as students with disabilities, not-so-talkative students, students that come from impoverished backgrounds and students that are on the verge of discontinuing their studies due to various personal and professional commitments (Fuller & Yu, 2014; Munoz Carril et al., 2013). Further, the content analysis emphasised that for educators to recognise the several contexts of their learners within the online class, it is vital to enhance communication throughout the learning process (pre-, during, and post).

On the other hand, classroom management is a skill educators need in online space (Albrahim, 2020). Therefore, educators must establish rules to maintain the classroom's code of conduct, monitor students' progress, manage time, and preserve mentorship and coaching.

The Role of Technology Within Education—Technological Skills

As an educator involved in online education, technological skills are vital (Guasch et al., 2010; Koehler et al., 2013). It has been further recognised that in terms of technological skills, basic IT and multimedia skills are vital (Bigatel et al., 2012; Kattoua et al., 2016). Output 1 also claims the importance of educators familiarising themselves with the interface of LMS and other online learning tools. The content analysis further addressed the need to handle unforeseen technological difficulties that could take place within online space and the need to prepare individualised experiences for the students in virtual space (Albrahim, 2020; Kattoua et al., 2016). In the context of DRR, Output 4 refers to various disruptive technologies (such as drones, virtual reality, and artificial intelligence) used within the DRR sector. The DRR educators must be up-to-date with these technological advancements. Output 1 report claims that one of the main challenges educators and learners face in online DRR education is related to relevant IT infrastructures such as internet connections and computers. Hence, the educators must intervene in his regard when needed.

The Role of the Institution in Online Education

The institutional setup plays a significant role in online education (Baran & Correia, 2014; Palloff & Pratt, 2011). The content analysis emphasises the importance of addressing privacy matters within the online space (Habibi, 2021), specifically in online assessment feedback (Bigatel et al., 2012; Farmer & Ramsdale, 2016). Further, the related studies also consider the need for proper legal and policy backgrounds to govern online learning and teaching (Bigatel et al., 2012; Habibi, 2021).

3.2.3 Attitudes

Given that attitudes are subjective and, hence, challenging to recognise categories of attitudes, the attitude domain was analysed based on the affective domain presented by Bloom in 1956 (Bloom et al., 1956). Bloom's taxonomy, a model that assists learners and educators in selecting relevant learning strategies, divides the educational objectives into the three categories of skills, knowledge and attitudes, out of which attitudes have been interpreted in the affective domain (Testa, Toscano, & Rosato, 2018). This domain consists of five levels, that has been presented from the lowest to the highest:

1. Receiving—Awareness of the necessity and preparedness to pay attention,
2. Responding—Active participation in the learning process,
3. Valuing—the ability to estimate the value of something,
4. Organisation—Comparison and arranging values, conflict resolution and establishing a unique value system, and
5. Characterisation—Establishing a universal, consistent, predictable value system to control learner behaviours. This requires practising professional ethics and updating strategies based on contextual differences (Wu et al., 2019).

The interviewees in the study believed that students' equal participation in online learning appertains to the educators' approach. They asserted that it is the educators' responsibility to ensure equal participation, and the onus is on them to improve the students' engagement in different online teaching strategies. UK_11 stated, "*It depends on how the lecturers deal with their students*". Stating a supporting view, UK_7 mentioned, "*I think the onus is on me to make sure that I have structured the questions that I ask them in such a way that I show them that we are getting the most out of them as opposed to putting the onus on them to answer the questions.*" And she further added "*This is all about relationship building, and so I try to encourage this group in this way. It encourages everyone to have something important to say and promotes that positively, and it shows openness on my part. Telling them when I screw up, and I think that allows them to feel that it is OK if they share their opinion or their position or argument.*" One aspect of educators ensuring equal participation is understanding the students' difficulties. For instance, UK_9 highlighted a question that educators should answer before designing an activity/ assignment/ homework: "*Do they have the kind of support, capability and confidence to do what you ask students to do in their own time?*".

The interviewees who agreed they are responsible for promoting students' equal participation mentioned some of their good practices. According to SW_1, "*You get students with visual impairments, non-native speakers. There is so many things that we must be careful of, even font size and font colour, use of complex languages*". For instance, UK_ 6 remarked that they had used software called Ally that allows students to change the language during an online lecture, which is very helpful for international students, as well as software that can read aloud written information and change the font to make the material more inclusive and accessible. UK__7, who strongly believes that educator has much to contribute to promoting equality and inclusivity in students' participation, explained one of her strategies: "*I start all classes telling them that no matter what anybody says, I am going to tell them that they are wrong. It does not matter what you say. I could agree with you 1000%. I could have the same position, but I will tell you might be wrong, and this is my tool to ensure you can respond to criticism. So, it is promoting critical analysis. So, I explain it to them as a pedagogical tool, and then they know they will be told they are wrong no matter what. So, it is OK if they get it wrong because no one will know they got it wrong.*"

UK__7, who supported the power of anonymity as a mechanism to give every student a voice, further stated, "*With the online, no one has to turn their camera on but me*". According to her, "*Some students are self-conscious of their living environment. Some students are doing their work from random places, so equal participation is, well, no one is going to be judged, and no one has to be worried about having all of that recorded as far as different perspectives.*" This is similar to UK__5's suggestion on adopting flexibility and fostering students to have more control over time and work at their own pace.

UK__3, as an educator representing a degree programme that offers to teach bilingually in a multilingual country, raised attention to the difficulties for monolingual students (students who need to speak/understand the languages the programme is being taught). He further stated the difficulty to find material in different languages: "*language barrier was there because we conduct lectures bilingually and if we upload a document documentary in English, it is very difficult for us to find its Sinhala or Tamil (majority of the students enrolled in the programme speak/understand these two languages) versions or Sinhala or Tamil documentaries in general*". However, it can be argued that this incident presents cultural impacts on education and, therefore, cannot be generalised. However, this brings attention to the idea that culture gaps are prominent, and they may also misinterpret student behaviour in classes. Although educators work hard to understand their students from different cultural backgrounds, these situations could occur until the educator-learner relationship matures. Therefore, openness is a crucial attribute to be practised from the initial days, and UK_7 presented her views on a healthy educator-learner relationship as presenting her concise post-structuralist views. UK_7 argued, "*So openness, conversation, and encouragement that everybody's position and argument is relevant.*". Meanwhile there were also concerns over the fact that the flexibility of online sessions would be abused by educators who, according to UK_8 "*do not want to come to work*" as it can be easier to work at home rather than "*coming up and facing the students and being challenged by the students*". Similarly, SW_9 highlighted the importance of adapting to different working environments, learning fast and being critical in performing better in the DRR discipline.

Typically, some DRR courses/programmes/modules are known to be a niche in higher education, particularly in some countries. Therefore, there are only a few experts capable of teaching those areas, and understandably, this could give rise to problems related to copyrights (for lecture material or recorded lecturers). Without a sufficient understanding of copyrights, some lecturers refused to allow their lectures to be recorded in a context where the recordings were the only choice for most students with internet-related issues. UK_3 described this experience as a head of the department: "*Some lecturers did not want to share other recordings because what they said was, what would happen if student uploads their recording illegally in other channels or YouTube and who is going to take that responsibility? So, in that context, some lecturers said they would conduct the live lecture but would not share the recording, and we were not in a position to force the lecturers to record*". This draws the attention to educating the lecturers on aspects like copyrights on material and exercising professional due diligence.

The above discussed results can be summarised under the following points:

- Willingness to adapt
- Flexibility to meet evolving needs and respond smoothly to unprecedented external changes
- Willingness to learn and self-evaluate
- Openness
- Growth mindset
- Observance and being responsible
- Being ethical
- Teamwork and networking.

Following the above findings of the project, the attitudes mentioned above were divided as per the levels of Bloom's affective domain to investigate the coherence of the said levels and the project findings (Table 1):

Based on the above table, the project findings on the DRR educators' attitudes complement the affective domain of Bloom's Taxonomy, which could be used as the basic categorisation of the domain of attitudes in the competency framework.

Table 1 The levels of Bloom's affective domain with project findings on educators' attitudes (*Source* Developed by the authors)

Receiving	Responding	Valuing	Organisation	Characterisation
Willingness to adapt	Teamwork and networking	Observance and being responsible	Willingness to learn and self-evaluate	Flexibility to meet evolving needs and respond smoothly to unprecedented external changes
Flexibility to meet evolving needs and respond smoothly to unprecedented external changes			Growth mindset	Being ethical
Openness			Being ethical	Teamwork and networking

3.3 *Development of the Competency Framework*

Following mapping the literature review findings with the content analysis results, the initial framework was developed. The development of the framework was based on five main principles:

3.3.1 Principle 1: Main Virtues

As this is the very first digital competency framework for DRR educators, the research team was mindful of the way forward of the developed framework. Hence, the framework was based on the following three primary virtues:

- Responsiveness: As per the findings discussed above, it is evident that DRR education and community have various needs and aspects. Hence, the framework will be responsive to the needs of DRR education and the diverse DRR community.
- Adaptability: The framework is developed in such a way that it can be adaptable to different educational contexts in Europe and the global setting.
- Flexibility: The lists mentioned in the framework are not exhaustive and limited to. The framework is sufficiently flexible for the prospective users to make necessary adjustments to suit their institutional and country-specific requirements.

3.3.2 Principle 2: Critical Dimensions of Digital Competencies

As per the findings discussed above, the overall dimensions of the framework will be based on the following:

- Knowledge

  - Pedagogical knowledge
  - Content knowledge
  - Technological knowledge (Koehler & Mishra, 2009)

- Skills

  - Digital pedagogical
  - Design
  - Content
  - Classroom management
  - Technological
  - Reflecting (Albrahim, 2020)

- Attitudes

  - Receiving
  - Responding
  - Valuing
  - Organisation
  - Characterisation (Wu et al., 2019).

3.3.3 Principle 3: Stages of Online Teaching

The reviewed literature also emphasised how the relevant digital competencies may differ based on the following stages of teaching:

a. Before teaching—preparing, planning, and designing.
b. During the teaching—facilitating, interacting, and providing and seeking feedback competencies.
c. After teaching—reflecting on teaching and feedback (Abdous, 2011).

Therefore, the overall recognised digital competencies for DRR educators were categorised based on the stages above, as addressing the overall teaching process is vital.

3.3.4 Principle 4: The Main Principles that Reimagine Online Distance DRR Education

Output 2 of the INCLUDE project was dedicated to developing a framework to reimagine online DRR education. The development of the said framework (Fig. 2) was based on the online DRR teaching and learning strategies and associated challenges recognised under Output 1. The said framework was based on the following principles:

- Inclusivity—promotion of inclusion education considering the diverse backgrounds of students such as impoverishments such as poverty, age, disabilities and gender,

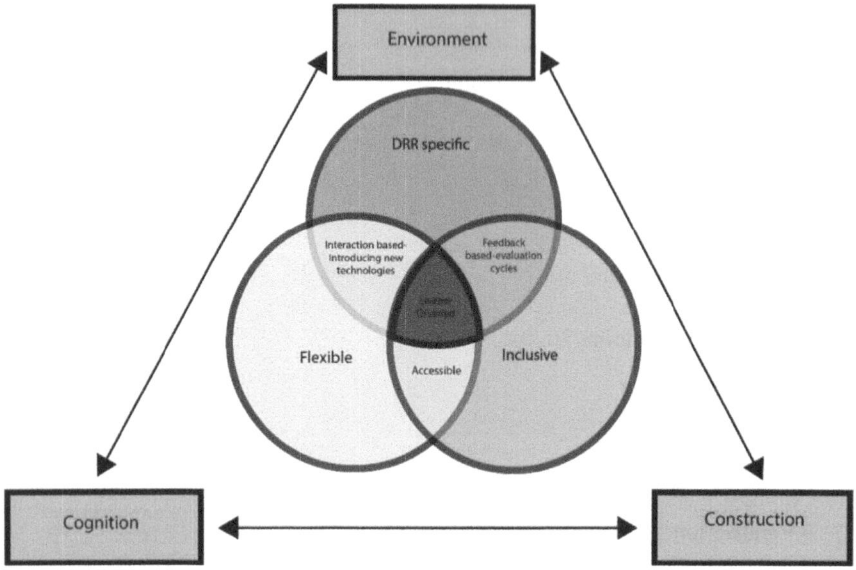

Fig. 2 INCLUDE Output 2 framework—framework to reimagine online DRR education (*Source* Developed by the authors)

- Flexibility—Keeping education up-to-date to accommodate the multi-disciplinary and diverse nature of DRR,
- Accessibility—Ensuring accessibility to learning content and classroom environments,
- Interactions—Developing an interactive learning environment,
- Feedback-based—providing feedback to students and getting feedback from the students on teaching,
- Use of technology—Using disruptive technologies related to DRR, such as Drones, virtual realities, artificial intelligence, and
- Privacy concerns—Considering the protection of user data.
- x Learner centred—Pleacing the learners at the core not just as receivers but as co-constructors

When considering these seven principles, it is evident that the abovementioned findings resonate well with them. In this context, it is evident that the categories recognised so far on digital competencies for DRR educators comply with the principles developed to reimagine digital DRR education.

Based on the above principles, the three main layers of the competency framework are knowledge, skills, and attitude. Under each dimension, the relevant competencies were derived based on the content analysis and the literature review. As the existing competency frameworks are based on digital education, the novelty of this framework is its dedication to online DRR education. Hence, the competencies were developed

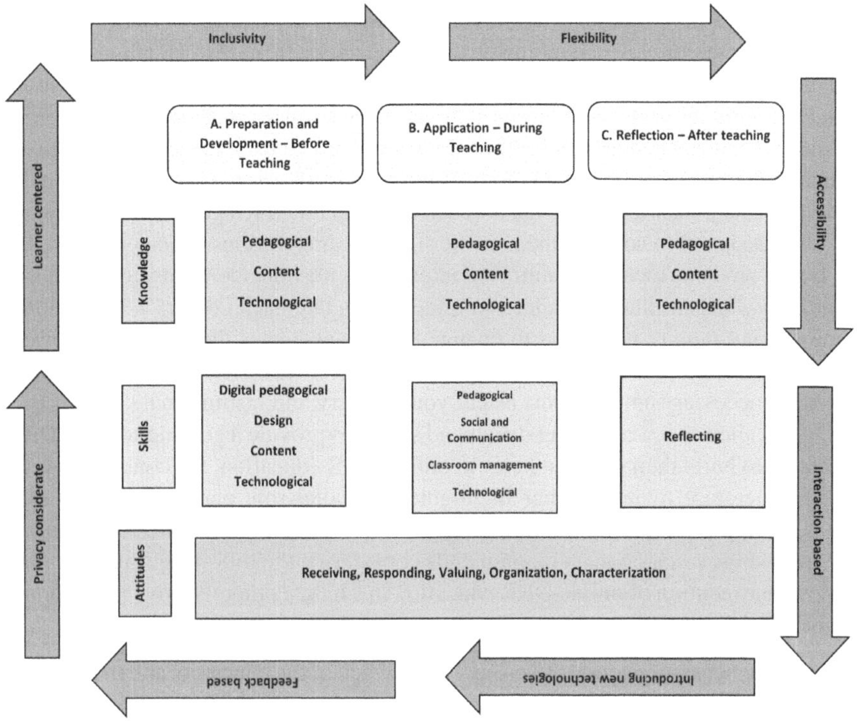

Fig. 3 Final version of the developed competency framework (*Source* Developed by the authors)

in considering the nature of DRR as a subject under the dimension of knowledge, skills, and attitudes.

Following the development of the initial competency framework, it was shared among the project partners for feedback. Following their feedback, the amended framework was utilised for external validation. It was conducted with the participation of seven DRR educators who represent various fields related to DRR based in the partner countries. Following this, the below final competency framework was developed (please refer to Annexure 1 for the detailed competency framework) (Fig. 3).

4 Conclusion and Way Forward

DRR education is multidisciplinary, ranging from crisis management and emergency medicine to cost control in disaster reconstruction. Hence, the role of a DRR educator is broad. This chapter presented the development of a competency framework that holds the digital competencies required by DRR educators. This framework is limited to the DRR educators' role in delivering educational content (teaching) compared to designing DRR programmes/ modules. The competencies required of DRR educators in different contexts vary depending on the context-specific needs/challenges. In such situations, this framework needs to be customised, and identify the context-specific mandatory and optional competencies. To do the needful, the following virtues were followed to make it convenient for the framework's users: responsiveness to changes, adaptability and flexibility. Even though this framework was developed based on the findings of Asia Pacific, Lithuania, Sweden and the United Kingdom, the framework has been designed to adapt to the context of the country various educators are based on. Based on the respective country's context, they may decide what is optional and mandatory. Further, the research team understands that each country's institutional contexts may require educators to change the competencies they have recognised. Hence, this framework has been developed upon the virtue of flexibility, where they can make necessary amendments as per your country and institutional contexts.

Despite the framework being developed solely to provide a guiding tool for DRR educators to build their competencies in online DRR education, the content analysis and the literature review further highlighted the vital role played by the institutional settings in providing online DRR education. As a way forward to this initial framework, the research team recommends considering the following aspects for the successful execution of online DRR education in a more comprehensive institutional setup:

1. Taking necessary actions to protect the privacy of the educators and the learners within an online space
    a. Institutional access
        i. Enrolment to courses
    b. Content securing
        i. Uploading content
        ii. Personal information that is required to enroll on the courses
    c. Course management
        i. Commenting
        ii. Assessments and feedback
2. Having a policy and regulatory framework in place in terms of the functionality of the platforms
3. Making sure that the relevant infrastructure is available for both educators and students
4. Having a sustainable strategy in terms of the maintenance of the platforms and facilities, specifically in the context of financial management
5. Taking necessary actions to protect the copyrights of the platform
6. Make sure relevant access is given for the online resources on DRR to both the educators and students following the applicable intellectual property guidelines
7. Assigning relevant roles in terms of the platform and training the appropriate parties
    a. Administrators
    b. Facilitators

   c. Students

8. Having a guideline and policy framework in place in terms of the online learning space within higher education and its way forward
9. Developing a guide that could be referred to by the users of the platform
10. Taking necessary steps to enhance the accessibility and the reachability of the platform through

    a. Fact access
    b. Multilinguality
    c. Scalability
    d. Cross-platform
    e. Good user interface and user experience.

Annexure

	A. Preparation and development—prior to teaching	B. Application—during teaching	C. Reflection—after teaching
Knowledge	***Pedagogical knowledge*** 1. Develop knowledge of appropriate* digital learning theories/pedagogical styles on online teaching strategies. E.g. Constructivist and cognitive learning theories *Appropriate to build a positive connection with different types of learners and circumstances ***Content knowledge*** 1. Plan the teaching content and assessments in a manner that caters for the needs of different types of learners (create inclusivity through content knowledge) 2. Update the knowledge on developments in the DRR content ***Technological knowledge*** 1. Plan the use of technology and tools for the online teaching	***Pedagogical knowledge*** 1. Promote collaboration 2. Enhance active learning/ student engagement 3. Handling distractions and manage dominating ideas ***Content knowledge*** 1. Improve the knowledge on area being taught ***Technological knowledge*** 1. Secure and responsible use of digital resources	***Pedagogical knowledge*** 1. Identify the existing challenges with previous online student cohorts 2. Self-evaluations to identify training/ learning needs 3. Peer feedback/review ***Content knowledge*** 1. Evaluating the information used to teach to make sure if the most up to date information has been used 2. Reflecting if the underrepresented and relevant information has been used in the teaching content ***Technological knowledge*** 1. Identify what facilitations to be proposed to the institution 2. Evaluating the technologies that were used in the teaching process
	Digital Pedagogical skills 1. Develop skills to utilize appropriate learning theories on online teaching strategies. E.g. Constructivist and cognitive learning theories 2. Develop skills on appropriate online techniques and learning strategies E.g. LMS, blended learning, only synchronous learning, only asynchronous learning, flipped classroom and live lectures 3. Explore under-presented areas 4. Promote interconnected thinking	***Digital Pedagogical skills*** 1. Use of appropriate learning theories into account in terms of deciding on the online teaching strategies. E.g. Constructivist and cognitive learning theories 2. Use of appropriate techniques and learning strategies. E.g. LMS, blended learning, only synchronous learning, only asynchronous learning, flipped classroom and live lectures 3. Ability to work with quality resources	

(continued)

(continued)

	A. Preparation and development—prior to teaching	B. Application—during teaching	C. Reflection—after teaching
Skills	***Design skills*** 1. Have a thorough preparation of the course in terms of its a. Learning outcomes b. Time frame c. Workload management of the educator and the learner 2. Design courses in way where teaching strategies could be diversified so that the content is flexible and adoptable as per the context of the course is offered. E.g. Use of both synchronous learning and asynchronous learning 3. Consider a wider audience in terms of designing the course. E.g. Geographical context, language and disabilities 4. Consider the practical aspect of Disaster Risk Reduction Education. E.g. filed visits, field work and placements 5. Consider the interdisciplinary aspect of Disaster Risk Reduction Education. E.g. natural sciences, engineering and Social Sciences 6. Consider the future career prospects of the learners 7. Design appropriate assessments that address diverse student needs and workable to differentiate machine (AI) generated work with human/student generated work	4. Evaluating the context of your online class when finalizing your pedagogical strategies. E.g. a. The level of students—undergraduate/post-graduate b. The past experiences of the students c. Motivation of the students based on their interactions and performance d. Other commitments of the students—employment and social responsibilities 5. Plan out to make sure that all types students are included within the learning process. E.g. Less vocal students, dominating students, students with physical and learning disabilities 6. Handle distractions that may occur online. E.g. Unauthorised admission and removal, unnecessary comments visa chat 7. Effective teaching in terms of a. Achieving intended learning outcomes b. Time managements c. Student performance in assessments d. Retrieval and retention of knowledge 8. Active learning strategies to keep students engaged. E.g. Zoom polls, chat, Spotify challenge, Kahoots 9. Flexible strategies that could accommodate in any of the changes of the above-mentioned facts 10. Use of anonymity as a benefit in the online space to get the less vocal students engaged. E.g. use of the chat option 11. Knowledge between the educator and the learner should be duly shared. E.g. Co-creation of knowledge through feedback	

(continued)

(continued)

	A. Preparation and development—prior to teaching	B. Application—during teaching	C. Reflection—after teaching
	Content skills 1. Content Available and prepared to be converted into various forms at a time of emergencies. E.g. content to be conveniently converted to online content 2. Reliable content in terms of the context the course is offered 3. Authentic content in terms of the context the course is offered 4. Up to date content in terms of latest of developments in the field of Disaster Risk Reduction 5. Address the interdisciplinary aspect of Disaster Risk Reduction 8. Address the practical aspect of Disaster Risk Reduction Education 9. The content should be accessible to a wider audience in terms of. E.g. Geographical context, Language, and Disabilities and other learning needs	***Social and communication skills*** 1. Resolve distractions and grievances that could take place within an online classroom. E.g. Unauthorised admission and removal, unnecessary comments visa chat 2. Gather information of the class to recognize the students who are affected by the digital divide and strategize teaching to get them involved 3. Make teaching material and lectures available in common and local languages 4. Ensure the equal participation within the class in terms of communication 5. Use of the benefit of anonymity within digital space to get more students engaged. E.g. use of the chat option 6. Recognize the students that are at the verge of discontinuing the course for various reasons and connect them with relevant authorities 7. Present the data and teaching content in a way that is accessible to everyone 8. Enhance communication and interactions during and after class via digital space to understand the contexts of the students 9. Gain feedback from students in terms of teaching and content for further development Giving feedback for students 10. Enhancing inclusivity within the online classroom in terms of a. Students with disabilities b. Students with professional commitments c. Students with emotional and health concerns d. Students that come from impoverishment backgrounds	***Reflecting skills*** 1. Reflecting on the feedback received from the students for further development 2. Self-evaluation 3. Peer evaluation

(continued)

(continued)

	A. Preparation and development—prior to teaching	B. Application—during teaching	C. Reflection—after teaching
	Technological skills 1. Relevant trainings and knowledge on the relevant digital literacy in terms of online teaching. E.g. a. Basic IT skills b. Multimedia skills 2. Relevant infrastructure is available and if not, to report to relevant authorities. E.g. a. Stable internet connection b. Computer c. Relevant software 3. Familiarizing with the interface and functions of the learning management systems and other online teaching tools 4. Familiarizing with new technological developments that are taking place in the Disaster Risk Reduction field. E.g. drones, AI and virtual reality	***Classroom Management skills*** 1. Set up the rules that should govern the classroom 2. Constant monitoring of students' progress 3. Efficient management of time 4. Maintain leadership, mentorship and coaching within the online classroom ***Technological skills*** 1. Use disruptive technologies within teaching to give students field work experience within a digital space. E.g.: Virtual reality glasses 2. Make note of the unforeseen technological difficulties that class face and take necessary steps 3. Use of various technological aspects to enhance the learner centric learning. E.g. cMOOCs. Repositories 4. Use of relevant technologies to develop an individualized experience to the students. E.g. cMOOCs	

(continued)

(continued)

	A. Preparation and development—prior to teaching	B. Application—during teaching	C. Reflection—after teaching
Attitudes	***Receiving*** • Openness • Growth mindset • Observance ***Responding*** • Teamwork and networking • Flexibility to meet evolving needs • Readiness and preparedness • Willingness to adapt • Swift responsiveness to unprecedented external changes ***Valuing*** • Inclusivity (respecting the diversity) • Willingness to learn and self-evaluate ***Organization*** • Adaptability ***Characterization*** • Being ethical • Responsibility and accountability		

References

Abdous, M. H. (2011). A process-oriented framework for acquiring online teaching competencies. *Journal of Computing in Higher Education, 23*(1), 60–77.

Albrahim, F. A. (2020). Online teaching skills and competencies. *Turkish Online Journal of Educational Technology-TOJET, 19*(1), 9–20.

Baartman, L. K. J., & de Bruijn, E. (2011). Integrating knowledge, skills and attitudes: Conceptualising learning processes towards vocational competence. *Educational Research Review, 6*(2), 125–134. https://doi.org/10.1016/j.edurev.2011.03.001

Baran, E., & Correia, A. P. (2014). A professional development framework for online teaching. *TechTrend, 58*(5), 95–101.

Baran, E., Bilici, C. S., Sari, A. A., & Tondeur, J. (2019). Investigating the impact of teacher education strategies on preservice teachers' TPACK abstract. *British Journal of Educational Technology, 50*(1), 357–370. https://doi.org/10.1111/bjet.12565

Bigatel, P. M., Ragan, L. C., Kennan, S., May, J., & Redmond, B. F. (2012). The identification of competencies for online teaching success. *Journal of Asynchronous Learning Networks, 16*(1), 59–78.

Bloom, B. S., Engelhart, M. D., Furst, E. J., Hill, W. H., & Krathwohl, D. R. (1956). *Taxonomy of educational objectives. The cognitive domain.* David McKay Co Inc.

Carril, P. C. M., Sanmamed, M. G., & Selles, N. H. (2013). Pedagogical roles and competencies of university teachers practicing in the elearning environment. *The International Review of Research in Open and Distance Learning, 14*(3), 462–487.

Centre for Research on the Epidemiology of Disasters (CRED). (2023). *2022 disasters in numbers.* CRED.

Craddock, J., & Gunzelman, R. (2013). Creating WOW: Characteristics of successful online instruction and facilitation. *The Journal of the Virtual Classroom, 26.*

Dobbin, E., Corrigan, P. M., Graham, C., Thomas, K. G., Freeburn, R. W., & Wheadon, H. T. (2009). PDGFRβ induces stem cell differentiation via the Ras/ERK and STAT5 signaling pathways. *Experimental Hematology, 37*(1), 111–121. https://doi.org/10.1016/j.exphem.2008.09.012

ElSayary, A. (2023). The impact of a professional upskilling training programme on developing teachers' digital competence. *Journal of Computer Assisted Learning.* https://doi.org/10.1111/jcal.12788

Farmer, H., & Ramsdale, J. (2016). Teaching competencies for the online environment. *Canadian Journal of Learning and Technology, 42*(3), 1–17.

Ferrari, A. (2013). *DIGCOMP: A framework for developing and understanding digital competence in Europe.* Joint Research Centre of the European Commission.

Ferrari, A., Punie, Y., & Redecker, C. (2012). Understanding digital competence in the 21st century: An analysis of current frameworks. In A. Ravenscroft, S. Lindstaedt, C. D. Kloos, & D. Hernández-Leo, *21st century learning for 21st century skills*, Berlin, Heidelberg.

From, J. (2017). Pedagogical digital competence-between values. Knowledge and skills. *Higher Education Studies, 7*(2), 43–50.

Fuller, P., & Yu, G. (2014). Lessons learned: Online teaching adventures and misadventures. *Journal of Social Sciences, 10*(1), 33–38.

Funkhouser, B. J., & Mouza, C. (2013). Drawing on technology: An investigation of preservice teacher beliefs in the context of an introductory educational technology course. *Computers & Education, 62*, 271–285. https://doi.org/10.1016/j.compedu.2012.11.005

Guasch, T., Alvarez, I., & Espasa, A. (2010). University teacher competencies in a virtual teaching/learning environment: Analysis of a teacher training experience. *Teaching and Teacher Education, 26*(2), 199–206.

Habibi, A. (2021). Book review: Education abroad bridging scholarship and practice. *Research in Comparative & International Education, 16*(3), 325–332. https://doi.org/10.1177/17454999211038078

Hall, R., Atkins, L., & Fraser, J. (2014). Defining a self-evaluation digital literacy framework for secondary educators: The DigiLit Leicester project. *Research in Learning Technology, 22*(0). https://doi.org/10.3402/rlt.v22.21440

Hämäläinen, R., Nissinen, K., Mannonen, J., Lämsä, J., Leino, K., & Taajamo, M. (2021). Understanding teaching professionals' digital competence: What do PIAAC and TALIS reveal about technology-related skills, attitudes, and knowledge? *Computers in Human Behavior, 117*, 106672. https://doi.org/10.1016/j.chb.2020.106672

Herdiansyah, H., Husein, S. I., Asrofani, F. W., Simamora, P. A. R., & Kholila, B. N. (2020). Disaster Awareness through Disaster Preparedness Education for Primary Schools. *IOP Confonference Series: Earth and Environmental Science, 519*, 012016. https://doi.org/10.1088/1755-1315/519/1/012016

Hew, K. F., & Brush, T. (2007). Integrating technology into K-12 teaching and learning: Current knowledge gaps and recommendations for future research. *Educational Technology Research and Development, 55*, 223–252.

Hoffmann, T. (1999). The meanings of competency. *Journal of european industrial training.*

Instefjord, E. J., & Munthe, E. (2017). Educating digitally competent teachers: A study of integration of professional digital competence in teacher education. *Teaching and Teacher Education, 67*, 37–45. https://doi.org/10.1016/j.tate.2017.05.016

Jena, P. K. (2020). Online learning during lockdown period for covid-19 in India. *International Journal of Multidisciplinary Educational Research, 9.*

Joo, Y. J., Park, S., & Lim, E. (2018). *Factors influencing preservice teachers' intention to use technology*

Kagawa, F. (2005). Emergency education: A critical review of the field. *Comparative Education Sciences, 41*(4), 487–503.

Kattoua, T., Al-Lozi, M., & Alrowwad, A. (2016). A review of literature on e-learning systems in higher education. *International Journal of Business Management and Economic Research, 7*(5), 754–762.

Kitagawa, K. (2021). Conceptualising 'disaster education'. *Education Sciences, 11*(5). https://doi.org/10.3390/educsci11050233

Koehler, M., & Mishra, P. (2009). What is technological pedagogical content knowledge (TPACK)? *Contemporary Issues in Technology and Teacher Education, 9*(1), 60–70.

Koehler, M. J., Mishra, P., & Cain, W. (2013). What is technological pedagogical content knowledge (TPACK)? *Journal of Education, 193*(3), 13–19. https://doi.org/10.1177/002205741319300303

Koohang, A., Riley, L., Smith, T., & Schreurs, J. (2009). E-learning and constructivism: From theory to application. In *Proceedings of the 2009 InSITE conference.* https://doi.org/10.28945/3321

Le Deist, F. D., & Winterton, J. (2005). What is competence? *Human Resource Development International, 8*(1), 27–46. https://doi.org/10.1080/1367886042000338227

Lyonga, N. A. N., Moluayonge, G. E., & Nkeng, A. J. (2021). A study of techno-pedagogical skills and teachers' performance in HTTTC Kumba, Cameroon. *European Journal of Education and Pedagogy, 2*(1), 46–50.

Mäkelä, T., Mehtälä, S., Clements, K., & Seppä, J. (2020). Schools went online over one weekend opportunities and challenges for online education related to the COVID-19 crisis.In *Paper presented at the EdMedia + Innovate Learning*

Malik, S. (2021). Learning theory of cognitivism and its implications on students' learning. *International Journal Peer Reviewed Journal Refereed Journal Indexed Journal, 7*(5), 67–71. www.wwjmrd.com

Mangione, G. R., Capuano, N., Orciuoli, F., & Ritrovato, P. (2013). Disaster Education: a narrative-based approach to support learning, motivation and students' engagement. *Journal of e-Learning and Knowledge Society, 9*(2), 129–152. https://doi.org/10.20368/1971-8829/837

Mota, F. P. B., & Cilento, I. (2020). Competence for internet use: Integrating knowledge, skills, and attitudes. *Computers and Education Open, 1*, 100015. https://doi.org/10.1016/j.caeo.2020.100015

Munoz Carril, P. C., Gonzalez Sanmamed, M., & Hernandez Selles, N. (2013). Pedagogical roles and competencies of university teachers practicing in the elearning environment. *The International Review of Research in Open and Distance Learning, 14*(3), 462–487.

Palloff, R. M., & Pratt, K. (2011). *The excellent online instructor: Strategies for professional development*. Wiley.

Perrenoud, P. (2005). La universitat entre la transmissió de coneixements i el desenvolupament de competències. *El debat sobre les competències a l'ensenyament universitari, ICE UB. Documents de Docència Universitària, 5*, 8–25.

Reid-Martinez, K., & Grooms, L. D. (2018). Online learning propelled by constructivism. In *Encyclopedia of information science and technology* (4th ed., pp. 2588–2598). IGI Global.

Salam, A., Mohamad, N., & Besar, M. N. A. (2011). Audit upon graduation: UKM house officers' competencies. *Procedia-Social and Behavioral Sciences, 18*, 460–463. https://doi.org/10.1016/j.sbspro

Scherer, R., Tondeur, J., Siddiq, F., & Baran, E. (2018). The importance of attitudes toward technology for pre-service teachers' technological, pedagogical, and content knowledge: Comparing structural equation modeling approaches. *Computers in Human Behavior, 80*, 67–80. https://doi.org/10.1016/j.chb.2017.11.003

Senanayake, A. C., Samarakkody, A., Malalgoda, C., Amaratunga, D., Haigh, R., Liyanage, C., Hamza, M., Kaklauskas, A., & Shaw, R. (2023). Towards an inclusive disaster education: The state of online disaster education from the learner's perspective. *Sustainability, 15*, 11042. https://doi.org/10.3390/su151411042

Septima, M. E., Wardoyo, M. A. I., Praptiwi, N. Y., Ashari, A. N. S., Ashari, A., Susanti, N. I., Jainudin;, Latifah, F., & Nugrahagung, P. P. (2019). Disaster education through local knowledge in some area of Merapi Volcano. *IOP Conference Series: Earth and Environmental Science, 271*, 012011. https://doi.org/10.1088/1755-1315/271

Shaw, R., Takeuchi, Y., Ru Gwee, Q., & Shiwaku, K. (2011). Disaster Education: An Introduction. In *Disaster Education* (pp. 1–22). https://doi.org/10.1108/s2040-7262(2011)0000007007

Siina, C. (2022). *Educators' digital competency framework*. UNICEF Regional Office for Europe and Central Asia.

Sopegina, V. T., Chapaev, N. K., & Simonva, M. V. (2016). Integration of pedagogical and technological knowledge in forming meta-competencies of a modern worker. *International Journal of Environmental & Science Education, 11*(5), 7836–7846.

Teodorescu, T. (2006). Competence versus competency: What is the difference? *Performance Improvement, 45*(10), 27–30. https://doi.org/10.1002/pfi.4930451027

Testa, S., Toscano, A., & Rosato, R. (2018). Distractor efficiency in an item pool for a statistics classroom exam: Assessing its relation with item cognitive level classified according to Bloom's Taxonomy. *Frontiers in Psycholog, 9*, 1585. https://doi.org/10.3389/fpsyg.2018.01585

The Internal Displacement Monitoring Centre (IDMC), & Norwegian Refugee Council (NRC). (2023). *GRID 2023: Internal displacementand food security*. IDMC.

Tuamsuk, K., & Subramaniam, M. (2017). The current state and influential factors in the development of digital literacy in Thailand's higher education. *Information and Learning Science, 118*(5/6), 235–251. https://doi.org/10.1108/ILS-11-2016-0076

TPACK, Teacher self-efficacy, and technology acceptance model. *Journal of Educational Technology & Society, 21*(3), 48–59. http://www.jstor.org/stable/26458506

van Laar, E., van Deursen, A. J. A. M., van Dijk, J. A. G. M., & de Haan, J. (2017). The relation between 21st-century skills and digital skills: A systematic literature review. *Computers in Human Behavior, 72*, 577–588. https://doi.org/10.1016/j.chb.2017.03.010

Voogt, J., Fisser, P., Pareja Roblin, N., Tondeur, J., & van Braak, J. (2013). Technological pedagogical content knowledge—A review of the literature. *Journal of Computer Assisted Learning, 29*(2), 109–121. https://doi.org/10.1111/j.1365-2729.2012.00487.x

Wu, W.-H., Kao, H.-Y., Wu, S.-H., & Wei, C.-W. (2019). Development and evaluation of affective domain using student's feedback in entrepreneurial massive open online courses. *Frontiers in Psychology, 10*(1109). https://doi.org/10.3389/fpsyg.2019.01109

Climate Resilience of Argo-Eco Systems

Importance of Sustainable Utilization Practices and Emerging Technologies Towards Food Security

Chathuri Maheshika Senanayake, Pasan Chinthana Bandara, Rumesh Liyanage, Ahinsa Anubhashani Kumari Lankanayaka, Piumali Jayethissa, and S. B. Navarathne

1 The Imperative of Achieving Food Security for a Sustainable Future

Meeting the global need for food security remains a prominent hurdle in the years ahead, especially considering the persistent rise in population, urbanization, changes in dietary preferences, and the impact of worldwide climate changes (Hazell & Wood, 2008). To appropriately tackle these concerns, the authors center their attention on the FAO's definition of food security expressed as, the state in which people consistently have dependable access to enough safe and nourishing food to fulfill their dietary requirements and promote good health (FAO, 1996). This extends beyond mere food availability and encompasses the capacity to secure it both physically and financially. Guaranteeing food security is essential for supplying sufficient nutrition, alleviating hunger, and lessening poverty. This notion encompasses four primary facets: availability, access, utilization, and stability (Fig. 1).

Each dimension guarantees that individuals and communities can obtain sufficient, secure, and nourishing food. The availability of sufficient access to a varied spectrum of food resources is imperative for addressing the dietary prerequisites of the population. It further encompasses variables including agricultural efficiency, complexities of supply chains, infrastructure for food preservation, and the dominant capacity for food production. Food access encompasses both the capability to

C. M. Senanayake (✉) · P. C. Bandara · R. Liyanage · A. A. K. Lankanayaka
Department of Biosystems Technology, Faculty of Technology, University of Sri Jayewardenepura, Pitipana, Homagama 10206, Sri Lanka
e-mail: chathurisnnk@sjp.ac.lk

P. Jayethissa · S. B. Navarathne
Department of Food Science and Technology, Faculty of Applied Sciences, University of Sri Jayewardenepura, Gangodawila, Nugegoda 10250, Sri Lanka

C. Malalgoda et al. (eds.), *Climate Change Adaptation in the Built Environment*,
https://doi.org/10.1007/978-3-031-75826-3_21

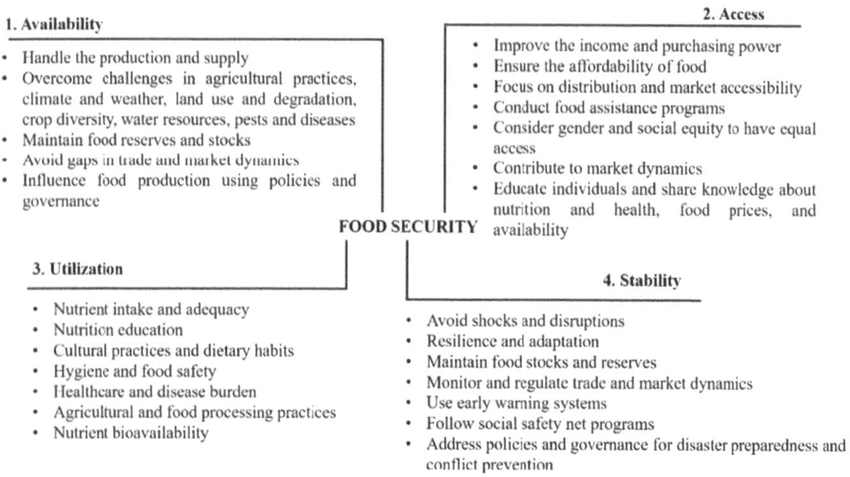

Fig. 1 Four primary pillars to achieve food security

procure sustenance and the absence of substantial impediments in doing so. Consistent and substantial increases in prices influence in dictating food accessibility. Even if food is present, its unattainability for a note-worthy segment of the populace undermines food security. Augmenting accessibility entails mitigating income inequalities, ensuring affordability, and resolving transportation and market-related challenges. Utilization refers to the process of converting food into nourishment. This covers factors such as diversity in the diet, food safety, cooking methods, and cultural preferences. It is vital to ensure that people consume a balanced and diverse diet to promote good health. Stability parameter refers to consistent access to food despite the events of extreme climate events, economic fluctuations, or conflicts. A stable food supply chain is resilient and can withstand various challenges without compromising the population's nutritional well-being. Long-term food security necessitates strategies that address both immediate and long-lasting challenges. Rather than increasing food production all four dimensions also address improving distribution systems, reducing inequalities, enhancing nutritional education, and promoting sustainable agricultural practices (Aborisade & Bach, 2014; FAO 2015).

The State of Food Security and Nutrition in the World report expresses, that at the end of 2021, the world faced high alerts regarding global food security. This was attributed to the enduring impacts of the COVID-19 pandemic, coupled with persistent conflicts and weather-related disruptions. In 2022, approximately 735 million people suffered from daily hunger measured by the prevalence of undernourishment. It can be observed that there was an increase of 122 million people experiencing hunger in 2022 compared to the situation in 2019 before the emergence of the pandemic (Fig. 2 and Table 1).

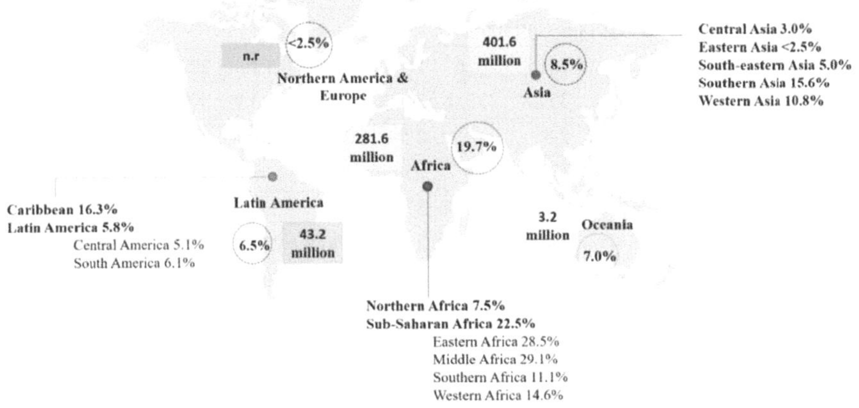

Fig. 2 Prevalence of undernourishment in the world in 2022 midrange (*Source* FAO, 2023)

The economic rebound witnessed in 2021 following the pandemic displayed a decreased pace in 2022. The escalation in costs of food, agricultural resources, and energy, intensified by the consequences of the conflict in Ukraine, had a detrimental effect on the restoration of jobs and earnings for the most susceptible populations. As a result, the reduction in hunger was impeded. It has been reported that rural areas generally experience a higher global food insecurity around 33.3%. Periurban areas and urban areas experience food insecurity as, 28.8, and 26.0% respectively. In every global region, women are disproportionately impacted by food insecurity compared to men. The gender gap diminished by approximately 2% points for severe food insecurity in Asia and around 1.3% points in Latin America and the Caribbean. On the other hand, in Africa and Northern America as well as Europe, the gap slightly widened for moderate or severe food insecurity, while it remained relatively stable for severe food insecurity (FAO, 2023).

An extensive analysis of global food insecurity involves a comprehensive evaluation of the prevailing conditions that affect the availability, accessibility, utilization, and stability of food resources on a worldwide scale. Evaluating global agricultural output, including crop yields and livestock production, helps determine the potential availability of food resources. Monitoring factors such as climate change impacts, natural disasters, and land degradation provides insights into the difficulties facing food production. Understanding the access aspect of food security necessitates a thorough evaluation of income levels and global economic imbalances. Analyzing trends in food prices, trade policies, and market access helps identify regions or populations at risk of limited access to affordable and nutritious food. Examining dietary patterns and nutritional intake across different regions provides insights into the utilization aspect of food security. This entails determining the prevalence of malnutrition, nutrient deficiencies, and dietary-related health problems. The socio-political environment must be understood in order to assess food security. Factors such as, conflicts, emigration, governance, and social safety nets may have a big impact on

Table 1 Number of undernourished population at the global and regional level from 2015 to 2022

Amount of undernourished people (millions)								
	2015	2016	2017	2018	2019	2020	2021	2022
Asia	357.8	336.0	319.3	325.2	343.9	396.2	414.1	401.6
Central Asia	2.8	2.6	2.5	2.2	2.0	2.5	2.4	2.3
Eastern Asia	104.2	n.r.	n.r.	n.r.	n.r.	n.r.	n.r.	n.r.
South-eastern Asia	47.9	41.6	37.4	36.5	35.0	35.2	36.0	34.1
Southern Asia	260.3	242.8	232.2	236.2	258.6	307.7	326.0	313.6
Western Asia	24.1	27.0	27.0	28.7	29.1	30.0	29.6	31.6
Africa	189.6	204.1	207.9	215.6	225.1	254.7	270.6	281.6
Northern Africa	12.3	13.4	14.4	14.6	14.4	15.1	17.6	19.5
Sub-Saharan Africa	177.3	190.7	193.5	201.0	210.6	239.6	253.0	262.0
Eastern Africa	96.8	106.1	108.6	110.8	116.9	126.4	131.2	134.6
Middle Africa	36.7	40.1	39.8	42.3	44.4	51.0	54.2	57.0
Southern Africa	5.9	5.3	5.1	5.1	5.5	6.4	6.8	7.6
Western Africa	37.9	39.2	40.1	42.9	43.8	55.8	60.8	62.8
Latin America and Caribbean	32.9	38.2	36.6	37.9	36.0	42.3	45.6	43.2
Latin America	27.3	32.4	30.9	31.8	29.8	35.6	39.1	36.0
Central America	11.2	10.5	10.4	10.5	9.0	8.5	8.9	9.1
South America	16.1	21.9	20.5	21.3	20.8	27.1	30.3	26.8
Caribbean	5.6	5.8	5.7	6.1	6.2	6.7	6.5	7.2
Oceania	2.5	2.5	2.5	2.7	2.8	2.7	2.9	3.2
Northern America and Europe	n.r.	n.r.	n.r.	n.r.	n.r.	n.r.	n.r.	n.r.
World	**588.9**	**586.4**	**571.8**	**586.8**	**612.8**	**701.4**	**738.9**	**735.1**

Note n.r.—not reported
Source FAO (2023)

food accessibility. Furthermore, it is important to study the ecological effects on food production such as; resource usage and potential environmental deterioration in establishing stability in food systems. A crucial component of the assessment is identifying groups and areas that are particularly susceptible to food insecurity owing to socioeconomic, geographic, or other factors. Understanding how efficiently emergency solutions address severe food insecurity requires evaluating the efficacy of international food aid programs and humanitarian actions. Assessing national and international trade, health, social welfare, and agricultural policies can assist in determining how they affect food security (McCarthy et al., 2018).

Tackling of global food insecurity
The evaluations assist in guiding policy choices, resource allocation, and focused interventions to solve global food insecurity issues. It's critical to remember that these evaluations are dynamic and take into account shifting global dynamics (international trade, agricultural policies), technological development, and changing socioeconomic conditions

2 Effect of Various Factors in Food Security

2.1 Impact of Climate Changes on Food Security

The increase in climate change poses a significant obstacle to attaining global food security, which has a major impact on various aspects of food production, distribution, and accessibility. The consequences of climate change on food security are intricate, causing both direct and indirect effects that affect the presence, accessibility, and efficient utilization of food resources. The influence of climate change on food security stems from changes in weather patterns, rising temperatures, and more frequent occurrences of extreme weather conditions (Wheeler & Von Braun, 2013) (Fig. 3).

According to the current status of food security, hunger is prevalent in countries of sub-Saharan Africa and South Asia due to the most alarming extreme conditions. However, it is crucial to acknowledge that, the number of undernourished people is seriously deficient in capturing the access, utilization, and stability of food security.

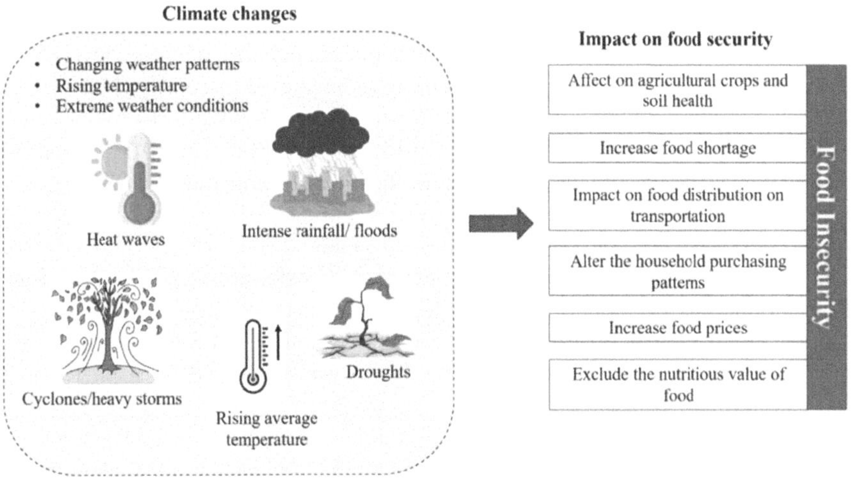

Fig. 3 Impact of climate changes on food security

As per the estimated results, climate changes adversely impact calorie deficits, omitting other nutritional insufficiencies and household or individual-level scarcities of food (Grebmer, 2012).

The effects of climate change on food production and availability are significant. These effects include changes in growing conditions, more frequent extreme weather events, lower crop yields, shifts in pest patterns, water scarcity issues, and disruptions in livestock and fisheries. These impacts disproportionately affect regions like sub-Saharan Africa and South Asia, highlighting vulnerabilities, typically about 10–20 million hectares of crop loss. To tackle these issues, immediate actions including the adoption of resilience measures, sustainable farming practices, and global collaboration efforts (Challinor et al., 2014; IPCC, 2014; Rosenzweig et al., 2014).

Climate change is increasing the frequency and severity of extreme weather events including storms, floods, and droughts. This phenomenon could lead to heightened risks of landslides and erosion, along with greater disparities in crop yields and local food accessibility. As a result, these elements may compromise food security and supply stability. Latin America, South Africa, and Southern Australia are agricultural regions that have traditionally had more natural disasters than the European region (Fischer et al., 2002). Sub-Saharan Africa has the majority of this land, while some regions of Asia are chronically undernourished as a result of the increased food volatility (Bruinsma, 2003).

As shown in the Intergovernmental Panel on Climate Change (IPPC) report 2007, the effectiveness of individuals in utilizing food will be influenced by climate change, as it changes the conditions for food safety while causing food and waterborne diseases. Higher temperature reports an increase in food poisoning in warmer regions of Israel, Australia, and the Pacific Islands. As well as, the floods resulting from heavy rainfall, can elevate the likelihood of the outbreak of water-borne diseases (Bruinsma, 2003; IPCC, 2014).

In the Sri Lankan context, agricultural and related fields are affected by climatic changes such as elevated temperatures, shifts in rainfall patterns, rising sea levels, and occurrences of severe weather events resulting in huge food insecurity (Mani et al., 2018). In particular, North, North Central, Western, and North Western provinces are the most negatively impacted areas (The World Bank, 2018). In late 2016, the nation faced its most severe drought period due to the intense temperature increase. As a result of that 40% reduction in rice cultivation during early 2017. Subsequently, rainfall worsened the situation for food crop yields. Therefore, rain-fed farmers found themselves in a more precarious situation. (Coslet et al., 2017). Additionally, Sri Lanka experiences 74% (UNDRR, 2019) of natural disasters like floods, droughts, landslides, and storms due to extreme weather conditions, and approximately US$ 380 million of average annual losses was accounted for from 1998 to 2012 (Siriwardana et al., 2018).

In African countries, climate changes including droughts, floods, and heat waves have a detrimental impact on crop yields and production as a result of subsequent increases in temperature and altered rainfall (Trisos et al., 2022). In sub-Saharan Africa, the occurrence of droughts has tripled from the 1970s to 2010s. Around 2.2

million people in Zimbabwe, Malawi, and Mozambique were affected by a devastating cyclone in 2018 and severe flooding in western and central Africa resulted in the loss of agricultural fields and livestock (Opoku et al., 2021; World Bank, 2021). However, according to the African Climate Policy Centre results in 2013, the annual toll of 1.7 million deaths in Africa is leading to food insecurity and malnutrition because of the climatic shifts.

Extreme climatic changes have a notable impact on supply shortages and price increases of food. Hence, these challenges can hinder individuals' ability to obtain sufficient and nutritious sustenance. Furthermore, disturbances in transportation and infrastructure caused by climate-related factors can hinder the movement of food, thereby restricting its availability in specific areas. This intricate relationship between climate change and food access highlights the pressing requirement for adaptive approaches and sustainable measures to guarantee food security for everyone. East Asia and the Near East/North African region have witnessed significant benefits attributed to lower food costs and substantial income expansion, as noted in the IPCC 2007 report.

The consequences of climate change on food security present an intricate and diverse obstacle. This phenomenon imperils crop yields, disturbs distribution networks, and amplifies pre-existing weaknesses in the food system. Swift actions are imperative to alleviate these effects by employing sustainable agricultural methods, implementing adaptive approaches, and fostering global collaboration.

2.2 *Problems in the Food Supply Chain*

Perishable and processed foods come to our dining table starting from agricultural farms, plantations or forests, and aquatic bodies such as the ocean. Foods from these sources move to the consumer in a chain-like fashion, hence called the "food supply chain". The food supply chain is a collection of processes such as food processing, distribution, retailing, and consumption (Bourlakis & Weightman, 2008). With the developed infrastructure for transportation and storage facilities our supply chains can be considered as global affairs where food commodities travel through the globe to reach its end user. In the simplest terms, a well-managed food supply chain would assure the food security and continuous food supply of a country, with minimum disturbances and minimum price fluctuations even when faced with external shocks (Bourlakis & Weightman, 2008; Davis et al., 2021).

However, it is not an exaggeration to state that the developing countries in Asia and Africa have fragile food supply chains in overall. Mainly, the front-end of the supply chain (customer demand) should be well coordinated with the back-end operations such as production and manufacturing (Aworh, 2021; Khan et al., 2022). Apart from challenges arising from the demand forecasts and infrastructural difficulties, climate change-driven natural disasters can and have affected many critical steps of supply chains. Unlike other challenges on the supply chain, the impact of climate change and related phenomena are considered to be more unpredictable and uncontrollable, making them a threat to any food supply chain (Escarcha et al., 2018). For example, climate change creates serious problems in agricultural sector. Untimely droughts and torrential floods can interfere with the conventional agricultural seasons and cycles destroying crops even before the harvesting stage (Davis et al., 2021; Hsiang et al., 2013). Large-scale supply chains are usually driven by forecasts of demand, however, lack of production creates an unnecessary leading time, resulting in food shortages and price fluctuations as experienced in the consumers' end (Davis et al., 2021).

Hurricane Maria (2017) was a significant natural disaster that created a sophisticated discussion about having shock-resilient supply chains. A Caribbean island and an unincorporated territory of the United States, Puerto Rico was in the receiving end of this natural disaster, where Hurricane Maria made landfall as a Category 5 hurricane. This led to a devastation of country's electricity and communication infrastructure for months. More importantly in the supply chain perspective, hurricane damaged the country's roads, bridges, and ports creating a huge issue related to food supply and food security as 85% of the country's food consumption was fulfilled by food imports. What resulted is an enormous effort in reconstructing the critical infrastructure of the country to ensure food and other supply chains. In a study conducted about the resilient model of a food supply chain that has been subjected to shocks, we are supposed to understand the supply chain nodes are interconnected rather than the conventional circular flow of products. Hence the availability of telecommunications and energy services are critical to continuing the food supply chain. In the case of Puerto Rico, they were too dependent on imports and policy-wise they did not show preparation to reduce the dependence on external nodes of the supply chain (Kim & Bui, 2019; Serra & Sanchez-Jauregui, 2022).

This is important to note, because even though a supply chain seems stable during normal operations, gaps and inconsistencies are more visible when they are subjected to unprecedented challenges like various natural disasters including pandemics. When considering the COVID-19 pandemic, it affected almost all the human-involved steps for a considerable amount of time. However, the food industry faced an immense pressure as some of the processing plants such as fish and meat processing plants became hotspots for spreading virus. However, since the food products are essential for continuing human lives, the food industry had to respond with crisis response plans for transportation, cleaning, sanitation, disinfection of facilities, COVID-19 testing, and screening of plant workers to continue the production processes and keep supply chain running (De Sousa Jabbour et al., 2020; Rizou et al., 2020).

Not only that, in the other end of the supply chain, when handling retail level of the supply chain, safety measures had to be taken to ensure social distancing when deliveries were done in communities with infected people which were again considered as hot spots. Here what mattered was the workers' medical condition, their personal hygiene (i.e. equipped with sanitizers and personal protective equipment), keeping working surfaces clean all the time, and keeping social distancing during any contact with consumers to make sure the supply chain workers are not affected by the situation (Aday & Aday, 2020).

Not only the developed countries, many developing countries also try to save the supply chains by importation at higher prices to fight food scarcity (Kummu et al., 2020). From the farmers' and food manufacturers' point of view, this result in a reduction of market share as their customer base is now replaced with much stronger and reliable suppliers. Nevertheless, dealing with this situation is regrettable, considering that our biosystems are endowed with diverse food resources and climates that enable us to cultivate our foods, rather than relying on climate-vulnerable supply chains.

Therefore, as a global community, we need to have plans to address excess and/or seasonal food commodity productions, especially for fruits and vegetables, which share a significant place in our diets. On a positive note, we see some individuals, investors, and researchers work on such resources at their capacity. Therefore, with careful planning, we have the potential to contribute a significant amount of regional resources to enhance global food security.

The approach to solving food supply chain problems needs to be much more trans-disciplinary as these systems are integrated in many aspects. The supply chain is more of a management affair however, to solve its issues and assure resilience, its dependence on agriculture techniques, manufacturing processes, transportation, and climate factors has to be incorporated into the broader picture. Each country needs to rethink its national and international policies to safeguard the activities of the supply chain making sure the food manufacturers can access the transportation sector, technology sector, market sector, and other relevant sectors to smoothly run the supply chain even during chaotic times (de Paulo Farias & dos Santos Gomes, 2020; Cappelli & Cinni, 2020, De Sousa Jabbour et al., 2020).

3 Measures in Achieving Food Security

3.1 Use of Indigenous Knowledge in Food Processing and Preservation

Indigenous knowledge is a collective system of knowledge, beliefs, traditions, and practices that are unique to a particular culture or region, and passed from generation to generation (Mapedza et al., 2022; Misty, 2009). Indigenous knowledge has been used for centuries in various aspects related to weather forecasting, water management, land preparation, pest and disease control, food production, processing, and preservation. The utilization of indigenous knowledge is a sustainable option for establishing food security. Food security is primarily threatened by natural disasters and extreme climate events. Indigenous knowledge has a vital contribution to identifying and preparing for disastrous climate changes and enforcing policies for disaster risk reduction (Zhang & Nakagawa, 2018).

In the Sri Lankan context, drought and prolonged periods of low rainfall impose a serious threat to the farmers in the dry zones to meet the continuous supply of food throughout the year. It has been reported that disastrous drought occurs once every 3–5 years (Panabokke et al., 2002). The village tank-cascade system stands as a classic example of utilizing indigenous knowledge, connecting a series of small tanks into a larger tank, primarily to reuse and recycle water. This strategy was important in establishing food security by tackling the disastrous drought conditions and as well as flood conditions.

Indigenous ancestors also known as the *Vedda* whose origin date back to the fifth and sixth century BC practiced numerous food processing and preservation activities.

The primary food source was derived from hunting forest animals using bows and arrows, as well as from aquatic fish. Fish and meat were prepared using direct roasting with a wood fire. The excess raw meat was preserved for a few days by immersing in bee honey, smoking, or sun drying. "*Perume*" is a preserved meat product developed by the ancestors that resemble the modern sausage (Panabokke et al., 2010). The world's first use of spices and herbs as a food flavoring agent dated back to the hunter's era, as they observed wrapping of meat in leaves enhanced the flavor. Another processed form of meat in ancient Sri Lanka was "*Mas guli*" made by mixing different spices and herbs and coated with rice flour. This was consumed after deep frying. The exclusive dependence on hunting was subsequently partially replaced with the commencement of crop cultivation. Yams and tubers (such as *Dioscorea pentaphylla*, *Dioscorea oppositifolia*), rice, finger millet, and maize were the main carbohydrate sources in the diet. Yams and tubers were often consumed in the roasted form. Whereas the grains and cereals were consumed after boiling (*Thalapa*) or preparing flatbread (*Rotti*) from their ground flour. A thick gravy (*Ānama*) prepared with smoked meat was a frequent side dish served along with flatbreads (Silva, 2015).

Establishing food security in African countries is critical as they have experienced severe drought and food shortages over the past few years. African people have their own traditional knowledge and traditional crops that could be used efficiently to mitigate climate-induced food insecurity. Their traditional crops such as; cassava, yam, plantain, sweet potato, millets, and sorghum serve as staples, soybeans, and groundnuts as oil crops, and a large variety of crops as fruits and vegetables, and green leafy vegetables (FAO, 1997). These traditional crops are often grown as rotation crops between the main crops. Resistance to drought, reducing soil erosion, providing resistance to pests and diseases, and controlling weed growth are some important benefits of traditional crop cultivation (Scott et al., 2020; Katz & Weaver, 2003). The most common processing methods of fruits and vegetables include; sun drying, boiling in salted water, and fermentation (FAO, 1997). Traditionally meat was consumed without processing. The most commonly consumed meat was from cattle, sheep, goats, and pigs. The meat was processed by smoking, salting, or sun drying to preserve it for a longer time (Tweheyo, 2018). Dried meat which was also known as *Nyirnyir* was fried with oil in the presence of garlic, dipped in camel ghee, and consumed along with tea (Tweheyo, 2018). In addition to meat, people from African countries, consume various insects including; caterpillars, crickets, and green bugs to fulfill their protein requirements. Similar to the processing of meat, insects were dried, roasted, and consumed as a snack or immersed in honey to preserve for a longer time (Katz & Weaver, 2003).

People living in the Arctic region experienced climate changes including; volcanic eruptions, stormy weather, glacial lake outburst floods, shorter snow cover seasons, and other changes that had a major impact on food security in that region (Thoman et al., 2020). Their traditional foods mainly come from hunting animals (Pacific walrus, Seals, bowhead whale, etc.) and harvesting marine sources (fishery species like Atlantic cod, Atlantic salmon) due to ecological reasons (Kozloy, 2004). The traditional foods were prepared of different parts of marine animals including, flesh, fat, tongue, entrail, etc. These traditional foods are composed of vital nutrients which are important to maintain the health and well-being of the people. Beluga whale is an important source of vitamin C, and Pacific walrus is a rich source of protein and fat (Kozloy & Zdor, 2003). People from Arctic community owned unique food processing and preservation practices. It was reported that some communities preserved meat in whey to survive in the winter. They used burned algae as a substitute for salt. The precious indigenous knowledge made people survive in extreme climate events for a very long time.

However, the rapid climate changes made communities less reliable solely on traditional practices. For instance, increased ocean warming is expected to cause an increase in the mortality of marine fish species in the Arctic region (Brattland et al., 2019). Due to the reduction of sea ice, people in northwest Greenland have to use skiffs to hunt walruses (AMAP, 2018; Hossain et al., 2018). Therefore, further advancements to indigenous knowledge and traditional practices are required to mitigate the risks imposed by climate change.

Packaging is an important operation to keep food products from contaminants, keep food safe during long-distance transportation, and maintain the freshness of the food product until consumption (Hotchkiss, 1997). In ancient times, natural materials including plant leaves, coconut shells, shells of shellfish, and animal skin were used as packaging materials. Pouches made from treated animal skins (leather) were used to store liquids (i.e., water and wine) and solids (i.e., flour and sugar). Pottery is considered one of the oldest human inventions that led to designing of earthenware (clay pots) which was used as a packaging material to store liquids and various solid foods (i.e., dried foods, fermented foods, honey, etc.). Those clay pots were sealed using corks, plant leaves, and wax to make them moisture and gas-proof, which increased the shelf life of the products to some extent. These packaging materials are environment-friendly and biodegradable, which has a minimum impact on the environment. However, with the growing population and their needs, these solutions alone would not be sufficient. Hence, the need of new technologies in the food packaging sector are required to increase efficiency and at the same time to have a minimum impact on the environment.

3.2 Popularizing Underutilized Food Crops

The need for large-scale food manufacturing and increased focus on socio-economic development of farmers have created a focus on a range of commercial food crops. These are grown also on much larger scales in agricultural lands involving conventional and advanced agricultural practices. Technological advancements in developing new crop varieties have led to the increased popularity of high-yield and stable cultivars among farmers worldwide, resulting in enhanced incomes. It is crucial to introduce crop varieties that demonstrate adaptability to local biosystems while exhibiting resilience to climate-related risks. This ensures a sustainable and robust agricultural system in the face of environmental challenges (Mayes et al., 2012; Varshney et al., 2009). However, we may have lost our interest in diversifying our crops due to our consumption and purchasing patterns. This has led to neglecting and underutilizing many plants and food commodities that are found regionally. It is worth to note this category of food crops represent an enormous diversity owing to their habitats and climatic zones indicating that they are resilient and adopted to their Biosystems very well (Mayes et al., 2012; Tafadzwanashe et al., 2017; Varshney et al., 2009).

On the other hand, Underutilized crops, with their rich history of providing nutritional and medicinal benefits over millennia, have unfortunately faced prolonged neglect in contemporary agricultural practices due to lack of awareness, low shelf-life, no or lack of manual cultivation, and restrictions in the supply chain. For the food crops that were used traditionally, processing or cooking recipes may have been transferred to younger generations orally, not necessarily in formal written forms (Malkanthi et al., 2014). This prolonged neglect has resulted in the oversight of certain edible crops, necessitating a rediscovery process as the initial step towards their popularization. (Padulosi et al., 2004).

Edible crops include all types of commodities such as cereals/seeds, green leaves, stem vegetables, roots and tuber crops, and fruits (Padulosi et al., 2004). In recent decades, numerous studies worldwide have been conducted to explore regionally found underutilized commodities and their potential as food sources. Due to the improvement and higher availability of characterization techniques and equipment, we can now examine a broader chemical profile of a food commodity. In the case of underutilized food commodities, they are rich in various functional chemicals in addition to the macronutrients displaying whole new perspectives of using them. Hence these research can also be understood as rediscoveries or scientific proof of traditional knowledge which can highly influential in getting attention from the consumers (Padulosi et al., 2004; Peduruhewa et al., 2021).

The prevailing research approach commonly involves collecting samples of the selected underutilized crop and extensively studying its nutritional assay, as exemplified by studies such as those conducted by Peduruhewa et al., in 2021, Nadeeshani et al., in 2018, and Jayasuriya and Rathnayaka in 2016. As mentioned in these studies, the raw materials are converted to a value-added food product (i.e. flour, solid food, ready-to-serve drink, etc.) and highlighted their sensory properties

(Peduruhewa et al., 2021; Weerasekara et al., 2012; Marimuthu et al., 2013). This is crucial as it offers hints into the potential commercialization of the studied underutilized crop. In further studies, advanced chemical assays, genome sequencing, market surveys, and feasibility studies can provide further insight (Marimuthu et al., 2013; Babarinde et al., 2019; Adhikari, Hussain & Rasul, 2017; Vu & Nguyen, 2017).

The notable advantage of this approach lies in the utilization of raw commodities readily available in a specific region to meet the daily food requirements of the local population and allowing a less complex supply chain that may survive in a pandemic situation as well. For example, depending on the climate zone, the availability of edible plants may differ. Leaf vegetables, flowering plants, ferns, and fruits can be understood through comprehensive studies and analyses. It is also noteworthy that regionally identified yet dormant delicacies can be further popularized for consumption. (Padulosi et al., 2004; Tafadzwanashe et al., 2017).

For example, Manuka honey can be identified as a monofloral honey which specifically produced by European honey bees (*Apis mellifera*) using the nectar of flowers of Manuka (*Leptospermum scoparium*) tree. Apart from its utilization as a traditional medicine for thousands of years, Manuka honey now has a significant place in modern pharmaceutical industry due to its proven antimicrobial properties (Alvarez-Suarez et al., 2014; Johnston et al., 2018).

On a similar note, Matsutake mushrooms (*Tricholoma matsutake*) can be identified some terrestrial fungi that grow in association with red pine trees *(Pinus densiflora*). Even though there are other terrestrial fungi found around the world, original wild type from Japan is expensive and preferred by the consumers (Yang et al., 2008).

With these examples, it is evident that the underutilized plants can be further studied to reveal various biosystem relationships with pollinators, other botanical and zoological symbioses, and region-specific properties to gain attention from consumers as well as food manufacturers. It is worth noting that such identified plants are raised from their underutilized status indicating their use has been improved and widened with a systematic approach making them more sellable in the local and international markets.

Once in a while availability of a large quantity of raw commodities can be a source for value-added food that can be commercialized after a comprehensive feasibility study. Underutilized crop varieties are mostly found as wild-grown and rarely seen in larger areas, especially since they inherit considerable economic importance. This seems to be important as some governments have taken some measures on conservation, utilization, and marketing of nutritious underutilized crops not only to establish food security in the country but also to gain economic benefit.

Such examples highlight the significance of a transdisciplinary approach in popularizing underutilized food commodities. To complement the biochemical analysis of these crops, food product development studies, botanical studies, biosystems studies including the study of pollinators of these plants could reveal highly commercial pathways for underutilized crops greatly. Feasibility studies focusing on the technical and market feasibility of the value-added products of underutilized raw materials can also help in this regard (Padulosi et al., 2004).

There are countries or even smaller communities that continue to utilize some of these crops to a considerable extent. Governments around the world, who are profoundly responsible for assuring food security should carefully look at these crops to analyze their potential and conserve them. This intervention from the government can be in the form of economic and policy changes allowing framework and financial backing for the conservation and sustainable use of underutilized crops. Quality of any national-level policy exploring and popularizing the underutilized food commodities can be improved by blending the regional knowledge of underutilized edible plants and food processing techniques. Furthermore, the level of intervention can be at various levels, even global considering the existence of global food supply chains that may be interested in the properties of underutilized crops (Galluzzi & López Noriega, 2014; Padulosi et al., 2004; Tafadzwanashe et al., 2017).

3.3 Home Gardening and Community Gardening

Home gardening has always been a solution to fight food insecurities. Essentially, living spaces encompass available vacant areas that can be transformed into home gardens. In every aspect, home gardens stand as a valuable source of crop diversity. These cultivated spaces not only enhance the aesthetic appeal of residences but also contribute significantly to promoting biodiversity, ensuring a sustainable supply of fresh produce, and fostering a closer connection between individuals and the food they consume. The crops are related not only to the climatic and geographical zones but also to the cultural and socioeconomic backgrounds of the garden owners. Home gardens usually consist of a mixed population of edible fruits, vegetables, nuts, leafy greens, medicinal plants as well as flowers and timbers. Plants are often grown in a multi-level arrangement especially due to the limitation of space. Home gardens are a good practice to maintain and conserve crop diversity as the heirloom species of plants can be grown without much effort in the garden system in a small scale. Predominantly home gardens are expected to fulfil the households' food demand entirely or partly while the excess can be shared or sold at local retail shops. As the materials needed for growing traditional species may not be produced commercially or in a large scale making small-scale home gardens to produce new generations of growing materials to other home gardeners as well (Galhena et al., 2013; Mohri et al., 2013; Taylor & Lovell, 2015).

A famous home gardening system exists in the mountain range of Sri Lanka dubbed as Kandyan Home Gardens. This region is mainly over laps with the wet zone providing a healthy environment for the growth of perennial crops. Conditions such as deep soil, year-round rainfall, and frequent dry periods helped a thriving culture of Kandyan Home Gardens (Sthapit et al., 2016).

As a common practice, governments should continuously invest in advertising and initiating programs leading to home gardens. Recent advances in agricultural technologies have enabled the establishment of home gardens even in small and cramped spaces within urban cities (Taylor & Lovell, 2015). However, the progression of such activities can be discontinued due to the busy lifestyles of gardeners, lack of consultation about soil preparation, plant care, and fertilizer application. This, unfortunately, can lead to the under-achievement of the objectives of home gardening.

In that sense, the introduction of community gardens can be used as a better alternative to assure food security regionally. Community gardens can be operated in easily accessible arable lands, possibly owned by the central government or local governments, and religious institutions. The workforce required to operate community gardens can be found among school children, senior citizens, and other interested people of the region. Consultation-wise, it would be easier, since agriculture consultants can focus their efforts with less traveling involved. Hence, the harvest can be shared among stakeholders or sold at a reasonable price without the influence of intermediaries and transportation costs. Still, the community gardens can be used to grow heirloom and underutilized crops in the region. With the support from more people, community gardens can adapt to a much systematic plant selection, nursery management, and cultivation practices (Corrigan, 2011; Furness and Gallaher, 2018).

3.4 Novel and Emerging Technologies in Food Processing and Food Preservation

Drying was the most practiced food preservation method since ancient times which was modified over the past century. Earlier the sole purpose of drying was to make available food for a long period and reduce the weight and the volume of the food product for the convenience of storage (Delagoda & Barbosa de Lima, 2016; Mujumdar, 2014). Solar energy was the major energy source utilized for many food processing and preservation practices, including drying. The efficiency of solar energy was sufficient to fulfill the requirements of a small group of people, however, as the population increased, people needed to discover novel and alternative energy sources. Traditional biomass fuels (solid fuels) were the first to arrive. Burning wood and crop residues were the major fuel options for the poorest communities whereas charcoal was a fuel option for the communities with a middle-income level (Kroon et al., 2013). Burning solid fuels caused several problems including, releasing smoke and toxic substances into the atmosphere, which ultimately increased air pollution (Zhang & Smith, 2003). This has led to the discovery of clean fuels including; kerosene, ethanol, methanol, natural gas, electricity, etc. (Kroon et al., 2013). Although fossil fuels satisfy 80% of the world's energy requirement, burning them causes a huge emission of carbon dioxide (CO_2) and nitrous oxide (N_2O) into the atmosphere. This in turn leads to disastrous effects on climate change, including; global warming due to the greenhouse effect, air pollution, increase in the acidity of oceans (Trop & Goricanec, 2016; Lucas et al., 2018). The discovery of renewable energy sources (solar, wind, biomass, geothermal, and hydro-power energy) has eliminated issues related to fossil fuels, as they served as sustainable energy sources

(Ashfaq & Ianakiev, 2018; Ludin et al., 2018). Also, this is crucial to meet the escalating energy demand by serving as a substitute for fossil fuels. To serve the needs of the increasing population, novel methods have emerged through modification of the current technologies to improve the characteristics that have not been achieved using the current methods. Novel processing technologies serve many functions including; maintaining freshness, and quality, and increasing the shelf life of food products. For instance, the primary purpose of drying was to remove moisture from foods and increase the shelf life of food (Jayas, 2016). In addition to the removal of water from the food product, the novel drying methods have also considered the chemical and biochemical changes in the food products. Therefore, a wide range of food products can be obtained (i.e., snacks, baked items) and their quality is increased.

Convective drying is the most popular drying method used to dry fruit and vegetables that applies the principle of convection, facilitating direct contact between the heating source and the food product for efficient moisture removal. (Artyukhov et al., 2019). The temperature of this process can be controlled which is advantageous compared to sun drying. However, still, the operating temperatures are too high which can cause deterioration of the organoleptic properties of foods. Freeze drying is introduced later to preserve the organoleptic properties of food products while drying them as this method employs low temperature and low pressure (Nowak & Jakubczyk, 2020). Spray drying is applied to dry liquid food products such as milk, fruit juices, etc., and is considered a feasible option to obtain powdered particles with a defined particle size through the process called atomization (Cal & Sollohub, 2010). Although the first discovery of spray drying dates back to 1860, over the past few centuries it has been modified to improve efficiency.

Following the drying process, packaging technology was implemented to extend the shelf life of the product further. The first use of glass containers as a food packaging material has been dated back to 3000 BC, prepared by pouring into molds of silica melted at high temperatures (Yaris & Sezgin, 2017). Other traditional packaging materials include paper and metals. Packaging materials have been modified and advanced over the past few decades to provide a higher degree of protection to food products, meet consumer requirements, and maintain freshness and quality during transportation and storage. Plastic (petroleum-based) also has been used as a packaging material for food products due to its high strength, durability, good moisture and air permeability, and lightweight, versatile nature (Marsh & Bugusu, 2007). A wide variety of plastics has evolved in the form of rigid and flexible packaging. However, due to its non-degradability, limited recycling capacity has caused potential risks to the environment and health. The emission of greenhouse gases contributing to global warming is linked to plastic production, as reported. In 2015, the combined effects of plastic production and incineration resulted in the release of 1.96 gigatons of greenhouse gases into the atmosphere (Sharma et al., 2023). Plastics contaminate aquatic bodies where macro and microplastics are found in a lot of marine species including the species that people consume too. The disposal of plastics on land releases many chemicals that impose toxic effects on both humans and the environment (Rhodes, 2018). These significant environmental and health concerns have spurred the search for novel materials in food packaging to address these issues. The discovery of edible packaging (edible coating and films) is a vital recent advancement in packaging technologies as it serves as a sustainable alternative to the widely used plastics. These types of packages are primarily made of byproducts generated in the food processing industry that are usually discarded as waste (Ncama et al., 2018). Edible coating and films provide good protection for highly perishable food commodities such as fruits and vegetables and enhance their sensory properties. In most instances, edible packaging is enriched with plant-essential oils which are rich in antioxidants and antimicrobial compounds, that increase the shelf life of food products and impose positive effects on health (Sharma et al., 2020). Therefore, the development and popularizing of edible packaging could be a promising solution to reduce the use of plastic polymers and decrease food by-products as waste.

4 Challenges and Recommendations

Food insecurity leads to multiple issues such as hunger, malnutrition, poverty, and political instability around the globe. Among the various contributing factors, climate change stands out as the foremost driver of global food insecurity. Establishing a resilience mechanism is thus important to achieve food security globally. The biggest challenge would be to manage the food requirements of the growing population with the available land. Ensuring a stable supply chain is paramount to maintaining a reliable food supply, especially in the face of extreme climate events. All the steps starting from the manufacturers to consumers should be thoroughly monitored, identify the gaps, and impose solutions. National-level programs should be implemented to promote traditional knowledge and emphasize the importance of the combination of traditional and new technologies in food processing and preservation. In these programs, it is also important to popularize the underutilized crops that are specific to each region even within the same country. Underutilized crops are the crops that have been neglected for so many years, however, they possess many nutritional and functional properties that may add a variety to the food plate. In addition, the use of underutilized crops in the diet eliminates the need for complex supply chains and ensures that the people in a particular region won't starve in situations like extreme climate events. Promoting home gardening is the best recommendation to popularize the underutilized crops. Encouraging people towards home gardening can be done through government-funded programs including, the distribution of seeds, providing consultations on planting, fertilizer application, etc. Community gardening is another approach towards establishing food security, where the major workforce is represented by senior citizens, school children, etc. Allocating sufficient lands for community gardening is thus important to establish food security at the global level.

5 Conclusion

Out of the 17 sustainable development goals (SDG), SDG 2 focuses on the complete elimination of hunger and malnutrition by 2030. Climate change has a major impact on food security. Approximately 821 million people globally are estimated to be chronically undernourished in the year 2017, as a result of severe drought and environmental degradation. In region wise, severe food insecurity is mostly observed in African regions. Climate change has severely affected the agriculture sector in Sri Lanka too. The country experienced prolonged droughts, flash floods, heat stress, and heavy rainfalls over the past few decades that declined crop production, caused deaths and displacements, and increased hunger and malnutrition among people. The existing policies or resilient systems consist of numerous gaps that need to be addressed promptly to establish food security globally. The absence of reliable food supply chains, inadequate preparedness for future climate variations, and lack of disaster risk reduction policy implementation are some of the gaps that need to be addressed.

Different communities own unique traditional food processing and preservation practices that have been used for ages to make food available in sufficient quantities at all times. These practices are eco-friendly and impose minimum or no adverse effects on the environment. However, urbanization and busy lifestyles of peoples have made people adopt new efficient methods of food processing and preservation. In the process of establishing food security, the integration of both traditional knowledge and systems with that of novel methods is vital to increase the accessibility to nutritious food sources by the growing population in a sustainable manner. Each region/country is comprised of a large number of crops that have been neglected or underutilized. These crops are widely grown in the regions and often show a higher adaptability to harsh climate conditions. Due to the high nutritional value of these crops, the incorporation of these crops into the diet is an effective way of eradicating hunger and malnutrition. Hence, popularizing the underutilized crops and disseminating knowledge regarding their nutritional properties is a timely requirement. Home gardening and community gardening is thus, an important measure to serve those purposes and achieve food security in a global context. Although governments have continuously invested in such programs, the progression of such activities has become less popular among people nowadays, due to the change in their lifestyles and lack of consultation on gardening practices. In the process of achieving food security, many challenges need to be addressed but not disregarded. As the global food demand is increasing with the growing population, the biggest challenge would be to increase food production with the available land for cultivation. Hence, the proposed strategies would be promising solutions to achieve food security globally.

References

Aborisade, B., & Bach, C. (2014). Accessing the pillars of sustainable food security. *European International Journal of Science and Technology, 3.*

Aday, S., & Aday, M. S. (2020). Impact of COVID-19 on the food supply chain. *Food Quality and Safety, 4*(4), 167–180. Available at: https://doi.org/10.1093/fqsafe/fyaa024

Adhikari, L., Hussain, A., & Rasul, G. (2017). Tapping the potential of neglected and underutilized food crops for sustainable nutrition security in the mountains of Pakistan and Nepal. *Sustainability, 9*(2), 291. Available at: https://doi.org/10.3390/su9020291

African Climate Policy Centre. (2013). *Climate change and health in Africa: Issues and options.*

Alvarez-Suarez, J. M., Gasparrini, M., Forbes-Hernández, T. Y., Mazzoni, L., & Giampieri, F. (2014). The composition and biological activity of honey: A focus on Manuka honey. *Foods, 3*, 420–432.

AMAP. (2018). Adaptation actions for a changing Arctic: Perspectives from the Baffin Bay/Davis Strait Region. *Arctic Monitoring and Assessment Programme (AMAP)*, Oslo, Norway, pp. 17–345.

Artyukhov, A., Artyukhova, N., Ostroha, R., Yukhymenko, M., Bocko, J., & Krmela, J. (2019). Convective drying in the multistage shelf dryers: Theoretical bases and practical implementation. In *Current drying processes*. IntechOpen.

Ashfaq, A., & Ianakiev, A. (2018). Features of fully integrated renewable energy atlas for Pakistan; wind, solar and cooling. *Renewable and Sustainable Energy Reviews, 97*, 14–27.

Aworh, O. C. (2021). Food safety issues in fresh produce supply chain with particular reference to sub-Saharan Africa. *Food Control, 123*, 107737.

Babarinde, G. O., Babarinde, S. A., Ojediran, T. K., Odewole, A. F., Odetunde, D. A., & Bamido, T. S. (2019) Chemical composition and toxicity of Jatropha curcas seed oil against Sitophilus zeamais Motschulsky as affected by pre-extraction treatment of seeds. *Biocatalysis and Agricultural Biotechnology, 21*, 101333. Available at: https://doi.org/10.1016/j.bcab.2019.101333

Bourlakis, M. A., & Weightman, P. W. H. (2008). *Food supply chain management*. Wiley.

Brattland, C., Eythórsson, E., Weines, J., & Sunnanå, K. (2019). Social-ecological timelines to explore human adaptation to coastal change. *Biodiversity Change and Human Adaptation, 48*, 1516–1529.

Bruinsma, J. (2003) World agriculture: Toward 2015/2030. In *A food and agriculture organization perspective*. Earthscan.

Cal, K., & Sollohub, K. (2010). Spray drying technique. I: Hardware and process parameters. *Journal of Pharmaceutical Sciences., 99*(2), 575–586.

Cappelli, A., & Cini, E. (2020). Will the COVID-19 pandemic make us reconsider the relevance of short food supply chains and local productions? *Trends in Food Science & Technology*, 99, 566–567. Available at: https://doi.org/10.1016/j.tifs.2020.03.041

Challinor, A. J., Watson, J., Lobell, D. B., Howden, S. M., Smith, D. R., & Chhetri, N. (2014). A meta-analysis of crop yield under climate change and adaptation. *Nature Climate Change, 4*(4), 287–291. Available at: https://doi.org/10.1038/nclimate2153

Corrigan, M.P. (2011) Growing what you eat: Developing community gardens in Baltimore, Maryland. *Applied Geography, 31*(4), 1232–1241. Available at: https://doi.org/10.1016/j.apgeog.2011.01.017

Coslet, C., Goodbody, S., & Guccione, C. (2017). *Special report: FAO/WFP crop and food security assessment mission to Sri Lanka*. FAO and WFP.

Davis, K. F., Downs, S., & Gephart, J. A. (2021). Towards food supply chain resilience to environmental shocks. *Nature Food, 2*, 54–65.

de Paulo Farias, D., & dos Santos Gomes, M. (2020). COVID-19 outbreak: What should be done to avoid food shortages? *Trends in Food Science & Technology*, pp. 291–292. Available at: https://doi.org/10.1016/j.tifs.2020.06.007

De Silva, A. (2015). *Indigenous Knowledge: An overview and case reviews.*

De Sousa Jabbour, A. B. L., Jabbour, C. J. C., Hingley, M., Vilalta-Perdomo, E., Ramsden, G., & Twigg, D. (2020). Sustainability of supply chains in the wake of the coronavirus (COVID-19/SARS-CoV-2) pandemic: Lessons and trends. *Modern Supply Chain Research and Applications, 2*(3), 117–122.

Delagado, J. M. P. Q., & de Lima, A. G. B. (2006). *Drying and energy technologies*. Springer International Publishing.

Escarcha, J. F., Lassa, J. A., & Zander, K. K. (2018). Livestock under climate change: A systematic review of impacts and adaptation. *Climate, 6*(3), 54. Available at: https://doi.org/10.3390/cli6030054

FAO. (1996). *Rome declaration on world food security and World Food summit plan of action*. World Food Summit.

FAO. (1997). Agriculture, food and nutrition for Africa. In *A resource book for teachers of agriculture*.

FAO. (2015a). FAO statistical pocketbook. In: NATIONS, F. A. A. O. O. T.U. (Ed.)

FAO. (2023). *FAOSTAT: Suite of food security indicators*. FAO.

Fischer, G., Shah, M., & van V. H. (2002). Climate change and agricultural vulnerability, a special report prepared as a contribution to the world summit on sustainable development. In *International Institute for Applied Systems Analysis*, Laxenburg, Austria.

Furness, W. W., & Gallaher, C. M. (2018). Food access, food security and community gardens in Rockford, IL. *Local Environment, 23*(4), 414–430. Available at: https://doi.org/10.1080/13549839.2018.1426561

Galhena, D. H., Freed, R., & Maredia, K. M. (2013). Home gardens: A promising approach to enhance household food security and wellbeing. *Agriculture and Food Security, 2*(8). Available at: https://doi.org/10.1186/2048-7010-2-8

Galluzzi, G., & López Noriega, I. (2014). Conservation and use of genetic resources of underutilized crops in the Americas—A continental analysis. *Sustainability, 6*(2), 980–1017. Available at: https://doi.org/10.3390/su6020980

Hazell, P., & Wood, S. (2008). Drivers of change in global agriculture. *Philosophical Transactions of the Royal Society B: Biological Sciences, 363*(1491), 495–515. Available at: https://doi.org/10.1098/rstb.2007.2166

Hossain, K., Raheem, D., & Cormier, S. (2018*) Food security governance in the Arctic-Barents region*. Springer.

Hotchkiss, J. J. (1997). Food-packaging interactions influencing quality and safety. *Food Additives Contaminants, 14*(6–7), 601–607.

Hsiang, S. M., Burke, M., & Miguel, E. (2013). Quantifying the influence of climate on human conflict. *Science, 341*(6151), 1235367.

Intergovernmental Panel on Climate Change. (2007). *Climate change: Impacts, adaptation, and vulnerability, contribution of working Group II to the fourth assessment report of the intergovernmental panel on climate change*. Cambridge University Press, Cambridge, UK.

Intergovernmental Panel on Climate Change. (IPCC). (2014). Climate Change 2014: Impacts, adaptation, and vulnerability. *Contribution of Working Group II to the Fifth Assessment Report of the IPCC*.

IPCC. (2014). *Climate Changes 2014: Impacts, adaptation, and vulnerability*.

Jayas, D. S. (2016) *Food dehydration, reference module in food science*. Elsevier

Jayasuriya, A., & Rathnayaka, U. (2016). Investigation of nutritional value of selected Sri Lankan traditional root and tubers. *Journal of Biosciences, 4*(8), 656–660. Available at: https://doi.org/10.21276/sajb.2016.4.8.10

Johnston, M., McBride, M. Dahiya, D., Owusu-Apenten, R., & Nigam, P. S. (2018). Antibacterial activity of Manuka honey and its components: An overview. *AIMS Microbiology, 4*(4), 655–664. Available at: https://doi.org/10.3934/microbiol.2018.4.655

Katz, H. S., & Weaver, W. W. (2003). *Encyclopedia of food culture* (Vol. 1).

Khan, S. A. R., Razzaq, A., Yu, Z., Shah, A., Sharif, A., & Janjua, L. (2022). Disruption in food supply chain and undernourishment challenges: An empirical study in the context of Asian countries. *Socio-Economic Planning Sciences, 82*, 101033. Available at: https://doi.org/10.1016/j.seps.2021.101033

Kim, K., & Bui, L. (2019). Learning from Hurricane Maria: Island ports and supply chain resilience. *International Journal of Disaster Risk Reduction, 39*, 101244. Available at: https://doi.org/10.1016/j.ijdrr.2019.101244

Kozlov, A. (2004). Impact of economic changes on the diet of Chukotka Natives. *International Journal of Circumpolar Health, 63*(3), 235–242. Available at: https://doi.org/10.3402/ijch.v63i3.17717

Kozlov, A. I., & Zdor, V. (2003). Whaling products as an element of indigenous diet in chukotka. *The Anthropology of East Europe Review, 17*(1), 127–138.

Kroon, B. V. D., Brouwer, R., & Van Beukering, P. J. H. (2013). The energy ladder: Theoretical myth or empirical truth? Results from a meta-analysis. *Renewable and Sustainable Energy Review, 20*, 504–513.

Kummu, M., Kinnunen, P., Lehikoinen, E., Porkka, M., Queiroz, C., Röös, E., Troell, M., & Weil, C. (2020). Interplay of trade and food system resilience: Gains on supply diversity over time at the cost of trade independency. *Global Food Security, 24*, 100360. Available at: https://doi.org/10.1016/j.gfs.2020.100360

Lucas, H., Pinnington, S., & Cabeza, L. F. (2018). Education and training gaps in the renewable energy sector. *Solar Energy, 173*, 449–455.

Ludin, N. A., Mustafa, N. I., Hanafiah, M. M., Ibrahim, M. A., Teridi, M. A. M., Sepeai, S., Zaharim, A., & Sopian, K. (2018). Prospects of life cycle assessment of renewable energy from solar photovoltaic technologies: A review. *Renewable and Sustainable Energy Reviews., 96*, 11.

Malkanthi, S. H. P., Karunaratne, A.S., Amuwala, S. D., & Silva, P. (2014). Opportunities and challenges in cultivating underutilized field crops in Moneragala District of Sri Lanka. *Asian Journal of Agriculture and Rural Development, 4*(1), 96–105. Available at: http://ageconsearch.umn.edu/rec-ord/198388

Mani, M., Bandyopadhyay, S., & Chonabayashi, S. (2018). *South Asia's hotspots: The impact of temperature and precipitation changes on living standards*. The World Bank Group.

Mapedza, E., Rashirayi, T., Xueliang, C., Haile, A. T., van Koppen, B., Ndiyoi, M., & Sellamuttu, S. S. (2022). Indigenous knowledge systems for the management of the Barotse Flood Plain in Zambia and their implications for policy and practice in the developing world. *Current Directions in Water Scarcity Research, 4*, 209–225.

Marimuthu, M., Venkatesh, M., & Poongodi, M. (2013). X-ray diffraction analysis of nano sized seed powder of white colored germplasm of Velvet Bean (*Mucuna pruriens*), an underutilized wild South Indian Food Legume. *Nano Biomedicine and Engineering, 5*(3), 121–126.

Marsh, K., & Bugusu, B. (2007). Food packaging—Roles, materials, and environmental issues. *Journal of Food Science, 72*(3), 39–55.

Mayes, S., Massawe, F. J., Alderson, P. G., Roberts, J. A., Azam-Ali, S. N., & Hermann, M. (2012) The potential for underutilized crops to improve security of food production. *Journal of Experimental Botany, 63*(3), 1075–1079. Available at: https://doi.org/10.1093/jxb/err396

McCarthy, U., Uysal, I., Badia Melis, R., Mercier, S., Donnell, C. O., & Ktenioudaki, A. (2018). Global food security—Issues, challenges, and technological solutions. *Trends in Food Science and Technology*. https://doi.org/10.1016/j.tifs.2018.05.002

Misty, J. (2009). Indigenous knowledges. In *International encyclopedia of human geography*, pp. 371–376.

Mohri, H., Lahoti, S., Saito, O., Mahalingam, A., Gunatilleke,N., Irham, Hoang, V. T., Hitinayake, G., Takeuchi, K., & Hearth, S. (2013) Assessment of ecosystem services in homegarden systems in Indonesia, Sri Lanka, and Vietnam. *Ecosystem Services, 5*, 124–136. Available at: https://ideas.repec.org/a/eee/ecoser/v5y2013icp124-136.html

Mujumdar, A. S. (2014) *Handbook of industrial drying* (p. 1348). CRC Press Taylor and Francis Group.

Nadeeshani, H., Wimalasiri, K. M. S., Samarasinghe, G., Silva, R., & Madhujith, T. (2018). Evaluation of the nutritional value of selected leafy vegetables grown in Sri Lanka. *Tropical Agricultural Research, 29*(3), 255–267. Available at: https://doi.org/10.4038/tar.v29i3.8265

Ncama, K., Magwaza, L. S., Mditshwa, A., Tesfay, S. Z. (2018). Plant-based edible coatings for managing postharvest quality of fresh horticultural produce: A review. *Food Packaging and Shelf Life, 16*, 157–167. Available at: https://doi.org/10.1016/j.fpsl.2018.03.011

Nowak, D., & Jakubczyk, E. (2020). The freeze-drying of foods—The characteristic of the process course and the effect of its parameters on the physical properties of food materials. *Foods, 9*(10), 1488.

Opoku, S. K., Leal Filho, W., Hubert, F., & Adejumo, O. (2021). Climate change and health preparedness In Africa: Analyzing trends in six African countries. *International Journal of Environmental Research and Public Health, 18*(9), 4672.

Padulosi, S., & Hoeschle-Zeledon, I. (2004). Underutilized plant species: What are they? *LEISA, 20*, 5–6.

Panabokke, C. R., & Punyawardena, B. V. R. (2010). *Climate change and rain-fed agriculture in the dry zone of Sri Lanka.* International Water Management Institute.

Panabokke, C. R., Sakthivadivel, R., & Weerasinghe, A. D. (2002). *Evolution, present status and issues concerning small tank systems in Sri Lanka.* International Water Management Institute.

Peduruhewa, P. S., Jayathunge, K. G. L. R., & Liyanage, R. (2021). Potential of underutilized wild edible plants as the food for the future—A review. *Journal of Food Security, 9*(4), 136–147. Available at: https://doi.org/10.12691/jfs-9-4-1

Rhodes, C. J. (2018). Plastic pollution and potential solutions. *Science Progress, 101*(3), 207–260.

Rizou, M., Galanakis, I. M., Aldawoud, T. M. S., & Galanakis, C. M. (2020). Safety of foods, food supply chain and environment within the COVID-19 pandemic. *Trends in Food Science & Technology, 102*, 293–299. Available at: https://doi.org/10.1016/j.tifs.2020.06.008

Rosenzweig, C., Elliott, J., Deryng, D., Ruane, A. C., Müller, C., Arneth, A., Boote, K. J., Folberth, C., Glotter, M., Khabarov, N., Neumann, K., Piontek, F., Pugh, T. A. M., Schmid, E., Stehfest, E., Yang, H., & Jones, J. W. (2014). Assessing agricultural risks of climate change in the 21st century in a global gridded crop model intercomparison. *Proceedings of the National Academy of Sciences, 111*(9), 3268–3273. Available at: https://doi.org/10.1073/pnas.1222463110

Scott, G. J., Rosegrant, M. W., & Ringler, C. (2020). Roots and tubers for 21st century: Trends, projections and policy options. 2020 vision for food. In *Agriculture and the environment. Discussion paper 31.*

Serra, K. L. O., & Sanchez-Jauregui, M. (2022). Food supply chain resilience model for critical infrastructure collapses due to natural disasters. *British Food Journal, 124*(13), 14–34. Available at: https://doi.org/10.1108/BFJ-11-2020-1066

Sharma, S., Barkauskaite, S., Jaiswal, A. K., & Jaiswal, S. (2020). Essential oils as additives in active food packaging: A review. *Food Chemistry, 343*, 128403.

Sharma, S., Sharma, V., & Chatterjee, S. (2023). Contribution of plastic and microplastic to global climate change and their conjoining impacts on the environment—A review. *The Science of the Total Environment, 875*, 162627.

Siriwardana, C. S. A., Jayasiri, G. P., & Hettiarachchi, S. S. L. (2018). Investigation of efficiency and effectiveness of the existing disaster management frameworks in Sri Lanka. *Procedia Engineering, 212*, 1091–1098. Available at: https://doi.org/10.1016/j.proeng.2018.01.141

Sthapit, B., Lamers, H. A. H., Ramanatha Rao, V., & Bailey, A. (Eds.). (2016). Tropical fruit tree diversity: Good practices for in situ and on-farm conservation. Issues in Agricultural Biodiversity, Oxon (UK) Earthscan for Routledge p. 396. ISBN: 978-1-138-78128-3.

Tafadzwanashe, M., Chimonyo, V. G. P., Chibarabada, T. P., & Modi, A. T. (2017) Developing a roadmap for improving neglected and underutilized crops: A case study of South Africa. *Frontiers in Plant Science, 8.* Available at: https://www.frontiersin.org/articles/https://doi.org/10.3389/fpls.2017.02143

Taylor, J., & Lovell, S. (2015). Urban home gardens in the Global North: A mixed methods study of ethnic and migrant home gardens in Chicago, IL. *Renewable Agriculture and Food Systems, 30*(1), 22–32. Available at: https://doi.org/10.1017/S1742170514000180

Thoman, R. L., Richter-Menge, J., & Druckenmiller, M. L. (2020). *NOAA Arctic Report Card 2020.*

Trisos, C. H., Adelekan, I. O., Totin, E., Ayanlade, A., Efitre, J., Gemeda, A., & IPCC. (2022). Africa. In: *Climate Change 2022: Impacts, adaptation, and vulnerability, Working Group II Contribution to the IPCC Sixth Assessment Report*. IPCC.

Trop, P., & Goricanec, D. (2016). Comparisons between energy carriers' productions for exploiting renewable energy sources. *Energy, 108*, 155–161.

Tweheyo, R. (2018). *Indigenous knowledge and food security: Enhancing decisions of rural farmers*. University of Groningen.

UNDRR. (2019). *Disaster risk reduction in Sri Lanka. United Nations office for disaster risk reduction,* Bangkok, Thailand.

Varshney, R. K., Nayak, S. N., May, G. D., & Jackson, S. A. (2009). Nextgeneration sequencing technologies and their implications for crop genetics and breeding. *Trends in Biotechnology, 27*(9), 522–530.

von Grebmer, K., Ringler, C., Rosegrant, M.W., Olofinbiyi, T., Wiesmann, D., Fritschel, H., Badiane, O., Torero, M., Yohannes, Y., Thompson, J., von Oppeln, C., Rahall, J. (2012). Global hunger index. The challenge of hunger: Ensuring sustainable food security under land, water, and energy stresses. *IFPRI*, Washington DC, USA, p. 67.

Vu, D. T., & Nguyen, T. A. (2017). The neglected and underutilized species in the Northern mountainous provinces of Vietnam. *Genetic Resources and Crop Evolution, 64*, 1115–1124. Available at: https://doi.org/10.1007/s10722-017-0517-1

Weerasekara, W. M. R. A. B., Rathnayaka, R. M. U. S. K., & Saranandha, K. H. (2012). Preparation of ready-to-serve beverage from tropical almond (Terminalia cattappa) fruit pulp. *Tropical Agricultural Research and Extension, 15*(4), 105–107. Available at: https://doi.org/10.4038/tare.v15i4.5260

Wheeler, T., & von Braun, J. (2013). Climate change impacts on global food security. *Science, 341*(6145), 508–513. Available at: https://doi.org/10.1126/science.1239402

World Bank. (2021). *Climate change adaptation and economic transformation in Sub-Saharan Africa*. World Bank.

Yang, X., He, J., Li, C., Ma, J., Yang, Y., & Xu, J. (2008). Matsutake Trade in Yunnan Province, China: An overview. *Economic Botany, 62*, 269–277. Available at: https://doi.org/10.1007/s12231-008-9019-6

Yarış, A., & Sezgin, A. (2017). *Food packaging: Glass and plastic, researches on science and art in 21st century.*

Zhang, H., & Nakagawa, H. (2018). Validation of indigenous knowledge for disaster resilience against river flooding and bank erosion. In *Science and Technology in Disaster Risk Reduction in Asia* (pp. 57–76).

Zhang, J., & Smith, K. R. (2003). Indoor air pollution: A global health concern. *British Medical Bulletin, 6*(1), 209–225. Available at: https://doi.org/10.1093/bmb/ldg029

Unveiling the Green Trail: Harnessing RFID Technology for Sustainable Vegetable Traceability

Shakila Dilshani Rukunayaka, Rumesh Liyanage, S. B. Nawaratne, Chathuri Maheshika Senanayake, K. G. L. R. Jayathunge, and B. A. V. P. N. Jayatissa

1 Introduction

In a world increasingly concerned about the far-reaching consequences of climate change and its profound impacts on global ecosystems and socio-economic systems, the need to embrace sustainable practices and adapt to these changes has never been greater. The places we live, work, and create—referred to collectively as the built environment—have become critical in this quest. People from numerous industries are looking for new ways to address climate change concerns (Baumann, 2021; IPCC, 2014). Within this global paradigm, the agro-food sector in Sri Lanka has received special attention as a key participant in reducing and adapting to climate change problems. Because of the country's unique geography and climate, it is subject to extreme weather, fluctuating rainfall, and rising temperatures. These changes are significant for the country's farming, food security, and rural populations (Sri Lanka Climate Resilience Program, 2020).

In light of the worldwide concern over climate change, adopting sustainable methodologies that are in accordance with the Sustainable Development Goals (SDGs) is imperative. The aforementioned necessity is notably conspicuous within the agro-food industry of Sri Lanka, owing to the country's vulnerability to the effects of climate change (Herath & Poon, 2021). By incorporating Sustainable Development Goals (SDGs) such as No poverty, Zero hunger, Good health and well-being, decent work and economic growth, responsible consumption and production, Climate

S. D. Rukunayaka · R. Liyanage (✉) · C. M. Senanayake · K. G. L. R. Jayathunge
Faculty of Technology, Department of Biosystems Technology, University of Sri Jayewardenepura, Dampe-Pitipana Rd, Colombo, Sri Lanka
e-mail: rumeshliyanage@sjp.ac.lk

S. B. Nawaratne · B. A. V. P. N. Jayatissa
Faculty of Applied Sciences, Department of Food Science and Technology, University of Sri Jayewardenepura, Nugegoda, Sri Lanka

C. Malalgoda et al. (eds.), *Climate Change Adaptation in the Built Environment*,
https://doi.org/10.1007/978-3-031-75826-3_22

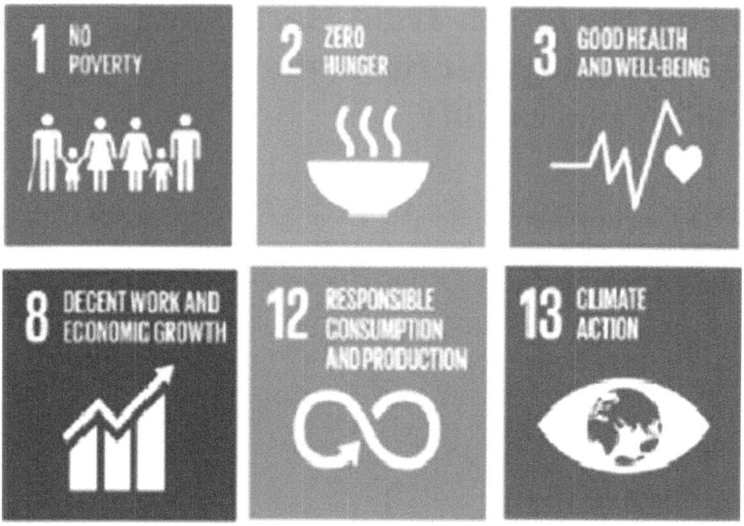

Fig. 1 SGD goals addressed by the current project (*Source* https://sdgs.un.org/goals)

action, the sector has the potential to enhance its ability to withstand extreme weather events, guarantee access to sufficient and nutritious food, and improve the well-being of rural communities (UN, 2015). Additionally, these efforts align with wider sustainable development goals (Fig. 1).

As the world's population grows and urbanization accelerates, the demand for food security and safety has reached an unprecedented surge. However, in order to protect the ecosystem for future generations, this quest for subsistence must be balanced with ecological protection. Sustainable agriculture has emerged as a convincing concept for striking a delicate balance between serving the nutritional needs of a growing population and protecting the environment (FAO, 2019). Vegetable traceability, as an essential component of sustainable agriculture, holds the promise of improving efficiency, transparency, and environmental stewardship in the agricultural supply chain. The seamless integration of Radio-Frequency Identification (RFID) technology is at the heart of this change (Wang et al., 2006). We hope to uncover the complex role of RFID in carving the "Green Trail" within Sri Lanka's agricultural supply chain by leveraging its strength.

The "Green Trail" represents a journey of accountability and transparency that begins with vegetable farming and ends with end-user consumption. Each stage of this trip is attentively documented using RFID-enabled traceability, allowing stakeholders to access essential information about where it came from, transportation, handling and storage of vegetables. Such real-time information gives crucial details that enable customers to make informed decisions, increasing their trust in the quality and reliability of the food they buy. Furthermore, RFID technology can help to streamline the agricultural supply chain, optimize logistics, and reduce waste. Monitoring and tracking vegetables at all stages, from harvest to distribution, promotes higher productivity and flexibility in the system of agriculture. RFID-driven traceability adds to savings in costs for growers, distributors, and consumers alike by lowering post-harvest losses and controlling spoiling, thereby improving the agricultural sector's overall sustainability.

In this chapter, a deeper look was taken at the relationship between sustainable vegetable traceability and climate change adaptation in the context of Sri Lanka. By shedding light on the many benefits and opportunities afforded by RFID technology, this chapter is expected to deliver perspectives that can help open pathways for a more versatile, environmentally friendly and sustainable future in Sri Lanka's agriculture industry and wider built environment.

1.1 Background

Agriculture is heavily reliant on the climate and natural resources; hence, climate change has a significant impact on agriculture. According to the FAO (2016), climate change has long-term consequences for agriculture and food security. Population growth and changes in feeding patterns will cause a 60% rise in global food demand by 2050 compared to 2006 levels, while climate change will continue to have an impact on global food systems. Agriculture, shifting land-use patterns in agriculture, food processing, supply chains, and consumption all amounted to 34% of global GHG emissions in 2015 (Crippa et al., 2021).

In the middle of incredible natural beauty, Sri Lanka has a distinctive combination of challenges as well as possibilities in the face of climate change and sustainable development. As a tropical island nation, the country is especially vulnerable to the negative effects of climate change, including as extreme weather events, changed rainfall patterns, and rising temperatures. These climate alterations not only endanger the country's biodiversity and ecosystems but also represent considerable hazards to the country's vital agricultural sector, on which the country's food security and economy rely heavily.

Sri Lanka is primarily an agricultural country, with small-scale farmers accounting for more than 80% of food producers. Climate change is already causing unpredictable rainfall and dramatic swings between droughts and intense rains in Sri Lanka, as predicted for South Asia (Chithranayana & Punyawardena, 2014; Esham et al., 2018). For example, between 2000 and 2018, the island was subjected to 30 large-scale flooding events, and in 2016–2017, the country experienced the worst drought in 40 years, affecting 20 districts and approximately 1.8 million people (Esham et al., 2018). According to FAO data, intense droughts that followed excessive precipitation have affected agricultural areas in Sri Lanka, endangering the food security of around 900,000 people (FAO, 2017). In this unpredictable climate change context, the need to implement sustainable agriculture strategies couldn't be more pressing (Fig. 2).

To decrease risks and adjust to changing climates, it is critical to reconcile productivity in agriculture while preserving the environment (Sri Lanka Climate Resilience

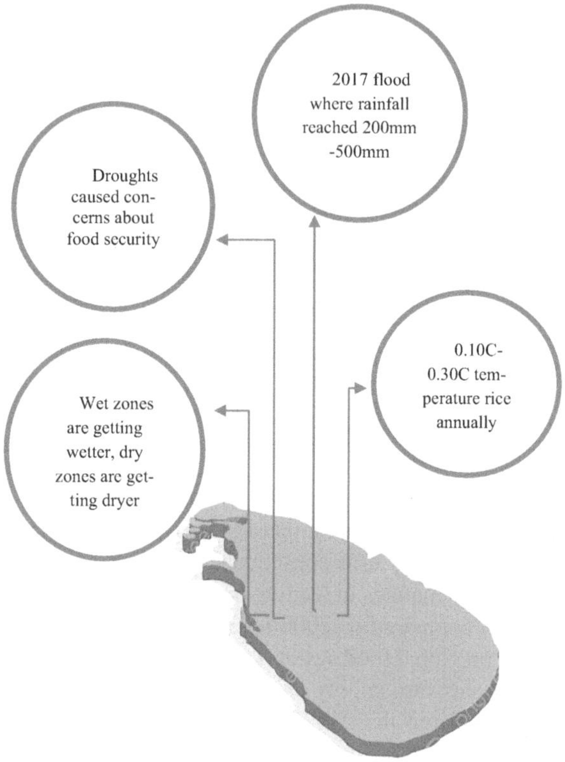

Fig. 2 Impact of climate change on Sri Lanka

Program, 2020). Sustainable agriculture methods highlight robustness, resource optimization, and ecological preservation with the goal of protecting the country's food security and people's well-being.

According to recent research, it has been observed that a considerable proportion, ranging from 20 to 46%, of vegetables obtained during harvesting in Sri Lanka undergo loss throughout the subsequent stages of post-harvest handling and storage (Sarananda et al., 2004; Wasala et al., 2014). This loss has significant economic implications, amounting to an estimated annual financial setback of around $300 million. The aforementioned losses not only have a detrimental impact on the state of food security, but they also serve to disrupt the income streams of farmers and hinder the advancement towards the attainment of Sustainable Development Goal 2, which aims to eradicate hunger (Zero Hunger) (UN, 2015).

The difficulty persists due to the insufficient storage and transportation facilities, which is further compounded by unpredictable meteorological conditions. It is imperative to prioritize urgent and focused endeavors, such as the adoption of sophisticated post-harvest technologies, enhancements to storage infrastructure, and the implementation of agricultural practices that are resilient to climate change. These measures are essential in reducing vegetable losses, promoting sustainable livelihoods, and mitigating the extensive impacts of climate change on the agro-food sector.

Traceability, a critical component of today's agricultural supply chains, is vital for sustaining the efficiency of processes and the safety of agricultural commodities. Traceability systems enable farmers and other interested parties to quickly identify, locate, and withdraw food lots in the case of a product recall, hence ensuring the safety of consumers (Andrade et al., 2019). Traditional communication methods such as phone and fax, have proven inadequate for efficient monitoring as well as data exchange between local vegetable growers and other involved parties in the supply chain. Introducing Radio-Frequency Identification (RFID) technology, in conjunction with a strong information infrastructure, has the opportunity to ensure complete traceability in the supply chain at a lower labor cost, benefiting small and medium-sized agricultural enterprises.

Regardless of the promise of sustained traceability, certain challenging circumstances remain, casting a pall over farmers' and the nation's economic well-being. Farmers are affected due to the current lack of a proper pricing structure for agricultural products, particularly vegetables, while consumers also endure the burden of paying outrageous costs for their fruits and vegetables (Sri Lanka Climate Resilience Program, 2020). This duality emphasizes the critical necessity for contemporary technologies to improve farmers' lives and strengthen the national economy. Furthermore, significant amounts of post-harvest losses in the vegetable industry, which now stand at 20–46%, worsen food insecurity and highlight the need to reduce these sorts of losses. Furthermore, the lack of a solid national vegetable distribution infrastructure, along with an absence of traceability systems, exposes the agricultural supply network to contamination & wastage possibilities with no effective methods of tracing or recall.

Effectively managing the supply chain for perishable goods, such as vegetables and fruits, presents unique challenges. This is mostly owing to the fact that the value of perishable items declines rapidly over time, setting it apart from standard supply chain tactics (Blackburn & Scudder, 2009). The utilization of advanced technologies such as blockchain and Internet of Things (IoT) enables efficient monitoring of fresh transportation chains by facilitating inventory and product quality monitoring. This effectively addresses the management challenge posed by perishability (Yang et al., 2017).

However, in Sri Lanka, the utilization of advanced traceability technology in the fruits and vegetables supply chain is significantly lower compared to the worldwide context. Consequently, Sri Lanka has the task of establishing an effective traceability system for fruits and vegetables. This obvious gap needs immediate action to reform the agricultural supply chain, increasing openness, accountability, and resilience in the face of possible disasters.

1.2 Purpose and Scope

The main aim of this chapter is to elucidate the revolutionary capacity of Radio-Frequency Identification (RFID) technology in changing the traceability of vegetables and enhancing sustainable agriculture in the distinctive built environment of Sri Lanka. The automated identification and data capture technology known as RFID has attracted considerable attention in several industries. This is primarily owing to its capacity to facilitate smooth and effective tracking and management of data. Within the realm of Sri Lanka's agriculture industry, radio frequency identification (RFID) emerges as a potent instrument for guaranteeing traceability, accountability, and sustainability across the entirety of the vegetable supply chain.

The current investigation, is aimed to examine the use of RFID technology in the context of sustainable vegetable traceability in Sri Lanka. The objective is to reveal the diverse benefits and potential prospects associated with this technology. This chapter will explore the functionalities of traceability systems that are enabled by RFID technology, specifically focusing on real-time monitoring and data collection. The aim is to provide stakeholders in Sri Lanka's agricultural sector with reliable and current information, enabling them to make decisions based on the RFID information. This topic has been previously discussed by Wang et al. (2006). RFID-enabled traceability systems establish an interlinked network for the interchange of data, so ensuring comprehensive documentation of crucial information pertaining to the source, handling, and processing of vegetables at every step of the supply chain (Golan et al., 2004). The transparency provided also enables consumers to make well-informed decisions, feeling assured about the safety and sustainability of the food they buy. Additionally, it cultivates a sense of responsibility among farmers and distributors. In addition, timely information regarding the status of vegetables can be accessed by farmers, distributors, and retailers. This access allows them to effectively manage logistics, limit wastage, and mitigate post-harvest losses. The aforementioned efficiency leads to financial benefits for individuals or groups with a vested interest in the agricultural sector, while also making a positive contribution to the long-term viability and environmental soundness of the industry as a whole.

In summary, the integration of RFID technology, sustainable vegetable traceability, and the built environment offers a promising approach to enhancing the resilience and sustainability of Sri Lanka's agricultural economy. In light of the pressing issue of climate change, sustainable agriculture has emerged as a promising solution for a successful and sustainable future. One key aspect that supports this approach is the utilization of RFID-enabled traceability, which enhances the potential for positive outcomes.

2 Understanding Sustainable Vegetable Traceability

2.1 Introduction

In a period characterized by swift urbanization, worldwide trade, and climate change, there has been a growing focus on examining the methods through which food is produced, distributed, and consumed. There is a growing global awareness among consumers regarding the implications of their dietary decisions on both personal well-being and the natural environment. Consequently, a significant transformation has occurred in consumer perspectives towards the consumption of food, characterized by an increasing focus on the aspects of food safety, quality, and sustainability. The dynamic nature of the agricultural sector has led to the emergence of sustainable vegetable traceability as an essential element in contemporary farming practices (Cruz Introini et al., 2018).

At its essence, vegetable traceability refers to the capacity to meticulously monitor and document the complete trajectory of vegetables, commencing from their cultivation on the farm and concluding with their consumption by individuals. According to ISO 22005 (2007) traceability is the ability to follow the movement of a feed or food through specified stage(s) of production, processing and distribution. The process entails the documentation and surveillance of vital data throughout every phase of the supply chain, which includes production, processing, packing, shipping, and retailing. According to Golan et al. (2004), the implementation of this extensive traceability system offers significant data points that enable stakeholders to ascertain the source, handling procedures, and potential hazards linked to every batch of vegetables.

Traceability refers to the capacity to track the historical progression, utilization, or whereabouts of an object or entity through the use of documented identifications. Furthermore, it can be broadly characterized as the capacity to systematically track and monitor the entire lifecycle of a product, encompassing its production, processing, and distribution stage. The concept of traceability can be categorized into three distinct categories, as identified by Perez-Aloe et al. (2007). The concept of back traceability, also known as supplier traceability, Internal traceability, sometimes referred to as process traceability, is a crucial aspect within the realm of academic research. Forward traceability, also known as client traceability, is a concept that refers to the ability to track and document the relationship between client requirements and the corresponding design and implementation decisions throughout the software development process.

The concept of end-to-end traceability comprises the three types of traceability, namely forward, backward, and lateral traceability. Given that traceability is applicable throughout all stages of a supply chain, traceability systems consist of three essential components. According to Bechini et al. (2005), firstly, there is the requirement for a unique identification system for units or batches of each component in a product. Secondly, there is the need to collect information regarding the time and location of each batch transfer or transformation. Lastly, there must be a mechanism in place to establish connections and relationships between this collected data. In a study conducted by Jansen-Vullers et al. (2003), the authors propose a framework consisting of four key components for effectively managing information related to manufacturing traceability. These components include: (1) ensuring physical lot integrity to establish the level of traceability resolution, (2) collecting tracing and process data, (3) establishing product identification and process linking mechanisms, and (4) facilitating reporting and system data retrieval.

2.2 The Importance of Vegetable Traceability

2.2.1 Improving Food Safety and Quality

The establishment of vegetable traceability systems plays a vital role in ensuring the preservation of food safety and quality standards. Consumers are expressing a growing need for enhanced confidence about the source and management of the food they consume, due to the escalating occurrence of foodborne illness outbreaks and incidences of contamination (Golan et al., 2004). Traceability systems enable consumers to obtain information pertaining to the manufacturing techniques, transit routes, and storage conditions associated with vegetables. The provision of transparency enables customers to make well-informed decisions, so fostering enhanced trust in the safety and quality of the agricultural products they purchase (Fig. 3).

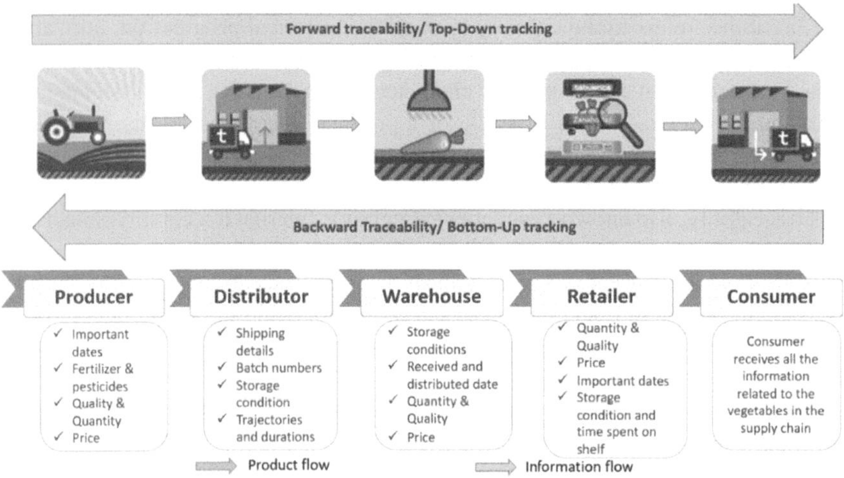

Fig. 3 Vegetable food supply chain in Sri Lanka and the details needed to be traced at each stage

2.2.2 Enhancing the Efficiency of the Supply Chain

An efficient supply chain is one that can be traced. The capacity to monitor the trajectory of veggies from their origin on the farm to their final destination on the table allows stakeholders to effectively identify and address obstacles, reduce unnecessary time lags, and enhance the efficiency of logistical operations. Enhanced efficiency not only mitigates wastage and spoilage but also yields financial benefits for farmers, distributors, and retailers. The utilization of technology such as Radio-Frequency Identification (RFID) enables the facilitation of real-time monitoring, which in turn allows for the prompt transmission of information. This capability enhances the ability to swiftly respond to potential disruptions, as highlighted by Onwude et al. (2020).

2.2.3 Promoting Sustainable Practices

The implementation of vegetable traceability holds significant potential in advancing the cause of sustainable agriculture. The identification of ecologically responsible and ethically sourced produce can be facilitated by stakeholders through an understanding of the production processes connected with each vegetable. Traceability systems can also contribute to the recognition of sustainable farming methods that effectively mitigate water use, minimize chemical inputs, and reduce greenhouse gas emissions. In light of growing consumer emphasis on sustainability, the utilization of traceability can emerge as a potent marketing strategy for companies who are committed to eco-friendly activities.

3 RFID Technology: An Overview

3.1 Introduction to RFID Technology

Radio frequency identification (RFID) has become an integral component of our everyday existence. The phrase "radio communication" is utilized to define any method that enables the exchange of information between a fixed unit and a mobile unit, or between multiple mobile units (Landt, 2005). At its core, RFID utilizes radio waves to automatically identify and track objects or individuals equipped with RFID tags. Simply put, on one end of the connection, there are less intricate devices known as tags, while on the other end, more sophisticated devices such as a multiprocessor unit can be observed. Tags are often compact and cost-effective devices that may be obtained in large quantities and conveniently affixed to the items intended for identification. Furthermore, they can be operated automatically (Fig. 4).

This advanced technology facilitates the smooth acquisition and transmission of data, offering immediate insights into the movement and condition of labeled objects.

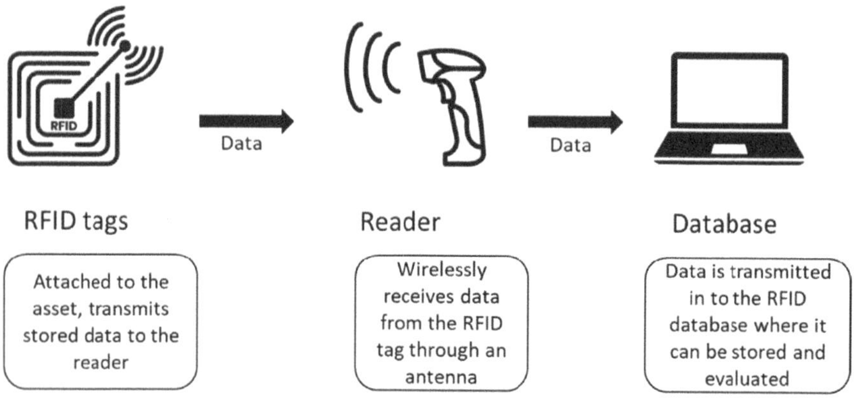

Fig. 4 Basic concept of RFID technology

When comparing RFID technology to standard barcode systems, it becomes evident that RFID has numerous notable advantages. These include the capability of non-line-of-sight communication, the capacity to scan several tags concurrently, and a higher resistance to environmental conditions such as dirt and moisture (Table 1).

3.2 Components of an RFID System

RFID tags, RFID readers, and a backend database or information system are the three main parts of an RFID system (Aguzzi et al., 2011; Roberts, 2006). For efficient data capture, storage, and retrieval, each component is essential.

3.2.1 RFID Tags

An antenna and a unique identification code (UIC) or integrated circuit (IC) enable RFID tags, which are small devices, to broadcast and receive data utilizing radio frequencies. Tags are capable of transmitting said information through a predetermined frequency that has been programmed onto the chip. The interpretation of radio frequency by readers enables the reading of data from tags, facilitating the calculation of their respective identities. Active and passive RFID tags are the two primary varieties. Active RFID tags can track large assets or moving vehicles because they have their own power source and can send signals over great distances. Instead of having an internal power source, passive RFID tags rely on the energy emitted by RFID readers to operate. In situations where short-range identification is sufficient, such as tracking agricultural products along the supply chain, passive tags are more cost-effective and often employed (Babaeian & Karmakar, 2018).

Table 1 Advantages of RFID tags over traditional tracing methods like barcodes and QR codes currently used in Sri Lanka

RFID tags	Barcodes and QR codes
RFID devices do not necessitate visual contact. RFID tags have the capability to be situated within various enclosures such as boxes, containers, or even inserted within objects or living organisms (Finkenzeller, 2004)	Requires visual contact
RFID technology has an extended reading range, enabling the scanning of RFID tags at distances exceeding 100 m (Jedermann et al., 2009; Ruiz-Garcia & Lunadei, 2011)	Reading range of barcodes using traditional laser scanners or linear image scanners is limited to a few inches to around two feet
According to Finkenzeller (2004), it is possible to achieve a reading rate of over 100 tags per second, indicating a significantly improved ability to interrogate or read tags at a faster pace. According to EPC Global (2008), RFID tags have the capability to write five tags per second, with active tags having a storage capacity of 1 MB and passive tags having a storage capacity of 4 kB	Barcodes can only be read individually (White et al., 2007)
The utilization of real-time monitoring, environmental sensing, as well as tracing and tracking techniques	No real time data monitoring can be done even though certain amount of information can be traced
Different types of sensors can be embedded with RFID Tags to collect information about various parameters like temperature, pH, concentration of gases, such as acetaldehyde or ethylene, shock and vibration, and humidity	Such sophisticated technologies do not exist with barcodes

Passive RFID tags can be categorized based on the frequency band employed, such as low frequency (LF), high frequency (HF), ultra-high frequency (UHF). UHF tags may experience periodic issues resulting in a decline in performance when exposed to materials that have the ability to absorb Radio Frequency (RF) energy, such as liquids and metals.

However, the design of certain UHF tags may effectively address these challenges, thereby establishing them as the optimal solution for item-level tracing systems throughout the whole supply chain (Zhang et al., 2017) (Fig. 5).

The long-range reading capability of RFID technology renders it a more favorable option than NFC for the purposes of managing the transportation and distribution of products.

In contrast, the widespread adoption of smartphones equipped with Near Field Communication (NFC) technology, the user-friendly nature of smartphone applications (Shobha et al., 2016) designed for both consumers and small-scale farm and food producers, the accessibility of cloud storage solutions, and the proliferation of Internet indicate the potential utilization of NFC-enabled smartphones for enhancing traceability within the food supply chain.

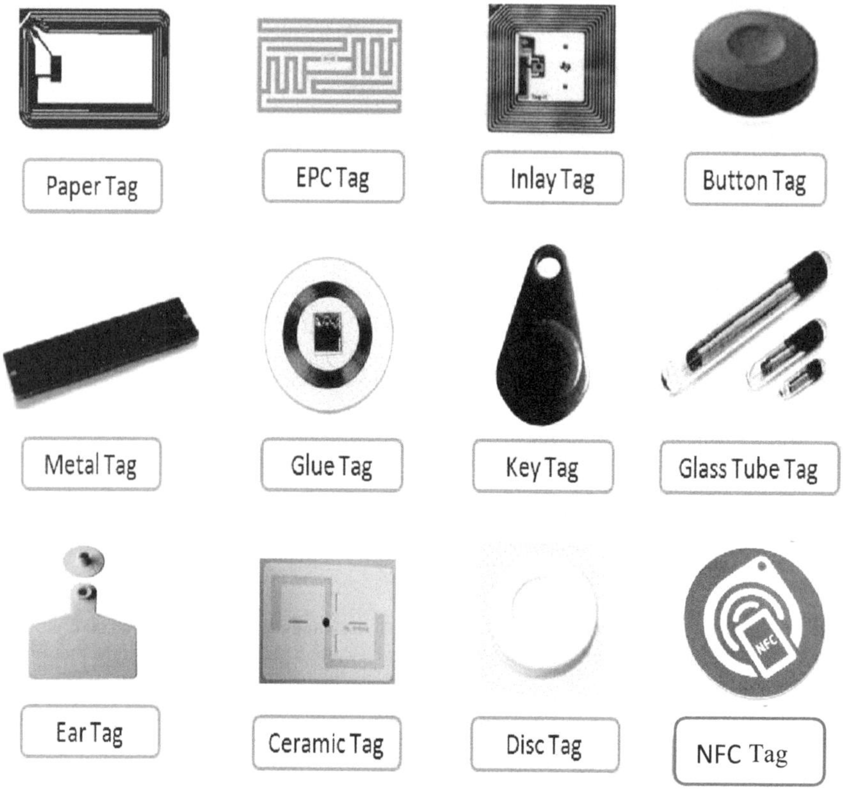

Fig. 5 Types of RFID tags (Dedy Irawan et al., 2018)

NFC tags
Near Field Communication (NFC) is a wireless technology derived from the Radio Frequency Identification (RFID) family of technologies. It operates at a frequency of 13.56 MHz, falling within the high-frequency (HF) range. It is a Near Field Communication (NFC) entities have the capability to exchange both power and data within a limited range of a few centimeters, specifically less than 10 cm (Myny et al., 2015). These devices possess the fundamental characteristics of RFID technology, which include operating in reader/writer mode with passive tags. However, they also possess the additional capability of sharing data among active devices that are powered
One notable distinction between NFC and RFID lies in the extensive integration of NFC into mobile devices, including smartphones, tablets, and notebooks. This facilitated the creation of apps specifically tailored to the needs of the stakeholders

3.2.2 RFID Readers

Radio waves are used by RFID readers, often referred to as interrogators, to connect with RFID tags. An RFID reader activates an RFID tag when it comes within read range of the reader, enabling the tag to communicate its unique identifying code back to the reader. To enable smooth data capture, RFID readers, which can be fixed or portable, are frequently positioned in strategic locations at important points throughout the supply chain. To track the movement and condition of vegetables in the context of sustainable vegetable traceability, RFID readers may be installed at many points, including farm entrances, packing facilities, distribution hubs, and retail outlets (Kelepouris et al., 2007).

In the case of NFC tags, the act of bringing the smartphone in close proximity to the object triggers the automatic activation of an application, which proceeds to retrieve and process all the data stored within the tag. The application possesses the capability to establish a connection with a website that serves as a repository for data. Additionally, it exhibits the ability to autonomously initiate more intricate operations on a smartphone, such as initiating phone calls, dispatching text messages, and transmitting GPS location information, among other functionalities.

3.2.3 Backend Database or Information System

The data collected by RFID readers is typically stored and managed in a backend database or information system. This system serves as a central repository for all RFID-related data, including information on the origin, handling, and processing of vegetables. Stakeholders within the supply chain can access and analyze this data to gain valuable insights into the flow of vegetables and identify potential areas for improvement in terms of efficiency, quality control, and sustainability (Wang et al., 2006).

3.2.4 RFID Software

The functionality of RFID technology extends beyond the mere existence of tags and readers, as it necessitates the utilization of specialized software to efficiently handle the information obtained from RFID devices. The utilization of software is essential for facilitating communication with RFID readers, as well as for the purpose of filtering and cleansing RFID data. Furthermore, software plays a crucial role in converting this data to be compatible with advanced applications, such as automatic data transformation and aggregation. The software can be utilized to encode RFID data into Physical Markup Language (PML) and subsequently transmit it to various destinations such as messages, streams via web services, Java Message Service, HTTP replies, or TCP/IP data packets (Bibi et al., 2017).

It is simpler in the case of NFC tags as any registered stakeholder along the supply chain can access the mobile application developed specifically for tracing to receive or input data. NFC works similarly to Wi-Fi or Bluetooth and works in a familiar manner to any small-scale farmers and distributors without any high-end technological knowledge.

3.3 RFID Tags and Their Applications in the Agri-food Sector

RFID tags have been widely utilized in many applications within the agriculture industry, providing innovative resolutions to enduring obstacles. RFID tags can be employed at many points throughout the agricultural supply chain to improve transparency and efficiency in the realm of sustainable vegetable traceability.

3.3.1 Traceability

The concept of traceability pertains to the capacity to identify the whereabouts of an animal, commodity, food product, or ingredient, and to track its historical trajectory along the supply chain, either in a forward direction (from source to consumer) or in a backward direction (from consumer to source) (Hobbs, Yeung, & Kerr, 2007). The utilization of RFID technology has proven to be effective in the implementation of traceability control. This is primarily attributed to its capacity to accurately identify, categorize, and efficiently manage the movement of commodities (Angeles, 2005; Attaran, 2007; Jones et al., 2004).

The utilization of RFID technology holds significant potential for achieving comprehensive traceability throughout the Agri-food supply chain. This is primarily due to its capability to provide unique identification for each individual product, regardless of the stage it is in (Bernardi et al., 2007). In the study conducted by Gandino et al. (2007) put forth a traceability system that utilizes RFID technology specifically designed for the purpose of tracking fruits within a warehouse setting. Hu et al. (2013) devised a system that aimed at enhancing traceability within the vegetable supply chain. The primary objective of this approach is to guarantee the transportation of vegetables that are both safe for consumption and free from pollution to the market.

This methodology involves the storage of vegetable cultivation information, such as harvesting date, harvest sequence number, and land code number, as well as vegetable processing information, from the RFID cards. By utilizing a unique identifier assigned to each batch of vegetables, it becomes possible to access traceability information pertaining to the vegetables. The fundamental details of the product are connected to a web platform, enabling accessibility for consumers through the utilization of RFID technology. The integration of Radio Frequency Identification (RFID) technology with the internet enhances the advantages of value addition, brand management, and logistics management through the enhancement of information transparency.

The integration of RFID tags with sensors enables the monitoring of environmental parameters, hence facilitating the adjustment of decisions in accordance with the real-time condition of the product. In their study, Wang et al. (2010) integrated RFID technology with sensor networks, GPS, and rule-based decision-making to effectively monitor temperature, humidity, and vibration levels within containers. If the environmental parameters are above the established safety threshold, the warning function will initiate an alert signal to notify the users. Furthermore, it offers significant insights in the event of product recall resulting from natural calamities.

3.3.2 Cold Chain Monitoring

The monitoring of the cold chain for perishable goods has a significant importance in order to fulfill the growing requirements for quality and safety imposed by customers and regulatory authorities. Refrigerated transportation is necessary for the transportation of perishable commodities, including vegetables, fruits, meat, and fish. The examination and evaluation of temperature gradients within refrigerated rooms, containers, and trucks is a primary focus within the industry due to the significant impact that temperature has on the longevity of perishable food items. In the conventional approach, temperature control is typically achieved through the utilization of thermometers or compact temperature loggers. This practice is commonly employed for a range of perishable food items, such as chilled fish and cold fish (Taokis & Labuza, 1989; Tinker et al., 1985), dairy products (Shellhammer & Singh, 1991), meat and poultry, as well as frozen fruit and vegetables. The current temperature tracking systems exhibit limitations in terms of convenience and precision due to the requirement for periodic temperature checks. Consequently, data collecting becomes a manual process, resulting in elevated labor costs. The utilization of Radio Frequency Identification (RFID) technology, in conjunction with temperature indicators (sensors), enables the systematic collection of time–temperature data at regular intervals. This integration enhances temperature management practices by autonomously capturing and storing temperature-related information. In addition, the utilization of RFID sensor tags has been found to be more cost-effective compared to the utilization of temperature loggers, as supported by Ogasawara and Yamasaki (2006) and Ursa (2006). In a study conducted by Gras (2006), the monitoring of a cold chain for frozen commodities was carried out with semi-passive and active RFID technology, which was equipped with temperature sensors. The utilization of real-time data on agricultural products enables users to expedite decision-making processes, thus mitigating potential losses within the cold chain. The integration of RFID technology alongside sensor networks guarantees the delivery of agricultural products to end users in optimal and pristine conditions.

3.3.3 Supply Chain Management

The management of perishable commodities within the supply chain presents significant importance and challenges due to several factors. These include the substantial variations in product characteristics, the stringent traceability requirements, the limited shelf life of food products and agricultural produce, and the necessity to maintain appropriate environmental conditions throughout the various stages from producer to end user (Bubny, 2000; Raman et al., 2001). According to Tajima (2007), the implementation of RFID technology holds the potential to enhance the efficiency of supply chains by mitigating inventory losses and reducing management costs. Additionally, it enables dynamic logistics planning, decreases labor expenses, and improves the efficiency of product flows. The utilization of RFID technology in the retail business has been shown to enhance inventory visibility due to the accurate and expedited collection of information (McMeekin et al., 2006; Raman et al., 2001). According to McFarlane and Sheffi (2003), radio frequency identification (RFID) enables a greater level of automated material inspection and handling. The implementation of automated data collecting through RFID technology has the potential to mitigate the issue of out-of-stock occurrences. The spoilage of perishable products constitutes a significant concern within the realm of supply chain management. The implementation of efficient stock rotation practices involves the retrieval of products from storage in a sequential manner based on their respective expiry dates. This practice ensures that products are sold prior to their deterioration, as emphasized by Shulman (2001). According to Bushnell (2000), the utilization of RFID technology has the potential to enhance operational savings by mitigating the labor expenses incurred in doing inventory counts of shelved merchandise.

The utilization of radio frequency identification (RFID) technology to gather data and store it in a centralized database has been found to enhance transparency among various stakeholders involved in the supply chain. These stakeholders can be classified as manufacturers, wholesalers, retailers, and customers. The research conducted by Hung and Sheng (2012) demonstrates that the implementation of RFID tags in plastic containers enhances the overall administration of the wholesale market, resulting in a concomitant decrease in human error, labor requirements, and costs. The utilization of RFID technology enables the tracking and tracing of objects, hence facilitating the acquisition of real-time information pertaining to particular goods. This information may be effectively disseminated across the various partners involved in the supply chain. By employing RFID technology, Dole Food Company, the leading global producer and distributor of fresh fruits and vegetables, successfully used a proactive measure to recall packed salads that were potentially contaminated with *E. coli* bacteria. This recall was initiated prior to any instances of consumer illness being recorded (Uldrich, 2007). In the context of climate-induced natural disasters, the implementation of product recalls can serve as a means to mitigate the adverse effects on perishable food items. With the availability of real-time data, decisions may be taken to effectively utilize them in a manner that maximizes profitability.

The discussion surrounding RFID technology and the establishment of sustainable vegetable traceability serves as a symbol of advancement within Sri Lanka. The utilization of RFID-enabled traceability systems those facilitates a comprehensive data-tracking process, which allows stakeholders to make informed decisions at various stages of the supply chain. This expands extensively, undergoing a metamorphosis into a panoramic representation of the optimization of resources and the achievement of sustainability within many sectors. The incorporation of RFID technology in the agricultural supply chain facilitates significant advancements, including several aspects such as climate change adaptation and enhanced food security. The potential for RFID technology to contribute to Sri Lanka's transition towards a more environmentally friendly and sustainable future is evident, highlighting the prospects of a prosperous trajectory ahead.

4 NFC-Enabled Vegetable Traceability System in Sri Lankan Fresh Agricultural Supply Chain

4.1 Implementing NFC in the Agricultural Supply Chain

This study presents a novel traceability system for the fresh veggies supply chain, which utilizes EPC (Electronic Product Code) technology to provide seamless tracking.

The proposed system for tracking and tracing items at the individual level is distinguished by its smooth integration of hardware and software subsystems in throughout the supply chain of fresh vegetables. This integration will ensure the continued involvement of stakeholders while also minimizing the expenses associated with implementing new technologies. The suggested system is a unique and cost-effective hybrid solution that guarantees seamless traceability. This will be achieved through the integration of many technologies like Android NFC mobile devices, NFC tags (specifically passive HF tags), and the more affordable data entry software technologies. The suggested solution will emphasize the importance of maintaining a distinct division between the logical EPC-based traceability architecture and the physical infrastructure. This separation is crucial in facilitating a flawless, uninterrupted, and adaptable product traceability process throughout the supply chain.

In addition, information sharing will be facilitated among various stakeholders in the agricultural sector, including farmers as well as, government institutions, authorities responsible for controls (such as Climate and geographical institutes, Payment Agencies and certification bodies), and final consumers. Throughout the production chain, data pertaining to earlier processes will be acquired by means of a mobile device held by the farmer and a tag affixed to the object. Subsequently, further information will be appended and stored in a cloud-based database, and subsequently inscribed onto another tag. The end customer has the ability to access comprehensive information regarding the supply chain by utilizing their smartphone to scan the tag affixed to the food goods, thereby gaining insight into the details presented by farmers and other stakeholders.

Moreover, the system is planned to exhibit the capability to provide support for the Integrated Administration and Control System (IACS) method. This is due to the fact that the information that is captured and generated by the system may be effectively utilized inside the IACS framework. Hence streamlining various administrative and control procedures. Public institutions may utilize the data collected for administrative verifications, while controllers can also make use of it to streamline the control process and minimize disruptions to farmers.

This system will facilitate the automatic time and geo referencing of data and actions. As a result, it guarantees prompt and precise trace-back when deemed required. The public is provided with information regarding the origin of vegetables. In addition, the real-time synchronization of the cloud database might be employed, for instance, to track the spread of pathogens in agricultural areas and promptly notify nearby farmers and relevant municipal authorities. The adoption of Near Field Communication (NFC) technology enhances food safety measures, benefiting both stakeholders involved in each stage of the supply chain and the end consumer. Furthermore, the application has the potential to enhance sustainable production practices and ensure product quality, ultimately leading to an enhancement in consumer trust and confidence.

4.2 *Proposed Methodology*

4.2.1 Identify the Gaps in the Vegetable Supply Chain in Sri Lanka

A comprehensive strategy will be employed to identify deficiencies in the vegetable supply chain in Sri Lanka. Prior to proceeding, a comprehensive examination of the available literature will be undertaken to evaluate prior studies, reports, and publications pertaining to the vegetable supply chain. This will offer valuable perspectives on prevalent problems and obstacles. In addition, government and business reports will be examined to obtain additional insights into supply chain inefficiencies. Stakeholder interviews will be carried out with essential participants in the vegetable supply chain, encompassing growers, wholesalers, retailers, and transportation providers. The interviews will yield qualitative data regarding topics such as transportation congestion, losses occurring after harvest, and control measures for ensuring quality. Field surveys will be conducted at several locations along the supply chain to directly observe activities and collect primary data. This task will need the documentation of procedures, identification of inefficiencies, and gathering of data pertaining to the handling, storage, and transportation of products. Afterwards, the gathered data will undergo analysis to identify prevalent deficiencies and inefficiencies in the vegetable supply chain. These problems will then be ranked according to their impact on the industry.

4.2.2 Design an RFID & GPS Vegetable Tracking System for Monitoring and Traceability

The development of the RFID and GPS vegetable tracking system will entail a sequence of deliberate and calculated measures. The selection of suitable RFID and GPS technologies for vegetable monitoring will mostly depend on criteria such as cost, coverage, accuracy, and compatibility with current infrastructure. Following that, a sturdy system architecture will be created, using RFID and GPS technology to provide instantaneous tracking and traceability. This will involve establishing data exchange protocols and communication routes to guarantee smooth operation. A prototype of the tracking system will be created to undergo testing and validation, guaranteeing its compatibility with certain vegetable kinds and supply chain players.

4.2.3 Develop a User-Friendly Mobile Application

The creation of a mobile application that is easy for users to navigate will adhere to a methodical process. The user requirements will be collected by conducting thorough interviews with stakeholders in the supply chain and end-users. This process will involve identifying crucial features, preferences for ease of use, and specific demands for data visualization. The application will be developed with a user-friendly and intuitive interface, which will be facilitated by the production of wireframes and user flows. After the design process, the mobile application will be created for both Android and iOS platforms and smoothly incorporated with the RFID and GPS tracking system.

4.2.4 Standardize the Developed Systems

The standardization efforts will consist of a methodical assessment of the created systems in relation to industry standards and legislation concerning traceability and technology. Necessary adjustments and alterations will be made to ensure compliance. Thorough documentation, consisting of user manuals and technical specifications, will be created for the systems. Additionally, training materials will be provided to stakeholders. A pilot test will be carried out to evaluate the functionality of the integrated system, obtaining feedback from a specific group of supply chain stakeholders in order to facilitate enhancements. In the end, the standardized solutions will be implemented on a broader scale through cooperation with government agencies and industry partners to successfully tackle deficiencies in the vegetable supply chain in Sri Lanka. During the research, we will consistently seek input and work closely with stakeholders to guarantee the generated systems are both relevant and effective.

4.3 Architecture of the Proposed System

The proposed system has been carefully designed to effectively incorporate the complexities of Sri Lanka's fresh vegetable market. In the context of this particular environment, vegetables obtained from both small-scale and large-scale agricultural producers undergo a process that encompasses the stages of harvesting, transportation, and temporary storage. The aforementioned veggies are thereafter distributed to small- to medium-scale retail establishments, while another portion is packaged and retailed in supermarkets.

The progression of data across this sequence manifests as a chronicle characterized by openness and the ability to be traced. Throughout the process of collecting and transporting vegetables, a significant amount of data is generated at each stage, encompassing crucial information regarding their source, quality, and handling. As the aforementioned vegetables are distributed to small- to medium-scale retail establishments, the trajectory of data pertaining to them continues to exhibit a dynamic nature. Data is collected and transmitted, by the integration of simple and profitable technologies facilitating a thorough comprehension of the trajectory of the produce.

The traceability information pertaining to vegetables will be transmitted to the system by each participant in the supply chain. During each stage of the supply chain, the respective stakeholder will acquire and incorporate data into the NFC tag, or introduce a new tag if deemed required. The subsequent procedures will be consistently executed at each stage of the process.

1. Initial data input into the NFC tag on the vegetable crate and update initial information into the primary cloud database
2. Next stakeholder reads information available on the NFC tag
3. Update the information in their relevant phase into the NFC tag
4. Introduce new NFC tag where necessary, including all the previous data obtained from the cloud database
5. Update all the information in the primary cloud database
6. Update all the information in the regulatory database.

The tag exclusively includes an URL to the data stored in the cloud database. When the object is transferred from one area to another, the tag functions as a supplementary document, allowing the information stored in the original database to be accessible in the subsequent database. The utilization of cloud storage facilities and services enables seamless information sharing between databases, even when they are managed by different software systems. Different software applications supporting Android devices and specified for each stage will be used in each phase to enter the information relevant to the specific stage (Fig. 6).

4.3.1 Farm

Farmers who have completed the registration process are granted privileges to utilize a specialized mobile application specifically designed for agricultural producers. This

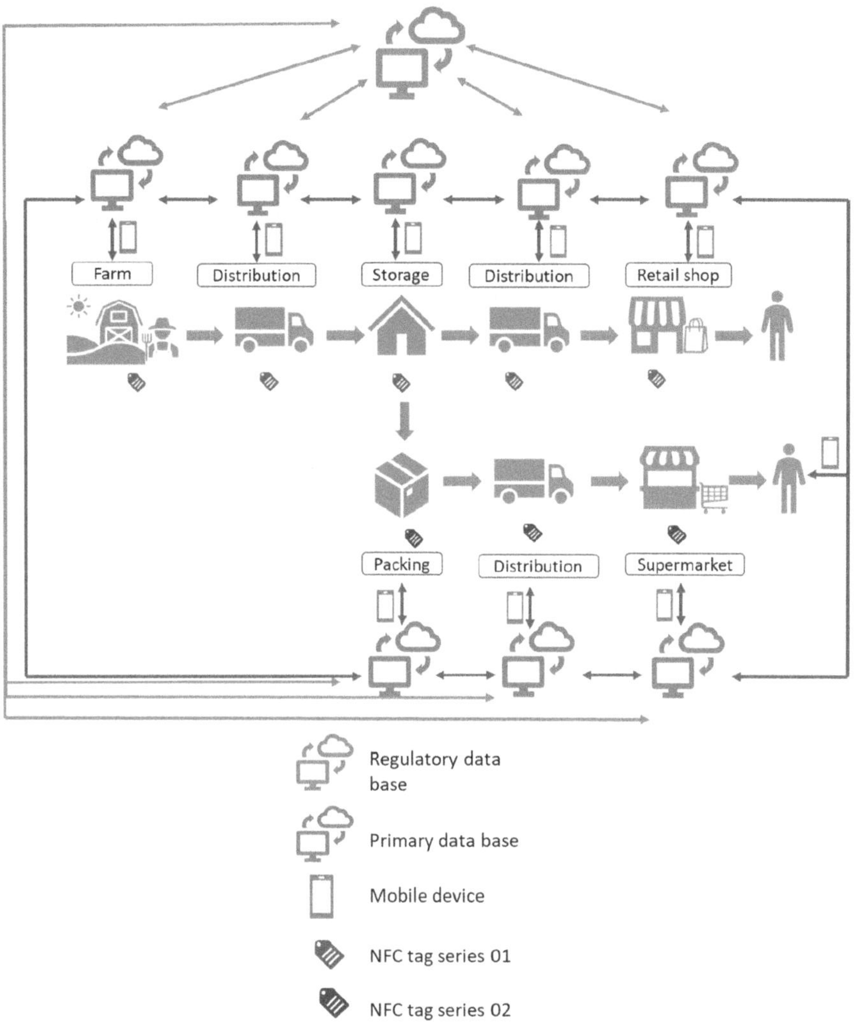

Fig. 6 The architectural design of the proposed system to trace fresh vegetables throughout the supply chain which leverages smartphones and NFC tags. The process of monitoring and tracking goods, as they move through the supply chain, involves the coordination of various entities and the utilization of multiple primary databases connected together and all connected together with one regulatory database

application facilitates seamless interaction between farmers and NFC tags, enabling remote scripting of the tags with relevant information. The entire process takes place exclusively within the application interface, allowing farmers to independently enter data that encompasses various activities linked to land management up until the final stage of harvesting. The NFC tags are affixed to particular crates (Fig. 7), which are allocated for specific varieties of vegetables.

Fig. 7 Vegetable crate with NFC tag on it

The dataset provided by agricultural practitioners encompasses a wide range of data points, which are thoroughly described in the accompanying Fig. 9a. After compilation, this dataset is smoothly integrated with the principal cloud storage system. As an integral component of an automated procedure, this consolidated data is later transferred to the regulatory database.

4.3.2 Distributors

Distributors are mandated to undergo a registration process in order to obtain authorization to utilize the platform. This registration process allows individuals to effortlessly access and modify pertinent information. Distributors may easily view the pre-existing data saved in the cloud database by scanning the NFC tags attached to the crates. The activation of the scanning process initiates a direct connection to a URL that provides distributors quick access to the primary cloud database, thereby providing them with useful information.

Distributors have the option to utilize a specialized mobile application that is customized to meet their specific requirements. This application functions as a central platform for inputting data pertaining to the transportation and distribution of vegetables. The recently acquired data has the potential to be smoothly incorporated into the same NFC tags that were utilized in the initial phases of the supply chain. Moreover, this approach allows for precise logging of any damage incurred by the produce during shipping. Figure 9b illustrates a list of data to be logged to the NFC tags by the distributors. The implementation of this complete strategy, which incorporates distributors into the data loop, guarantees the continuous maintenance of current and precise information across all stages of the process.

4.3.3 Storage and Intermediaries

Within the specific setting of Sri Lanka, it is common for the distribution of vegetables to undergo an intermediary phase, wherein the product is temporarily stored prior to being directed towards shops. The intermediary stage may include several activities like grading and packing, in addition to the direct transit of goods to merchants. It is worth mentioning that Sri Lanka has a considerable population of small-scale vegetable vendors, so enhancing the variety of the distribution network throughout the nation.

In this particular paradigm, intermediaries assume a crucial role as key players in the traceability system that is being suggested. After the registration procedure, intermediaries participate by conducting scans of NFC tags that are attached to crates. This operation initiates the inclusion of crucial information directly into these tags. Then vegetables are either advanced to the stage of packaging and selling or are directly allocated to retailers. Figure 9c shows the list of data entered by the intermediaries.

4.3.4 Grading and Packing

Fresh vegetables undergo a grading and packing process in certain situations, particularly when they are destined for retail sale in supermarkets. In these particular cases, it is imperative for grading and packing stakeholders to be registered within the system. This registration enables individuals to integrate the relevant information into NFC tags.

It is crucial to acknowledge that, while dealing with the evaluation and repacking of materials, a unique methodology is necessary. The process of transitioning veggies from crates to newly packed packets require the placement of new Near Field Communication (NFC) tags. This requires a systematic series of steps: those involved in this phase must initially extract data from the pre-existing NFC tags attached to the crates. Following this, the aforementioned information is then revised and consolidated, incorporating the supplementary data obtained from the packing centers. A potential list of information is mentioned in the Fig. 9d. These revised details are then transcribed into the newly implemented Near Field Communication (NFC) tags, which are affixed to the recently packaged vegetable packets. The rigorous procedure guarantees the uninterrupted dissemination of precise and current data, even during the progression of vegetables through the steps of grading and packing.

4.3.5 Retail Shops and Supermarkets

The study's focus spans into two main categories of retail facilities, namely small-scale retail shops and supermarkets. In the context of small-scale retail establishments, it is customary for shopkeepers to display vegetables in stacked formations for the purpose of sale. In this configuration, the NFC tag linked to a particular crate should be conspicuously exhibited in close proximity while the corresponding batch of fruit is being offered for sale. This exhibit provides consumers with the opportunity to acquire extensive information regarding the entire journey of the veggies, starting from the farm and destined at the shop. Through the utilization of a mobile application (tailored for customers), individuals are able to promptly obtain access to the aforementioned information by scanning the NFC tag.

Figure 8 illustrates a visual representation of this configuration. The implementation of exhibiting tags in close proximity to vegetable heaps establishes a straightforward and transparent linkage between customers and the traceability data. In certain supermarkets as well, a comparable methodology can be employed, wherein NFC tags are also prominently displayed in close proximity to stacks of vegetables. However, when addressing the issue of packaged veggies sold in supermarkets, a distinct approach must be employed. Within this particular scenario, a novel Near Field Communication (NFC) tag is attached to every individual packet containing vegetables. The recently implemented NFC tag effectively encompasses all relevant information pertaining to the whole supply chain, from the farm to the shop. There are some details that has to be entered by the retailer as well in Fig. 9e.

The suggested system facilitates the active participation of farmers, distributors, intermediaries, and retailers, customers as well as the regulatory authorities in a comprehensive data-sharing network. The process of traceability encompasses various stages, including seedling identification, crop monitoring, and the crucial phases of harvesting, distribution, and retail. These stages are interconnected, forming a cohesive web of traceability. The use of this holistic approach holds the potential to mitigate inefficiencies, enhance the allocation of resources, and diminish post-harvest losses. Moreover, the system's adaptability caters to a wide

Fig. 8 NFC tags displayed in a fruit stack

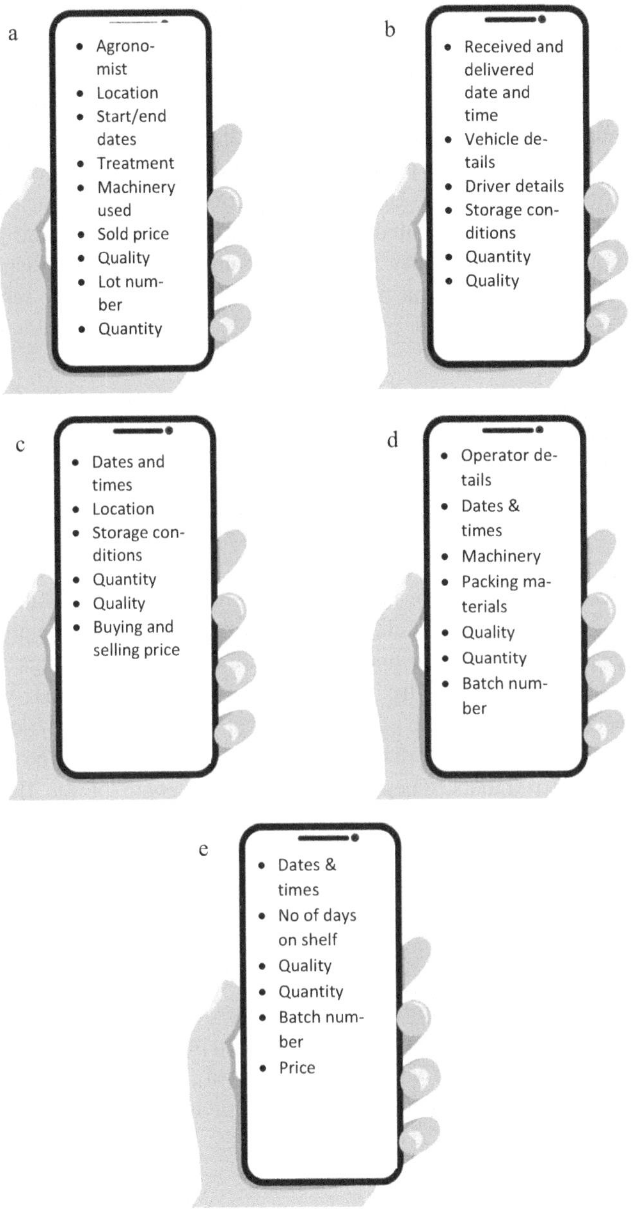

Fig. 9 Sets of data to be uploaded by each stakeholder in their corresponding phases. **a** Farmer, **b** distributor, **c** storage and intermediaries, **d** packers, **e** retail and supermarkets

range of situations, including both small-scale retail establishments and larger supermarket settings, thereby guaranteeing that consumers are equipped with comprehensive information while making their buying choices. This system is grounded in the fundamental principles of sustainability, resilience, and adaptability to meet consumer demands. Through the facilitation of real-time data collecting, this technology enables stakeholders to make well-informed decisions, fosters consumer trust, and promotes the adoption of sustainable consumption patterns. In conclusion, the adoption of this integrated approach not only facilitates the adaptation of Sri Lanka's agricultural industry to climate change but also establishes a foundation for a vegetable distribution network that is both ecologically sensitive and consumer-oriented.

5 Challenges and Limitations

5.1 *Environmental Effect*

The performance of RFID systems can be influenced by several environmental variables, such as the presence of dust, dampness, severe temperatures, metallic objects, and high-water content in tagged materials. According to the findings of Dobkin and Weigand (2005), the readability of tags at a frequency of 915 MHz was significantly influenced by the presence of metal and water. In their study, Fletcher et al. (2005) conducted an analysis to determine the factors influencing the strength of radio frequency (RF) signals. They found that the strength of the RF signal is depends on both the quantity of water it traverses and the period of the sinusoidal wave. The majority of food products contain a significant amount of water, and their dielectric characteristics undergo changes when subjected to an electromagnetic field (Nelson & Bartley, 2000). The presence of external signal sources, such as mobile phones and barcode scanners, has the potential to produce errors throughout the reading process. According to Amador et al. (2010), the majority of tags employ direction-sensitive antennas, meaning that the position of the tag can impact its readability. Collisions on the air interface and transmission problems have been identified as potential causes of false negative reads (Brusey et al., 2003). The disposal of non-biodegradable RFID tags has been identified as a potential environmental issue (Li et al., 2006).

5.2 *Operator Acceptance of New Technologies*

Extensive research has been conducted in the academic literature about the investigation in utilizing a combination of several RFID technologies and standards to enhance supply chain management (Guido et al., 2012). The application of these methods has also been successful in the agro-food sector (Trebar et al., 2011; Catarinucci & Cuinas, 2011).

Nevertheless, the establishment of a comprehensive and seamless traceability system, spanning from the initial production phases to the final consumption by end users, is currently in its nascent phase, with numerous unresolved challenges.

Many scholars work and put forth ideas that are seen overly intrusive and, as a result, are not readily embraced by those responsible for their implementation. An illustrative instance pertains to the utilization of Wireless Sensor Networks (WSN) within greenhouses with the aim of attaining precision agriculture (Keshtgari & Deljoo, 2012; Roy & Bandyopadhyay, 2008). Despite the numerous potential advantages offered by this technology, its implementation remains significantly constrained due to the reluctance of experienced agronomists who assert that no sensor node can ever supplant their expertise. Hence, a crucial element in the process of reengineering is the comprehensive comprehension of the proposed solution by the operators prior to its acceptance and implementation.

5.3 *Data Management*

The data management systems face numerous obstacles due to the diverse features of RFID data. The data generated by RFID technology is characterized by its large volume, time-dependent nature, and dynamic changes, which contribute to the heightened complexity of data management. Errors in the data may necessitate the implementation of filtering or pre-processing techniques on the raw data. According to Wang and Liu (2005), RFID applications have the capability to generate data in real-time regarding changes in location and environmental elements. Therefore, it is crucial to employ a comprehensive data model that is well-suited for application-level interaction, encompassing activities such as tracking and monitoring. Laddhad (2006) provides a comprehensive analysis of different methodologies employed in data warehousing, such as the ER (entity relationship) model and RFID-Cuboids. Academic researchers are currently engaged in the development of methodologies and models aimed at enhancing the efficiency of data management and the process of cleaning and filtering RFID data.

5.4 *Collisions*

When two items with tags that are equidistant from the reader are positioned next to each other, both tags will become energized upon receiving the signal from the reader and broadcast their response to the reader simultaneously. In this scenario, the data from both tags would be overlaid, leading to a collision between the tags (Want, 2006). Moreover, in situations where many readers are present, the interrogation signals emitted by each reader can potentially interfere with one another, leading to reader collision. Hence, anticollision techniques are to be employed to prevent collisions between multiple tags and readers (Want, 2006). The anti-collision protocol facilitates the establishment of a transmission schedule for the tags, ensuring that each tag is able to successfully connect with the reader.

5.5 *Virus Attacks*

According to a source in the field of RFID technology (RFID Update, 2006), data obtained via RFID and afterwards handled by a database management system is susceptible to computer viruses. According to Rieback et al. (2006), the data obtained through RFID technology is vulnerable to both nefarious intentions and viral attacks. A substantial quantity of data produced by the RFID tags is kept within a database. In order to mitigate the risk posed by viruses, it is imperative to remove any characters or strings from the provided data that could potentially be exploited for malicious purposes. Additionally, it is recommended to employ Secure SQL (structured query language) Coding techniques.

6 Conclusion

There are multiple traceability systems available in the industry at present. However, the complexity of enabling smooth information flow among the different stakeholders involved in the process poses a significant issue, given the diverse range of potential implementations. Paper-based solutions are frequently utilized to ensure traceability in many situations. However, this methodology is distinguished by a slow distribution of information and a lack of ability to utilize data for improving the efficiency of fresh vegetable processing and, subsequently, the final outcome. This study proposes a comprehensive traceability system that utilizes NFC (Near Field Communication) identification, specifically evaluated in the context of the fresh vegetable supply chain, as a response to the aforementioned difficulties.

RFID technology is widely utilized across the food supply chain. However, the novelty of this research resides in its proposal of an integrated traceability framework that establishes a connection between end customers and the system via cellphones and NFC technology. The incorporation of automated tracing methods for Small and Medium-sized Enterprises (SMEs) operating in the agricultural sector is made possible by the user-friendly and cost-effective characteristics of the accompanying application and hardware.

While the integration of RFID technology into the food supply chain has previously taken place, this concept still remains new to the food and agricultural sector in Sri Lanka. The greater readability of NFC tags in various visibility circumstances highlights their advantages over QR codes.

Furthermore, the NFC-based traceability system exhibits potential applications that transcend normal supply chain management. It holds significant promise in addressing the repercussions of abrupt climatic variations and natural calamities. During periods of environmental disruption, such as instances of severe weather or unanticipated emergencies, the utilization of NFC technology's real-time monitoring capabilities can be of paramount importance in safeguarding the integrity and reliability of food safety and security measures. By efficiently documenting the trajectory of newly harvested veggies from their origin to the final destination on consumers' tables, this system has the potential to offer vital observations regarding the safety of the produce in the face of various challenges. The application provides valuable data that can assist various stakeholders, including authorities, relief organizations, and consumers, in evaluating the feasibility of the food supply during periods of interruption. The increased transparency can enhance the process of making informed decisions and allocating resources efficiently, so guaranteeing that necessary sustenance is provided to people who require it, while also limiting potential health hazards. Therefore, the emergence of the NFC-based traceability system is not only beneficial for improving routine operations but also plays a crucial role in times of crisis, thereby highlighting its significance in protecting food security and public health.

References

Aguzzi, J., Sbragaglia, V., Sarria, D., Garcia, J. A., Costa, C., del Rio, J., Manuel, A., Menesatti, P., & Sarda, F. (2011). A new laboratory radio frequency identification (RFID) system for behavioural tracking of marine organisms. *Sensors (Basel), 11*(10), 9532–9548.

Amador, C., & Emond, J. P. (2010). Evaluation of sensor readability and thermal relevance for RFID temperature tracking. *Computers and Electronics in Agriculture, 73*(1), 84–90.

Andrade, R. G., Batistella, M., Pires, P. P., Visoli, M. C., & Amaral, T. B. (2019). Geotraceability: An Innovative Strategy for Extraction of Information and an Aid for Sustainable Cattle Raising. *International Journal of Advanced Engineering Research and Science, 6*(5).

Angeles, R. (2005). RFID technologies: Supply-chain applications and implementation issues. *Information Systems Management, 22*(1), 51–65.

Attaran, M. (2007). RFID: An enabler of supply chain operations. *Supply Chain Management: An International Journal, 12*(4), 249–257.

Babaeian, F., & Karmakar, N. C. (2018). Hybrid chipless RFID tags-a pathway to EPC global standard. *IEEE Access, 6*, 67415–67426.

Baumann, F. (2021). The next frontier—human development and the anthropocene: UNDP human development report 2020. *Environment: Science and Policy for Sustainable Development, 63*(3), 34–40.

Bechini, A., Cimino, M. G., Lazzerini, B., Marcelloni, F., & Tomasi, A. (2005, January). A general framework for food traceability. In *Symposium on Applications and the Internet Workshops*, IEEE, pp. 366–369.

Bernardi, P., Demartini, C., Gandino, F., Montrucchio, B., Rebaudengo, M., & Sanchez, E. R. (2007). Agri-food traceability management using a RFID system with privacy protection. *International Conference on Advanced Networking and Applications* (AINA-07), 68–75.

Bibi, F., Guillaume, C., Gontard, N., & Sorli, B. (2017). A review: RFID technology having sensing aptitudes for the food industry and their contribution to tracking and monitoring of food products. *Trends in Food Science & Technology, 62*, 91–103.

Blackburn, J., & Scudder, G. (2009). Supply chain strategies for perishable products: The case of fresh produce. *Production and Operations Management, 18*(2), 129–137. https://doi.org/10.1111/j.1937-5956.2009.01016

Bottani, E., Montanari, R., & Rizzi, A. (2009). The impact of RFID technology and EPC system on stock-out of promotional items. *International Journal of RF Technologies, 1*(1), 6–22.

Brusey, J., Floerkemeier C., Harrison, M., & Fletcher, M. (2003). Reasoning about uncertainty in location identification with RFID. In *IJCAI-2003 Workshop on Reasoning with Uncertainty in Robotics, Acapulco, Mexico.*

Bubny, P. (2000). Keeping cool with RFIDS. *Supermarket Business, 55*(5), 95.

Bushnell, R. (2000). RFID's wide range of possibilities. *Modern Materials Handling, 55*(1), 37.

Catarinucci, L., Cuinas, I., Expósito, I., Colella, R., Fernandez, J. A. G., & Tarricone, L. (2011). RFID and WSNs for traceability of agricultural goods from Farm to Fork: electromagnetic and deployment aspects on wine test-cases. *Proceedings of the 19th International Conference on Software, Telecommunications and Computer Networks* (SoftCOM '11), pp. 12–15.

Chithranayana, R. D., & Punyawardena, B. V. R. (2014). Adaptation to the vulnerability of paddy cultivation to climate change based on seasonal rainfall characteristics. *Journal of the National Science Foundation of Sri Lanka, 42*(2), 119–127.

Crippa, M., Solazzo, E., Guizzardi, D., et al. (2021). Food systems are responsible for a third of global anthropogenic GHG emissions. *Nat Food, 2*, 198–209. https://doi.org/10.1038/s43016-021-00225-9

Cruz Introini, S., Boza, A., & Alemany Díaz, M. D. M. (2018). Traceability in the Food Supply Chain: Review of the literature from a technological perspective. *Dirección y Organización, 64*, 50–55.

Dedy Irawan, J., Adriantantri, E., & Farid, A. (2018). RFID and IOT for attendance monitoring system. In *MATEC web of conferences.*

Dobkin, D. M., & Weigand, S. M. (2005). Environmental effects on RFID tag antennas. In *IEEE MTT-S International. Microwave Symposium Digest*

E. P. C. Global. (2008). *EPCTM radio-frequency identity protocols Class-1 Generation-2 UHF RFID Protocol for Communications at 860 MHz–960MHz.*

Esham, M., Jacobs, B., Rosairo, H. S. R., & Siddighi, B. B. (2018). Climate change and food security: A Sri Lankan perspective. *Environment, Development and Sustainability, 20*(3), 1017–1036.

FAO. (2016). *The state of food and agriculture: climate change, agriculture and food security.* FAO of UN, Rome, Italy.

FAO. (2017). *Special report on FAO/WFP crop and food security assessment mission to Sri Lanka.* http://www.fao.org/3/i7450e/i7450e.pdf

FAO. (2019). *The state of food security and nutrition in the World 2019. Safeguarding against economic slowdowns and downturns.* Food and Agriculture Organization of the United Nations. http://www.fao.org/3/ca5162en/ca5162en.pdf

Finkenzeller, K. (2004). *Radio-frequency identification fundamentals and applications* (p. 2). Wiley.

Fletcher, R., Marti, U., & Redemske, R. (2005). *Study of UHF RFID signal propagation through Complex Media'.* IEEE Antennas and Propagation Society International Symposium.

Gandino, F., Montrucchio, B., Rebaudengo, M., & Sanchez, E. R. (2007). Analysis of an RFID-based information system for tracking and tracing in an agri-food chain. In *Proceedings of the First Annual RFID Eurasia, Turkey*, pp. 1–6.

Golan, E., Krissoff, B., Kuchler, F., Calvin, L., Nelson, K., & Price, G. (2004). *Traceability in the U.S. food supply: Economic theory and industry studies.* U.S. Department of Agriculture, Economic Research Service. https://www.ers.usda.gov/publications/pub-details/?pubid=43519

Gras, D. (2006). RFID based monitoring of the cold chain. In *2nd International Workshop on Cold Chain Management, Bonn, Germany.*

Guido, L. M., & Patrono, P. (2012). Evaluating potential benefits of the use of RFID, EPC global, and ebXML in the pharmaceutical supply chain. *International Journal of Healthcare Technology and Management, 13*(4), 198–222.

Herath, G., & Poon, W. C. (2021). Climate change, hunger and food security in Asia with special reference to Sri Lanka: Can the sustainable development goals be achieved by 2030. In *Exploring synergies and trade-offs between climate change and the sustainable development goals*, pp. 37–54.

Hobbs, J. E., Yeung, M. T., & Kerr, W. A. (2007). *Identification and analysis of the current and potential benefits of a national livestock traceability system in Canada*. http://www4.agr.gc.ca/resources/prod/doc/pol/pub/nltsc-sntac/pdf/nltscsntac_e.pdf

Hu, J., Zhang, X., Moga, L. M., & Neculita, M. (2013). Modeling and implementation of the vegetable supply chain traceability system. *Food Control, 30*(1), 341–353.

Hung, C. C., & Sheng, C. T. (2012). Application of plastic containers with RFID in the marketing of fruit and vegetables in Taiwan. *International Journal of Agricultural and Biological Engineering, 5*(1), 35–42.

IPCC. (2014). *Climate Change 2014: Synthesis report. Contribution of Working Groups I, II, and III to the Fifth Assessment Report of the Intergovernmental Panel on Climate Change*. IPCC.

ISO. (2005). *ISO 22000:2005, Food safety management systems—Requirements for any organization in the food chain.*

Jansen-Vullers, M. H., van Dorp, C. A., & Beulens, A. J. (2003). Managing traceability information in manufacture. *International Journal of Information Management, 23*(5), 395–413.

Jedermann, R., Ruiz-Garcia, L., & Lang, W. (2009). Spatial temperature profiling by semi passive RFID loggers for perishable food transportation. *Computers and Electronics in Agriculture, 65*(2), 145–154.

Jones, P., Clarke-Hill, C., Shears, P., Comfort, D., & Hillier, D. (2004). Radio frequency identification in the UK: Opportunities and challenges. *International Journal of Retail and Distribution Management, 32*(3), 164–171.

Kelepouris, T., Pramatari, K., & Doukidis, G. (2007). RFID-enabled traceability in the food supply chain. *Industrial Management & Data Systems, 107*(2), 183–200.

Keshtgari, M., & Deljoo, A. (2012). A wireless sensor network solution for precision agriculture based on ZigBee technology. *Wireless Sensor Network, 4*, 25–30.

Laddhad, K. (2006). RFID data management. *M.Tech thesis submitted to IIT-Bombay*, India.

Landt, J. (2005). The history of RFID. *IEEE Potentials, 24*(4), 8–11. https://doi.org/10.1109/MP.2005.1549751

Li, S., Visich, J. K., Khumawala, B. M., & Zhang, C. (2006). Radio frequency identification technology: Applications, technical challenges and strategies. *Sensor Review, 26*(3), 193–202.

McFarlane, D., & Sheffi, Y. (2003). The impact of automatic identification on supply chain operations. *International Journal of Logistics Management, 14*(1), 1–17.

McMeekin, T. A., Baranyi, J., Bowman, J., Dalgaard, P., Kirk, M., Ross, T., Schmid, S., & Zwietering, M. H. (2006). Information systems in food safety management. *International Journal of Food Microbiology, 112*(3), 181–194.

Myny, K., Tripathi, A. K., van der Steen, J. L., & Cobb, B. (2015). Flexible thin-film NFC tags. *IEEE Communications Magazine, 53*(10), 182–189.

Nelson, S. O., & Bartley, P. G. J. (2000). Measuring frequency- and temperature-dependent dielectric properties of food materials. *Transactions of the ASAE (American Society of Agricultural Engineers), 43*(6), 1733–1736.

Ogasawara, A., & Yamasaki, K. (2006). A temperature-managed traceability system using RFID tags with embedded temperature sensor. *NEC Technical Journal, 1*(2), 82–86.

Onwude, D. I., Chen, G., Eke-Emezie, N., Kabutey, A., Khaled, A. Y., & Sturm, B. (2020). Recent advances in reducing food losses in the supply chain of fresh agricultural produce. *Processes, 8*(11), 1431.

Perez-Aloe, R., Valverde, J. M., Lara, A., Carrillo, J. M., Roa, I., & Gonzalez, J. (2007). Application of RFID tags for the overall traceability of products in cheese industries. In *1st Annual RFID Eurasia* (pp. 1–5). IEEE.

Raman, A., Dehoratius, N., & Zeynep, T. (2001). Execution: The missing link in retail operations. *California Management Review, 43*(3), 136–152.

RFID Update. (2006). Study: RFID vulnerable to viruses and worms. *RFID Update,* March 15. www.rfidupdate.com/articles/home.php. Accessed 13 March, 2021.

Rieback, M. R., Simpson, P. N. D., Crispo, B., & Tanenbaum, A. S. (2006). RFID malware: Design principles and examples. *Pervasive and Mobile Computing, 2*(4), 405–426.

Roberts, C. M. (2006). Radio frequency identification (RFID). *Computers & Security, 25*(1), 18–26.

Ruiz-Garcia, L., & Lunadei, L. (2011). The role of RFID in agriculture: Applications, limitations and challenges. *Computers and Electronics in Agriculture, 79*(1), 42–50.

Sarananda, K. H., Balasuriya, S. T., & Ganeshalingam, K. (2004). Quality of papaya variety 'Rathna' as affected by postharvest handling. *Tropical Agricultural Research and Extension, 7*, 72–78.

Shellhammer, T. H., & Singh, R. P. (1991). Monitoring chemical and microbial changes of cottage cheese using a full history time-temperature indicator. *Journal of Food Science, 56*(2), 402–405, 410.

Shobha, N. S. S., Aruna, K. S. P., Bhagyashree, M. D. P., & Sarita, K. S. J. (2016). NFC and NFC payments: A review. In *2016 International Conference on ICT in Business Industry & Government* (ICTBIG) (pp. 1–7). IEEE.

Shulman, R. (2001). Perishable systems take center stage. *Supermarket Business, 56*(4), 47–48.

Siuli Roy, D., & Bandyopadhyay, S. (2008). Agro-sense: precision agriculture using sensor-based wireless mesh networks. In *Proceedings of the 1st ITU-T Kaleidoscope Academic Conference, Innovations in NGN: Future Network and Services* (K-INGN '08), pp. 383–388.

Sri Lanka Climate Resilience Program. (2020). Climate resilient infrastructure mainstreaming in Sri Lanka. *Overview of Policies and Initiatives.* http://www.climatechange.gov.lk/

Tajima, M. (2007). Strategic value of RFID in supply chain management. *Journal of Purchasing and Supply Management, 13*(4), 261–273.

Taokis, P. S., & Labuza, T. P. (1989). Reliability of time-temperature indicators as food quality monitors under non-isothermal conditions. *Journal of Food Science, 54*(4), 789–792.

Tinker, J. H., Salvin, J. W., Learson, R. J., & Empola, V. B. (1985). Evaluation of automated time-temperature monitoring system in measuring the freshness of chilled food. *IIFIIR Commissions C2 D3, 4*, 286–290.

Trebar, M., Grah, A., Melcón, A., & Parreno, A. (2011). Towards RFID traceability systems of farmed fish supply chain software. In *Proceedings of the 19th International Conference on Software, Telecommunications and Computer Networks* (SoftCOM '11), pp. 6–11.

Uldrich, J. (2007). Dole: Let us thank RFID technology. *The Motley Fool.* www.fool.com/investing/value/rfidsaves-the-dole.aspx. Accessed 13 April, 2016.

United Nations. (2015). Transforming our world: The 2030 Agenda for Sustainable Development (A/RES/70/1). United Nations.

Ursa, Y. (2006). *Cold-trace: A mobile-based traceability solution rendering fleet management more effective.* E-Challenge.

Wang, F., & Liu, P. (2005). Temporal management of RFID data. In *Proceeding of Very Large Data Bases Conferences* (VLDB), 1128–1139.

Wang, L., Kwok, S. K., & Ip, W. H. (2010). A radio frequency identification and sensor-based system for the transportation of food. *Journal of Food Engineering, 101*(1), 120–129.

Wang, N., Zhang, N. Q., & Wang, M. H. (2006). Wireless sensors in agriculture and food industry - Recent development and future perspective. *Computers and Electronics in Agriculture, 50*(1), 1–14.

Want, R. (2006). An introduction to RFID technology. *IEEE Pervasive Computing, IEEE CS and IEEE ComSoc, 5*(1), 25–33.

Wasala, C. B., Dissanayake, C. A. K., Dharmasena, D. A. N., Gunawardane, C. R., & Dissanayake, T. M. R. (2014). Postharvest losses, current issues and demand for postharvest technologies for loss management in the main banana supply chains in Sri Lanka. *Journal of Postharvest Technology, 2*(1), 80–87.

White, G., Gardiner, G., Prabhakar, G., & Abd Razak, A. (2007). A comparison of barcoding and RFID technologies in practice. *Journal of Information, Information Technology and Organizations, 2*, 118–131.

Zhang, J., Tian, G. Y., Marindra, A. M., Sunny, A. I., & Zhao, A. B. (2017). A review of passive RFID tag antenna-based sensors and systems for structural health monitoring applications. *Sensors, 17*(2), 265.
Yang, Z., Zheng, K., Yang, K., & Leung, V. C. (2017). A blockchain-based reputation system for data credibility assessment in vehicular networks. In 2017 IEEE 28th annual international symposium on personal, indoor, and mobile radio communications (PIMRC) (pp. 1–5). IEEE.
Dedy Irawan, J., Adriantantri, E., & Farid, A. (2018). RFID and IOT for attendance monitoring system. In MATEC Web of Conferences.

The Resilience of Agro Ecosystems Through Higher Education; Identification of Key Thematic Areas

Asitha De Silva, Dilanthi Amaratunga, Kinkini Hemachandra, I. U. Rodrigo, G. Y. Jayasinghe, U. T. G. Perera, and D. M. N. A. Dissanayake

1 Introduction

Agro ecosystems are natural ecosystems that have been modified to produce food while retaining many of the characteristics of natural ecosystems (Hodson, 2012). Nevertheless, Agro ecosystems are threatened by climate change and other anthropogenic factors, for example, shifting of job markets, urbanization, and labour shortages in agricultural production across the world (Hoy, 2015; Liebman et al., 2015). Hence, agro ecosystem resilience has been identified as a global priority for assuring global food security. For example, Agro ecosystems' resilience links the Sustainable Development Goals (SDGs) across several goals including goal 2, 3, 11, 13 and 15 (United Nations, 2015). Further, assuring agro-ecosystems' resilience plays a major role in these planned development targets.

However, the increasing decline in agro-ecosystems have presented urgency for new ways of managing agricultural and natural systems. Dussi et al. (2019) have identified the importance of higher education as a major factor increasing agro-ecosystem's resilience. Although the previous studies have developed several frameworks for agro-eco system's resilience (Cabell et al., 2012), the role of the higher education institutes on agro-eecosystems' resilience has not been properly discussed in the current literature. Antoni et al. (2019) also explain this issue concerning the tropical agro-ecosystems in the world which have been severely impacted in the last few decades.

G. Y. Jayasinghe · U. T. G. Perera (✉)
University of Colombo, Colombo, Sri Lanka
e-mail: thisara@geo.cmb.ac.lk

A. De Silva · D. Amaratunga · K. Hemachandra · U. T. G. Perera
University of Huddersfield, Huddersfield, UK

I. U. Rodrigo · G. Y. Jayasinghe · D. M. N. A. Dissanayake
University of Ruhuna, Mapalana, Sri Lanka

C. Malalgoda et al. (eds.), *Climate Change Adaptation in the Built Environment*,
https://doi.org/10.1007/978-3-031-75826-3_23

Therefore, it is essential to understand the role and the perspectives of higher education in agro-ecosystems' resilience, and this perspective assumes a different course of action in management. The existing knowledge on tropical agro-ecosystems could be further enhanced by identifying alternative approaches from the global context. Moreover, higher education institutions could also contribute to agro-ecosystems' resilience through their expertise and capacities in conducting research. Thus, a framework will facilitate evaluating opportunities and challenges associated with higher education institutions for contributing to agro-ecosystems' resilience.

2 Methodology

This research aims to develop a framework to understand the key concepts of agro-ecosystems' resilience through the higher education system. Under the research aims, this study cover the objective of identifying the global context of tropical agro-ecosystems' resilience. To achieve this objective, this study adopted a qualitative method as the study's methodology. An extended literature review was conducted to cover the global context of tropical agro-ecosystems' resilience. The literature review used academic databases and policy documents. Academic data was extracted as a combination of systematic and snowball methods. Accordingly, three databases were used for conducting the extended literature review; Scopus, Science Direct and Emerald. The snowball method was adopted to investigate further the background and the practices related to Agro ecosystems resilience where it was necessary. In order to achieve research objectives, search strings were developed, identifying the key terms and their related terms. Accordingly, the following search strings were used to conduct the systematic literature review.

"Agro ecosystems" OR "Tropical Agro ecosystems" OR "Rice-based Agro ecosystems" AND "Resilience" OR "Ecosystem resilience" OR "Agro ecosystems resilience" AND "Higher education" OR "Higher education institutes" OR "Higher education teaching".

To limit the number of documents used in the review, several inclusion and exclusion criteria were adopted. The inclusion criteria were set as follows;

- Research papers, book chapters, journal articles
- Search terms either in the title, abstract or the keywords
- Papers are written in English and
- Articles related to one or more domains of Agro ecosystems and higher education.

In contrast, the exclusion criteria were set:

- Reviews, meta-analyses and commentaries
- Articles not written in English.

After applying the inclusion and exclusion criteria, 88 papers were shortlisted for review from all three databases. Next, the repetitive articles among the three databases were manually excluded, and finally, 46 documents were used for conducting the systematic literature review.

Regarding policy documents, four global policy documents that were related to Agro ecosystems resilience were studied. The documents are the 2030 Agenda for Sustainable Development, the Sendai Framework for Disaster Risk Reduction, the Paris Agreement and the New Urban Agenda.

The finalised data sources were analysed using thematic analysis to identify key themes. Accordingly, the following section presents the key thematic areas identified by the systematic literature review. One hundred fifty-one concepts were identified from the literature review and grouped into 21 categories. The 21 categories were further grouped into five key areas of resilience: Agro ecosystems, environment, policy and governance, and higher education. After the development of the initial framework, a questionnaire survey and a round table discussion were planned to collect and validate the framework. The initial framework comprises three layers; the first layer consists of five key areas, the second layer consists of 21 concepts, and the third layer consists of another 125 concepts as extensions from the second layer.

3 Results and Discussion

As explained above, the results were illustrated in the framework in Figs. 1 and 2. A bottom-up approach was adopted to identify the key areas of the framework. Accordingly, Fig. 1 demonstrates the complete BRIATE framework with 151 concepts and all three layers while Fig. 2 demonstrates the finalized BRITAE Framework with the first two layers, According to Fig. 1, the five thematic areas in the first layer were presented in detail in the following section.

3.1 Resilience

Resilience is the first thematic area identified by the study. Resilience refers to the capacity to recover. However, under the BRITAE framework, it is mainly targeted at agro ecosystems resilience. Four key areas were identified as important in making agro ecosystems resilience through higher education; socio-ecological resilience, sustainability, disaster risk, and climate change.

3.1.1 Socio-ecological Resilience

Socio-ecological resilience refers to the socio-ecological systems' ability to cope with the changes (Cote & Nightingale, 2012) caused by climate change, natural hazards, or even man-made hazards (Liu et al., 2016). The environmental system approach could be integrated for socio-ecological resilience. This system understands

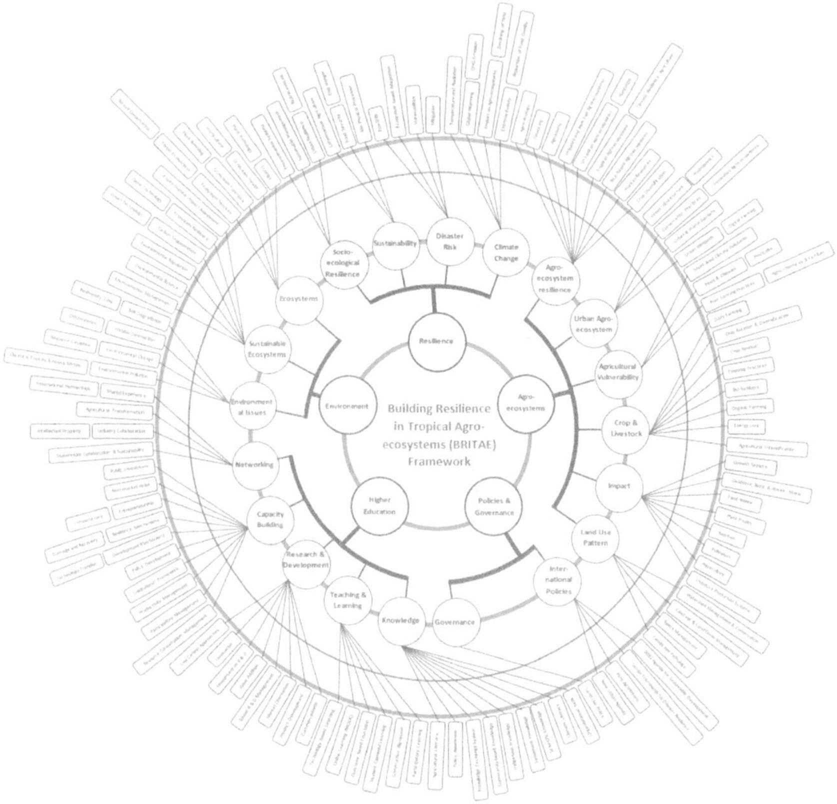

Fig. 1 Completed BRIATE framework: three tiered framework with 151 concepts

the earth as a system consisting of interconnected microsystems (Bennett & Chorley, 2015). Similarly, sustainable management was identified as another contributor to socio-ecological resilience. It integrates concepts of sustainability and management (Tregidga & Milne, 2006). Sustainable management concerns conservation and preservation of process or the system while increasing its productivity and efficiency (Guthrie et al., 2010).Urban resilience is targeted through green and blue infrastructure initiatives, which focus on restoration and preservation of natural environmental processes for the survival of humans (Zuniga-Teran et al., 2020) and structural engineering methods and ecosystems based adaptation measures towards urban resilience (Brink et al., 2016).

3.1.2 Sustainability

Sustainability focuses on managing and improving human development while sustaining the natural systems, ecosystems and natural resources by conservation

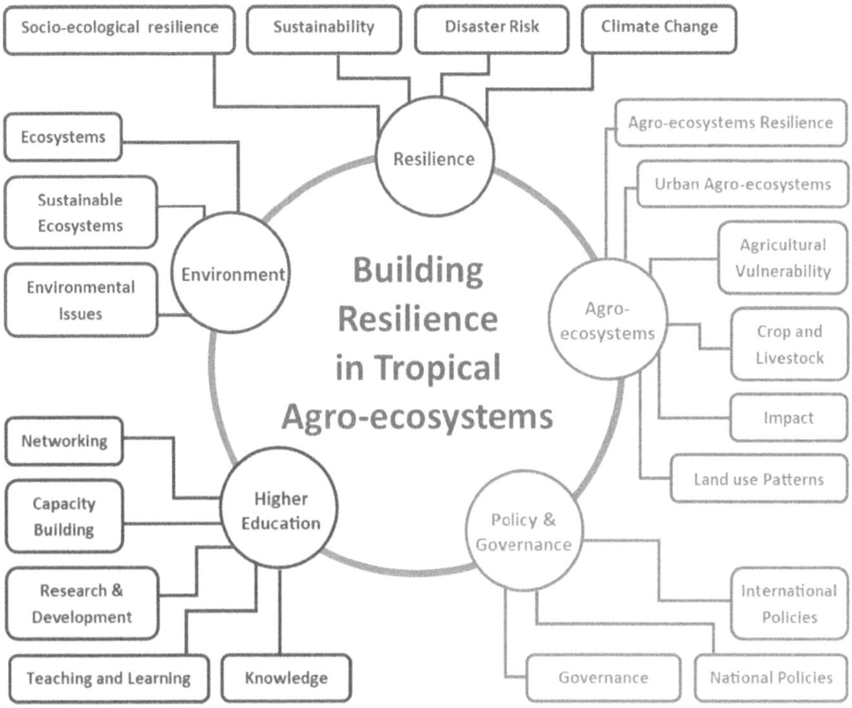

Fig. 2 Key thematic areas of the BRITAE framework

and preservation (Werbach, 2011). Conservation agriculture is one such approach that focuses on preserving and conserving natural resources, biodiversity, soils, and water while increasing the productivity of the harvest (Hobbs et al., 2008). Agriculture is significant due to its connectivity with food security and hunger. Hence, Current sustainable development agendas search for new approaches and methods to increase food security through smallholder farming practices (Mango et al., 2014), household garden practices (Galhena et al., 2013), and even urban agricultural practices (Opitz et al., 2016). Biological and physical processes are key ingredients for sustainability (Pickett et al., 2014).

3.1.3 Disaster Risk Reduction

Another significant concept related to socio-ecological resilience uncovered from the literature review was disaster risk reduction (DRR). Agro ecosystems resilience has functioned as a disaster risk mitigation measure combining with ecosystems based adaptation measures (Ayyam et al., 2019). For example, introducing seasonal crops into a steep slope not be the best solution for landslide mitigation. It may need a hybrid approach between soft and hard engineering mechanisms (Onuma & Tsuge,

2018). Similarly, eco-DRR is another approach for disaster risk reduction and mitigation mechanisms (Gupta & Nair, 2012). Creating space for carbon sequestration, integrating water features to reduce urban heat, public open green spaces, and urban wetlands are some of the measures adopted and Eco-DRR and green and blue infrastructure perspectives (Andersson et al., 2019). However, disasters risks could not be fully mitigated (Stein & Stein, 2014). Agro ecosystems resilience could be introduced as a disaster risk mitigation measure with a combination of ecosystem-based adaptations (Ayyam et al., 2019).

3.1.4 Climate Change

The fourth concept related to resilience is climate change. Climate change is the long term change in the existing weather patterns of the world (Turner & Annamalai, 2012). Agro ecosystems are highly sensitive to climate change (Hamouda & Elhabil, 2017; Mera et al., 2006) and become one of the key concerned areas of the decision-makers (Várallyay, 2007). Therefore, current agriculture practices focus on reducing GHG emissions and strengthening Agro ecosystems resilience (Falkenmark et al., 2007). Because greenhouse gases affect the Agro ecosystems through temperature and radiation. Temperature and radiation levels affect plant growth stages and declining yields (Craparo et al., 2015). Daily temperature fluctuations in tropical and subtropical regions are directly responsible for yield loss (Ashraf & Michalopoulos, 2015). It is clear that climate change directly threatens global food security (Brown & Funk, 2008).

3.2 Agro Ecosystems

The second key thematic area identified within the first layer of the framework is Agro ecosystems. The term Agro ecosystems is derived from ecosystems primarily focus on a coherent unit of agricultural activities in a specific location (Oliver et al., 2015). Agro ecosystems are among the key topics covered in sustainable development to achieve food security and end hunger (Khumairoh et al., 2012). Therefore, many development planners concern about Agro ecosystems resilience, which has been threatened by climate change, population growth, global warming, and urbanization (El-Sharkawy, 2014). On top of that, increasing disasters due to climate change and global warming have also adversely affected existing Agro ecosystems (Cabell & Oelofse, 2012).

3.2.1 Agro Ecosystems Resilience

For a resilient Agro ecosystems, biological diversities and richness have become primary factors (Wania et al., 2006). Among such practices, agroecology assures the best use of nature and its services for food production without harming the natural resources (Altieri, 2002). Agroecological concepts are applicable in some farming practices, such as organic, integrated, conventional, intensive or extensive (Wezel et al., 2009). For agoecosystems resilience, agronomy is similarly important. Agronomy uses science and technology to produce food, fuel, fibre, and land restoration (Sumberg, 2017). In addition to the use of science and technology in agriculture, the human resource also plays a major role in Agro ecosystems resilience. For developing human resources related to Agro ecosystems, the focus should be given to basic education in agriculture, agricultural extension and training, and technical and professional education in agriculture (Spielman et al., 2008).

Agricultural activities are conducted based on irrigated and/or rain-fed Agro ecosystems (Thebo et al., 2014). For example, rain-fed agriculture is commonly used in paddy cultivation in Asia, based on the two monsoon rainfalls (Ismail et al., 2013). In Asia, rice-based Agro ecosystems are prominent (Settele et al., 2018). South Asia, East Asia and Southeast Asia are the largest rice produces in the world (Silva et al., 2015). Asia being one of the most vulnerable regions for climate change, the impact on rice-based Agro ecosystems is significantly high (Alam et al., 2011). Under Agro ecosystems resilience, plantations are another sector resilience to be introduced because of their risk due to climate change (Fonta et al., 2011). However, plantations are more industrial oriented due to their mono-cropping system and become highly sensitive to climate change (Fonta et al., 2011). Besides, tropical Agro ecosystems are another agricultural output with more diverse and rich agricultural ecosystems (Andres & Bhullar, 2016) and hence considered as the global climate engine (Laliberté et al., 2015).

3.2.2 Urban Agro Ecosystems

Rapid urbanization is visible in every part of the world due to population growth (Buhaug & Urdal, 2013). Due to urbanization, many cultivated lands are being lost for agriculture, threatening global food security (Pandey & Seto, 2015). By incorporating green and blue infrastructure, urban Agro ecosystems have revolutionized the traditional thinking of agriculture where everyone can productively involve in agricultural production (Lin et al., 2017). Green infrastructure methods have been used as a climate change mitigation and disaster risk reduction strategy in urban areas (Staddon et al., 2018). Green walls, green roofs, wildlife overlaps biodiversity parks, and fish ladders are some of the practices adopted under green and blue infrastructure (Manual et al., 2015). Advancing from traditional practices, vertical gardening (Specht et al., 2014), hydroponic methods (Rufí-Salís et al., 2020), indoor cultivation techniques (R Shamshiri et al., 2018) are introduced to make urban Agro ecosystems more resilience. Urban gardens and home gardens have become popular among communities recently due to the Covid pandemic situation, focusing on growing their food (Lal, 2020). Among such novel farming methods, hydroponic is identified as a technology or horticulture practice with no soils (Jones, 2016). Plants are grown with only roots exposed to water where the nutrients are added (Naik et al., 2015).

Wetlands are also useful for flood protection, water quality enhancement, food chain support and carbon sequestration (Verhoeven & Setter, 2010). When it comes to urban wetlands, they are significantly important with disaster risk reduction, especially in flood mitigation (Furuta, 2019).

3.2.3 Agricultural Vulnerability

Due to climate change, many Agro ecosystems have become vulnerable directly and indirectly (Mougou et al., 2011). Short-lived climate pollutants is one such condition making agricultural vulnerabilities. Due to progressive increases of ozone and black carbon layers in the atmosphere, agricultural production is affected through direct effects on crops and indirect effects on climate (Fuhrer & Booker, 2003). Rising temperature, the number, and insect pests' appetite increase Agro ecosystems vulnerabilities (Tubby & Webber, 2010). Similarly, rising temperature and climate change also increase the occurrence, prevalence, and severity of plant diseases (Duku et al., 2016). Therefore, pesticides have become much popular since it controls pest infestations (Sattler et al., 2007). However, the use of agrochemicals, pesticides and fertilizers, have harmful effects on human health (Aktar et al., 2009) and the environment (Mahmood et al., 2016). These pesticides will eliminate the pests and all living organisms, which are very important for cultivation (Chowdhury et al., 2008).

3.2.4 Crop and Livestock

Crop and livestock farming was identified as a key aspect under the Agro ecosystems' resilience theme. Crop and livestock combine animal and crop farming as agricultural production (Wright et al., 2012). Dairy farming is part of agricultural production (Rosati & Aumaitre, 2004) and is significantly important to community wellbeing (Alvarez-León et al., 2006). Developing rural households practice crop and livestock mixed methods to make their lives more resilient for survival (Sempore et al., 2016). For resilience Agro ecosystems, crop rotation and diversification have been considered important (Bowles et al., 2020). Monoculture practices are considered as poor farming practices since they degrade topsoils and soil nutrients (Feintrenie et al., 2010). Where there are diversified crops, the nutrient balance will be preserved (Li et al., 2019).

Cropping practices/cropping systems have also been practiced for long ago (Borchers et al., 2014). The cropping practices focus on crop types, crop rotation and diversification, residue management, crop nutrient, and water management on agricultural land (Niblack & Chen, 2004). Adopting good cropping practices positively impacts food security and environmental sustainability (Matthews et al., 2013). Another strategy for increasing agricultural resilience is agricultural intensification. It refers to increases in agricultural production per unit of inputs (Rudel et al., 2009).

Biofertilizers are a natural way of feeding plants and currently a much popular agricultural practice worldwide (Akram et al., 2020).

3.2.5 Impact on Agro Ecosystems

Agro ecosystems directly linked with global food security (Hillel & Rosenzweig, 2011). Agro ecosystems are directly affected by changing temperature and precipitation levels and indirectly by natural hazards, as stated earlier. Such changes affect the growth seasons (Mohanraj et al., 2016; Vlam et al., 2014), which are significantly important in cultivating seasonal crops (Cornelissen et al., 2018). In addition, plants are subjected to environmental stresses that reduce and limit crops' productivity (Ahmad et al., 2009). Two types of environmental stresses; biotic stress and abiotic stress (Rejeb et al., 2014). Biotic stresses are the biological factors that reduce the agricultural productivity by various pathogens such as fungi, bacteria, oomycetes, nematodes and herbivores (Gull et al., 2019), while abiotic stresses are the physical conditions that effects on agricultural productivity such as radiation, salinity, floods, drought, extremes in temperature, heavy metals, etc. (Cramer et al., 2011).

Climate change also affects the pollination process (Memmott et al., 2007). Up to 35% of the global crop production depends on pollinator-dependent crops (Potts et al., 2016). Pollination occurs in a mutualistic relationship between an animal and a plant. Climate change further affects livestock systems mainly through rising temperatures and changes in precipitation too (Rojas-Downing et al., 2017).

The ultimate goal of Agro ecosystems is to contribute to global food security. Nevertheless, some countries report food waste while others suffer from food shortages (Hegnsholt et al., 2018; Loewenberg, 2015). Even within countries, such disparities could be evident (Saxena, 2018). Foods are wasted during production (Hafid et al., 2017), process, retailing (Swenson, 2008) (Messner et al., 2020), distribution (Plazzotta et al., 2017) and consumption (Reynolds et al., 2019). It is estimated that global food waste is more than one-third of global food production (Lipinski et al., 2013).

3.2.6 Land Use Patterns

Land patterns are changing rapidly all over the world due to development activities (Waddell, 2002). Changing land-use patterns sometimes disrupt the balance of natural environmental processes, resulting in disasters (Khan et al., 2015). Under the land-use changes, degradation of wetlands (Jiang et al., 2017), settlement congestion in flood plains (Taş et al., 2013), reduction of rain forests (Chaudhary et al., 2016), are some of the burning issues faced by many countries.

Therefore, watershed management and conservation have been identified as significant for improving water quality and other natural resources (Alemayehu et al., 2009). This will allow anyone to use the land and water resources comprehensively while conserving for the future. Similar to watershed management, land use management also plays a major role in Agro ecosystems. Land use is the way the land has been utilized for anthropogenic activities. Properly planned land use is key to sustainable agriculture (Schilling et al., 2008). Land use and land cover management practices will develop the best practices for sustainable development, including agriculture and disaster resilience (Banba & Shaw, 2017

In the field of agriculture, the geotagging mechanism is adopted. Geotagging refers to adding geographical identification markers like latitude and longitudes for agricultural lands (Kanchi et al., 2019). This will support decision-makers to find a wide range of location-specific information without going to the fields (Kanchi et al., 2019). Satellite technology is another strategy introduced into agriculture to increasing productivity and resilience to changing world (Teixeira et al., 2016). It has been used for detecting and controlling pests and disease, understanding water and nutrient status, planning crop nutrition programmes, informing in-season irrigation, predicting yields and estimating harvest timing (Onojeghuo et al., 2018).

3.3 Environment

Under the five key areas of the first layer of the BRITAE framework, the environment was identified as the third key thematic area. This is because Agro ecosystems threaten the natural environment due to deforestation and environmental pollution caused by agricultural practices (Olanipekun et al., 2019). Accordingly, the following related sub-elements were identified within the concept environment.

3.3.1 Environmental Issues

Development has led to rapid environmental pollution leading to climate change, global warming, and sea-level rise as consequences (Shukla et al., 2017). Deforestation is one such anthropogenic activity challenging the environment. During the development processes, deforestation occurs primarily to obtain land for agriculture, construction, or wood for timber or fuel (Geist & Lambin, 2001). Agricultural deforestation is significant, especially in developing nations where rural farmers encroach on forests and shrubs for their cultivation. Hence, Agro ecosystems resilience focuses on habitat restoration, preservation and conservation based on flora, with an agricultural value (Sajjad et al., 2018). Agriculture has also become a key contributor to biodiversity loss by converting natural habitats to intensely managed systems and releasing pollutants, including GHGs (Dudley & Alexander, 2017). Such impact of agriculture finally affects large scale environmental change (Parr et al., 2003). With the population rise, the environmental change was further accelerated (Li & Ma, 2014).

To fulfil the agricultural output for the rising population, increasing agricultural lands' productivity becomes a major concern among farmers. Thus, farmers use more fertilizers and pesticides, which causes environmental pollutions again by adding toxicity content in the soil (Gomiero et al., 2011). Even though farmers had traditional knowledge on environmentally friendly mechanisms to control pests and increase agricultural productivity (Dubey et al., 2010), with technological advancement, that knowledge has been depleted (Reyes-García et al., 2014).

3.3.2 Ecosystems

Another key concept identified under the environment theme is the ecosystem. Ecosystem refers to the extent of space or a geographical area where animals, plants and other biotic and abiotic factors function together for survival (Begon & Townsend, 2020). Ecosystems services provide services for human survival (Daily & Matson, 2008). Nutrient recycling, photosynthesis, and soil formation are some of the supportive services provided by the ecosystems (Eldridge & Delgado-Baquerizo, 2017).

Ecosystems, including forest ecosystems, are currently under the threat of degrading primarily due to anthropogenic activities (Führer, 2000). Agriculture is also a key contributor to the degradation of forest ecosystems, especially in tropical regions (Boerner et al., 2007). Hence, sustainable agricultural practices focus on nature conservation through low carbon emission, reduction of agro-chemicals, reduction of tillage, preserve biodiversity, mix cultivation practices, use of organic matter (Hobbs et al., 2008). Such practices help secure ecosystems health, which is directly linked with ecosystems, ecosystems services, natural capital, and biodiversity.

3.3.3 Sustainable Ecosystems

Many international policies have emphasised the significance of sustainable ecosystems to maintain and balance anthropogenic activities in the development process (Primmer et al., 2015). Specifically, ecosystems resilience has become significant for Agro ecosystems (Cabell & Oelofse, 2012). Ecosystems resilience refers to ecosystems' capacity to deal with disturbances, either by resisting them, recovering from them, or adapting to them, whilst retaining their ability to deliver services and benefits now and in the future. Hence, environmental scientists have pointed out the need to balance the inputs and outputs, which is recognised as the environmental balance. However, the environmental balance is disrupted by development activities and natural hazards (Paul, 2011).

Therefore, to restore the environmental balance, environmental regulations are introduced. Such regulations are adopted internationally, regionally, and local to protect the environment. There are several types of environmental regulations; command and control, economic instruments, co-regulation and voluntary initiatives, self-regulation, and education and information (Haq & Schwela, 2008). Environmental Impact Assessment is another mandatory requirement of all development activities to assess the environmental consequences of a plan, policy, program, or actual project before deciding to move forward with the proposed actions (Glasson & Therivel, 2013). In addition to these regulatory mechanisms, some environmental practices have been introduced. Clean technology/green technology is one such mechanism introduced. It refers to a set of technologies that either reduce or optimize natural resources whilst simultaneously reducing the negative effect that technology has on the planet and its ecosystems (Acemoglu et al., 2016).

3.4 Policies and Governance

Agricultural growth and comprehensive food and nutrition security could not be achieved without increasing resilience with legislative and policy environment support. The following subsections explain the international, national and governance related to Agro ecosystems resilience.

3.4.1 International Policies

Proper guidance is a much-needed aspect when developing sustainable agricultural practices and Agro ecosystems that are resilient to adverse impacts. 2030 Agenda for Sustainable Development is committed to achieving sustainable development in its three dimensions, economic, social and environmental, in a balanced and integrated manner. Food and agriculture are key to achieving the entire set of SDGs. Investing in agriculture crops, livestock, forestry and fisheries—and rural development are powerful tools to end poverty and hunger and open the door to sustainable development (United Nations, 2015). The Sendai Framework for Disaster Risk Reduction (SFDRR) 2015–2030 is another global policy that governs global affairs against disaster risks. The framework is recognized as a human-based and bottom-up approach that encourages related institutions to develop and implement sustainable policies that support evaluating disaster risks and disaster monitoring and implementation (Busayo et al., 2020).

The third global policy relevant to Agro ecosystems is the Paris agreement. It is a legally binding international treaty on climate change to prevent GHG emissions and global warming. Changing climate due to GHGs and global warming has devastative impacts on Agro ecosystems and global food production (Guiot & Cramer, 2016). The agreement consists of two sections focused on food security and ending hunger to build community resilience. The New Urban Agenda also represents pivotal importance towards bridging urban and rural communities in the planning and development of food systems (FAO, 2020).

3.4.2 Governance

Governance is essential when it comes to the implementation and continuation of the tasks proposed by decision-makers. Therefore, policies and laws are applicable to assure governance within the governing process. Most Sri Lankans depend on agriculture for their livelihood. Improper land use could lead to various human, biological and environmental problems. Hence, land use planning should be the prime concern to solve many land use-related issues. According to available statistics, nearly 44% of agricultural lands have been subject to land degradation. A large number of fragmented agricultural lands, which are unproductive, have contributed to this land degradation. Similarly, a large extent of agricultural land in plantations areas are underutilized. Therefore, the National Land Use Policy established in 2007 by the Ministry of Lands focused on rational utilization of lands as a resource to ensure food security, a high quality of life, equity and ecological sustainability (National Land Use Policy, 2007). Therefore, the National Land Use Policy could be considered as a significant legislative measure securing the governance in the agricultural policy.

3.5 Research in Higher Education Institutions

Higher/tertiary education is vital for knowledge creation and dissemination. Hence, for building Agro ecosystems resilience, higher education plays a major role. Thus, the following sub-themes were identified as relevant concepts for higher education in Agro ecosystems resilience.

3.5.1 Knowledge

According to the literature review, the knowledge of climate science literacy, agricultural literacy, and indigenous knowledge is vital to building resilience in tropical Agro ecosystems. Climate literacy is an understanding of the influence of humans on climate and climate's influence on humans and society. A climate-literate person understands the essential principles of Earth's climate system, how to assess scientific climate information, how to communicate about climate and climate change in a meaningful way, and how to make informed and responsible decisions (Miléř & Sládek, 2011). Similar to climate literacy, agricultural literacy is further significant for Agro ecosystems knowledge. The knowledge and/or understanding of our food and agriculture is referred to as agricultural literacy. (Frick et al., 1991). Another aspect of knowledge is traditional or indigenous knowledge, the knowledge systems embedded in the cultural traditions of regional, indigenous, or local communities. Traditional knowledge includes types of knowledge about traditional technologies of subsistence, for example, tools and techniques for hunting or agriculture. By intentionally connecting academic standards to the real world, higher educational institutes could lower the gap between knowledge and action that can contribute to the community (Melaville et al., 2006).

Scientists and researchers around the world believe that science and research can influence sustainability through expertise, know-how, and abilities. In this way, science and research could help feed, cultivate, shelter, educate, and employ the world's growing human population while preserving the world's essential life-support and biodiversity structures (Strigl, 2003). This could be aided by systemic knowledge, which is the holistic interpretation of interpersonal expectations or norms, technological structure, and relationships (Sheffield, 2005). Furthermore, awareness of government policies is critical in higher education because it can have a huge impact on the capacity building of higher educational institutes relevant to building resilience in Agro ecosystems.

3.5.2 Teaching and Learning

Several methods of teaching and learning are found in the literature and covered in the framework. Participatory learning is one such teaching method where Collaboration is used as a technique within participatory learning. Collaboration promotes working effectively and respectfully with diverse teams, exercising flexibility and a willingness to make compromises to accomplish a common goal. It assumes shared responsibility as a principle for collaborative work while valuing an individual's contributions (Seel, 2011). In contrast, student-centred learning is recognised as a teaching approach that substitutes active learning and incorporates self-paced learning programs and cooperative community scenarios. Hence, learners develop higher levels of critical thoughts, problem-solving, enhancing learning habits, and increasing overall participation (Overby, 2011).

Nevertheless, the present teaching and learning activities have taken an unprecedented shift, reducing one-on-one traditional classroom activities to online teaching and learning methods using technology. Among such online teaching and learning methods, Massive Open Online Courses (MOOCs) have gained considerable attention from the academic community over the past years (Gaebel, 2014). In addition, the new ICT-based tools such as Zoom, MS Teams, Google Meet are largely used for teaching and learning activities throughout the world, especially due to the current COVID-19 pandemic.

3.5.3 Research and Development

Universities, technical colleges, research institutions, vocational training institutions and other higher education institutions constantly generate new knowledge by conducting research and development studies. A country's commitment to research and development is measured as a percentage of the Gross Domestic Product (GDP) allocated to research and development (R&D). However, in Sri Lanka, the present allocation of GDP on research and development is only 0.11% of GDP, far below that of other countries in the region and the world (The World Bank, 2021). As a result, new product development has received relatively little attention in most of the higher educational institutes in Sri Lanka, leading to a knowledge gap in policymaking and academic research (Durkin & McKenna, 2011).

As a solution, a market orientation approach is encouraged, with the goal of developing and implementing effective organisational partnership strategies (Hemsley-Brown & Oplatka, 2010). Market orientation could help higher education institutions address many of their challenges (Akonkwa, 2009). This approach has proven to be effective in developing countries. They regard the commercialization of their research as the foundation for the increment base (Elmi, 2009). This has been regarded as the added value of research and development. As a result, value addition has been widely promoted as a strategy for gaining a competitive advantage and creating consumer value in a hostile market environment (Nagarajan et al., 2013). In this process, it has become increasingly important for companies/organizations to optimize the effectiveness of research and development, specifically in engineering and technology-centred fields.

3.5.4 Capacity Building

Due to rising student enrolment, HEIs face the challenge of managing institutional resources to meet the rising demand. (Sempebwa, 2011). Specifically, to improve research capacities within universities, supporting infrastructure facilities become vital. Research centres and IT facilities are major infrastructure facilities applicable for developing the capacity building in HEIs (Aithal, 2016). However, due to the increasing and competing demand for public investments, universities receive limited allocations for capacity building. Hence, additional funds are being sourced through project proposals and other direct funding strategies (Pruvot et al., 2015).

Capacity building is further related to performance and productivity. Measuring productivity in the higher education sector is not straightforward. Some researchers argue that productivity within higher education is related to student's quality (Lewis & Dundar, 1999). One way to elevate productivity in the higher education sector is through administrative and legislative procedures. The development of policies is an essential aspect when defining organizational goals and providing guidance to achieve them, ensuring the organization's processes, improving overall organization's performance and efficiency and clarifying what needs to be done (Canada Mortgage & Housing Cooperation, 2018).

Participatory management is another strategy to enhance capacity building and productivity through employee's expertise and experience (Shagholi & Hussin, 2009). Hence, many organizations have shifted to practice Human Resource approaches and participatory management concepts in enhancing productivity in HEIs. In addition, many policies are introduced to improve institutional frameworks as a strategy to ensure academic quality and teaching excellence (Dill & Beerkens, 2013).

Several studies have shown how development mechanisms can be linked to the capacity building of HEIs. Capacity development in higher education institutions could be targeted to help achieve development goals. Capacity-building programmes, for example, could be targeted to contribute to the implementation of Clean Development Mechanisms as a strategy to reduce global warming (Subbarao & Lloyd, 2011). In this context, technology transfer aids in the dissemination of basic understanding, information, and innovation from scientific and technological research conducted by a university, institute, or government laboratory to a broader audience, which includes individuals or firms in the private and quasi-private sectors (Liu & Jiang, 2001). Resilience, likewise, has emerged as a viable framework for integrating humanitarian and long-term development initiatives (FAO, 2018). When it comes to connecting development goals, resilience is crucial.

As a result of capacity building programmes conducted by HEIs, entrepreneurs can be generated. Entrepreneurship generates new value or economic gain. Specifically, agro entrepreneurs initiate, take risks, and identify opportunities in uncertain agricultural business environments (Tripathi & Agarwal, 2015). Addo (2018) emphasised that the participation of young graduates in agriculture as agro-preneurs is a must when addressing technological and digital revolution in agro food systems, changes in food consumption patterns, and environmental changes in the agricultural sector. In addition to developing agropreneurs, HEIs contribute to valuing non-market services, such as valuing ecosystems services. Ecosystems provide services that are difficult to measure in monetary terms and are referred to as non-market services (Costanza et al., 1997).

3.5.5 Networking

Networking is the exchange of information and ideas among people with a common profession or special interest, usually in an informal social setting. Among such networks, university-industry links are prominent. Different dimensions of these university-industry links where individuals and teams from universities and industries work together on specific projects and produce common outputs. Interaction of higher education system and industry with the main aim to boost knowledge and technology exchange is known as the university-industry collaboration (Ankrah & Omar, 2015).

The primary goal of international partnerships by universities is to elevate the impact of research and development activities; by joint research, co-authoring of research papers, resource sharing through joint research grants and research equipment sharing, and strengthening students' education. (Hamdullahpur, 2020). A sound legal background should assure the ownership of innovative and creative work for stronger international partnerships. Intellectual property rights play a major role in this regard (Haugen, 2020). In addition, experience sharing is also important for stronger collaborations and networking (de Wit-de Vries et al., 2019).

4 Conclusions

Agro-ecosystems are significant due to many reasons, as explained in the introduction. Nevertheless, they face many challenges within the rapid globalisation, rising population and changing climatic context. The increasing challenges demand innovative practices for securing agro-ecosystems' resilience. The paper provides an initial framework to support agro-ecosystems' resilience identifying the key concepts where strategies could be introduced. The framework has pointed out the prominent links to tropical agro-ecosystems' resilience. The initial framework will be tested by conducting a survey among high-education institutions in Sri Lanka and later on will be validated by conducting a roundtable discussion. Accordingly, the next step would be to evaluate the status of Sri Lankan agro-ecosystems' resilience, particularly within the higher education institutions in Sri Lanka. Therefore, the framework will guide the HEIs in assessing their capacities, strengths and challenges for contributing to the agro-ecosystems' resilience. In addition, the framework further provides inputs for developing a novel curriculum targeting building resilience in tropical agro-ecosystem in Sri Lanka as a joint programme.

Acknowledgements This work is funded by the Erasmus +Programme of the European Commission under the call on Education, Audiovisual and Culture Executive Agency (International Capacity Building—Joint Projects). The grant No. 610012-EPP-1-2019-1-LK-EPPKA2-CBHE-JP). The European Commission support for the production of this publication does not constitute an endorsement of the contents which reflect the views only of the authors, and the Commission cannot be held responsible for any use which may be made of the information contained therein.

References

Acemoglu, D., Akcigit, U., Hanley, D., & Kerr, W. (2016). Transition to clean technology. *Journal of Political Economy, 124*(1), 52–104.

Acquaah, G. (2008). *Horticulture: Principles and practices*. Pearson education, Inc.

Addo, L. K. (2018). Factors influencing Agripreneurship and their role in Agripreneurship Performance among young Graduate Agripreneurs. *International Journal of Environment, Agriculture and Biotechnology, 3*(6), 268289.

Agrios, G. (2012). *Plant pathology*. Elsevier.

Ahmad, A., Diwan, H., & Abrol, Y. P. (2009). Global climate change, stress and plant productivity. In *Abiotic stress adaptation in plants* (pp. 503–521). Springer.

Aithal, P. S. (2016). How to increase research productivity in higher educational institutions–SIMS Model. *International Journal of Scientific Research and Modern Education (IJSRME), 1*, 447–458.

Akcali, B. Y., & Sismanoglu, E. (2015). Innovation and the effect of research and development (R&D) expenditure on growth in some developing and developed countries. *Procedia-Social and Behavioral Sciences, 195*, 768–775.

Akonkwa, D. B. M. (2009). Is market orientation a relevant strategy for higher education institutions? Context analysis and research agenda. *International Journal of Quality and Service Sciences.*

Akram, M. S., Cheema, M. A., Waqas, M., Bilal, M., & Saeed, M. (2020). *Role of bio-fertilizers in sustainable agriculture.*

Aktar, W., Sengupta, D., & Chowdhury, A. (2009). Impact of pesticides use in agriculture: Their benefits and hazards. *Interdisciplinary Toxicology, 2*(1), 1–12.

Alam, M., Siwar, C., Islam, R., Toriman, M. E., & Basri, T. (2011). Climate change and vulnerability of paddy cultivation in north-west Selangor, Malaysia. *Academic Series of Universiti Teknologi MARA Kedah, 6*(1), 45–56.

Alemayehu, F., Taha, N., Nyssen, J., Girma, A., Zenebe, A., Behailu, M., Deckers, S., & Poesen, J. (2009). The impacts of watershed management on land use and land cover dynamics in Eastern Tigray (Ethiopia). *Resources, Conservation and Recycling, 53*(4), 192–198.

Alhorr, Y., Eliskandarani, E., & Elsarrag, E. (2014). Approaches to reducing carbon dioxide emissions in the built environment: Low carbon cities. *International Journal of Sustainable Built Environment, 3*(2), 167–178.

Altieri, M. A. (2002). Agroecology: The science of natural resource management for poor farmers in marginal environments. *Agriculture, Ecosystems and Environment Natural Resources, 93*(1–3), 1–24.

Alvarez-León, E.-E., Román-Vinas, B., & Serra-Majem, L. (2006). Dairy products and health: A review of the epidemiological evidence. *British Journal of Nutrition, 96*(S1), S94–S99.

Andersson, E., Langemeyer, J., Borgström, S., McPhearson, T., Haase, D., Kronenberg, J., Barton, D. N., Davis, M., Naumann, S., & Röschel, L. (2019). Enabling green and blue infrastructure to improve contributions to human well-being and equity in urban systems. *BioScience, 69*(7), 566–574.

Ankrah, S., & Omar, A. L. T. (2015). Universities–industry collaboration: A systematic review. *Scandinavian Journal of Management, 31*(3), 387–408.

Antoni, C., Huber-Sannwald, E., et al. (2019). Socio-ecological dynamics of a tropical agricultural region: Historical analysis of system change and opportunities. *Land Use Policy, 81*, 346–359. https://doi.org/10.1016/j.landusepol.2018.10.028

Ashraf, Q., & Michalopoulos, S. (2015). Climatic fluctuations and the diffusion of agriculture. *Review of Economics Statistics, 97*(3), 589–609.

Ayyam, V., Palanivel, S., & Chandrakasan, S. (2019). Natural disasters and coastal agro-ecosystems. In *Coastal ecosystems of the tropics-adaptive management* (pp. 175–198). Springer.

Banba, M., & Shaw, R. (2017). *Land use management in disaster risk reduction*. Springer.

Barrow, C. (2006). *Environmental management for sustainable development*. Routledge.

Begon, M., & Townsend, C. R. (2020). *Ecology: From individuals to ecosystems*. Wiley.

Bennett, R. J., & Chorley, R. J. (2015). *Environmental systems: Philosophy, analysis and control*. Princeton University Press.

Biggs, J. B. (2011). *Teaching for quality learning at university: What the student does*. McGraw-Hill Education (UK).

Blum, A. (2018). *Plant breeding for stress environments*. CRC Press.

Boerner, J., Mendoza, A., & Vosti, S. A. (2007). Ecosystem services, agriculture, and rural poverty in the Eastern Brazilian Amazon: Interrelationships and policy prescriptions. *Ecological Economics, 64*(2), 356–373.

Bolay, J.-C., Schmid, A., Tejada, G., & Hazboun, E. (2012). *Technologies and innovations for development: Scientific cooperation for a sustainable future*. Springer Science & Business Media.

Borchers, A., Truex-Powell, E., Wallander, S., & Nickerson, C. (2014). *Multi-cropping practices: Recent trends in double-cropping*.

Bowles, T. M., Mooshammer, M., Socolar, Y., Calderón, F., Cavigelli, M. A., Culman, S. W., Deen, W., Drury, C. F., y Garcia, A. G., & Gaudin, A. C. (2020). *Long-term evidence shows that crop-rotation diversification increases agricultural resilience to adverse growing conditions in North America*. One Earth.

Brink, E., Aalders, T., Ádám, D., Feller, R., Henselek, Y., Hoffmann, A., Ibe, K., Matthey-Doret, A., Meyer, M., & Negrut, N. L. (2016). Cascades of green: A review of ecosystem-based adaptation in urban areas. *Global Environmental Change, 36*, 111–123.

Bronson, K. (2019). Looking through a responsible innovation lens at uneven engagements with digital farming. *Journal of Life Sciences, 90*, 100294.

Brown, M. E., & Funk, C. C. (2008). *Food security under climate change.*

Buhaug, H., & Urdal, H. (2013). An urbanization bomb? Population growth and social disorder in cities. *Global Environmental Change, 23*(1), 1–10.

Busayo, E. T., Kalumba, A. M., Afuye, G. A., Ekundayo, O. Y., & Orimoloye, I. R. (2020). Assessment of the Sendai framework for disaster risk reduction studies since 2015. *International Journal of Disaster Risk Reduction, 50*, 101906.

Cabell, J. F., & Oelofse, M. (2012). An indicator framework for assessing agro ecosystems resilience. *Ecology Society, 17*(1).

Canada Mortgage and Housing Cooperation. (2018). *Why are policies important*? Retrieved January 13 from https://www.cmhc-schl.gc.ca/en/maintaining-and-managing/managing-affordable-housing/housing-management-hub/policy-information/why-are-policies-important

Cao, W., Zhu, Y., Tian, Y., Yao, X., & Liu, X. (2006). Research progress and prospect of digital farming techniques. *Scientia Agricultura, 39*(2), 281–288.

Chang, G.-C. (2006). *Strategic planning in education: Some concepts and steps.* United Nations Educational, Scientific and Cultural Organization.

Chaudhary, R. P., Uprety, Y., & Rimal, S. K. (2016). Deforestation in Nepal: Causes, consequences and responses. In *Biological and Environmental hazards, risks, and disasters* (pp. 335–372).

Chersich, M. F., Wright, C. Y., Venter, F., Rees, H., Scorgie, F., & Erasmus, B. (2018). Impacts of climate change on health and wellbeing in South Africa. *International Journal of Environmental Research Public Health Nutrition, 15*(9), 1884.

Chowdhury, A., Pradhan, S., Saha, M., & Sanyal, N. (2008). Impact of pesticides on soil microbiological parameters and possible bioremediation strategies. *Indian Journal of Microbiology, 48*(1), 114–127.

Cote, M., & Nightingale, A. J. (2012). Resilience thinking meets social theory: Situating social change in socio-ecological systems (SES) research. *Progress in Human Geography, 36*(4), 475–489.

Cornelissen, G., Nurida, N. L., Hale, S. E., Martinsen, V., Silvani, L., & Mulder, J. (2018). Fading positive effect of biochar on crop yield and soil acidity during five growth seasons in an Indonesian Ultisol. *Science of the Total Environment, 634*, 561–568.

Costanza, R., d'Arge, R., De Groot, R., Farber, S., Grasso, M., Hannon, B., Limburg, K., Naeem, S., O'Neill, R. V., & Paruelo, J. (1997). The value of the world's ecosystem services and natural capital. *Nature, 387*(6630), 253–260.

Cramer, G. R., Urano, K., Delrot, S., Pezzotti, M., & Shinozaki, K. (2011). Effects of abiotic stress on plants: A systems biology perspective. *BMC Plant Biology, 11*(1), 1–14.

Crampton, J. W., Graham, M., Poorthuis, A., Shelton, T., Stephens, M., Wilson, M. W., & Zook, M. (2013). Beyond the geotag: Situating 'big data'and leveraging the potential of the geoweb. *Cartography and Geographic Information Science, 40*(2), 130–139.

Craparo, A. C. W., Van Asten, P. J., Läderach, P., Jassogne, L. T., & Grab, S. W. (2015). *Coffea arabica* yields decline in Tanzania due to climate change: Global implications. *Agricultural and Forest Meteorology, 207*, 1–10.

Dabbert, S., Haring, A. M., & Zanoli, R. (2004). *Organic farming: Policies and prospects.* Zed Books.

Daily, G., & Matson, P. (2008). Ecosystem services: From theory to implementation. *Proceedings of the National Academy of Sciences, 105*(28), 9455–9456.

de Wit-de Vries, E., Dolfsma, W. A., van der Windt, H. J., & Gerkema, M. P. (2019). Knowledge transfer in university–industry research partnerships: A review. *The Journal of Technology Transfer, 44*(4), 1236–1255.

Dill, D. D., & Beerkens, M. (2013). Designing the framework conditions for assuring academic standards: Lessons learned about professional, market, and government regulation of academic quality. *Higher Education, 65*(3), 341–357.

Dissanayake, P., Hettiarachchi, S., & Siriwardana, C. (2018). Increase in disaster risk due to inefficient environmental management, land use policies and relocation policies. Case studies from Sri Lanka. *Procedia Engineering, 212*, 1326–1333.

Drexler, J. Z., Krauss, K. W., Sasser, M. C., Fuller, C. C., Swarzenski, C. M., Powell, A., Swanson, K. M., & Orlando, J. L. (2013). *A long-term comparison of carbon sequestration rates in impounded and naturally tidal freshwater marshes along the lower Waccamaw River*, South Carolina, Wetlands. https://doi.org/10.1007/s13157-013-0456-3

Dubey, N. K., Shukla, R., Kumar, A., Singh, P., & Prakash, B. (2010). Prospects of botanical pesticides in sustainable agriculture. *Current Science, 98*(4), 479–480.

Dudley, N., & Alexander, S. (2017). Agriculture and biodiversity: A review. *Biodiversity, 18*(2–3), 45–49. https://doi.org/10.1080/14888386.2017.1351892

Duku, C., Sparks, A. H., & Zwart, S. J. (2016). Spatial modelling of rice yield losses in Tanzania due to bacterial leaf blight and leaf blast in a changing climate. *Climatic Change, 135*(3–4), 569–583.

Durkin, M., & McKenna, S. (2011). Informing the marketing of higher education to younger people. *Irish Marketing Review, 21*(1), 41–47.

Dussi, M. C., Flores, L. B., Barrionuevo, M. E., & Dussi, S. E. (2019). Agroecology in higher education: a multidimensional vision as a resilience strategy to climate change. *Acta Horticulturae, 1258*, 79–86. https://doi.org/10.17660/ActaHortic.2019.1258.11

El-Sharkawy, M. A. (2014). Global warming: Causes and impacts on Agro ecosystems productivity and food security with emphasis on cassava comparative advantage in the tropics/subtropics. *Photosynthetica, 52*(2), 161–178.

Eldridge, D. J., & Delgado-Baquerizo, M. (2017). Continental-scale impacts of livestock grazing on ecosystem supporting and regulating services. *Land Degradation & Development, 28*(4), 1473–1481.

Elmi, M. (2009). *Commercialization methods of research findings* (pp. 6–10). Publishing Trade Promotion Organization of Iran.

Falkenmark, M., Finlayson, M., Gordon, L. J., Bennett, E. M., Chiuta, T. M., Coates, D., Ghosh, N., Gopalakrishnan, M., de Groot, R. S., & Jacks, G. (2007). *Agriculture, water, and ecosystems: Avoiding the costs of going too far*.

Fan, S., & Pandya-Lorch, R. (2012). *Reshaping agriculture for nutrition and health*. The International Food Policy Research Institute.

Fankhauser, S., & Schmidt-Traub, G. (2011). From adaptation to climate-resilient development: The costs of climate-proofing the Millennium Development Goals in Africa. *Climate Development, 3*(2), 94–113.

FAO. (2018). *The state of food and agriculture migration, agriculture and rural development*. http://www.fao.org/3/I9549EN/i9549en.pdf

FAO. (2020). *FAO, WMO and partners call for early warning and early action to avoid disasters*. United Nations. Retrieved 07th January from http://www.fao.org/news/story/en/item/1313921/icode/

Feintrenie, L., Schwarze, S., & Levang, P. (2010). Are local people conservationists? Analysis of transition dynamics from agroforests to monoculture plantations in Indonesia. *Ecology Society, 15*(4).

Feng, D., & Zhao, G. (2020). Footprint assessments on organic farming to improve ecological safety in the water source areas of the South-to-North Water Diversion project. *Journal of Cleaner Production, 254*, 120130.

Fonta, W., Ichoku, H., & Urama, N. (2011). Climate change and plantation agriculture: A Ricardian analysis of farmlands in Nigeria. *Journal of Economics Sustainable Development, 2*(4).

Frick, M. J., Kahler, A. A., & Miller, W. W. (1991). A definition and the concepts of agricultural literacy. *Journal of Agricultural Education, 32*(2), 49–57.

Führer, E. (2000). Forest functions, ecosystem stability and management. *Forest Ecology and Management, 132*(1), 29–38.

Fuhrer, J., & Booker, F. (2003). Ecological issues related to ozone: Agricultural issues. *Environment International, 29*(2–3), 141–154.

Furuta, N. (2019). Role of Wetlands for disaster risk reduction. In *International symposium on the conservation of the coastal areas of the Bay of Bengal*

Gaebel, M. (2014). *MOOCs: Massive open online courses*. EUA Geneva.

Galhena, D. H., Freed, R., & Maredia, K. M. (2013). Home gardens: A promising approach to enhance household food security and wellbeing. *Agriculture & Food Security, 2*, 1–13.

Geist, H. J., & Lambin, E. F. (2001). What drives tropical deforestation. *LUCC Report Series, 4*, 116.

Glasson, J., & Therivel, R. (2013). *Introduction to environmental impact assessment*. Routledge.

Gomiero, T., Pimentel, D., & Paoletti, M. G. (2011). Environmental impact of different agricultural management practices: conventional vs. organic agriculture. *Critical Reviews in Plant Sciences, 30*(1–2), 95–124.

Guiot, J., & Cramer, W. (2016). Climate change: The 2015 Paris Agreement thresholds and Mediterranean basin ecosystems. *Science, 354*(6311), 465–468.

Gull, A., Lone, A. A., & Wani, N. U. I. (2019). Biotic and abiotic stresses in plants. In *Abiotic and biotic stress in plants*. IntechOpen.

Gupta, A. K., & Nair, S. S. (2012). *Ecosystem approach to disaster risk reduction*. National Institute of Disaster Management New Delhi.

Guthrie, J., Ball, A., & Farneti, F. (2010). *Advancing sustainable management of public and not for profit organizations*. Taylor & Francis.

Hafid, H. S., Nor'Aini, A. R., Mokhtar, M. N., Talib, A. T., Baharuddin, A. S., & Kalsom, M. S. U. (2017). Over production of fermentable sugar for bioethanol production from carbohydrate-rich Malaysian food waste via sequential acid-enzymatic hydrolysis pretreatment. *Waste Management, 67*, 95–105.

Hamdullahpur, F. (2020). Global citizens for the twenty-first century: The role of international partnerships in university education. In *Successful global collaborations in higher education institutions* (pp. 23–30). Springer.

Hamouda, Y. E., & Elhabil, B. H. (2017). Precision agriculture for greenhouses using a wireless sensor network. In *2017 Palestinian International Conference on Information and Communication Technology (PICICT)*

Handeland, S. O., Imsland, A. K., & Stefansson, S. O. (2008). The effect of temperature and fish size on growth, feed intake, food conversion efficiency and stomach evacuation rate of Atlantic salmon post-smolts. *Aquaculture, 283*(1–4), 36–42.

Haq, G., & Schwela, D. (2008). *Air quality governance and policies*. https://doi.org/10.13140/RG.2.1.1292.5288

Haugen, H. M. (2020). The UN Declaration on Peasants' Rights (UNDROP): Is Article 19 on seed rights adequately balancing intellectual property rights and the right to food? *The Journal of World Intellectual Property*.

Hayward, M. D., Bosemark, N. O., & Romagosa, T. (2012). *Plant breeding: principles and prospects*. Springer Science & Business Media.

Hegnsholt, E., Unnikrishnan, S., Pollmann-Larsen, M., Askelsdottir, B., & Gerard, M. (2018). *Tackling the 1.6-billion-ton food loss and waste crisis*. The Boston Consulting Group, Food Nation, State of Green.

Hemsley-Brown, J., & Oplatka, I. (2010). Market orientation in universities. *International Journal of Educational Management*.

Hillel, D., & Rosenzweig, C. (2011). *Handbook of climate change and Agro ecosystems: impacts, adaptation, and mitigation* (Vol. 1). World Scientific.

Hobbs, P. R. (2007). Conservation agriculture: What is it and why is it important for future sustainable food production? *Journal of Agricultural Science, 145*(2), 127.

Hobbs, P. R., Sayre, K., & Gupta, R. (2008). The role of conservation agriculture in sustainable agriculture. *Philosophical Transactions of the Royal Society B: Biological Sciences, 363*(1491), 543–555.

Hodson, E., Niggli, U., Kitajima, K., Lal, R., & Sadoff, C. (2021). *Boost nature positive production.*
Holdenrieder, O., Pautasso, M., Weisberg, P. J., & Lonsdale, D. (2004). Tree diseases and landscape processes: The challenge of landscape pathology. *Trends in Ecology & Evolution, 19*(8), 446–452.
Hoy, C. W. (2015). Agro ecosystems health, agro ecosystems resilience, and food security. *Journal of Environmental Studies and Sciences, 5*(4), 623–635.
Huang, J.-K., Jiang, J., Wang, J.-X., & Hou, L.-L. (2014). Crop diversification in coping with extreme weather events in China. *Journal of Integrative Agriculture, 13*(4), 677–686.
Ifejika Speranza, C. (2010). Resilient adaptation to climate change in African agriculture. *Studies.*
IPCC. (2014). *Climate change: Impacts, adaptation and vulnerability—Summary for policymakers.*
Ismail, A. M., Singh, U. S., Singh, S., Dar, M. H., & Mackill, D. J. (2013). The contribution of submergence-tolerant (Sub1) rice varieties to food security in flood-prone rainfed lowland areas in Asia. *Field Crops Research, 152*, 83–93.
Jiang, Q.-H., & Wu, Y.-B. (2016). Lack of space: The development of traditional Wushu of the "pain". *Journal of Harbin Sport University, 1*, 13.
Jiang, W., Lv, J., Wang, C., Chen, Z., & Liu, Y. (2017). Marsh wetland degradation risk assessment and change analysis: A case study in the Zoige Plateau, China. *Ecological Indicators, 82*, 316–326.
Jones, J. B. (2016). *Hydroponics: A practical guide for the soilless grower*. CRC Press.
Kahrl, F., Li, Y., Su, Y., Tennigkeit, T., Wilkes, A., & Xu, J. (2010). Greenhouse gas emissions from nitrogen fertilizer use in China. *Environmental Science & Policy, 13*(8), 688–694.
Kanchi, R. R., Sreeramula, V. P., & Palle, D. V. (2019). *Implementation of smart agriculture using CloudIoT and its geotagging on android platform.*
Kaphle, A., Navya, P. N., Umapathi, A., & Daima, H. K. (2018). Nanomaterials for agriculture, food and environment: Applications, toxicity and regulation. *Environmental Chemistry Letters, 16*(1), 43–58.
Khan, M. M. H., Bryceson, I., Kolivras, K. N., Faruque, F., Rahman, M. M., & Haque, U. (2015). Natural disasters and land-use/land-cover change in the southwest coastal areas of Bangladesh. *Regional Environmental Change, 15*(2), 241–250.
Kharb, S., & Singh, V. (2004). Nutriceuticals in health and disease prevention. *Indian Journal of Clinical Biochemistry, 19*(1), 50.
Khumairoh, U., Groot, J. C., & Lantinga, E. A. (2012). Complex agro-ecosystems for food security in a changing climate. *Evolution Trends in Ecology, 2*(7), 1696–1704.
Kirono, D., Kent, D., Hennessy, K., & Mpelasoka, F. (2011). Characteristics of Australian droughts under enhanced greenhouse conditions: Results from 14 global climate models. *Journal of Arid Environments, 75*(6), 566–575.
Lal, R. (2020). Home gardening and urban agriculture for advancing food and nutritional security in response to the COVID-19 pandemic. *Food Security Agronomy for Sustainable Development*, 1–6.
Laliberté, F., Zika, J., Mudryk, L., Kushner, P., Kjellsson, J., & Döös, K. (2015). Constrained work output of the moist atmospheric heat engine in a warming climate. *China Earth Sciences, 347*(6221), 540–543.
Lavoie, J. R., Daim, T., & Carayannis, E. G. (2020). Technology transfer evaluation: Driving organizational changes through a hierarchical scoring model. *IEEE Transactions on Engineering Management*.
Leake, A. (2003). Integrated pest management for conservation agriculture. In *Conservation agriculture* (pp. 271–279). Springer.
Lewis, D. R., & Dundar, H. (1999). Costs and productivity in higher education: Theory, evidence, and policy implications. In *Higher education: Handbook of theory and research* (pp. 39–102). Springer.
Li, J., Huang, L., Zhang, J., Coulter, J. A., Li, L., & Gan, Y. (2019). Diversifying crop rotation improves system robustness. *Agronomy for Sustainable Development, 39*(4), 38.

Li, S., & Ma, Y. (2014). Urbanization, economic development and environmental change. *Sustainability, 6*(8), 5143–5161.
Liang, D., Wood, C. L., McQuerry, K. J., Ray, D. L., Clark, J. D., & Bewley, J. M. (2013). Influence of breed, milk production, season, and ambient temperature on dairy cow reticulorumen temperature. *Journal of Dairy Science, 96*(8), 5072–5081.
Liebman, M., Schulte, L. A., Kapuscinski, A. R., Salvador, R. J., & Locke, K. (2015). Enhancing agro ecosystems performance and resilience through increased diversification of landscapes and cropping systems: Diversity affects agroecosystem performance and resilience. *Elementa: Science of the Anthropocene, 3.*
Lin, B. B. (2011). Resilience in agriculture through crop diversification: Adaptive management for environmental change. *BioScience, 61*(3), 183–193.
Lin, B. B., Philpott, S. M., Jha, S., & Liere, H. (2017). Urban agriculture as a productive green infrastructure for environmental and social well-being. In *Greening cities* (pp. 155–179). Springer.
Lindell, M. K. (2013). Recovery and reconstruction after disaster. *Encyclopedia of Natural Hazards, 8*, 12–824.
Lipinski, B., Hanson, C., Lomax, J., Kitinoja, L., Waite, R., & Searchinger, T. (2013). Reducing food loss and waste. *World Resources Institute Working Paper, 1*, 1–40.
Liu, H.-L., Willems, P., Bao, A.-M., Wang, L., & Chen, X. (2016). Effect of climate change on the vulnerability of a socio-ecological system in an arid area. *Global Planetary Change, 137*, 1–9.
Liu, H., & Jiang, Y. (2001). Technology transfer from higher education institutions to industry in China: Nature and implications. *Technovation, 21*(3), 175–188.
Liu, X., Zhang, X., Wang, Y., Sui, Y., Zhang, S., Herbert, S., & Ding, G. (2010). Soil degradation: A problem threatening the sustainable development of agriculture in Northeast China. *Plant, Soil and Environment, 56*(2), 87–97.
Livesey, S., Hartman, C., Stafford, E., & Shearer, M. (2009). Performing sustainable development through eco-collaboration: The Ricelands habitat partnership. *The Journal of Business Communication (1973), 46*(4), 423–454.
Lloyd, S. J., Kovats, R. S., & Chalabi, Z. (2011). Climate change, crop yields, and undernutrition: Development of a model to quantify the impact of climate scenarios on child undernutrition. *Environmental Health Perspectives, 119*(12), 1817–1823.
Loewenberg, S. (2015). Conflicts worsen global hunger crisis. *The Lancet, 386*(10005), 1719–1721.
Louwagie, G., Gay, S. H., Sammeth, F., & Ratinger, T. (2011). The potential of European Union policies to address soil degradation in agriculture. *Land Degradation & Development, 22*(1), 5–17.
Mahmood, I., Imadi, S. R., Shazadi, K., Gul, A., & Hakeem, K. R. (2016). Effects of pesticides on environment. In *Plant, soil and microbes* (pp. 253–269). Springer.
Mango, N., Nyikahadzoi, K., Makate, C., Dunjana, N., & Siziba, S. (2015). The impact of integrated agricultural research for development on food security among smallholder farmers of southern Africa. *Agrekon, 54*(3), 107–125.
Manion, P. D. (2003). Evolution of concepts in forest pathology. *Phytopathology, 93*(8), 1052–1055.
Manual, D., Infiltration-Trench, B., & Well, D. (2015). *Green infrastructure.*
Masudian, P., Farhadpoor, M. R., & Ghashgayizadeh, N. (2013). Commercializing university research results: A case study by Behbahan Islamic Azad University. *Library Philosophy and Practice (e-Journal)*, 2–26.
Matthews, R. B., Rivington, M., Muhammed, S., Newton, A. C., & Hallett, P. D. (2013). Adapting crops and cropping systems to future climates to ensure food security: The role of crop modelling. *Global Food Security, 2*(1), 24–28.
McCauley, L. A., Jenkins, D. G., & Quintana-Ascencio, P. F. (2013). Isolated wetland loss and degradation over two decades in an increasingly urbanized landscape. *Wetlands Ecology Management, 33*(1), 117–127.
McClain, M. E. (2013). Balancing water resources development and environmental sustainability in Africa: A review of recent research findings and applications. *Ambio, 42*(5), 549–565.

Mechler, R., Calliari, E., Bouwer, L. M., Schinko, T., Surminski, S., Linnerooth-Bayer, J., Aerts, J., Botzen, W., Boyd, E., & Deckard, N. D. (2019). Science for loss and damage. Findings and propositions. In *Loss and damage from climate change* (pp. 3–37). Springer.

Meerow, S., & Newell, J. P. (2019). Urban resilience for whom, what, when, where, and why? *Urban Geography, 40*(3), 309–329.

Melaville, A., Berg, A. C., & Blank, M. J. (2006). *Community-based learning: Engaging students for success and citizenship.*

Memmott, J., Craze, P. G., Waser, N. M., & Price, M. V. (2007). Global warming and the disruption of plant–pollinator interactions. *Ecology Letters, 10*(8), 710–717.

Mendelsohn, R., & Neumann, J. E. (2004). *The impact of climate change on the United States economy.* Cambridge University Press.

Mera, R. J., Niyogi, D., Buol, G. S., Wilkerson, G. G., & Semazzi, F. H. (2006). Potential individual versus simultaneous climate change effects on soybean (C3) and maize (C4) crops: An agrotechnology model based study. *Global Planetary Change, 54*(1–2), 163–182.

Messner, R., Johnson, H., & Richards, C. (2020). From surplus-to-waste: A study of systemic overproduction, surplus and food waste in horticultural supply chains. *Journal of Cleaner Production, 278*, 123952.

Meštrović, T. (2018). What are nutraceuticals? *News-Medical.* Retrieved 09 December from https://www.news-medical.net/health/What-are-Nutraceuticals.aspx

Miléř, T., & Sládek, P. (2011). The climate literacy challenge. *Procedia-Social and Behavioral Sciences, 12*, 150–156.

Mohanraj, I., Ashokumar, K., & Naren, J. (2016). Field monitoring and automation using IOT in agriculture domain. *Procedia Computer Science, 93*, 931–939.

Mori, A. S., Furukawa, T., & Sasaki, T. (2013). Response diversity determines the resilience of ecosystems to environmental change. *Biological Reviews, 88*(2), 349–364.

Mougou, R., Mansour, M., Iglesias, A., Chebbi, R. Z., & Battaglini, A. (2011). Climate change and agricultural vulnerability: A case study of rain-fed wheat in Kairouan. *Central Tunisia. Regional Environmental Change, 11*(1), 137–142.

Mulvaney, R. L., Khan, S. A., & Ellsworth, T. R. (2009). Synthetic nitrogen fertilizers deplete soil nitrogen: A global dilemma for sustainable cereal production. *Journal of Environmental Quality, 38*(6), 2295–2314.

Musa, M. F., & Ahmad, Z. (2012). Higher education physical assets and facilities. *Procedia-Social and Behavioral Sciences, 50*, 472–478.

Nagarajan, V., Mohanty, A. K., & Misra, M. (2013). Sustainable green composites: Value addition to agricultural residues and perennial grasses. *ACS Sustainable Chemistry & Engineering, 1*(3), 325–333.

Naik, P., Swain, B., & Singh, N. (2015). Production and utilisation of hydroponics fodder. *Indian Journal of Animal Nutrition, 32*(1), 1–9.

Nanney, B. (2004). *Student-centered learning.* Retrieved November 30, 2012.

Niblack, T. L., & Chen, S. Y. (2004). *Cropping systems and crop management practices.*

Nieto-Romero, M., Oteros-Rozas, E., González, J. A., & Martín-López, B. (2014). Exploring the knowledge landscape of ecosystem services assessments in Mediterranean Agro ecosystems: Insights for future research. *Environmental Science & Policy, 37*, 121–133.

Olanipekun, I. O., Olasehinde-Williams, G. O., & Alao, R. O. (2019). Agriculture and environmental degradation in Africa: The role of income. *Science of the Total Environment, 692*, 60–67.

Oliver, T. H., Heard, M. S., Isaac, N. J., Roy, D. B., Procter, D., Eigenbrod, F., Freckleton, R., Hector, A., Orme, C. D. L., & Petchey, O. L. (2015). Biodiversity and resilience of ecosystem functions. *Trends in Ecology, 30*(11), 673–684.

Onojeghuo, A. O., Blackburn, G. A., Huang, J., Kindred, D., & Huang, W. (2018). Applications of satellite 'hyper-sensing' in Chinese agriculture: Challenges and opportunities. *International Journal of Applied Earth Observation and Geoinformation, 64*, 62–86.

Onuma, A., & Tsuge, T. (2018). Comparing green infrastructure as ecosystem-based disaster risk reduction with gray infrastructure in terms of costs and benefits under uncertainty: A theoretical approach. *International Journal of Disaster Risk Reduction, 32*, 22–28.
Opitz, I., Berges, R., Piorr, A., & Krikser, T. (2016). Contributing to food security in urban areas: Differences between urban agriculture and peri-urban agriculture in the Global North. *Agriculture and Human Values, 33*, 341–358.
Overby, K. (2011). Student-centered learning. *Essai, 9*(1), 32.
Pandey, B., & Seto, K. C. (2015). Urbanization and agricultural land loss in India: Comparing satellite estimates with census data. *Journal of Environmental Management, 148*, 53–66.
Parr, T. W., Sier, A. R. J., Battarbee, R. W., Mackay, A., & Burgess, J. (2003). Detecting environmental change: Science and society—Perspectives on long-term research and monitoring in the 21st century. *Science of the Total Environment, 310*(1–3), 1–8.
Paul, B. K. (2011). *Environmental hazards and disasters: Contexts, perspectives and management*. Wiley.
Perry, D. A., Oren, R., & Hart, S. C. (2008). *Forest ecosystems*. JHU Press.
Peters, D. (2020). *Building an Institutional Framework (WD)*. Retrieved January 17 from https://sswm.info/water-nutrient-cycle/water-distribution/softwares/creating-and-enabling-environment/building-an-institutional-framework-%28wd%29
Pickett, S. T., McGrath, B., Cadenasso, M. L., & Felson, A. (2014). Ecological resilience and resilient cities. *Building Research Information Processing in Agriculture, 42*(2), 143–157.
Plazzotta, S., Manzocco, L., & Nicoli, M. C. (2017). Fruit and vegetable waste management and the challenge of fresh-cut salad. *Trends in Food Science & Technology, 63*, 51–59.
Post Disaster Needs Assessment Guidelines—Agriculture, Livestock, Fisheries and Forestry. (2014). https://www.gfdrr.org/sites/default/files/publication/pdna-guidelines-vol-b-agriculture-livestock-fisheries-forestry_0.pdf
Potts, S. G., Imperatriz-Fonseca, V., Ngo, H. T., Biesmeijer, J. C., Breeze, T. D., Dicks, L. V., Garibaldi, L. A., Hill, R., Settele, J., & Vanbergen, A. J. (2016). The assessment report on pollinators, pollination and food production: Summary for policymakers. In *Secretariat of the intergovernmental science-policy platform on biodiversity...*
Primmer, E., Jokinen, P., Blicharska, M., Barton, D. N., Bugter, R., & Potschin, M. (2015). Governance of ecosystem services: A framework for empirical analysis. *Ecosystem Services, 16*, 158–166.
Pruvot, E. B., Claeys-Kulik, A.-L., & Estermann, T. (2015). *Designing strategies for efficient funding of universities in Europe* (pp. 20–22). European University Association.
R Shamshiri, R., Kalantari, F., Ting, K., Thorp, K. R., Hameed, I. A., Weltzien, C., Ahmad, D., & Shad, Z. M. (2018). *Advances in greenhouse automation and controlled environment agriculture: A transition to plant factories and urban agriculture*.
Rejeb, I. B., Pastor, V., & Mauch-Mani, B. (2014). Plant responses to simultaneous biotic and abiotic stress: Molecular mechanisms. *Plants, 3*(4), 458–475.
Reyes-García, V., Aceituno-Mata, L., Calvet-Mir, L., Garnatje, T., Gómez-Baggethun, E., Lastra, J. J., Ontillera, R., Parada, M., Rigat, M., & Vallès, J. (2014). Resilience of traditional knowledge systems: The case of agricultural knowledge in home gardens of the Iberian Peninsula. *Global Environmental Change, 24*, 223–231.
Reynolds, C., Goucher, L., Quested, T., Bromley, S., Gillick, S., Wells, V. K., Evans, D., Koh, L., Kanyama, A. C., & Katzeff, C. (2019). Consumption-stage food waste reduction interventions—What works and how to design better interventions. *Food Policy, 83*, 7–27.
Rojas-Downing, M. M., Nejadhashemi, A. P., Harrigan, T., & Woznicki, S. A. (2017). Climate change and livestock: Impacts, adaptation, and mitigation. *Climate Risk Management, 16*, 145–163.
Rosati, A., & Aumaitre, A. (2004). Organic dairy farming in Europe. *Livestock Production Science, 90*(1), 41–51.

Rosmiza, M. Z., Davies, W. P., Cr, R. A., Jabil, M. J., & Mazdi, M. (2016). Prospects for increasing commercial mushroom production in Malaysia: Challenges and opportunities. *Mediterranean Journal of Social Sciences, 7*(1 S1), 406–406.

Rudel, T. K., Schneider, L., Uriarte, M., Turner, B. L., DeFries, R., Lawrence, D., Geoghegan, J., Hecht, S., Ickowitz, A., & Lambin, E. F. (2009). Agricultural intensification and changes in cultivated areas, 1970–2005. *Proceedings of the National Academy of Sciences, 106*(49), 20675–20680.

Rufí-Salís, M., Calvo, M. J., Petit-Boix, A., Villalba, G., & Gabarrell, X. (2020). Exploring nutrient recovery from hydroponics in urban agriculture: An environmental assessment. *Resources, Conservation Recycling, 155*, 104683.

Rui, J., Peng, J., & Lu, Y. (2009). Succession of bacterial populations during plant residue decomposition in rice field soil. *Applied and Environmental Microbiology, 75*(14), 4879–4886.

Sajjad, M., Li, Y., Tang, Z., Cao, L., & Liu, X. (2018). Assessing hazard vulnerability, habitat conservation, and restoration for the enhancement of mainland China's coastal resilience. *Earth's Future, 6*(3), 326–338.

Sattler, C., Kächele, H., & Verch, G. (2007). Assessing the intensity of pesticide use in agriculture. *Agriculture, Ecosystems & Environment, 119*(3–4), 299–304.

Saxena, N. C. (2018). Hunger, under-nutrition and food security in India. In *Poverty, chronic poverty and poverty dynamics* (pp. 55–92). Springer.

Schilling, K. E., Jha, M. K., Zhang, Y. K., Gassman, P. W., & Wolter, C. F. (2008). Impact of land use and land cover change on the water balance of a large agricultural watershed: Historical effects and future directions. *Water Resources Research, 44*(7).

Schulze, E.-D., & Mooney, H. A. (2012). *Biodiversity and ecosystem function.* Springer Science & Business Media.

Schwarz, E. J., Wdowiak, M. A., Almer-Jarz, D. A., & Breitenecker, R. J. (2009). *The effects of attitudes and perceived environment conditions on students' entrepreneurial intent.* Education + Training.

Sebatta, C., Wamulume, M., & Mwansakilwa, C. (2014). Determinants of smallholder farmers' access to agricultural finance in Zambia. *Journal of Agricultural Science, 6*(11), 63.

Seel, N. M. (2011). *Encyclopedia of the sciences of learning.* Springer Science & Business Media.

Sempebwa, J. (2011). Evaluating the utilisation of resources in higher education institutions: The case of teaching space at a Ugandan university. *Evaluation, 17*(3), 247–259.

Sempore, A. W., Andrieu, N., Le Gal, P.-Y., Nacro, H. B., & Sedogo, M. P. (2016). Supporting better crop-livestock integration on small-scale West African farms: A simulation-based approach. *Agroecology; Sustainable Food Systems, 40*(1), 3–23.

Settele, J., Heong, K. L., Kühn, I., Klotz, S., Spangenberg, J. H., Arida, G., Beaurepaire, A., Beck, S., Bergmeier, E., & Burkhard, B. (2018). *Rice ecosystem services in South-East Asia.* Springer.

Shagholi, R., & Hussin, S. (2009). Participatory management: An opportunity for human resources in education. *Procedia-Social and Behavioral Sciences, 1*(1), 1939–1943.

Sheffield, J. (2005). Systemic knowledge and the V-model. *International Journal of Business Information Systems, 1*(1–2), 83–101.

Shukla, J. B., Verma, M., & Misra, A. K. (2017). Effect of global warming on sea level rise: A modeling study. *Ecological Complexity, 32*, 99–110.

Silva, F., Stevens, C. J., Weisskopf, A., Castillo, C., Qin, L., Bevan, A., & Fuller, D. Q. J. P. O. (2015). *Modelling the Geographical Origin of Rice Cultivation in Asia Using the Rice Archaeological Database. 10*(9), e0137024.

Spady, W. G. (1994). *Outcome-based education: Critical issues and answers.* ERIC.

Specht, K., Siebert, R., Hartmann, I., Freisinger, U. B., Sawicka, M., Werner, A., Thomaier, S., Henckel, D., Walk, H., & Dierich, A. (2014). Urban agriculture of the future: An overview of sustainability aspects of food production in and on buildings. *Agriculture Human Values, 31*(1), 33–51.

Spielman, D. J., Ekboir, J., Davis, K., & Ochieng, C. M. (2008). An innovation systems perspective on strengthening agricultural education and training in sub-Saharan Africa. *Agricultural Systems, 98*(1), 1–9.

Staddon, C., Ward, S., De Vito, L., Zuniga-Teran, A., Gerlak, A. K., Schoeman, Y., Hart, A., & Booth, G. (2018). Contributions of green infrastructure to enhancing urban resilience. *Environment Systems Decisions, 38*(3), 330–338.

Stein, S., & Stein, J. (2014). *Playing against nature: Integrating science and economics to mitigate natural hazards in an uncertain world*. Wiley.

Strait, J. E., Dawson, P., Walther, C. A. P., Strait, G. G., Barton, A. K., & McClain, M. B. (2020). Refinement and psychometric evaluation of the executive skills questionnaire-revised. *Contemporary School Psychology, 24*(4), 378–388.

Strigl, A. W. (2003). Science, research, knowledge and capacity building. *Environment, Development and Sustainability, 5*(1–2), 255–273.

Subbarao, S., & Lloyd, B. (2011). Can the clean development mechanism (CDM) deliver? *Energy Policy, 39*(3), 1600–1611.

Sumberg, J. (2017). *Agronomy for development: The politics of knowledge in agricultural research*. Taylor & Francis.

Swenson, D. (2008). *Estimating the production and market value-based impacts of nutritional goals in ne Iowa*.

Taş, M., Taş, N., Durak, S., & Atanur, G. (2013). Flood disaster vulnerability in informal settlements in Bursa, Turkey. *Environment and Urbanization, 25*(2), 443–463.

Teixeira, A. H. d. C., Leivas, J. F., & Bayma-Silva, G. (2016). *Options for using Landsat and rapid eye satellite images aiming the water productivity assessments in mixed agro-ecosystems*.

Thebo, A., Drechsel, P., & Lambin, E. (2014). Global assessment of urban and peri-urban agriculture: Irrigated and rainfed croplands. *Environmental Research Letters, 9*(11), 114002.

Tregidga, H., & Milne, M. J. (2006). From sustainable management to sustainable development: A longitudinal analysis of a leading New Zealand environmental reporter. *Business Strategy the Environment, 15*(4), 219–241.

Tripathi, R., & Agarwal, S. (2015). Rural development through Agripreneurship: A study of farmers in Uttar Pradesh. *Global Journal of Advanced Research, 2*(2), 534–542.

Trivedi, P., Wallenstein, M. D., Delgado-Baquerizo, M., & Singh, B. K. (2018). Microbial modulators and mechanisms of soil carbon storage. In *Soil carbon storage* (pp. 73–115). Elsevier.

Tubby, K., & Webber, J. (2010). Pests and diseases threatening urban trees under a changing climate. *International Journal of Forest Research, 83*(4), 451–459.

Tůmová, E., & Gous, R. M. (2012). Interaction of hen production type, age, and temperature on laying pattern and egg quality. *Poultry Science, 91*(5), 1269–1275.

Turner, A. G., & Annamalai, H. (2012). Climate change and the South Asian summer monsoon. *Nature Climate Change, 2*(8), 587–595.

United Nations. (2015). *Sustainable development goals*. United Nations Retrieved May 19, from https://www.un.org/sustainabledevelopment/sustainable-development-goals/

Vandermeer, J. (2011). *The ecology of agro ecosystems*. Jones & Bartlett Learning.

Várallyay, G. (2007). Potential impacts of climate change on agro-ecosystems. *Agriculturae Conspectus Scientificus, 72*(1), 1–8.

Verhoeven, J. T., & Setter, T. L. (2010). Agricultural use of wetlands: Opportunities and limitations. *Annals of Botany, 105*(1), 155–163.

Vlam, M., Baker, P. J., Bunyavejchewin, S., & Zuidema, P. A. (2014). Temperature and rainfall strongly drive temporal growth variation in Asian tropical forest trees. *Oecologia, 174*(4), 1449–1461.

Waddell, P. (2002). UrbanSim: Modeling urban development for land use, transportation, and environmental planning. *Journal of the American Planning Association, 68*(3), 297–314.

Wania, A., Kühn, I., & Klotz, S. (2006). Plant richness patterns in agricultural and urban landscapes in Central Germany—Spatial gradients of species richness. *Landscape Urban Planning, 75*(1–2), 97–110.

Weichselgartner, J. (2001). Disaster mitigation: the concept of vulnerability revisited. *Disaster Prevention Management: An International Journal.*

Werbach, A. (2011). *Strategy for sustainability.* Strategic Direction.

Wezel, A., Bellon, S., Doré, T., Francis, C., Vallod, D., & David, C. (2009). Agroecology as a science, a movement and a practice: A review. *Agronomy for Sustainable Development, 29*(4), 503–515.

Wright, I. A., Tarawali, S., Blümmel, M., Gerard, B., Teufel, N., & Herrero, M. (2012). Integrating crops and livestock in subtropical agricultural systems. *Journal of the Science of Food, Agriculture & Food Security, 92*(5), 1010–1015.

Zell, C., & Hubbart, J. A. (2013). Interdisciplinary linkages of biophysical processes and resilience theory: Pursuing predictability. *Ecological Modelling, 248*, 1–10.

Zuniga-Teran, A. A., Gerlak, A. K., Mayer, B., Evans, T. P., & Lansey, K. E. (2020). Urban resilience and green infrastructure systems: Towards a multidimensional evaluation. *Current Opinion in Environmental Sustainability, 44*, 42–47.

Green Consumer Behavior Index (GCBI): Evaluation of Consumer's Green Behaviour Towards Sustainable Consumption and Eco-conscious Society

H. A. C. Priyankara, U. Premathilaka, and C. M. Navaratne

1 Introduction

Climate change is a global issue seeking coordinated worldwide efforts to counteract its adverse outcomes (Desai, 2022). The concept of sustainable consumption is a key concept that shows the direct relationship of society and the environment towards sustainable development. Furthermore, sustainable consumption is aligned with economic growth and environmental damages which involve various behavioural disciplines and ecological principles (Vargas-Merino et al., 2023). A rising trend of green consumption can be observed in both industrialized nations' and developing countries' economies with the influence of social and environmental behaviours of the customer achieving consumer well-being (Haba et al., 2023).

The rapid population growth accelerates not only ecological deterioration but also energy crisis which highlights the importance and urgency of green consumption through energy conservation, green development, sustainable consumption patterns and pollution reduction to meet the basic needs of people (Khalil & Zaidi, 2014). As a result of this situation, consumers are changing their consumption patterns and purchasing habits to achieve a more sustainable environment through sustainable consumption (Barbarossa & Pastore, 2015). The consumer's present choices and

H. A. C. Priyankara (✉)
Faculty of Technology, Department of Biosystems Technology, University of Ruhuna, Matara, Sri Lanka
e-mail: chamilahacp@gmail.com

U. Premathilaka · C. M. Navaratne
Faculty of Agriculture, Department of Agricultural Engineering and Environmental Technology, University of Ruhuna, Matara, Sri Lanka
e-mail: champa@agricc.ruh.ac.lk

H. A. C. Priyankara
International Program in Hazardous Substance and Environmental Management (IP-HSM), Graduate School, Chulalongkorn University, Bangkok, Thailand

C. Malalgoda et al. (eds.), *Climate Change Adaptation in the Built Environment*,
https://doi.org/10.1007/978-3-031-75826-3_24

actions have long-term impacts on future generations; adaptation to a sustainable lifestyle is crucial to protect the ecosystem and preserve natural resources for the future (Handayani et al., 2017).

The green consumption behavior allows people to participate in environmental protection, which is important in saving our planet from pollution (Yue et al., 2020). Green consumerism is a continuation of global consumerism action, which began with consumer knowledge of their rights to receive appropriate, safe and environmentally friendly product (Handayani et al., 2017). Green consumers are essential to green innovation and green consumption since it not only benefits the environment, society and economy, but it is also necessary for long-term development. Green innovation will lower environmental risks, such as Carbon Dioxide (CO_2) emissions, other effects of climate change and most of firms struggle to save resources to protect the environment because of consumer behavior, which is responsible for the change in environmental challenges (Yue et al., 2020).

Identification of drivers that affect green purchasing behavior is important to increase purchasing of green products and green consumerism to increase market shares of green products in the global market. Therefore, the main scientific contribution of this study is to develop a tool to evaluate the green behavior of a consumer to maintain a sustainable lifestyle. This study focuses on development of a "Green Consumer Behavior Index-(GCBI)" as a new tool to evaluate consumer behavior towards promoting sustainable consumption. The index will be enabling the empirical following of the consumption patterns of the consumer and as a result policy makers will be properly informed and stakeholder discussion about sustainable consumption will be encouraged.

1.1 Objective of the Study

The socio-economic, technical and environmental trends are considered as the main trends that affect society and individuals causing inequality, globalization, environmental deterioration and demographic diversity (Chekima et al., 2015).The consumer, who has a positive attitude towards the environment, does not always translate their green purchasing into actual purchasing behavior (Luthra, 2022). Identification of factors related to sustainability is crucial for developing a responsible, sustainable citizen; however, it can be hard to evaluate the behaviour of someone based on a wide variety of sustainability indicators. Therefore, identifying behavioral indicators to include in an index would help to distinguish individuals who follow a sustainable lifestyle and the areas that need to be adopted towards sustainable consumption.

This research develops a composite index to provide regular quantitative measures of consumer behavior by identifying the individual influencing factors for green behavioral change. Furthermore, the development index aims to promote sustainable consumption by increasing consumer awareness and providing individual reference points for consumers to compare their own consumption patterns.

2 Literature Review

2.1 *Sustainable Consumption*

Sustainability is a concept related to the development of products, goods, and services which is significant because humans place a great deal of strain on the environment and its finite resources by consuming beyond their fundamental needs such as larger homes, luxury automobiles, the latest technology and appliances, clothing and entertainment.

The term "sustainable consumption" which is considered as the main driver of sustainable development has different labels by different authors (Quoquab & Mohammad, 2020). It can be referred to as avoiding the excessive consumption of goods and services while minimizing the negative impact of consumption on the environment (Handayani et al., 2017). The primary goal of sustainable consumption is not to deteriorate environmental quality through increased production of goods and services (Jain et al., 2021), nor is it to reduce consumption, but rather to reduce its environmental impact (Tripathi & Singh, 2016). The production and consumption of retail goods like clothing, electronics, and other consumables account for one-fourth of all worldwide emissions. Although consumers are becoming more aware of the negative consequences of certain purchasing habits, over-consumption is still prevalent in some product categories (Poore & Nemecek, 2018). Building collaboration between a wide range of stakeholders and across sectors in all nations is necessary to accomplish this through sustainable consumption and production (Abdulrazak & Quoquab, 2017). Sustainable consumption and production are supported by key elements including waste management, sustainable resource management, design for sustainability, cleaner production and resource efficiency, sustainable transport, eco-labelling and certification, sustainable procurement, sustainable marketing, and promoting sustainable lifestyles, all contributing to a comprehensive approach (Kazi & Mehta, 2016).

2.2 *Green Consumer Behavior*

Green consumer behavior is a form of consumption that protects the environment for both the present and future generations. Green consumption practices aim to maintain the ecological environment and reduce the harmful effects of consumption on the environment throughout the acquisition, usage, and after-use of commodities (Barbarossa & Pastore, 2015). The responsible consumption of consumers who preserve their own health and individual interests while considering environmental protection, resource conservation, pollution prevention and social responsibility can be refer to as green consumption. Green consumption behavior is a crucial prerequisite to ensure long-term sustainability whereby customers attempt to conserve the ecological environment and minimize the detrimental effects of consumption (Rahnama & Rajabpour, 2017).

2.3 Green Products, Green Market and Consumer Awareness

To attain sustainable consumption, one of the key strategies is to buy more ecologically friendly goods. Utilizing environmentally friendly items helps ensure that the planet's natural resources, like plastic, metal and water are protected (Foo et al., 2021). Green products, also known as environmentally friendly items, are those that can be recycled or conserved, and do not pollute the environment or deplete natural resources and are also beneficial for human health (Jain et al., 2021). Green products serve customer requirements while protecting the environment and promoting a more sustainable society. Reducing waste and maximizing resource efficiency are the primary objectives of these goods. Their production is linked to a product development process which is designed in such a way that the environmental impact is considered throughout the product's life cycle (Luthra, 2022).

Even though customers are acting positively toward buying green items, it was also shown that when it comes to making actual purchases, pricing, environmental concern, and quality are still the main variables influencing consumer choice (Luthra, 2022). It is revealed that cost is the most critical consideration and the biggest barrier to purchasing environmentally friendly items and pricing and consumer's green purchasing behavior are significantly correlated (Joshi & Rahman, 2015). If the cost of the green product is kept low, consumers will be motivated to purchase it (Barbarossa & Pastore, 2015).

In green marketing, the knowledge of the benefits and drawbacks of green products are important for both consumers and seller because they can enhance sales while avoiding over-consumption (Jain et al., 2021). Green advertisements, eco-label, eco-brand, environment friendly packaging are some of the green marketing tools that encourage consumers to purchase environmentally friendly products and application of these tools is crucial for changing consumer behavior in purchasing as it reduces the harm that synthetic items do to the environment (Nica, 2020). The cost reduction, brand image, legal compliances and environmental friendliness are identified as the main four benefits of eco-friendly packaging on sustainable supply chain management. If manufacturing companies do not consider employing eco-friendly packaging, it will affect the consumer's intention to purchase a competitor's product that uses eco-friendly packaging (Wang et al., 2021).

Awareness of green products can have a positive relationship with purchasing intentions. As consumers become more familiar with green products, they are subsequently become aware of their existence, which may have an impact on their intents to make green purchasing and subsequent actions (Rahnama & Rajabpour, 2017). Utilizing green products will assist people in transitioning to a green lifestyle, however, only a small portion of the population is aware of the use of green products. The public must be made aware of the benefits of using green products and how to recognize such features in the goods they buy, and this would contribute to preserving the ecosystem for future generations.

2.4 Theories Related to Green Behaviour

To understand the context of green consumer behaviour, there are several theories that explain the reasons and factors for people engaging in environmentally perceived actions which refer to their individual actions and behaviors. The different theories have different perspectives when explaining the factors that influence green behaviour and what motivates individuals' action towards conserving the environment (Trong Nguyen et al., 2023). Fishbein and Azjen combined two psychological theories of behavior change to create the Theory of Reasoned Action (TRA) and Theory of Planned Behavior (TPB), which aims to explain and predict consumer behavior (Madden et al., 1992).

2.4.1 Theory of Planned Behavior

This theory is widely used in consumer behavior studies and according to this theory, the green behavior of an individual is driven by three factors. Individuals show their intention to engage in actions towards green consumption with the influence of their attitude behavior, social pressure/subjective norms and perceived behavioral control. A positive intention to engage in green behaviour is led by favorable attitudes which are reflected as positive outcomes associated with their green actions. Subjective norms explain how social factors influence green behavior. Individual perceptions of how important to others (family, friends etc.,) is aligned with this as normative belief and be motivated with the perceived social pressures. The behavioral intention will transfer the actions towards the actual green behavioral translation (Charlotte Nickerson, 2023). The following theories that are relevant to consumer behavior were used for the study.

2.4.2 Theory of Reasoned Action Behavior (TRA)

According to theory of reasoned action behavior (TRA), people have higher intentions and are more likely to carry out the suggested conduct if they perceive the behavior as positive and believe that others want them to undertake the behavior (subjective norm). The Theory of Reasoned Action has four main terms: subjective norm, belief, intention, and attitude. Related to this study, subjective norms are the total number of significant individuals in a consumer's life and their perception of whether they would want them to perform the green behavior. The main points of this model are that attitudes are function beliefs. If consumers think that buying environmentally friendly products will lead to reduce accumulation of toxic particles to the soil, they will have a positive attitude towards it (Wilson & Edelyn, 2022).

2.4.3 Theory of Green Purchasing Behavior (TGPB)

Another theory that is involved in this study is the Theory of Green Purchasing Behavior (TGPB), which adequately and clearly describes consumer purchasing patterns for eco-friendly products. TGPB united the components from social psychology theories based on pro-social motivations, image of green purchasing, attitude toward green purchasing and social norms (Han, 2020). This theory is associated with developing the Green Consumer Behavior Index as it considers the role of the variables discovered through the qualitative process in analyzing consumer green purchase behaviour.

2.4.4 Theory of Consumer Buying Behavior

The theory of consumer buying behavior examines how consumers decide what to buy and enables companies and marketers to profit from these behaviors by predicting how and when consumers will make a purchase (Chawla & Naved Khan, 2015). It contends that a consumer's preferred brand decision is influenced by their motivations, alternative options and decision mediators that link their motivations to those alternatives such as whether consumers believe wooden bottles are better than single-use plastic bottles to reduce the harmful effects for living organisms in the soil (Soomro et al., 2020).Therefore it is possible to identify a need and produce something that fills it by comprehending these mediators, competing brands and consumer brand awareness.

2.5 Identification of the Factors that Effect on Green Consumer Behaviour and Sustainable Consumption

. There has been a lot of prior research on environmentally friendly consumer behavior and identifying the factors influencing green behavior, which was the main reason for those investigations (Ezzati & Yani, 2017). When selecting the main categories for the GCBI, the carbon emissions from different activities that were recorded in the previous studies along with the consumers' consumption patterns were considered with the theories of consumer studies (Koide et al., 2021). Figure 1 shows the individual's carbon emission contribution resulting from different daily activities.

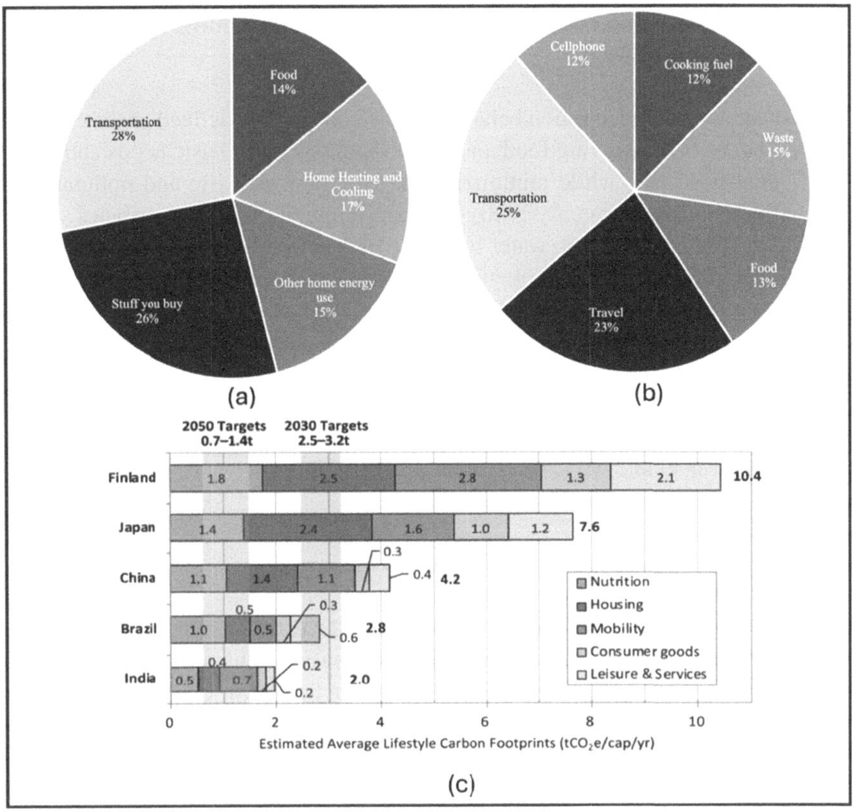

Fig. 1 Emission comes from different activities of a consumer. **a** Gaps between the current lifestyle carbon footprints of average citizens in countries and 2030–2050 targets (Koide et al., 2021). **b** Emissions come from five key areas of our lives; where the average American's carbon emission comes from Union of Concerned Scientists (2012). **c** Break up of average carbon footprint (per capita) for rural households when wood is considered CO_2 neutral (Bhoyar et al., 2014)

By considering the above findings, the GCBI considered the following behavioral actions as main categories and marks were distributed by considering the previous studies mentioned.

2.5.1. Food and water consumption behaviour
2.5.2. Product selection and purchasing behaviour
2.5.3. Household waste management practices
2.5.4. Energy conserving behaviour of the consumer.

2.5.1 The Sustainable Food and Water Consumption Behaviour of the Consumer

The food and water consumption behavior of the consumer is defined as the efficient use or practices of conserving food and water that respond to basic needs and bring a better quality of life, while minimizing the emission of waste and pollutants to the environment (Vermeir et al., 2020). The consumption of food contributes significantly to ecological footprint, water footprint, and carbon footprint. Therefore, it is crucial to comprehend the psychological-social factors that influence food and water consumption behavior to improve the efficacy of resources conservation measures that resulting environmental benefits (Hartmann & Siegrist, 2017). Furthermore, the food and water consumption behavior include the food selection, storage, preparation, consumption, cleaning up of food and way of disposing and reducing waste by avoiding excessive packaging, less meat consumption results more environmental benefits (Kasavan et al., 2017; van Bussel et al., 2022).

2.5.2 Product Selection and Purchasing Behavior

When increasing world population, with an elevated level of consumption resulting in resources shortage and product selection and purchasing behavior of the consumer plays a key role due to the high volume of production (Ran et al., 2022). When compared to similar competing goods or services, environmentally responsible purchasing, also known as green purchasing, refers to choosing goods and services that have less impact on both humans and the environment. Environmental sustainability emerged from a change in views regarding reaction on conserving environment, which has changed customer demand and user behavior (Fernando et al., 2020). A relationship has been established between the environment and the qualities of green products due to the prioritization process within consumer behavior. Utilizing eco-friendly products instead of exclusive marketing techniques further connects these items (Quoquab & Mohammad, 2020).

2.5.3 Energy Conserving Behavior

Sustainability of energy use can be defined as reducing energy usage and improving energy efficiency using energy efficient products (Rajmohan & Weerahewa, 2010). The residential and service sectors have seen the most significant growth in electricity usage during the previous few decades. CO_2 emissions have increased because of this rising electricity use and to reduce CO_2 emission and local air pollution, the globe must quickly switch to low-carbon energy sources like nuclear and renewable technology (Van Acker et al., 2015). The way energy is used in houses is influenced by several factors, including energy prices, household and dwelling features, climate, energy sources, the types and efficiency of home appliances (Tufaner & Turker, 2016). It is difficult to comprehend how a variety of intricate factors shape attitudes and behaviors that conserve energy. Therefore, more knowledge is needed regarding the factors influencing consumption and household lifestyle to promote changes in behavior that reduce energy use. Consumers play a key role in energy conservation by purchasing green label products, recycling, lowering electricity usage and using energy efficient household appliances. Installing energy-saving appliances minimizes energy waste and helps to promote low-carbon economies (Wijekoon & Sabri, 2021). The factors that impact on Consumer's Energy conserving behavior are included in Table 1.

2.5.4 House-Hold Waste Management Practices

Sustainable management of waste produced by humans has an impact on the air, water and land. The term "waste management" refers to the appropriate procedures for gathering, moving, managing, and monitoring various waste products in a way that does not harm the environment and humans (Almasi et al., 2019). Utilizing the most environmentally friendly methods, sustainable waste management guides waste on its path to become a resource recovery and towards circular emissions as well (Barloa et al., 2016).

Ruediger et al. (2012), stated that the requirement of a sustainable index of green consumer behavior and its both quantitative and qualitative testing requirements to address the missing links with previous studies. The study suggested development of sustainable consumer index as concrete measures of consumer green behavior. Lack of green communication tool, requirement of alternative conceptual studies are highlighted in the context of assessing consumers green bahavior as stated in Paço et al. (2019). Most of the studies focused on the specific behaviors of green consumers with cross-sectional data of personal traits and making casual interpretations of findings. Moreover, none of the studies separate the actual green self-image into individuals' utility functions and motivating the consistent behavior (Binder & Blankenberg, 2017). Individual value orientation and actual observed behaviors need to quantify to find how the green lifestyle translates in to action where this value action gap shows 72% of the respondents in their intentions and actions (Binder & Blankenberg, 2017). Ruediger et al. (2012) and Testa et al. (2020), the findings of studies provide the framework for further development to an index, and both provide only the hypothesized index development and demonstrating the factors. Carbon Footprint (CFP) calculators, ecological footprint calculators and water footprint calculators also can be identified as important tools to evaluate green consumers. These tools provide good insight for individuals to identify their actions contributing adverse effects to the environments where provide quantifications information. CFP calculators are popular tools for educating individuals about the impacts of daily/short term or long-term habits and stimulate the interest in users and motivate their commitment towards emission reduction as well (Burgui-Burgui & Chuvieco, 2020; Loyarte-López et al., 2020). However, these tools are limited and subjective to few individual actions and the tools are unable to provide the compressive assessments

Table 1 Identification of the factors that effect on green consumer behaviour and sustainable consumption

Main category	Study	Factors identified in the study
1. Food and water consumption behavior	de Boer et al. (2016)	Meat consumption: discuss transitioning to low-carbon society with the support of consumers food-related options and diet change
	Shafiee-Jood and Cai (2016)	Food saving behavior: focus on enhancing food security and environmental sustainability by reducing food losses and food waste
	Nica (2020)	Organic food consumption: Buying organic food as sustainable consume decision making behavior
	Singha and Eljamal (2020)	Water conservation behavior: promoting conservation and consumption behaviour and related issues
2. Product selection and purchasing behavior	Joshi and Rahman (2015)	Price of green product: Factors affecting green purchasing behavior and future research direction
	Yadav and Pathak (2017)	Purchasing intention: Determinants of consumer's green purchase behaviour in a developing nation, applying and extending the theory of planned behaviour
	Yadav and Pathak (2017)	Environment concern: Green purchasing behavior; Green products and Ajzen's Theory planned behavior
	Zhang and Zhang (2015)	Brand image: The impact of brand image on consumer behavior
3. Energy conserving behavior	Abdel Wahed Ahmed and Abd El Monem (2020)	Mode of transportation: sustainable and green transportation for better quality of life
	Van Acker et al. (2015)	Travel behavior: travel behavior, lifestyle, and sustainable urban mobility is concerned in the study

(continued)

Table 1 (continued)

Main category	Study	Factors identified in the study
	Dutta and Hwang (2021)	Environment concern when purchasing vehicle: green electric vehicle purchase intention of consumer
	Laicane et al. (2015)	Appliances handling and electricity saving: Evaluation of household electricity savings and analysis of household electricity demand profile and user activities
	Sangroya and Nayak (2017)	Buying behavior of green energy: Factors influencing buying behavior of green energy consumer
4. Household waste management	Kang et al. (2020)	E-waste disposing behavior: Electronic waste collection system and household electronic waste management
	Chikowore (2020)	The household waste disposing method: Factors that influence on household waste management practices
	Mcneill and Moore (2015)	Disposing behavior of old cloths: fashionable consumers and attitudes to sustainability in clothing choice, and sustainable fashion consumption and the fast fashion conundrum
	Nunkoo et al. (2020)	Household food waste management: best practices, motivation, attitudes and barriers

of behaviors of the consumer with attitudes and their norms and behavioral transformations into actions. On the other hand, these calculators lack adequate information to provide context for their results, or this information is not effectively conveyed.

Furthermore, according to the above explained context, there is a requirement for developing a comprehensive tool that can capture the multidimensionality of consumer green behavior, including the cognitive, emotional, and social dimensions. Therefore, this study was conducted to construct a GCBI as a novel measure to evaluate consumer green behavior and it will provide an individual reference point for consumers to compare their own consumption patterns and motivate changing behaviors toward reducing impacts to the environment.

3 Methodology

In this section, the steps and stages outline the systematic approach that followed in the research to investigate and understand and measure consumers' green behavior. The followed methodology and framework are described in detail below.

3.1 Development an Index to Evaluate Green Behavior and Sustainable Consumption: Green Consumer Behavior Index (GCBI)

The study was conducted according to the design framework shown in Fig. 2. The study followed the different processes in various stages to develop the GCBI including the factor identification, development of GCBI, conducting the survey, data analysis, and updating the GCBI according to the comments. Furthermore, validation of the index was done after addressing feedback receiving for the initial survey by presenting their own individual scores.

3.1.1 Development of the GCB Index

The GCBI was developed concerning four major categories and the factors identified in each major category that mentioned in Sect. 2.5. The index was constructed as a questionnaire based on the theories mentioned in Sect. 2.4, and the identified factors

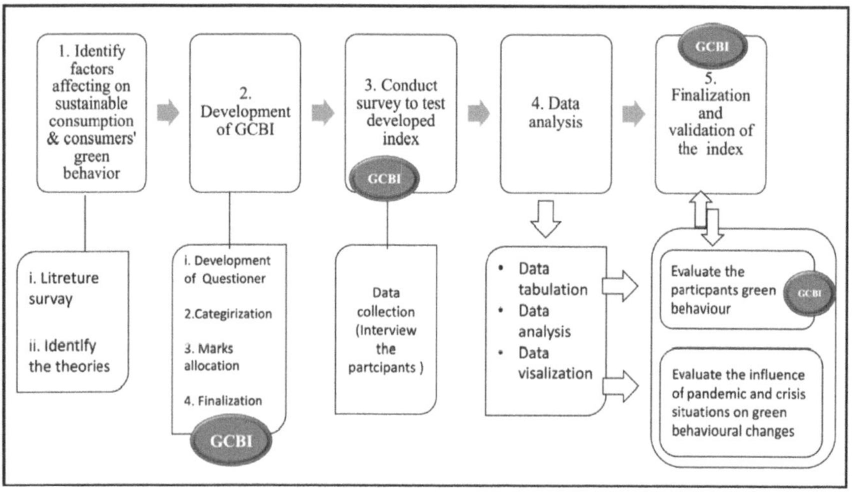

Fig. 2 The framework of the study: development process of GCBI

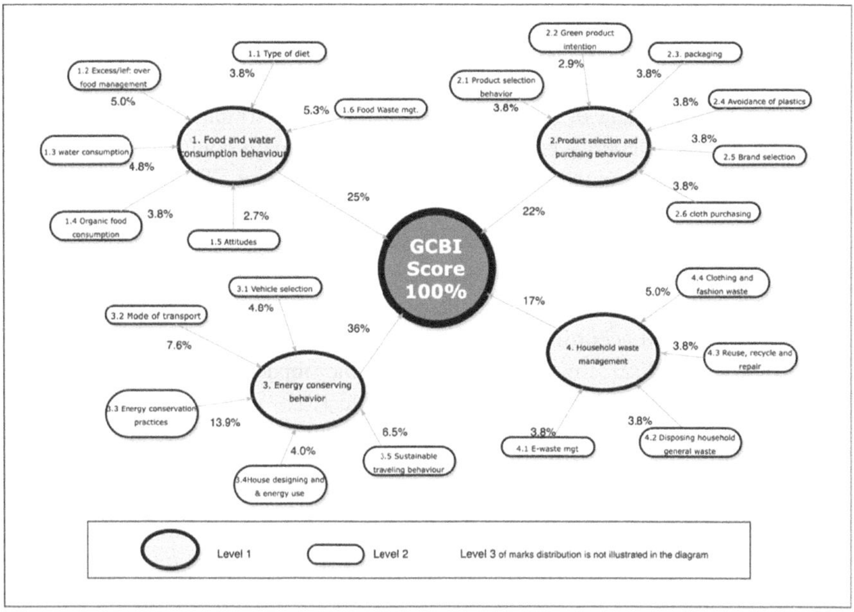

Fig. 3 Marks distribution among major categories of GCBI

in Sect. 2.5. The final marks/scores were allocated to each question that represents the behavioural activities, and allocated marks were distributed by concerning their positive and negative impacts to the environment. Furthermore, the actions/activities/behavioral are aligned with the different subgroups. The allocations of the marks in various categories are mentioned in Fig. 3.

Most human activities revolve around the necessities of life such as food, transportation, clothing and housing which have resulted in significant carbon dioxide emissions. The factors were identified in four major areas: food and consumption behavior, Product selection and purchasing behavior, energy conserving behavior and house-hold waste management practices. The household and individual activities have direct and indirect effects on energy use and related carbon dioxide emission. Personal travel and residential energy use are direct factors, while the operations of housing, transportation and food consumption are considered indirect influences on energy consumption and CO_2 emission (Barbarossa & Pastore, 2015). This research also focused on examining those diverse green behaviours within the context of the social and psychological bases of sustainable lifestyles.

The factors belonging to main consumption categories were evaluated in the index's first level as behavioral activities that consumers engage in today. These behavioral activities were categorized according to their relativity and the distribution of marks in second level represented the sub-categories of the index. The food and water consumption behaviour, product selection and purchasing behavior, energy conserving behavior and house-hold waste management practices are the major categories of the index which are consisted with subcategories while the distribution of marks in third level represented the major categories of the index.

The higher percentage of marks (36%) was allocated to the energy-conserving behavior category as energy is considered one of the main contributors to carbon emission which destroys the quality of the environment. The second highest percentage of marks (25%) in the index were allocated to the food and water consumption behaviour category. The product selection and purchasing behaviour category allocated 22% of marks as the third category. As the final category (4th category) 17% of marks are allocated for household waste management related activities. The marks were allocated according to the findings of previous studies (Bhoyar et al., 2014; Koide et al., 2021; Testaet al., 2020) and the impacts to the environment as shown in Fig. 1a–c and studies mentioned in Table 1.

Level three of marks distribution assesses the behaviours of the consumer based on behavioral actions, attitudes and norms based on the theories of green consumerism explained in Sect. 2.4.

As mentioned in the theories in Sect. 2.4, used to select the 3rd level factors for the index including motivation factors, social norms, belief, intention, attitude, self-images and market-based factors. Furthermore, the findings form studies in Sect. 2.5 adapted to 3rd level criteria selection as shown in Table 2.

The total marks were categorized into a scale with five different categories to evaluate individual behavior. Consumers who show an individual score of less than 40 are categorized as consumers with poor green behavior and consumers who have scored 41–54 belong to the unsatisfied green behavior category. For the satisfied green behavior category represents the consumers with a 55–64 score, score 65–79 is categorized as very satisfied green behavior. Finally, the consumers who achieved an 80–100 score were identified as those with outstanding behavior. Furthermore, the marks of the categories reflect the consumers' behavior and the opinions regarding the scale and the score values were evaluated during the validation.

3.2 *Data Collection and Data Analysis*

To assess the consumers' green behavior using selected factors based on theories and the previous studies explained in Sect. 2.4 and 2.5, a questionnaire-based survey was conducted to collect information. The questionnaire was developed based on categories and subcategories explained in the mark's allocations Sect. 3.1 to collect the data for GCBI. The questionnaire that was used in this study was divided into five sections. The first section collects the demographic information/socio-economic characteristics including gender, age, education, income and other four sections consist with 25 questions to access four major categories (see the Sect. 3.1). The questions provided under each subcategory in the questionnaire included level three activities explained in Table 2.

Table 2 Marks distribution of each subcategories as Level 3

Lv2	Lv3 Activity	Type of scale	Marks for activity/bahvior
1.1	a. Meet consumption (Regularly, seldom, fish and eggs, eggs, Vegetarian) b. Meet type (Beef, lamb/mutton, pork, chicken, other)	LS LS	0–4 0.2–0.8
1.2	Leftover/excess food management (throw as waste, feed animals, share with other, keep for further use, use for biogas production, no excess only prepare for family requirement)	MO	0–1.75 Maximum: cumulative value 5.25
1.3	Water consumption Use mindfully and encourage family to save water, concern on water saving, reuse when possible, concern if someone remind, expect to change habits to save water)	LS	5–1
1.4	Organic food consumption (grow, grow and some buy, buy market, consume own grown inorganic, buy inorganic, doesn't care the type)	LS	5–1
1.5	Attitude: prefer to consume products Better for environment, based on health benefits, quality, eco Fredy products more expensive, less trust/ trust issues)	LS	2.8–0
1.6	Actions to minimize waste (avoid bulk purchase, concern expire date, use preservation techniques separate storage,, choose less perishable and consume first, use natural preservatives,)	MO	1.4–0.2 Maximum: cumulative value 5.6
2.1	Purchase green products (even higher price, only affordable, when discount, when cheaper, not concern)	LS	3–0
2.2	Green Product intention (with eco label, no eco.lable but not harmful to env. Recyclable, reusable, less harmful to env, no concern)	LS	4–0
2.3	Packaging (excess packaging, less package, recyclable, compostable, reusable, no packaging)	LS	0–4

(continued)

Table 2 (continued)

Lv2	Lv3 Activity	Type of scale	Marks for activity/bahvior
2.4	Avoid plastic: (carry own bags, select reusable bags, use cloth bag, try to use minimum bags, purchase bags when shopping)	LS	0–4
2.5	Brand selection (Sustainable brands only, sustainable when affordable, try to purchase of sus. Brand on friend recommendations, concern eco-labels, brand priority over sustainability, Don't concern on sus. Brands)	LS	0–4
2.6	Purchasing cloths (latest fashion, Less ironing, match with many items, suitable to many occasions, durability, ecofriendly materials)	LS	0–4
3.1	Vehicle selection factors (price, brand/design/model, durability/long term use/fuel efficiency/env. friendliness)	MO	0–2 Maximum: cumulative value 5
3.2	Mode of transport (activities: commuting, trips, others Modes: foot, cycle, public transport, taxi, privet vechicle)	MO	0–2 Maximum: cumulative value 8
3.3	Energy conserving practices (in lightings, use of refrigerators, cooking, use of air conditioning, avoid anergy intensive hobbies, use renewable energy sources, no practices)	MO	0–2 Maximum: cumulative value 6.4
3.4	House design and energy use (design, space utilization, water system, air quality mgt., construction materials, vegetation, no considerations)	MO	0–2 Maximum: cumulative value 4.2
4.1	E-waste (reuse/repair, donate, sell, store in home, dumping)	LS	4 -0
4.2	Daily waste (composting, give to waste truk after separation, sell to collectors, dumping, buring)	LS	4 -0

(continued)

Table 2 (continued)

Lv2	Lv3 Activity	Type of scale	Marks for activity/bahvior
4.3	Waste product use (food waste-biogas, wood: heating, use for recreations, dumping)	LS	4–0
4.4	Manage old cloths and fashions (resell, donate, wear home, use other purposes, upcycling, no use)	MO	0–1.75 Maximum: cumulative value 4.2

MO Multiple use, *LS* Licket scale/lener scales values with one option

The developed index was distributed initially among 131 consumers who were selected by purposive sampling technique. The participants were selected according to Testaet al., 2020 (study: Green consumer behavior: insights from survey and experiments), by considering the age groups in demographics. The consumers aged above 16 were selected as the target group for interviewing, testing the index and collecting feedback for further improvement. Since this study targets secondary level school children, the age groups were slightly adjusted where that study considered participants from 18 to 70 and this study considers four age groups as 16–25, 26–35, 36–45 and 45–70. The questionnaire used different scale formats in different questions such as Likert scales, semantic differential scales and multiple selections option. The collected data were analyzed using Microsoft Excel and IBM SPSS Version 26 software to identify the factors influencing sustainable consumption. Consumer green behavior was assessed based on the criteria outlined in Sect. 2.5. Descriptive statistics including bar charts and pie charts were used to explain and compare the individual performances and the overall performances with different category levels. Furthermore, correlation analysis was conducted between demographic variables and GCBI scores. The validation of the index was conducted with a different questionnaire as explained in the Sects. 3.3.

3.3 Validation the Developed GCBI

The developed index was validated to justify the composite measure resulting from the previous phases. The interviews and questionnaire survey (validation questionnaire) were conducted to test the index validity in the second round. The thirty consumers were selected by purposive sampling technique among those who previously participated in the study. Participants were selected according to explain 3.2 by considering age and education and other demographic factors. After distributing their individual scores among different levels of the index, the participants evaluated their personal standing points on the identification of behavioral aspects that need to improve, index value reflection of sustainable consumption behavior, satisfaction and further improvements of behavioral actions and their recommendation for further use of GCBI. The validation results are explained in Sect. 4.4.

4 Results and Discussion

Within this section, the key findings of the study are presented and organized according to the research objectives and the framework shown in Fig. 2 in the Methodology section. Through a systematic analysis of data collected from a diverse and representative sample, the study attempted to identify the individual behaviors, preferences, and perceptions that shape consumer choices in the context of sustainable consumption and environmental consciousness. The factors identified through the literature survey that influence green consumer behavior and sustainable consumption were described in Sect. 2.5.

4.1 Demographic Information of the Participated Consumers

The demographic information of the respondents, including gender, age, education level and income level were inquired at the first part of the questionnaire. The overview of the demographic characteristics of the study participants is essential for understanding the composition of the sample and its relevance to the research objectives and the findings. Table 3 shows the demographic features of respondents who participated in the questionnaire survey conducted to investigate the GCB index.

The study group consisted of 131 people who filled in the questionnaire. There was approximately equal participation from all four age groups for the questionnaire. The majority of participants had completed their higher studies (38%) and a minority (30%) of the respondents had completed only their junior secondary education. The minority (9%) of respondents represented the higher income levels category with an income greater than LKR.65000. 19% of the respondents were reported to belong to the category of unemployed.

Table 3 Results of demographic features of respondents

Variable	Group	%	Variable	Group	%
Gender	Male	43	Income level	< Rs.15,000	12
	Female	57		Rs. 15,000–Rs 30,000	18
Age	16–25	23		Rs. 30,000–Rs. 45,000	22
	26–35	26		Rs. 45,000–Rs 65,000	20
	36–45	26		< Rs. 65,000	9
	46–65	24		Unemployed	19
Education level	Higher education	38			
	Senior secondary education	32			
	Junior secondary education	30			

4.2 *Score Distribution in the Sample and Comparison of Achieved Score Against Total Score in Different Levels of Index*

To accurately present the realistic self-image and the value gap of individual without bias while ensuring the evaluation process does not merely reflect the average all consumers, it is crucial to use random consumer example as the means of explaining the evaluation method demonstrate in the study. As given in Fig. 4, any individual can compare his/her achieved score with total score. From this the consumer can identify each category and subcategories which he/she has to improve to be a successful green consumer.

The distribution of achieved scores of each category in the study group against the total allocated standard score is given in Fig. 4. According to the gathered data the categories, consumption behavior, purchasing behavior, energy conserving behavior and household waste management behavior must improve by 24.45, 43.77, 23.33 and 32.94% respectively to reach standard allocated score for sustainable lifestyle. According to the chart, the above population has less concern for the environment in the aspect of product selection and purchasing. The studied population achieved the highest score in energy-conserving behavior, which signifies consumers have positive attitudes and actions towards energy-conserving activities in daily life.

A breakdown of achieved scores of energy conserving behavior in relative to standard allocated score is given in Fig. 5. According to gathered data, mode of transport was identified as the least sustainable energy conserving behavior, and it needs to be improved by 93% to reach the sustainability level. Using zero emission methods including walking and cycling for short distance travelling, switching to practices like carpooling, using public transportation and using maritime and river transport

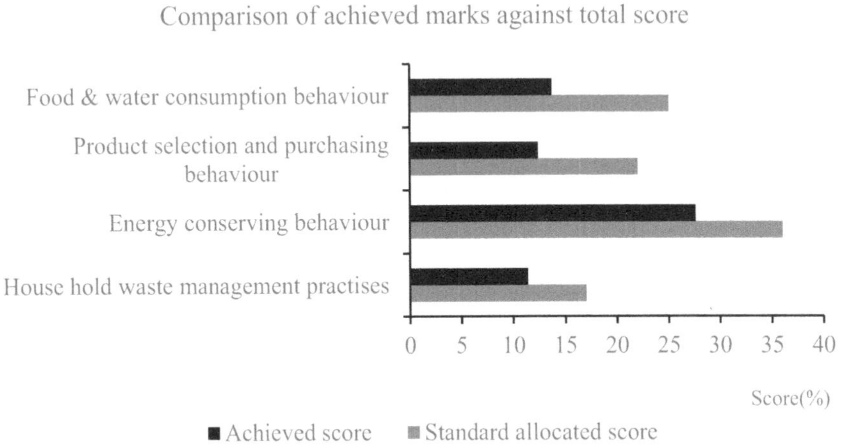

Fig. 4 Distribution of GCBI scores in the sample

instead for air trade can reduce the energy usage while reducing GHG emissions. Ride sharing apps could be developed and publicized to connect potential carpooling partners. Telecommunications methods should be introduced and improved for remote work rather than travelling longer distances to work.

The attitude on greenhouse (building) designing should be improved by 70% to achieve the target score of sustainable behaviour. The remaining 52% of the vehicle purchasing behaviour can be improved by choosing EV and hydrogen vehicles and vehicles with more fuel efficiency and durability. Sustainable travelling and energy conserving habits need to develop by 20 and 14.2% respectively to reach the target score. (Fig. 5a).

Any person can change his/her behaviour into a sustainable household waste management practice without any cost. It just requires a change in attitude in day today life. The practice of reusing waste items has least score in household management, and it need to be improved by 47.36% to reach sustainable behaviour. According to Fig. 5b majority of the population had scored in waste management practices which means majority of them dispose waste properly, reduce the amount of waste by recycling and up cycling. (Fig. 5b).

To achieve the sustainable behavior, the behaviors of water consumption, attitude on organic food, organic food consumption, avoidance of food waste, excess food management and type of diet need to improve by 33.33, 25.95, 42.01, 54.71, 56 and 55.26% respectively (Fig. 5c).

Excess food waste can be managed by avoiding bulk buying, sharing with neighbors, preserving for future use, putting in biogas plant and by providing as feed for animals. Choosing vegetarian, organic food, with less meat can be considered as a more sustainable type of diet. The protein requirement can be fulfilled by choosing sustainable seafood, plant-based meat and soya bean milk over meat and cow milk. Using local and seasonal foods also contributes a lot to reducing emissions. Organic home gardening is a particularly good practice to fulfill household nutrient requirements in a more sustainable manner (Fig. 5c).

As in Fig. 5d the respondents have paid less attention to selecting green products and brand image of the product. While doing shopping, most consumers use plastic bags and have poor intention on buying products with green packages.

The GCB Index was scaled into five levels namely, outstanding, very satisfactory, satisfactory, unsatisfactory and poor. The majority of the respondents (58%) belonged to the unsatisfactory level in GBCI scale while only minority (11%) belong to the outstanding level of the GBCI scale. The outstanding behavior was showed by only 8% of population and they had less concern on food and water consumption behavior compared to other three categories which means when buying food, and using water resources they deviate slightly from following sustainable behaviors. The consumers who showed unsatisfactory green behavior had more unsustainable behaviors in product selection and purchasing compared to the other three categories.

The marks distribution of GCBI is allocated the highest number of marks for the energy consumption behavior although the consumers who showed poor green behavior (< 40) had the least concern on sustainable energy conserving behavior. Meanwhile, they have shown more sustainable behaviors related to the food and water

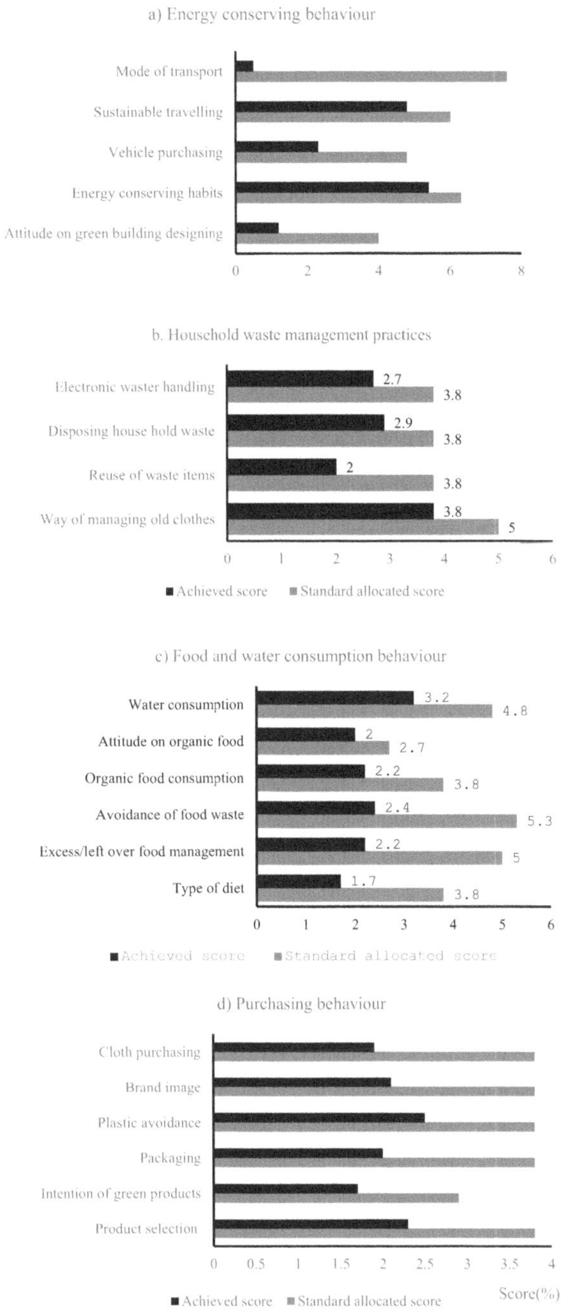

Fig. 5 Distribution of scores in subcategories of main categories

Fig. 6 Behavior analysis of poor, unsatisfactory and outstanding levels of sample

consumption category compared to the other three categories. The consumers who showed unsatisfactory green behavior had more unsustainable behaviors in product selection and purchasing compared to the other three categories. The outstanding behavior was shown by only 8% of the population and they had less concern about food and water consumption behaviour compared to the other three categories which mean when buying food, and using water resources they deviate slightly from following sustainable behaviors as shown in Fig. 6.

4.3 Correlation Between GCBI and Demographic Variables

The demographic features gender, age, number of family members and level of education did not show any relationship with the GCBI. The income level and education level were positively correlated with GCBI (Table 4).

It is crucial for policymakers to understand how green behavior of female impact on sustainability. Researchers has found that women do key role in sustainability than men. Women make more impact on others in society since they stay at home and are closer to family members. Women make most decisions on the purchases

Table 4 Correlation between GCBI and demographic variables

Variable	Coefficient value	*P* value	Null hypothesis
ETA coefficient value			
Gender	0.065	0.465	Accepted
Age	0.115	0.905	Accepted
Pearson coefficient value			
Family members	– 0.006	0.944	Accepted
Education	0.385	0.021	Rejected
Spearman coefficient value			
Income	0.248	0.004	Rejected

of goods for the house (Kawgan-Kagan, 2020). Based on Fig. 7a, also, it revealed, higher number of female participants showed outstanding green behavior compared to male participants. In consumer behavior, gender and age have a significant effect according to previous studies. There was a clear distinction between men's and women's view about purchasing environmentally friendly goods and claimed that gender is determined by the biological differences between men and women which manifest in their varying personalities (Pudaruth et al., 2015). According to the studies, Women are more likely than males to purchase environmentally friendly items and to be concerned about their safety and health (Kawgan-Kagan, 2020).

The sample data was not strong enough to conclude that a relationship between the categorical variable exists in the population (Fig. 7b). However, based on the results of a recent study it was found that age showed a positive relationship with pro-environmental behavior such as avoiding waste, reducing use. It revealed that old workers in the organization performed more of environmentally friendly behavior and they had stronger tendency to perform resources conserving behavior (Wang et al., 2021).

According to Fig. 7d, consumers who had shown outstanding green behavior, had average income level. According to study, consumer income and its relation to sustainable food consumption, which was conducted in Hungary, they revealed that ecological footprint of food consumption was not equally distributed among the income group and middle upper income group had showed unsustainable behavior related to food consumption (Veto and Né Mózner, 2014).

The consumers who had higher education qualifications were more aware about green purchasing and resources conserving behavior compared to consumers who had lower education as shown in Fig. 7c. Identify the influence of education in pro-environmental behavior in Europe which revealed that education causes consumers to be more concerned with social welfare and pro-environmental behaviors (Meyer, 2015).

4.4 Validation Test Results and the Participants' Feedback

The index was distributed as a questionnaire among 30 participants with the purpose of validation. The participants were selected via purposive sampling technique as mentioned in Sect. 3.3. The index was applied to analyze the negative and positive green behaviors of the sample and comments were given according to their index value. The participated consumers were interviewed to get their opinions on index with respect to their behavior. The mean index score of the validation sample was 57.9%. Figure 8 reflects the distribution of the index score of the validated sample and average index was 58% which score by validation sample for GCBI.

Most of them agreed with their index marks related to their general behavior and they could identify their unfavorable behaviors which impact on the environment negatively. The participants were satisfied with the evaluation of their lifestyle's sustainability with GCBI. They would like to use it to do change their current behavior

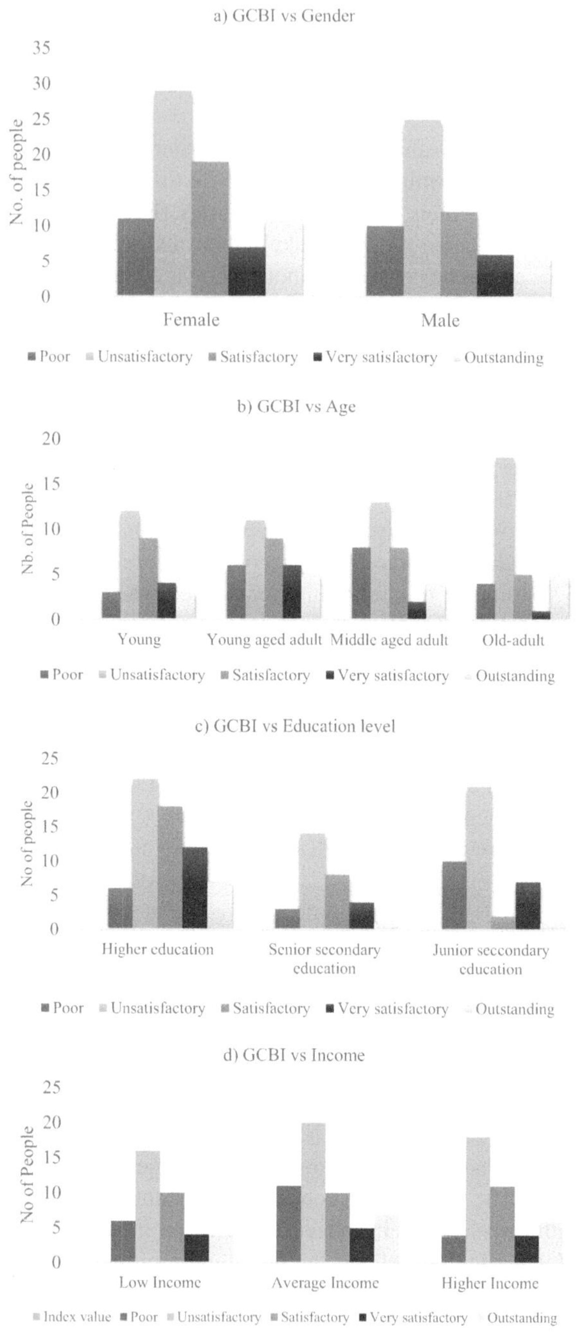

Fig. 7 Relationship between GCBI and demographic variables

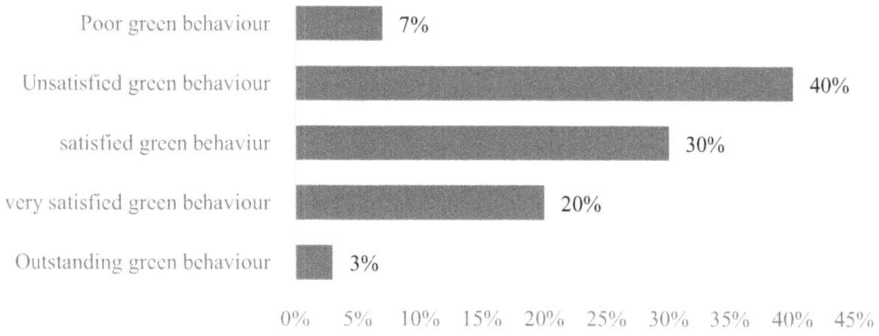

Fig. 8 The distribution of participants of validation test results under the GCB Index scale

Table 5 Feedback of the consumers towards GCBI during the validation

Feedback of the participants	Mean	Scale
I identified behavioral aspects that need to improve	2.13	Agree
My index value reflects my behavior on sustainable consumption	2.1	Agree
I am satisfied with my current sustainable consumption behavior	3.6	Disagree
I expect to change my current behavior to live sustainably	1.5	Strongly agree
I recommend GCBI to further use	1.8	Strongly agree

sustainably. The majority of the participants volunteered to introduce the GCBI to their colleagues to ensure sustainable consumption behavior of the community (Table 5).

5 Summary of the Results and Discussion

The general understanding of facts, ideas, and relationships pertaining to the environment and ecosystems is called environmental knowledge. The Green Consumer Behavior Index evaluates on consumer's environmental knowledge and attitudes on how products are manufactured, their effects on the environment, eco-labeling and eco-packaging of products, organic food consumption and energy-conserving practices. Consumers' green consumption behavior must be able to maintain environmental quality because the individual consumption pattern in their modern lifestyle contributes to high environmental damage. Effective consumer behavior measurement helps companies to develop their marketing strategies and consumers can focus more on improving their sustainable lifestyle. Reducing individual resource use and improving corporate environmental performance are both benefited by the consumer's GCBI. As an increasing number of organizations focus on product and processes meant to minimize environmental harm, understanding the impact of green consumer values on purchase behavior is crucial. The findings of the index will assist consumers, producers, policymakers and academics in facilitating or encouraging sustainable consumption behavior through government legislation, product development or other sustainability initiatives. In contrast to other environmental surveys, this study focuses on actual behavior and material lifestyles rather than just opinions and attitudes and this includes metrics like house energy use, travel patterns and the relative penetration of green items compared to standard products.

5.1 Limitations of the Study

Green consumerism is a complex concept that is defined with various factors and interactions. Due to the complexity of access, the GCBI required large data including information on individual consumption, norms, attitudes. The time duration for interviews extended beyond the expected time frame due to complexity, the higher number of questions and some questions required further explanations. It may affect to interest of the consumer's to complete the assessment and observe the results The index may complex to the aged below 16 and some areas were not appropriate to properly evaluated some age groups since they have not control of some factors including household operations. Some selected factors are linked with the economic state of the consumers where hinder the individuals' actual green behaviors and intentions. Consequences can arise due to different scales of mark allocations. Cultural background and the behaviors link with religions that affect to green behavior not properly assed in the index. Since the index is newly proposed, to calculate the index values and present the results to participants required additional time. These limitations highlighted the requirement of programmable or web-based applications to present the results upon completion of the assessment. These limitations highlighted the requirement of programmable or web-based applications to present the results upon completion of the assessment.

5.2 Recommendations and Further Direction of the Study

The Green Consumer Behavior Index provides an opportunity to identify the drawbacks of individual consumer behavior and track their behavioral changes over the period of time. By raising consumer awareness and giving them a standard score against which to evaluate their own consumption habits, the Green Consumer Behavior Index aims to promote sustainable consumption. As highlighted in Sect. 4.7 it can be developed into an application, tool or web-based program which will enable easy access to the wider audience. The results can be stored in data base and can be used for further development of consumer behavior models with different techniques like machine learning application for green behavior identification. The index has to be developed into an education and awareness rising tool on green consumer behavior and can be used in school, education programs, awareness campaigns. The multidimension approach when used in the development of the index provides many directions for further development that effectively encourage individual's sustainable consumption. The index will provide the valuable insight about the consumers preference and behaviors related to suitability that can be used for marketers and business to promote and tailor their services, products and comminutions towards brand loyalty. Furthermore, this index can be used for benchmarking consumer actions and evaluate the progress towards green consumerism. The GCBI, can be used for quantifying and tract green consumer behaviors, where serves as a critical tool in the collective effort to move towards a more sustainable and environmentally friendly society. Since there less applications and studies on quantification of behaviors and behavioral intentions this study will be a better baseline study that can be expanded according to the future assessments and studies.

6 Conclusion

The Green Consumption Behavior Index is a composite measure of consumer's environmentally sustainable behavior which developed to evaluate consumer's behavior in four broad areas; product selection and purchasing behavior, food and water consumption behavior, sustainable waste, management practices, and energy conserving behavior. The results showed an association between demographic factors and the green behavior index value and one of the key findings was the positive relationship between education level and index value. The majority of the respondents of the selected sample (58%) belonged to the unsatisfactory level in GBCI scale while only minority (11%) belong to the outstanding level of the GBCI scale. The outstanding behavior was showed by only 8% of population and they had less concern on food and water consumption behavior compared to other three categories. Majority of participants agreed with their GCBI value which reflects their environmentally conscious behavior. GCBI was identified by participants as a tool to identify their individual green behavior. Further, using it as a reference point, they can modify their behavior in future. The index findings assist consumers, producers, policymakers and academics in facilitating or encouraging sustainable consumption behaviour through government legislation, product development or other sustainability initiatives.

References

Abdel Wahed Ahmed, M. M., & Abd El Monem, N. (2020). Sustainable and green transportation for better quality of life case study greater Cairo. *Egypt, 16*, 17–37. https://doi.org/10.1080/16874048.2020.1719340

Abdulrazak, S., & Quoquab, F. (2017). *Exploring consumers' motivations for sustainable consumption: A self-deterministic approach, 30*, 14–28. https://doi.org/10.1080/08961530.2017.1354350

Almasi, A., Mohammadi, M., Azizi, A., Berizi, Z., Shamsi, K., Shahbazi, A., & Mosavi, S. A. (2019). Assessing the knowledge, attitude and practice of the kermanshahi women towards reducing, recycling and reusing of municipal solid waste. *Resources, Conservation and Recycling, 141*, 329–338. https://doi.org/10.1016/J.RESCONREC.2018.10.017

Barbarossa, C., & Pastore, A. (2015). Why environmentally conscious consumers do not purchase green products: A cognitive mapping approach. *Qualitative Market Research, 18*, 188–209. https://doi.org/10.1108/QMR-06-2012-0030/FULL/PDF

Barloa, E. P., Lapie, L. P., Paul, C., De La Cruz, P. (2016). *Knowledge, attitudes, and practices on solid waste management among undergraduate students in a Philippine State University* (Vol. 6).

Bhoyar, S. P., Dusad, S., Shrivastava, R., Mishra, S., Gupta, N., & Rao, A. B. (2014). Selection and peer-review under responsibility of the Organizing Committee of Understanding the impact of lifestyle on individual carbon-footprint Selection and peer-review under responsibility of the Organizing Committee of. *Procedia-Social and Behavioral Sciences, 133*, 47–60. https://doi.org/10.1016/j.sbspro.2014.04.168

Binder, M., & Blankenberg, A. K. (2017). Green lifestyles and subjective well-being: More about self-image than actual behavior? *Journal of Economic Behavior & Organization, 137*, 304–323. https://doi.org/10.1016/J.JEBO.2017.03.009

Burgui-Burgui, M., & Chuvieco, E. (2020). Beyond carbon footprint calculators. New approaches for linking consumer behaviour and climate action. *Sustainability, 12*(16), 6529. https://doi.org/10.3390/su12166529

Chawla, M., & Naved Khan, M. (2015). Online buying behaviour: A brief review and update. *Aima Journal of Management & Research, 9.*

Chekima, B., Chekima, S., Syed Khalid Wafa, S. A. W., Igaua, O. A., & Sondoh, S. L. (2015). Sustainable consumption: The effects of knowledge, cultural values, environmental advertising, and demographics. *International Journal of Sustainable Development & World Ecology, 23*, 210–220. https://doi.org/10.1080/13504509.2015.1114043

Chikowore, N. (2020). Factors influencing household waste management practices in Zimbabwe. *Journal of Material Cycles and Waste Management, 23*, 386–393. https://doi.org/10.1007/s10163-020-01129-9

de Boer, J., de Witt, A., & Aiking, H. (2016). Help the climate, change your diet: A cross-sectional study on how to involve consumers in a transition to a low-carbon society. *Appetite, 98*, 19–27. https://doi.org/10.1016/J.APPET.2015.12.001

Desai, B. H. (2022). Regulating global climate change: From common concern to planetary concern. *Environmental Policy and Law, 52*, 331–347. https://doi.org/10.3233/EPL-219050

do Paço, A., Shiel, C., & Alves, H. (2019). A new model for testing green consumer behaviour. *Journal of Cleaner Production, 207*, 998–1006. https://doi.org/10.1016/J.JCLEPRO.2018.10.105

Dutta, B., & Hwang, H. G. (2021). Consumers purchase intentions of green electric vehicles: The influence of consumers technological and environmental considerations. *Sustainability, 13*, 12025. https://doi.org/10.3390/SU132112025

Ezzati, N. A. A. N., Yani, A. (2017). *The determinant factors of green consumption behavior.*

Foo, M.Y., Kanapathy, K., Zailani, S., Shaharudin, R., Meow, Y., Foo, K., Kanapathy, S., Zailani, M., & Rizaimy, S. (2021). Green purchasing: Capabilities, practices and effects on firms' triple bottom line performance. *Studies of Applied Economics, 39*. https://doi.org/10.25115/EEA.V39I3.4160

Haba, H. F., Bredillet, C., & Dastane, O. (2023). Green consumer research: Trends and way forward based on bibliometric analysis. *Cleaner and Responsible Consumption, 8*, 100089. https://doi.org/10.1016/J.CLRC.2022.100089

Han, H. (2020). Theory of green purchase behavior (TGPB): A new theory for sustainable consumption of green hotel and green restaurant products. *Business Strategy and the Environment, 29*, 2815–2828. https://doi.org/10.1002/BSE.2545

Handayani, W., Prayogo, R., & Prayogo, R.A. (2017). Green consumerism: An Eco-friendly Behaviour Form Through the Green Product Consumption and Green Marketing. *Sinergi: Jurnal Ilmiah Ilmu Manajemen, 7*. https://doi.org/10.25139/SNG.V7I2.364

Hartmann, C., & Siegrist, M. (2017). Consumer perception and behaviour regarding sustainable protein consumption: A systematic review. *Trends in Food Science & Technology, 61*, 11–25. https://doi.org/10.1016/J.TIFS.2016.12.006

Jain, V. K., Dahiya, A., Tyagi, V., & Dube, P. (2021). Predicting factors affecting green purchase intentions towards green products. *World Review of Science, Technology and Sustainable Development, 17*, 360–374. https://doi.org/10.1504/WRSTSD.2021.117893

Joshi, Y., & Rahman, Z. (2015). Factors affecting green purchase behaviour and future research directions. *International Strategic Management Review, 3*, 128–143. https://doi.org/10.1016/J.ISM.2015.04.001

Kang, K. D., Kang, H., Ilankoon, I. M. S. K., & Chong, C. Y. (2020). Electronic waste collection systems using Internet of Things (IoT): Household electronic waste management in Malaysia. *Journal of Cleaner Production, 252*, 119801. https://doi.org/10.1016/J.JCLEPRO.2019.119801

Kasavan, S., Mohamed, A. F., Halim, S. A. (2017). Sustainable food waste management in hotels: case study langkawi unesco global geopark. *Planning Malaysia, 15*. https://doi.org/10.21837/PM.V15I4.317

Kawgan-Kagan, I. (2020). Are women greener than men? A preference analysis of women and men from major German cities over sustainable urban mobility. *Transp Res Interdiscip Perspect, 8*, 100236. https://doi.org/10.1016/J.TRIP.2020.100236

Kazi, S. M., & Mehta, T. (2016). *Demystifying sustainable consumption & production*. TERI [www document]. https://www.teriin.org/opinion/demystifying-sustainable-consumption-production. Accessed 18 November, 2023.

Khalil, H. B., & Zaidi, S. J. H. (2014). Energy crisis and potential of solar energy in Pakistan. *Renewable and Sustainable Energy Reviews, 31*, 194–201. https://doi.org/10.1016/J.RSER.2013.11.023

Koide, R., Lettenmeier, M., Akenji, L., Toivio, V., Amellina, A., Khodke, A., Watabe, A., & Kojima, S. (2021). Lifestyle carbon footprints and changes in lifestyles to limit global warming to 1.5°C, and ways forward for related research. *Sustainability Science, 16*, 2087–2099. https://doi.org/10.1007/S11625-021-01018-6/FIGURES/5

Laicane, I., Blumberga, D., Blumberga, A., & Rosa, M. (2015). Evaluation of household electricity savings. Analysis of household electricity demand profile and user activities. *Energy Procedia, 72*, 285–292. https://doi.org/10.1016/J.EGYPRO.2015.06.041

Loyarte-López, E., Barral, M., & Morla, J. C. (2020). Methodology for carbon footprint calculation towards sustainable innovation in intangible assets. *Sustainability, 12*, 1629. https://doi.org/10.3390/SU12041629

Luthra, C. (2022). Determinants of green purchase behaviour—a literature review on specific product categories. *Academy of Marketing Studies Journal, 26*.

Madden, T. J., Ellen, P. S., & Ajzen, I. (1992). A comparison of the theory of planned behavior and the theory of reasoned action. *Personality and Social Psychology Bulletin, 18*, 3–9 https://doi.org/10.1177/0146167292181001

Mcneill, L., & Moore, R. (2015). Sustainable fashion consumption and the fast fashion conundrum: Fashionable consumers and attitudes to sustainability in clothing choice. *International Journal of Consumer Studies, 39*, 212–222. https://doi.org/10.1111/ijcs.12169

Meyer, A. (2015). Does education increase pro-environmental behavior? Evidence from Europe. *Ecological Economics, 116*, 108–121. https://doi.org/10.1016/J.ECOLECON.2015.04.018

Nica, E. (2020). Buying organic food as sustainable consumer decision-making behavior: Cognitive and affective attitudes as drivers of purchase intentions toward environmentally friendly products. *SHS Web of Conferences, 74*, 04018. https://doi.org/10.1051/SHSCONF/20207404018

Nickerson, C. (2023). Theory of reasoned action (Fishbein and Ajzen, 1975) [www document]. https://www.simplypsychology.org/theory-of-reasoned-action.html. Accessed 17 September, 23.

Nunkoo, R., Bhadain, M., & Baboo, S. (2020). Household food waste: Attitudes, barriers and motivations. *British Food Journal, 123*, 2016–2035. https://doi.org/10.1108/BFJ-03-2020-0195/FULL/PDF

Poore, J., & Nemecek, T. (2018). Reducing food's environmental impacts through producers and consumers. *Science, 1979*(360), 987–992. https://doi.org/10.1126/SCIENCE.AAQ0216/SUPPL_FILE/AAQ0216_DATAS2.XLS

Pudaruth, S., Juwaheer, T. D., & Seewoo, Y. D. (2015). Gender-based differences in understanding the purchasing patterns of eco-friendly cosmetics and beauty care products in Mauritius: A study of female customers. *Social Responsibility Journal, 11*, 179–198. https://doi.org/10.1108/SRJ-04-2013-0049/FULL/PDF

Quoquab, F., & Mohammad, J. (2020) A review of sustainable consumption (2000 to 2020): What we know and what we need to know. *Journal of Global Marketing, 33*, 305–334. https://doi.org/10.1080/08911762.2020.1811441

Rahnama, H., & Rajabpour, S. (2017). Identifying effective factors on consumers' choice behavior toward green products: The case of Tehran, the capital of Iran. *Environmental Science and Pollution Research International, 24*, 911–925. https://doi.org/10.1007/S11356-016-7791-X

Rajmohan, K., & Weerahewa, J. (2010). Household Energy Consumption Patterns in Sri Lanka. *Sri Lankan Journal of Agricultural Economics, 9*, 55. https://doi.org/10.4038/SJAE.V9I0.1833

Ran, Y., Nilsson Lewis, A., Dawkins, E., Grah, R., Vanhuyse, F., Engström, E., & Lambe, F. (2022). Information as an enabler of sustainable food choices: A behavioural approach to understanding consumer decision-making. *Sustainable Production and Consumption, 31*, 642–656. https://doi.org/10.1016/J.SPC.2022.03.026

Ruediger, H., Fateh Ali Khan, M., & Article, Y. (2012). Standard-Nutzungsbedingungen. *Amfiteatru Economic Journal.*

Sangroya, D., & Nayak, J. K. (2017). Factors influencing buying behaviour of green energy consumer. *Journal of Cleaner Production, 151*, 393–405. https://doi.org/10.1016/J.JCLEPRO.2017.03.010

Shafiee-Jood, M., & Cai, X. (2016). Reducing food loss and waste to enhance food security and environmental sustainability. *Environmental Science and Technology, 50*, 8432–8443. https://doi.org/10.1021/ACS.EST.6B01993/ASSET/IMAGES/LARGE/ES-2016-01993W_0003.JPEG

Singha, B., & Eljamal, O. (2020). A review on water conservation and consumption behavior: Leading issues, promoting actions, and managing the policies. *Proceedings of International Exchange and Innovation Conference on Engineering & Sciences (IEICES), 6*, 171–178. https://doi.org/10.5109/4102484

Soomro, R. B., Mirani, I. A., Sajid Ali, M., & Marvi, S. (2020). Exploring the green purchasing behavior of young generation in Pakistan: Opportunities for green entrepreneurship. *Asia Pacific Journal of Innovation and Entrepreneurship, 14*, 289–302. https://doi.org/10.1108/APJIE-12-2019-0093

Testa, F., Iraldo, F., Bianchi, G., Iorio, V. di, Iovino, R., & Vizzoto, F. (2020). *Green consumer behaviour: From survey and experiments.* https://doi.org/10.13140/RG.2.2.27204.14727

Tripathi, A., & Singh, M. P. (2016). Determinants of sustainable/green consumption: A review. *International Journal of Environmental Technology and Management, 19*, 316–358. https://doi.org/10.1504/IJETM.2016.082258

Trong Nguyen, L., Nguyen, T. H., Ngoc Nguyen, H., Dai Nguyen, L., Thi Thu Nguyen, D., Duy L. E. (2023). *Determinants of green consumer behavior: A case study from Vietnam.* http://www.editorialmanager.com/cogentbusiness. https://doi.org/10.1080/23311975.2023.2197673

Tufaner, M. B., & Turker, T. (2016). The importance of the sustainable development in terms of quality of life of the future generations. *Journal of Social and Administrative Sciences, 3.*

Union of Concerned Scientists. (2012). *Cooler smarter: Geek out on the data*! | [www document]. https://www.ucsusa.org/resources/cooler-smarter-geek-out-data#.W-4zk-hKjD5. Accessed 18 November, 2023.

Van Acker, V., Goodwin, P., & Witlox, F. (2015). Key research themes on travel behavior, lifestyle, and sustainable urban mobility. *International Journal of Sustainable Transportation, 10*, 25–32 https://doi.org/10.1080/15568318.2013.821003

van Bussel, L. M., Kuijsten, A., Mars, M., & van 't Veer, P. (2022). Consumers' perceptions on food-related sustainability: A systematic review. *Journal of Cleaner Production, 341*, 130904. https://doi.org/10.1016/J.JCLEPRO.2022.130904

Vargas-Merino, J. A., Rios-Lama, C. A., & Panez-Bendezú, M. H. (2023). Sustainable consumption: Conceptualization and characterization of the complexity of "being" a sustainable consumer—A systematic review of the scientific literature. *Sustainability, 15*, 8401. https://doi.org/10.3390/SU15108401

Vermeir, I., Weijters, B., De Houwer, J., Geuens, M., Slabbinck, H., Spruyt, A., Van Kerckhove, A., Van Lippevelde, W., De Steur, H., & Verbeke, W. (2020). Environmentally sustainable food consumption: A review and research agenda from a goal-directed perspective. *Frontiers in Psychology, 11*, 520238. https://doi.org/10.3389/FPSYG.2020.01603/BIBTEX

Veto, Z., & Mózner, N. (2014). Sustainability and consumption structure: Environmental impacts of food consumption clusters. A case study for Hungary. *International Journal of Consumer Studies, 38*, 529–539. https://doi.org/10.1111/IJCS.12130

Wang, Y., Hao, F., & Liu, Y. (2021). Pro-Environmental behavior in an aging world: Evidence from 31 countries. *International Journal of Environmental Research and Public Health, 18*, 1–13. https://doi.org/10.3390/IJERPH18041748

Wijekoon, R., & Sabri, M. F. (2021). Determinants that influence green product purchase intention and behavior: A literature review and guiding framework. *Sustainability, 13*, 6219. https://doi.org/10.3390/SU13116219

Wilson, N., & Edelyn (2022). Extending the Theory of Reasoned Action (TRA) model to predict consumers' intention to purchase green cosmetics. *Jurnal Manajemen Bisnis dan Kewirausahaan, 6*, 565–570. https://doi.org/10.24912/JMBK.V6I5.18746

Yadav, R., & Pathak, G. S. (2017). Determinants of consumers' green purchase behavior in a developing nation: Applying and extending the theory of planned behavior. *Ecological Economics, 134*, 114–122. https://doi.org/10.1016/J.ECOLECON.2016.12.019

Yue, B., Sheng, G., She, S., & Xu, J. (2020). Impact of consumer environmental responsibility on green consumption behavior in China: The role of environmental concern and price sensitivity. *Sustainability, 12*, 2074. https://doi.org/10.3390/SU12052074

Zhang, Y., & Zhang, Y. (2015). The impact of brand image on consumer behavior: A literature review. *Open Journal of Business and Management, 3*, 58–62. https://doi.org/10.4236/OJBM.2015.31006

Risk and Vulnerability of Food System to Climate Change: A Case Study of Upcountry Vegetable Supply Chains in Sri Lanka

M. Esham, A. W. Wijeratne, I. J. A. Ruhunuge, and N. Jayathilake

1 Introduction

In Sri Lanka, there is considerable apprehension regarding how climate change might affect food production, distribution, food security, and the well-being of local communities. In the 2021 Notre Dame-Global Adaptation Index (ND-GAIN, a measure illustrating a country's current vulnerability to climate disruptions), Sri Lanka secured the 104th position out of 185 countries with a score of 40.6 in the ND-GAIN Country Index. Additionally, its vulnerability score stood at 0.468, placing it at 124th, while the readiness score was 0.399, ranking 99th globally.

Among South Asian countries, in Sri Lanka, the intake of fruits and vegetables lags behind the recommended levels established by the World Health Organization (WHO) (Jayawardena et al., 2020). Despite vegetables constituting 8% of the total household food expenditure on average (CBSL, 2019b), vegetable consumption remains notably low in Sri Lanka (Jayatissa et al., 2014). A nutritional survey has drawn attention to a substantial segment of the Sri Lankan population who do not adhere to a balanced diet, as their daily intake of fruits, vegetables, and dairy products falls considerably below the dietary guidelines established by the government (Jayawardena et al., 2013). Several factors contribute to this low daily consumption

M. Esham (✉) · A. W. Wijeratne
Department of Agribusiness Management, Faculty of Agricultural Sciences, Sabaragamuwa University of Sri Lanka, Belihuloya, Sri Lanka
e-mail: esham@agri.sab.ac.lk

I. J. A. Ruhunuge
Data Management Division, Hector Kobbekaduwa Agrarian Research and Training Institute, Colombo, Sri Lanka

N. Jayathilake
International Water Management Institute, Colombo, Sri Lanka

I. J. A. Ruhunuge
Faculty of Graduate Studies, Sabaragamuwa University of Sri Lanka, Belihuloya, Sri Lanka

C. Malalgoda et al. (eds.), *Climate Change Adaptation in the Built Environment*,
https://doi.org/10.1007/978-3-031-75826-3_25

of fruits and vegetables, including limited purchasing power, seasonal fluctuations in market prices, and a lack of awareness of nutritious value (Jayawardena et al., 2013).

In developing countries, there is evidence of a negative correlation between the cost of food and dietary diversity (Brinkman et al., 2010). Consequently, some researchers found that income played a pivotal role in determining fruit and vegetable consumption levels in Sri Lanka (Weerahewa et al., 2013). This implies that individuals with lower incomes are often unable to afford fruits and vegetables, which tend to be relatively expensive. The cost of fruits and vegetables typically increases during adverse weather conditions, posing a particular challenge for low-income individuals. For example, heavy rains have a significant impact on vegetable cultivation in up-country regions, resulting in a scarcity of vegetables and subsequent price increases (CBSL, 2018). Extensive evidence suggests that extreme weather events occurring in regions where vegetables are cultivated contribute to a rise in vegetable prices (Esham et al., 2018). The lack of emphasis on climate change adaptation within the vegetable sector can be attributed to limited research on climate change impacts on upcountry vegetable supply chains.

In comparison to other South Asian regions, historic climatic data for Sri Lanka indicate warmer extremes (Sheikh et al., 2015a). Most of the recent studies done in Sri Lanka exhibited an alarming trend in temperature associated with more frequent droughts (Naveendrakumar et al., 2019). As a result, climate change will persist in placing significant stress on regional food systems. Climate shocks have extensive consequences across the entirety of the food supply chain, influencing various phases including food production, processing, distribution, activities in the wholesale and retail sectors, and ultimately, the act of consumption. The consequences of climate change also reach into the realm of food security, underscoring the necessity to consider the four critical dimensions of food security: the presence of food, its accessibility, stability, and proper utilization (Esham, 2021). Furthermore, vulnerability is typically related to climatic factors such as rainfall variability and temperature variations which raise the risks to food security in the event of a sudden shock.

There is a lack of research that has focused on the transmission of environmental shocks through food supply chains and their consequential impact on food security. This gap in knowledge hinders our understanding of the effects of environmental shocks on the food systems (Davis et al., 2021). The existing research on mitigation and adaptation in the food system is predominantly focused on food production, there is a dearth of information and specific guidance regarding how to mitigate and adapt to climate change at other stages in the value chain (Fanzo et al., 2018). There is an emphasis on the need for studies to understand the environmental and social systems' impact on food systems and the dynamics of the propagation of environmental shocks across supply chains through empirical research to ensure effective interventions (Davis et al., 2021). This chapter attempts to contribute to the gap in empirical studies by quantifying the impact of climate change on supply chain nodes of the upcountry vegetable supply chain in Sri Lanka.

The study presented in the chapter was done to capture the risks and vulnerability associated with upcountry vegetable supply chains due to climate variability in upcountry vegetable growing regions as well as along transport routes. The specific goals of this study were; (1) To assess the hazard, vulnerability, and risk scores (2) To identify instances of climate-related hazard events occurring in production regions and along transportation routes (3) To quantify the influence of climate hazards on collectors and wholesalers based in Dedicated Economic Centres (DECs) (4) To quantify the wastage of vegetables at DECs resulting from climate-related hazards.

1.1 Climate in Sri Lanka

Sri Lanka, a tropical island in the Indian Ocean north to the equator, is particularly vulnerable to the adverse impacts of climate change due to its geographical positioning. The distribution of rainfall in Sri Lanka is significantly shaped by its topography and the influence of two seasonal wind patterns (Chandrapala, 1996). The country features a tropical monsoon climate characterized by two well-defined monsoon seasons, each with its distinct weather patterns. Sri Lanka experiences two distinct monsoon seasons. The southwest monsoon (SWM) occurs from May to September, while the northeast monsoon (NEM) takes place from December to February (Ahmed & Suphachalasai, 2014). Additionally, there are two inter-monsoon seasons: the first inter-monsoon (FIM), which spans from March to April, and the second inter-monsoon (SIM), occurring from October to November (Zubair, 2002). Sri Lanka's geographical features result in distinctive rainfall patterns, exhibiting significant spatial variation despite the country's relatively small size. Sri Lanka's climatic zones can be categorized into three main zones namely wet, intermediate, and dry. The wet zone, situated in the southwest, experiences an annual rainfall of over 2500 mm and maintains high humidity levels, primarily influenced by the southwest monsoon. Conversely, the dry zones, located in the southern and north-western parts of the country, receive less than 1750 mm of annual rainfall. The intermediate zones, found in the eastern and central regions, receive rainfall ranging from 1750 to 2500 mm, primarily due to the influence of the northeast monsoon.

1.2 Observed Climate Change in Sri Lanka

Temperature and rainfall are the two key climatic factors that exert a substantial influence on food production in Sri Lanka. The historical climate data demonstrates a consistent trend of both extreme and notable systematic warming (Sheikh et al., 2015b). When examining rainfall data from the past century, it becomes evident that there has been a diminishing trend in up-country regions situated at an elevation of 900 m above mean sea level. However, there are no discernible patterns in dry zone rainfall (Mani et al., 2018). Comparable warming patterns are noticeable in the seasonal average temperatures during the critical agricultural phases of the Yala season spanning from April to September and Maha from October to March

(Jayawardena et al., 2013). Rainfall variability plays a crucial role in crop production since the timing of cropping seasons relies heavily on rainfall patterns. Notably, in contrast to earlier decades, there has been a recent increase in the inter-decadal variability of rainfall (Ranasinghe et al., 2014).

Throughout the past three decades, there has been observable regional and temporal variability in rainfall patterns, as detected through an analysis of rainfall data (Wickramagamage, 2015). Specifically, the dry zone has witnessed an uptick in rainfall variability and extreme weather events in recent years (Panabokke & Punyawardena, 2010). The significant fluctuations in rainfall have led to frequent episodes of both floods and droughts, constituting the two most prevalent climatic hazards impacting agricultural output in Sri Lanka. Recent studies have disclosed a rising tendency for intense rainfall events within the dry zone during both cultivation seasons. This trend has led to situations where excessive moisture becomes a challenge during the reproductive phase of crops.

1.3 Climate Projections for Sri Lanka

According to regional climate model predictions, the temperature will rise by 2–3 °C by the end of the twenty-first century (Zubair et al., 2015). An additional projection for the end of the twenty-first century, utilizing the General Circulation Model (GCM), indicates an expected temperature increase of 2.84 °C (Zubair et al., 2015). In line with the Special Report on Emissions Scenarios (SRES), downscaled scenarios indicate that by the year 2080, temperatures are expected to increase by approximately 2.5–4.5 °C under the A2 scenario and approximately 2.5–3.25 °C under the B2 scenario (Cline, 2007). A rise in temperature of 1.1–2.4 °C and 1.5–2.8 °C for the two major crop seasons will occur in 2025 (Jayawardhana & Warnakulasooriya, 2020). According to the CCKP model, by the 2090s, the average temperature in Sri Lanka is estimated to increase by approximately 3.2 °C. Anticipated temperature trends indicate that maximum and minimum temperatures will experience accelerated increases compared to historical norms, yet they are projected to remain below global averages. Based on statistically downscaled predictions from the KNMI Climate Explorer, which employ a lower geographic resolution, we can expect a rise of approximately 3.5 °C under the given scenario by the 2090s, with a deviation of 1.2 °C below the global average.

1.4 Climate Change, Food System, and Food Security Nexus

The long-term consequences of climate change carry substantial implications, particularly for agriculture and food security. This situation is further compounded by two interrelated factors: the growth of the population and shifting dietary preferences. It is projected that these factors will lead to a 60% surge in global food consumption by

2050 when compared to levels in 2006. Concurrently, the persistent impact of climate change on global food systems continues to be a pressing concern, as highlighted by the FAO in 2006. In this context, Sri Lanka stands out as a particularly vulnerable nation, as it grapples with heightened exposure to climate change, increased sensitivity to its effects, and limited adaptability. Sri Lanka exhibits a heightened sensitivity to multiple aspects of climate change, including increasing temperatures, alterations in rainfall patterns, sea-level elevation, and the intensification of extreme weather events (Paltan et al., 2018). Conversely, Sri Lanka's hidden climatic hotspots represent significant vulnerabilities to the repercussions of climate change, with these hotspots concentrated in crucial agricultural areas. Notably, the North, North Central, Western, and North Western provinces of Sri Lanka find themselves on the frontline of challenges posed by climate change, bearing the brunt of the resulting damages (Yatawara, 2005).

It appeared that authorities and policymakers in Sri Lanka were either inadequately informed about international sustainability initiatives or were hesitant to transition from traditional agriculture practices to more sustainable approaches. The overarching agricultural policy seemed to give priority to traditional farming practices while placing a strong emphasis on export-oriented agriculture (Ministry of Agriculture, 2019). Given that over 80% of the population resides in rural areas (CBSL, 2019a), policies related to food security should take precedence. Currently, Sri Lanka fulfills over 80% of its food requirements through domestic production, while essential imports such as wheat, sugar, fish, and dairy products constitute approximately 65% of the total food imports (World Food Programme, 2017). Sri Lanka enjoys a substantial degree of self-sufficiency in rice production, with over 75% of other food crops cultivated domestically. Nonetheless, climate change stands as a considerable threat to local food production and the broader food security situation in Sri Lanka. In 2022, Sri Lanka held the 79th position among 113 countries in the Global Food Security Index, achieving an overall 61% (FAO, 2016).

Food security is a focal point on the developmental agendas of numerous developing nations, with its core concern revolving around the sustainability of food production and the assurance of secure food trade. Myriad factors come into play when assessing food security, ranging from population growth to climate change, urbanization, industrialization, shifts in land use, water scarcity, income growth, nutritional trends, and the dynamics of global energy supply and food trade (Wheeler & Braun, 2013). Significantly, among these factors, climate change emerges as a profound influencer of food system activities and various dimensions of food security (Wheeler & Braun, 2013).

Food systems can be understood as the dynamic interactions occurring within and between both the natural and human environments, ultimately encompassing a spectrum of activities that span from food production to consumption (Gregory et al., 2005). These food systems are shaped by an array of determinants rooted in the biophysical, social, economic, and political realms, collectively referred to as food system drivers. The ultimate outcomes of these interactions encompass food security, societal well-being, and environmental assets (Ericksen, 2008).

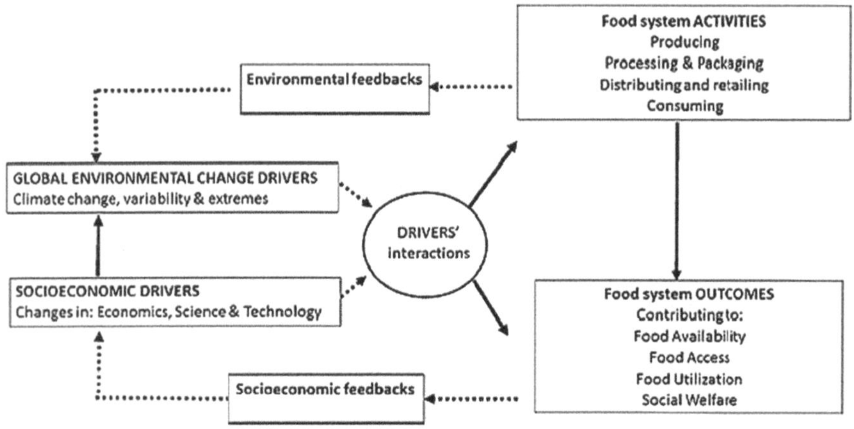

Fig. 1 A conceptual framework that establishes connections between the drivers of climate change and various activities and outcomes within the food system

On a macro scale, the efficiency of a food system hinges on its ability to adapt and align supply with the demand for various food items. When food systems face stressors, it puts food security at risk. These stressors can emerge from a multitude of origins, incorporating climate variations and a diverse range of other social and ecological influences (Gregory et al., 2005). Climate change, specifically, will exert a complex influence on every facet of the food system, covering everything from production to consumption. This influence will manifest in various forms, impacting production, markets, food prices, and various aspects of the supply chain and infrastructure. While a significant portion of existing research has primarily concentrated on the implications of climate change on food production and productivity, there is a shortage of information concerning its effects on other essential food system functions, such as postharvest storage, processing, distribution, and consumption.

1.5 Climate Change Impact on Upcountry Vegetable Supply Chains and Food Security

Climate change exerts a profound influence on agriculture, the primary source of sustenance and income in rural areas. Alterations in temperature and precipitation patterns have the potential to disrupt ecosystems, leading to changes in ecosystem services, which in turn have far-reaching implications for agricultural production and food security.

This comprehensive concept encompasses four fundamental components: food availability, which involves the entire food production and distribution process; food access, incorporating factors such as affordability, allocation, and personal food choices; food utilization, which addresses aspects related to nutritional value, societal significance, and food safety; and stability which refers to the stability of other three dimensions (Fig. 1). Crucially, it is imperative to recognize that the food system plays a pivotal role in influencing all these facets of food security (Fig. 2) (Wheeler & Braun, 2013).

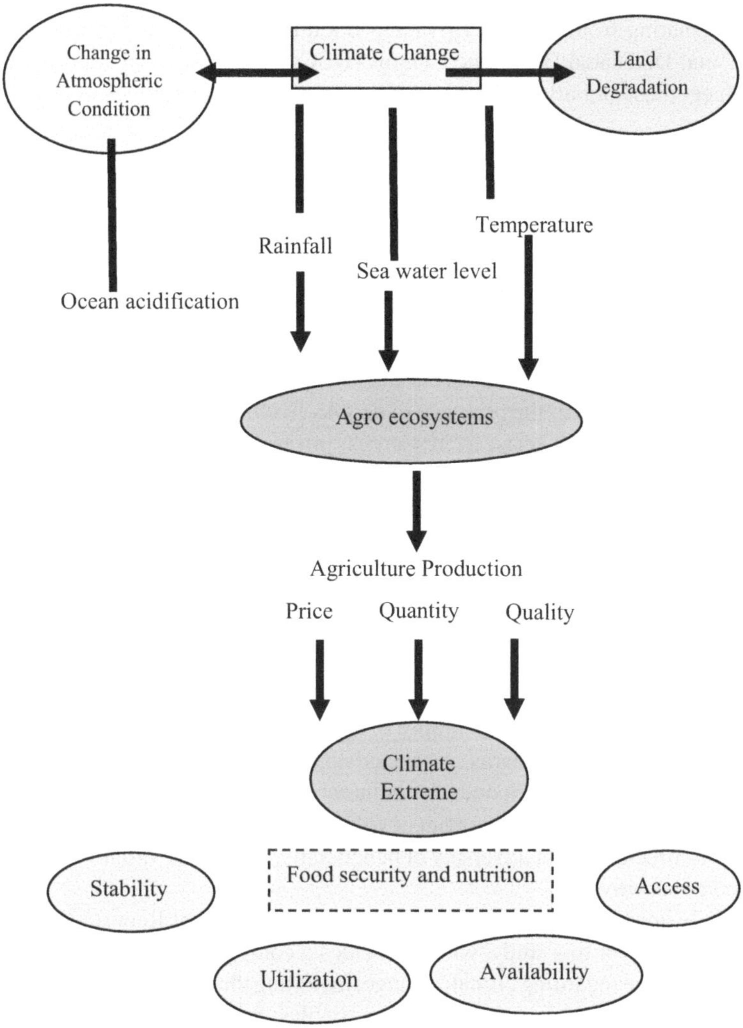

Fig. 2 Cascading effects of climate change impacts on food security and nutrition. *Source* FAO (2016)

2 Methodology

2.1 Study Areas

The focus of the study is the Colombo City Region and the food sheds that feed Colombo. However, this study focused primarily on upcountry vegetable supply chains originating from Nuwara Eliya and Badulla. Vegetable supermarket chains, retailers, and Dedicated Economic Centres (DECs) including Keppetipola, Nuwara Eliya, Meegoda, Dambulla and Embilipitiya were used as the sources to collect data.

2.2 Risk and Vulnerability Assessment Framework

The definition of **Hazard** is the potential occurrence of a natural or human-induced physical event, trend, or physical impact that results in loss of life, injury, or other health impacts, as well as damage and loss to property or infrastructure. Hazard is an external factor that represents stress caused by climate change.

Exposure is defined as the presence of people, livelihoods, species or ecosystems, environmental functions, services, resources, infrastructure, and economic, social, and cultural assets in potentially impacted places and settings. It can be quantified in terms of the percentage or amount of assets exposed, such as land, stakeholders, and the volume of local output.

Vulnerability is defined as a system's inherent property. The degree to which a system is sensitive to, or unable to cope with, the detrimental consequences of climate change, including climatic variability and extremes, is characterized as vulnerability. Vulnerability is made up of two parts: adaptive capacity and sensitivity. These are acknowledged as inherent characteristics of a system that incline it to be negatively affected by any injury. The ability of systems, institutions, humans, and other species to react to possible damage, take advantage of opportunities, or respond to consequences is the first component of vulnerability.

Sensitivity is the second component of vulnerability. The degree to which a system or species is affected, either adversely or beneficially, by climate variability or change is defined as sensitivity.

The study approach outlined in the IPCC Fifth Assessment Report (AR5) serves as the foundation for this study, which provides a comprehensive assessment of the state of knowledge regarding climate change, including the risk assessment approach. Risk (R) is defined as a function of three variables: hazard (H), exposure (E), and vulnerability (V). Vulnerability, exposure, and hazard interact to create risk where; $R = f(H, E, V)$.

The risk assessment process involved the identification of elements such as hazards, exposure, and vulnerability. These elements were determined through a thorough analysis of content, both from secondary sources and primary data, for each node within the value chain. Scores were assigned to the nodes in the supply chain for which data was accessible. This numerical approach allowed for the measurement of the degree of risk and vulnerability at various nodes. The dataset included a combination of content analysis, secondary data, and primary data. The information was then processed to construct indices that reflect the levels of risk and vulnerability associated with individual nodes within the supply chain.

2.3 Data Collection Approach

A mixed-method approach was used, incorporating various research techniques to investigate the impact of climate change. The methods involved a comprehensive online search of media coverage related to climate hazards over six years (2015–2020), the compilation of secondary data involved a thorough examination of literature and documents, and expert consultations. A comprehensive profile of the food system and refinement of relevant variables for vulnerability assessment was undertaken. For the development of vulnerability indices and metrics secondary data gathered through content analysis of published data sources were used. Additionally, primary data were obtained from stakeholders through questionnaire surveys to validate climate-related risks, exposure levels, sensitivity, and adaptive capacity, which subsequently aided in the development of vulnerability indices. Expert input and validation were sought to verify the robustness of these vulnerability indices and measures.

2.4 Content Analysis

Data over six years was used for online content analysis to gather climatic events reported in Sri Lanka's major electronic media. The individual events that were identified served as the basis for characterizing the nature of hazards, exposure, and vulnerability. Meteorological data supported the identification of hazards, while climate data and other pertinent secondary data supported the assessment of exposure and vulnerability. The information obtained through web-based content analysis played a crucial role in identifying several key aspects, including:

1. *Identification of Climate Hazards* which enabled the recognition of climate-related risks
2. *Exposure Assessment* which involved determining the extent to which relevant nodes within the system were exposed to climate hazards.
3. *Impact Assessment* in which the primary focus was on assessing the implications of climate events on diverse aspects, encompassing production, assets, infrastructure, and socio-economic conditions. This assessment aimed to determine both the sensitivity of these elements and their ability to withstand the effects of these hazards.

2.5 Assessment of Risk and Vulnerability

When evaluating the value chain nodes, our attention was directed toward multiple aspects of risk assessment. These encompassed an analysis of hazards, exposure, and vulnerability, which in turn included considerations of sensitivity and capacity. This comprehensive evaluation relied on a combination of content analysis and the utilization of secondary and primary data sources. For supply chain nodes where data was accessible, specific scores were assigned. The numerical scales provided a means to measure the magnitude of risk and vulnerability. For illustration, a value of zero was suggested when no impact was observed, while a higher number was recommended when the extent of damage was more severe. This approach allowed for a more precise measurement of the intensity of risk and vulnerability associated with each node in the supply chain. The assessment involved a careful examination of the scores assigned to supply chain nodes for up-country vegetables, conducted on a seasonal basis (Yala and Maha) spanning from 2015 to 2020. These scores were instrumental in the development of risk and vulnerability indices. As a benchmark, we identified the best seasons characterized by the least hazard impact to assess the risk and vulnerability levels associated with different supply chain nodes.

In line with the theoretical framework in the AR5 (Fifth Assessment Report), the following operationalization process for the quantitative assessment of climate hazards' influence on Climate-Resilient Food Systems (CRFS) was proposed:

$$H_{ijk} \in \mathrm{Z}^{-} : H_{ijk} \in [0, -4]$$

where H_{ijk} represents an integer (Z) ranging from 0 to − 4, each assigned value carries the following significance: it characterizes the magnitude of the hazard within the jth sector at the ith point in time for the kth node within the food supply chain. Exposure scores showed; no exposure (0), moderate exposure (− 1), strong exposure (− 2), very strong exposure (− 3), and population, assets, and infrastructure were severely exposed (− 4). The sensitivity score was determined by aligning the attributes of the supply chain node with the impact of the hazard. Scale for measuring sensitivity denoted: no sensitivity (0), mild sensitivity (− 1), moderate sensitivity (− 2), high sensitivity (− 3), and very high sensitivity (− 4). Concurrently, the adaptive capacity score was determined through a comprehensive evaluation that incorporated content analysis, secondary data, and primary data sources. The culmination of this assessment involved the combination of sensitivity and adaptive capacity ratings, resulting in the calculation of the overall vulnerability score. Vulnerability scores were calculated by independently assigning values to the two fundamental components of vulnerability, which are adaptive capacity and sensitivity. These values were determined based on information obtained from the content analysis, as well as from secondary and primary data sources (Fig. 3).

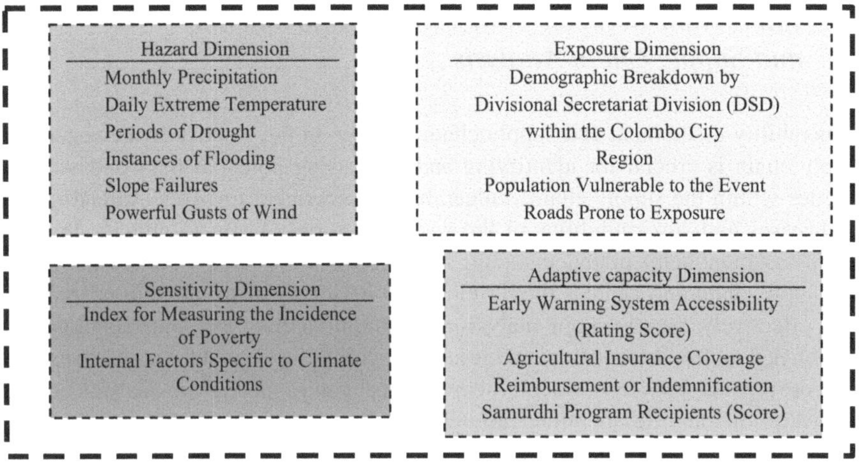

Fig. 3 Variables used for scoring different dimensions. *Source* Authors

2.6 Data Processing and Analysis

The dataset covered a variety of sources, including content analysis, secondary data, and primary data. These sources consisted of published reports, a comprehensive literature review, and an analysis of publicly available data. For both secondary and primary data, two separate databases (DBs) were established. Standard coding procedures were employed when inputting the data into these digital databases. Prior to conducting the analysis, a rigorous data-cleaning process was undertaken to ensure accuracy and consistency. Subsequently, this information was subjected to further processing to create risk and vulnerability indices tailored to specific supply chain nodes (Fig. 4). These indices were instrumental in determining the levels of risk and vulnerability associated with each supply chain node.

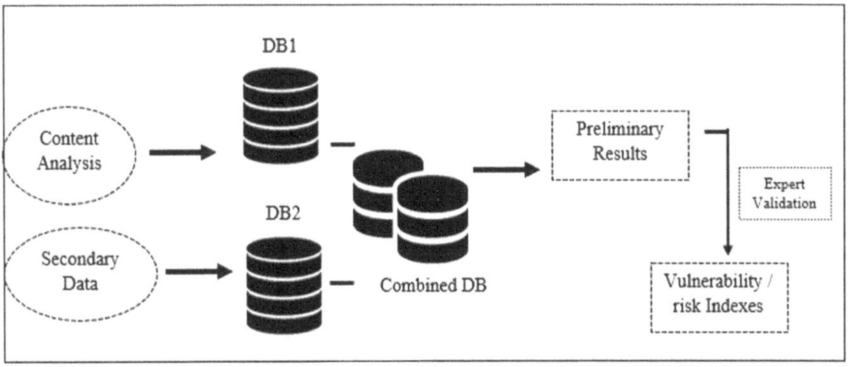

Fig. 4 Flowchart of data processing and analysis. *Source* Authors

2.7 Interaction Between Vulnerability Assessment and Supply Chain Analysis

Vulnerability assessment and supply chain analysis in the context of the vegetable supply chain is crucial for identifying and addressing potential risks and vulnerabilities within the supply chain. Vulnerability assessment involves evaluating the weaknesses and susceptibilities of the vegetable supply chain to climate change. This assessment helps in understanding the potential impacts of climate hazards on the supply chain and enables the development of strategies to mitigate or manage them effectively. Supply chain analysis focuses on mapping and understanding the flow of vegetables, information, and resources within the vegetable supply network. The convergence between vulnerability assessment and supply chain analysis occurs when the vulnerabilities identified through the assessment are analyzed in the context of the supply chain structure and dynamics. This analysis helps in identifying the hotspots or nodes within the supply chain that are most vulnerable to disruption or have the potential to propagate the impact along the supply chain. The basis for building a resilient food supply chain is to identify key processes and actors in food supply chains and understand their interaction with and exposure to environmental variability and the economy (Davis et al., 2021).

3 Results and Discussion

3.1 The Influence of Climatic Hazards on Upcountry Vegetable Supply Chains

The primary climate-related challenges affecting vegetable production in upcountry regions include wind, landslides, flooding, and heavy rainfall. These climate hazards also have affected the intermediaries within the vegetable supply chain in upcountry areas. Up-country DEC operations at interim points are particularly vulnerable to heavy rainfall and strong winds. The effects of climate-related risks are ultimately passed on to consumers in the form of increased prices as they flow through the supply chain. Traditional vegetable supply networks reliant on DEC contribute to roughly 85% of vegetable deliveries to the Colombo metropolitan region. By creating direct links between vegetable farmers within supply chains and supermarkets via a network of collection centers, it becomes possible to optimize the supply chains of supermarkets. The primary impediments affecting the affordability of vegetables for lower-income communities are the impacts of climate hazards on production and significant post-harvest losses throughout the supply chain.

3.2 Overview of Upcountry Vegetables

The majority of upcountry vegetables are primarily cultivated in the Nuwara Eliya and Badulla regions. In 2019, the Nuwara Eliya District played a significant role in the national production of various vegetables, contributing 57% of cabbage, 75% of carrots, 86% of leeks, and 37% of beans. Conversely, the Badulla District also made substantial contributions, accounting for 22% of cabbage, 23% of carrots, 10% of leeks, and 35% of beans in the national output. In contrast, the Colombo District had a relatively limited contribution to the overall national vegetable production (Table 1). As a result, the majority of upcountry vegetables are delivered to Colombo from these two regions with relatively large food miles.

3.3 Climate Hazards Related to the Production of Upcountry Vegetables

The principal climatic challenges faced by the Nuwara Eliya and Badulla regions include wind, landslides, floods, and heavy rainfall. During the period spanning from 2015 to 2020, Nuwara Eliya has reported 50 extreme events including 8 (16.0%) heavy rain events, 22 (44.0%) flood events, 22 (44.0%) landslide events, 30 (60.0%) wind events, and 2 (4.0%) droughts. Badulla has reported 36 extreme events including 8 (22.2%) heavy rain events, 7 (19.4%) flood events, 8 (22.2%) landslide events, 25 (69.4%) wind events, and 5 (13.9%) drought events. Rathnapura (transport route) has reported 21 extreme events including 15(71.4%) heavy rain events, 7 (33.3%) flood events, 8 (38.1%) landslide events, 4 (19%) wind events, and 4 (19%) droughts. Kandy (transport route) has reported 6 extreme events including 4 (66.7%) heavy rain events, 3 (50%) flood events, 3 (50%) landslide events, 2 (33.3%) wind events, and 1 (16.7%) drought. Kegalle (transport route) has reported 10 extreme events including 8 (80%) heavy rain events, 5 (50%) flood events, 6 (60%) landslide events, and 3 (30%) wind events.

In Badulla District, wind was the most prevalent climatic hazard, although severe rainfall and flooding also posed challenges to upcountry vegetable production. Climate hazards like heavy rainfall, flooding, landslides, and strong winds in districts like Ratnapura, Kandy, and Kegalle along transportation routes have impacted the transportation of vegetables from the Nuwara Eliya and Badulla districts to the Colombo city region.

Table 1 Major upcountry vegetable production from 2015 to 2022

Year	Vegetables	District production (MT)		National production (MT)
		Nuwara Eliya	Badulla	
2015	Tomatoes	22,775 (29)	13,880 (17)	79,553
	Cabbage	61,785 (53)	28,116 (24)	117,095
	Carrot	48,258 (71)	16,369 (24)	67,673
	Beans	19,719 (30)	25,297 (38)	66,163
	Leeks	35,263 (85)	4472 (11)	41,434
2016	Tomatoes	33,723 (36)	15,073 (16)	92,748
	Cabbage	69,704 (57)	26,132 (21)	123,365
	Carrot	57,974 (71)	21,635 (26)	82,037
	Beans	32,520 (39)	32,315 (39)	83,366
	Leeks	55,152 (90)	4580 (7)	61,122
2017	Tomatoes	25,881 (32)	15,155 (19)	80,839
	Cabbage	48,487 (49)	26,610 (27)	99,466
	Carrot	49,702 (69)	19,338 (27)	71,607
	Beans	28,152 (32)	32,515 (37)	87,385
	Leeks	51,195 (91)	3354 (6)	56,045
2018	Tomatoes	34,146 (34)	19,293 (19)	101,404
	Cabbage	54,097 (49)	26,933 (24)	111,141
	Carrot	51,697 (73)	16,405 (23)	71,051
	Beans	28,802 (34)	34,296 (41)	83,966
	Leeks	44,486 (87)	5393 (11)	51,330
2019	Tomatoes	17,869 (23)	18,003 (23)	77,858
	Cabbage	65,917 (57)	27,101 (23)	116,577
	Carrot	60,577 (75)	17,706 (22)	80,766
	Beans	24,146 (37)	22,719 (35)	65,450
	Leeks	27,326 (86)	3159 (10)	31,879
2020	Tomatoes	18,661 (21)	19,409 (21)	90,507
	Cabbage	64,659 (52)	30,371 (24)	125,745
	Carrot	65,553 (73)	20,896 (23)	90,224
	Beans	29,438 (35)	27,813 (34)	82,973
	Leeks	30,162 (84)	3745 (11)	35,695
2021	Tomatoes	20,795 (23)	68,324 (20)	89,668
	Cabbage	76,450 (59)	25,008 (19)	129,106
	Carrot	73,833 (77)	18,266 (19)	95,416
	Beans	34,656 (45)	22,292 (29)	76,864
	Leeks	38,575 (86)	4783 (11)	44,733
2022	Tomatoes	28,063 (32)	10,270 (12)	88,081

(continued)

Table 1 (continued)

Year	Vegetables	District production (MT)		National production (MT)
		Nuwara Eliya	Badulla	
	Cabbage	77,693 (67)	14,477 (13)	116,662
	Carrot	62,540(81)	12,063 (16)	77,178
	Beans	29,188 (43)	23,734 (35)	68,181
	Leeks	43,755 (84)	7045 (14)	51,804

Note Percentage contribution in parentheses. production values are rounded to the nearest integer
Source Department of Census and Statistics (DCS) (2020)

3.4 Consequences of Climate Hazards on Conventional Vegetable Supply Channels

In Colombo City, currently, the new Manning Market Complex in Peliyagoda serves as the main distribution center for vegetables. In November 2020, it moved from Pettah to Peliyagoda. There are 1192 stores in the new Manning Market Complex in Peliyagoda. These stores are run by commission salespeople who buy from farmers and collectors and resell to wholesalers and retailers. The distribution of vegetables within the Colombo city region is significantly influenced by Dedicated Economic Centers (DECs).

There are 14 DECs in the country, 5 of which are found in the Colombo city region including Narahenpita, Welisara, Meegoda, Piliyandala, and Ratmalana whereas, within the boundaries of the Colombo city limits, there are 18 other rather big markets mostly engaged in the retailing of vegetables.

These markets are situated in Maradana, Grandpass, Dematagoda, Borella, Wellawatta, Kirulapona, and Narahenpita. There are various additional vegetable merchants running stores, including sporadic marketplaces called Pola. According to approximate estimates, traditional supply channels provide roughly 85% of the vegetable requirements of the Colombo city region. For a better understanding of climatic risks affecting upcountry vegetables including handling postharvest losses, a survey of commission agents at DECs was done covering 77 commission agents from Dambulla, Keppetipola, Nuwara Eliya, Embilipitiya, and Meegoda. Considering the influence of climate-related risks on the activities of collectors and wholesalers situated in DECs revealed that 56 and 44% were affected by heavy rain and drought respectively in Dambulla. In Keppetipola, 67 and 33% were affected by heavy rain and wind while in Nuwara Eliya, 100% of them were affected by heavy rain.

The survey results showed that different geographical areas have different climatic hazards that influence DECs. Based on feedback from respondents in the upcountry DECs located in Keppetipola and Nuwara Eliya, it was revealed that heavy rainfall and strong winds were the main challenges impacting their operations. Rain, drought, and high temperatures hamper operations of vegetable handling leading to wastage at Dambulla, Meegoda, and Embilipitiya DECs.

Table 2 Climate-induced factors leading to vegetable wastage at DECs

Produce	Dambulla		Keppetipola		Meegoda	
	High temperature	Heavy rain (%)	High temperature	Heavy rain (%)	High temperature	High rain (%)
Tomato	25.00%	38.33	No impact	–	No impact	
Carrot	–	23.33	–	10.00	–	20.00
Leek	–	25.00	–	20.00	–	17.50
Cabbage	25.00%	39.00	–	41.00	–	18.33
Beans	–	46.88	–	10.63	–	15.00
Knolkhol	–		–	50.00	–	–

Source Authors' survey

Tomatoes, cabbage, and radishes are the vegetables that are most commonly damaged when handled during high temperatures, according to commission agents in Dambulla DEC. High rainfall during handling has an impact on most vegetable varieties. No vegetables were harmed by high temperatures at Keppetipola or Meegoda DECs. Heavy rainfall, however, had an impact on the majority of upcountry vegetables (Table 2).

The availability of some vegetables is severely impacted by climatic risks. At Keppetipola, supplies of tomatoes, potatoes, leeks, and brinjal considerably decrease as a result of heavy rains, while supplies of beans and brinjals are negatively impacted by high wind (Table 2). The drought has an impact on Dambulla's supplies of tomatoes, pumpkins, and cucumbers. Floods had little impact on the supplies going to Dambulla and Keppetipola (Table 2).

3.5 Supermarket Vegetable Supply Chains

Supermarkets are crucial as middlemen in the delivery of vegetables to the Colombo city region. Regarding the delivery of vegetables to the Colombo city region, the two prominent supermarket chains, Keels and Cargills, collectively represent approximately 10% of the market share. There are 411 Cargills outlets in the country of which more than 300 are in the Colombo area. The collecting centers, which connect to around 10,000 out-grower farmers, handle more than 95% of procurement. Cold trucks are used to transport the product obtained through collecting centers to centralized fruit and vegetable processing facilities. These units dispatch products to retail outlets. Heavy rain and drought were the main climatic risks that had an impact on distribution and retail operations.

There are more than 100 Keels outlets throughout the island, with roughly 60 of them in the Colombo district. All vegetables were purchased through their collectors, operating from collecting centers located in the production regions. Produce was moved from production areas to retail locations using cold trucks. Prices typically rise when there is a shortage of supplies due to rain or a drought, and supplies of specific vegetables, such as beans and green leaves, are severely impacted. Produce losses as a result of harsh weather are considerable and range from 10 to 20% depending on the kind of vegetable, for example, salad leaves 16–17%, carrots 8–9%, and potatoes 3–4%. None of the mentioned supermarket chains utilize weather information or early warning systems in their decision-making process. As an adaptation measure, they use alternative routes to reach Colombo when transit routes are impassable due to severe weather. This always contributes to an increase in the prices of vegetables that enter the Colombo city region.

3.6 *Climate Change Impact on Vegetable Prices*

Except for certain vegetable varieties, such as leeks, for which a correlation exists between rainfall patterns and the retail prices in Colombo, there is insufficient empirical evidence to establish a direct correlation between rainfall patterns and the retail price of vegetables from upcountry regions. The price of leeks rises when there is a significant increase in rainfall with a reasonable lag time. High postharvest losses (spoilage) of leeks as a result of persistent rain in the production regions are mostly to blame for this. Figure 5 shows the seasonal trend of price variation, with prices rising during the holiday months of April and December. It was impossible to understand the link between output and price without access to monthly information on vegetable production.

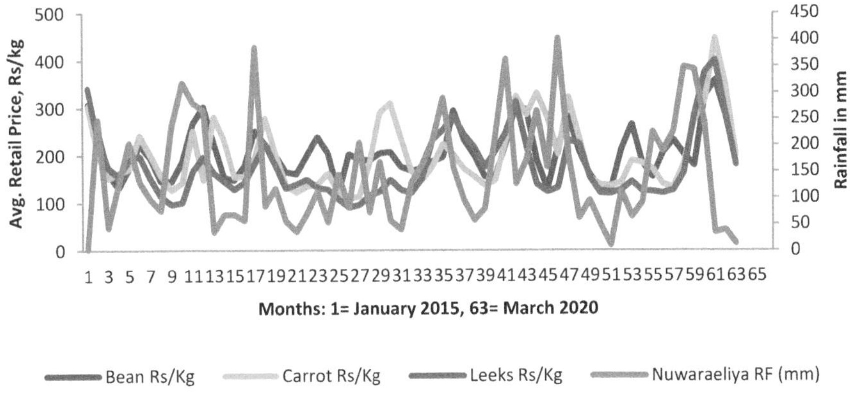

Fig. 5 Variations in monthly precipitation in Nuwara Eliya and retail prices of upcountry vegetables in Colombo. *Source* Authors

3.7 Possible Implications of Climate Hazards on Food Security and Nutrition

The consequences of climate impacts on production extend beyond the immediate agricultural context, affecting both the economy and society. The effects of climate change manifest through various channels, leading to a range of social and economic implications. These implications can encompass shifts in agricultural earnings, alterations in food markets, changes in pricing and trade dynamics, and shifts in investment patterns. Furthermore, they can have the potential to reduce agricultural income, impact the availability of physical capital, force farmers to sell productive assets like livestock to cope with income fluctuations, and even hinder the capacity for future investments.

These effects directly impact agricultural households, constraining their ability to cover other essential expenses such as healthcare and education. On a national scale, they may lead to price escalation of agricultural products, which can affect the economic and social well-being of the entire population. This is particularly pronounced in countries where a significant portion of family budgets is allocated to food expenses. Climate change impacts can have both quantity and price effects, leading to increased market volatility with profound implications on the availability, accessibility, and stability of food security dimensions.

Sri Lanka faces a significant challenge in achieving adequate per capita vegetable consumption. Only a mere 3.5% of individuals manage to meet the daily recommended intake of five servings of fruits and vegetables, as reported in Ingram (2011). Over the period spanning 2013–2017, the annual per capita availability of vegetables for consumption fluctuated between 134 and 146 kg, according to data from the Department of Census and Statistics in 2019. However, the actual per capita vegetable consumption remains considerably low at just 22 kg (equivalent to 60 g per day) (Ingram, 2011). This translates to a daily vegetable intake ranging from 225 to 300 g, falling short of the recommended daily amount for a balanced diet. Unfortunately, a substantial 70% of Sri Lankan adults tend to overconsume starchy foods, surpassing the recommended starch consumption guidelines, as highlighted in research (Jayawardena et al., 2013).

There are several factors contributing to the comparatively low vegetable consumption levels. Notably, seasonal price fluctuations in the vegetable market and the limited purchasing power of the most economically disadvantaged segments of the population are major contributors. As depicted in Table 3, the present consumption rates would need to quadruple or quintuple to meet the recommended dietary intake. Nevertheless, considering the expenses associated with vegetables, particularly given their price fluctuations, it appears unlikely that the most impoverished communities in the Colombo city region will be able to attain the necessary quantities of vegetables for a well-balanced diet.

Evidence from previous studies indicates that climate change disruptions to food supply chains and consequent price volatility compromise the food security and nutrition of poorer and vulnerable communities (Niles et al., 2017; Tirado et al.,

Table 3 Household expenditure on vegetables in Colombo District, per capita consumption, and recommended intake

Item	Monetary income per month per household (LKR)	Household food expenditure share	Monthly household expenditure (LKR)	Current per capita consumption	Recommended per capita
Vegetables	65,264	8.2%	1805	65 g	225–300 g

Note Household (HH) size = 4 persons
Source Household Income and Expenditure Survey (HIES) (2019)

2013). Drawing from these insights, it is rational to conclude that the primary factors contributing to the inaccessibility of vegetables for lower-income dwellers stem from the adverse effects of climate-related hazards on production and the substantial post-harvest losses that occur throughout the supply chain. Furthermore, there are notable concerns related to the safety of upcountry vegetables. Farmers frequently harvest vegetables shortly after pesticide treatment, disregarding the essential safety interval between the application of agrochemicals and the harvest, as highlighted in research (Silva & Dematawewa, 2021). This practice is becoming increasingly prevalent, particularly in response to irregular rainfall patterns in regions where upcountry vegetables are cultivated. Hence, it poses risks to the well-being and safety of consumers in the Colombo city region who rely on these vegetables for sustenance. To mitigate the hazard of consuming vegetables contaminated with agrochemicals, some upcountry vegetable farmers opt to cultivate vegetables for personal consumption on separate plots and refrain from consuming produce from their commercial plots.

These consequences are expected to evolve gradually in response to the changing impacts of climate change. They are likely to differ among various regions and populations. Evaluating the consequences of the increased frequency of extreme weather events, along with unpredictable climatic patterns, is of utmost importance. Additionally, it is crucial to anticipate both the immediate and long-term outcomes resulting from substantial alterations in global warming trends. Estimating the economic impacts of long-term global warming poses challenges, primarily because it relies on the outcomes of global climate modeling studies, which are run at much coarser spatial resolutions (Padmajani et al., 2014). Additionally, these effects are greatly influenced by the actions that individuals and communities take to adapt and mitigate negative effects and promote positive outcomes, which can be unpredictable.

In summary, the impact of climate change on different aspects of society and the economy is complex and diverse, and these effects will continue to transform as the climate undergoes alterations. Evaluating and tackling these consequences necessitates a multidisciplinary and adaptable strategy to adequately counter and adapt to the challenges presented by climate change.

4 Conclusions and Recommendations to Build Climate-Resilient Food Systems

The current study is an attempt to deviate from predominately crop production-focused risk and vulnerability assessments towards supply chain-focused risk and vulnerability assessment to provide insights into compounding and cascading risks of climate change on food supply chains (Davis et al., 2021). The Colombo city region primarily receives the majority of upcountry vegetables from the Nuwara Eliya and Badulla districts. These regions face significant climatic challenges, with wind, landslides, floods, and heavy rains being the most prominent hazards. Wind, in particular, is the prevailing hazard in both Nuwara Eliya and Badulla, with moderate intensity. The impact of these climatic hazards, especially wind and high rainfall can be highly detrimental to vegetable production in the Nuwara Eliya and Badulla areas. Given the high-risk nature of the upcountry vegetable supply chain, it is evident that the impacts of climatic hazards can swiftly translate into significant price increases for the end customer.

Heavy rain and strong winds emerged as the primary threats impacting the operations of upcountry DECs, including Keppetipola and Nuwara Eliya, positioned at intermediate points within the supply chain. Climate-related hazards also led to vegetable spoilage both at the DECs and during transit. Unfortunately, early warnings and climate forecasts are not given due consideration by wholesalers and commission agents at these DECs when making operational decisions. Supermarkets serve as pivotal intermediaries within the vegetable supply chain to the Colombo city region. The two largest supermarket chains jointly command roughly 10% of the market share in the provision of vegetables to the Colombo city region. Surprisingly, none of the supermarket chains have acknowledged the use of early warning systems or weather-related information in their procurement decisions, even though such information is crucial for minimizing the impact of climate hazards.

The primary reasons behind the inaccessibility of vegetables for the less affordable communities within the Colombo city region are the adverse effects of climatic risks on production and postharvest losses all along the supply chain. Additionally, climate change has raised concerns regarding food safety. The irregular and excessive rainfall patterns have compelled upcountry vegetable producers to increase their use of agrochemicals, including fungicides and insecticides, in an attempt to protect their crops. Given the limited attention to the issue of quantifying risk and vulnerabilities in the food supply chain it hinders the endeavors to minimize and mitigate risks along the food supply chains (Rathore et al., 2017; Zhao et al., 2022). This study contributes to understanding food supply chain risks and vulnerabilities, and the interplay between risks and vulnerabilities of vegetable supply chains in Sri Lanka. This study also provides insights to understand how environmental disruptions such as climate change impacts transmit across local food supply chains.

5 Recommendations for Building Resilient Upcountry Vegetable Supply Chains

Addressing the impact of climate change on upcountry vegetable growers and supply chain actors indeed requires effective adaptation strategies. Here are some key considerations and strategies for building climate-resilient supply chains:

Infrastructure Investment: Investing in cold chain facilities, adequate storage space, and essential amenities is crucial. This infrastructure can help in preserving the quality of vegetables, reducing postharvest losses, and ensuring a consistent supply of produce. Cold chain facilities, such as cold storage, refrigerated trucks, and processing facilities, play a crucial role in preserving the freshness and nutritional value of vegetables from the farm to the consumer.

Capacity Expansion at DECs: Expanding the capacity of DECs is essential for accommodating increased demand and managing supply effectively. Upgrading DECs can enhance the efficiency of the supply chain.

Decision Support Systems (DSS): The implementation of a decision support system can facilitate the real-time matching of vegetable supply and demand in the Colombo city region by incorporating climate and weather data. This technology-driven approach enables the optimization of logistics, waste reduction, and proactive management during climate extremes. By integrating climate and weather data, a DSS can gather and analyze current information on vegetable supply and demand, empowering stakeholders to make well-informed decisions.

Vulnerability Assessment: It's important to acknowledge that from upstream (farmer end) to downstream (consumer end), the vegetable supply chain is vulnerable to the impacts of climate hazards. Therefore, it is necessary to develop tailor-made adaptation strategies to address risks and vulnerabilities across supply chains. Risk and vulnerability assessments help identify potential risks and vulnerabilities for supply chain actors in the food supply chain, ensuring that customers have access to safe and high-quality food products. By assessing and addressing vulnerabilities, such as contamination risks or improper handling practices, customers can have confidence in the safety and integrity of the food they consume. On the farmers' side, risk and vulnerability assessments help farmers and intermediaries in the supply chain to identify potential risks and vulnerabilities in their operations. By understanding risks and vulnerabilities assessments appropriate adaptation measures can be supported. This can include measures to address climate-related risks, pest and disease management, or infrastructure vulnerabilities.

Policy Actions: Strengthening policies that support climate-resilient agriculture and supply chains is essential. This includes providing incentives for sustainable practices, supporting risk management mechanisms, and promoting climate-smart agriculture. Strengthening policies that support climate-resilient agriculture and supply chains can help supply chain actors adapt to these changes and minimize the negative effects on agricultural productivity food losses and wastage. Hence policymakers need to collaborate with relevant stakeholders, including farmers, industry representatives, and environmental experts, to develop and implement effective policies that address the specific challenges faced by Sri Lanka's agriculture sector in the context of climate change.

Public Awareness: Raise awareness among farmers, supply chain stakeholders, and consumers about climate change impacts and adaptation measures. Encourage the adoption of sustainable and climate-resilient practices and sustainable consumption.

In summary, building resilience in upcountry vegetable supply chains requires a comprehensive approach that involves infrastructure development, technological advancements, policy support, and stakeholder collaboration. Implementing these strategies can effectively reduce the adverse effects of climate change on the vegetable supply chain and guarantee a consistent and secure food supply for consumers. These measures will enhance the resilience of vegetable supply chains while concurrently shielding consumers from the risks of price fluctuations and sudden price increases.

6 Competing Interests

The authors have no conflicts of interest to declare that are relevant to the content of this chapter.

Acknowledgements The authors wish to acknowledge the Sabaragamuwa University of Sri Lanka's research grants and the International Water Management Institute for funding this research.

References

Ahmed, M., & Suphachalasai, S. (2014). *Assessing the costs of climate change and adaptation in South Asia.*

Brinkman, H. J., De Pee, S., Sanogo, I., Subran, L., & Bloem, M. W. (2010). High food prices and the global financial crisis have reduced access to nutritious food and worsened nutritional status and health. *The Journal of Nutrition, 140*(1), 153S-161S.

CBSL. (2018). *Annual report.*

CBSL. (2019a). *Annual report 2019.*

CBSL. (2019b). *Economic and social statistics of Sri Lanka.*

Chandrapala, L. (1996). Long-term trends of rainfall and temperature in Sri Lanka. In *Climate variability and agriculture*, pp. 153–162.

Cline, W. R. (2007). *Global warming and agriculture: Impact estimates by country.* Peterson Institute. DCS. Agricultural Household Survey 2016/17, Department of Census and Statistics.

Davis, K. F., Downs, S., & Gephart, J. A. (2021). Towards food supply chain resilience to 1073 environmental shocks. *Nature Food, 2*(1), 54–65.

Ericksen, P. J. (2008). Conceptualizing food systems for global environmental change research. *Global Environmental Change*, 234–245. https://doi.org/10.1016/j.gloenvcha.220.09.02

Esham, M., Jacobs, B., Rosairo, H. S. R., & Siddighi, B. B. (2018). Climate change and food security: A Sri Lankan perspective. *Environment, Development and Sustainability, 20*(3), 1017–1036.

Esham, M. (2021). *Towards food security in Sri Lanka: A framework to strengthen the adaptive capacity of the food system to climate change*. Unpublished manuscript.

Fanzo, J., Davis, C., McLaren, R., & Choufani, J. (2018). The effect of climate 1085 change across food systems: Implications for nutrition outcomes. *Global Food Security, 18*, 12–19.

FAO. (2016). *Working document. Food and Agriculture Organization of the United Nations, City region food system situation analysis*, Colombo, Sri Lanka.

Gregory, P. J., Ingram, J. S. I., & Brklacich, M. (2005). Climate change and food security. *Philosophical Transactions of the Royal Society of London B. Biological Sciences, 360*(1463), 2139–2148. https://doi.org/10.1098/rstb.2005.1745

Ingram, J. (2011). A food systems approach to researching food security and its interactions with global environmental change. *Food Security, 3*(4), 417–431.

Ivanov, D., Sokolov, B., Solovyeva, I., Dolgui, A., & Jie, F. (2015). Ripple effect in the time-critical food supplychains and recovery policies. *IFAC-PapersOnLine, 48–3*, 1682–1687.

Jayatissa, R. L. N., Wickramasinghe, W. D., & Piyasena, C. (2014). *Food consumption patterns in Sri Lanka*. Hector Kobbekaduwa Agrarian Research and Training Institute.

Jayawardena, R., Byrne, N. M., Soares, M. J., Katulanda, P., & Hills, A. P. (2013a). Food consumption of Sri Lankan adults: An appraisal of serving characteristics. *Public Health Nutrition, 16*(4), 653–658.

Jayawardena, R., Jeyakumar, D. T., Gamage, M., Sooriaarachchi, P., & Hills, A. P. (2020). Fruit and vegetable consumption among South Asians: A systematic review and meta-analysis. *Clinical Research & Reviews*.

Jayawardhana, M., & Warnakulasooriya, B. (2020). Impact of problems associated with supply chain management practices of wholesalers on their business performance in the coconut industry in Sri Lanka with special reference to Kurunegala District. *Vidyodaya Journal of Managment, 06*(1), 59–87.

Mani, M., Bandyopadhyay, S., & Chonabayashi, S. (2018). *South Asia's hotspots: The impact of temperature and precipitation changes on living standards*. The World Bank Group.

Ministry of Agriculture. (2019). *Sri Lanka's overarching agricultural policy*. Ministry of Agriculture. http://www.agrimin.gov.lk/web/images/Information_Act/Development/2019_08_19_Draft_OAP.pdf

Naveendrakumar, G., Obeysekera, J., Vithanage, M., & Center, S. L. S. (2019). *Journal of Dry Zone Agriculture, 5*(2), 38–47. Faculty of Agriculture, University of Jaffna, Sri Lanka.

Niles, M. T., Ahuja, R., Esquivel, J. M., & Mango, N., Duncan, M. (2017). *Climate change and food systems: Assessing impacts and opportunities*. Meridian Institute.

Padmajani, M. T., Aheeyar, M. M. M., & Bandara, M. A. C. S. (2014). *Assessment of pesticide usage in up-country vegetable farming in Sri Lanka*. HARTI Research Report No: 164, Hector Kobbekaduwa Agrarian Research and Training Institute. https://doi.org/10.3390/su12125012

Paltan, H., Allen, M., Haustein, K., Fuldauer, L., Dadson, S. (2018). Global implications of 1.5 and 2 °C warmer worlds on extreme river flows. *Environmental Research Letters, 13*. https://iopscience.iop.org/article/10.1088/1748-9326/aad985/pdf

Panabokke, C., & Punyawardena, B. (2010). Climate change and rain-fed agriculture in the dry zone of Sri Lanka. In *Paper presented at the Proceedings of the national conference on water, food security and climate change in Sri Lanka* (Vol. 2). *Water quality, environment and climate change*, Colombo, June 9–11, 2009.

Ranasinghe, E. M. S., De Silva, C. S., & Piyadasa, R. U. K. (2014). Inter-decadal variability of seasonal rainfall in Sri Lanka within the period of 1951–2010. In *Paper presented at the Annual academic session 2014 Open University of Sri Lanka*, Colombo, Sri Lanka.

Rathore, R., Thakkar, J. J., & Jha, J. K. (2017). A quantitative risk assessment methodology and evaluation of food supply chain. *International Journal of Logistics Management, 28*(4), 1272–1293.

Sheikh, M. M., Manzoor, N., Ashraf, J., Adnan, M., Collins, D., Hameed, S., & Shrestha, M. L. (2015a). Trends in extreme daily rainfall and temperature indices over South Asia. *International Journal of Climatology, 35*(7), 1625–1637. https://doi.org/10.1002/joc.4081

Sheikh, M. M., Manzoor, N., Ashraf, J., Adnan, M., Collins, D., Hameed, & Shrestha, S. (2015). Trends in extreme daily rainfall and temperature indices over South Asia. *International Journal of Climatology, 35*(7), 1625–1637. https://doi.org/10.1002/joc.4081

Silva, G. L. L. P., & Dematawewa, C. M. B. (2021). Livestock and poultry to assure sustainability in the food system. In Weerahewa, J., Dandeniya, W., & Marambe, B. (Eds.), *Agricultural research for sustainable food systems in Sri Lanka: Volume 1: A historical perspective.* Springer. https://doi.org/10.1007/978-981-15-2152-2_13.

Tirado, M. C., Crahay, P., Mahy, L., Zanev, C., Neira, M., Msangi, S., & Müller, A. (2013). Climate change and nutrition: Creating a climate for nutrition security. *Food and Nutrition Bulletin, 34*(4), 533–547.

Weerahewa, J., Rajapakse, C., & Pushpakumara, G. (2013). An analysis of consumer demand for fruits in Sri Lanka 1981–2010. *Appetite, 60*, 252–258.

Wheeler, T., & Von Braun, J. (2013). Climate change impacts on global food security. *Science, 341*(6145), 508–513.

Wickramagamage, P. (2015). Spatial and temporal variation of rainfall trends of Sri Lanka. *Theoretical and Applied Climatology*, 1–12.

World Food Programme (2017). *Sri Lanka country strategic plan (2018–2022).* WFP.

Yatawara, R. A. (2005). *Poverty and trade issues in Sri Lanka.* Centre for Poverty Analysis.

Zhao, G., Olan, F., Liu, S., Hormazabal, J. H., Lopez, C., Zubairu, N., & Chen, X. (2022). Links between risk source identification and resilience capability building in agri-food supply chains: a comprehensive analysis. *IEEE Transactions on Engineering Management.*

Zubair, L. (2002). El Nino–southern oscillation influences on rice production in Sri Lanka. *International Journal of Climatology, 22*(2), 249–260.

Zubair, L., Nissanka, S. P., Weerakoon, W. M. W., Herath, D. I., Karunaratne, A. S., Prabodha, A. S. M., & Vishwanathan, J. (2015). *Handbook of climate change and agroecosystems: The agricultural model intercomparison and improvement project integrated crop and economic assessments* (Part 2).

Building Resilience for Adaptation to Climate Change in High Value Marine Fisheries: A Case on Lobster Fisheries, Sri Lanka

B. M. R. L. Basnayake, D. A. M. De Silva, and U. S. P. K. Liyanage

1 Introduction

As a global phenomenon in recent past years the sea level has increased by an average of 3.1 mm/year (Dangendorf et al., 2017) as a result of increased temperature the dissolved oxygen levels decrease, and oxygen minimum zones in the oceans have expanded in both coastal and offshore areas over the last decades. Moreover, water acidity has increased by 26% due to absorption of increasing amounts of anthropogenic CO_2 by the ocean (Barange et al., 2018).

According to the United Nations Food and Agriculture Organization (FAO), fisheries sector is providing the primary protein source for many of the developing countries and in 2020, global capture fisheries production recorded as 90.3 million tons, with an estimated value of USD 141 billion, including 78.8 million tons from marine waters and 11.5 million tons from inland waters and in the same year, an estimated 58.5 million people were as full-time or part-time fishers and out of that 21% are women fishers (Agriculture Organization of the United Nations, 2022).

Changes in the climate is affecting many ways of marine fishery resources such as productivity and distribution (Chávez & García-Córdova, 2012) and it is the one of key reasons for the drastic decline of marine populations over the past 30 years (WWF, 2015). The vulnerability has defined by the Intergovernmental Panel on Climate Change's (IPCC), it can define as "The degree to which a system is susceptible to, or unable to cope with the adverse effects of climate change, including climate variability and extremes. Vulnerability is a function of the character, magnitude, and

B. M. R. L. Basnayake (✉) · D. A. M. De Silva
Department of Agribusiness Management, Faculty of Agricultural Sciences, Sabaragamuwa University of Sri Lanka, PO Box 02, Belihuloya, Sri Lanka
e-mail: ruwinibasnayake@gmail.com

U. S. P. K. Liyanage
Regional Research Center, National Aquatic Resources Research and Development Agency, Colombo, Sri Lanka

C. Malalgoda et al. (eds.), *Climate Change Adaptation in the Built Environment*,
https://doi.org/10.1007/978-3-031-75826-3_26

rate of climate variation to which a system is exposed, its sensitivity, and its adaptive capacity" [6]. Increased frequency of extreme conditions such as floods, droughts, and windy storms, has impact on continuing the fishing activities and causes more damage and disruption of infrastructures, and the coastal zone (Chávez & García-Córdova, 2012). Mainly these impacts could be dividing as direct and indirect impacts on fish stocks. As direct impact; effects act on physiology and behavior and alter growth, reproductive capacity, mortality and distribution of fish stocks and as indirect impact; alter the productivity, structure and composition of the marine ecosystems on which fish depend for their food (Brander, 2010). Any kind of long-term effects of climate change can be having a huge impact on economically important, commercial level fisheries if the environmental conditions and water quality become unsuitable for high value species and range shifts lead to a decrease in local abundance (Fogarty et al., 2007).

Lobster fisheries is considering as highly valued seafood delicacy all over the world and with the high global demand, the world lobster trade is recorded as more than doubled over the past 20 years period (Boavida-Portugal et al., 2018). Further, due to ocean warming conditions and changing in ocean currents, several lobster stocks have already affected in the world and long larval phase of lobsters causes for more vulnerable for climatic changes. These changes are expected in to continue in the future as well (Caputi et al., 2013; Cockcroft et al., 2008; Pecl et al., 2009; Wahle et al., 2015).

Sri Lanka has 1760 km long coastal belt which is recorded as the 3rd longest costal line in South Asia and which contains unique ecosystem with lagoons, estuaries, mangroves, salt marshes, sand dunes, beaches and coastal marshy wetlands and different water bodies (Nianthi & Shaw, 2015). The major occupation of the coastal community is fisheries, which accounts for almost three quarters of the total fish production in Sri Lanka (Athulathmudali et al., 2011). Moreover, it is a well-known fact that Sri Lanka earns a higher income from its fisheries sector and in 2022 it is contributing 1% of the GDP in 2022 and export earning recorded as 26,749 Mt, with a value of Rs. Mn. 63,222 (USD Mn. 318.04) in 2021 (Central Bank Report, 2022). Among the high value marine coastal fishing activities lobster fisheries is considering as a good income source which is export oriented and household consumption is negligible (Koralagama et al., 2007). Major exporting countries are Japan, Thailand, Hong Kong, UK, Singapore, and Korea. According to export development board data in 2021, the lobster exporting value recorded as approximately USD 2,710,270.80. (Fig. 1) In recent past years due to COVID-19 impacts, the major income source of tourism in coastal area was totally collapsed and on the other hand, world market demand increased due to lack of supplies from major players. With these conditions, majority of non-fishers and fishers were adopted in lobster diving which is having high economic value that accelerated extra pressure on lobster stocks and over exploitation.

As shown in the Fig. 1, the exports value earned is showing improvement mainly due to growing demand in the foreign market. Which ultimately affecting on increase the competition among the export companies and middlemen in the landing sites for limited production and over exploitation.

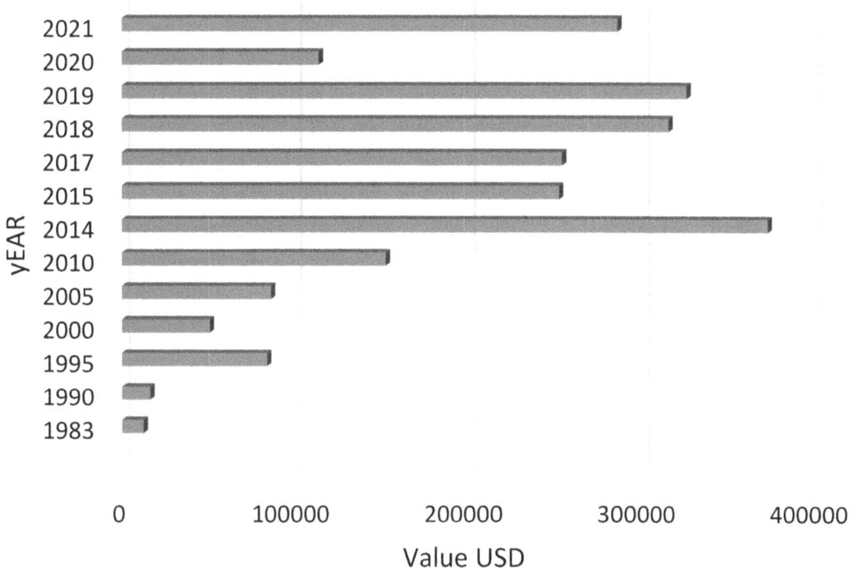

Fig. 1 Lobster exports earnings from 1990–2021. *Source* Sri Lanka Export Development Board

According to the Global Climate Change Models (GCCM), as a tropical island, Sri Lanka is expected to be affected by adverse impacts of climate change especially sea level rising due to the global warming impact (Gornitz, 1995; Munasinghe, 1998). Further, the sea level rise has intensified coastal erosion, siltation of estuaries salt water intrusion into rivers and salination of coastal plains and this increased siltation and reduced the fresh water discharge into the coastal zone that enhanced sedimentation on the coastal zone (Athulathmudali et al., 2011). Overexploitation, greenhouse gas emission, land use changes, sand mining like anthropogenic activities also causing for accelerate the slow processes of climate change (Nianthi & Shaw, 2015).

The climate change impact for lobster fisheries in Sri Lanka is also high specially affecting on their growth, advection, size at maturity and usual habitats appear to be sensitive to specially the changes in sea water temperature (Rajasuriya et al., 2002). Further, already, one species of lobster *Panulirus polyphagus* has completely disappeared from Sri Lankan coastal areas mainly due to overexploitation (Liyanage & Long, 2019). Since their main habitats of lobster juveniles based on seagrass beds, coral reefs, lime stones and sand stones it is obviously vulnerable to the climate change impacts (Long et al., 2010). The climate literacy level of fishermen is in low level and according to the recent research done by Sandaruwan et al. (2022), out of 200 sample, 67% of fishermen are highly exposed to climate-change induced hazards and extreme weather events and 61% of them are sensitive to those hazards however, only 34% have taken measures to adopt those hazards due to lack of knowledge and awareness. Further, in climate adoptive measures, it has identified many gaps and barriers such as knowledge and skills, technologies, information, and livelihood assets and thus, there is a crucial requirement to bolster the adaptive capacity of the country towards successful climate change adaptation (Dasandara et al., 2021).

Therefore, it is very much important to a adapt and build resilience in coastal communities who are engaging in lobster fishing who has already identified as particularly vulnerable to the climate change (Greenan et al., 2019). Further, empowering lobster fishery community and authorities and recognition of the risks associated with climate change could be the initial steps to mitigate the impact. With that, the present study eyed to identify the vulnerability of lobster fishery, to determine the climate impacts in lobster fishery and to identify community-based adaptation measures in Sri Lanka.

2 Literature Review

As shown in the above Table 1, The impact of climate change could vary depending on the fishing effort, type of fish, and the awareness level of fishermen. The majority of the literature highlights that sea level rise, temperature rise, ocean acidification, salinity, and high wind conditions are key climate change factors they experience. These impacts mainly affect their income, fishing efforts, fishermen's health, and lead to a decrease in catch volume or production.

Further, as adaptation strategies; community-based fisheries management, improving management of coastal habitats, improving management of coastal fisheries resources and policies and regulations are the common findings in the literature survey.

3 Methodology

3.1 Study Location

The study was conducted in South and west cost of Sri Lanka by mainly covering the Hambanthota, Matara, Galle, Puttalam and Chillaw fisheries districts (Fig. 2).

Hambanthota district considered as major spiny lobster producing district in Sri Lanka which is contributing 60% to the national production (Liyanage & Sharma, 2022). As shown in the Table 2, more active fishermen were recorded in these coastal areas and where highest lobster catch is recorded.

Moreover, both south and west coastal areas are found as vulnerable to climate change. As shown in the following map (Fig. 3) of vulnerability of marine fisheries sector based on sea level rise exposure both study location areas are highly susceptible for climate change impacts. As it is shown, the study locations are the most vulnerable

Table 1 Impact and adaptation measures of climate change in fisheries sector

Author	Topic	Objectives	Key findings
Mohammed and Uraguchi (2013)	Impacts of climate change on Fisheries: implications for food security in Sub-Saharan Africa	To examine the contribution of fisheries to poverty reduction and food security, and portray the potential impacts of climate change on the already strained resource in Sub-Saharan African (SSA) countries	Water Surface Temperature Rise: Such changes in ocean temperatures can change the dynamics of aquatic environments of the region. Changes in ocean dynamics could lead to changes in migration patterns of fish and possibly reduce fish landings, especially in coastal fisheries Sea Level Rise: In coastal areas, sea level rise may alter the salinity of estuarine habitats, inundate wetlands, and reduce or eliminate the abundance of submerged vegetation, adversely affecting those species which rely on these coastal habitats for reproduction and recruitment Ocean Acidification: it could potentially slow the growth of plankton and invertebrates that are at the bottom of the food chain. Thus, acidification can alter the productivity at certain trophic levels, thereby disrupting the complex food chain of aquatic ecosystems with effects on the productivity of fisheries
Timmers, (2012)	Impacts of climate change and variability on fish value chains in Uganda	To obtain information on climate change exposure, sensitivity, and adaptive capacity of activities along the major fish value chains	Storms and high winds on the lakes are dangerous for fishers and result in input, infrastructure and gear destruction Destruction of inputs and gear increased danger to boat crew and Fishers Fishing communities have limited awareness of climate change and weak adaptive capacity Fishers and fish farmers are largely unaware of the scientific concept of climate change, though they are known to be adapting in response to extreme wet and dry seasons, or climate variability. Fisher folk and farmers have little to no resources to effectively adapt to climate change, and there is no policy to guide adaptation Specific to fish production

(continued)

Table 1 (continued)

Author	Topic	Objectives	Key findings
Shaffril et al., (2011)	The potential impact of cli mate change environmental hazards on quality of life of fishermen community in Malaysia	To discover the climate change impact on the fishermen's quality of life	Affect fishermen health; they will be frequently exposed to disease such as fever, flu and cough resulting in lesser time for them to go out to the sea, thus lower their income
Abu Samah et al., (2019)	Factors affecting small-scale fishermen's adaptation toward the impacts of climate change: reflections from Malaysian fishers	To examine the influence of individual differences on the small-scale fishermen's climate change adaptation practices toward climate change	Small vessels are vulnerable to extreme winds and waves, which calls for a delay or cancelling of their fishing operation. Fewer fishing days mean less income Performing fishing operations with smaller vessels and smaller engine capacity could expose them to more threats and as the majority of them just carry a mobile phone as a basic communication tool, it hinders them from being able to seek for help should any emergency happens The rising sea levels are causing coastal erosions and as most of the SSFM's houses are located near shore, they needed to be transferred to a safer place Strong winds and waves causes damage to the SSFM's physical assets such as the jetty, vessels, and their fishing gears. As most of their houses are built from wood, such house design cannot withstand the threats posed by climate change

(continued)

Table 1 (continued)

Author	Topic	Objectives	Key findings
Najim (2012)	Impact of climate change on culture-based fisheries of seasonal reservoirs in Sri Lanka and resilience capacities of rural communities	To identify the changes of reservoir filling patterns in the scenario of climate change and to identify the water retention period of seasonal reservoirs in Sri Lanka	Increased flooding due to increasing number of high rainfall events will affect inland aquaculture and capture fishery Due to pollution, sedimentation and any adverse changes in water quality parameters of surface water bodies (mainly tanks) that sustain this fishery Drought would lead to lower yields in seasonal tanks, and thereby have severe impacts on the inland fishery as investments for aquaculture may not yield adequate returns Reduced production from the inland fishery would affect rural nutrition and incomes for those dependent on this activity Decreased primary productivity caused by temperature anomalies can affect the abundance of species higher up in the food chain in the long-term Rising temperatures may also lead to greater evapotranspiration and evaporation in irrigation tanks which may have impacts on the inland fishery and aquaculture **Impacts on inland fish stocks**: - A temperature rise of about 2 °C may have substantial impacts on the distribution, growth and reproduction of fish stocks – in both marine and inland waters - Rising temperatures may also lead to changes in spawning areas and distribution patterns of commercially important fish stocks - Decreased primary productivity caused by temperature anomalies can affect the abundance of species higher up in the food chain in the long-term - Rising temperatures may also lead to greater evapotranspiration/evaporation in irrigation tanks which may have impacts on the inland fishery and aquaculture

(continued)

Table 1 (continued)

Author	Topic	Objectives	Key findings
Macfadyen and Allison (2009)	Climate change, fisheries, trade and competitiveness: understanding impacts and formulating responses for commonwealth small states	To study impact of climate change on the fisheries sector in small developing commonwealth States, with a particular emphasis on trade and competitiveness	**Adaptation** Decommissioning, introduction of fishing rights, improved monitoring control and surveillance Provision of safe havens/harbours for the fishing sector to protect vessels can ensure continuing levels of catches available for trade; while climate-proofing any such developments through careful siting and appropriate engineering will be increasingly important Managing declining incomes if fish catches fall due to lower catches, and efforts aimed at diversification and fostering alternative livelihood activities Enabling exit from the fishery sector in response to downturns, by providing training in alternative occupations or through general investments in skills and capabilities, Dealing with fisheries' status as a 'safety net' activity. In many countries, climate change impacts on livelihoods in other sectors **Policy-level:** Invest in improved weather information and storm warnings Ensure that the fishery sector is included in national disaster preparedness and response planning Ensure that fisheries line management agencies are familiar with disaster preparedness and response procedures **Mitigation** Policy options to contribute towards climate change mitigation could however include taxation, regulation and incentives related to: Fuel use and/or vessel engine emissions; Technical innovations to reduce fuel usage and emissions in fishing vessel engines (e.g. more efficient vessel design reducing engine size requirements); Promotion of fuel-efficient fishing methods Research into support for low impact aquaculture Improvements in building design and handling practices to reduce energy requirements and increase energy efficiency e.g. through better insulation in ice plants, freezing plants, cold stores and chill stores

(continued)

Table 1 (continued)

Author	Topic	Objectives	Key findings
Shyam et. al. (2014)	Vulnerability of coastal fishery communities to climate change in Kerala state	To determine the vulnerability of the fisheries sector, in particular the fishing communities under consideration to changes in the biophysical system	-Change in monsoon patterns-Decrease in occurrences of coastal upwelling -Increased sea surface temperature -Increased sea level rise -Increased salinity Due to increase in sea surface temperature increase in fast growing, smaller, short lived and low value fish Increase distribution of pelagic fish Sharks and catfish are declining due to decrease in frequency and intensity of monsoon rainfall, less opportunities to go out fishing- decrease in catch, decrease in income
Roy et. al.(2019)	Consequences of climate change on fish diversity in Dekhar Haor Bangladesh	To study the implications of climate change on fish biodiversity as experienced by the local people of Dekhar Haor To expressed the local community's perceptions about the manifestations and consequences of climate change	-**Delay spawning** During high temperatures, the brood fish stay at the bottom of the water body to keep their body temperature in normal condition. This condition hampers the spawning of brood fish. During low temperature, fish body metabolism becomes slow - **Reduce body immunity** Temperature fluctuation reduces body immunity. Different unwanted diseases may take place. As a result, fish mortality may increase -**Stress of mother/brood fish** Fluctuation of temperature increase fish mobility and thus may cause serious stress of brood fish -**Suffocation of brood fish** Temperature change causes loss of dissolved oxygen and creates suffocation of brood fish. Brood fish seem to gulp at the surface of the pond - **Fermentation of fish food** During high temperatures, Some microorganisms may grow. They are responsible for the fermentation of fish food. Thus, fish nutrients may reduce in the ponds. As a result, brood fishes face malnutrition problems. This reduces the spawning rate

(continued)

Table 1 (continued)

Author	Topic	Objectives	Key findings
Lam et al. (2020)	Climate change, tropical fisheries and prospects for sustainable development	To effects of climate change on tropical marine fisheries, highlighting the socio-economic impacts to both tropical and extratropical nations, and discuss potential adaptation measures	Coral reefs, mangroves and sea grasses - which provide feeding areas limits, and the median return time between pairs of severe-coral-bleaching events has diminished steadily The aerobic capacity of fishes and invertebrates is also affected by decreasing water oxygen content and its interactions with ocean warming. Ocean deoxygenation can limit important metabolic functions that impair growth and reproduction in fishes and invertebrates, as well as reducing their temperature tolerances and, thus, geographic ranges. Ocean warming and deoxygenation also reduces the body size of some marine fishes and invertebrates, particularly in the tropics
Salgueiro-Otero (2022)	Transformation in times of climate change: what makes a fisher diversify livelihoods?	To shed light on factors that help to explain the transformative behavior of small-scale fishers when faced with climate change impacts	Flexibility (current income diversification level and diversity of target fishery resources) has a strong, positive relationship with livelihood diversification responses Social-ecological system knowledge, communication with different fishing groups (bridging communication), trust in institutional actors, and gender (female) were positively related to livelihood diversification when faced with climate impacts. Fishing experience, communication within fishing groups (bonding communication), and trust in other fishing groups (bridging trust) were negatively related to diversification
Villasante et al. (2022)	shar	To investigate individual and household-level adaptive responses to climate change among fishers	The fishers' vulnerability decreases as the size of households increases, while fishers who pay a mortgage and who live in households with fewer active members tend to be more vulnerable Galician shell fishers have developed a wide range of adaptation strategies to anticipate and respond to climate change impacts, namely harvesting pricier and more abundant species, reducing household expenses and increasing social involvement in shellfishery associations

Fig. 2 Study location map

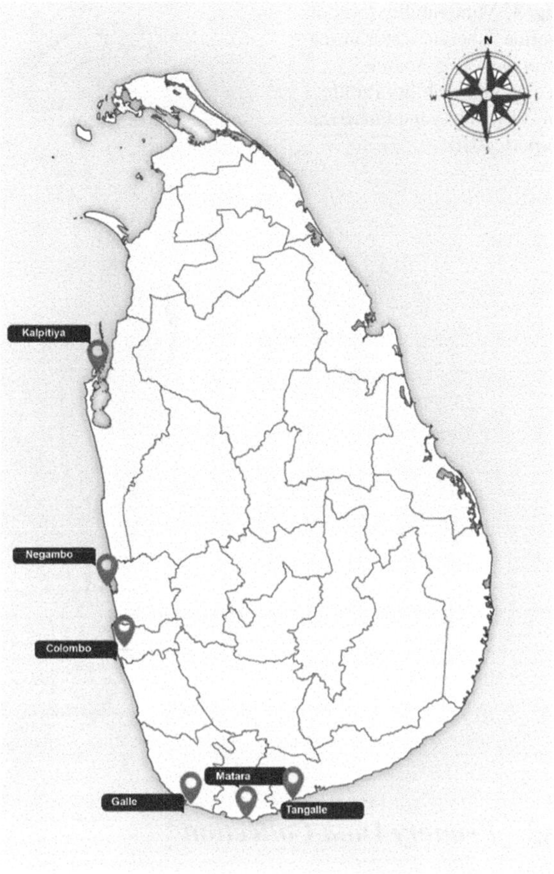

Table 2 Active fishermen and Lobster production by districts in 2019

District	Active fishers	Lobster production (Mt.)
Galle	11,390	10
Matara	14,840	10
Hambanthota	13,780	160
Puttalam	16,860	130
Chillaw	9890	10

Source Ministry of Fisheries, 2020

areas to climate hazards and the study locations were selected purposely and priority has given to the severity of the climate change impacts or disaster incidents.

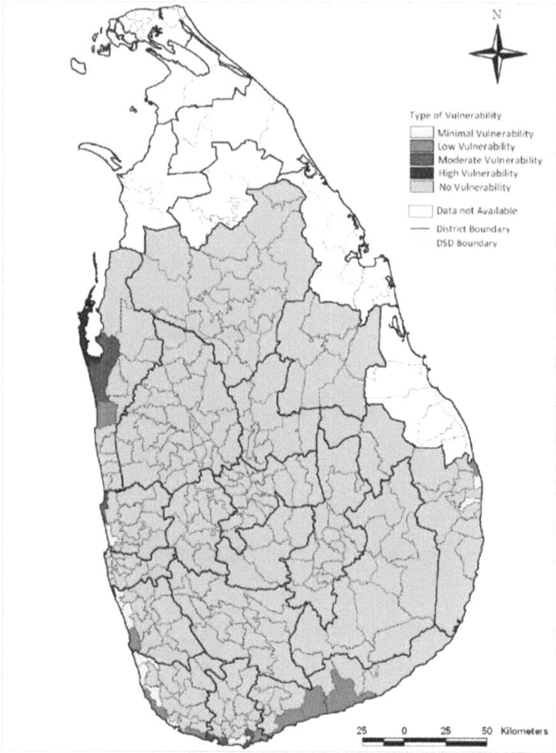

Fig. 3 Vulnerability level of marine fisheries sector in sea level exposure. *Source* Sector Vulnerability Profile on Agriculture and Fisheries Report, 2010

3.2 Primary Data Collection

A field survey guided by a questionnaire was conducted to collect data from a randomly selected sample of 60 lobster fishermen and 12 collectors across both Provinces. The primary aim was to assess the current status of the lobster industry and the impact of climate change hazards.

In addition to the questionnaire, key informant interviews and field observations were instrumental in gathering data from various stakeholders. The sample profile included 5 fisheries inspectors, 5 district fisheries officers, 10 exporters, 2 scientists, 2 customs officials, and 6 hoteliers. During these interviews and discussions primarily focused on the vulnerability level of the industry, the impacts of climate change, and existing and potential adaptive measures. The expertise of the participants, coupled with their experiences and knowledge, provided valuable insights into these crucial aspects.

This multi-faceted approach helped to comprehensively understand the challenges facing the lobster industry and to explore potential solutions.

3.3 Secondary Data Collection

Secondary data were obtained from reports published by the Ministry of Fisheries, National Aquatic Resources Research and Development Agency, Export Development Board, Department of Census and Statistics, as well as published journal articles on climate change impact of fisheries industry. These sources were crucial in providing a broader context for the current research.

The primary use of secondary data was to complement and strengthen the results of the primary survey conducted. By integrating information from these reputable sources, the study was able to enrich its analysis and validate the findings. This approach ensured a comprehensive understanding of the lobster industry's current status and its vulnerability to climate change impacts.

3.4 Data Analysis

The study utilized both qualitative and quantitative data. Quantitative data obtained during the survey were analyzed using the Microsoft Excel 2019 package, providing a structured and systematic approach to understanding the numerical aspects of the research.

Additionally, a prioritizing recommendations matrix was employed to graphically analyze the possible and potential adoption measures. This matrix facilitated a visual representation of the recommendations, allowing for a clearer understanding of the most important and feasible actions to be taken. The combination of quantitative analysis and graphical representation enhanced the study's ability to derive meaningful insights and actionable recommendations for the lobster industry the against climate change impacts.

4 Results and Discussion

The results revealed that, the typical supply chain of the lobster fisheries consists with fishermen, collector, distributor, local hotels and exporters.

As shown on the above Table 3 all the fishermen were men and majority of them were belongs to 30–50 age category while Buddhist and catholic participants were high. There was one woman collector in southern province and rest of the collectors of the sample were male. Same as fishermen, Buddhist and catholic participants are high in collector level as well. Both majority of fishermen and collectors were having primary or secondary education. Exports dominated by male counterparts and one female exporter recorded from western province. Moreover, its despite that, exporters are enjoying higher revenue comparing to the poor fishermen. While fishermen are selling their harvest, live animals for USD 6–15, and dead animals for USD 3–6 and the exporters are selling live animals for USD 46–62 and dead animals for USD 12–19. Of the sample of lobster supply chain 70% were practiced skindiving and scuba diving while 30% were using nets to catch the lobsters.

Further, the market for lobster fisheries is mainly organized for exports (95%) and remaining moves to local market. Main buyers were tourist hotels and restaurants. Quality concerns are very much important for exports while second grades, low quality died animals were directing towards to the local tourist hotels. The product are hole Lobster, Lobster Tail, Lobster Meat and Lobster Claw.

4.1 Climate Change Vulnerability of Lobster Fisheries

The vulnerability level was measured based on two criteria: stock status and socio-economic exposure (Greenan et al., 2019). In Sri Lankan costal water, we can find spiny lobsters mainly in south coast area than west coast or north coast areas. Usually, shallow waters less than 30 m in depth is ideal to find lobsters and they mainly live in Seagrass beds, Coral reefs, Lime stones and Sand stones. Further, six species of spiny lobster occur along the Sri Lankan costal area.

From the sample, majority of fishermen (95%) were aware on climate hazards and perceived that, over the past 10 years, climate change may have an impact and threatened to the final lobster harvest. Main lobster fishing season commence from August up to March. Moreover, February, September and October months are restricted for lobster fishing as a precaution for overexploitation and avoiding egg bearing period.

As shown in the following Fig. 4, the climate change vulnerability level differs from upper stream to downstream. Both Fishermen and collectors are more vulnerable for depletion of species and over exploitation of lobster while exporter is more vulnerable for over exploitation.

Further, 88.3% agreed to the statement of changes of lobster volume and 85% agreed that the diversity of catch is also changed within past 10 years period (Fig. 5). moreover, the fishermen have pointed out that they have to spend more time to catch lobsters due to low volume.

Based on the literature, there are six lobster species have identified in Sri Lanka and they are belonging to the family *Palinuridae* which is available in most of the Asian region. However, one of these species (*Panulirus polyphagus*) already disappeared from Sri Lankan coastal waters (Liyanage & Long, 2019).

As shown in the Table 4, the highest catch is recorded in *P. homarus* and all composition of each species have changed significantly in past years while *P. ornatus*, the most valuable species composition declined drastically. Percentage abundance of the *P. longipes* and *P. penicillatus* in same for a long period and also no any individual of *P. polyphagus* fond remain in the landings. After the introduction of the 1996 Southern Coastal Lobster Fisheries Regulations there was a trend of increasing *P. homarus* in the catch. The mud spiny lobster *P. polyphagus* is completely exploited

Table 3 Demographic factors of the respondents

Criteria	Southern province			Western province		
Participants	Men	Women	Total	Men	Women	Total
Fishermen	30	0	30	30	0	30
Age under 30	4	0	4	3	0	3
30–50	19	0	19	19	0	19
over 50 years	7	0	7	8	0	8
Religion Buddhist	22	0	22	8	0	8
Catholic	8	0	8	22	0	22
Muslim	0	0	0	0	0	0
Education primary	19	1	20	21	0	21
Secondary	9	1	10	9	0	9
Higher	0	0	0	0	0	0
Monthly income level > 25,000LKR	4	0	4	6	0	6
25,000–50,000LKR	18	0	18	13	0	13
50,000–1,00,000LKR	8	0	8	7	0	7
< 1,00,000LKR	0	0	0	4	0	4
Collectors	4	1	5	7	0	7
Age under 30	0	0	0	1	0	1
30–50	3	1	4	5	0	5
over 50 years	1	0	1	1	0	1
Education primary	1	1	2	5	0	5
Secondary	3	0	3	2	0	2
Higher	0	0	0	0	0	0
Monthly income level > 25000LKR	0	0	0	1	0	1
25,000–50,000LKR	4	1	5	4	0	4
50,000–1,00,000LKR	0	0	0	2	0	2
< 100000LKR	0	0	0	0	0	0
Exporters	3	1	4	5	1	6
Age under 30	0	0	0	1	0	1
30–50	1	0	1	5	0	5
over 50 years	0	1	1	4	0	4
Education primary	0	0	0	2	0	2
Secondary	1	1	2	4	0	4
Higher	0	0	0	4	0	4
Monthly income level > 25000LKR	0	0	0	0	0	0
25,000–50,000LKR	0	0	0	0	0	0
50,000–1,00,000LKR	0	0	0	2	0	2
< 1,00,000LKR	1	1	2	7	1	8

Source Survey data, 2022

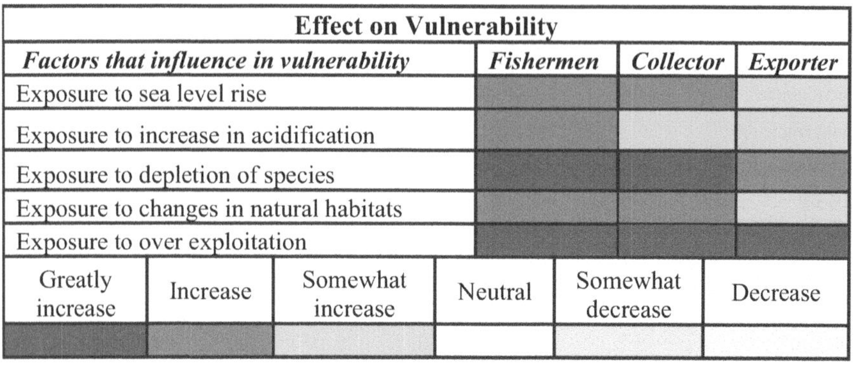

Fig. 4 Effect of Vulnerability in lobster fisheries

Fig. 5 Negative effects on lobster species due to the climate changes

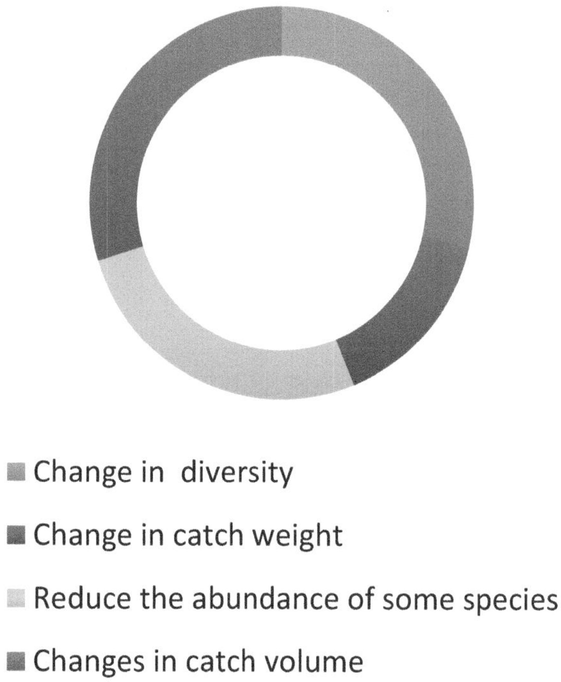

and was absent from the catch from January 2006 until the present (Liyanage & Sharma, 2022).

This declining mainly because of over exploitation of resources without considering the legal barriers. And habitat changes due to climate hazards also have an impact on declining of lobster species (Sandaruwan et al., 2022). With reference to the filed discussions, incensement of disasters such as flood, storm surges conditions may affect in near shore habitats and associated fishery resources and near shore

Table 4 Changes in Species wise catch

Species	English name	Percentage of catch								
		1989	1996	1998	2006	2015	2017	2018	2019	2020
P. homarus	Scalloped spiny lobster	74	70	83	86	86	91	72	85	73
P. ornatus	Ornate spiny lobster	8	10	5	3	13	3	7	1	0
P. penicillatus	Prong horn or variegated spiny lobster	6	8	5	3	< 1	1	3	2	8
P. longipes	Japanese spiny lobster	4	6	4	2	< 1	1	3	2	1
P. versicolor	Painted spiny lobster	6	5	3	6	< 1	4	15	10	18
P.polyphagus	Mud spiny lobster	2	1	0	0	0	0	0	0	0

Source Liyanage and Sharma (2022) and Jayakody (1997)

reefs which causing for more serious coastal erosion and saline intrusion due to sea level rise and ultimately threaten for survival of lobster species.

Table 5 is explaining the existing areas of available natural habitats and impact of human activities and climate change in Sri Lankan coastal belt.

Another key cause for lobster stock changes is habitat changes. According to Jayakody 1997, *P. versicolor*, *P.longipes* and and *P. penicillatus* are occupied in the coral reef areas and *P. homarus* occupied in several habitats such as coral reef, sand stone, lime stone reef, rock pools, and crevices as their typical habitats. The damage to the coral reef could be accelerate due to storm surges, coastal flooding, coastal current waves changes and coastal erosion like climate change incidents and limit the lobster population growth (Sandaruwan et al., 2022) (Table 6).

Table 5 Extent of marine habitats

Habitat	Area (ha)	Causal agent	Effect
Estuaries and lagoons	> 1,29,075	Beach sand mining	Increased erosion, reduce the water quality
Mangroves	6,080	Improperly sited groins, harbours, revetments and jetties	Increased erosion and increased acidifications
Salt Marshes	23,819	removal of coastal vegetation	Erosion and retreat
Coral Reefs	68,000	reef breaking or dynamiting for mining or fishing	Increased wave energy on beaches resulting in erosion

Source Coast Conservation Department Sri Lanka, 2007

Table 6 Projected average rising temperature

Year	Sri Lanka
2025	0.4 °C
2050	0.9 °C
2075	1.6 °C
2100	2.4 °C

Source Department of Meteorology

Rising temperature could be cause to bleach and impoverish the coral reef and these changes are adversely affecting on distribution and growth of lobsters. Further, sand and sediments deposit in coral also could effect on habitat changes of lobsters.

Moreover, acidification and increased atmospheric CO_2 could be effect on dissolve the external eggs and crust of lobsters. A temperature rise of about 2^0C may have substantial impacts on the distribution, growth and reproduction of stocks and reduce the productivity as well (Liyanage & Sharma, 2022).It is proven that, from 2010–2019 the lobster catch volume has drastically reduced (Fig. 6).

According to the literature, Changes in natural habitats due to higher usage of boats and anchorage of boats near coral reefs, due to climate change associated natural disasters and dragging of nets causing to damage the feeding and breeding habitats of lobsters and could be affect on low productivity records (Chávez & García-Córdova, 2012; Sandaruwan et al., 2022). Another issue is catching egg bearing female animals, which adversely affecting on lobster production in the country. It is a major cause of stock over exploitation. Poor awareness of fishermen and loop holes of existing regulations, mainly causing for this issue. Peak breeding period of the

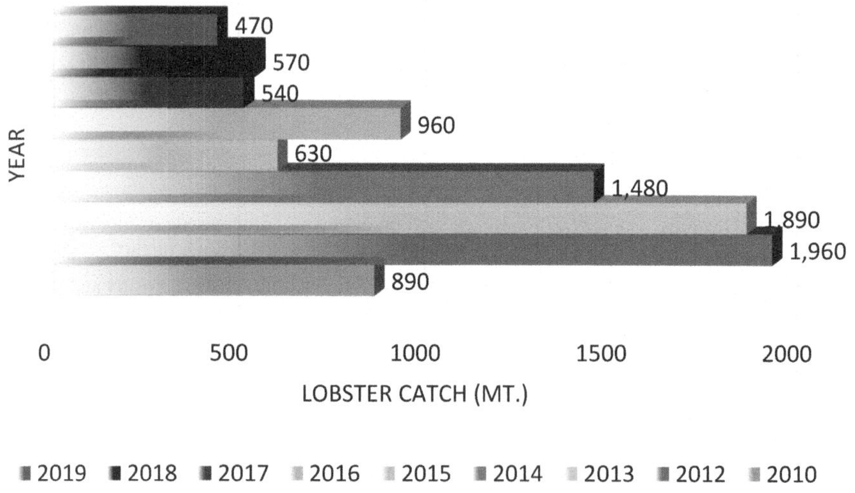

Fig. 6 Lobster catch volume (Mt.). *Source* Ministry of Fisheries

lobsters are mainly determined by the oceanographic conditions such as temperature, salinity and length of photo period in their environment (Sachlikidis et al., 2005). In the Indian ocean which located around Sri Lanka, reversal South-west and north-east monsoonal wind is playing vital role influencing oceanographic changes which are inducing biological productivity of the ocean ecosystem (Vinayachandran et al., 2004; Yapa, 2000). Therefore, breeding cycles of the many ocean creatures is couple with these seasonal changes (Vinayachandran et al., 2004) (Fig. 7).

The following Table 7 describes the undersized egg bearing lobsters catch comparison in 2019 and 2020 which is directly impact on depletion of the stocks and threatening to the sustainable production.

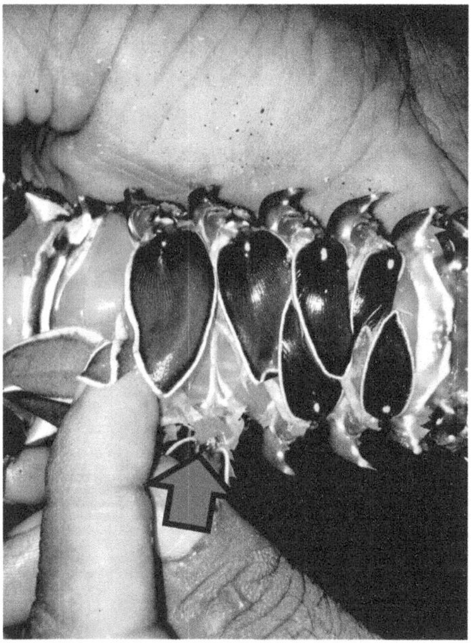

Fig. 7 External egg bearing female animal. *Source* Survey data, 2022

Table 7 Caught under sized and egg-bearing lobster stock status

Criteria	2019	2020
P. homarus	2.53%	0.8%
P. ornatus	0.74%	0.27%
P. penicillatus	0.6%	5.35%
Undersized males	3.47%	0.57%
Undersized females	2.35%	2.84%
Egg bearing females	48.53%	47.03%

Source Liyanage and Sharma (2022)

4.2 Impact of Climate Changes

Of those interviewed, about 80 per cent perceived that there is a climate change impact on their daily fishing activities. According to the survey conducted among lobster fishermen, they have identified that as main climate impacts for the lobster fishing are; wind changes (92%), changes in rainfall (87%), unpredicted weather conditions (83%), storm factor (61%), sea level rise (58%) and increase in temperature (51%) (Fig. 8). During the in-depth interviews, the expertise has mentioned as costal erosion, costal current waves pattern changes, changes in acidification, temperature changes are directly affecting on lobster growth and distribution in Sri Lankan costal water. These climate changes have proved in meteorological records as; a temperature increase of 0.01–0.036 °C y^{-1}, rainfall had decreased by 10–35 mm, Precipitation, evaporation and transpiration patterns are not stable.

or predictable during the last 30 years across the country (Roy et al., 2019).

The rapid urbanization of coastal areas can see due to unplanned coastal development and infrastructure, poor governance and more intensive industrial activity while housing for biologically rich coastal environment with unique ecosystems. It's recorded as more than 65% of the total country's population resident in coastal area and total population in the coastal belt is projected to be 8.4 million in 2010 as the negative effect (Lam et al., 2020). Livelihood due to the climate changes which shown in following Fig. 9, the fishermen have stated that; decrease monthly income, loss the employments, reduced their loss of fishing days and damage to boats and other infrastructure as main concerns.

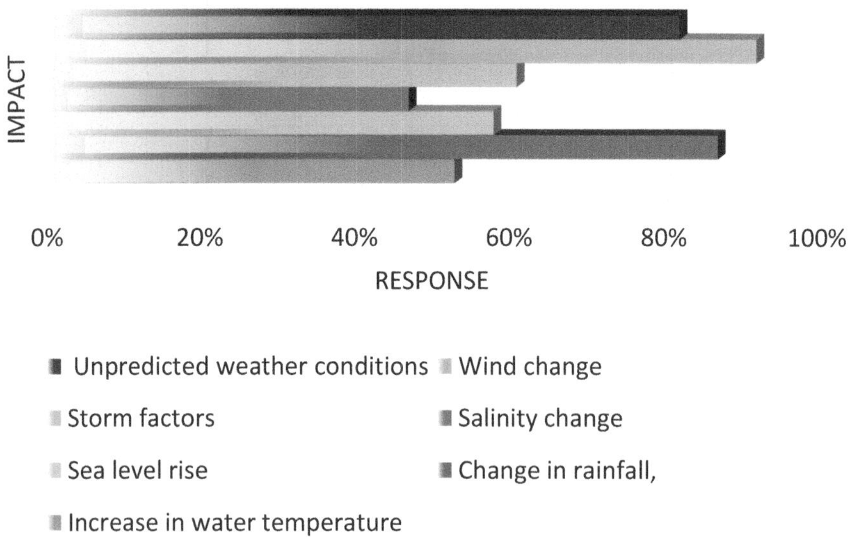

Fig. 8 What are the climate changes impacts that you have face during past years? *Source* Survey data, 2022

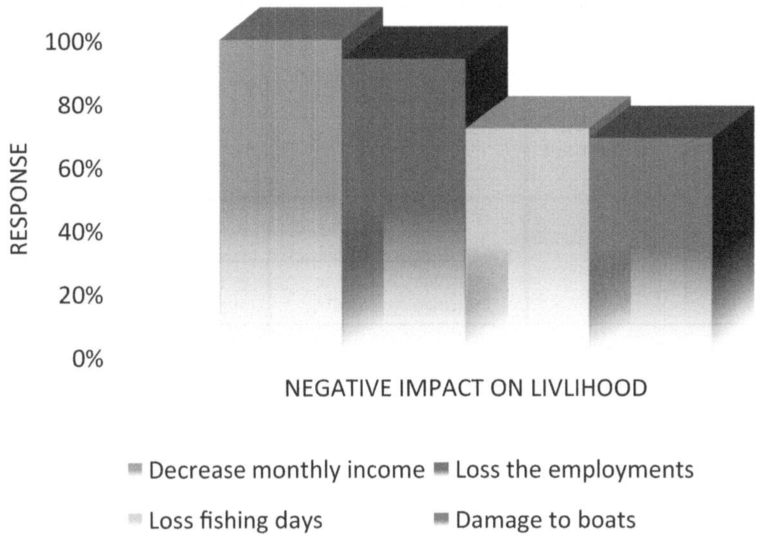

Fig. 9 What are the negative effects on your livelihood due to climate change? *Source* Survey data, 2022

Further, they are using radio, GPS, mobile and past experience to predict weather conditions daily and they mainly based on alerts issued by the government or navy for their day today fishing activities.

The study conducted by Nianthi and Shaw (2015) shows that there is a significant economic impacts of climate change, specially from sea level rise. According to the results they have stated that economic impact has identified in following sectors; land resources (an estimated amount of 1242 million rupees), tourism sectors (affected cost: 201 million rupees and replacement cost of 1174 million rupees in the study area) and industries (small industries impact 152 million rupees).

4.3 Adaptation Measures for Climate Change Impact

The adoption measures and recommendations to mitigate climate change impact have analyzed by using prioritizing recommendations matrix (Fig. 10) for the lobster fisheries. As shown there, the recommendations and adoption measures have organized based on its impact and difficulty level of implementation. The color gradient expresses the time-bound of each recommendation and adaptation measures. These measures and potential strategies were identified during in-depth interviews and the survey. Moreover, literature also based on develop each criteria.

As the measures to build resilience, educating and conducting community-based arrangement of resources management is essential. It is important to educate fishermen specially on conservation and management of the lobster stocks and the

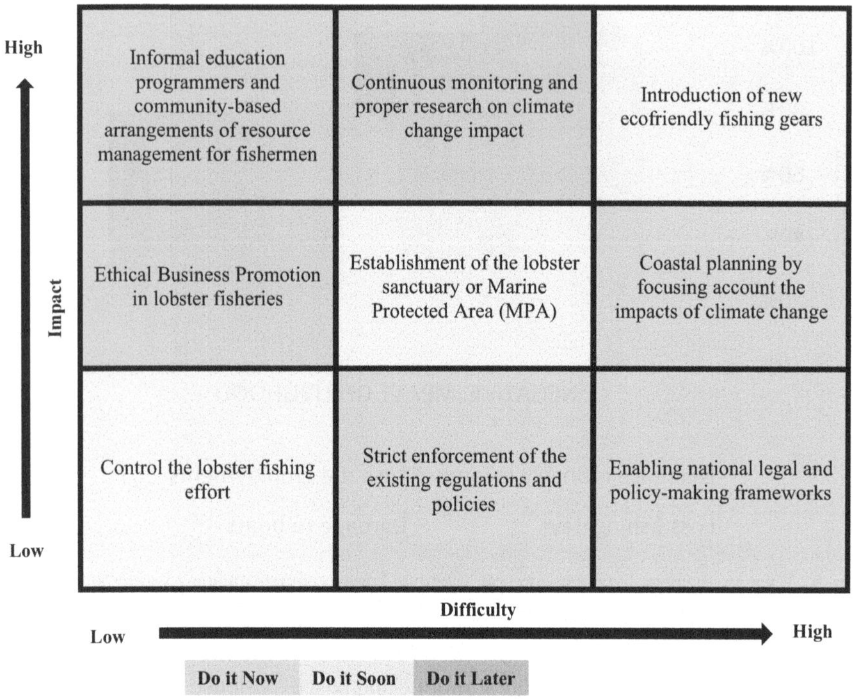

Fig. 10 Prioritizing recommendations matrix. Adopted from Dent and Collins, 2021

ecosystem in collaboration with the resource users is essential to mitigate the issue. Further, aware on impact of over exploitation of lobster stocks could be threaten to their income also essential as an adoptive measure. The second immediate action which can take is the continuous monitoring and research on climate change impact. In here, with proper monitoring, can collect information on lobster population specially in species wise in each costal area, climate change incidents, habitats change in location wise etc. which could use for future planning and could use to evaluate progress in moving towards management goals. Through researches, can identify loopholes in excising climate hazards adoptive measures, can develop new knowledge which may be useful in revising assumptions, ideas, model structures. Ethical business promotion is another immediate action which can take to mitigate over exploitation and selling egg bearing or removed female animals. Since the harvesting of berried females adversely influence on the growth of depleted lobster stock, we can aware local hoteliers, high-end restaurants on closing seasons of lobsters and importance of protecting and conservation lobster stocks.

The second category of adoptive measures are recognized as "do it soon". Where the actions should be taken in immediately. These actions also prioritizing to implement soon. The first action as mentioned in the recommendation matrix is control the lobster fishing efforts that could be support to manage and protect natural habitats.

Because during lobster fishing season, fishermen in eastern province are moving in to southern and west costal area. With that huge damage could happening accelerate the destroy of natural habitats especially coral reef which are housing for lobsters. To control the fishing effort, it is necessary to introduce a new license for the lobstermen in district wise is essential and regular monitoring is also had to implement. Establishment of the lobster sanctuary or marine protected area (MPA) is another adoptive measure could take for conserve the lobster species. This is currently practising for sea turtles and can apply same strategy to lobsters as well. There was a suggestion from the scientist during the in-depth interviews to implement this lobster sanctuary in great basses' reef in south coastal area which is located about 16 km away from the land and less fishing pressure. Lobster breeding and releasing programmes also could be started in parallel with the implementation of lobster sanctuary same as other countries are practicing. Third adoptive measure and recommendation is Introduction of new eco-friendly fishing gears again to reduce over exploitation and ecosystem conservation. Since 95% of fishers are using bottom set gill nets, its causing to damage coral reef. Therefore, its essential to introduce new gear to minimize the number of immature lobsters represent in the catch (Minimize the growth overfishing).

There are some adoptive measures which we have to address in national level and taking time to implement. Strict enforcement of the existing regulations and policies is one of those adoptive actions. Especially on catching berried females in large scale, removal eggs, complete prohibition of catching females at least one year or revision of the minimum legal size at least more 2 cm, and control through the specific license for lobster fishing and number of fishing gear units per crafts. Further, enabling national legal and policy-making frameworks special focusing in emergency response and disaster risk reduction could be used as an instrument to up lift fisheries community's livelihood. Moreover, building capacity for climate change adaptation through a policy-making framework could be a practical solution to mitigate the issue. Coastal planning by focusing account the impacts of climate change is another essential step to adopt climate change impacts. In here can focus on sea level rise, higher temperatures, prolonged droughts, severe rainfall, cyclones and storm surges such common incidents and can target the sustainable utilization of the resource. further, getting support of the coastal communities in planning, management, monitoring and evaluation with the support of government agencies could be more effective.

5 Conclusion

The continuous global demand of lobster market valued at USD6.1 billion in 2022 and forecasted USD12.43 billion value in 2029 results overexploitation of resources and meanwhile in Sri Lanka, the stocks have depleted within past years. In both globally and locally there is increasing concern over the consequences of climate change in many sectors and specially on fisheries production. Climate change is an additional pressure on top of the many (overfishing and habitat degradation) which existing lobster fisheries sector already experience. Global warming leading to climate change and sea level rise, impacting on coastal ecosystem health, livelihoods and the increased frequency and severity of natural hazards crippled the lobster fishery All the spiny lobster species are fully exploited and *P.polyphagus* is completely disappeared from Sri Lankan coastal waters. The survey conducted shows that 88.3% of the sample are experiencing changes in lobster volume, with 85% of respondents noting changes in the diversity of catches within the past 10 years. Further, changes in temperature, ocean acidity and oxygen level are directly affecting on change the water chemistry and causing change animal habitats, influencing on reproductive cycles and speed of sextual maturity (larvae and juvenile) which ultimately reduce the catching stock. The impact of climate change mainly revolves around creating economic risk with fewer fishing days. This study highlights the importance of proper planning and the implementation of a strict regulatory system. It integrates information on lobster vulnerability and provides a framework and data for further in-depth analyses by climate change adaptation planners. The implementation of integrated and adaptive participatory management, along with improved early warning systems and education, will be essential to mitigate the existing situation in Sri Lanka.

References

Abu Samah, A., Shaffril, H. A. M., Hamzah, A., & Abu Samah, B. (2019). Factors affecting small-scale fishermen's adaptation toward the impacts of climate change: Reflections from Malaysian fishers. *SAGE Open, 9*(3), 2158244019864204.

Agriculture Organization of the United Nations. Fisheries Department. (2022). The State of World Fisheries and Aquaculture, 2022 (Vol. 3). Food & Agriculture Organization

Athulathmudali, S., Balasuriya, A., & Fernando, K. (2011). *An exploratory study on adapting to climate change in coastal areas of Sri Lanka.* Centre for Poverty Analysis, Colombo.

Barange, M., Bahri, T., Beveridge, M. C., Cochrane, K. L., Funge-Smith, S., & Poulain, F. (2018). Impacts of climate change on fisheries and aquaculture. *United Nations' Food and Agriculture Organization, 12*(4), 628–635.

Boavida-Portugal, J., Rosa, R., Calado, R., Pinto, M., Boavida-Portugal, I., Araújo, M. B., & Guilhaumon, F. (2018). Climate change impacts on the distribution of coastal lobsters. *Marine Biology, 165*, 1–7.

Brander, K. (2010). Impacts of climate change on fisheries. *Journal of Marine Systems, 79*(3–4), 389–402.

Caputi, N., Lestang de, S., Frusher, S., & Wahle, R. A. (2013). The impact of climate change on exploited lobster stocks. In *Lobsters: biology, management, aquaculture and fisheries*, pp. 84–112.

Central Bank Report. (2022). https://www.cbsl.gov.lk/en/publications/economic-and-financial-reports/annual-reports/annual-report-2022

Chávez, E. A., & García-Córdova, E. A. (2012). Effect of climate change on the Caribbean lobster fisheries. *Proceedings of the 64th Gulf and Caribbean Fisheries Institute*, Puerto Morelos, Mexico.

Cockcroft, A. C., Van Zyl, D., & Hutchings, L. (2008). Large-scale changes in the spatial distribution of South African West Coast rock lobsters: An overview. *African Journal of Marine Science, 30*(1), 149–159.

Dangendorf, S., Marcos, M., Wöppelmann, G., Conrad, C. P., Frederikse, T., & Riva, R. (2017). Reassessment of 20th century global mean sea level rise. *Proceedings of the National Academy of Sciences, 114*(23), 5946–5951.

Dasandara, S. P. M., Kulatunga, U., Ingirige, M. J. B., & Fernando, T. (2021). Climate change challenges facing Sri Lanka: A literature review. In Sandanayake, Y. G., Gunatilake, S. and Waidyasekara, K. G. A. S. (eds), *Proceedings of the 9th World Construction Symposium, Sri Lanka* (pp. 183–195). https://doi.org/10.31705/WCS.2021.16

Fogarty, M., Incze, L., Wahle, R., Mountain, D., Robinson, A., Pershing, A., Hayhoe, K., Richards, A., & Manning, J. (2007). *Potential climate change impacts on marine resources of the Northeastern United States*. Report to Union of Concerned Scientists.

Gornitz, V. (1995). Sea-level rise: A review of recent past and near-future trends. *Earth Surface Processes and Landforms, 20*(1), 7–20.

Greenan, B. J., Shackell, N. L., Ferguson, K., Greyson, P., Cogswell, A., Brickman, D., Wang, Z., Cook, A., Brennan, C. E., & Saba, V. S. (2019). Climate change vulnerability of American lobster fishing communities in Atlantic Canada. *Frontiers in Marine Science, 6*, 579.

Jayakody, D. S. (1997). The spiny lobster fishery in relation to bottom conditions on the south coast of Sri Lanka. *Journal of National Aquatic Resources and Research Agency, 35*, 1–10.

Koralagama, D. N., Amaraslnghe, O., & Jayakody, S. (2007). Export market of lobsters in Sri Lanka, with special reference to slipper lobsters. *Proceedings of the Fourth Academic Sessions, Technical Session, III*, 84–90.

Lam, V. W., Allison, E. H., Bell, J. D., Blythe, J., Cheung, W. W., Frölicher, T. L., Gasalla, M. A., & Sumaila, U. R. (2020). Climate change, tropical fisheries and prospects for sustainable development. *Nature Reviews Earth & Environment, 1*(9), 440–454.

Liyanage, U., & Long, B. (2019). *Status of the South Coast Lobster Fishery*. National Aquatic Resources Research and Development Agency.

Liyanage, U. S. P. K., & Sharma, R. (2022). *Assessment of the South coast spiny lobster fishery*. National Aquatic Resources Research and Development Agency

Long, G., Amarasiri, C., Rajasuriya, A., Dissanayake, D. C. T., Liyanage, U. S. P. K., Jayasinghe, R. P. P. K., Athukoorala, A. A. S. H., Karunathilake, K. M. B. C., Fernando, H. S. G., & Fernando, W. V. A. T. D. (2010). Status of resources, fisheries management initiatives on sea cucumber, chank, lobster, shrimp, and marine aquarium fish in the northwest, south, and east coast of Sri Lanka. *Sri Lanka Fisheries Atlas* (Vol. I). National Aquatic Resources Research and Development Agency, Sri Lanka.

Macfadyen, G., & Allison, E. (2009). Climate change, fisheries, trade and competitiveness: understanding impacts and formulating responses for Commonwealth small states.

Mohammed, E. Y., & Uraguchi, Z. B. (2013). *Impacts of climate change on fisheries: Implications for food security in Sub-Saharan Africa* (pp. 114–135). Nova Science Publishers Inc.

Munasinghe, M. (1998). Climate change decision-making: Science, policy and economics. *International Journal of Environment and Pollution, 10*(2), 188–239.

Najim, M. M. M. (2012). Impact of climate change on culture-based fisheries of seasonal reservoirs in Sri Lanka.

Nianthi, K. W. G. R., & Shaw, R. (2015). *Climate change and its impact on coastal economy of Sri Lanka*. The Global Challenge, pp. 1–21.

Pecl, G., Frusher, S., Gardner, C., Haward, M., Hobday, A., Jennings, S., Nursey-Bray, M., Punt, A., Revill, H., & Van Putten, I. (2009). *The east coast Tasmanian rock lobster fishery–vulnerability to climate change impacts and adaptation response options*. Report to the Department of Climate Change. Department of Climate Change, Canberra.

Rajasuriya, A., Zahir, H., Muley, E. V., Subramanian, B. R., Venkataraman, K., Wafar, M. V. M., Khan, S., & Whittingham, E. (2002). Status of coral reefs in South Asia: Bangladesh, India, Maldives, Sri Lanka. In *Proceedings of the Ninth International Coral Reef Symposium* (Vol. 2, pp. 841–845), Bali, 23–27 October 2000.

Roy, N. C., Sen, R. C., & Chowdhury, M. A. (2019). Consequences of climate change on fish diversity in Dekhar Haor Bangladesh. *Int. J. Fish. Aquat. Stud, 7*(6), 118–124.

Sachlikidis, N. G., Jones, C. M., & Seymour, J. E. (2005). Reproductive cues in Panulirus ornatus. *New Zealand Journal of Marine and Freshwater Research, 39*(2), 305–310.

Salgueiro-Otero, D., Barnes, M. L., & Ojea, E. (2022). Transformation in times of climate change: What makes a fisher diversify livelihoods? *Frontiers in Marine Science, 9*, 888288.

Sandaruwan, K. P. G. L., De Silva, D. W. L. U., & Liyanapathirana, N. (2022). *Assessment of vulnerability of climate change on coastal fishery of Sri Lanka and adaptation strategies*. National Aquatic Resources Research and Development Agency (NARA), Scientific Sessions 2022

Shaffril, H. A. M., Samah, B. A., Uli, J., & D'Silva, J. L. (2011). The potential impact of climate change environmental hazards on quality of life of fishermen community in Malaysia. *Australian Journal of Basic and Applied Sciences, 5*(7), 507–515.

Shyam, S. S., Kripa, V., Zacharia, P. U., Mohan, A., & Ambrose, T. V. (2014). Vulnerability assessment of coastal fisher households in Kerala: A climate change perspective. *Indian Journal of Fisheries, 61*(4), 99–104.

Timmers, B. (2012). Impacts of climate change and variability on fish value chains in Uganda.

Villasante, S., Macho, G., Silva, M. R., Lopes, P. F., Pita, P., Simón, A., Balsa, J. C. M., Olabarria, C., Vázquez, E., & Calvo, N. (2022). Resilience and Social Adaptation to Climate Change Impacts in Small-Scale Fisheries. *Frontiers in Marine Science, 9*, 802762.

Vinayachandran, P.N., Chauhan, P., Mohan, M., & Nayak, S. (2004). Biological response of the sea around Sri Lanka to summer monsoon. *Geophysifcal Research Letters, 31*(1).

Wahle, R. A., Dellinger, L., Olszewski, S., & Jekielek, P. (2015). American lobster nurseries of southern New England receding in the face of climate change. *ICES Journal of Marine Science, 72*(suppl_1), i69–i78.

WWF. (2015). Living blue planet report species, habitats and human well-being.

Yapa, K. K. (2000). Seasonal variability of sea surface chlorophyll-a of waters around Sri Lanka. *Journal of Earth System Science, 109*(4), 427–432.

Economics, Marketing and Value Chain Enhancement of the Cottage Industry Based in Highly Vulnerable Agro-Ecosystem: A Case of Kithul (*Caryota urens*)

D. Achini M. De Silva, B. M. R. L. Basnayake, S. K. Gunathilake, and M. A. E. K. Jayasinghe

1 Introduction

1.1 Value Chains, Agricultural Value Chains and Artisanal Agricultural Value Chains

A value chain is a conceptual business model describing the full chain of a business's activities in the creation of a product or service starting from input supplies then production, processing, marketing and reach to end user or consumer (Porter, 1985). The value chain framework is made up of primary activities (inbound operations, operations, outbound logistics, marketing and sales, service) and secondary activities (procurement and purchasing, human resource management, technological development, and company infrastructure) (Cuddeford, 2014). Value chain analysis is key to identify the performance in each point or node of the value chain. Further, it helps to identify the flow of products and information, linkages, governance mechanism, income distribution, dependencies, and level of sustainability at each point.

Agricultural value chain begins with the producers; the farmers who grow crops and raise animals and ended up with the consumers who eat, drink, wear and use the final products. And in the middle are many thousands of men and women, and small and large businesses. Each person and each business perform one small step in the

D. A. M. De Silva (✉) · B. M. R. L. Basnayake · M. A. E. K. Jayasinghe
Department of Agribusiness Management, Faculty of Agricultural Sciences, Sabaragamuwa University of Sri Lanka, Belihuloya, Sri Lanka
e-mail: desilva.achini@yahoo.co.uk

S. K. Gunathilake
Department of Natural Resources, Faculty of Applied Sciences, Sabaragamuwa University of Sri Lanka, Belihuloya, Sri Lanka

C. Malalgoda et al. (eds.), *Climate Change Adaptation in the Built Environment*,
https://doi.org/10.1007/978-3-031-75826-3_27

chain, and each adds value along the way—by growing, buying, selling, processing, transporting, storing, checking, and packaging (Dong, 2021). Agriculture industry will require profound changes in order to fulfill its multiple functions in a context of harsher environmental conditions and demographic and market transformations (Trienekens, 2012). Global agri-food systems are undergoing a rapid transformation towards higher-value products, stricter food standards, and closer international and vertical integration. Agriculture value chains are going through a paradigm shift, from efficiency-driven industrial agriculture to resilience-focused eco-friendly agriculture (Akyüz et al., 2023). Building resilient agricultural value chains received prime attention from policy making to implementation crews. Establish agricultural value chains of commercial crops, livestock, and fish, were able to change its structure, functions, and roles to into dynamic market requirements. In contrast, artisanal agricultural value chains of many developing nations facing difficulties in system adjustment.

Agriculture value chains in developing countries often are characterized by dual value chains operating in parallel for the same product: one informal or traditional and the other formal or modern (Agricultural value chains: A game changer for small holders., 2014). Small holder agricultural producers are frequently involved in informal chains that deliver products to local middlemen, local consumers, primary processors and then to small local stores. Establish or formal value chains can deliver the same product, usually in better or more uniform quality, from larger farms or more organized groups of small farmers to more commercial wholesalers and from there to supermarkets or exporters. Duality of agricultural value chains has been accentuated by the explosive growth of retail chains. Informal agricultural value chains characterized by low-quality products, and low prices and low returns for value chain actors. Therefore, research attention needs to be focus on to find ways to integrate small producers into formal modern value chains, both domestic and export-oriented.

Value chain encompasses the creation of value along with flow of products, knowledge and information, income, services, payments, and the social capital needed to organize producers and communities. Information is especially important to all value chain actors and flows in two directions: markets inform producers of price, quantity and quality needs, product handling and technology options, while producers inform processors and markets on production quantities, locations, timing, and production issues (Gunaratne et al., 1996). Information asymmetry is common evil among small holder agricultural value chains, especially on artisanal value chains. This chapter aims to investigate the value chain of Kithul Industry, one of the oldest Agri-food value chain managed by rural farmers.

1.2 Context of Kithul Value Chain

Kithul (*Caryota urens*) is a plant belonging to the family Palmae, primarily found in the wet zone of Sri Lanka. Native rain forest species of tropical Asia mainly found in India, Myanmar, and Malaysia (Ethnobiol, 1995; Kithul 2021). The kithul palm is known by several names; include solitary fishtail palm, wine palm and jaggery palm (Ranawake & Pathirana, 2024). Kithul (*Caryota urens*) value chain (KVC) structure can be described as farmers, collectors/traders, processors, exporters, wholesalers, retailers, and consumers are analyzed from physical, financial, infrastructure and relational perspectives, the influence of the external environment, both micro and macro within which value chains operate must be expressed and understood. Therefore, exploration of value chains begins with environmental scanning. The KVC's micro environmental components are local institutions, stakeholders, local markets, financial institutions, and distributors. Macro environment consists with marketing, economic, social, political, technological, environmental, and legal factors that influences the value chain operations The socio-cultural environment of the KVC derived from rural villages located in wet zone of Sri Lanka. Marketing environment is twofold, informal rural markets and formal or organized markets catering to both urban markets locally and international markets.

1.3 Geographical Distribution of Kithul Value Chain

Biophysical environment explains the status of Kithul growing areas. In general, Kithul trees are growing in Tropical rain forests, buffer zones, and home gardens of the wet zone of Sri Lanka (Fig. 1). Annual bad weather events of South-West monsoons directly hit the production and processing. Upstream of the KVC spread over 12 wet zones districts of Sri Lanka while downstream of the KVC spread all over the country as well as beyond the national boundaries Scoping study on KVC helps to identify the socio-cultural and economic status of the families, communities and small and medium businesses derived in KVC.

Geographical distribution of Kithul trees (Fig. 1) highlighted that wet and intermediate zone are prominent in Kithul cultivation. Main growing areas are centered in Kandy and Rathnapura districts. Kithul tappers distributed in all Kithul growing areas as this is practiced as cottage industry. Tapper concentration is high in Rathnapura, Kandy, Badulla, Nuwara Eliya, Badulla, Galle and Matara districts compared to other areas and this figure is positively correlated with extend of Kithul plantations.

2 Methodology

The study was carried out in Kalawana and Weddagala of Rathnapura district of Sabaragamuwa province, Elpitiya of Galle district and Deniyaya of Matara District of the Southern Province of Sri Lanka where the Kithul industry is established traditionally. The study relied on both quantitative and qualitative data and the research approach is deductive. Case Study Approach used to collect qualitative data from 40 respondents and respondent groups were Kithul tappers, processors and traders

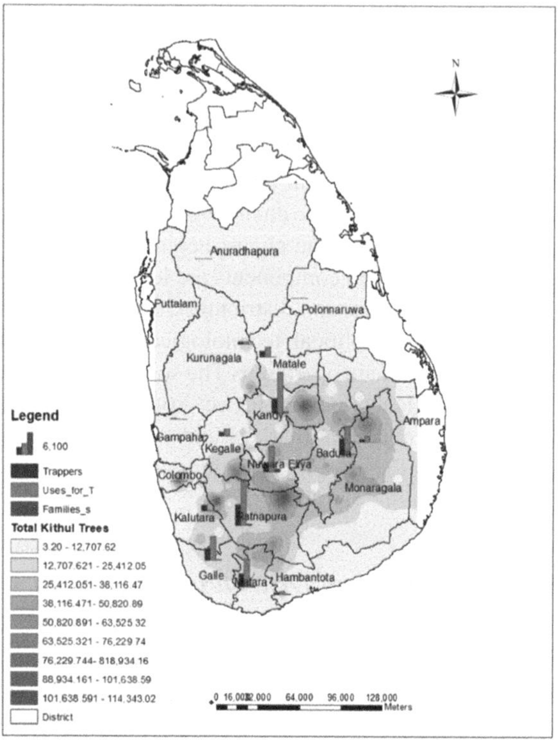

Fig. 1 Geographical distribution of KVC

who are involving in the Kithul upstream of the value chain. Exporters, retailers and consumers were considered from downstream of the value chain. Non availability of data base on Kithul value chain and registered producers pushed towards to use non probability sampling techniques. Both purposive and snowball sampling methods were instrumental to identify the value chain actors and collect the primary data. Data collection begins with storytelling exercise with selected Kithul value chain actors. Open ended discussions were conducted individually and audio records were used to store the information shared. Further, in-depth interviews were performed with same value chain actors with an interview guide. Key thematic areas of questions were on product portfolio, volume and values, cost of production, processing knowhow, and functions and roles of value chain actors. Moreover, field observations were used to monitor the work of each value chain actor, identify the technology they used, processing methods, issues, problems, and coping strategies to meet the market requirements.

3 Economics of Kithul Industry

Kithul palms are an economic windfall for their owners (Karunaratne & Akurugoda, 2022). Kithul trees are naturally growing in Tropical rain forest highlands, and home gardens and the contribution of Kithul trees on rural households are significant (Everett, 1995). Figure 2 explains the price distribution among value chain members based on different value-added products.

The income is highly varied from tapper to trader as well as growing region, stats of the tree, and skills, experience, and knowledge of tapper (Seneviratne & Dissanayake, 2017). Direct marketing is common for Kithul, jaggary and treacle usually the tapper/processor sells to the collector or directly sells to the customers. In general, they produce about 25 bottles per month and production volume depends on the based on the demand. Moreover, loyal local consumers maintain close relationships with primary producers, and they can buy Kithul products from the farm gate with the expected quality.

Kithul flour, one of the common rural desert and breakfast delicacy is famous for its health benefits. Production of Kithul flour is limited to few locations and production process followed traditional technology. Excess production of Kithul flour directly sell to intermediaries of the area.

Kithul toddy is having its own loyal customer base it is selling one bottle as 0.6–1US$) And they produce 5–10 bottles per day. Kithul vinegar production is also practicing in very low level. They found the raw materials without spending much cost such as wood from the forest, coconut shells and used glass bottles from neighbours or sellers. The initial expenses must spend as for ropes: 2–3.3 US$, Clay pot 4 US$, Knife 8US$, cost of labour cost 5–7 US$ and wood if bought 11 US$.

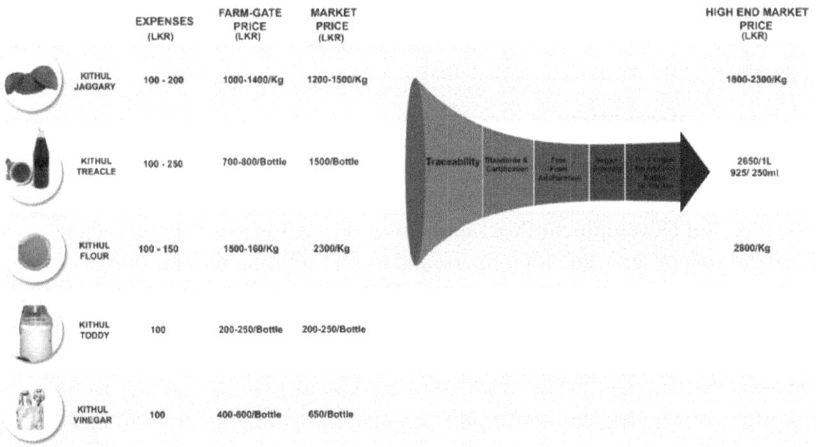

Fig. 2 Economics of KVC: price distribution along the KVC for 5 main products. *Source* Field survey

Large scale manufactures producing Kithul value added products for high end market follow various strategies with their own trained tappers and maintain the safety and quality standards. They follow value pricing strategy and price is based on market segmentation. Digital marketing methods are prominent among them. Websites, mobile marketing, social media, e-business platforms are frequently used to promote their products. Further, their products come with attractive packaging, labelling, guaranteed high quality and ultimately enjoying higher profits.

4 Value Chain Analysis of Kithul

4.1 Value Chain Environment: Institutional Analysis

Policies, legal framework, institutional arrangements, and initiatives are key components of the value chain environment. Toddy is the main targeted unprocessed product from Kithul tree and that belongs to the alcoholic beverage category. Therefore, legal aspects are managed by the Excise Department of Sri Lanka. License for manufacturing & selling processed/ Molasses/ Palmyra/ Special/ Coconut Arrack, and /or for manufacture & issue to the wholesale outlets, license to manufacture Vinegar, toddy, license to collect toddy and license for production and sale of fermented toddy. Amendments to the Exercise act, the Excise (Amendment) Act, No. 4 of 2018, focused more on safety and quality aspects; no tree producing toddy, other than Kithul tree shall be tapped, and no toddy shall be drawn or lowered from any tree other than the Kithul tree. Further, excise ordinance notification no.04/2021 highlighted that the maximum amount of toddy transported by an individual has been increased to three litres from one and a half litres and the hours for toddy sale has been extended to 9:00 p.m. Regulations provide limits to operate and aimed to manage the natural resources in sustainable way.

Government Intervention:

Ministry of Rural Economics Affairs, later ministry of primary industries and plantation crops introduced several programs to strengthen the Kithul industry, where National Kithul Development Programme of 2011–2014 period. Government initiatives on training programme for tappers, and master training programmes to increase the number of tappers and impart tapping skills to young generation. Provision of safety kits for Kithul tappers (Safety jacket, helmet, belt, knife etc.) and introduction of insurance schemes to mitigate the risk of climbing related accidents and life threats of Kithul tappers. About 1500 of safety kits were issued among Kithul value chain. Moreover, introduction of traditional and Kasper technology to increase the sap yield and bring unproductive trees into the tapping process and increase the sap yield.

Quality improvement for Kithul treacle and jaggery (GMP, SLS certification, cost effective stoves, brix meters, PH meters and devices to assess the level adulteration) to enable it to compete with other sweeteners such as maple syrup of Canada (20 GMP Certifications). Moreover, registration and issue of identity cards to recognize Kithul producers and overcome legal impediments. Facilitation to link the Kithul producers with large scale exporters including buy-back arrangements, placing the Kithul products in city cops, supermarkets and franchising in the private sector and establish model sales outlets with high quality Kithul products. Moreover, nursery development and expansion efforts to the Kithul plantations through increase the number of Kithul trees in the country and convert it as a commercial operation. Further, institutional mechanism facilitates to develop user-friendly tapping system to reduce the risk of climbing or suitable elevators for climbing Kithul trees. Further, job security assured through introduced insurance schemes for Master Trainers, trainees and Kithul tappers. Value addition promote through available facilities, Promote, demonstrate, and encourage Kithul products such as jaggery and treacle and Kithul based value added food, beverage and handicraft products through exhibitions, trade fairs and other modalities. Further, government initiatives facilitate the exports of value added Kithul products through participation to foreign trade fairs, export promotion as well manage the participation of trade fairs.

Governance:

The shared and rent methods in resource management are more common in the Kithul industry. Informal arrangements were more prominent mechanism for manage the Kithul industry, If the tappers shared the tree with the owner, they share the tree as every other day. if the tree shared by two tappers, they share the tree by adopting their own time schedule. If one tapper is collected the sap in morning, the other tapper will collect the sap in the evening with mutual understanding. In rent method, usually the owner rent the tree at least 06 months period and as the rent the tapper must provide half amount of the sap to the owner in daily. Rent method use for mobile trappers or outside tappers. Table 1 highlight the key regulatory mechanism govern the value addition in KVC.

Governance mechanism of KVC enrich with the rules operating in KVC, system of coordination, regulation, and control. Further, governance mechanism of KVC refers to both official rules and community rules. Official governance mechanism implies that interacts between KVC actors and official representatives of local and central

Table 1 Rules and standards affecting KVC

Type	Example	Enforcement & sanction
Official or legal standards	Adulteration (% of Kithul should be in treacle, jaggery)	Ban non-compliant products from export markets penalties
Voluntary standards	Organic, Fair Trade	Ineligibility for certification or value-added labelling
Commercial requirements	Colour, composition, taste, packaging, shelf-life, storage requirements, etc	Rejection of products from consumer, retailer, etc. Price based on quality Reduced price by retailer Limit market entry

government's responsible organizations. In general, official governance arrangements in line with legal requirements and that aim to protect user rights, consumers, and natural resources. Community rules or un-official governance mechanisms were imposed by the user to protect the industry from mal practices, protect primary producer communities, and cater to consumer demand through ethical supplies. Certificates and standards on KVC are an important governance arrangement made by importing countries, importers to ensure safe products to their clientele. Moreover, maintain standards add extra cost to the producers where small-scale producers unable to meet the requirements. Safety and quality management compliance is practised exporters to meet the market entry requirements of their buyers.

4.2 Value Chain Map: Value Chain Actors, Functions, Roles, Input–Output Relations

Value chain mapping follows a specific product through the various steps of the chain, from production to consumption. Value chain mapping process maps all actors involved in the process, their interactions, and power relations. Further, value chain mapping from a gender perspective provides insights on the gender division of the labour along the chain. Figure 3 discuss the Kithul value chain, its macro and micro environment, core value chain process, value chain actors and primary and supportive activities.

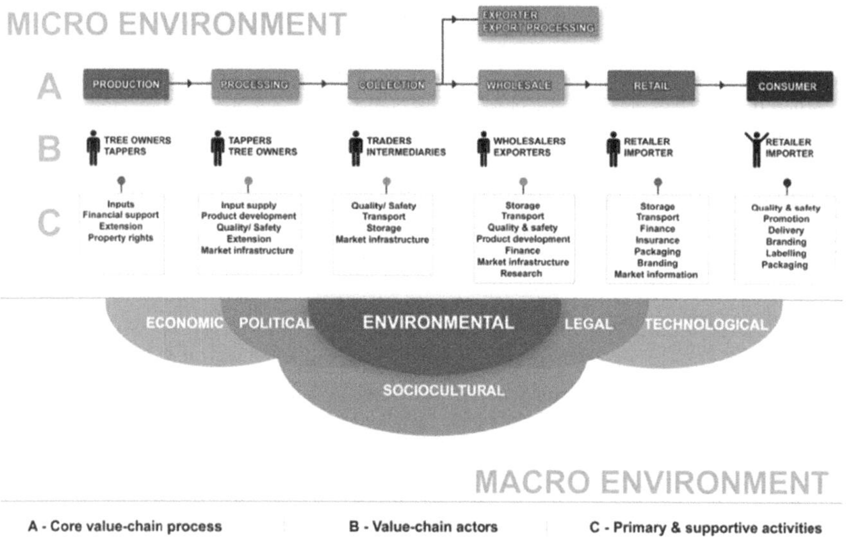

Fig. 3 Kithul value chain map embedded in its environment

Upstream of the KVC

Producers: Growers and tappers

Kithul industry is practiced in 12 Districts of Sri Lanka (Karunaratne & Akurugoda, 2022). In each District there are villages where treacle and jaggery are produced and sold for livelihood of tappers and their families. Kithul growing villages of wet zone of Sri Lanka have been defined as villages where either treacle or jaggery or both are produced for marketing purposes, in addition to their own consumption. Traditionally, villagers used naturally grown Kithul trees of forest areas, buffers zones or home gardens for the tapping. Growing market demand for Kithul products places special growing arrangements of Kithul plantations. Area based plant management arrangements can be observed among different Kithul growing districts. Further, it has been observed that practices related to obtaining sap, such as tapping methods, vary from District to District. In addition, the number of Kithul growing villages also varies from region to region (Seneviratne et al., 2005; Zoysa, 1992). Kandy District has the highest number (106), whereas Kalutara has the lowest number of Kithul villages (16) (Seneviratne et al., 2005). Since the number of tappers per village is same in each district and the number of villages being varied from District to District (Collins et al., 2015).

Traditionally, tapping was a major source of income for many people in the highlands of wet zone of Sri Lanka, particularly in villages bordering the forest. Many tappers traditionally gleaned their toddy from palms inside the forest. Present day tapping inside the Sinharaja, and other forest reserves has been described recently (IFAD Strategic Framework 2015) and the forests are government owned and as large areas are classified as reserves, would-be palm tappers must have a license. Culture of tapping Kithul trees varied across the regions. Typically, there will be several people in each village known as tappers who tap all the palms in their neighbourhood. Kithul trees grown in private home gardens are owned by the land owner. When a palm approaches flowering, the owner notifies a tapper. In general, from the time of the first sweet toddy production, the tapper and the tree's owner then share the yield equally by alternating days. The process of tapping varies in some specifications among individual tappers and between regions. Figure 4 showcase the preparation of Kithul inflorescence for tapping.

Downstream of the KVC

Wholesalers

Downstream of KVC composed of two main nodes, wholesalers, and retailers. Wholesalers playing bridging role between primary processors and retailers (Fig. 5). Wholesalers used to collect Jaggery and treacle from primary processors of their own network of suppliers. Established connections with primary processors ensure regular supplies in expected quality and quantity. Wholesalers are common in informal markets and sometimes they engage in retailing too. In contrast, export oriented firms maintain their own production process, and export products to retail chains and specialty stores for ethnic products.

Retailers

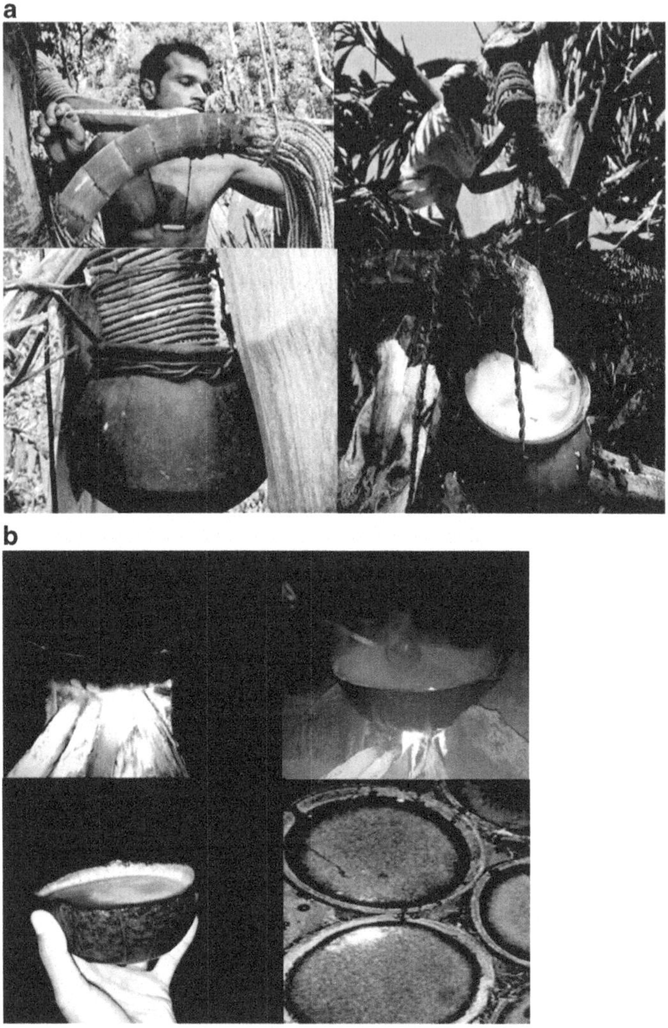

Fig. 4 A **a.** Process of Tapping the the Kithul Flower: Selection of suitable inflorescence, prepare flower for tapping, collecting the sap into clay pot **b.** Process of primary processing of of Kithul Jaggary: in traditional way; Boiling the sap, poring boiled thick treacle into coconut shells, and finished jaggery in coconut shells

There are different retail formats identified in KVC and most common types are mobile retailers of village fairs, grocery stores, and retail chains (Fig. 5). Female retailers are common in village fairs of rural areas, and they bring their own production to the marketplace. Further, village level collectors also engage retailing in urban and peri-urban areas where female retailers are prominent. All retail chains facilitate the retailing of Jaggery, Treacle and flour under their own brands and producer

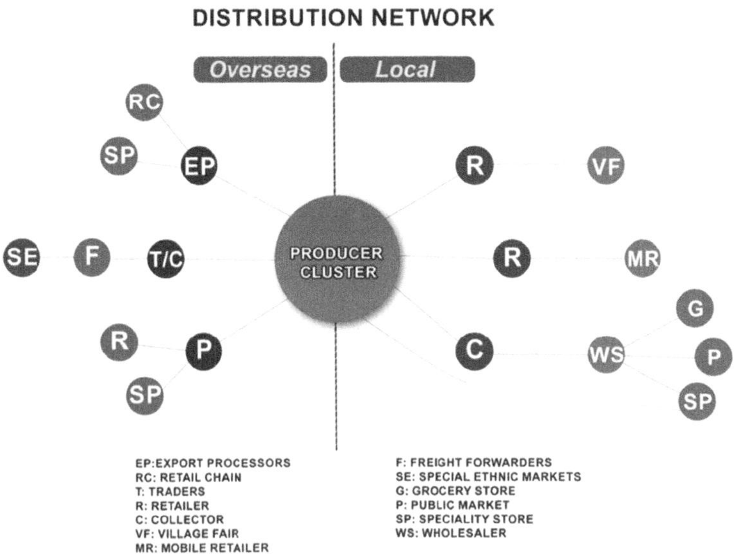

Fig. 5 Distribution networks of both local and overseas markets for kithul products

brands. Loyal consumers always going with their trusted brands and trust based on quality and authenticity of the products. Moreover, export-oriented processors and processors of branded Kithul products maintain safety and quality management procedures.

Consumers

Agricultural value chains are defined as interconnected actors who engage in sequence of processes that produce raw materials, add value using services and information which end users will pay for the benefits. Consumers govern modern value chains through the expectations; needs and wants, perceived values and willingness to pay for the created value. Toddy, treacle, jaggery and flour, most common products from Kithul tree are popular delicacies of rural villages of Sri Lanka which rural consumers test the quality through traditional taste and colour. Popular local brands based on regional names or names of area of production are based on unique taste and flavour of the local produce.

Key role of the KVC and its actors is creating and delivering value to the consumers. Identify the attributes of consumer value is vital to develop sustainable KVC. Identify the attributes of consumer value is vital to develop sustainable KVC. Most importantly find out the attributes of consumer value and convey the message to primary producers, tappers and processors will lead to cater the needs and wants of end users while developing sustainable KVC. In general, small holder Kithul tappers and processors who produce range of primary Kithul based products for rural and regional markets beyond their villages have little knowledge on type of consumers who consumer their products (users) and mode of consumption (uses).

On the other hand, consumers have no clear idea on primary producers, their socio-cultural background, living standards and especially the families depend on income from primary products.

Demographic (age, gender, income, ethnicity, religion, and stage of family life cycle, etc.), geographic (region, country, city, etc.), behavioural (knowledge, experience, attitude, usage, response, etc.), psychographic (attitudes, values, opinions. etc.) and benefits (perceived benefits and actual benefits) directly linked with consumer behaviour, purchase intention and the value consumer put on product (Wijesinghe et al., 2015). Details of market segmentation for Kithul product explain in Fig. 6 Geographic, demographic, behavioural and psychological clusters were prominent for Kithul products. Geographic segment is mainly based on ethnic markets established in EU, USA and Canada, Middle East, Australia, and East Asian markets. Mid to old age females of Asian origin belongs to upper middle class representing the main demographic segment. Health concern families were more into the behavioural segment while lifestyle and values were more considered in psychological market segment.

Product portfolio of the Kithul products varied from food products, beverages, to timber. Figure (7) explains the diversity of the Kithul based products produced and marketed in various marketplaces Kithul palm yields a rich, dark syrup with a highly valued, distinctive taste (Karunaratne & Akurugoda, 2022). In comparative testing of raw sugars Kithul jaggery ranked among those of highest quality (Gómez et al., 2020). Its export potential is being tested with success in the Australian consumer market (Karunaratne & Akurugoda, 2022). Kithul products move to the market via various channels; mainly informal channels handle biggest share of the jaggery and treacle market. Formal value chains govern by the export-oriented firms and distribution channels eyeing on export markets.

Kithul flour production, a very traditional cottage practice converted into a business practice or value chain. Health concern segment of both high-end local market and ethnic marketplaces demand on healthy Kithul flour. Kithul flour has huge potentials of being used to food industry to provide better gelling property with other functional needs. Based on this study Kithul flour can be promoted because of its positive

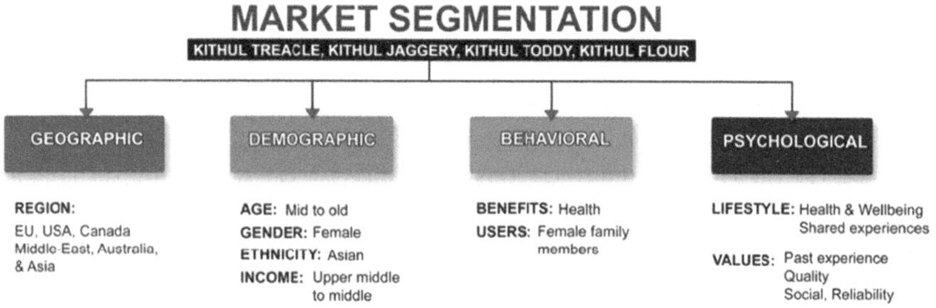

Fig. 6 Market segmentation for Kithul products; Treacle, Jaggery, Toddy & flour

Fig. 7 Kithul Products in Local Market Place (Jaggery, Flour, Treacle and Toddy

nutritional and Physico-chemical properties for the enhancement of different food products (International Labor Organization (ILO) 2021).

Kithul toddy, one of the earliest alcoholic beverages produced in rural areas of Sri Lanka. Toddy market is concentrated mainly in rural areas and rarely reached into urban markets. Kithul vinegar, another important product produced traditionally for household use and rarely enter local markets. In general, primary producers use traditional knowledge and technology in their production process. In high end market, large scale manufactures producing Kithul value added products with their own trained tappers. Their value pricing is very high and most of them are using websites, mobile marketing, social media, e-business platforms to promote their products. Further, their products come with attractive packaging, labelling, guaranteed high quality and ultimately enjoying higher profits.

Income Distribution Along the Value Chain

Income distribution in value chains highlight the top gainers and losses of value chain. In general agriculture value chains act favorably on downstream, especially for intermediaries. The Upstream, primary producers and end-users receive less favorable returns. Kithul value chain also behave like other small scale agricultural value chains except decision making and governance. Figure 8 brings the share of primary producers and intermediaries of the Kithul value chain.

High income variability is common among the value chain actors, and which depends on production capacity and quality of output. In general, Jaggery and treacle value chains are informal, short and target local markets except export-oriented chains. Tappers and or primary processors directly sell their products to consumers. Year-round production varied with bad weather and demand fluctuations based on seasonal demand. Especially, traditional ethnic and religious festivals place huge demand on both treacle and jaggery as essential ingredient on traditional sweet meat production.

Gender Relations in Value Chain

Gender in value chains, especially their role, responsibility and functions are essential to smooth flow of value chain. Agricultural value chains are heavily dependent on nature and people's participation. Gender participation in KVC explains in Fig. 9.

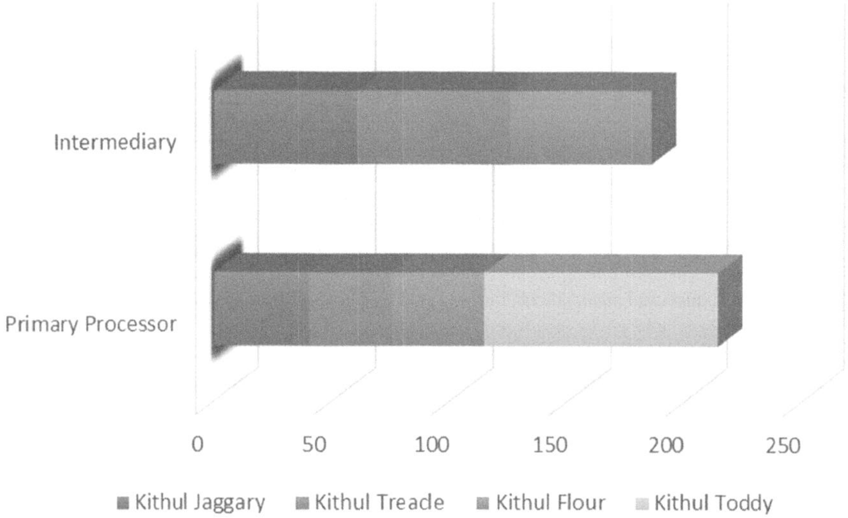

Fig. 8 Income share of the primary producers and intermediaries along the various Kithul value chains

Upstream functions of the KVC managed by the rural households and adult male members of the family playing key role in preparation of Kithul inflorescence for tapping, daily tapping operations, collecting the sap and primary processing (Table 2). In each activity of the value chain there is higher involvement of male than female. Generally, a tapper must spend 1–2 weeks to prepare the tree for the first tapping after identifying the matured flower of a Kithul tree. They have divided their time schedule as 6.00 a.m. for the first tapping, 12.00 p.m. for 2nd tapping, 6.00 p.m. fir third tapping and if they practice 4th tapping it is doing between 10–11 p.m. within a day.

The higher women involvement can observe in the Kithul products processing specially in Kithul flour production. And low involvement in Kithul toddy and vinegar production. All the trading related activities and decisions on pricing, selling point, selling amount etc. are taken by men. Female members are playing supportive roles in primary processing of treacle, jaggery and flour. Moreover, female members engage in retailing jaggery and treacle in village fairs, roadside stores, and mobile vending.

Gender backed constraints limit the participation of female members to access to productive resources involved in decision making and control of value chain functions (IFAD Strategic Framework 2015). Females playing key role in managing household economy through the income generated from value chain participation mainly and other sources while performing their other domestic roles. Gender relations determine access to assets and resources, participation in decision making, etc. all of which have a direct impact on the performance of the value chain (IFAD Strategic Framework 2015). Typical women's product in KVC, i.e., Jaggery, Treacle, flour in small quantities and produced in traditional way and target local market might

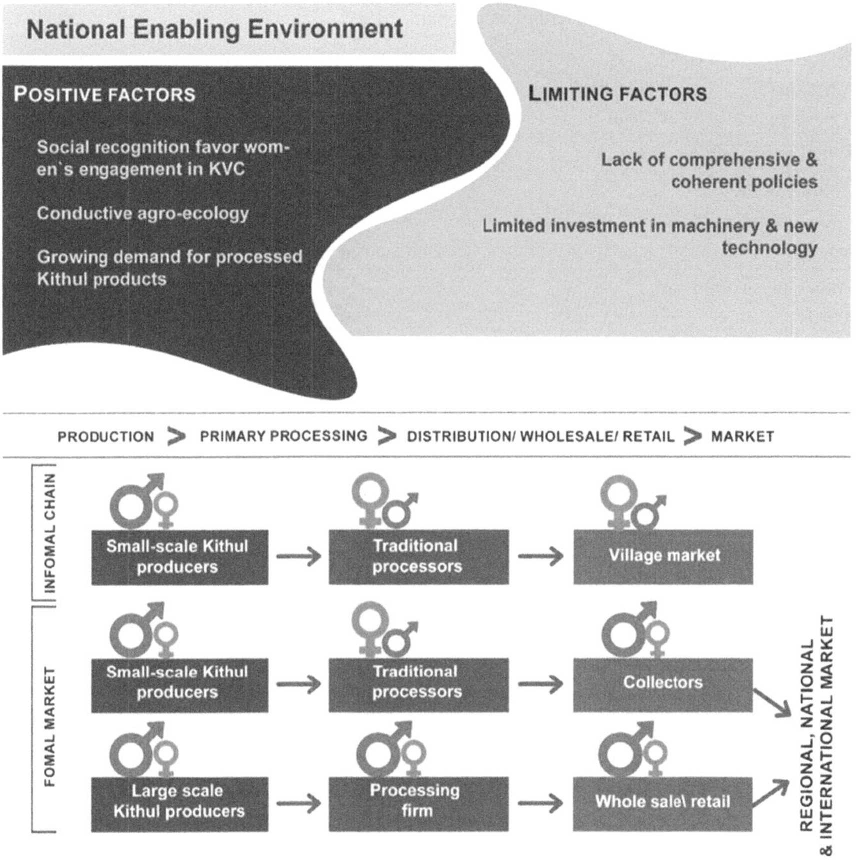

Fig. 9 Gender sensitive mapping of the Kithul value chain

offer limited opportunities for profits, but those tiny ventures, provide strategic entry points for enhancing women's participation in KVC bypassing their traditional roles and importing their interactions with supplies and service providers.

5 Value Chain Development

Establishing linkages, building structures, coordinating mechanisms, and developing capacities of institutions and value chain actors are key considerations in value chain development. Value Chain Development, in turn, refers to an approach which takes a product, service or commodity as the basis for analysis (Wijesinghe et al., 2015). Value chain analysis helps to understanding nature of value chain

Table 2 Gender involvement in Kithul value chain

Matrix roles and tasks				
Activities	No. of trees tapping per day/Per Person	Time (in hours or days)	Men	Women
Kithul tree tapping	2–6	3–4 times per day	XXX	
Collecting the sap		2–3 times per day	XXX	
Processing: Kithul jaggery		5–6 h after one collecting	XX	XX
Processing: Kithul treacle		2–4 h after one collecting	XX	XX
Processing: Kithul flour			X	XXX
Processing: Kithul toddy			XXX	
Processing: Kithul vinegar		1–2 Weeks after one collecting	XX	X
Trading			XXX	X

Source Field Survey

which can generate growth, job creation and poverty reduction. There were countable number of studies focused eye on Kithul industry and this is first involvement on analysis of KVC. KVC analysis aimed to identify constraints to and opportunities for increasing value chain performance, so that interventions can be designed and implemented to address constraints and improve outcomes. KVC development can contribute to decent work by: creating more equal opportunities for productive work for women and men, increasing their incomes, providing household income security, enhancing social integration through networking, dialogue and cooperatives, providing better prospects for entrepreneurial development, expose to learning new skills, improving occupational safety and health as well as good manufacturing practices with socially responsible production, providing appropriate incentives in the form of new livelihood opportunities to both men and women, and conserve nature.

Especially, value chain development for Kithul, as a smalls scale cottage industry established in rural areas of the country needs to identify its context; people, nature, and economics of it. Inclusive development approach is vital to address the issues of three pilers and ensure balanced development ensuring greater benefits for poor and excluded groups from economic growth and development. Value chain development aims for 'win–win' outcomes such as improved performance in production and processing of Kithul products and growth of income at the same time as positive impacts for the poor and vulnerable groups of the rural areas. Enterprising tappers and primary processors will ensure better jobs and returns, household wellbeing, along with social upgrading.

Increasing visibility of women's productive role in KVC is fundamental for ensuring women's role is recognized and valued, and that women are considered legitimate value chain actors and gain access to the resources and services they need to become more efficient and competitive. Further, recognizing women's domestic help as work is also essential for increasing their socio-economic status within households and communities and dismantling rigid gender stereotypes about women's and men's expected roles and responsibilities.

Poor financial literacy of women is one of the important aspects to address in most of the rural settings. In general, income generated used for daily priorities, such as food, educating children, communication, and medical expenses. Only handful of them engage in savings and invest in income generating activities. Therefore, awareness on household level financial handling, savings and investing in income generating activities are vital to strengthen the women's position in value chain.

5.1 Participatory Value Chain Development

Value chain development process essentially need to be interconnected with communities, local stakeholder groups, and macro level institutions. Institutional analysis confirmed that institutions responsible for agriculture, small scale industries, financial institutions, local governments, regulatory bodies, research institutions and markets are functioning separately, institutional mandates are overlapping, poor policy directions, and weak or fragile intuitional networks. Government initiatives on KVC development was unsuccessful due to the weak institutional capacities, interventions, and poor participation of communities. Therefore, identifying target communities, find out their real problems and issues, design and develop community based appropriate solutions will earn expected returns and enhance the KVC.

KVC analysis help to identify its structure, functions and people involved in each node. Value chain empowerment should be based on needs analysis on each node. There may be node specific requirements which cannot be generalize along the chain. In general, aging tappers, youth migration from Kithul industry to another industry, tapping is not attracted by youth anymore, etc. are visible problem s which need to be addressed to ensure supply base of the KVC. Mechanizations options along with safe working environment and better returns will ensure labour retention. Further, development of appropriate technology for tapping, collecting sap and processing will make their lives easier and comfortable. Moreover, empower women and children through both formal and informal education, enterprising the youth, market development and access to better marketplaces and financial literacy will ensure value chain empowerment.

5.2 Promotion and Upgrading Kithul Value Chain

Promotion and upgrading KVC essentially needs to investigate its inherited problems. This study finds out that lack of government intervention, poor institutional involvement, high labour cost, aging tappers, limited number of experienced tappers, poor or bad weather conditions, limited access to technology, and wild animal threat as critical issues which cripple the industry. Small scale processors operating individually, producing limited quantities, and less access to markets are more vulnerable compared to those who linked to the markets or buyers. Therefore, establishment of marketing, channels, allow small payers to enter the market systems, facilitate marketing process will reduce their vulnerabilities.

Moreover, research intervention is crying need to enhance the KVC. Explore health benefits of Kithul products, product design and development, identify potential consumers, overseas markets, proper packaging methods, processing and preservation technologies, storage and shelf-life studies, product and process innovations will ensure future market potential of KVC.

6 Summary

Kithul one of the important palm trees growing in tropical rain forests, especially the wet zone of Sri Lanka employing thousands and securing their livelihoods. Kithul treacle and jaggery considered as healthy delicacy which consumes generations. Toddy, alcohol beverage produced and consumed locally received much attention due to its taste. In general, two types of value chains were observed; informal chains ended up in producer regions, limited value addition and cater the need of local consumers. Formal lengthier chains are organized, catering to the needs of urban markets of the country as well as overseas markets. Formal value chains are much focused on food safety and quality management in production and processing, packaging, labelling, branding, storage and shelf-life, and product innovations. Private sector involvement is high where weak government interventions marginally facilitate the functioning of value chain. Building a competitive KVC is essential to ensure sustainable returns to its membership from tapper to consumer. Priority should focus on understanding the nature of products demanded by local, regional, and international markets and consumer segmentation. Build investment capacity on factor inputs, services, and research to meet the requirements of the prospective markets will secure the entry routes. Further, improve business enabling environment; policy and institutional environment will facilitate the value chain development process.

Acknowledgements Special thanks to W.M.T.B. Weddagala for his invaluable contributions in figure design, as well as the dedicated assistance provided by students during the data collection process.

References

Akyüz, Y., Salali, H. E., Atakan, P., Günden, C., Yercan, M., Lamprinakis, L., Kårstad, S., Solovieva, I., Kasperczyk, N., Mattas, K., Lazaridou, D., Yener, G., Alayidi, A., Kunchulia, I., Basilidze, L. and Knez, M. (2023). Case Study Analysis on Agri-Food Value Chain: A Guideline-Based Approach. *Sustainability, 15*(7), 6209 [online]. https://doi.org/10.3390/su15076209

Collins, R. C., Dent, B., & Bonney, L. B. (2015). *A guide to value-chain analysis and development for Overseas Development Assistance projects.* Australian Centre for International Agricultural Research: Canberra, ACT. https://www.CollinsDentBonney2015Aguidetovalue-chainanalysisanddevelopmentforOverseasDevelopmentAssistanceprojectsACIAR.pdf

Cuddeford, V. (2014). *An introduction to agricultural value chains.* https://idl-bnc-idrc.dspacedirect.org/bitstream/handle/10625/52685/IDL-52685.pdf

De Zoysa, N. (1992). Tapping patterns of the kitul palm (*Caryota urens*) in the Sinharaja area, Sri Lanka. *Principles, 36*, 28–33.

Dong, L. (2021). Toward resilient agriculture value chains: Challenges and opportunities. *Production and Operations Management, 30*(3), 666–675.

Ethnobiol, J. (1995). The Kitul palm: Ethnobotany of caryota urens L. in highland Sri Lanka. *Winter, 15*(2), 161–176.

Everett, Y. (1995). The Kitul palm: Ethnobotay of *Caryyota urenus L.* in highland Sri Lanka, *Journal of Ethnobiology, 15*(2):161–176.

Gómez, M. I., Meemken, E., & Verteramo Chiu, L. J. (2020). Agricultural value chains and social and environmental impacts: Trends, challenges, and policy options. *FAO eBooks.* https://doi.org/10.4060/cb0715en

Gunaratne, W. D. L., Premakumara, K. V. S., & Heenkenda, A. P. (1996). Kitul (Caryota urens L.): A popular multipurpose eplant among the rural community. In, *Proceedings of the Seventh Multipurpose Tree Species Workshop* (pp. 226–233). https://www.devex.com/news/agricultural-value-chains-a-game-changer-for-small-holders-83981

IFAD Strategic Framework: 2011–2015 https://www.ifad.org/documents/38714170/40264252/Scaling+up+note+-+Sustainable+inclusion+of+smallholders+in+agricultural+value+chains.pdf

International Labor Organization (ILO). (2021). *Value Chain Development for Decent Work A systems approach to creating more and better jobs*, 3rd ed. ILO. Available at: https://www.ilo.org/wcmsp5/groups/public/---ed_emp/documents/publication/wcms_766283.pdf

Karunaratne, M., & Akurugoda, I. (2022). Protecting special wild tree spe-cies and traditional knowledge towards securing livelihoods of rural com-munities: A study on Kithul *(Caryota urens)* industry in Sri Lanka. XV World Forestry Congress.

Kithul. (2021). A Treacle Tale. *The Sunday Observer.* https://archives1.sundayobserver.lk/2021/10/03/junior/kithul-treacle-tale

Norton, R. (2014). Agricultural value chains: A game changer for small holders. [online] Devex. Available at: https://www.devex.com/news/agricultural-value-chains-a-game-changer-for-small-holders-83981#:~:text=Improving%20agricultural%20value%20chains%20can

Porter, M. E. (1985). *Competitive Advantage: Creating and Sustaining Superior Performance.* Simon and Schuster, New York. ISBN 9781416595847.

Ranawake, A. L., & Pathirana, R. (2024). Indigenous knowledge in the Kithul (Caryota urens L.) industry of Sri Lanka and its scientific basis. *Genetic Resources and Crop Evolion.* https://doi.org/10.1007/s10722-023-01847-7

Seneviratne, M. A. P. K., & Dissanayake, D. M. D. K. S. (2017). A study on the practice of adulteration of Kitul (*Caryota urens*) sap products in Sri Lanka. *Sri Lanka Journal of Social Sciences, 39*(2), 69. https://doi.org/10.4038/sljss.v39i2.7441

Seneviratne, M. A. P. K., Wanigasundara, W. A. D. P. & Wijeratne, M. (2005). A study on the practice of adulteration of Kitul (*Caryota urens*) sap products in Sri Lanka. *Tropical Agricultural Research, 17*, 264–270 [online]. Available at: https://dl.nsf.gov.lk/server/api/core/bitstreams/a0f0fc5b-f59d-4e4f-a59d8c57f63fa146/content

Trienekens, J. H. (2012). *3. Value Chains in Developing Countries A Framework for Analysis* (pp. 43–68). Amsterdam University Press eBooks. https://doi.org/10.1515/9789048514991-003

Wijesinghe, C., Wicramasinghe, I., & Saranandha, K. H. (2015). Kithul Flour (*Caryota urens*) as a Potential Flour Source for Food Industry. *American Journal of Food Science and Technology, 3*(1), 10–18. https://doi.org/10.12691/ajfst-3-1-2

The manufacturer's authorised representative in the EU is Springer Nature Customer Service Centre GmbH, Europaplatz 3, 69115 Heidelberg, Germany. If you have any concerns regarding our products, please contact ProductSafety@springernature.com

Printed and bound by CPI Group (UK) Ltd, Croydon, CR0 4YY
01/10/2026
02225938-0011